# Meteorology

*Eric W. Danielson*
Hartford College for Women, University of Hartford

*James Levin*
Pennsylvania State University

*Elliot Abrams*
AccuWeather, Inc.

Boston, Massachusetts Burr Ridge, Illinois Dubuque, Iowa
Madison, Wisconsin New York, New York San Francisco, California St. Louis, Missouri

**WCB/McGraw-Hill**

*A Division of The* **McGraw-Hill** *Companies*

METEOROLOGY

This book is printed on recycled, acid-free paper containing 10% postconsumer waste.

1 2 3 4 5 6 7 8 9 0 QPD/QPD 9 0 9 8 7

ISBN 0-697-21711-6

Vice president and editorial director: *Kevin T. Kane*
Publisher: *Edward E. Bartell*
Sponsoring editor: *Lynne M. Meyers*
Developmental editor: *Daryl Bruflodt*
Marketing manager: *Lisa L. Gottschalk*
Project manager: *Gloria G. Schiesl, Janice M. Roerig-Blong*
Production supervisor: *Sandra Hahn*
Designer: *K. Wayne Harms*
Interior and cover designer: *Rokusek Design*
Cover Image: © *A. & J. Verkaik/The Stock Market.*
Illustrations: *AccuWeather, Precision Graphics*
Photo research coordinator: *Lori Hancock*
Art Editor: *Renee A. Grevas*
Compositor: *Interactive Composition Corporation*
Typeface: *10/12 Cochin*
Printer: *Quebecor Printing*

**Library of Congress Catalog-in-Publication Data**

Danielson, Eric William, 1940-
Meteorology / Eric W. Danielson, James Levin, Elliot Abrams.
p. cm.
Includes index.
ISBN 0-697-21711-6
1. Meteorology. I. Levin, James, 1946- . II. Abrams, Elliot, 1947- . III. Title
QC861.2.D36 1998
551.5--dc21

97-27826
CIP

www.mhhe.com

# Meteorology

# Dedication

*We dedicate this book to*
*Esther, Alice, Seth and Mary,*
*Andy, Heidi and Sarah,*
*Bonnie, Mike and Randy —*
*for their support, encouragement, and understanding, and*
*for weathering this project with patience and good humor.*

# Brief Contents

# Contents

# *Unit 5*

## *Are There Limits to Atmospheric Predictability?*

### Chapter 11

### *Circulations Large and Small*



### Chapter 12

### *Thunderstorms and Tornadoes*



### Chapter 13

### *Hurricanes*



### Chapter 14

### *How Stable Is Earth's Climate?*

# Preface

Meteorology! Meteorology is the branch of science devoted to the study of the atmosphere.

It is an exciting time to study meteorology. We are witnessing great technical advances from the installation of a national Doppler radar network to the nearly instant availability of weather data and charts via the Internet. We are seeing atmospheric issues in the headlines, from global warming and ozone depletion to devastating storms in places as far apart as Florida and the Bay of Bengal. And we are benefiting from theoretical advances that are leading to improvements in forecasting events as different as tornadoes and global temperature patterns. We welcome you to this exciting arena of study and hope you find it as fascinating as we do.

## Science Content, Science Process

Textbooks are conveyors of information. Whether the subject is science, history, or business, the textbook must convey basic facts and concepts of its discipline. We have strived to present the facts and concepts of atmospheric science in a way that is clear to the nonspecialist. In places where we felt that clear explanation rested on an understanding of more basic scientific concepts, we provided that background as well.

But a textbook that limits itself to presenting facts presents only one dimension of the subject it seeks to portray: the result bears as little resemblance to its subject as a stuffed animal does to its living counterpart.

From the first days of this project, we have sought to produce a book that more faithfully reflects the many facets of its subject. To us, meteorology is far more than a collection of facts and concepts, a collection of answers. Meteorology is a human activity; it is a quest for understanding pursued by thousands of women and men every day. As we attempt to puzzle out the unknown, we do much more than collect and analyze the data, which is the stereotypical image of the scientist at work. In addition to analyzing data, atmospheric scientists are engaged in a wide variety of activities such as imagining explanations of puzzling phenomena, creating experiments and inventing instruments to test their proposed explanations, brainstorming with colleagues to get fresh ideas, following hunches, writing about their work, presenting it to others—in short, employing a whole range of skills and procedures. This collection of activities, when employed according to certain rules commonly called "the scientific method," has been spectacularly successful in advancing our understanding of the natural world. We feel that by including scientific process in this textbook, we can convey to our readers a true sense of what it means to know something in science; the roles played by facts, laws, models, and theories; the limitations of science; and the sense in which scientific knowledge is tentative. We also feel that this is the only way to show the excitement of following a career in science.

To maintain a focus on process, we have grouped the 14 chapters into five units on the basis of "unit questions," each one an issue of interest to the meteorological community and the public. The chapters within one unit present relevant scientific concepts; the skills, strategies, and instruments one needs to work on the problem; its scientific and cultural context; and a review of what we know about its solution.

This attention to scientific process gives the book a somewhat unconventional feel. For example, we devote more space to topics such as causality, predictability, and the uses and limitations of scientific models than is customary in most introductory meteorology texts. We also include the names of a variety of present-day atmospheric scientists, some famous, others not, to emphasize the human face of the scientific enterprise.

## The AccuWeather Connection

One of the unique features of this textbook is its relation to AccuWeather, Inc. AccuWeather has supported the project in various ways, including granting the authors access to its AccuData database during development and rendering the artwork. Furthermore, AccuWeather has agreed to provide adopters of the textbook with 12 free hours of access time to the complete AccuData database. In this way, AccuWeather has enhanced considerably the opportunity for students to become actively involved in science process and to apply concepts studied in the text to real-time weather conditions. At the end of each chapter, we list resources available at AccuWeather, as well as elsewhere on the World Wide Web, related to that chapter's reading. Additional links and resource suggestions are available online at our book's Internet home page at www.mhhe.com/sciencemath/geology/danielson. We strongly encourage you to take advantage of these opportunities.

## Pedagogy

It is well-known that active participation promotes effective learning. Although options for participation while reading are limited, they do exist. We make a point of asking the reader questions, many questions. Frequently, we direct the reader to key data in tables and figures, ask at various points to consider how the discussion has developed and where it is headed, and in general try to provoke the reader into conversation with us. You will notice that our writing style is somewhat more conversational and less didactic than in other texts.

We believe in some aspects of the currently popular constructivist learning models. For example, we accept the notion that teaching is not akin to pouring information into the container of the student's mind but is instead a process of assisting every learner to mentally construct his or her own understanding based on firsthand observations and experiences, as far as possible. To support this construction process, we have adopted several approaches and techniques. For example, we have made a point of developing concepts from the specific to the general, thus equipping the reader with factual building blocks that can then be arranged to form relational structures. We offer help in structure building by presenting "concept maps" at frequent points. A concept map charts the relationships between different pieces of information and can assist learners in constructing such structures in their own minds. In some places, we challenge readers to construct their own concept maps, thereby developing a deeper understanding of the material presented.

Structure building is a challenging task. It requires concentration and necessitates carefully following the discussion. It is primarily for this reason that our text does not contain "boxes," "focus items," or other special-interest features scattered through each chapter. Although these devices can add interest, we feel they are a significant distraction to the reader.

The more connections the student can make between new mental structures and previous experience, the more robust the new structures will be. Thus, in presenting new ideas, we offer students many links to related everyday experiences. We also return frequently to concepts previously presented, revisiting them in new contexts.

(We do not subscribe, incidentally, to the more radical constructivist idea that all knowledge is "socially constructed" by groups of people. We feel that the history of science is replete with examples of individuals who made great scientific advances on their own, advances eventually adopted by the scientific community on the basis of their superiority to other explanations of nature.)

## To the Instructor

The book is intended for use in a one-semester introductory meteorology course. We have written it with a variety of readers in mind: nonscience majors seeking to fulfill a science course requirement, students majoring in subjects such as environmental science or geography for whom a meteorology course is required or recommended, and those who just want to indulge their curiosity or fascination with weather and the atmosphere.

The order of presentation of material is fairly conventional. After a one-chapter introduction to daily weather and synoptic weather systems, we move from energy concepts to humidity, clouds, and precipitation, then to dynamic meteorology and weather forecasting, and conclude with a chapter on climate. The early chapters contain many basic science concepts used in later chapters; therefore, we recommend moving through the text in the sequence presented.

Authors and instructors alike struggle with the depth-breadth dilemma in introductory science courses. Given limited time and an ever-expanding body of material to cover, is it better to "cover the field" to limited depth or to go deep in places and leave other topics untouched? We chose to narrow the scope somewhat by scaling back or eliminating discussion of some topics, such as the upper atmosphere (except as regards stratospheric ozone), the sun as an energy source, and atmospheric optics. While we regret the omission of these topics, doing so has given us room to explore subjects such as global warming, atmospheric thermodynamics, and weather forecasting to greater depths. Instructors not wishing to devote so much time to these latter topics can skip later sections of the relevant chapters without loss of continuity.

A word about the "unit question": Its purpose is to provide context and direction through an extended segment of the book, as well as to sustain an atmosphere of inquiry. The question itself is not intended to be the single "goal" for a given unit or even the most important question that might be asked. In some places, the unit question is very much in the foreground; elsewhere, it is much less visible. We feel there is no need to impose its presence on every lesson; instead, we offer it for use when and where it fits the purposes of your course.

## To the Student

### Meteorology and Math

The first question on many meteorology students' minds is not about tornadoes; it is not about hurricanes; it is not about global warming. Most often, question number 1 is, "Is there a lot of math in meteorology?"

Our response is, "Luckily, yes." The fact that many meteorological phenomena can be expressed in numerical terms has much to do with our success in understanding the atmosphere as well as we do. On the other hand, we recognize that mathematics is a language that few nonscience students speak fluently. As a result, we have assumed little mathematical prowess on the part of the reader and have not built explanations on mathematical reasoning where it was possible to do otherwise. However, you will be employing basic mathematics to make numerical computations throughout the course. And in a few places, the use of symbolic mathematical reasoning cannot be avoided, so we plunge in. Do not fear, however: you can do it, and you will. In many years of teaching this course, we have seen that students do not find the math level a serious problem.

### Study Aids

At the beginning of each chapter, you will find a list of chapter goals. We suggest you refer to this list periodically to check on your progress through the chapter.

Key words are identified in bold print. These are the basic terms you need to speak the language of the atmospheric scientist. Key words are listed at the end of each chapter and defined in the glossary.

At several points in each chapter you will encounter Checkpoints. Checkpoints provide a brief summary of the material just presented, along with a few practice questions with which you can gauge your understanding.

Many Checkpoints include diagrams known as concept maps. A concept map is a chart consisting of boxes connected by arrows and illustrating relations between concepts. (For an example, turn to the concept map on page 15.) Creating concept maps of your own is an excellent way of reviewing and memorizing material, as well as organizing it in your own mind.

Each chapter ends with a summary of the entire chapter, a list of the chapter's key words, Exercises (shorter, easier), Problems (longer, harder), Explorations (suggestions for research projects), and Resource Links (connections to additional information available both in printed form and electronically, such as on the World Wide Web).

At the end of the text, a series of appendices provides tables, charts, and other reference information useful to the student of meteorology.

### Suggestions for Reading This Book

Here are a few suggestions for reading your text:

1. Do your reading in small- to medium-sized doses. In general, the material from one Checkpoint to the next within a chapter is a reasonable amount for one sitting. Then take a short break or change subjects for a while before continuing.
2. Begin by previewing the section to get an idea of the territory ahead. Read only the headings and subheadings, look at the figures, and read their captions. Read the section review and the questions at the Checkpoint at section end.
3. Next, write down a few questions you expect to have answered or hope to find answers to during your reading. Your questions might be ones that occurred to you while you previewed the section, or you might create them from the section's headings and subheadings. For example, from the subheading "Land and Sea Breeze Circulations," your question might be simply, "What are land and sea breeze circulations?" or "How do they form?" As you read, look for answers to your questions.
4. As you read, strive to be actively engaged with the section by writing marginal notes, underlining, searching for answers to questions you asked, writing notes on separate paper, explaining concepts to a classmate, drawing diagrams, constructing outlines or concept maps, and so forth. The very process of performing these steps improves considerably your retention of the material. On the other hand, coloring whole sections of text with highlighting pens is of dubious value; it requires too little mental input to be a useful learning activity.
5. Use the chapter objectives, Checkpoint materials, and the chapter-end summary and questions as checks on your learning.
6. Enjoy yourself! To us, your authors, the atmosphere is a source of endless interest and inspiration. We hope to convey to you some of the excitement we feel in watching the wind come up, studying a fresh satellite image, or seeing a thunderstorm blossom on a summer afternoon. Some people claim that scientific understanding removes the mystery and excitement from natural phenomena. We feel just the opposite—that with some scientific understanding of a natural event, one can appreciate it far more deeply. We hope that our book heightens your appreciation of our atmosphere and the events that occur within it.

## Acknowledgments

We would like to express our thanks to the following reviewers for their thoughtful and thorough responses to various drafts:

David J. Berner *Normandale Community College*
Mark S. Binkley *Mississippi State University*
Dr. Michael Fitzwater *California State University-Sacramento*
Drannan Hamby *Linfield College*
C. Woodbridge Hickcox *Emory University*
Rudi Kiefer *UNC-Wilmington*
Daniel J. Leathers *University of Delaware*
Chris Mantzios *Doane College*
Jonathan Merritt *Penn State University*
Dr. David W. Miller *Wesleyan MetroPark Nature Center*
Dr. Rich Miller *Milwaukee Area Technical College*
Joseph Pifer *Bloomsburg University*
James L. Robinson *Aims College*
Peter J. Robinson *University of North Carolina*
Roger Sandness *South Dakota State University*
Thomas W. Schmidlin *Kent State University*
Stephen J. Stadler *Oklahoma State University*
Jerry L. Steffens *San Jose State University*
Harold E. Taylor *The Richard Stockton College of New Jersey*

Joel Myers, president of AccuWeather, Inc., offered many useful comments on various drafts. All of these reviewers have improved the text immeasurably. Alice and David Veazey and Seth Danielson of the University of Alaska at Fairbanks have participated in many helpful discussions, through all stages of the book's development, concerning its meteorology content and its pedagogical approach. Louise Loomis of Hartford College for Women assisted with early conceptualization of the pedagogy.

We would also like to thank our colleagues at Hartford College for Women, Pennsylvania State University and AccuWeather for their unflagging support; the Everett MacKinnon and Bertram Briand families of Cape Breton Island for logistical support; Donald P. LaSalle and Thomas Alena of the Talcott Mountain Science Center for providing access to weather data; and Brett Colton of AccuWeather for graphics support.

Our families, to whom this book is dedicated, helped in countless ways, from sketching figures and proofreading to shouldering additional household duties so that we could write.

Finally, we acknowledge with pleasure and gratitude all members of our WCB/McGraw-Hill book team, particularly Lynne Meyers, Daryl Bruflodt, Janice Roerig-Blong, Gloria Schiesl, Lori Hancock, and Renee Grevas, for their skill and care in turning our writing into a book.

# Exploring the Atmosphere: What Lies Ahead?

Science is a voyage of discovery. For some scientists, such as Fridtjof Nansen, shown here, a voyage is literally involved; for others, the voyage is only a metaphoric one. In either case, however, it is useful at the outset for the explorer to have a rough sense of the territory ahead.

The first Unit (Chapters 1 and 2) is shaped around the general question "What lies ahead?" for the student beginning a course in meteorology. The unit deals with such questions as: "What is the atmosphere made of? How far upward does it extend? What are the patterns into which it is organized?" With answers to these questions, you will be oriented and prepared for the work ahead.

Another question, a question of a different kind, provides a unifying theme for this and all later units. It is, "How does one go about finding answers to questions like those posed above?" Interestingly, the answer in part is, "By asking more questions." This unit introduces you to the art of asking and answering scientific questions, to the processes of observing, hypothesizing, testing, and publishing—the process known as science.

# Chapter 1

# Introduction to Meteorological Inquiry

## A First Look at Weather Systems

# Chapter Goals and Overview

Welcome to **meteorology,** the study of the earth's atmosphere!

The **atmosphere,** viewed from space in the photograph on the facing page, is the thin layer of gases that surrounds our planet and is known informally as air. The earth's blue skin, the atmosphere sustains all life on earth, protects us from meteorites from space and cancer-causing radiation from our sun, and serves as the medium in which the variety of events called weather take place.

Atmospheric scientists devote their professional energies to asking questions about the atmosphere's structure and behavior and to seeking answers to those questions. Some questions are obvious and persistent, such as "What will tomorrow's weather be?" Others are less obvious: "How do snowflakes form? How can you tell if a cloud-seeding method has any effect? What will the earth's climate be like 100 years from now?" Still others are more dramatic, because they deal with events that control the destinies of individuals and whole societies: "What is causing the thinning of the ozone layer, which threatens life on this planet? How can we predict turbulence of the sort that downs aircraft on take-off or landing, or blizzards that strand people for days, or cyclones that drown hundreds of thousands of people in south Asia, or storms like hurricane Andrew in 1992, which caused $30 billion in losses?" To study meteorology is to learn how to pose such questions and to search for their answers—in short, to learn the methods of scientific inquiry. Even if you do not intend at this point to become a scientist, understanding the process of scientific inquiry is important. Today's citizenry faces a steady stream of issues that have scientific dimensions. Understanding those dimensions means understanding scientific process as well as scientific facts.

We want to begin this book with a topic with which you are familiar; therefore, we have chosen as our first problem one dealing with simultaneous reports of radically different weather conditions at two locations in the Midwest. Expressed informally, the question is, "How can the weather simultaneously be autumn-like in Chicago, Illinois, and summer-like just 500 kilometers (300 miles) away in Evansville, Indiana?"

## Chapter Goals

In the process of investigating this question, you will learn how to:

make a weather observation;
interpret weather observations on a weather map;
recognize patterns in the apparently random jumble of weather map data; and
employ a model to explain a variety of daily weather events.

To master these skills, you will need to absorb a considerable amount of factual information. You will learn:

names and meanings of the variables that comprise a meteorological observation;
names and uses of a number of meteorological instruments;
weather map features such as highs, lows, and fronts; and
a model that relates map features to weather conditions you observe.

You will also become acquainted with the concept map, a powerful tool for organizing and remembering course content.

## A Meteorological Puzzle

Figure 1.1 tabulates the weather occurring at Chicago and Evansville at the same time on a summer afternoon. Spend a moment examining these two reports. First, imagine yourself in Chicago at the time of the observation. You notice the sky is covered with low, dark-gray clouds, you feel a brisk wind coming off Lake Michigan, and the temperature of 61°F makes it feel more like October than July. Now, picture yourself in Evansville: here, the temperature is 91°F, the sun is blazing down through nearly clear skies, and you begin to wilt in the muggy, oppressive air of a hot July afternoon.

From your own experience, you recognize that the situation presented in Figure 1.1 is unusual, and it provokes some obvious questions. For example, "Why is it sunny in Evansville, while Chicago's sky is completely overcast? Why are the winds at the two sites coming from opposite directions? Why do the temperatures differ by 30 degrees?" And, more generally, "Why would two places so close to each other be experiencing weather that's so different?" A more fundamental question is, "How do you go about answering the previous questions?" One focus of this chapter is to seek answers to these questions.

You may be wondering how you can investigate atmospheric problems "without knowing anything yet." Do not be concerned. You will acquire the relevant concepts and facts in context, as you need them. Furthermore, we believe that you know a great deal about the atmosphere's behavior already. Whether or not you can verbalize your knowledge, you have developed an extensive, partly unconscious sense of the atmosphere's behavior from your own experience as an inhabitant and observer of the earth's atmosphere. You have already observed many of the atmospheric phenomena you will be studying in this course, from cloud formation on a summer afternoon to—possibly—Atlantic hurricanes, drought conditions in the Southwest, or Midwest tornadoes; you may have experienced meteorological events that we have not. In addition, you have probably viewed many weather presentations on television. Thus, in a sense, you have 18 or more years of experience in this subject. We respect that and encourage you to put your own meteorological knowledge and experience to work for you throughout this course.

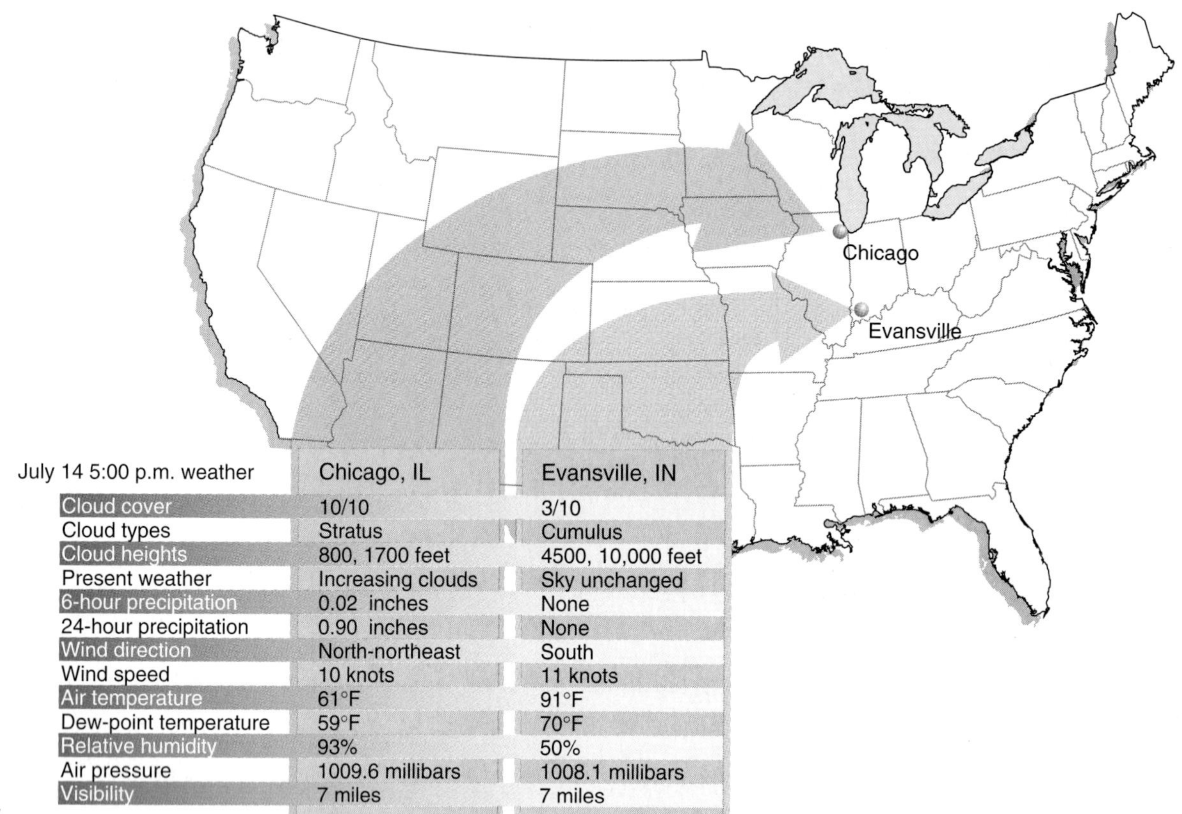


| July 14 5:00 p.m. weather | Chicago, IL | Evansville, IN |
|---|---|---|
| Cloud cover | 10/10 | 3/10 |
| Cloud types | Stratus | Cumulus |
| Cloud heights | 800, 1700 feet | 4500, 10,000 feet |
| Present weather | Increasing clouds | Sky unchanged |
| 6-hour precipitation | 0.02 inches | None |
| 24-hour precipitation | 0.90 inches | None |
| Wind direction | North-northeast | South |
| Wind speed | 10 knots | 11 knots |
| Air temperature | 61°F | 91°F |
| Dew-point temperature | 59°F | 70°F |
| Relative humidity | 93% | 50% |
| Air pressure | 1009.6 millibars | 1008.1 millibars |
| Visibility | 7 miles | 7 miles |

Figure 1.1 Weather conditions prevailing in Chicago, Illinois, and Evansville, Indiana, at 5:00 P.M. on July 14,1992. Note the remarkable difference in conditions, despite the two locations' proximity.

## Searching for an Explanation

How do you find the answer to a scientific question? Unfortunately, there is no single scientific method that investigators follow each day in their work. However, one strategy that is basic to scientific inquiry is to hypothesize an answer to your question and then attempt to prove its validity beyond a reasonable doubt.

But where do you go to find proof? Scientific proof comes from observing nature, measuring and recording relevant natural phenomena (often, but not always, in experimental tests), and then comparing your data with the assertions of the hypothesis. If your hypothesis fits the facts, you may accept it; if not, you must reject it. Thus nature—and not existing theory or the opinions of other scientists—is the ultimate authority, the supreme court of scientific proof.

If your hypothesis survives extensive testing and you accept it as true, you have found one answer to the question. If not, then you have eliminated one possible explanation, and you repeat the process, beginning with a new hypothesis, new testing, and so on, until eventually the hypothesis you make holds up to scrutiny. Because one's first guess rarely is the correct one, this method often becomes one of systematically eliminating wrong answers and thereby narrowing the range of possible explanations until at last you make the correct hypothesis.

How do you know what hypothesis to make first? There are no rules about this. As in guessing the winner of the Kentucky Derby or the Super Bowl, it helps to have information, experience, and luck. (But it is said that "fortune favors the prepared mind." Many an apparently lucky guess is actually the product of complex and subtle reasoning combined with a thorough understanding of the question.)

Experience has shown that it is best to begin with simple explanations of limited range. Thus, one might first want to verify that the observations are valid, that is, to eliminate the possibility that the weather did not really differ as much as the data in Figure 1.1 suggest. If the data are flawed, then we are wasting our time seeking an explanation for an event that never occurred. A reasonable starting hypothesis, therefore, might be, "There was no such glaring contrast between Chicago's and Evansville's weather on July 14; the apparent differences resulted from differences in observing instruments or technique."

To test the hypothesis that the data in Figure 1.1 somehow are flawed, we need to know more about how meteorological observations are made and what they mean. We turn to these topics now.

# Weather Observations and Instruments

In meteorology, the word *observation* often refers to a rather specific set of measurements that together define the state of the atmosphere at some time and place. The reports in Figure 1.1 are two such standard surface observations, measurements of conditions within a meter or two of the earth's surface. You are probably familiar with many of the components of the standard meteorological observation from watching weather forecasts on television, hearing them on radio, or reading them in the newspaper.

You probably also understand that the modern meteorologist uses a variety of sophisticated instruments to observe the weather. While instruments are important, the observer can learn a great deal about the atmosphere's state by simply going out and looking, without any instruments except the human body. Notice in the following discussion how much of the observation you can complete without instruments. We encourage you to get in the habit of making such observations yourself. Notice the clouds, the temperature, and the wind whenever you are outside. The air is as much a part of your surroundings as the clothes you wear and the ground you walk on; it is closer, in fact, than either of them.

We will now examine each element of the standard meteorological observation to explore more completely each element's meaning and to evaluate the hypothesis that the data in Figure 1.1 are erroneous. If the observations were consistent except for a few of the more subjectively determined components of the observation (such as cloud type) or if the differences were momentary, then the impulse to accept the hypothesis (i.e., that one or the other observation is in error) would be strengthened. On the other hand, if several or many of the data differ dramatically and remain different over several hours or more, we would suspect that the differences were real and would therefore reject the hypothesis.

## Observation Times

A few minutes before each hour, observers at thousands of weather stations throughout the world make, record, and disseminate hourly weather observations. (For a refresher on worldwide times and on the geographical coordinate system based on latitude and longitude, see Appendices G and F.) As meteorological observations are exchanged among stations in many different time zones, times reported in forms such as "5:00 P.M." become problematical: 5:00 P.M. in Chicago is 6:00 P.M. in New York and 3:00 P.M. in Los Angeles. To avoid confusion, all observation times refer to a single reference time, which is that kept at the Greenwich, England, Observatory and is abbreviated UTC (from "coordinated universal time") or simply Z. In July, local time in Chicago and Evansville is five hours earlier than UTC. Therefore the 5:00 P.M. Central time observations are 10:00 P.M. UTC. Actually, UTC times are given not in terms of A.M. and P.M. but in terms of a 24-hour clock; thus, 10:00 P.M. becomes 22:00 UTC, or simply 22 Z.

## Observing Clouds and Visibility

### *Cloud Cover*

The observers at Chicago and Evansville made the cloud cover observation in the same way you would: by going outside to a spot that affords an unobstructed view of the whole sky and then estimating the fraction of the sky that is covered by clouds. Standard observing procedure is to estimate cloud coverage to the nearest tenth.

The Chicago observer reported sky cover of 10/10. As you would surmise, 10/10 means that the sky was completely cloudy, the meteorological term for which is **overcast.** At Evansville, the sky cover was 3/10, the general term for which is **scattered.** Scattered clouds correspond to a sky cover between 1/10 and 5/10. Other terms are **broken,** which means coverage of 6/10 to 9/10, and **clear,** which means 0/10 of cloud coverage. (Note that the term **clear** refers only to cloud cover, not to the clarity of the air. You will find more detailed cloud information in Chapter 6.) If the sky cannot be seen because of haze, fog, or precipitation or some other obstruction, the observer indicates that the sky is **obscured.**

Is it possible that the cloud cover observations are in error? It is difficult to imagine an experienced observer confusing scattered with overcast conditions. Perhaps one observer accidentally recorded the wrong value for cloud cover for that one hour? Table 1.1 shows each station's July 14 observations from 21 Z to 23 Z (4:00 to 6:00 P.M.). The observed differences in cloud cover persist throughout the period, making a careless error improbable and increasing the likelihood that the differences are real and our hypothesis is false.

### *Cloud Types*

For the moment, we will work with just a few major cloud types, as illustrated in Figure 1.2. Cloud genera are not distinct entities

*Table 1.1* **Chicago and Evansville Weather from 4:00–6:00 P.M. (21–23 Z), July 14, 1992**

| Location | Time | Cloud Cover | Visibility | Present Weather | Wind Dir./Sp. | Air Temp. | Dew Point | Air Pressure |
|---|---|---|---|---|---|---|---|---|
| Chicago | 4:00 | Overcast | 7 mi. | Inc'g clouds | NE/11 | 62°F | 60°F | 1008.9 mb |
| | 5:00 | Overcast | 7 mi. | Inc'g clouds | NNE/10 | 61°F | 59°F | 1009.6 mb |
| | 6:00 | Overcast | 7 mi. | Inc'g clouds | N/13 | 61°F | 59°F | 1009.6 mb |
| Evansville | 4:00 | Scattered | 8 mi. | Sky unchanged | SW/18 | 90°F | 70°F | 1008.1 mb |
| | 5:00 | Scattered | 7 mi. | Sky unchanged | S/11 | 91°F | 70°F | 1008.1 mb |
| | 6:00 | Scattered | 7 mi. | Sky unchanged | SW/8 | 90°F | 70°F | 1007.4 mb |

A Stratus: Flat, sheetlike

B Cumulus: Heaped up

C Cirrus: Wispy

D *Nimbo-*, or *-Nimbus:* Prefix or suffix meaning "Producing precipitation."

Figure 1.2 Basic cloud forms. Figure 6.1 on page 156 provides a more extensive atlas of cloud types. Note: Nimbus is not a separate type but is used in combination with cumulus and stratus (i.e., cumulonimbus, nimbostratus).

like genera and species of living things. Instead, a variety of hybrid cloud forms, such as stratocumulus and cirrocumulus, often occur.

The reported height of a cloud is that of the cloud's base above the ground (not sea level). An observer measures cloud heights in any of a number of ways, shown in Figure 1.3, including simply estimating the height if other means are not available.

Note that distinctly different cloud types prevail at the two locations and that the Chicago observer measured the height of the overcast stratus cloud deck to be considerably lower than the scattered cumuli reported at Evansville. The fact that both cloud height and cloud types differ from Chicago to Evansville casts doubt on the hypothesis that the differences are due to a careless observational error.

### *Visibility*

The **prevailing visibility** is a measure of the atmosphere's horizontal transparency. Specifically, it is the farthest distance you can see along at least half of the horizon. Figure 1.4 shows how to determine the prevailing visibility when it varies in different directions. For many people, the difficult part of making a visibility measurement is finding a location affording an unobstructed view to the horizon. (This is one reason that airports make excellent observation sites. To ensure safety, obstructions such as trees and tall buildings are usually not found immediately adjacent to airports. The same absence of obstructions that permits safe takeoffs and landings affords weather observers the best opportunity to see the entire sky.) Once you have managed that, the rest is easy. Simply scan the horizon, noting the distance to the farthest objects you can see over at least half the horizon.

If the visibility is less than 7 miles, the observer must report the restriction (snow, rain, haze, etc.) as some form of "present weather," which is explained in the next section. The reported visibility of 7 miles at both Chicago and Evansville was considered unrestricted. Thus, the visibility values show no significant difference between the two weather reports.

Visibility is a crucially important variable for aircraft taking off and landing. At many airport weather stations, instruments

A Aircraft

B Balloon

C Ceilometer

D Estimation

Figure 1.3 Four methods of determining the heights of cloud bases above the ground. (*A*) Airplane pilots often report heights. (*B*) Helium-filled balloons rise at nearly constant rates. By timing the balloon's flight until entering the cloud, the observer can determine cloud base height. (*C*) The ceilometer reflects light off the cloud base. The observer can find cloud height by measuring time elapsed from transmission to reception.

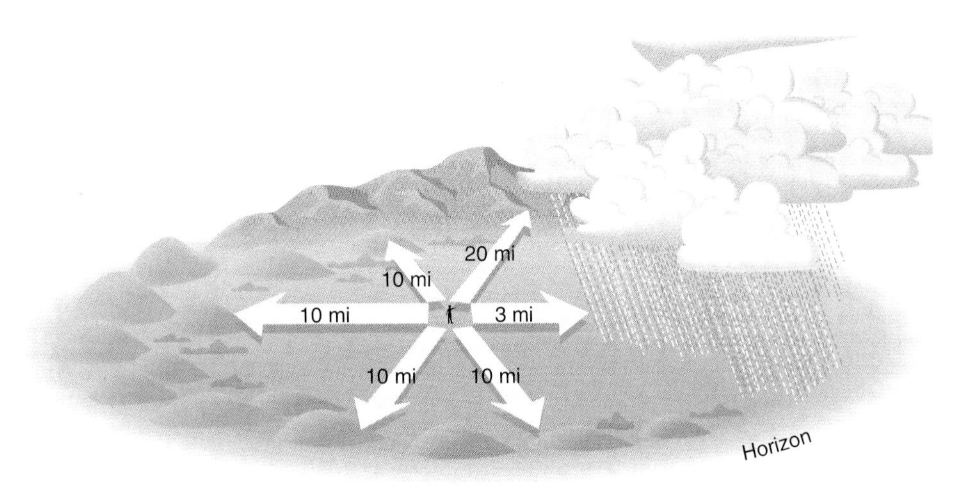


Figure 1.4 Observing visibility. The observer would report the visibility to be 10 miles, because that is the greatest distance she can see over at least half of the horizon.

called **transmissometers** measure visibility along runways. A transmissometer fundamentally is a combination of light source and receiver. The receiver measures the fraction of light that arrives from the source compared to the intensity emitted and translates that fraction into visibility units, in miles or kilometers.

## Observing Present Weather and Precipitation

### *Present Weather*

The Chicago observer reported "increasing or thickening clouds" for **present weather.** Present weather is a description of (generally) visible events occurring in the atmosphere; loosely, present weather describes "what it is doing outside." The observer chooses a descriptor from a list of choices, shown in Appendix H. This process removes much of the subjectivity in reporting present weather. Present weather encompasses phenomena such as rain, thunder, and haze but not descriptors such as "hot," "windy," and so on, because the latter are covered more precisely elsewhere in the weather observation.

The "sky unchanged" report for Evansville means that no precipitation was falling, no phenomena such as lightning, fog, or haze were observed, and the extent of cloud cover was much the way it had been over the past three hours.

Although the present weather differed from Chicago to Evansville, the difference is not a startling one, and it neither supports nor refutes the hypothesis that the observations are flawed.

### *Quantity of Precipitation*

Precipitation amount is the depth of water substance that has fallen over a given period; it is measured with a rain gauge or a snow gauge. The six- and 24-hour totals listed for Chicago and Evansville cover the periods ending at 00 Z on July 15 (7:00 P.M. on July 14).

In theory, a tin can is a suitable rain gauge. In practice, however, it is useful to funnel the collected water into a narrow gauge, because most precipitation amounts are small. (It is difficult to measure three-one hundredths of an inch of rain in the bottom of a container. Therefore, rain from a wider area is funneled into the container, making it easier to measure; then the measuring gauge is adjusted to compensate for the extra rain collected by the funnel.) For snow, both the depth on the ground and the melted equivalent are recorded. Figure 1.5A presents some typical instruments used to measure liquid precipitation. Location is a critical issue for a precipitation gauge. A precipitation gauge installed too close to trees or buildings gives unrepresentative readings.

## Observing Wind

### *Wind Speed*

Wind is nothing more than air in motion. Although you cannot see the wind directly, it is easy enough to observe its impact on objects it encounters. Then you can use the adaptation of the **Beaufort wind scale** in Table 1.2 to translate your observation into an estimate of wind speed. With practice, a person can become an accurate estimator of wind speed using the Beaufort scale. Beaufort (pronounced "*bow*-fort," as in *bow tie*) was a 19th-century British naval officer who developed a method of estimating wind speed according to the appearance of the ocean surface (wave heights, etc.). In Beaufort's time, shipping ran on wind power, so maritime wind data were of crucial importance. In Table 1.2, Beaufort's original scale has been expanded to include effects observed on land.

Wind speed is measured and reported in units called **knots.** A knot is defined as 1 nautical mile per hour. A nautical mile is about 15% (800 feet) longer than the 5280-foot statute mile; therefore, the 10-knot wind speed reported in Chicago is equivalent to 11.5 mph. As with visibility and present weather, wind speed data at the two stations are quite similar and are consistent over a three-hour period; thus, they offer no evidence in support of the hypothesis that the data are erroneous.

### *Wind Direction*

You can observe the wind direction simply by standing in an open area and turning slowly until you feel the wind blowing directly on your face. The direction you are facing is the wind direction. (A moist finger held in the wind will do also.) Note that the direction is reported as that from which the wind is blowing. Thus, a report of a north wind means it is coming from the north and moving toward the south. Terms like *northerly* and *easterly* also mean that the wind is coming from those directions. (Ocean currents, however, are described in the opposite sense. For example, a current running from south to north is called a north current.)

The **anemometer** and **wind vane** measure wind speed and direction, respectively. Sometimes the two instruments are combined into a single device. Figures 1.5B and C illustrate models typical of those used at weather observing stations.

Note that wind speed is one of the few weather elements for which the Chicago and Evansville observers recorded nearly identical values (10 and 11 knots, respectively). The wind directions at the two locations are diametrically opposed, however, and remain that way from at least 21 Z to 23 Z. The consistency of each set of wind direction observations increases the likelihood of its accuracy.

## Observing Temperature and Humidity

An observant, trained person could proceed this far through a weather observation with no measuring equipment beyond a tin can, funnel, and ruler for measuring rainfall. From this point on, however, special instruments are required: the human body simply is not a very sensitive instrument for making unassisted estimates of the air's temperature, its humidity, and especially its pressure. We need to extend our senses through the use of instruments, to make reliable observations of these elements.

### *Temperature*

Air temperature is most commonly measured with a thermometer. As in the case of precipitation measurement, the thermometer's location is more important than its type: thermometers exposed to

A

B

C

D

Figure 1.5 (*A*) Precipitation measuring devices. The funnel magnifies the depth of precipitation collected; to compensate, the gauge's scale must be expanded by the same factor. (*B*) and (*C*) Typical wind measuring instruments. Note that the tail on the wind vane keeps the pointer heading into the wind; hence, wind direction refers to the direction *from which* the wind is blowing. (*D*) The standard instrument shelter supports thermometers and (often) humidity gauges a uniform distance above the ground and protects them from direct sunlight and precipitation.

*Table 1.2* **Beaufort Wind Scale**

| Beaufort number | General description | Land and sea observations for estimating wind speeds | Wind speed knots* |
|---|---|---|---|
| 0 | Calm | Smoke rises vertically. Sea like mirror. | < 1 |
| 1 | Light air | Smoke, but not wind vane, shows direction of wind. Slight ripples at sea. | 1–3 |
| 2 | Light breeze | Wind felt on face, leaves rustle, wind vanes move. Small, short wavelets. | 3–6 |
| 3 | Gentle breeze | Leaves and small twigs moving constantly, small flags extended. Large wavelets, scattered whitecaps. | 7–10 |
| 4 | Moderate breeze | Dust and loose paper raised, small branches moved. Small waves, frequent whitecaps. | 11–15 |
| 5 | Fresh breeze | Small leafy trees swayed. Moderate waves. | 16–20 |
| 6 | Strong breeze | Large branches in motion, whistling heard in utility wires. Large waves, some spray. | 21–26 |
| 7 | Near gale | Whole trees in motion. White foam from breaking waves. | 27–32 |
| 8 | Gale | Twigs break off trees. Moderately high waves of great length. | 33–40 |
| 9 | Strong gale | Slight structural damage occurs. Crests of waves begin to roll over. Spray may impede visibility. | 41–47 |
| 10 | Storm | Trees uprooted, considerable structural damage. Sea white with foam, heavy tumbling of sea. | 48–55 |
| 11 | Violent storm | Very rare; widespread damage. Unusually high waves. | 56–64 |
| 12 | Hurricane | Very rare; much foam and spray greatly reduce visibility. | 65 and over |

Source: *Smithsonian Meteorological Tables*, 1966. Washington, D.C.: Smithsonian Institution.

*1 knot = 1.15 mph

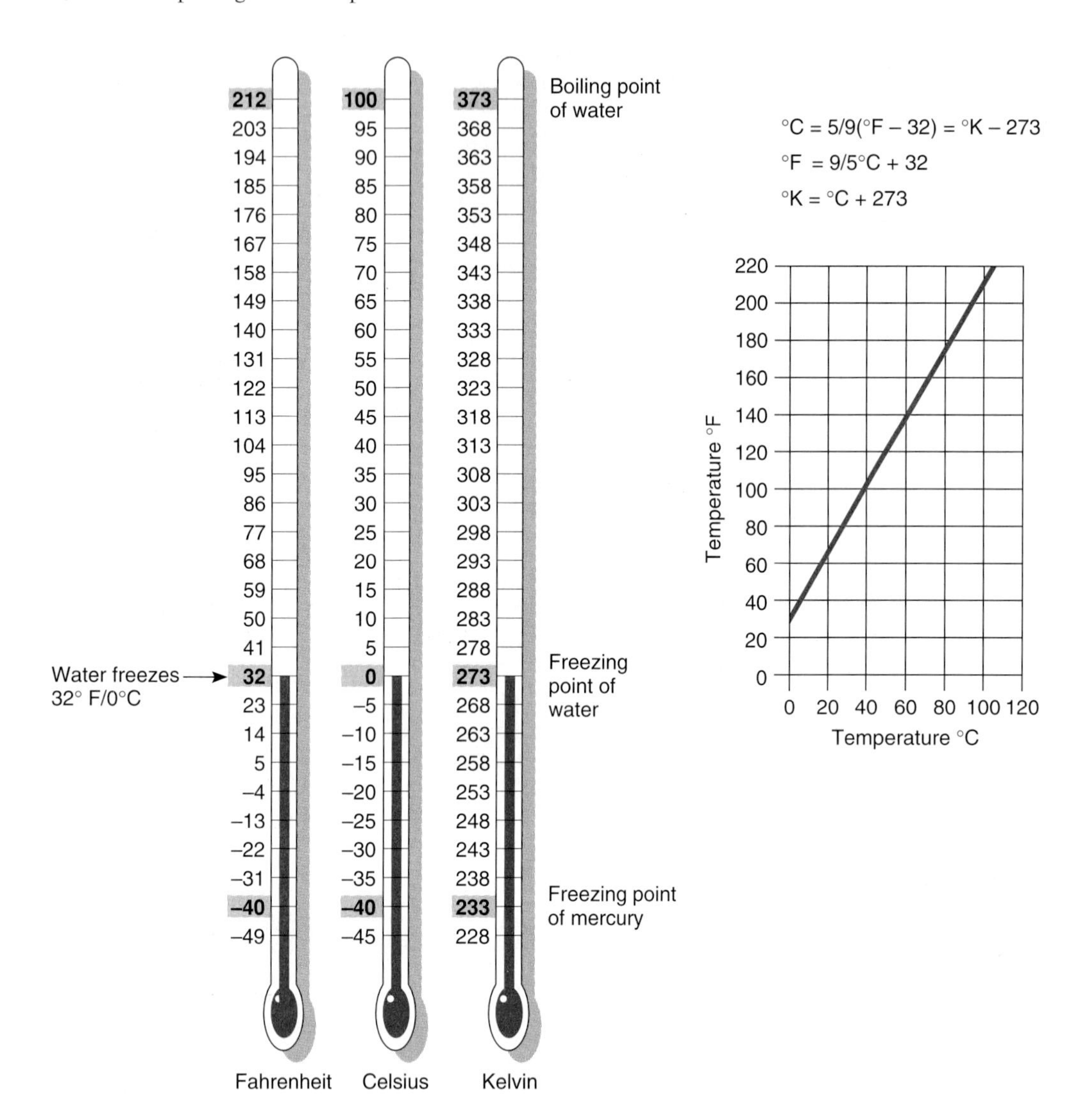


Figure 1.6 Fahrenheit (F), Celsius (C) and Kelvin (K) temperature scales.

direct sunlight or placed on the ground on hot blacktop, for example, can register extraordinary and meaningless temperatures as far as the atmosphere is concerned. Therefore, the thermometer is installed in a shelter such as the one pictured in Figure 1.5C, 1.5 meters (5 feet) above the ground. Three different thermometer scales are used in meteorology: **Fahrenheit, Celsius,** and **Kelvin.** For the moment we need concern ourselves with only the Fahrenheit and Celsius scales, which are shown in Figure 1.6.

Considering again the disparity between the Chicago and Evansville weather conditions, it is difficult to imagine that experienced observers misread the thermometer to such a degree. Furthermore, reference back to Table 1.1 shows that the reported differences persisted over several hours. Thus, an errant observer would have had to repeat the error several times in succession, which is highly improbable. At this point, the "observer error" hypothesis seems doubtful.

## *Humidity*

The term **humidity** refers to the atmosphere's water vapor content. Water vapor is water in its gas phase and as such is invisible. In Chapter 5, you will study atmospheric humidity in detail. For now, we will briefly introduce two of the most widely used humidity variables, the dew-point temperature and relative humidity.

The **dew point,** or dew-point temperature, is the temperature at which water vapor begins to condense into dew (or fog) if the air were cooled but otherwise unchanged. To understand the concept of dew-point temperature, imagine the following, illustrated in Figure 1.7: You are in Evansville and, seeking relief from the heat, you reach for a pitcher of water someone has poured and left on a table some time ago. The water has adjusted to the air temperature, which is 91°F, so you add ice, stirring as you do so. The water temperature in the pitcher cools, as does the pitcher and the air in contact with it. You continue adding ice and stirring, and the temperature of ice water, pitcher, and air in contact continues to fall. When the temperature reaches 70°, you notice a film of tiny water droplets beginning to form on the outside surface of the pitcher. You have cooled the air touching the pitcher to its dew point. The procedure just outlined is not just a thought experiment; some dew point instruments employ essentially the same procedure.

Now suppose you perform the same experiment in Chicago. This time, the pitcher sitting out at air temperature initially is at 61°F. You add ice and notice almost immediately, after the temperature has fallen just 2 degrees to 59°F, that dew has formed. Apparently the air in Chicago is much closer to its dew point than in Evansville, where it had to cool 21 degrees before condensation occurred.

There are at least two ways of interpreting these observations. One is by comparing the dew-point temperatures at the two stations: 70°F in Evansville, 59°F in Chicago. As you will see in Chapter 5, a higher dew-point temperature means a greater amount of water vapor present in a given amount of air. Thus, more water vapor is present in the Evansville air than in Chicago's.

From another point of view, however, the Chicago air can be considered more humid than Evansville's. Notice that in Chicago, the difference between temperature and dew point (the "temperature dew-point spread") is only 2°F (61°F 2 59°F), compared to a 21°F spread (91°F 2 70°F) in Evansville. Thus, in Chicago a much smaller temperature drop would be required to initiate the condensation of water vapor as dew or fog than in Evansville. **Relative humidity** is a term that reflects the significance of the temperature-dew point spread. You will learn a more precise definition of relative humidity in Chapter 5. For now, think of relative humidity as an index whose value decreases as the temperature-dew point spread increases. Thus, Chicago, with a 2-degree temperature-dew point spread, has a higher relative humidity (93%) than does Evansville (21-degree spread and relative humidity of 50%).

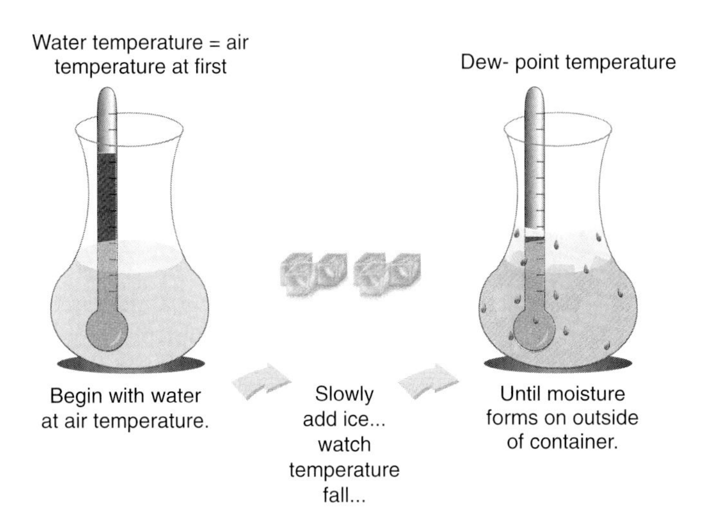


Figure 1.7 Lowering the water temperature will cause moisture to condense on the outside surface of the pitcher. The temperature at which this occurs is called the dew-point temperature. The higher the dew-point temperature, the more moisture in the air to begin with.

## Observing Air Pressure

Our difficulties in sensing values of temperature or humidity unaided by instruments are minor compared to the problem of sensing air pressure. The only time you are liable to be aware of air pressure at all is when it changes in some drastic fashion, such as when you rise 30 floors in an elevator, descend into a coal mine, take off in an airplane, or drive up a mountain. Such motions might cause a pressure change of a few percent, which you might experience as a popping in your ears. On the other hand, inspection of the 22 Z data in Figure 1.1 shows that the air pressure values at Chicago and Evansville differ by just 1 millibar out of roughly 1009, which is less than one-tenth of 1%. No one can sense such small pressure changes.

(At this point you may be asking, "Then why bother with air pressure at all, if we never feel it?" The reason will become clear later in this chapter; keep the question in mind as you continue your reading.)

What is air pressure? As with humidity, we will spend considerable attention on this important subject later in this book. For now, we simply state that **air pressure** is the force exerted by the air on a given area. And what, then, is a force? A force is a push or pull. What, exactly, does air push on? On everything it comes in contact with but particularly, for purposes of this course, on other air. If the air pressure varies from place to place, then the air within that region is subject to unbalanced forces, as you can see in Figure 1.8. Such unbalanced forces may cause the air to go into motion. We recognize that motion as wind.

To return to the question of how air pressure is measured: the **barometer** is the instrument that extends the observer's

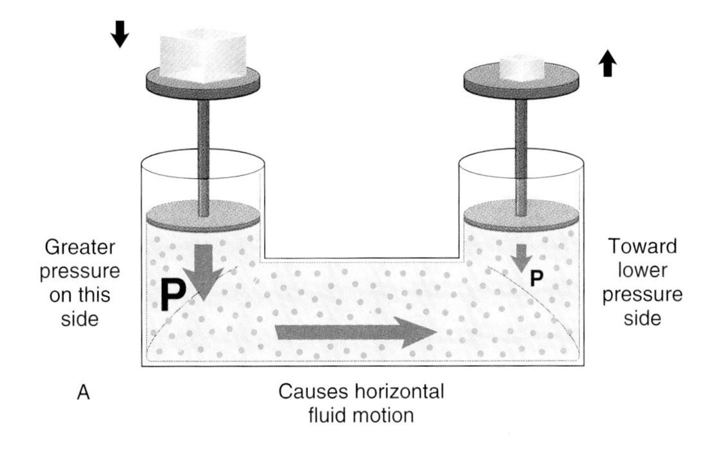


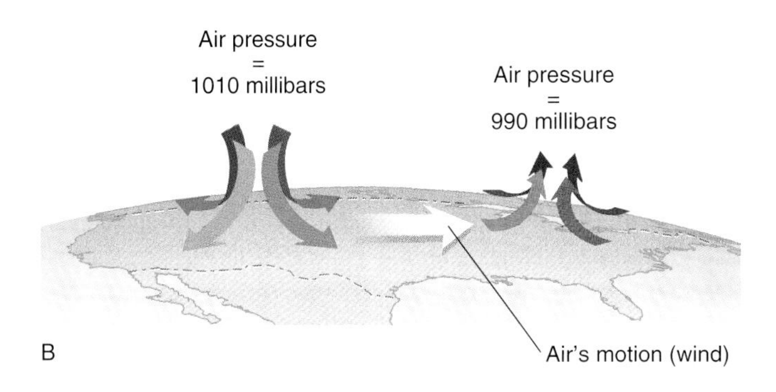


Figure 1.8 Air in the region indicated is subject to a greater pressure force from the high pressure in the west than from the lower pressure to the east. In response to the unbalanced pressure forces, the air will begin moving eastward.

senses into the world of air pressure. The two basic kinds of barometer are illustrated in Figure 1.9. Units of air pressure measurement are numerous. They include inches of mercury, millimeters, millibars, and Pascals. In this chapter, we will restrict ourselves to using **millibar** units. Typical air pressure values near sea level are slightly over 1000 millibars.

You were reminded a few paragraphs ago about the great variations in air pressure that occur as you move vertically. Isn't it necessary to compensate for the different altitudes of reporting stations? It is, indeed. It is standard practice to adjust the measured surface air pressure to eliminate effects of different station altitudes. Such adjusted pressures are called "sea-level pressures." Unless explicitly stated otherwise, stations' surface air pressures (including those at Chicago and Evansville in this example) are sea-level values (more on this topic in Chapter 2).

Looking back at the data in Table 1.1, you can see that Chicago and Evansville reported very similar sea-level air pressures. The absence of significant difference neither supports nor refutes the hypothesis that one or both reports are flawed.

## Evaluating the First Hypothesis

Now you are informed about weather observing practices and are in a better position to accept or reject the hypothesis that the 5:00 P.M. observations for Chicago and Evansville differ because of observational errors. You have seen, for example, that observing instruments are located in standard ways, so we should not expect local idiosyncrasies to explain the observed differences. You have also learned that observers use similar instruments and make reports according to a standard set of procedures (think of observing visibility or present weather), so the differences are not likely to be the result of different observing styles. Furthermore, the major differences in cloud cover, precipitation, wind direction, temperature, dew point, and humidity readings, and their persistence over several hours, tend to rule out the possibility that random observer errors somehow created the apparent differences. It seems appropriate, therefore, despite some similarities in the two stations' weather, to reject the hypothesis that one or both reports are flawed and that there is no "real" difference in the two reports. (Note, however, that we have not proved that the reports are completely error-free. We have simply determined it is likely that real differences existed in the weather at the two sites.)

Thus, we shall assume that the differences in Chicago's and Evansville's weather were real and that further inquiry is necessary to understand those differences. We shall go on to compare the observations with those made before and after 22 Z, and with those made at other nearby stations, after a review of this section.

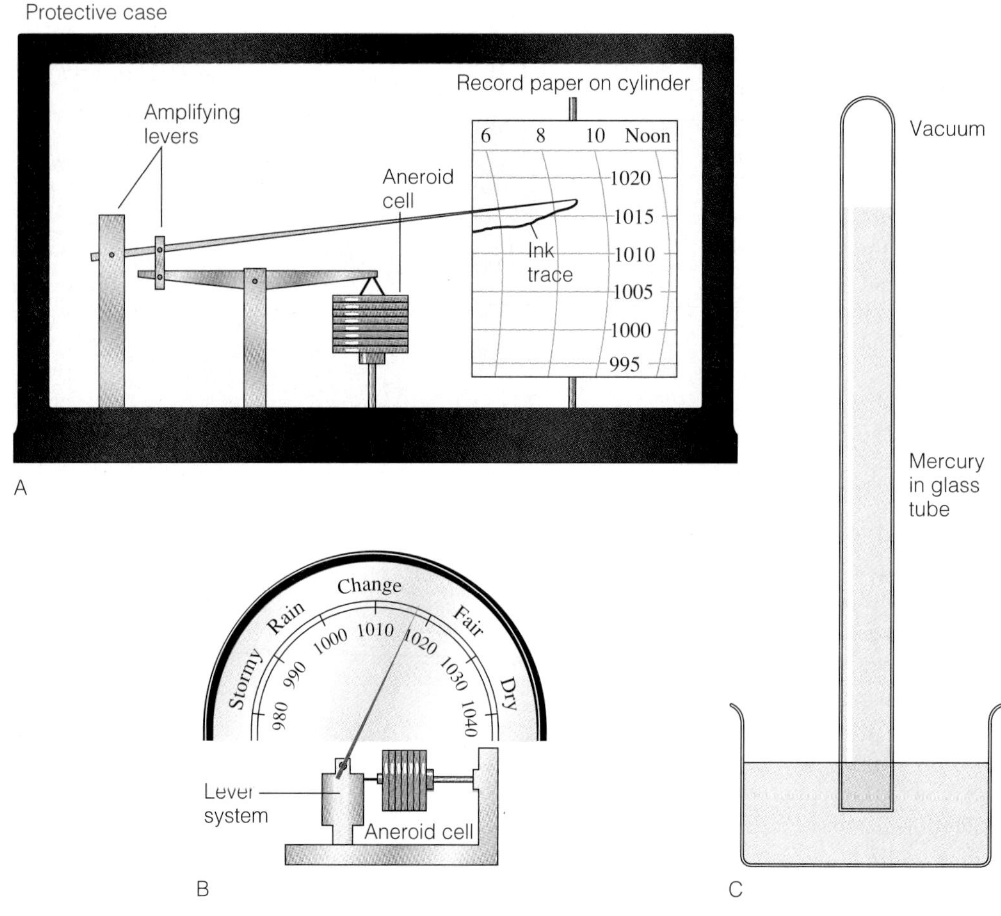


Figure 1.9 The aneroid (*A* and *B*) contains a partially evacuated, air-tight cell. Increases and decreases in the outside air pressure cause the aneroid cell to contract and expand, respectively. These changes are indicated on a graph (*A*) or on a dial (*B*). The mercury barometer consists of a tall (36 inch) tube from which all air has been removed and which stands in a reservoir of mercury. Mercury rises in the tube to a height at which its pressure on the reservoir's surface equals that of the air. The height of the mercury column, then, is a measure of the air's pressure.

# *Checkpoint*

***Review***

The **concept map** in Figure 1.10 summarizes this section's information about weather observations and instruments.

Read the map by tracing your way through various paths, working generally downward. Note that concepts, ideas, and objects appear in boxes; the boxes are connected by lines that indicate the relationships

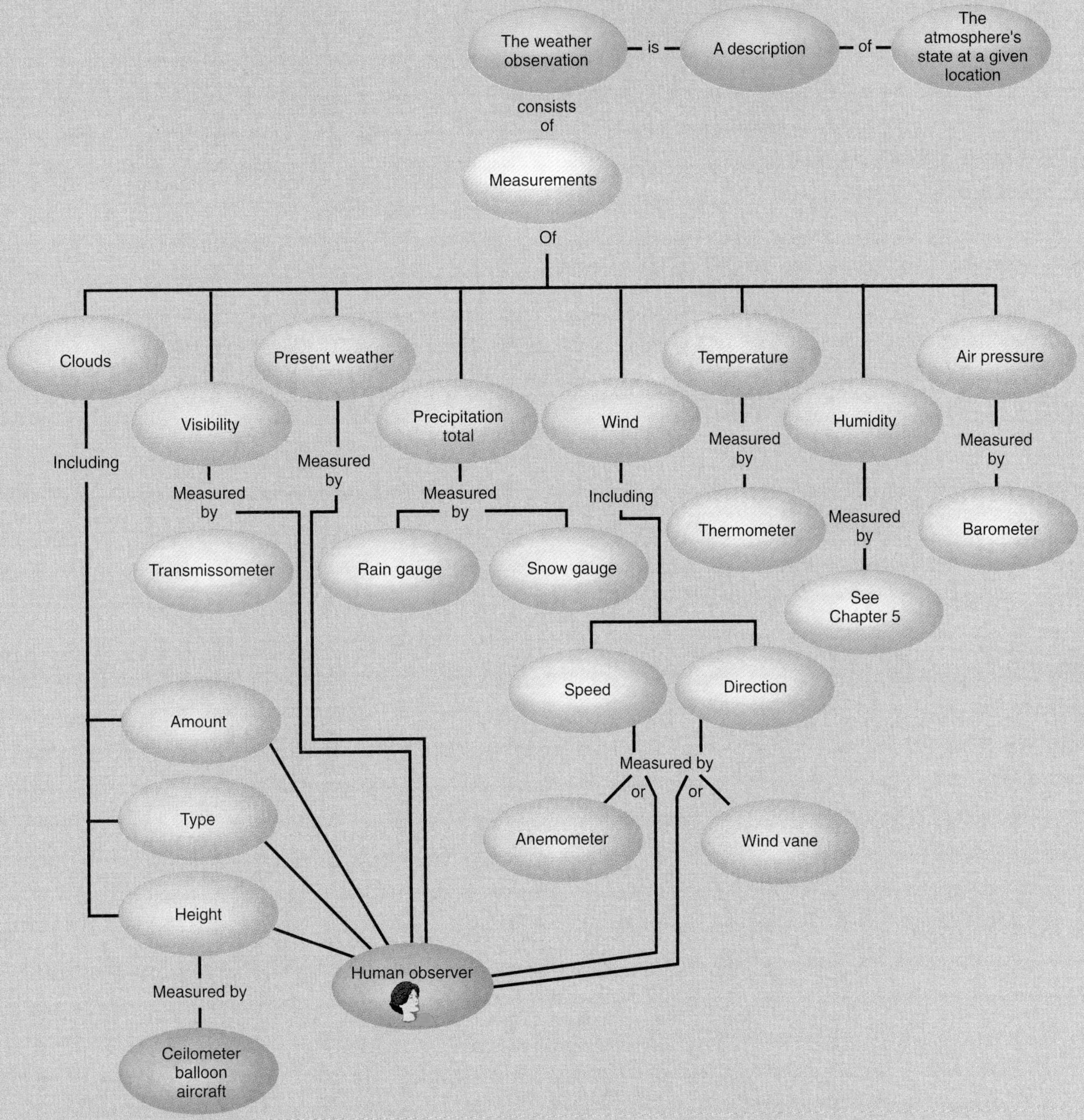

Figure 1.10 Concept map: The weather observation.

between objects. From the concept map, you see that the weather observation consists of measurements of eight variables, several of which have two or three components (e.g., wind consists of direction and speed). Thus, the relationship between the "Measurement" box and the eight weather elements below it is one of parts to the whole. Note that the map provides a convenient way of identifying various meteorological instruments with the variables they measure. The concept map is a powerful tool for organizing and remembering information. We will use them frequently throughout this text.

***Questions***

1. A weather station reports broken clouds. What fraction of the sky is cloudy?
2. Does a north wind mean (a) the wind is coming from the north or (b) going toward the north?
3. Suppose you observe continuous, light drizzle falling. What present weather category would you select to describe the condition? (See Appendix H.)
4. Name two variables that describe the atmosphere's water vapor content. How do these variables differ?

## Weather Observations in Space and Time

Now that we have established that the Chicago-Evansville weather differences appear to be real, it would be useful to know how extensive a phenomenon we are dealing with. Did the differences persist for hours, days, months? Are they permanent differences? Knowing the scale of the event will help us form hypotheses that match the problem.

### The Context in Time

We know already from the previous section that the discrepancies persisted for at least two hours. Table 1.3(A and B) provides us with more extensive data. You can see that the differences existed for at least 12 hours. On the other hand, notice that the average July data (Table 1.3B) indicate that long-term average conditions at the two locales are very similar. This line of inquiry has narrowed our search. We know that whatever caused the disparity maintained its influence for a number of hours but then began to relinquish its hold.

### The Context in Space

What is the spatial context for these observations? Did other observers near Chicago report conditions like the Chicago data? Is Evansville's observation representative of conditions over a larger area? An answer of no to either question would suggest that that location's weather was subject to strong local influences. (For example, you will see later that an observing station in a deep valley may experience weather conditions very different from nearby, nonvalley locations.) Figure 1.11, which illustrates temperatures recorded at a number of sites at 22 Z, reveals that similar temperatures tended to occur near one another, forming regions of like values. At this time, Chicago lay in a very different temperature zone from Evansville.

**Table 1.3A Chicago and Evansville Weather from 11:00 A.M. to 11:00 P.M., July 14, 1992 (16 Z on 7/14 to 03 Z on 7/15)**

| Location | Time | Cloud Cover | Visibility (mi.) | Present Weather | Wind Direction/Speed | Air Temperature (F) | Dew Point (F) | Air Pressure (mb) |
|---|---|---|---|---|---|---|---|---|
| Chicago | 11:00 | Overcast | 7 | Light rain | NE/8 | 61 | 60 | 1010.3 |
| | 1:00 | Overcast | 9 | Light rain | NE/5 | 62 | 61 | 1009.9 |
| | 3:00 | Overcast | 4 | Shower, fog | NNE/12 | 62 | 61 | 1009.6 |
| | 5:00 | Overcast | 7 | Inc'g clouds | NNE/10 | 61 | 59 | 1009.6 |
| | 7:00 | Overcast | 2 | Drizzle, fog | N/10 | 60 | 60 | 1010.3 |
| | 9:00 | Overcast | 1/2 | Drizzle, fog | NE/12 | 60 | 60 | 1009.0 |
| | 11:00 | Overcast | 3/8 | Drizzle, fog | NE/10 | 61 | 61 | 1009.1 |
| Evansville | 11:00 | Clear | 6 | Sky unchanged | S/12 | 87 | 73 | 1012.4 |
| | 1:00 | Scattered | 7 | Sky unchanged | SSW/16 | 89 | 72 | 1011.2 |
| | 3:00 | Scattered | 7 | Sky unchanged | SSW/15 | 91 | 72 | 1008.7 |
| | 5:00 | Scattered | 7 | Sky unchanged | S/11 | 91 | 70 | 1008.1 |
| | 7:00 | Scattered | 7 | Sky unchanged | SW/8 | 88 | 69 | 1007.4 |
| | 9:00 | Scattered | 7 | Inc'g clouds | WNW/14 | 76 | 69 | 1007.8 |
| | 11:00 | Overcast | 5 | Fog | SW/8 | 71 | 70 | 1008.5 |

**Table 1.3B** Average July Data, Chicago and Evansville

| | Chicago | Evansville |
|---|---|---|
| CLOUD COVER | 5/10 | 6/10 |
| WIND | SW/8 | SSW/7 |
| DAILY HIGH TEMPERATURE (°F) | 85 | 89 |
| DEW POINT (°F) | 66 | 70 |
| MONTHLY PRECIPITATION (INCHES) | 3.37 | 3.00 |

## A New Hypothesis

The patterns revealed in Figure 1.11 do not "explain" the temperature differences between Chicago and Evansville. They do inform us of the spatial extent of the temperature differences, however. It is clear that the differences are not highly local events but that each station is part of a larger pattern. This insight leads directly to the next question: Why do regions show such temperature contrasts? In a sense, we have moved to a larger, more general, and perhaps more difficult question to answer. But the revelation that such temperature patterns exist over large areas also gives us added confidence that the problem is "real" and not some accident of measurement.

The knowledge that the Chicago-Evansville temperature differences are part of a much larger pattern suggests that differences in cloud cover, wind direction, and so forth also may be part of larger patterns. We will make that assumption (which we will check) and hypothesize that the temperature and other patterns are components of a single, large-scale weather system that was causing the differences between Chicago's and Evansville's weather. To test this new hypothesis, however, we need to present some weather map plotting techniques.

## A Mapping Aid: The Plotting Model

It would be useful to see the complete 22 Z weather reports for each observing station, not just temperatures. However, if we proceeded by listing the reports next to each station, as we did in Figure 1.1, very few stations' data would fit. Furthermore, a map should be a visual aid, whereas a list of data is not. We need a more concise and visual way to represent each station's report.

Figure 1.12A shows Chicago's 22 Z weather, plotted according to a station plotting model used to display observed weather. (This model is somewhat simplified from the standard plotting

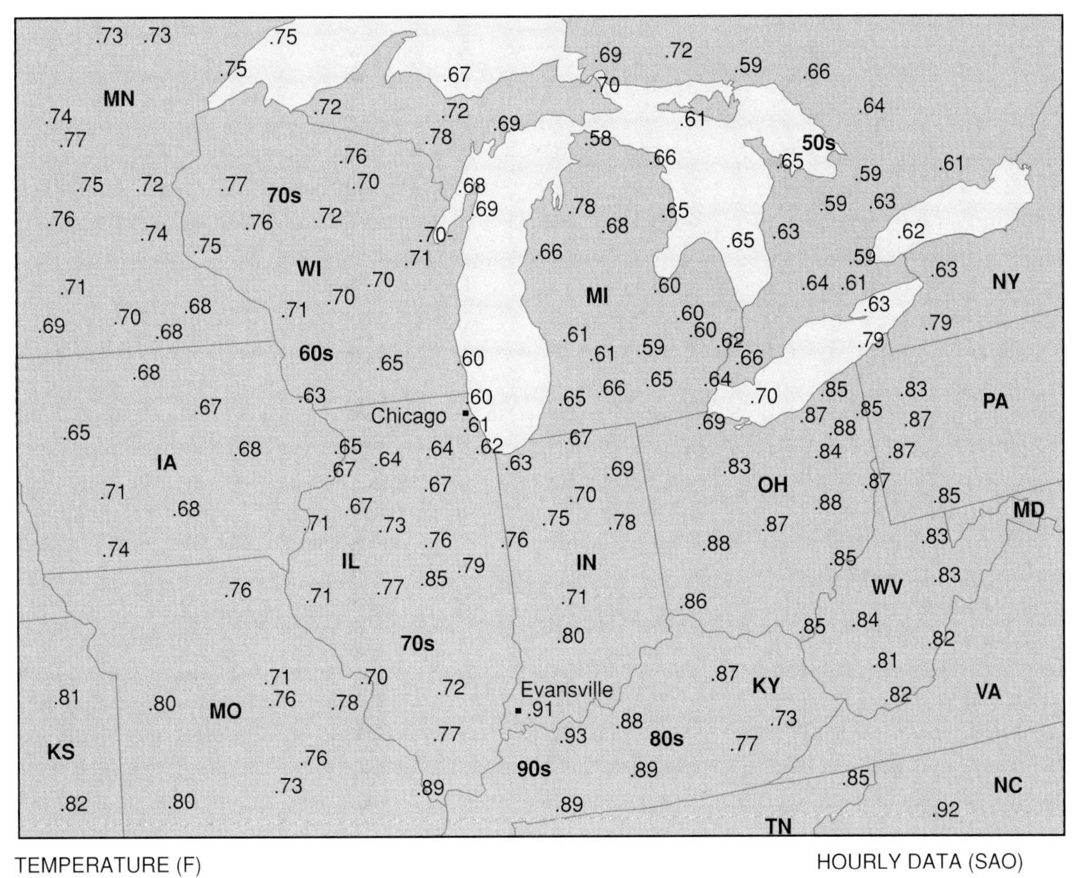


Figure 1.11 Surface temperatures at 22 Z on July 14. Note the tendency for similar temperatures to cluster in zones oriented east to west.

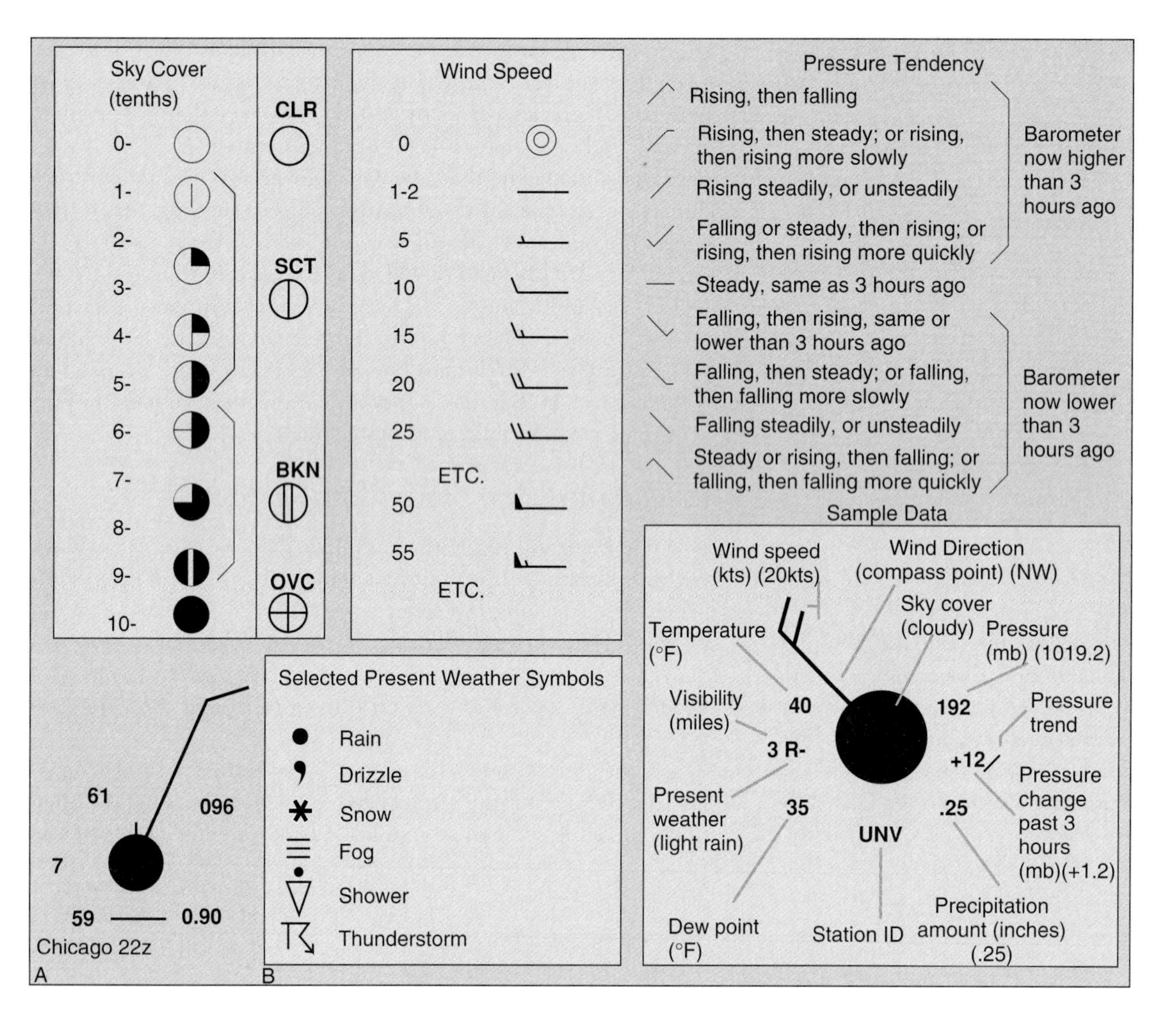


Figure 1.12 (*A*) Station plot for Chicago's 22 Z weather. The long bar below the cloud circle signifies status clouds. The short vertical mark at the cloud circle's top is the present weather symbol for "increasing clouds." Other symbols should be clear from the text and from comparison with the data in Figure 1.1. (*B*) Some symbols used in the station plotting model. For a more complete set of tables, see Appendix H.

model in which additional data are plotted for each station.) The circle at the center locates the station on the map and indicates the total amount of cloud cover according to the system shown in Figure 1.12. The arrow shows wind speed, rounded off to the nearest 5 knots (5 knots for a short feather, 10 knots for each long feather), and wind direction (the arrow travels with the wind, feathers at the back). Each of the other variables has a particular position in which it is plotted. Thus, temperature always appears above and left of the cloud circle, air pressure above and right. Because temperature is always plotted in the same position, it is not necessary to label it as temperature or to specify its units. This procedure reduces clutter on the plotted map.

Air pressure data are plotted without the leading 9 or 10 and without the decimal point. Thus, pressure of 1029.7 millibars is plotted as 297; 987.6 mb would be plotted as 876.

How can you tell whether a given plotted pressure is missing a leading 9 or a 10? The only sure answer is, "By comparing it with data at other stations in the vicinity." This will become second nature once you have studied a number of weather maps in detail. In the meantime, here is some guidance: Most sea-level pressure measurements are within 10 or 20 millibars of 1000 millibars; therefore, in reconstructing the original pressure value, affix whichever value (9 or 10) gives a pressure closest to 1000 mb. Thus, 923 would decode to 992.3 mb—not 1092.3, which would be a world record. More specifically, you can usually assume that for plotted values in the range 000 to 500, a leading 10 has been omitted, and for values from 500 through 999, a leading 9 has been omitted.

Notice that Figure 1.12 gives symbols for plotting the three-hour **pressure tendency.** The pressure tendency is the amount and nature of the change in air pressure over the past three hours. For example, a pressure tendency of +14/ means that the air pressure has risen 1.4 millibars over the past three hours, and the rise has been a steady one. Pressure tendency information is particularly useful for determining how pressure systems are changing and moving, as you will see in a later section.

## Weather Map Patterns on July 14

Now that you are familiar with the standard plotting model, we can consider our latest hypothesis, that the weather differences from Chicago to Evansville arise from the presence of a large-scale weather system passing through the Midwest. We begin evaluating this possibility by turning to Figure 1.13, in which the 22 Z, July 14 weather is plotted for a number of locations in the region.

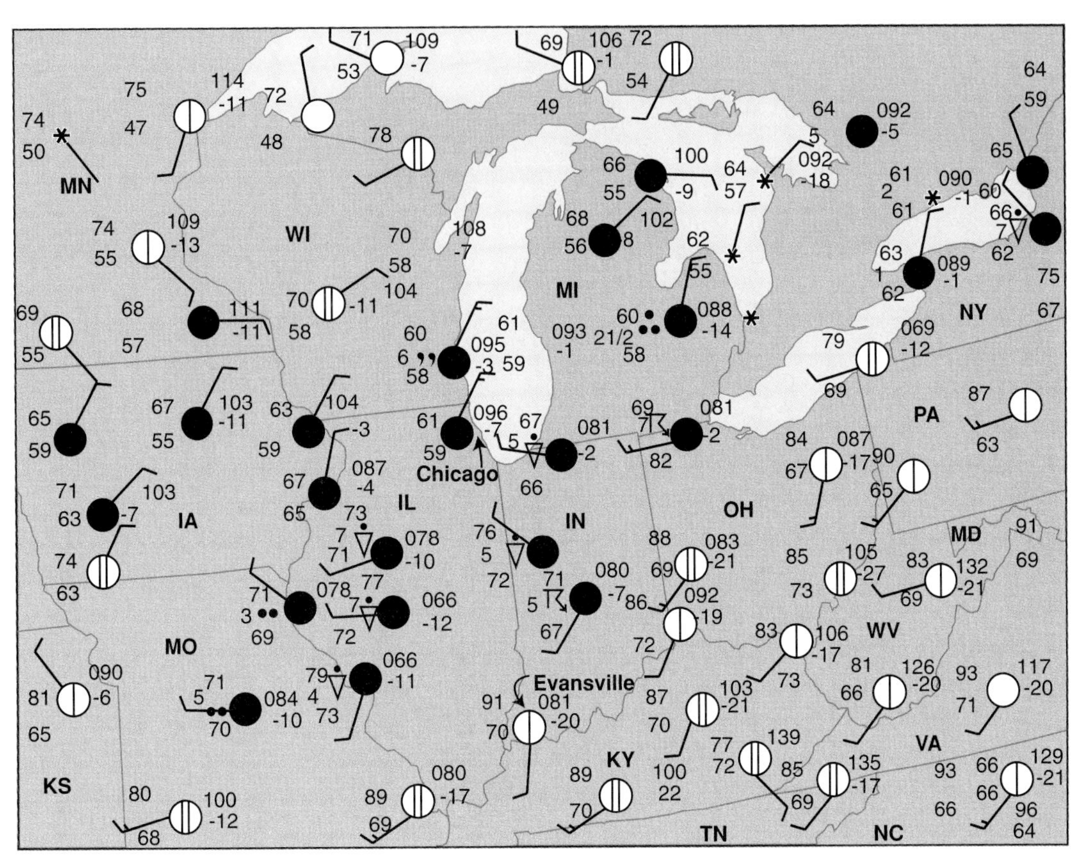


Figure 1.13 Station plots of weather conditions at 22 Z on July 14. Note the tendency for high temperatures to occur at stations with southwest winds and low temperatures where winds are northeast and skies are overcast.

## *Patterns of Clouds, Precipitation, and Air Pressure*

Study Figure 1.13 for a moment, looking for patterns in other weather elements besides temperature. You may notice that cloudiness is concentrated in a broad zone running from west to east across the map. In contrast, locations in the northwest and south sections of the map are reporting nearly clear skies. Chicago lies in the cloudy band, Evansville in the partly sunny southern zone. Note also that precipitation, in the form of showers and thunderstorms (shown by the $\dot{\nabla}$ and ⊼ symbols), is falling in the cloudy region, particularly in Illinois and Indiana.

If you examine the values of air pressure, you find a pattern that correlates with those of cloud cover and precipitation: higher pressures are reported in the relatively clear skies to the northwest and south, especially the southeast; lower pressures occur along the cloudy zone. Two observers in central Illinois, in the region of greatest shower activity, reported the very lowest pressure, 1006.6 millibars (066 on the map).

## *The Air Pressure Pattern in More Detail*

The focus on air pressure in the discussion about patterns leads us to examine the pressure pattern in more detail, in Figure 1.14. The center of lowest pressure is labeled with a large L and is referred to informally as a **"low."** A second, smaller low lies over western New York and is similarly labeled. The two regions of high pressure **("highs")** are labeled with an H. These features, as you probably have assumed, are the familiar "highs" and "lows" of the daily weather map. In addition, the pattern of isobars in Figure 1.14 makes the pressure pattern more evident. **Isobars** are lines of equal air pressure. Each isobar represents the locations of some particular value of air pressure reduced to sea level. On one side of the isobar, pressure is higher; on the other side, it is lower. For example, follow the isobar labeled 1008 that encircles the Illinois low in Figure 1.14. Notice it passes directly through two stations reporting air pressures of exactly 1008.0 millibars (080 on the map). Within the loop formed by the isobar, air pressures all are lower than 1008.0 millibars; outside of the loop, all are higher. Similarly, the 1010.0 isobar defines locations having a pressure of 1010.0 millibars. Within the isobar's loop, pressures are lower than 1010.0; outside the loop, they are higher.

Note that the northeast-southwest orientation of the low pressure region and the presence of the two highs are all much more obvious in Figure 1.14 than in Figure 1.13. Isobar patterns provide the meteorologist a way of seeing the field of air pressure at a glance, instead of interpreting numerical values plotted at each station.

Next, examine the wind flow in Figure 1.14. Remember that the plotted wind arrows point with the wind into each station's

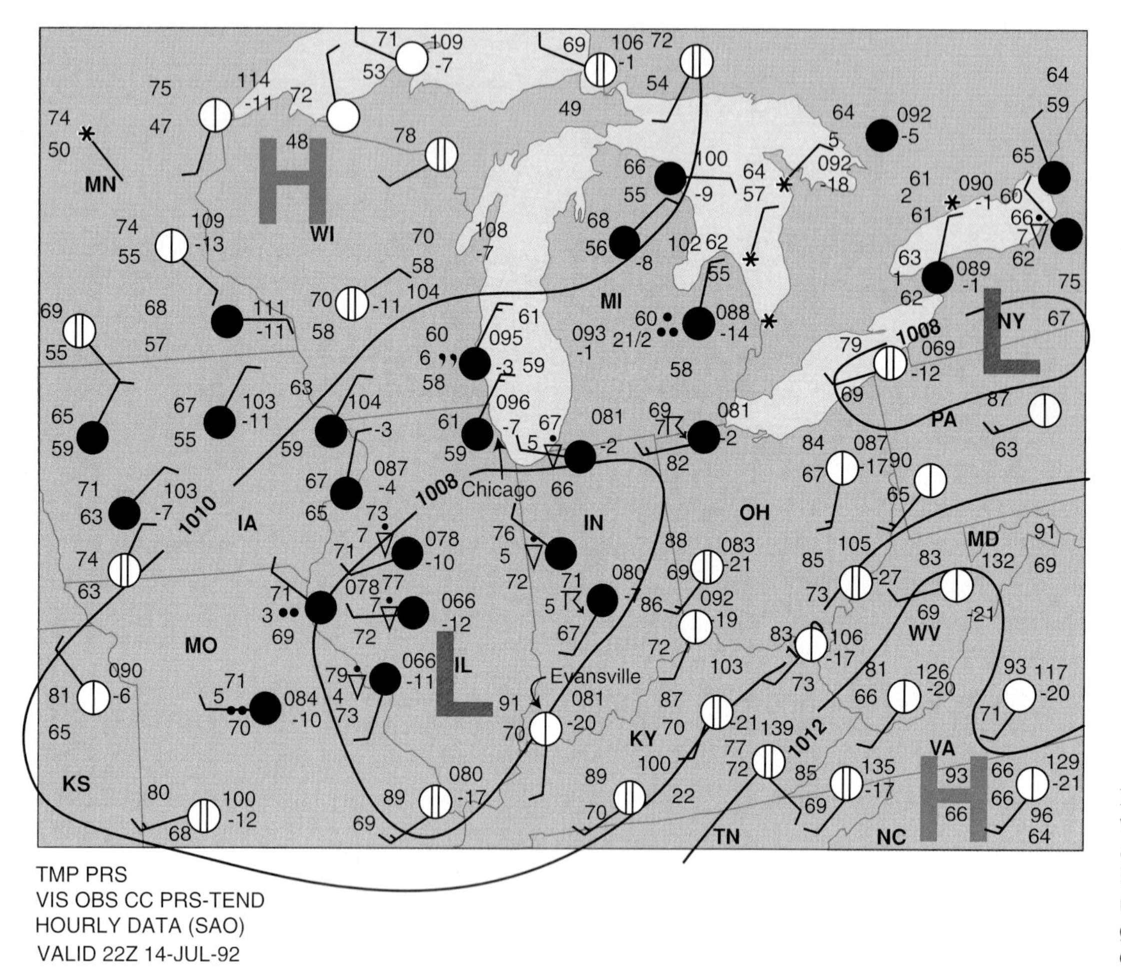


**Figure 1.14** The surface weather map, showing high and low pressure centers and isobars. The isobar pattern makes evident the low's elongation in a northeast-southwest direction.

cloud circle, with the feathers at the rear of the arrow. Notice that the wind to the north of the low pressure region is mainly from the northeast, while south of the lows it tends to come from the southwest. This pattern results in a counterclockwise flow of air around the lows, more or less along the isobars. On the other hand, the wind flow tends to be clockwise around the high pressure center in the northwest corner of the map.

Closer inspection of the wind flow (at Evansville and Chicago, for example) reveals that at most stations, it is not simply parallel to the isobars but tends to cross them as well, from higher toward lower pressure. The overall pattern in Figure 1.14, then, is one in which winds tend to spiral clockwise and outward from centers of high pressure but counterclockwise and slightly inward toward low pressure centers. (You will see in later chapters that these rotations are reversed in the Southern Hemisphere; winds blow clockwise and inward around lows, counterclockwise and outward around highs.)

Thus, the underlying assumption for the latest hypothesis seems true: a number of weather variables are organized into large-scale patterns. Therefore, we now consider the hypothesis itself: that the patterns, and therefore the Chicago-Evansville weather differences, are caused by a passing weather system.

## Evaluating the Weather System Hypothesis

Returning to the latest hypothesis that the Chicago-Evansville weather differences are the result of patterns created by a transient, large-scale weather system, you probably recognize there is a great deal of evidence in its favor. Chicago lies just to the north of a low pressure center, which is bringing that city north winds, clouds, and therefore low temperatures. Evansville lies southeast of the low and as a result is experiencing south winds, clear skies typical of the high to its east, and a very warm (91°) temperature. The presence of only scattered clouds at Evansville, which is not far from the low pressure center, is somewhat inconsistent with the present hypothesis, but in all, you have made good progress in understanding the July 14 weather differences. Although we will need to look further before accepting the hypothesis, there is no reason to reject it at this point. (See checkpoint on facing page)

# Checkpoint

***Review***

Trace your way through the various branches of the concept map in Figure 1.15. Note that the plotting model itself is indicated as an important visual aid. Practice visualizing the wind flow pattern on the weather map (Figure 1.14) by concentrating on the plotted wind arrows. Note from the concept map how much information contour lines such as isobars add to the map.

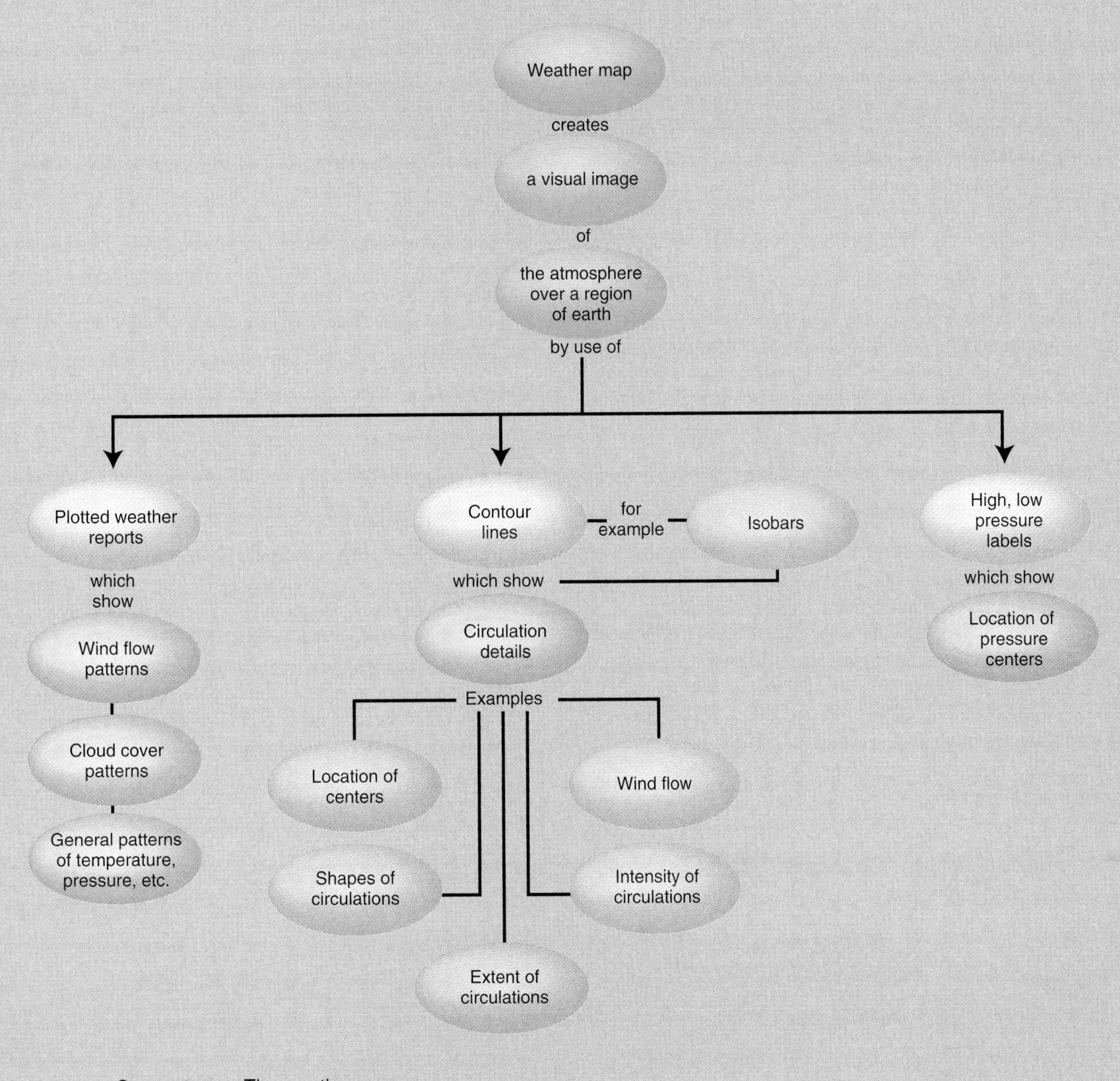


Figure 1.15 Concept map: The weather map.

***Questions***

1. You observe a cloud cover of 5/10 and wind from the west at 25 knots. How would you plot these conditions on a weather map?
2. In Figure 1.14, Dayton, Ohio, reports a coded air pressure of 083. What is the actual pressure in millibars?
3. What relation is there in Figure 1.14 between wind directions and isobars?

# Basic Models of High and Low Pressure Systems

Weather systems like those revealed in the last section are common and important features of the atmospheric landscape. Frequently, they are the dominant cause of weather changes from place to place and from day to day. It will be useful, therefore, to develop a rather detailed descriptive model of these large-scale, or **synoptic scale,** weather systems. A synoptic scale weather system is one several hundred to 1000 or so miles in diameter and whose lifetime is roughly one week.

## A Preliminary Model

Assume for the moment that Figure 1.14 portrays typical synoptic scale weather system behavior. To summarize from this figure and from the related discussion in the last section, the following features are evident:

1. The pattern of air pressure is the main organizing structure.
2. Winds circulate around Northern Hemisphere low pressure centers (Lows) in a counterclockwise and slightly inward sense. A low pressure system and its attendant circulation, such as the one in Figure 1.14, is known as a **cyclone.** (Many people mistakenly equate *cyclone* with *tornado.* As you can see from the definition above, *cyclone* is a more general term. Any low pressure system and its circulation, regardless of its size or intensity, is considered a cyclone. Thus, tornadoes, hurricanes, and larger, generally less destructive examples, like the circulation we are investigating at present, are all cyclones.)
3. Clouds and precipitation are located commonly in regions of low pressure.
4. Wind flow in the vicinity of high pressure centers (highs) is just the opposite of the low pressure case. Here, winds circulate clockwise (in the Northern Hemisphere) and slightly outward. Fair skies are the rule in high pressure regions. High pressure systems and their circulations are called **anticyclones,** since their rotation is opposite of that in cyclones.
5. Temperature patterns result to a considerable extent from the combined effects of latitude, wind flow, and cloud cover. Higher temperatures generally are found in the south and in regions having south winds. It is colder in the north and where wind flow is from the north. Daytime temperatures tend to be higher on clear days than on cloudy days, especially in summer.

These five observations are incorporated into the three-dimensional model depicted in Figure 1.16. Notice that the wind flows clockwise and outward from the highs and "feeds into" the low. Where does this air go, as it moves into the low center? According to the model, the air rises. You will see in later chapters that when air rises, it cools rapidly; and you know from the discussion about dew points that cooling air eventually reaches its dew-point temperature, causing its water vapor to begin to condense. Thus, the model suggests that lows, as regions of rising air, will be regions of cloudiness, which is borne out on the July 14 weather map. Conversely, the model suggests that air must undergo **subsidence,** the meteorological term for sinking air, in the vicinity of highs. Subsidence warms the air, thereby increasing the temperature-dew point spread and lowering the relative humidity. Thus, the model predicts clear skies in highs, which again is consistent with the July 14 data.

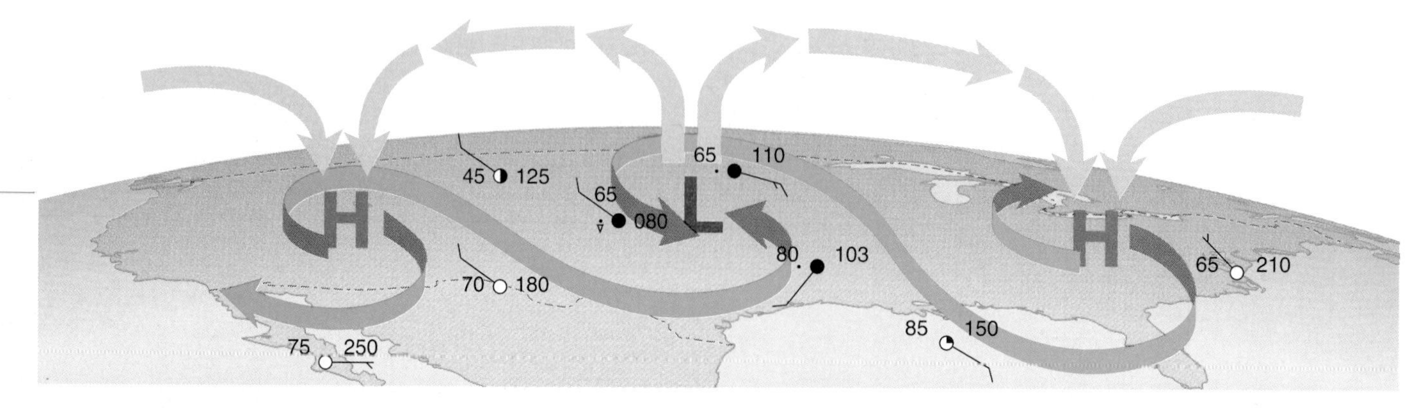


**Figure 1.16** A preliminary 3-dimensional circulation model. Note how the air circulates between the high and the low pressure centers. The plotted temperatures, pressures, and so on are merely plausible values and are given to show general relations, not actual values.

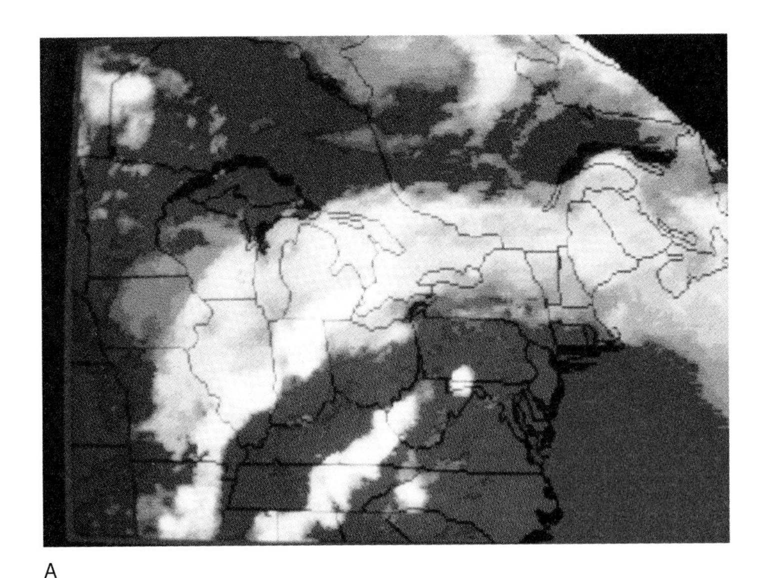
A

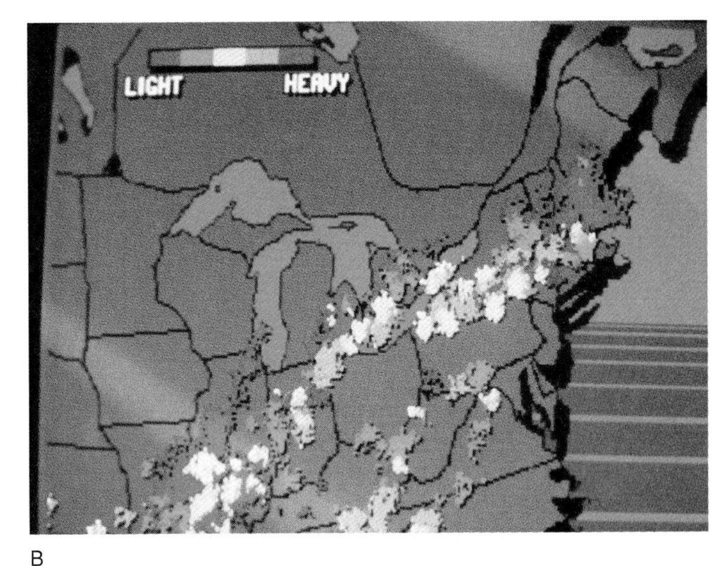

B

Figure 1.17 (*A*) The satellite image of the 22 Z weather shows cloudiness prevailing in the regions of lowest pressure. (*B*) Weather radar image for 22 Z. Note that the line of precipitation from New England through the Midwest lies along the axis of lowest pressure in Figure 1.14.

Figure 1.16 indicates that at higher altitudes, a reverse flow occurs: air converges over highs to replace the subsiding air, whereas over lows rising air eventually diverges. The result is a continuous circulation system. You will recognize aspects of this circulation model in many different atmospheric phenomena, from the July 14 highs and lows to phenomena as varied as ocean breezes and hurricanes.

We remind you, however, that this is a simple and preliminary model, and not the last word. In later chapters, you will see it refined and adjusted in various ways as it is applied to cyclones of different scales.

## Evaluating the Model: How Well Does It Work?

The model of Figure 1.16 needs to be tested further before we accept it. Does the model remain credible as we examine new data? We consider this question briefly in this section.

Figure 1.17A depicts an image made by weather satellite at the same time as the surface observations of Figure 1.14. In some ways this is a far more complete view of cloud cover than surface observations from the relatively few sites in Figure 1.14. In a sense, the satellite view fills in between surface reporting stations. Compare the two figures. Note that the satellite image indicates that clouds predominate in regions of low pressure. High pressure regions tend to be clear. Further data are presented in Figure 1.17B, which shows regions in which precipitation was occurring at 22 Z. Again, observe the tendency for precipitation to occur in low pressure regions and the continuity of the precipitation pattern.

While granting the circulation model in Figure 1.16 a certain amount of validity, you probably also recognize that it has left unanswered some important questions. For example, we noted before that the model fails to explain why Evansville, which is nearly as close to the low pressure center as Chicago and has lower air pressure, has clear skies. Other, more general questions remain: Why are clouds and precipitation distributed so irregularly within the region of low pressure and mainly to the north side of the low? Why do wind directions change so abruptly in certain regions of the map? Why do large regions have nearly the same temperatures, while other, smaller areas exhibit large temperature change? The model of Figure 1.16 is a great simplification of the atmosphere's actual behavior and has no answers for such questions.

Re-examine for a moment the data plotted in Figure 1.13 with the model of Figure 1.16 in mind. This time, look for observations that conflict with the model, rather than support it. You can see that the problems mentioned in the last paragraph (sudden wind shifts and temperature changes, for example) are evident at a number of locations. The next section presents an important improvement, the addition of fronts. Other refinements of the model will be made in later chapters. Even then, however, for all its usefulness and accuracy, the model will have its imperfections. After all, it is only a model, a human simplification of the atmosphere's complexity.

## Adding Fronts to the Model

Figure 1.18 presents the July 14 map once again but with a new feature, a red and blue line running through the low pressure zone. South of the line, notice that weather conditions are rather uniform: nearly every station is reporting winds from the southwest, partly cloudy skies, and warm temperatures. North of the line, conditions are dramatically different: stations are reporting north or northeast winds and much lower temperatures. Clouds and precipitation are concentrated mainly in a region just north of the line.

The line we have added to the map in Figure 1.18 is called a **front.** For now, we may define a front as a narrow zone of division between air masses of different characteristics. The front in

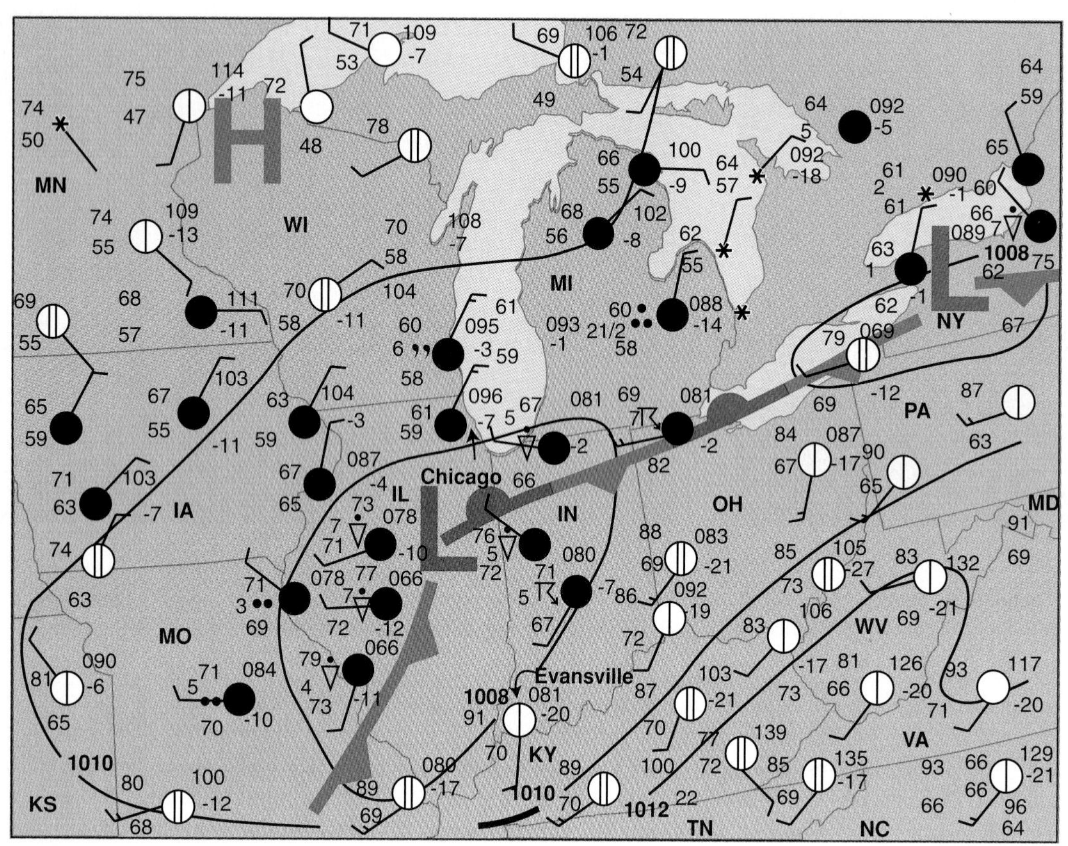


Figure 1.18 The 22 Z surface weather map, with fronts added. The fronts mark the line of division between two different air masses. North of the front, winds are predominantly from the northeast; south of the front, from the southwest. The front marks the line along which these air flows meet.

Figure 1.18 marks the line along which warm air from the southeastern states meets cooler air spilling southwestward from Wisconsin and Minnesota.

Meteorologists recognize four basic types of fronts, distinguished mainly by their motions. A **cold front** is one in which colder air advances to replace warmer air. In a **warm front** the opposite occurs: the colder air recedes, allowing warmer air to replace it. A **stationary front,** as you would surmise, is one showing no evident motion. The fourth type, called an **occluded front,** is a hybrid and occurs when a cold front overtakes a warm front. Figure 1.19A shows weather map symbols representing different frontal types.

What weather conditions prevail along a front, where air streams from the two different highs meet? Figure 1.19B shows a vertical cross section through both the cold front and the warm front. Note that the warmer air, being lighter, rises over the cool air. (The same principle is at work in a hot air balloon. The hot air inside the balloon is lighter than the unheated air surrounding the balloon, causing the air to rise, carrying the balloon with it. Chapter 2 covers these processes in more detail.) As the warm air rises above the front, it cools and eventually reaches its dew point; clouds and precipitation form in the rising air. The structure shown in Figure 1.19B implies that we should expect clouds and precipitation to develop primarily on the cold air side of the front's surface position. It also suggests that temperature and wind changes occur rather abruptly at the front, rather than gradually as one might expect with a simple, circular model such as that in Figure 1.16.

Notice also in Figure 1.19B that a typical cold front has a steeper slope than the warm front. These slope differences are related to important differences in cloudiness and precipitation. Cold fronts are associated with more violent, showery, shorter-duration precipitation primarily from cumuliform clouds; warm frontal clouds more typically are stratiform (i.e., of the stratus type), and precipitation is steadier and of longer duration but less violent.

Synoptic scale low pressure circulations with fronts are called, appropriately, **frontal cyclones.** The idealization of a typical frontal cyclone in Figure 1.19C gives a more detailed depiction of the system's fronts, associated highs, and attendant weather. Of course, actual values of pressure, temperature, and so on will vary markedly by time of day, season of the year, and location; therefore, do not concern yourself with memorizing the actual weather conditions in Figure 1.19C. Instead, concentrate on the overall features of the model, such as circulation directions around highs, lows, and fronts, and distribution of cloud cover, precipitation, temperatures, and so forth.

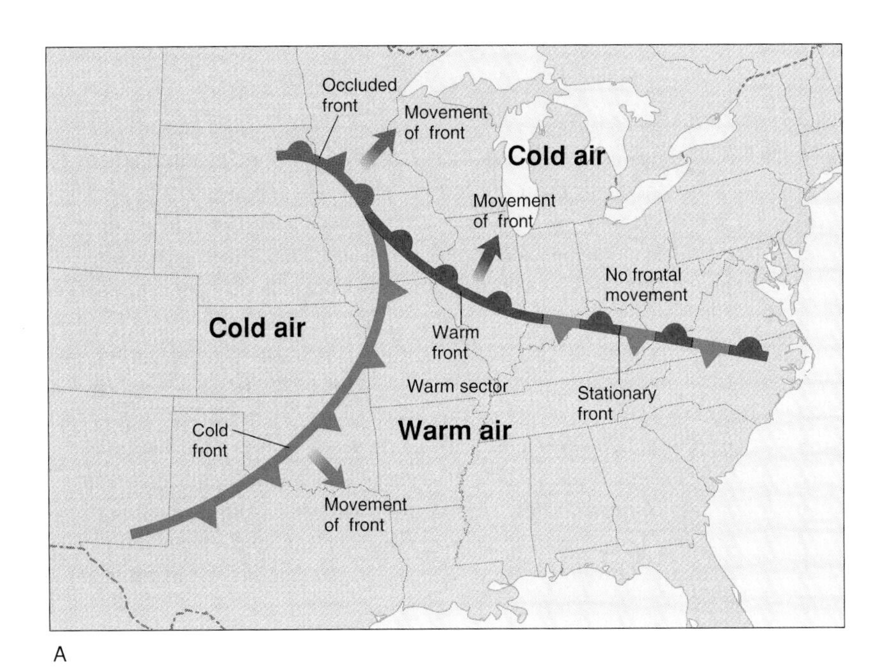


A

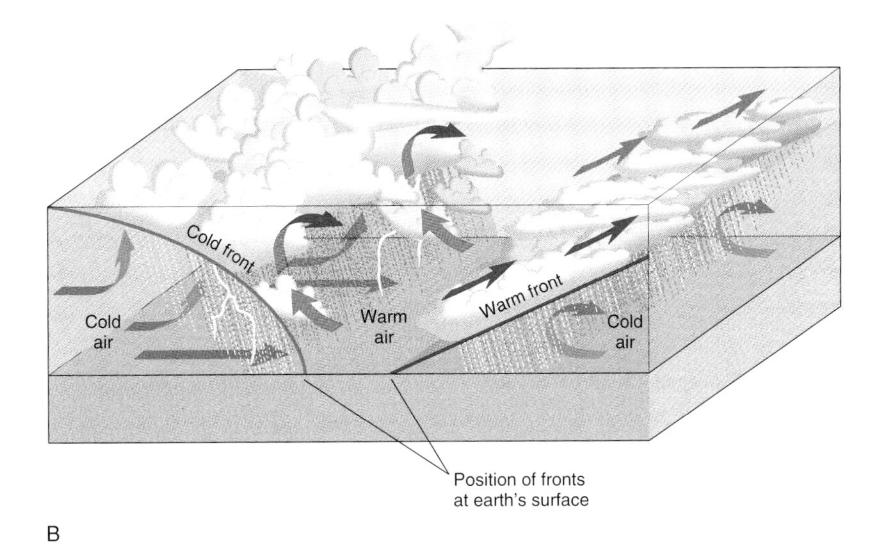


B

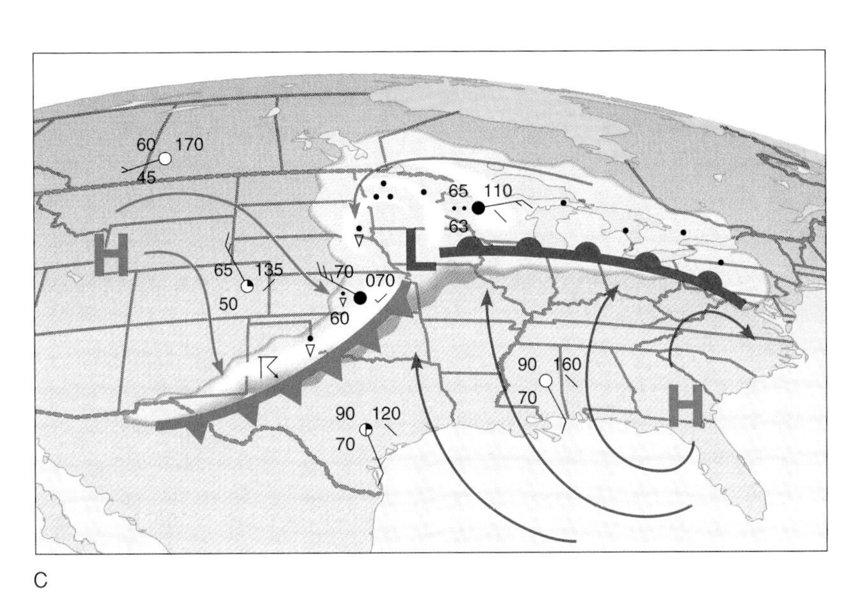


C

Figure 1.19 (*A*) Plotting symbols for four types of fronts. Notice that the triangles and semicircles are placed on the forward side of cold, warm, and occluded fronts. (*B*) Cross-sectional view of a cold front and warm front. Notice the differences in slope, cloud type and precipitation type between cold and warm fronts. (Note: The slopes of both fronts are greatly exaggerated, for clarity. A typical front slopes upwards less than one part in 100. (*C*) The frontal cyclone model.

Analyzing the July 14 weather in terms of the frontal cyclone model shown in Figure 1.19 substantially improves our ability to understand the weather patterns on that day. The front's presence explains the observation that most of the cloudiness and precipitation occur in a similar band running slightly to the north of the zone of lowest pressure (as shown in Figure 1.17) and the large temperature and wind changes often seem to occur over small distances along this same band.

More specifically, we are now able to answer the question with which we opened the chapter: Why are Chicago and Evansville experiencing such different weather? Our answer, as illustrated in Figure 1.18, is that a weak low pressure center and associated front lie between the two stations. Chicago, lying to the north of the low and the front, is experiencing frontal clouds and the cool northeast winds originating from a high pressure center in Wisconsin. Evansville lies south of the low and front, in the system's warm sector. A warm sector is the wedge of air lying to the warm side of a frontal cyclone between its cold front and warm front and typically is a region of fair weather (see Figure 1.19A). Evansville's air originates in a high, centered to the southeast, which brings it high temperatures and high dew points.

## How Do Frontal Cyclones Move and Evolve?

Two important aspects of the frontal cyclone model remain to be considered: First, how do these features move, if at all? Anyone making a prediction of tomorrow's weather will want specific information about the direction and speed of motion of nearby highs, lows, and fronts. In Figure 1.20, you can see the region's weather at 5:00 A.M. on July 15 (10 Z), just 12 hours later than the depiction in Figure 1.14. Notice that most features have shifted eastward somewhat. This behavior is typical. There are many exceptions, but generally weather systems in middle latitudes travel with some west-to-east component.

Pressure tendencies often give useful clues about the movements of highs, lows, and fronts. Pressures fall in advance of an approaching low or front and rise in advance of a high. Thus, the pressure tendency alone can be a useful indicator of tomorrow's weather.

The second question is, what sort of life cycle do frontal cyclones experience? More specifically, how do they form, how do they grow, how do they dissipate? These are important

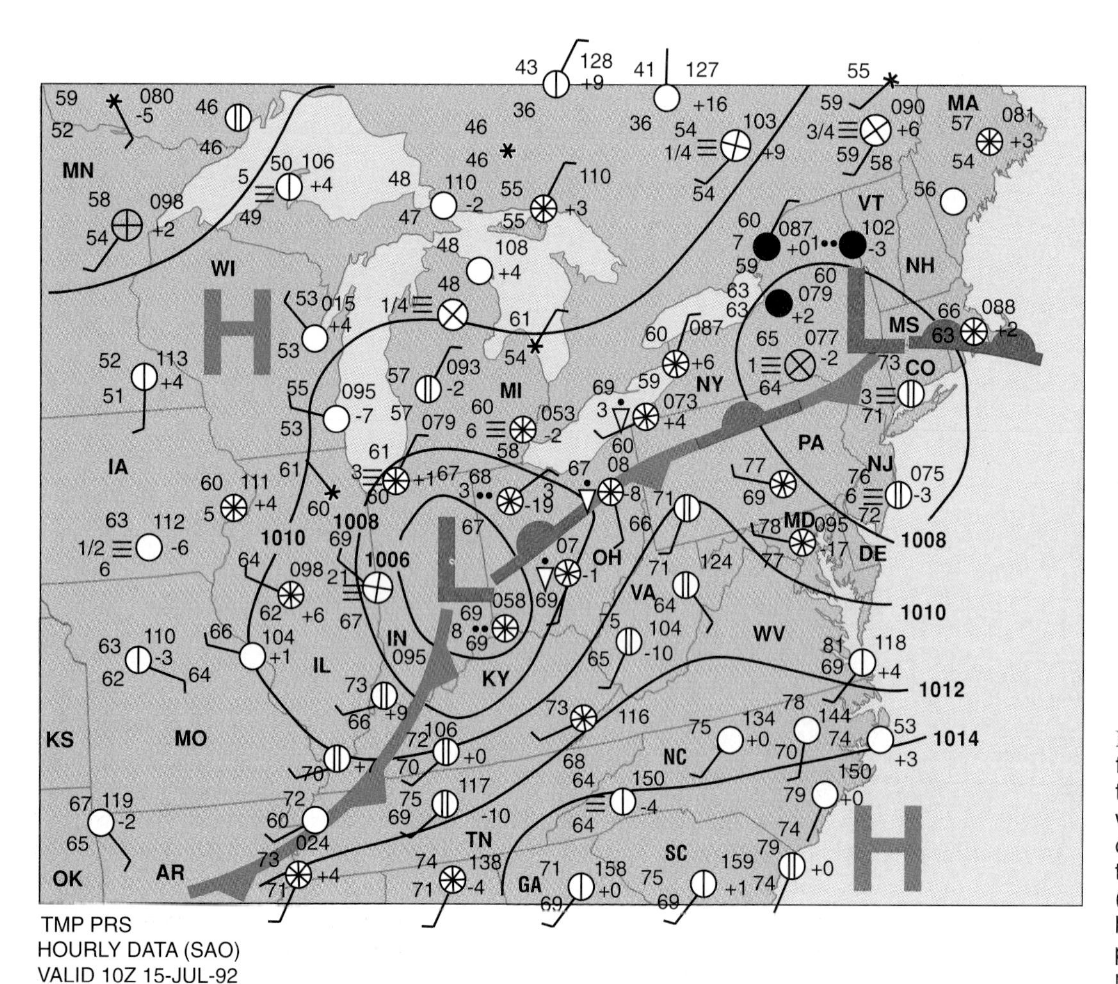


**Figure 1.20** The weather map for 10 Z, July 15 (5:00 A.M. Central time, July 15). Note the eastward movement of the pressure centers and fronts compared to the 5:00 P.M., July 14 positions (Figure 1.14). Also note that both low centers have lower central pressures than on the previous map.

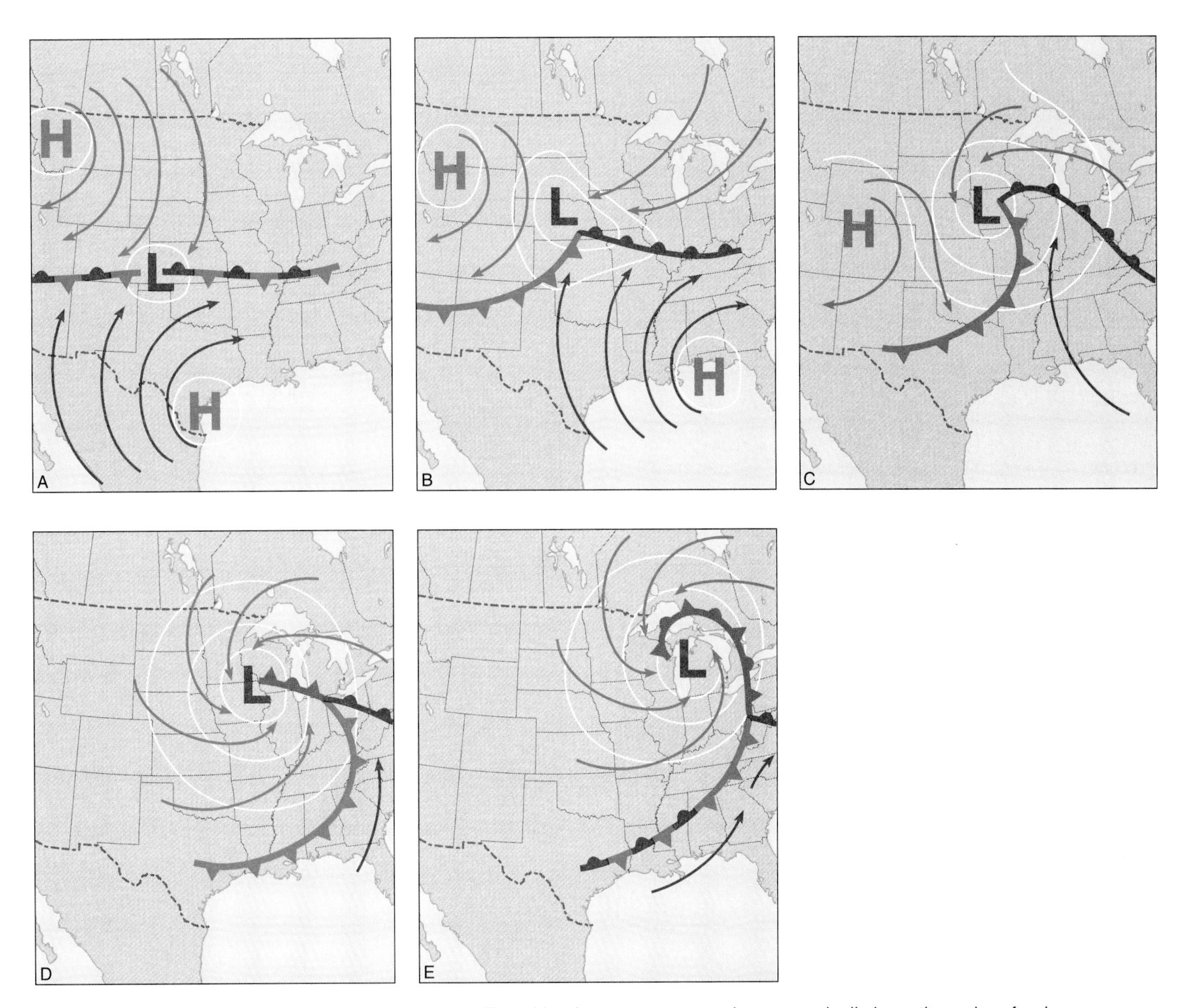


Figure 1.21 Stages of frontal cyclone development. Transition from one stage to the next typically is on the order of a day.

and difficult questions, ones we will consider in more depth later on. For now, we will continue to restrict ourselves to a purely descriptive model, as shown in Figure 1.21.

What is the life span of a typical frontal cyclone? Examine Figure 1.21 and notice that each stage is said to last a day or so. Thus, a typical system persists for roughly a week. (Highs may persist for a good deal longer.) Also notice from Figure 1.21A that the low tends to form on a stationary front, along which already exists a counterclockwise wind direction pattern. As the low develops, the circulation shapes the front into a wavelike pattern, shown in Figure 1.21B. In advance of the low (i.e., to its east), the front tends to move northward into colder air as a warm front, due to the wind circulation around the low. West of the low, the situation is reversed: winds around the low carry the front southward, bringing colder air to regions previously warm. Hence, this section of the front is behaving as a cold front. Note, however, that both warm and cold fronts are segments of the same front; the designators "warm" and "cold" refer only to the temperature changes caused by the front's movement.

Over the next two or three days, the typical low intensifies, that is, its central pressure becomes lower (Figure 1.21C). Everything about the low—its cloud cover, precipitation intensity, temperature contrast, wind speeds—becomes more intense. The fronts continue to turn counterclockwise around the low center, forming an increasingly pronounced backward-wave structure. The cold front generally travels faster than the warm front and eventually overtakes it, forming an occluded front, as shown in Figure 1.21D.

Generally, the low reaches its greatest intensity at the onset of occlusion. Thereafter, as the wave pattern seems to break and fall backwards (Figure 1.21E), attendant weather events be-

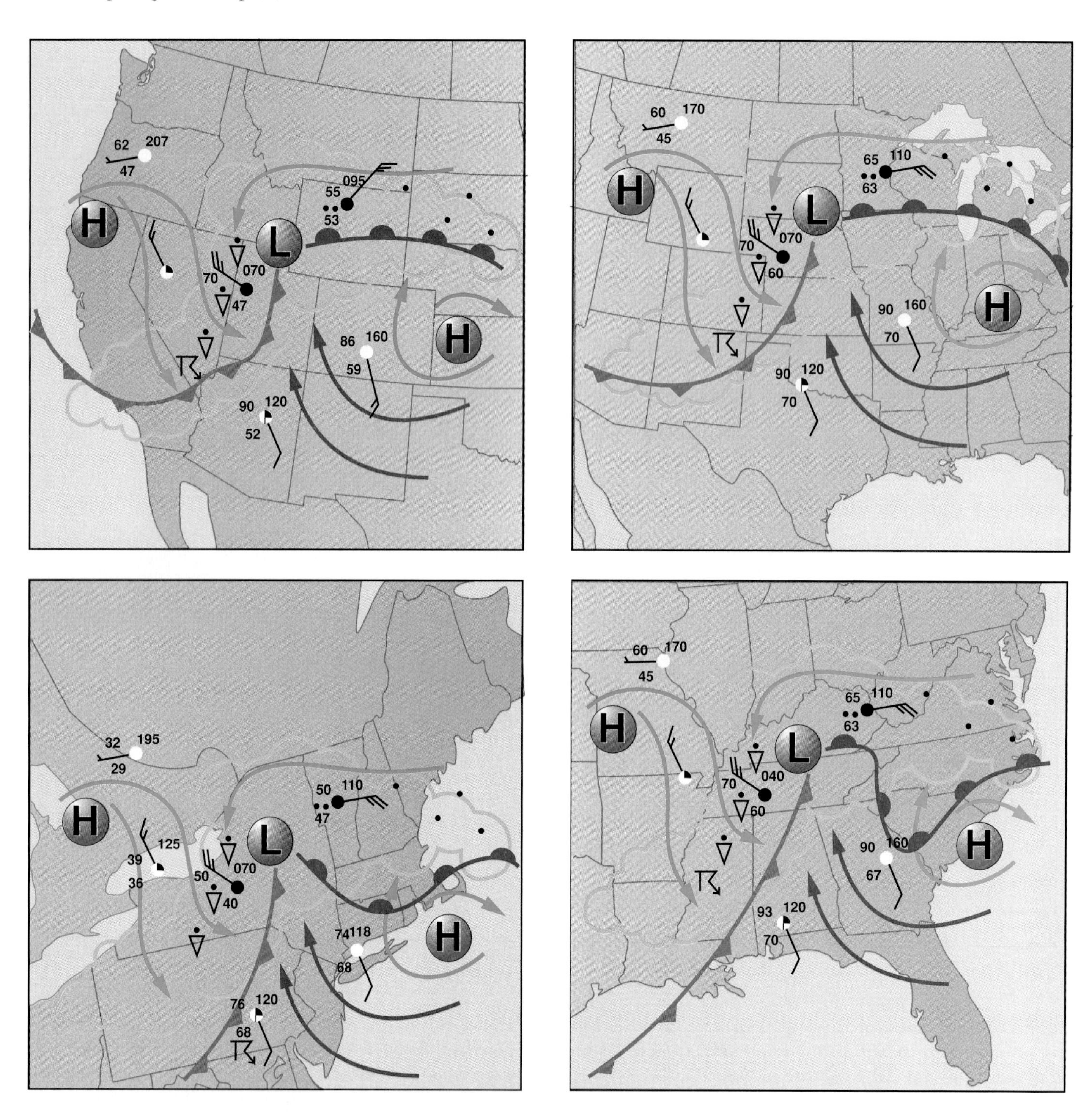


**Figure 1.22** Examples of frontal cyclones. Compare these examples with the stages of cyclone development in Figure 1.21.

come less intense. In a few more days, the low's circulation weakens and dissipates.

## Evaluating the Frontal Cyclone Model

Once again we return to assess the validity of a proposed explanation. In this case, we consider whether the presence of a passing frontal low pressure system of the type described above accounts satisfactorily for the Chicago-Evansville weather differences.

We have already examined the specific meteorological data in terms of the frontal model and have found that the correspondence between the observed data and the model is quite satisfactory. For the model to be generally useful, however, we ought to be able to recognize its presence on other days. If the model fits only the July 14 weather, then it is of limited value and doubtful validity even on that day. Figure 1.22 presents simplified weather maps for four other days, chosen more or less at random. In each case, it is relatively easy to find patterns consistent with the model.

The fact that it is a regular feature adds substantially to its usefulness.

A more satisfactory and instructive way to verify the model's validity would be to take regular weather observations throughout this course and compare them with real-time weather images available online at several sites on the World Wide Web, such as those listed in the Resources Links Section at the end of this chapter. Consistency between your local observations and large-scale weather map features would strengthen both your confidence in the model and your skill in applying it.

Thus, the frontal cyclone model seems to provide a plausible and rather effective basis for understanding the July 14 weather conditions in the Midwest. We accept the hypothesis that a passing weather system (a frontal cyclone, as it turns out) is responsible for the sharp differences between Chicago's and Evansville's weather.

## Checkpoint

***Review***

The concept map in Figure 1.23, a somewhat more involved one than previous examples, summarizes the major features of the frontal cyclone model. Notice that fronts are associated with cyclonic circulations, not with anticyclones (highs).

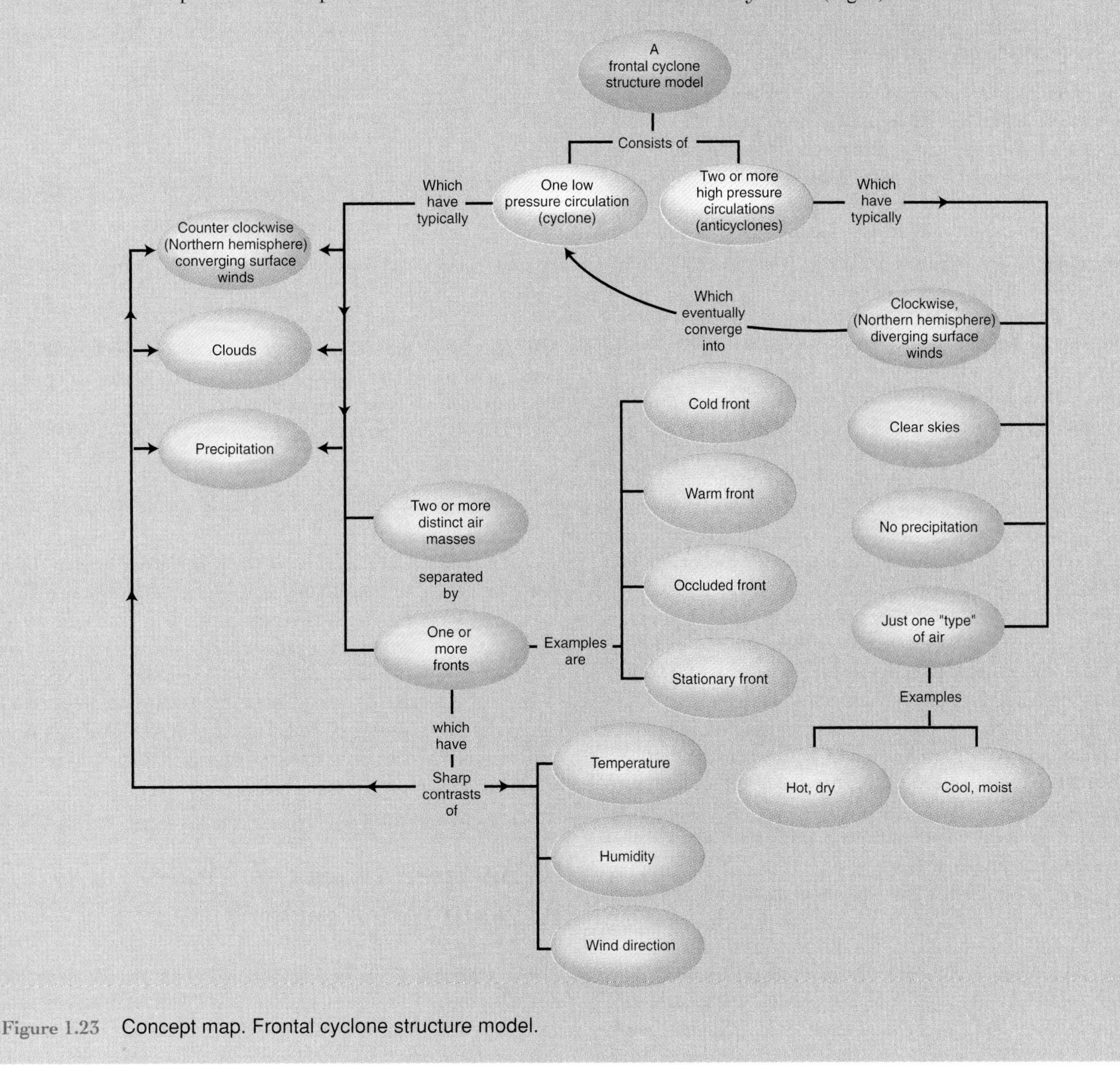


Figure 1.23 Concept map. Frontal cyclone structure model.

*Questions*

1. In what direction do winds circulate around lows? Around highs?
2. What is a front? What is the difference between a cold front and a warm front? What is an occluded front?
3. Why are clouds and precipitation generally found on the cold air side of a front's surface position?
4. What is wrong with the expression (sometimes uttered by the meteorologically uninitiated) that "a high front is approaching"?

# Applying the Frontal Cyclone Model

Many times during this course you will find the frontal cyclone model a useful tool. In this section, we will consider two examples of its utility.

## Using the Model to Forecast Local Weather

Suppose you make the following weather observation: clear skies, wind from the south at 15 knots, air pressure at 1010 millibars and falling steadily, temperature of 70°F. Assume that these conditions are due in some part to the influences of synoptic scale cyclones and anticyclones, and fronts (often a reasonable assumption for many mid-latitude locations). Can you use the cyclone model in conjunction with your observation to estimate future conditions?

Although the model alone cannot produce a definitive forecast, it can provide some insight. Return to Figure 1.19C for a moment. Where in the model do you find weather similar to the conditions you have just observed?

Clearly, there is no exact match between the sample conditions shown in Figure 1.19C and your observation. But conditions south of the low center in the model resemble your report fairly closely. Matching your observation to the model in this manner suggests that you are located in a warm sector. Your placement also implies that there is a low center to your north or northwest and a cold front to your west.

What will happen next? How will your weather change? Since you are reporting clear skies, no drastic change seems imminent. However, the model suggests the presence of a cold front to your west, and you know that frontal systems generally have a tendency to move eastward. Furthermore, you reported falling pressure, and fronts are located in regions of lower pressure. All of this suggests a cold frontal passage within the next day. Figure 1.19C shows that as the front passes, you can expect a brief period of showery weather, then clearing and cooler with northwest winds and rising pressure.

Reconsider, for a moment, the process of matching your observation to the model in Figure 1.19. The most useful piece of data was wind direction. Knowing wind direction alone, in fact, you can make a general estimate of your location within the high-low model. A French navigator named Buys Ballot noticed this fact and summarized it in a convenient "law" that now bears his name: Buys Ballot's law states that for observers in the Northern Hemisphere, if you stand with your back to the wind, low pressure will be to your left (and slightly ahead). Figure 1.24 illustrates Buys Ballot's law in practice.

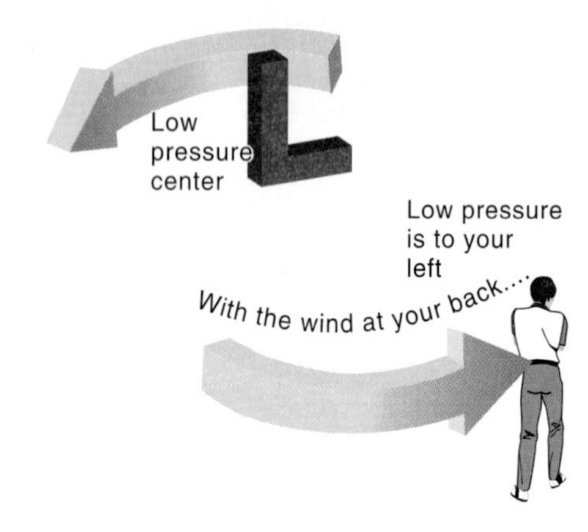


Figure 1.24 Buys Ballot's law. Standing with your back to the wind, lower pressure lies to your left, higher pressure to your right.

## Using the Model to Estimate Weather Conditions Elsewhere

Sometimes it is useful to reverse the process outlined in the previous section. That is, if we know the configuration of highs, lows, and fronts on a weather map, we can estimate weather conditions at a location not reporting the weather at all. Such a process is known as interpolation. In general, interpolation is the process of estimating unspecified data based on values that are given at surrounding locations (or times, etc.). You will use interpolation often as you work with weather charts and other graphs in this text.

# Reflection: Does the Model Explain the Weather?

Clearly, the model we have developed is a powerful tool for visualizing the atmosphere's structure and behavior on the synoptic scale. It is invaluable for summarizing a great deal of disparate information in a single image. It also is a powerful tool for showing relations between weather conditions at different places and for predicting conditions at locations for which no observations exist. Can we go farther than that and say that the model explains weather conditions?

Questions of this sort are "essay questions": they have no single correct answer. On the one hand, when we voice statements such as, "The reason our wind is from the south is that we are located in the warm sector of the low," we imply that our warm sector location explains the fact that our wind is southerly. But how do we know we are in the warm sector? Well, the warm sector is a region where the wind is from the south (among other things). So the wind is from the south because we are in the warm sector, and we know we are in the warm sector because the wind is from the south! Clearly, A can't be the cause of B while B simultaneously is the cause of A.

Suppose we say instead, "The reason we have a south wind is that the air is moving in response to pressure differences within the weather system." As we mentioned earlier, pressure is a force, and forces can cause motion; so this explanation feels more satisfactory. But to fully accept this explanation, we need to know specifically how pressure differences cause wind. Thus, we must move to another level of explanation, relating local weather conditions to basic physical principles such as the laws of motion, gravity, and so on. Atmospheric scientists have been extraordinarily successful over the past century in learning how to apply these basic principles to explain and predict local weather conditions. Their success is a recurrent theme of this book.

But if an explanation of local weather rests on these basic physical principles, don't we need to understand what the physical principles themselves rest on before we can feel the we have a true explanation of the weather? In other words, don't we need to know answers to such questions as " Why are there laws of motion, why is there gravity, why is there space?" At this point, the only answer to such questions seems to be "that's just the way the universe is made," which of course is no explanation at all. Thus, we follow a series of "why-because" exchanges to arrive finally at the limits of our current knowledge. We find there is no ultimate scientific explanation or understanding, only reference to the next higher level of generality.

This example is a useful reminder that the goal of scientific inquiry is not to find out why, in any absolute sense. The goals are considerably less ambitious: they are to find out what. The aim of scientific inquiry is description, including construction of models and principles, like those in this chapter, that show relations between different objects and phenomena, and can be used to make predictions. Whenever we speak of explanation, we must remember our explanations are of a limited nature. We must recognize an explanation in science is, in essence, a connection to a more general model or rule, which itself has no ultimate scientific explanation.

# Study Point

## Chapter Summary

In this first chapter, you have been instructed in the use of a number of tools employed in pursuing atmospheric investigations. Some of these are tools in the literal sense: they include thermometers, barometers, and transmissometers, used to measure various properties of the air. Other tools are more abstract or metaphorical: these include the station plotting model, a tool used to picture a single location's weather observation; and the frontal cyclone model, a useful tool in making sense of the array of data you see on a typical weather map. This first chapter has also introduced you to several key ideas concerning scientific inquiry. The most important is that science is a never-ending process of activity that includes noticing, asking questions, posing explanations in the form of hypotheses, testing and then rejecting or accepting one's hypotheses, raising new questions, and so forth in a continuing cycle. The chapter's meteorological content included an introduction to the basic variables that comprise the standard weather observation: clouds, visibility, present weather, precipitation amounts, wind, temperature, humidity, and air pressure. We also discussed techniques and instruments involved in measuring these weather elements and for plotting the reports on weather maps. We presented circulations, fronts, and the frontal cyclone model, a comprehensive scheme for visualizing relations among the various weather elements at different locations. We concluded with a reflection on the meaning of scientific explanation.

## Key Words

| | |
|---|---|
| meteorology | atmosphere |
| overcast | scattered clouds |
| broken clouds | clear |
| obscured | prevailing visibility |
| transmissometer | present weather |
| Beaufort wind scale | knot |
| anemometer | wind vane |
| Fahrenheit | Celsius |
| humidity | dew point |
| relative humidity | air pressure |
| force | barometer |
| millibar | concept map |
| pressure tendency | low |
| high | isobar |
| synoptic scale | cyclone |
| anticyclone | subsidence |
| front | cold front |
| warm front | stationary front |
| occluded front | frontal cyclone |

## Review Exercises

1. List eight weather elements that are included in a normal weather observation.
2. For each of the weather elements in question 1, name the instrument used to measure that variable.
3. The highest air temperature ever officially recorded was 136°F, in El Azizia, Libya. What is this temperature in degrees Celsius?
4. What would you estimate the air pressure to be in central Ohio in Figure 1.18?
5. Describe the weather associated with the passage of a typical cold front. Do the same for a typical warm front and an occluded front. (Why do we not ask the question for a stationary front?)
6. Suppose you drove from Flint, Michigan, to St. Louis, Missouri, on July 14, 1992. Describe the weather you would have experienced.
7. You go out to make your daily observation and notice that along one-fourth of the horizon, you can see 20 miles, but along the remainder of the horizon, you can see only 5 miles due to haze. What do you report for the visibility?
8. Construct a concept map for different types of fronts.

## Problems

1. Your sister in Atlanta calls and says the temperature at her house was 120 yesterday afternoon. The newspaper says the official high was "only" 90. Pose a reasonable hypothesis to explain the discrepancy.
2. The weather in Dallas one recent afternoon is shown below. Explain what might have caused the change.

| Time | Temperature | Wind | Clouds and Weather |
|---|---|---|---|
| 1 P.M. | 90 | S/10 | 2/10 |
| 2 P.M. | 91 | S/11 | 3/10 |
| 3 P.M. | 77 | NW/15 | 10/10, shower |
| 4 P.M. | 74 | NW/15 | 4/10 |
| 5 P.M. | 74 | NW/12 | 2/10 |

3. An observer reports the weather at her location to be overcast, winds from the southeast at 3 knots, visibility 3 miles due to fog, temperature of 51°F, and air pressure at 1006.5 millibars. (a) Plot her report according to the plotting model. (b) The dew-point temperature was not

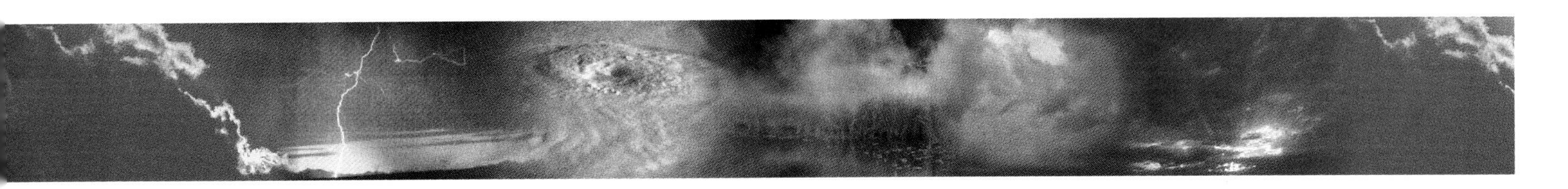

reported for the station. Considering the other observed conditions, what is a likely value for the dew point? Explain your answer.

4. Why is it colder, typically, to the east of a high pressure center than to its west?
5. It is a sunny winter morning. The wind is from the southeast at 15 knots, the temperature is 20°F, the dew point is 15°F, and the barometer is 1020.3 millibars and falling steadily. What will your weather be like for the next 24 hours? Explain your forecast.
6. Re-state Buys Ballot's law to apply to high pressure instead of low pressure. Make a sketch illustrating your law.
7. In what direction do typical highs and lows move? Suppose they moved the opposite way. For a viewer in Washington, D.C., how would TV weather presentations change? Do you think forecasts there would be more accurate than now or less? Why?

## Explorations

1. What sort of regularities exist in your weather? How does it differ from that at the nearest meteorological observing station? You can answer these questions by maintaining a daily record of weather observations. You can measure and record cloud information, visibility, present weather, and wind speed and direction with no instruments beyond your five senses. If you have access to a thermometer, barometer, and humidity gauge, record the readings from these items also. If you have a personal computer, store your data in a computer spreadsheet. Then use the spreadsheet's graphing capabilities to display your data.
2. To what extent are weather map features reflected in your local weather? Make a practice of checking at the Weather Channel, NOAA Weather Radio station, weather sites on the World Wide Web (see suggestions in Resource Links below), local weather forecasting programs, or newspaper weather forecasts. Relate weather conditions you observe to the highs, lows, and fronts shown on the forecast map.
3. Log onto the Internet or connect to a weather service such as AccuWeather and collect surface maps six hours apart in time. Do these maps tend to confirm the basic components of the circulation model presented in this chapter? Thus, do weather systems tend to move eastward, do winds circulate around highs and lows as expected, is fair weather associated with highs and not lows or fronts, and so forth? (See Resource Links below for suggested Internet addresses.)

## Resource Links

The following is a sampler of print and electronic references for topics discussed in this chapter. We have included a few words of description for items whose relevance is not clear from their titles. More challenging readings are marked with a C at the end of the listing. For a much extensive and continually updated list, see our World Wide Web home page at www.mhhe.com/sciencemath/geology/danielson.

Watts, A. 1994. *The Weather Handbook.* Dobbs Ferry, NY: Sheridan House. 187 pp. A nontechnical introduction to weather and forecasting.

Middleton, W. E. K. 1969. *Invention of the Meteorological Instruments.* Baltimore: Johns Hopkins Press. 362 pp. C

University of Michigan's Weathernet at http://cirrus.sprl.umich.edu/wxnet/ is a rich source of weather information and links to other sources.

Pennsylvania State University's hourly U.S. Weather Statistics at http://www.ems.psu.edu/cgi-bin/wx/uswxstats.cgi is an excellent source for current weather conditions.

AccuWeather's home page at http://www.AccuWeather.com is an extensive source of current weather conditions including worldwide satellite images, Doppler radar images, current and forcast surface weather maps, current weather condition, and weather discussions.

For readers with access to AccuData, AccuWeather's real time database, SFC gives worldwide surface weather observations. RAD provides radar images. DIS gives discussions of current weather conditions. SATMAP gives satellite images. SURMAP gives surface and forecast maps. SURA enables the user to plot and analyze weather data for any location.

For users with access to AccuWeather's AccuData, SURMAP 1 chart gives positions of highs, lows fronts, and precipitation for current conditions. The Weather Channel© has a similar chart at http://www.weather.com/images/curwx.gif. Other relevant AccuData charts products include HRD xxx, for latest data at station *xxx*, and PLS *mmm*, where mmm is a given background map.

# Chapter 2

# *Is the Atmosphere's Composition Changing?*

# Chapter Goals and Overview

Is the composition of earth's atmosphere changing? Measurements of the atmosphere's nitrogen, oxygen, and argon content (which together comprise over 99.9% of the composition of dry air) indicate no significant change in these gases over the period meteorologists have made such measurements (roughly the past 100 years). On the other hand, the photo on the facing page is convincing proof that other processes occur that obviously alter the air's composition.

How is this paradox resolved? Is the influence of factors such as air pollution and volcanoes limited to 0.1% of the atmosphere's composition? Is that 0.1% changing? Is it important, despite its modest size? Are there other processes that remove the injections of volcanic and human-made matter? Are the changes so slow that they don't appear in measurement trends? These are some of the issues that will occupy us in this chapter.

The topic for investigation for this chapter has a very different character from that of the last. In Chapter 1, you dealt with atmospheric behavior on a large scale, and you employed a model appropriate for this range of behaviors. That model was a holistic one: you saw how atmospheric conditions in different places were part of much larger patterns. At times, it would have been helpful to position yourself high above the atmosphere to look down on it all. In this chapter, you will move in the opposite direction: you will want to be able to visualize smaller and smaller scales of reality, right down to the world of atoms and molecules. The first tasks, therefore, will be to review basic concepts about the structure of matter and to develop a model appropriate for dealing with problems on this scale. Then you will employ the model to investigate atmospheric composition and its changes.

## Chapter Goals

By chapter's end, you should be able to:

describe, using words and/or diagrams, a model for conceptualizing matter as an assemblage of atoms and molecules;

distinguish among atoms, molecules, elements, compounds, and mixtures;

explain the concepts of temperature, pressure, density, and phase change in terms of the molecular model and/or kinetic theory of gases;

name the five most common gases in the earth's atmosphere and their approximate concentrations;

name several important air pollutants, their sources, and strategies for controlling or reducing their effects; and

explain how ozone is created and destroyed in the atmosphere, the role of humans in these processes, and some of the consequences of ozone pollution and depletion.

# Properties of Matter

Consider for a moment the properties all objects from the world of matter hold in common. For example, what is it about this textbook that makes it a "real" object and not an image, an idea, or a memory? A standard response, which will serve us well, is that the book, like all matter, has mass and takes up space. But exactly what does that mean? Although these are familiar concepts, it is useful to define them precisely at this point.

## Large-Scale Properties

Large-scale properties of matter include its mass, volume, and density.

### *Mass*

**Mass** is a measure of an object's resistance to a change of motion. As you can see in Figure 2.1A, the more massive an object, the harder it is to alter its course—or to start it moving. Mass is an innate property of an object and is constant regardless of the object's location. Mass is usually measured in units of grams (g) or kilograms (kg). Some students confuse mass and weight. Unlike mass, weight is a force; it is a measure of earth's (or another body's) gravitational pull on an object and depends on both the object's mass and the strength of gravity. Thus, if gravity changes, your weight changes. On the moon, for example, you would weigh only one-sixth as much as on earth, although your mass is the same in both places.

### *Volume*

Matter's second basic property, that it takes up space, is related to the concept of **volume**. Volume is simply a measure of the amount of space occupied by an object (Figure 2.1B). The fact that matter occupies volume means that you cannot put two objects in the same space. In meteorology, volume is given in units of cubic meters ($m^3$) or cubic centimeters ($cm^3$). (For a refresher on exponents and powers-of-10 notation, see Appendix B.)

### *Density*

**Density** is the amount of mass contained in a given volume. Since every physical object has mass and volume, it also has a certain density, which is its mass divided by its volume. In formula form, $\rho = M/V$, where density is represented by $\rho$, the Greek letter *rho*, and M and V are mass and volume, respectively (see Figure 2.1C). Units of density are grams per cubic centimeter ($g\ cm^{-3}$), kilograms per cubic meter ($kg\ m^{-3}$), and so on.

## Atomic Structure of Matter

### *Atoms*

Imagine doing the following thought experiment: you capture a piece (an atom) of argon and place it under a fictitious, extremely powerful measuring instrument with which you can

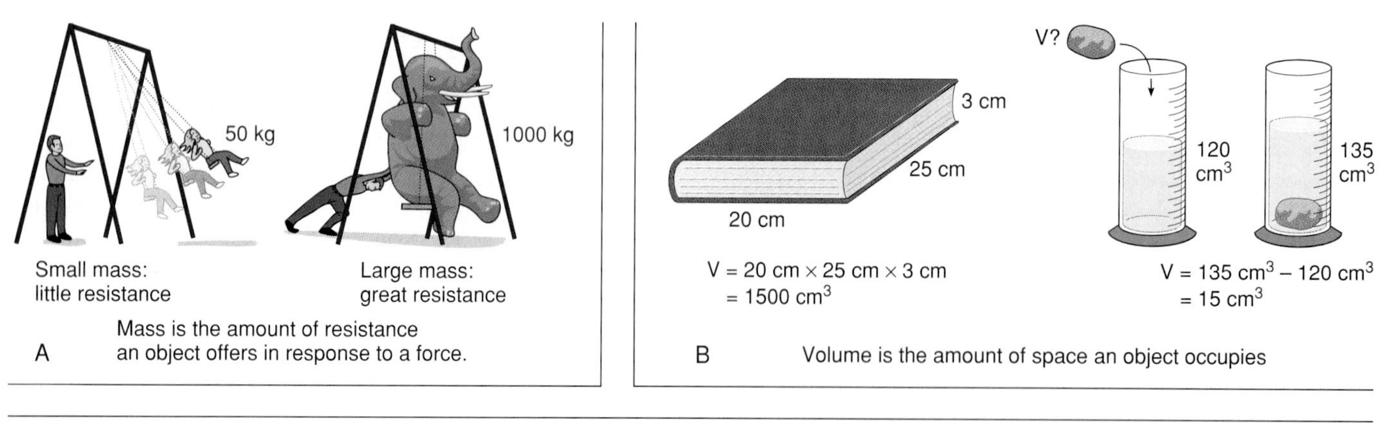


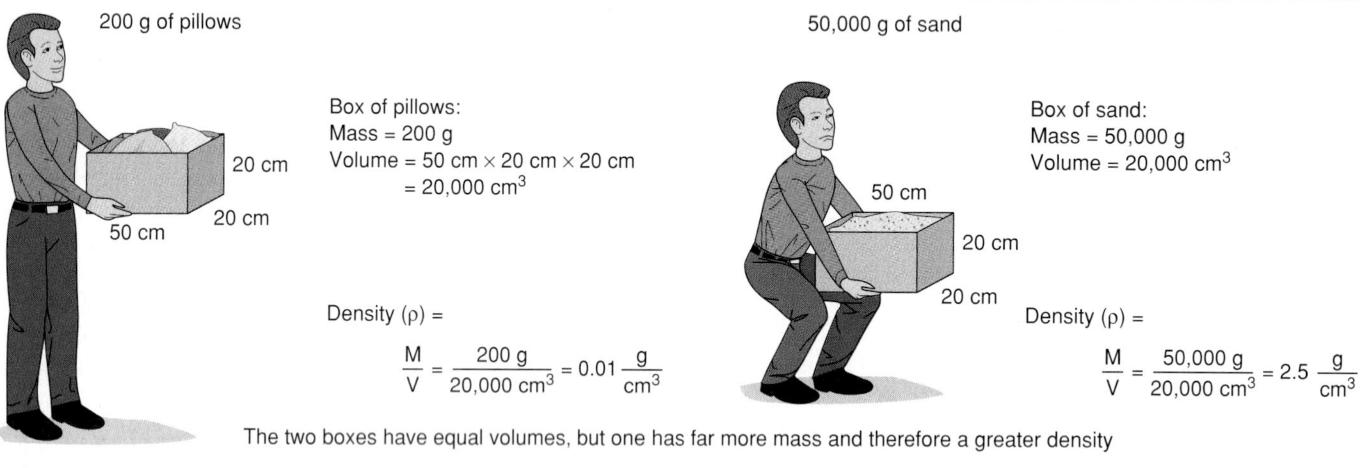


Figure 2.1 Some basic physical properties of matter.

determine all manner of the particle's properties, including its shape, size, and structure (if any), its mass, and its electric charge. Figure 2.2 shows your view of the argon through this hypothetical instrument, according to currently accepted atomic theory.

You can see in Figure 2.2 that your single argon particle is not an undifferentiated lump of "argonness" but has a structure of its own. Notice that it consists of a central **nucleus** of 40 smaller particles. Eighteen of these smaller particles, named **protons,** have identical, positive electric charges and masses. Each of the other 22 nuclear particles is just slightly more massive than a proton but has no charge at all; being electrically neutral, it is called a **neutron**. Surrounding this 40-particle nucleus in a haze-like series of orbits (called "orbitals") are 18 incredibly tiny particles, called **electrons,** each carrying identical electric charges of exactly the same size as the proton's charge but opposite in sign: negative charges. The overall configuration of nuclear protons and neutrons, surrounded by electrons, is known as an **atom**. Thus, your argon particle is more properly called an argon **atom**.

The electrons' distance from the argon atom's nucleus is immense compared to the size of the nucleus itself: a model of the argon atom using a bunch of grapes to represent the protons and neutrons in the nucleus would require the outermost electrons to be located 5 miles from the nucleus. In a sense, atoms and all physical objects are mostly empty space: if you could tightly pack the particles comprising a person's body, eliminating all the space between nuclei and electrons, you could fit the population of New York City into a sphere of incredible density with the same diameter as the period at the end of this sentence.

Are all of these tiny subparticles—that is, the protons, neutrons, and electrons of argon atoms—also made of argon? Suppose you could split the nucleus, a process known as nuclear fission, carving one proton from the nucleus, as shown in Figure 2.3. Testing these two separate pieces, you would find that neither one exhibits the properties of an argon atom any longer. In fact, the single proton exhibits the properties of a hydrogen atom, while the other nucleus, with the remaining 17 protons, resembles the element chlorine! Recombining the proton with the chlorine nucleus, you would find you have an argon atom once more. (The process of combining nuclei is called nuclear fusion; fusion is the opposite process from fission.)

These considerations suggest that argon is not some ultimate, indivisible substance, but exists as the result of a specific, atomic configuration of protons, neutrons, and electrons. Confirmation of this hypothesis might involve further fissions and

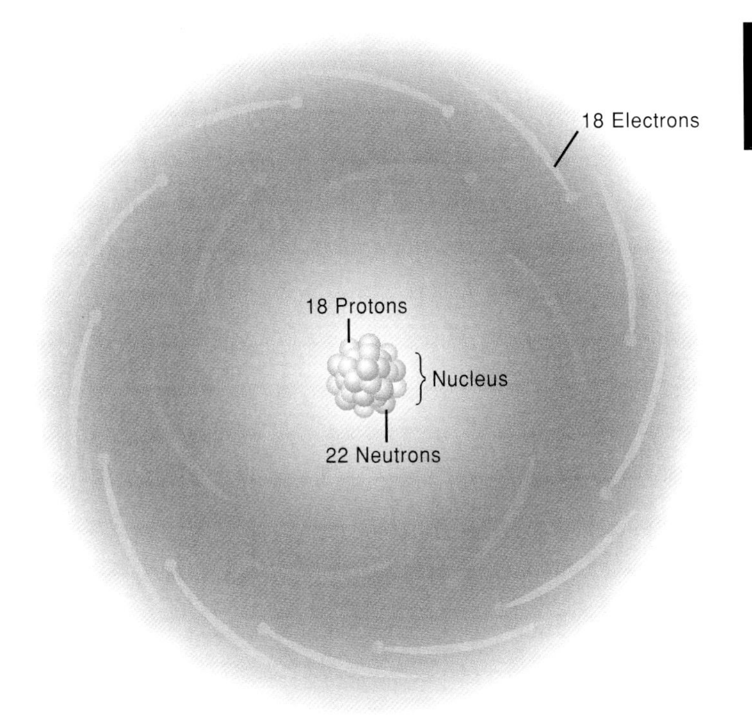


Figure 2.2 An atom of argon. Note that the distance between the atom's nucleus and its electrons is far greater than shown (see the text). As indicated—except that it should show a greater space between nucleus and electron "haze"; it would be good not to show electrons as points, or as occupying orbits like planets.

**Table 2.1 Experiments on an Argon Nucleus**

| Trial Number | Proton Count (Atomic number) | Neutron Count | Electron Count | Resultant Element |
|---|---|---|---|---|
| STARTING CONDITIONS: | | | | |
| 1 | 18 | 22 | 18 | Argon |
| REMOVE ONE PROTON, NEUTRON, ELECTRON: | | | | |
| 2 | 17 | 22 | 18 | Chlorine |
| 3 | 18 | 21 | 18 | Argon |
| 4 | 18 | 22 | 17 | Argon |
| REMOVE TWO PROTONS, NEUTRONS, ELECTRONS: | | | | |
| 5 | 16 | 22 | 18 | Sulfur |
| 6 | 18 | 20 | 18 | Argon |
| 7 | 18 | 22 | 20 | Argon |
| ADD ONE PROTON, NEUTRON, ELECTRON: | | | | |
| 8 | 19 | 22 | 18 | Potassium |
| 9 | 18 | 23 | 18 | Argon |
| 10 | 18 | 22 | 19 | Argon |
| ADD TWO PROTONS, NEUTRONS, ELECTRONS: | | | | |
| 11 | 20 | 22 | 18 | Calcium |
| 12 | 18 | 24 | 18 | Argon |
| 13 | 18 | 22 | 20 | Argon |

18 Protons
22 Neutrons
Begin with an argon nucleus...
... split off one proton, and end up with...
1 Proton
17 Protons
22 Neutrons
Hydrogen and Chlorine

Figure 2.3 Fission of one proton from the argon atom's nucleus yields nuclei of hydrogen and chlorine.

fusions of the argon atom. Systematically adding and removing different numbers of protons, neutrons, and electrons and observing the results would yield the data presented in Table 2.1.

### *Elements*

Examine the data in Table 2.1; notice that all of the argon atoms contain 18 protons. The number of neutrons or electrons may vary, but the proton count, known as the **atomic number,** does not. Note also that atoms with atomic number 17 are chlorine atoms and those with atomic number 19 are potassium. An atom's atomic number determines whether it is an atom of argon, chlorine, or some other substance.

A substance composed of atoms of the same atomic number is known as an **element**. Thus, argon is an element, composed entirely of atoms with atomic number 18; chlorine is an element composed of atomic number 17 atoms; and so forth. On earth, elements of atomic number 1 through 92 occur naturally; they are listed for your reference in Appendix E. Note the abbreviations given to the different elements in that table: Ar for argon, O for oxygen, and so on. You will use these abbreviations frequently in this and later chapters.

### *Isotopes*

The data in Table 2.1 illustrate several other useful concepts concerning the structure of matter. For example, notice that although an argon atom's atomic number must be exactly 18, its neutron count may vary considerably. An atom with a given number of neutrons is called an **isotope**. Eight different isotopes of argon occur naturally, having neutron counts ranging from 17 to 24. Different isotopes are typically indicated by their **atomic mass number,** which is the total number of protons plus neutrons in the nucleus. Thus, the most common isotope of argon, with 18 protons and 22 neutrons, is known as argon-40, or simply Ar-40. An argon atom with just 17 neutrons would be designated Ar-35.

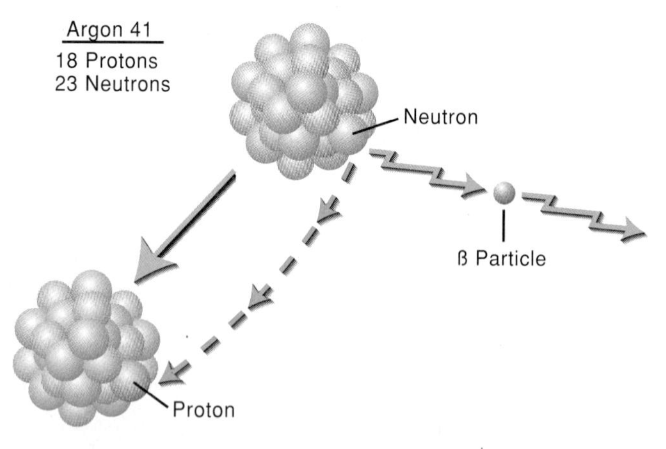


**Figure 2.4** An argon-41 nucleus undergoes radioactive decay by emission of a "beta particle" (an electron) and change of a neutron into a proton. The result is a potassium-41 nucleus.

### *Radioactive Decay*

If you could observe a single atom of the isotope argon-41 continuously over a period of several hours, you probably would witness the remarkable event shown in Figure 2.4: at some moment, one of the atom's neutrons likely would change spontaneously into a proton, accompanied by ejection of a particle from the atom's nucleus. With this additional proton, the nucleus would no longer be argon, whose atomic number is 18, but an atom of atomic number 19, which is listed in Appendix E as potassium.

This process is an example of **radioactive decay**. Radioactive decay is the transmutation of a nuclear particle into one or more other particles plus energy. Many isotopes are unstable, meaning that they decay radioactively. Later in this chapter, you will consider pollution caused by radioactive decay of the unstable gas radon-222.

## Molecules, Compounds, and Mixtures

### *Molecules*

The atoms of many substances do not occur singly but **bond,** or join chemically, in pairs (see Figure 2.5). Thus, most nitrogen

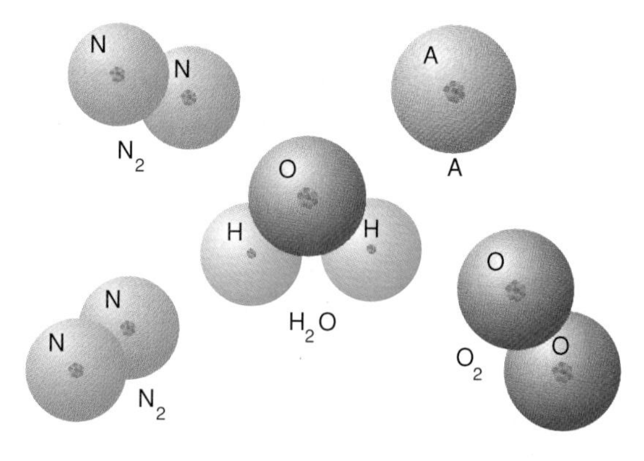


**Figure 2.5** Most atoms in the air do not appear singly, but are bonded into molecular configurations.

**Table 2.2 Properties of Atmospheric Constituents**

| Substance | | Atomic Number | Atomic Mass Number | Common Form | Molecular Mass |
|---|---|---|---|---|---|
| Nitrogen | (N) | 7 | 14 | diatomic | 28 |
| Oxygen | (O) | 8 | 16 | diatomic | 32 |
| Argon | (A) | 18 | 40 | monatomic | 40 |
| Water vapor | ($H_2O$) | 1+1+8 | 1+1+16 | triatomic | 18 |
| Carbon dioxide | ($CO_2$) | 6+8+8 | 12+16+16 | triatomic | 44 |
| Neon | (Ne) | 10 | 20 | monatomic | 20 |
| Helium | (He) | 2 | 4 | monatomic | 4 |

atoms in the lower atmosphere are bonded in pairs. The same is true of oxygen atoms. Water vapor consists of a cluster of three bonded atoms, two hydrogens and an oxygen. In fact, only argon, of the half-dozen or so most common ingredients of the lower atmosphere, appears in atomic form, that is, as single atoms.

### *Compounds*

Two or more atoms chemically bonded, such as the two nitrogen atoms or the two hydrogen plus one oxygen atom in water, comprise a **molecule**. The substance formed by a single kind of molecule ($H_2O$, for example) is called a **compound** (water, in this case). Most of the substances we encounter in our environment are, in fact, compounds and not pure, elemental substances. Incidentally, note the parallel between atoms and elements, on the one hand, and molecules and compounds. Atoms are the fundamental building blocks of the elements, as molecules are for compounds; splitting an atom or breaking a molecule changes that substance into another.

A molecule such as an oxygen-oxygen pair is called a **diatomic** molecule, because it is composed of two atoms. It is designated $O_2$. Similarly, diatomic nitrogen is abbreviated $N_2$. The substance **ozone,** an important atmospheric gas, is a **triatomic** (three atom) molecule of oxygen; its label is $O_3$. Table 2.2 summarizes some of the key atomic-scale properties of the most common atmospheric gases. A substance composed of individual atoms not bonded to each other is referred to as **monatomic** (one atom).

### *Mixtures*

Suppose you could count and catalog a sample of 1 million air particles. Despite the large number of particles involved, physically the sample would be a very small one: if you lined up those 1 million molecules in single file, they would form a line just one-tenth of a millimeter long, roughly the thickness of one page of this book.

Table 2.3 shows typical results of such a count, indicating that air exists in the form of molecules of $N_2$, $O_2$, $H_2O$, atoms of Ar, an occasional molecule of $CO_2$, and, much more rarely, some other atom, molecule, or larger particle—a clump of molecules, such as a water droplet. Note that the units in Table 2.3 are "parts per million," because your sample size was 1 million. See the discussion of these units at the bottom of the table.

These data show that air cannot be called an element or a compound because it is not composed of a single kind of atom or molecule. Instead, air is said to be a **mixture**. A mixture is a collection of different atoms, molecules, and/or larger particles in physical proximity to each other but not bound chemically.

*Table 2.3*

**Typical Results of Counting 1 Million Particles of Air**

| Name of object | | Count |
|---|---|---|
| ATOMS AND MOLECULES: | | |
| Nitrogen | ($N_2$) | 774,500 |
| Oxygen | ($O_2$) | 207,500 |
| Argon | (A) | 9,250 |
| Water vapor | ($H_2O$) | 8,365 |
| Carbon dioxide | ($CO_2$) | 352 |
| Neon | (Ne) | 18 |
| Helium | (He) | 5 |
| Methane | ($Ch_4$) | 2 |
| Krypton | (Kr) | 1 |
| Hydrogen | (H) | 1 |
| OTHER OBJECTS: | | |
| Dust particle | | 4 |
| Water droplet | | 1 |
| Pollen grain | | 1 |
| ATMOSPHERIC SUBSTANCES NOT APPEARING IN SAMPLE OF 1 MILLION: | | |
| Dinitrogen oxide | ($N_2O$) | |
| Carbon monoxide | (NO) | |
| Xenon | (Xe) | |
| Ozone | ($O_3$) | |
| Ammonia | ($NH_3$) | |
| Nitrogen dioxide | ($NO_2$) | |
| Nitric oxide | (NO) | |
| Sulfur dioxide | ($SO_2$) | |
| Hydrogen sulfide | ($H_2S$) | |

A note on terminology: A scientist might refer to these counts as "parts per million" because each is based on an imaginary sample of 1 million members. Thus, the argon count is said to be 9340 parts per million (abbreviated as 9340 ppm), the $CO_2$ count is 360 ppm, and so on. For very small concentrations, you may have to use a larger sample to count many objects of interest; then the reference you use might be "parts per billion," or ppb. Counts based on samples of 1000 items are described as parts per thousand, as you would guess, which is abbreviated ppt. Finally, counts compared to a sample of 100 items are called percent (per 100), which is represented by the symbol %, but you knew that one already.

Of course, you can scale a count upward or downward from one of these bases to another, just by multiplying or dividing by 10, 100, and so on. Thus, the $CO_2$ count of 360 ppm is the same as 0.360 ppt or 0.036%.

# Checkpoint

***Review***

We have presented a model of matter's structure based on the concept of the atom, along with a number of terms that help describe that structure. Although the experiments conducted here were only thought experiments, a wide variety of actual experiments provide evidence that the atomic model is a valid one. Remember also that models are human inventions, human attempts

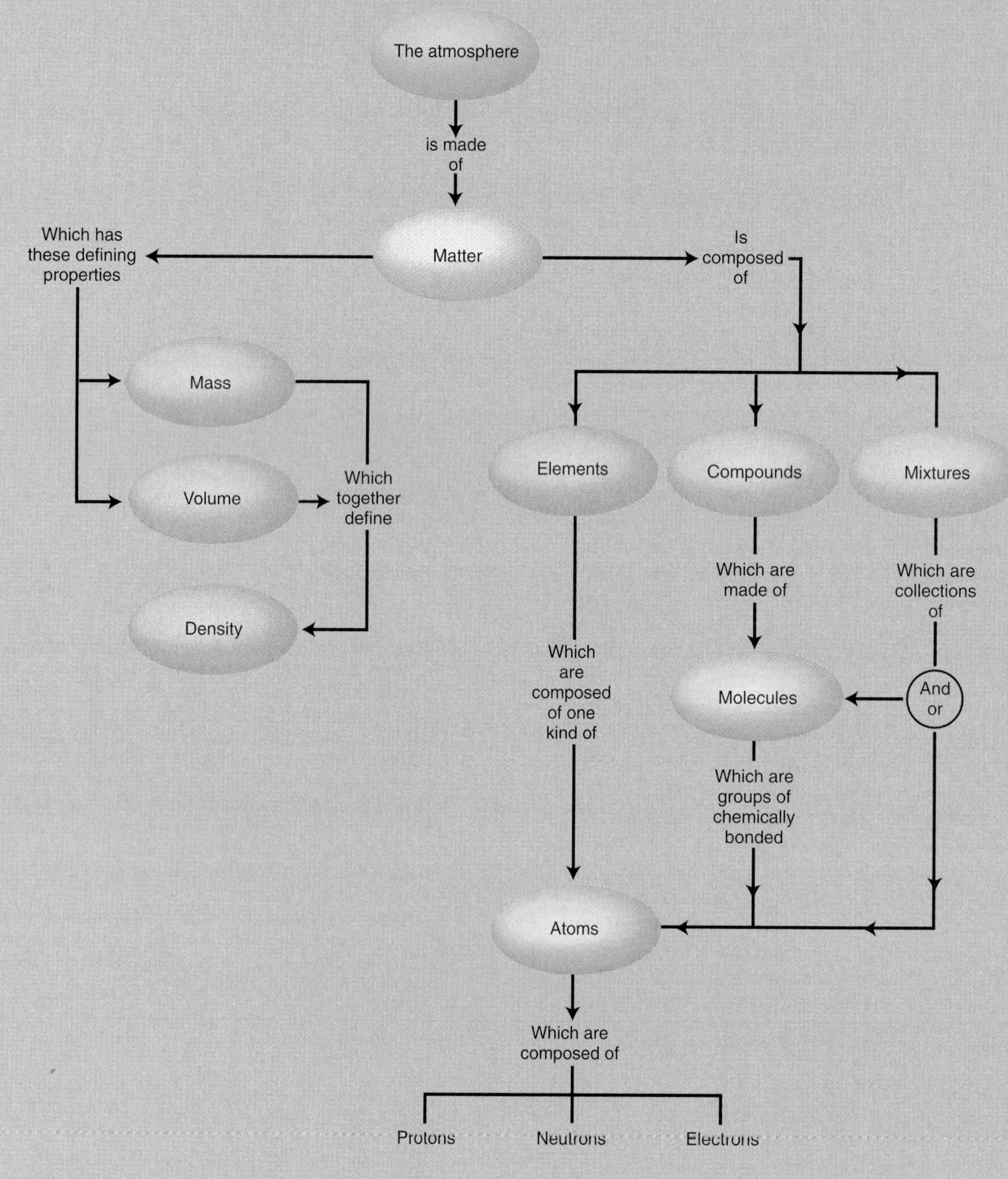


Figure 2.6 Concept map: Composition of Matter.

to describe nature. Thus, no model is literally perfect or true; but a model need not be perfect to be useful.

For review, trace through the concept map in Figure 2.6, making sure you understand the relations different objects bear to one another. As an additional exercise, you might extend the diagram to include properties of protons, neutrons, and electrons, and the process of radioactive decay.

***Questions***

1. Describe how you could determine the density of this textbook; of yourself; of a mountain made of granite.
2. Draw a sketch of a single oxygen atom, showing correct numbers of protons, neutrons, and electrons. (Use Appendix E for data.) Do the same for a hydrogen atom.
3. What is the difference between an atom and a molecule? Between an element, a compound, and a mixture?
4. Suppose you removed a particle containing two protons and two neutrons from the nucleus of an iron atom. What kinds of atoms would the resultant nuclei be? (Again, refer to Appendix E.)

# Molecules in Motion: Kinetic Theory and the Atomic Model

In science, a theory is a set of assumptions and principles that apply to a wide variety of phenomena and have withstood repeated tests. The central concept of the last section, that matter is composed of atoms, which in turn are composed of smaller particles, is known as the atomic theory of matter. This section deals with a theory that relates molecular motion and heat.

## The Well-Mixed Atmosphere

Figure 2.7A shows a mixture of materials of differing densities. From everyday experience you know that such mixtures tend to "settle out," with the densest particles falling and accumulating at the bottom. Air, which as you have seen is composed of particles of different densities, doesn't seem to settle out in this fashion, however. You might wonder, "What keeps the mixture called air mixed? Why aren't we, at the earth's surface, wading around in a layer of $CO_2$ and argon, the densest gases, as fantasized in Figure 2.7B?"

Consider how you might keep the material in Figure 2.7A mixed: perhaps by shaking vigorously. As long as the particles are moving fast enough, they do not settle out but remain more or less evenly distributed throughout the container. Could this explanation apply to the air as well? That is, could the air particles be moving sufficiently rapidly to keep themselves from settling out? In fact, molecules of air do move at tremendous speeds, often faster than the speed of sound, which is approximately 350 meters (1150 feet) per second. Furthermore, this motion is neither smooth nor organized but occurs in random directions. A typical molecule collides with neighbors many millions of times per second, changing direction and speed with each collision. Such vigorous collisional activity ensures that the different molecules remain well mixed. (Note the difference between these random, high-speed molecular motions and the much slower motion known as wind, in which air molecules move more or less together from one location to another.)

The portrayal of air as an assembly of innumerable tiny particles in constant and rapid collisional motion is commonly a starting assumption of a theory known as the **kinetic theory of gases**. (The word *kinetic* means "pertaining to motion.") Kinetic theory is central to understanding the atmosphere's behavior; therefore, we shall develop this important theory further.

## Kinetic Theory and Temperature

What causes air molecules to move so rapidly? Consider the everyday observation that heating water to a boil causes the water to move vigorously, while freezing it seems to stop all visible motion. Guessing that the air might respond similarly to temperature differences, you might hypothesize that adding heat to a gas sample increases molecular motion: the more heat added, the faster air molecules move. To test this hypothesis, examine the data given in Table 2.4.

Is the hypothesis confirmed? How, if at all, does the average molecular motion vary with temperature? Does the speed at a particular temperature vary in any systematic way with the gas molecules' molecular mass?

Inspection should tell you that the hypothesis is indeed confirmed: the speed does indeed vary with temperature, and hotter gas molecules move faster, on the average. Notice also that at a given temperature, less massive molecules have greater mean speeds.

## Phases of Matter

Everyday matter exists in three phases: as solids, as liquids, and as gases. Molecular motions within solids, liquids, and gases differ from one another in important ways. For example, consider the water molecules in Figure 2.8. In solid phase, the molecules move only by vibrating "in place"; they are not free to move about with respect to each other. An input of energy causes the molecules to move more vigorously and freely—while still remaining "in contact" with each other—a process we identify with melting, or changing phase to liquid. Further energy input frees the molecules from each other entirely, allowing them to expand indefinitely, their volume constrained only by the dimensions of their container. This change of phase is known as **evaporation** and represents the transition from liquid to gas phase. The reverse process, the transition from gas to liquid phase, is called **condensation.**

A

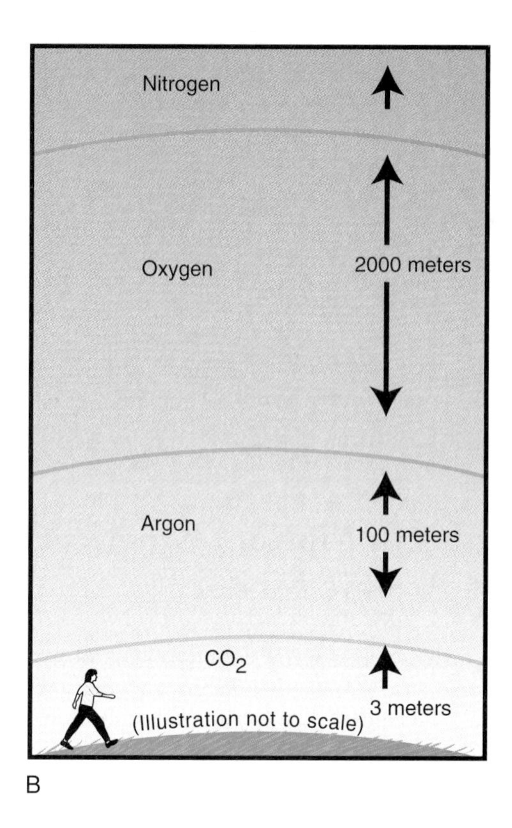


B

Figure 2.7 (*A*) Left unperturbed, the densest objects will settle to the bottom of the chamber, leaving lighter (less dense) ones above. (*B*) If the atmosphere settled out according to the density of its component gases, it would look like this. What problems does this suggest?

Table 2.4 The Effect of Temperature of the Mean Speeds of Air Molecules

| Gas | Molecular Mass | Temperature (Celsius) −20 | 0 | 20 | 40 | 60 | 80 | 100 |
|---|---|---|---|---|---|---|---|---|
| Argon | 40 | 366 | 380 | 394 | 407 | 420 | 432 | 444 |
| Carbon dioxide | 44 | 349 | 362 | 375 | 388 | 400 | 412 | 424 |
| Helium | 4 | 1157 | 1202 | 1245 | 1287 | 1328 | 1367 | 1405 |
| Hydrogen | 2 | 1637 | 1700 | 1761 | 1820 | 1878 | 1933 | 1987 |
| Mercury | 202 | 163 | 169 | 175 | 181 | 187 | 192 | 198 |
| Nitrogen | 28 | 437 | 454 | 471 | 486 | 502 | 517 | 531 |
| Oxygen | 32 | 409 | 425 | 440 | 455 | 469 | 483 | 497 |
| Water Vapor | 18 | 546 | 567 | 587 | 607 | 626 | 644 | 662 |

Average speeds in meters per second

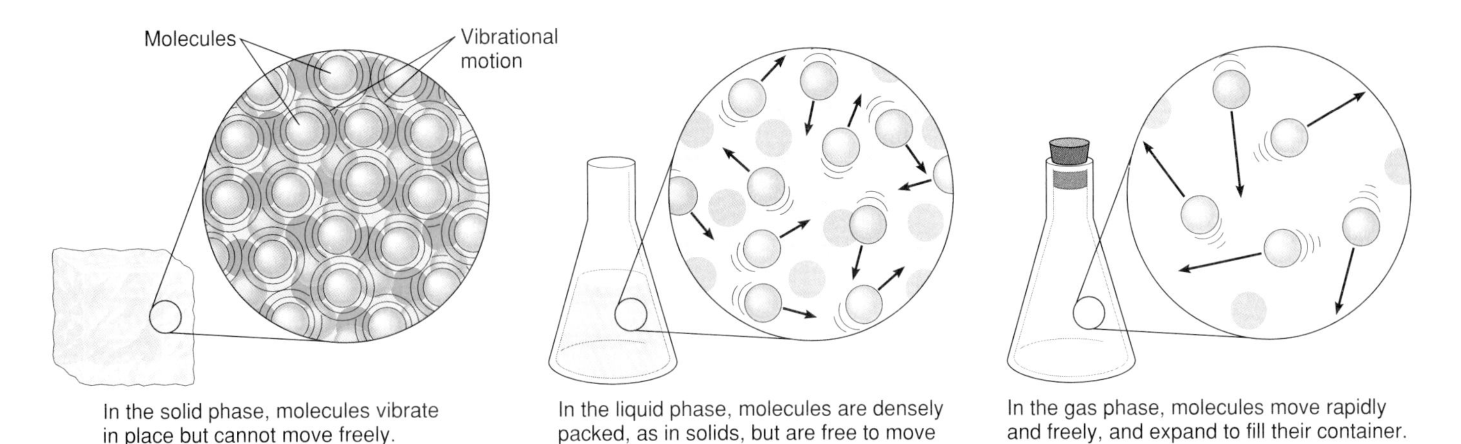


Figure 2.8 Phases of matter.

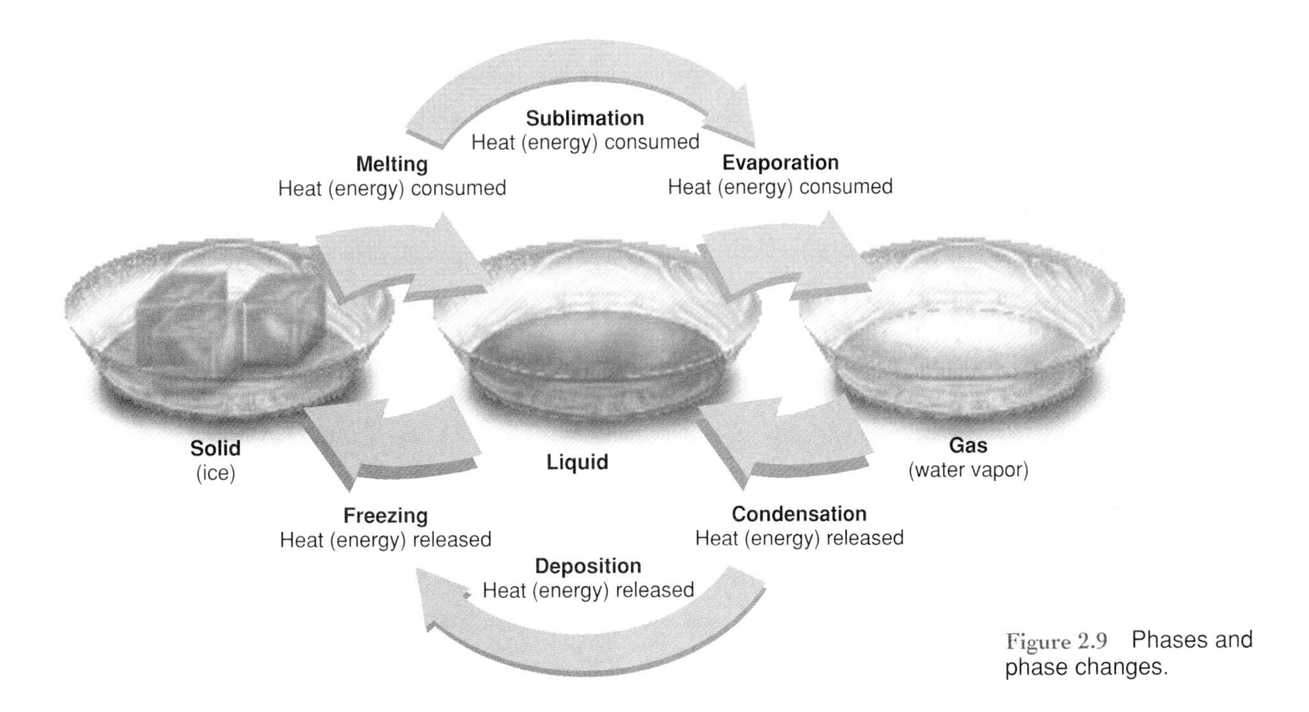


Figure 2.9 Phases and phase changes.

Figure 2.9 summarizes various phase changes. Note that matter may change directly from solid to gas, a process known as **sublimation,** and may change from gas directly to solid, which is known as **deposition**. The appearance and subsequent disappearance of frost on a window pane on a very cold morning provides an example of deposition followed by sublimation. If the temperature remains below freezing, then the water skips the liquid phase as it passes from water vapor to solid frost and then, later, back again to water vapor.

## Kinetic Theory and Air Pressure

A force, as you read in Chapter 1, is a push or a pull. A persistent force may result from a continuous push or pull, as in Figure 2.10A, or from the collective effect of a great many impacts, as in Figure 2.10B. Now, suppose the missiles in the figure are not tennis balls but countless numbers of air molecules moving at the speed of sound; also, cover the tennis racket with something more resistant to air, as in Figure 2.10C. The collective impact of millions and millions of air molecules striking the tennis racket cover creates a sizable force on each square centimeter of the racket cover. The average force on each square centimeter is just the quotient of the total force divided by the number of square centimeters and is known as the pressure. Since the pressure in this case is caused by the air, it is called air pressure.

Thus, air pressure can be explained in terms of the kinetic theory of gases: rapidly moving air molecules exert a pressure on any object with which they collide.

The amount of pressure the air exerts under normal circumstances may surprise you. Near sea level, in a typical weather

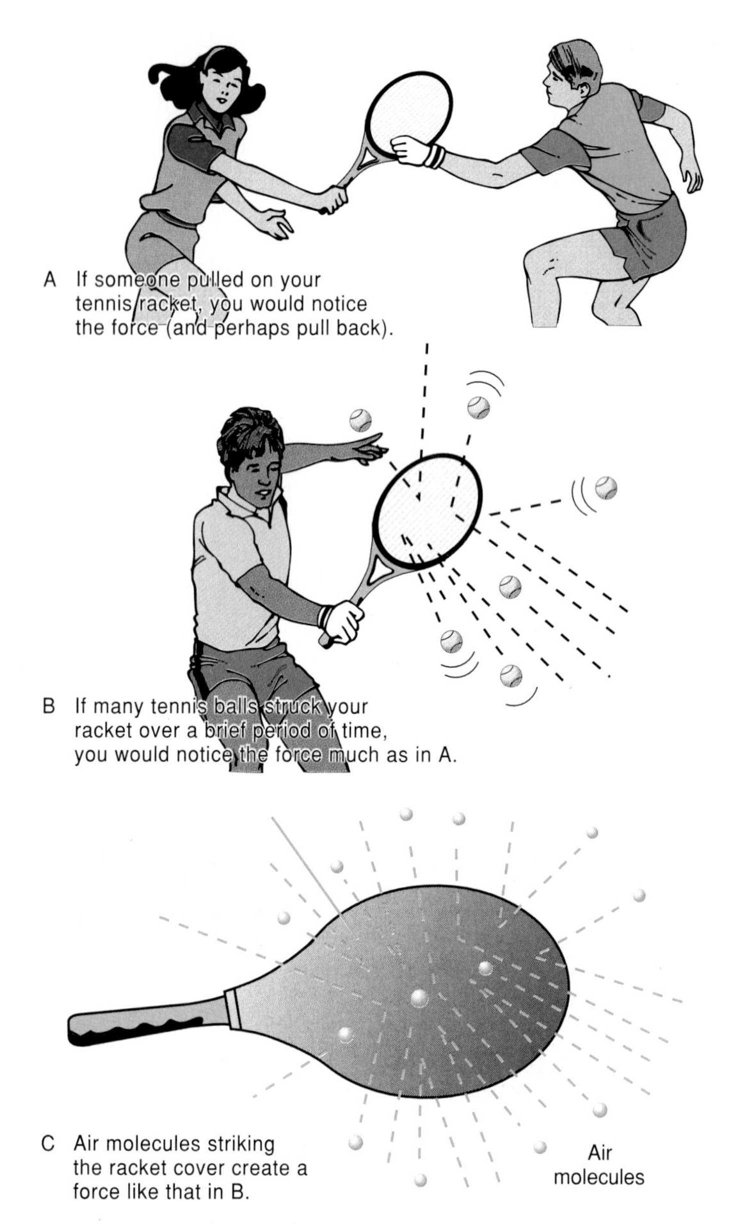


Figure 2.10 Impacts of particles create pressure forces.

system with barometer readings of around 1000 millibars, the pressure amounts to roughly 14 pounds per square inch. If you assume that the tennis racket cover in Figure 2.10 has a surface area of about 200 square inches, the total force comes to 200 × 14 pounds of force, or a total of 2800 pounds of force due to air pressure! What keeps the racket from flying off at the speed of sound, pushed away by all that air pressure? Only the fact that the other side of the cover is experiencing the same amount of force in the opposite direction. Because air molecules move randomly in all directions, the pressure they exert is virtually the same on each side of the racket. The two forces are equal in magnitude but opposite in direction, so their effects cancel out and no motion occurs.

## Pressure, Temperature, and Density

Consider the tennis ball model of pressure in Figure 2.10 once more. If you hit the racket with twice as many tennis balls in the same amount of time, it is evident you will double the pressure on the racket. The same reasoning applies at the molecular level: all other things being equal, doubling the density of gas molecules results in a doubling of the number of impacts and therefore a doubling of the gas pressure.

Suppose you increase the speed at which the incoming tennis balls hit the racket. How will that affect the pressure? To what would this correspond at the molecular level? You should recognize that increasing the tennis balls' speed (which is analogous to increasing the temperature of a gas) would increase the force on the racket (analogous to an increase of gas pressure).

## The Gas Law

You have seen in the last two sections that kinetic theory provides a model for visualizing and understanding the behavior of molecules making up a gas, through the concepts of temperature, pressure, and density. These three variables are known as the **variables of state,** because they define the state, or basic properties, of a gas. You have seen also that the theory implies certain relationships among these variables of state. For example, if density is unchanged, then increasing the temperature implies that the molecules move more rapidly, thereby increasing the pressure. And increasing the density without altering the temperature increases the number of molecular impacts, thereby also increasing the pressure. These relations are summarized in a formula known as the **gas law** (sometimes referred to as the ideal gas law, or the equation of state): for a given sample of gas,

$$\text{pressure} = \text{constant} \times \text{density} \times \text{temperature}.$$

In mathematical shorthand, the gas law is written $P = C \rho T$, where P is pressure, C the constant, T temperature, and $\rho$ (the Greek letter *rho*) the density.

The "constant" in this formula is a constant for a particular gas sample but may vary from sample to sample. For a **monatomic** gas (a gas composed of single atoms), its value is different from that for diatomic gas, and so on. The constant also depends, as you might expect, on the units chosen for the variables of state. If you are measuring pressure in millibars, for example, the constant will have to be a different value in order for the formula to work than if your pressure is measured in pounds per square inch. In fact, the "constant" is the gas law's most challenging concept; otherwise, the expression is quite straightforward.

For an application of the gas law, consider the demonstration in Figure 2.11. The air molecules in the pressure cooker move more rapidly as they become hotter, according to kinetic theory. You also know that their faster movement means they collide with greater force; thus, the pressure also increases as a result of the heating. The gas law says the same thing: a rise in

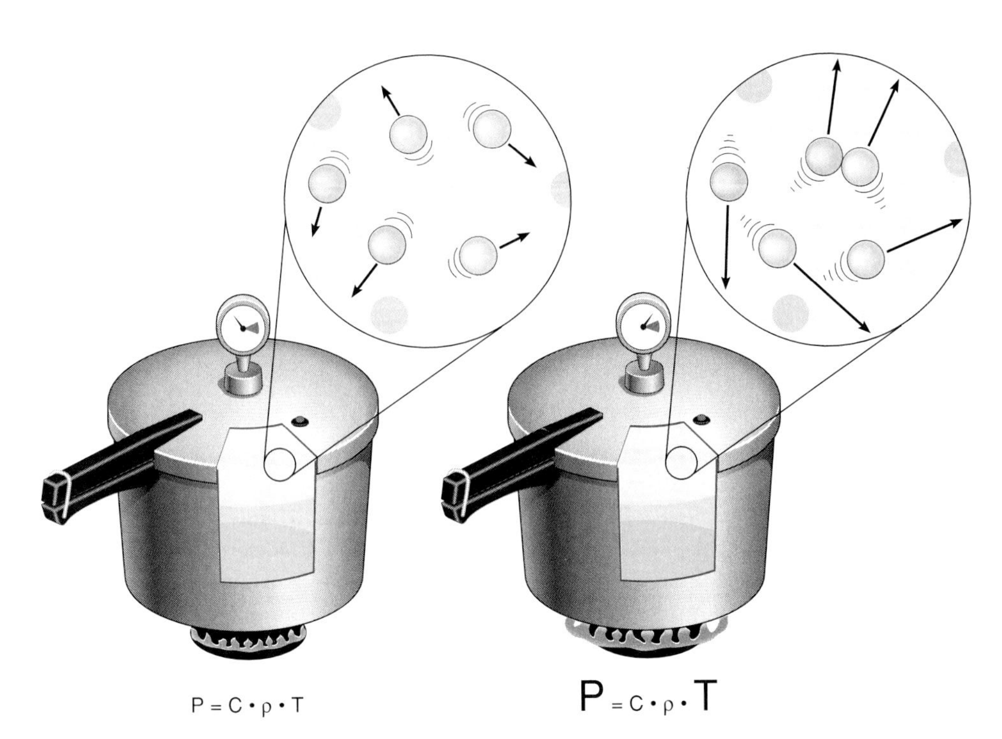


Figure 2.11 As the temperature rises, the pressure must rise to match it. Density does not change at first, because the mass of the air and the volume of the container remain constant.

temperature means the right side of the equation is larger; to balance, the left side must increase by the same amount if the density remains constant. Thus, the pressure must rise in reaction to the temperature increase.

In a real pressure cooker, the air density does not remain constant. You add water to the pot before you seal it, and some of this water evaporates as it is heated. This water vapor adds molecules to the air in the cooker, causing the air's density to increase, as well as its temperature. Both density and temperature, therefore, act to increase the air pressure in the cooker. (Refer to the gas law to see why this is true.) Eventually, the pressure force may be great enough to lift the weight that covers the safety valve, causing a small amount of hot air to escape with a hissing sound. As the air escapes, the air density in the cooker decreases; the pressure, therefore, falls, as you can see from the gas law. The weight clinks back down onto the safety valve, and temperature and pressure begin to increase once again. At high temperatures, the nearly constant hissing of escaping air and clinking of the safety valve's weight provide a commentary of the gas law in action.

Of course, the atmosphere is not a closed container like the pressure cooker. We could simulate actual atmospheric conditions more realistically by redoing the demonstration with the pressure cooker's top removed. This case is considerably more complex, as now the air's volume and density, as well as its temperature and pressure, are all free to change. In later chapters, we will explore this situation in some detail, as we consider such problems as the formation of clouds and the development of storms.

## Checkpoint

***Review***

Note in the concept map in Figure 2.12 that the molecules of a gas are in rapid, random motion and that these speeds depend on the gas's temperature. This leads to a molecular-level explanation of pressure and connects the behavior of molecules to the gas law, which is one of the meteorologist's most useful tools of inquiry.

***Questions***

1. You have two air samples, one at 0°C, the other at 30°C. If you could see the molecules in these two samples, how would they differ?
2. At a temperature of 20°C, what is the average speed of an oxygen molecule in the air? A helium atom? (See Table 2.4.)
3. What is the cause of air pressure, according to the kinetic theory of gases?
4. Suppose you have a pressure cooker with a plugged valve; no air can enter or leave the cooker. If you put it in the freezer, what will happen to the pressure of the air inside the cooker? To its temperature? To its density?

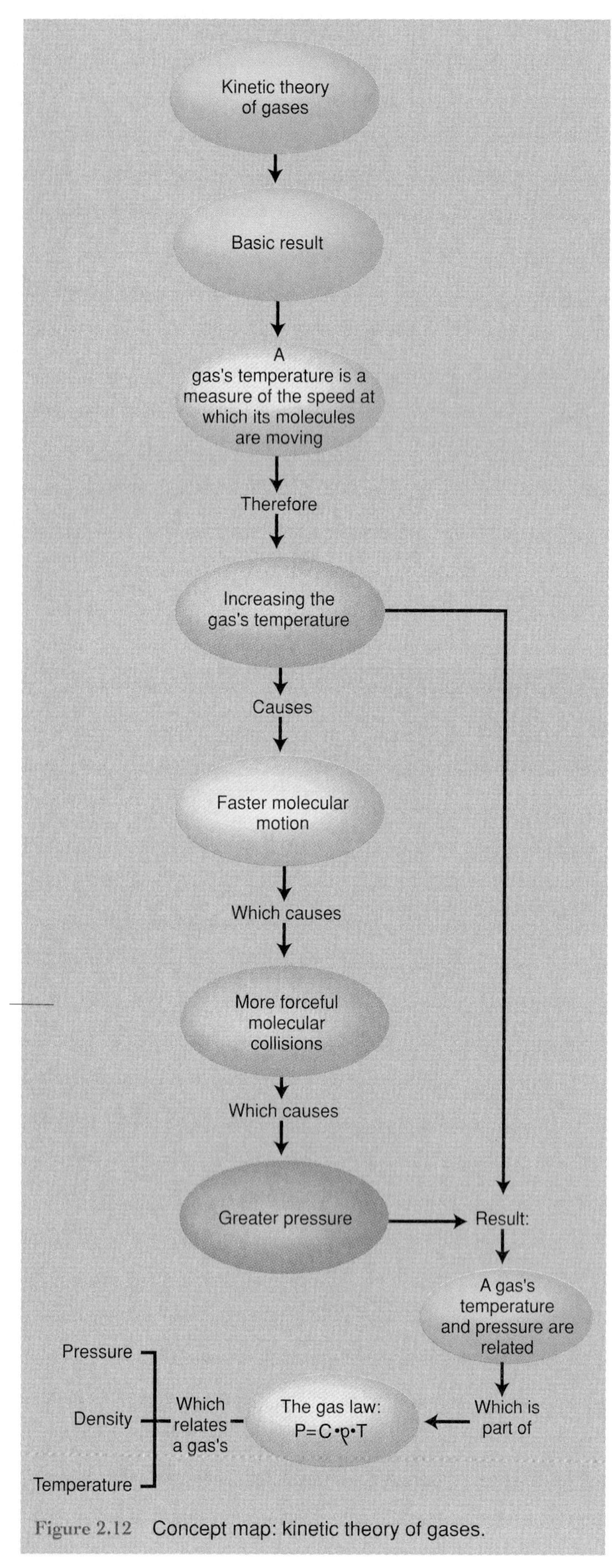


Figure 2.12 Concept map: kinetic theory of gases.

# Composition Changes in the Lower Atmosphere

Equipped with the atomic model of matter and the kinetic theory of gases, including the gas law, at last you are prepared to consider the atmosphere's composition and its changes in some detail. This section is an inventory of atmospheric substances and their changing presence. As the title indicates, we deal here with the "lower" atmosphere. For now, you can take this to mean the air from the earth's surface up to an altitude of 10 to 15 kilometers (6 to 10 miles). Later, we will be more specific.

Many of the substances you will investigate are considered **air pollutants**: that is, they are present in the air in amounts sufficient to cause injury to life or the environment. Like many labels, however, this one must be used with discretion. As you will see, some substances are pollutants in some circumstances but not in others. Others are pollutants to some organisms but important resources to others.

## Nitrogen, Oxygen, and Argon

In the opening paragraph of this chapter, you learned that the proportions of nitrogen, oxygen, and argon have remained nearly constant for the past century or so. Later in the chapter, we will discuss changes in major atmospheric constituents such as oxygen that have occurred over the billions of years of the earth's evolution. Now, however, we move on to less common substances whose variations are more significant in the short term.

## Water Vapor and the Hydrologic Cycle

The downpour pictured in Figure 2.13 serves as a vivid reminder that the concentration of $H_2O$ can vary dramatically and suddenly. These large variations are due largely to the ease with which water changes phase in the earth's atmosphere. For example, in locations of high temperature and abundant water supply, such as the tropical oceans, water evaporates readily and may comprise as much as 4% of the mass of the atmosphere. In the very coldest places, on the other hand, few $H_2O$ molecules exist in gas form; here, as little as 0.01% of the air's mass may be water vapor. As air masses travel across the earth's surface and through weather systems of the sort discussed in Chapter 1, the air's temperature, pressure, and access to water varies, leading to a constant exchange of $H_2O$ between the atmosphere and the earth's surface. This exchange of water through processes of phase change, precipitation, transportation, and runoff is known as the **hydrologic cycle** and is shown in brief in Figure 2.14. Notice that the units are in thousands of cubic kilometers of water. Thus, for example, the atmosphere is shown to contain 13 thousands of cubic kilometers of water (i.e., 13,000 $km^3$).

When the different flows in a cycle diagram are labeled with values, as they are in Figure 2.14, the diagram may be called a budget. Several terms are often used in discussions of cycles and budgets: **sources** are processes that increase the amount of substance under study ($H_2O$ here) at a given point, **sinks** decrease the amount of the substance, and **fluxes** are the rates of flow of the substance. **Storage,** as the name suggests, refers to the amount of the substance contained (stored) in various locations. Applying these terms to Figure 2.14, the flux of 423 units (i.e., 423,000 $km^3$) per year in the form of precipitation represents a sink, or loss, to the atmosphere but a source (a gain) to the earth's surface. Note the vast storage of water in the oceans (no surprise!); also, observe that the atmosphere's 13 units of water vapor stored at a given time are far less than the annual flux of 423 units that passes through the atmosphere via evaporation and precipitation.

What do you conclude from the fact that the flux entering the atmosphere via evaporation, 423 units, exactly equals the precipitation flux out (324 + 99)? The implication is that the storage of water in the atmosphere is not changing from year to year. If the incoming flux were greater, the amount in storage would have to increase. If the sinks were greater, the amount in storage would be decreasing.

In theory, the budget is a particularly useful tool for investigating this chapter's question: "Is the atmosphere's composition changing?" To see if the concentration of a given substance is changing, you can draw up a budget for that substance, showing all sources and sinks. If the sources' fluxes exceed those of the sinks, then the component is increasing its concentration. If the sinks exceed the sources, the substance is becoming more scarce. A practical limitation to this approach is that often the magnitudes of the sources and sinks are not well known. Thus, we face a dilemma: although the budget represents the exact accounting sequence we need to perform, we may not know what specific values to enter into the calculation. This very limitation has its own value, however. It has stimulated and continues to stimulate a great amount of research, all part of the ongoing evolution of scientific inquiry.

Figure 2.13 On a local scale, the atmosphere's composition can change suddenly and dramatically. Imagine the number of water molecules leaving the atmosphere per second in this downpour.

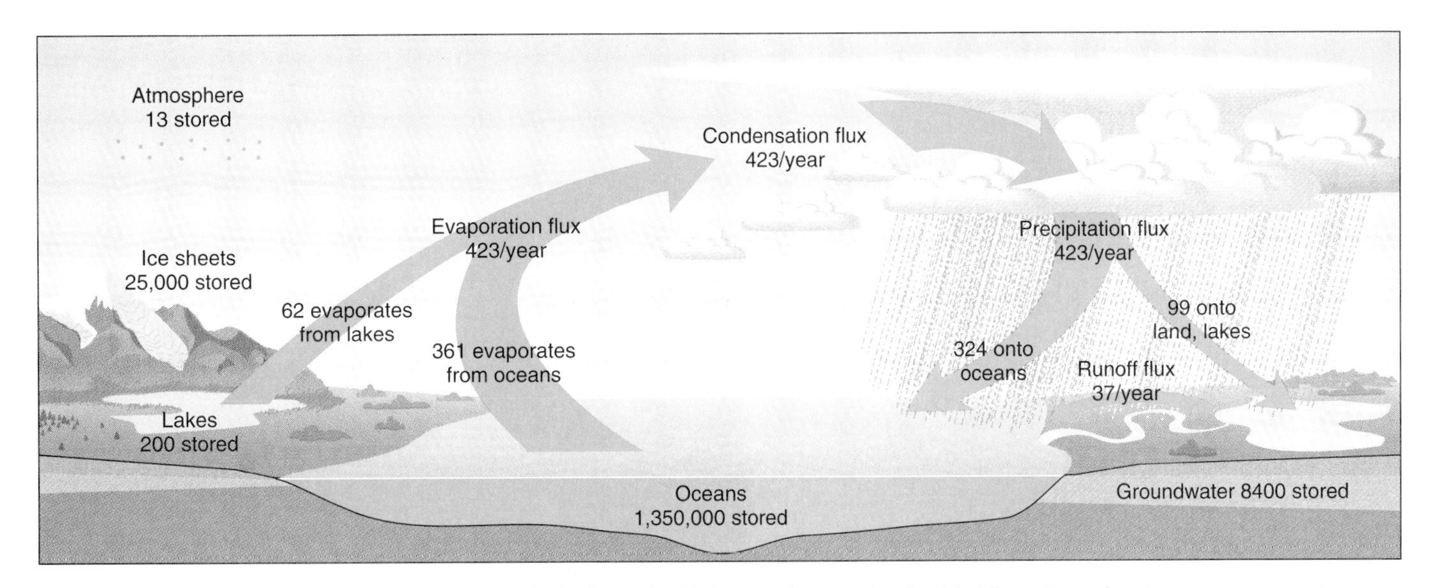


Figure 2.14 A simplified diagram of the earth's hydrologic cycle. Units are thousands of cubic kilometers of water. Source: Based on data from J. Peixoto and A. Oort, *Physics of Climate*, 1992, American Institute of Physics, New York, NY.

On a more immediate note, cycle and budget diagrams are also useful study tools. Note their resemblance to concept maps.

## Carbon Dioxide

Figure 2.15A shows the Mauna Loa Observatory in Hawaii. At an altitude of 3000 meters (11,000 feet) above sea level on an island far from the world's population centers, the observatory is an ideal location from which to monitor the atmosphere's composition unaffected by local pollution sources. Over the past 40 years, scientists at Mauna Loa have collected a valuable set of data on atmospheric carbon dioxide.

In Figure 2.15B, you can see the Mauna Loa carbon dioxide record. Study this graph for a moment.

Note from the figure that carbon dioxide makes up slightly under four-one hundredths of 1 percent of the atmosphere: that is, 0.036%, or about 360 parts per million. Compared to $O_2$, $N_2$, Ar, and $H_2O$, this is a trivial amount. However, note also the graph's overall upward trend, an indication that the atmosphere's $CO_2$ concentration has increased through the period shown. In addition, the series of small ripples suggests a cyclic pattern that exists along with the long-term increase.

How great was the increase of $CO_2$ over the period of time plotted? The peak value in 1958, on the left edge of the chart,

A

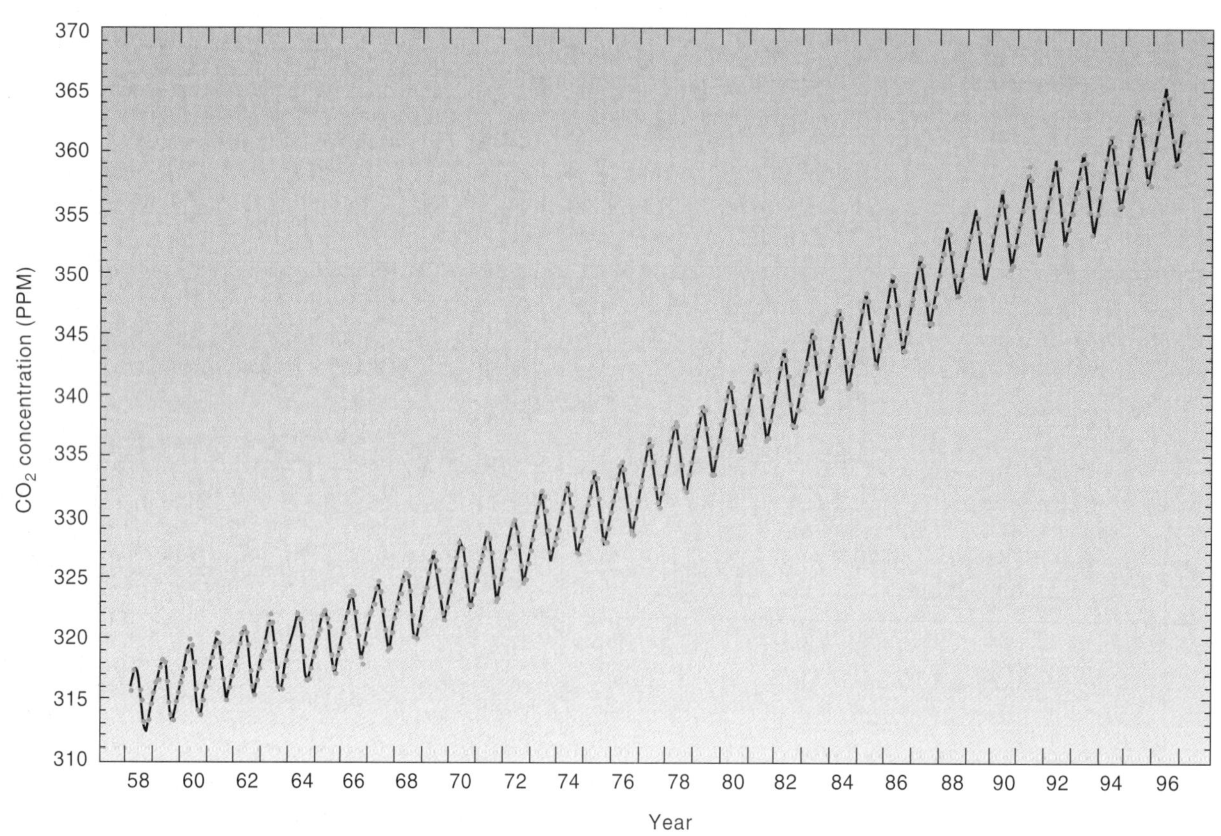


B

**Figure 2.15** (*A*) Mauna Loa Observatory is an important site for air quality measurements. (*B*) Monthly average concentrations of atmospheric $CO_2$ in parts per million by volume of dry air (ppm) versus time in years, observed at Mauna Loa Observatory, Hawaii with a continuously recording nondispersive infrared gas analyzer. The smooth oscillating curve is a fit of the data to a trend plus the annual cycle which increases linearly with time, and does not try to connect all the monthly data points, shown as dots.

was 318 parts per million (ppm); on the right side, in 1996, the peak value had increased to 366 ppm. Thus, the increase was 48 ppm. By itself, this number is not particularly revealing. More important is the fractional or percentage increase 48 ppm represents. Based on the 1958 value of 318, the fractional increase is (366–318)/318, or nearly 15%. Expressed in these terms, the $CO_2$ increase appears significant indeed.

What might have caused this steady increase in atmospheric $CO_2$? You may recall from an earlier science course that plants and animals have a good deal to do with $CO_2$. To refresh your memory (or to fill in), a plant's leaves combine $CO_2$ from the air, water from the soil, and sunlight—in a process called **photosynthesis**—to make plant tissue. We can represent photosynthesis schematically as follows:

$$6CO_2 + 6H_2O \text{ — sunlight} \rightarrow C_6H_{12}O_6 + 6O_2.$$

The first term, $6CO_2$, means "six molecules of $CO_2$"; thus, it represents six carbon atoms and 12 oxygen atoms (since one $O_2$ is just two oxygen atoms). Similarly, $6H_2O$ represents six water molecules, or 12 hydrogen plus six oxygen atoms. The molecule $C_6H_{12}O_6$ on the right side of the reaction is a form of sugar, called glucose. With this explanation, you can read the photosynthesis expression above as: "Six molecules of $CO_2$ combine with six molecules of $H_2O$ with energy from sunlight to produce one molecule of glucose plus six molecules of $O_2$."

Compare the number of carbon atoms on each side of the reaction. Notice they are the same (six on each side). The same is true for oxygen (18 on each side) and hydrogen (how many?). These equalities are not coincidences; instead, they illustrate a basic rule of chemical reactions: the atoms involved are "conserved," that is, the same atoms appear on each side of the → symbol. The changes are only in how the atoms are arranged to form molecules.

From the photosynthesis formula above, you can see that as plants photosynthesize, they remove $CO_2$ from the atmosphere and store the carbon (the C in $CO_2$) in the form of sugar. Therefore, growing plants are a sink of atmospheric $CO_2$. At the same time, plants return the $O_2$ from the carbon dioxide to the atmosphere, thereby enriching the atmosphere with oxygen.

The respiration of animals operates in the opposite sense from the point of $CO_2$ and $O_2$. With every breath you take, you alter the atmosphere's composition by taking many millions of $O_2$ molecules, combining them with C atoms to create $CO_2$, and exhaling this $CO_2$ into the atmosphere. (Of course, your individual $CO_2$ contribution is trivial compared to the amount of $CO_2$ in the atmosphere already, so don't hold your breath.) Plants also respire, consuming oxygen and returning $CO_2$ to the atmosphere. The formula for respiration is just the reverse of that for photosynthesis:

$$C_6H_{12}O_6 + 6O_2 \rightarrow 6CO_2 + 6H_2O + \text{energy}$$

Compare this expression with the one for photosynthesis. Notice that in the case of respiration, the reaction does not require energy but releases it instead.

Thus, the plant and animal kingdoms form a remarkable symbiotic relationship in which the waste gas for one is a needed resource for the other. Although there are other sources and sinks in the atmosphere's $CO_2$ budget, as you will see, over the years this plant-animal relationship has had a major impact on the atmosphere's $CO_2$ levels.

Reflecting on the plant-animal symbiosis, you might hypothesize that the $CO_2$ increase is the result of an increase of animal life or a decrease in plant life over the past 100 years. Although the human species certainly has multiplied during that period, the increase is not great enough to account for the rise shown in Figure 2.15B. Nor can a change in plant population explain the change.

The explanation lies in what happens at the end of a plant's life. Normally, when a plant dies it decomposes rather quickly, within a few months or years. Decomposition processes release the plant's carbon reserves, built up during its life, back to the atmosphere in the form of $CO_2$. Thus, the plants' removal of $CO_2$ is only temporary, persisting only as long as the plant is not decaying.

Some plants, however, do not decay the moment they die. In marshy or swampy areas, dead plant matter may accumulate in the water, prevented from decomposing by the absence of free (atmospheric) oxygen. Over the course of the earth's long history, a vast amount of plant matter has been interred in this manner, its carbon kept from the atmosphere. Such materials are known as fossil fuels: we know them more familiarly as coal, oil, and natural gas.

Burning fossil fuel (or wood) releases the former plant's carbon. The chemical reaction is the same as that for respiration, with the result that $CO_2$ is released into the air. Thus, burning fossil fuels releases millions of years' worth of stored carbon into the atmosphere. This carbon appears to be responsible for the $CO_2$ buildup shown in Figure 2.15B. In Chapter 4, you will learn of the role that increasing atmospheric $CO_2$ may be playing in the long-term warming of our planet.

Return to Figure 2.15B for a moment and examine the short-term oscillations of $CO_2$. How long is the period of oscillation, the time from one maximum to the next? If you count the number of peaks over the graph, you find that there are as many peaks as there are years: thus, the period of oscillation is one year. When does the peak occur? Close inspection of the curve reveals that concentrations of $CO_2$ reach their maximum early each year (in February or March) and their minimum in August or September.

Because most of the world's land masses, and therefore most forests, are found in the Northern Hemisphere, the lowest annual $CO_2$ readings occur each September, at the end of the Northern Hemisphere's growing season. Highest $CO_2$ values occur in late winter (Northern Hemisphere) months because photosynthesis is at a minimum then; the fact that it is summer in the Southern Hemisphere and trees there are busily removing $CO_2$ from the air does not offset the absence of photosynthesis in the north. Thus, the annual $CO_2$ oscillations are mainly a result of growing plants' uptake of $CO_2$ during the growing season.

The world's oceans are another important part of the $CO_2$ cycle. Vast quantities of $CO_2$ are dissolved in sea water, and some of the $CO_2$ added to the atmosphere by combustion and

Figure 2.16 These trees are victims of acid rain.

other processes eventually is absorbed into the oceans. The role of the oceans in absorbing and releasing $CO_2$ is an active subject of research.

## Sulfur, Nitrogen, and Acid Rain

You have probably heard a good deal already about **acid rain** and **acid fog**—precipitation and fog made acidic through the addition of sulfuric and nitric acids, among others. Acid rain and acid fog have become major problems in locations whose air bears the residue of high-temperature combustion and sulfur emissions. Figure 2.16 shows examples of this damage.

### *Sulfur Compounds*

Roughly 1 percent of plant matter is sulfur. As plant tissue decays or is burned, its sulfur (S) combines with oxygen ($O_2$) to form sulfur dioxide, $SO_2$, which enters the atmosphere. Thus, wood and the fossil fuels—coal, oil, and natural gas—are all reservoirs of potential atmospheric sulfur. Coal is particularly rich in sulfur; over half of humankind's sulfur contribution to the atmosphere comes from coal burning.

Sulfur dioxide itself is highly toxic; it can cause human death at concentrations over 500 parts per million. Furthermore, atmospheric $SO_2$ eventually combines with $O_2$ and $H_2O$ to form sulfuric acid, $H_2SO_4$. A single molecule of sulfuric acid does not remain single for long; it soon combines with other like molecules and attracts water vapor molecules, forming tiny liquid droplets called **aerosol particles.** Aerosol particles are microscopic solid or liquid particles of either human or natural origin that are suspended in the atmosphere. Polluted aerosol particles carry a large portion of the airborne sulfuric acid. Their presence in the air often is evident as visibility-restricting haze. It may also be evident as irritation to eyes and lungs; people report greater respiratory difficulties when acid aerosol particles are present. Under extreme conditions, such air pollution can be fatal to humans. In London in 1952, thousands of people died from respiratory infections aggravated by sulfur-based pollutants. In Donora, Pennsylvania, several dozen people succumbed due to a 1948 air pollution episode.

Is human activity causing increases in the concentrations of atmospheric sulfur? If so, then air samples hundreds or thousands of years old should show lower sulfur levels than current

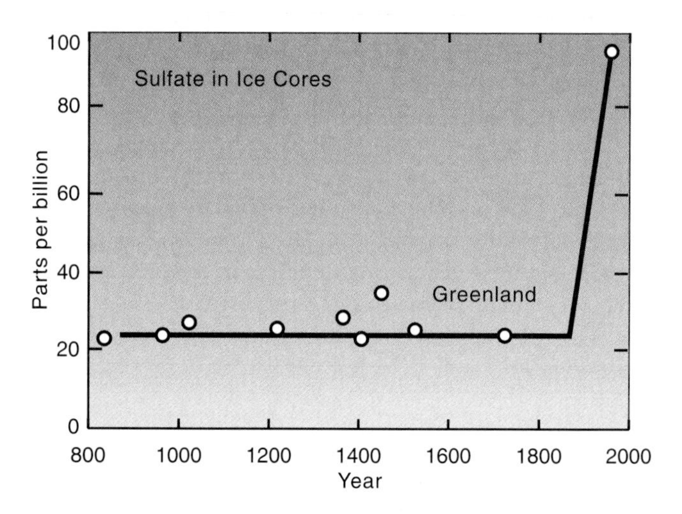


Figure 2.17 Atmospheric sulfur, captured in the Greenland ice cap, shows dramatic increases after 1800, according to the research of M. M. Herron.

samples. But where would one find 1000-year-old air? One location is deep in the Greenland ice cap. Greenland's ice has existed for many thousands of years. Each year, as new snow accumulates and gradually solidifies to ice, it traps tiny air bubbles. These air bubbles are the air samples scientists need. Figure 2.17 shows the long-term history of sulfur compounds in the atmosphere, based on air samples from the Greenland ice cap. Note the recent increase.

The concept map in Figure 2.18A traces the formation of acid haze and acid rain from sulfur emissions.

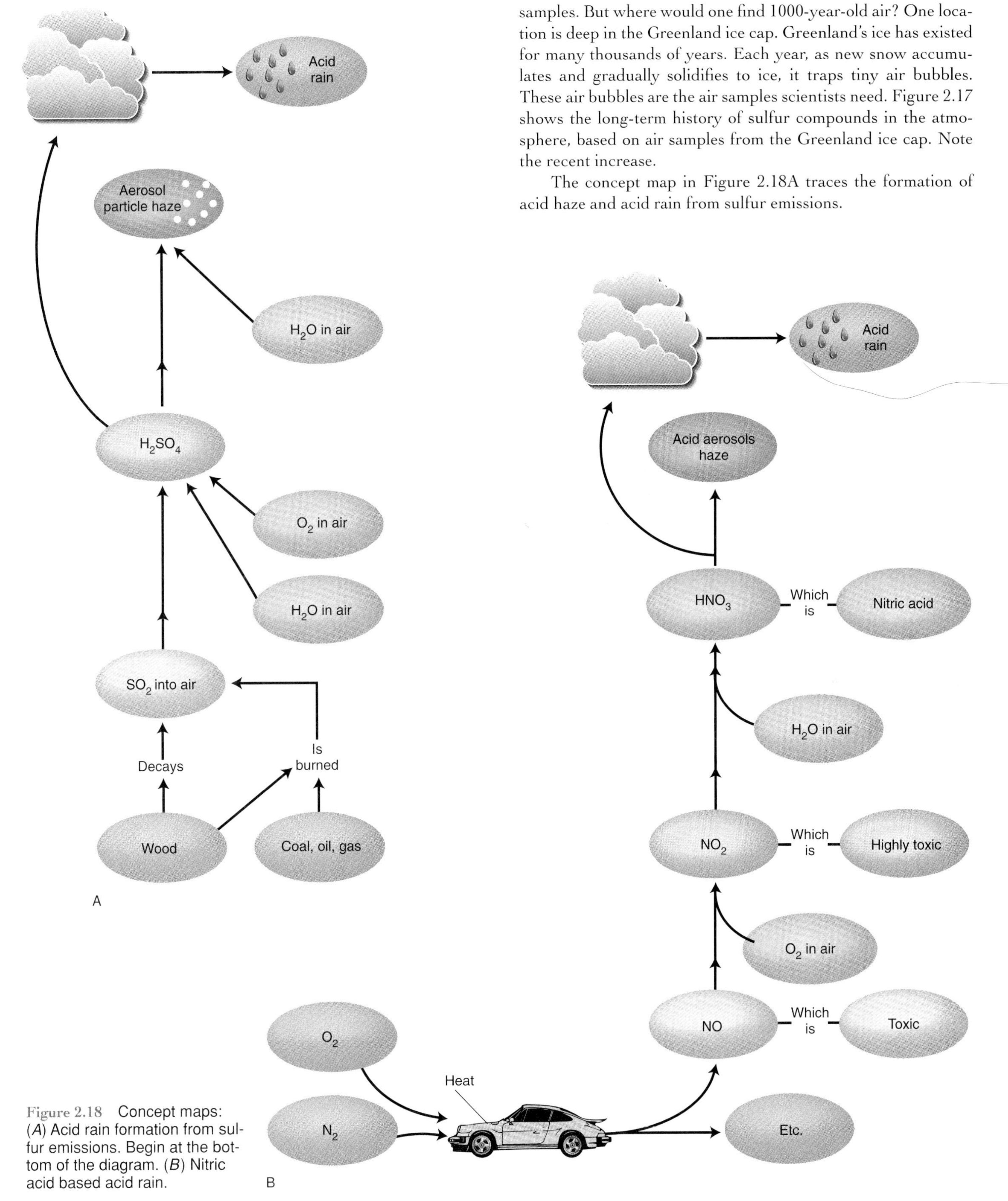


Figure 2.18 Concept maps: (*A*) Acid rain formation from sulfur emissions. Begin at the bottom of the diagram. (*B*) Nitric acid based acid rain.

### *Nitrogen Compounds*

As you know, diatomic nitrogen and oxygen ($N_2$ and $O_2$) make up nearly 99% of the atmosphere. When the air is heated to great temperatures, as in burning wood, coal, and so on, the nitrogen and oxygen combine to make nitric oxide, known as NO:

$$N_2 + O_2 \text{ heat} \rightarrow NO + NO$$

Nitric oxide reacts further with oxygen and water to make nitrogen dioxide ($NO_2$) and nitric acid ($HNO_3$). NO is somewhat toxic, $NO_2$ is highly toxic, and $HNO_3$ is an acid that forms aerosols and causes the same problems attributed to sulfuric acid.

Notice the source of nitric acid in the concept map in Figure 2.18B. It is not released into the atmosphere as sulfur is. Instead, normal atmospheric nitrogen is altered through the heat of combustion to create NO, which then further reacts to form $NO_2$ and nitric acid. It is interesting and ironic that the two most common substances in the atmosphere, $N_2$ and $O_2$, combine under heating to form such powerful toxins. In one form, oxygen and nitrogen are harmless, even valuable components of the atmosphere. In another form, they are deadly pollutants.

## Ozone

Someone once described *pollutant* as a resource out of place. Examples of this description come readily to mind: acid in the rain, instead of in car batteries; snow falling on a highway instead of on a ski slope; trash along roadsides instead of in recycling bins. Atmospheric ozone exemplifies the description also. Near the earth's surface, ozone is a pollutant because it is highly toxic and corrosive and because it may contribute to climate change. On the other hand, it is an unmitigated resource high in the atmosphere because there, it shields life on earth from lethal radiation. In this section, you will investigate ozone the pollutant in the lower atmosphere; in a few pages, you will examine ozone the resource in the upper atmosphere.

If you glance back at Table 2.3, you will see that $O_3$ comprises a tiny portion of clean air: only one oxygen molecule out of every 10 million is triatomic. If the air is polluted with other substances, however, $O_3$ levels may rise much higher. For example, if the air contains nitrogen dioxide, $NO_2$, and the sky is sunny, the $NO_2$ molecules will absorb sunlight that may break the $NO_2$ bonds, as follows:

$$NO_2 + \text{sunlight} \rightarrow NO + O$$

This process is an example of **photochemical dissociation;** *photochemical* because it requires sunlight to proceed and *dissociation* because the reaction separates (dissociates) an O from the parent $NO_2$ molecule.

Monatomic oxygen (O) does not remain dissociated in the lower atmosphere; instead, it recombines very rapidly with the first oxygen molecule it encounters, which is most likely to be an $O_2$:

$$O + O_2 \rightarrow O_3$$

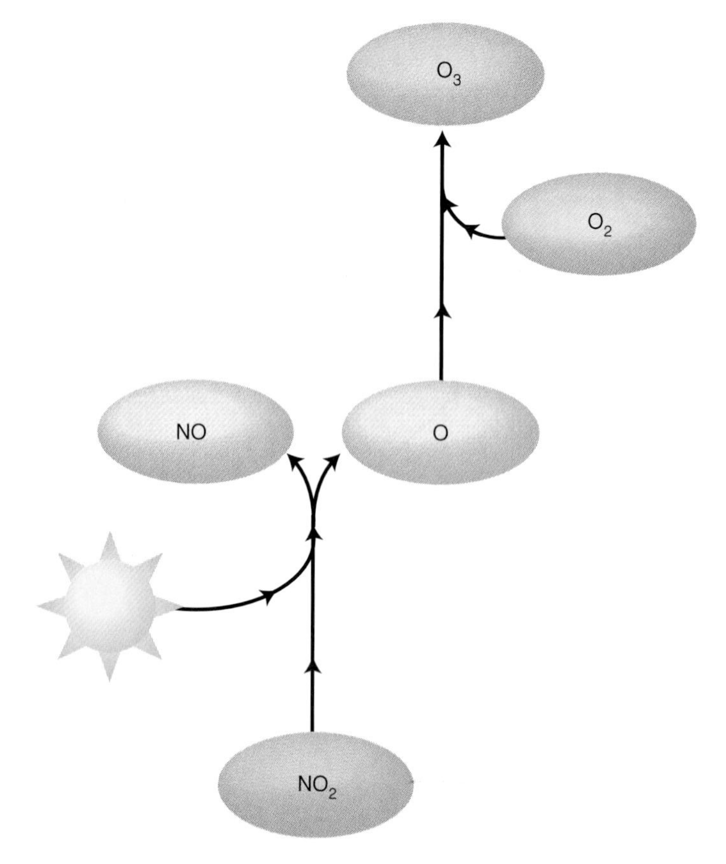


Figure 2.19 Concept map: Ozone production from nitrogen dioxide.

The result, as you see, is the formation of an ozone molecule. Figure 2.19 summarizes this ozone formation process.

If sufficient amounts of ozone are created, a weather event called **photochemical smog** occurs. Smog, originally coined as a combination of the words *smoke* and *fog*, now generally refers to a visible haze or fog polluted with ozone, $NO_2$ (which gives the smog its brown appearance), and other pollutants.

Ozone is considered a pollutant for a number of reasons: to cite just three examples, it damages plant leaves, causes eye and respiratory irritations, and damages materials such as rubber. Its effects can be so severe that an ozone report is now a regular part of many weather forecasts. (In weather forecasts, ozone and related compounds are often termed *oxidants*.) At times of high $O_3$ levels, ozone can cause eye irritation and respiratory ailments; therefore, people are advised to limit strenuous outside activities or to remain indoors. Table 2.5 shows the impact that $O_3$ can have on humans and our environment. Note the very small background value for $O_3$ and the relatively high values it can reach during brief periods (up to a few days) of severe pollution, called episodes.

What conditions would be conducive to high ozone levels? First, nitrogen dioxide must be present in large amounts; therefore, an urban setting, one with great numbers of automobiles for example, is the most likely site. Second, as you learned above, sunlight is required; therefore, episodes are most likely in high

**Table 2.5** Ozone Concentrations (Parts per Billion)

| Concentration | Description |
|---|---|
| 10–20 | Average value in lower atmosphere |
| 25 | "Good" air quality index |
| 45 | "Moderate" air quality index; some damage to materials, vegetation |
| 100 | Maximum allowable industrial exposure |
| 120 | EPA limit: not to be exceeded more than once a year |
| 135 | "Very unhealthful" air quality index; some people feel ill |
| 225 | "Hazardous" air quality index; most people feel symptoms |
| 400 | Values recorded in ozone episodes |
| 1000 | Extensive damage to vegetation |
| 10,000 | Values reached in upper atmosphere |

pressure regions. High centers are favored also because winds there tend to be light, preventing the ozone from dissipating. Highs also tend to keep air from mixing vertically, as you will see later; this effect adds to the observed correspondence between high pressure and ozone episodes. The Los Angeles region is particularly well suited for $O_3$ episodes, given its great numbers of automobiles, frequent high pressure weather, surrounding mountains that act as a barricade for free air movement, and surface cooling provided by the ocean, plus the presence of warm air aloft, which prevents the air from mixing vertically (see Figure 2.20). But Los Angeles is not unique. Each summer, the U.S. Environmental Protection Agency regularly reports dozens of cities, large and small, in violation of ozone pollution standards. During the hot, sunny summer of 1988, New York City's air exceeded the ozone limits on 42 days—nearly one day in two!

Figure 2.20 A morning of intense photochemical smog in Los Angeles.

## Other Atmospheric Constituents

Many trace substances (substances with very low concentrations) are under close scientific scrutiny because their presence in the atmosphere is suspected to be increasing and, despite their minuscule presence, they may significantly affect human health or the earth's climate. Indeed, one of the key lessons of this section is the fact that trace elements, in portions measured in parts per million or parts per billion, can have such a potent impact on life and the environment. A great number of different substances fall into this category. We will touch on just a few that are particularly representative or important.

### *Methane*

You may already have a warm association with methane: methane is the chief component of natural gas. Informally known as marsh gas, methane is released into the atmosphere as plants decay. Vast supplies of methane lie beneath the earth's surface, the decompositional results of countless plants of times past. Rice paddies and cattle are significant present-day methane sources.

Methane's chemical formula is $CH_4$, which makes it a **hydrocarbon,** a compound formed from hydrogen and carbon atoms. Figure 2.21 depicts methane's changing presence in the atmosphere. Although the absolute amount of methane is small compared to that of $CO_2$, methane has the potential, like $CO_2$, to warm the earth's climate. Thus, its recent steady increase is of some concern. In Chapter 4, you will consider methane's climate-altering potential in more detail.

### *Carbon Monoxide*

Like its cousin $CO_2$, carbon monoxide (CO) is a product of combustion. However, unlike $CO_2$, carbon monoxide is a lethal threat to human health; CO poisoning is solely responsible for a number of human deaths every year. It is not a gas to treat lightly.

Carbon monoxide is a common emission product from gasoline engines. A car running without pollution control devices emits more than a pound of CO for every tank of gas burned. Carbon monoxide is colorless and odorless, which contributes to its danger.

What makes CO so lethal is its strong tendency to combine with hemoglobin, the compound in blood that carries oxygen to the body's cells. Carbon monoxide is several hundred times more effective than $O_2$ in combining with hemoglobin; thus, in a CO-rich environment, no $O_2$ combines with hemoglobin, and no $O_2$ reaches the cells. The result is the same as if no air at all were present: the organism dies of asphyxiation. At CO levels greater than 100 ppm, people are subject to headaches and nausea due to oxygen deprivation.

Although global CO levels do not appear to be increasing, local values can far exceed the 100 ppm limit under circumstances such as those in Figure 2.22.

### *Heavy Metals*

You are probably aware that heavy metals—elements such as lead and mercury—are extremely toxic even in tiny concentrations. It is speculated that lead from pipes poisoned ancient Romans and that Isaac Newton accidentally poisoned himself experimenting with mercury. Modern sources of heavy metal pollution are numerous, including lead-based paints and leaded gasoline. Atmospheric concentrations of lead increased rapidly during the mid-20th century due to leaded gasoline emissions. There is good news, however. Recent studies indicate that lead levels have dropped by more than 85% since the mid-1960s. The decrease is attributed to the decline in use of lead as a gasoline additive, an important example of how society can respond to solve a serious environmental problem.

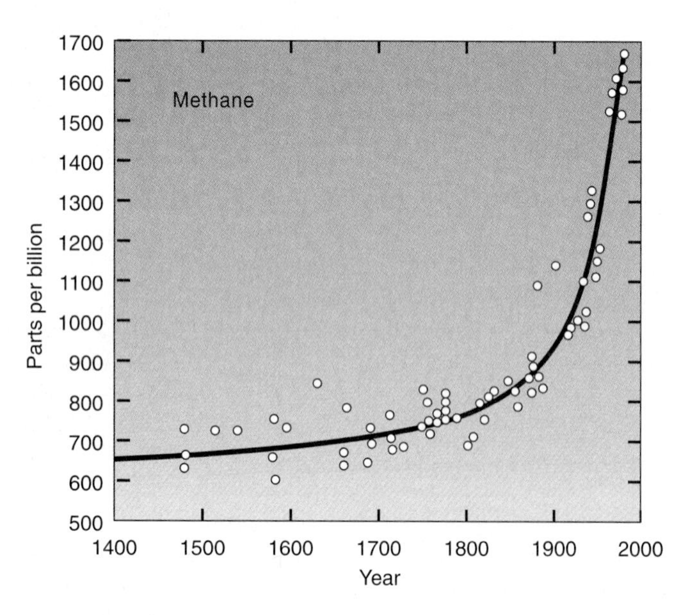


Figure 2.21 Atmospheric methane measurements, from ice cores (prior to 1950) and from direct measurement of the atmosphere (after 1950).

Figure 2.22 Carbon monoxide levels in stagnant air can exceed safe levels and lead to illness or death.

### *Particles*

In August 1992, the Anchorage, Alaska, International Airport was closed for two days because of an accumulation of volcanic dust and ash that fell from the atmosphere and piled up to a depth of 1 centimeter (1/2 inch). Small solid particles of dust and dirt, both human-made and natural, are common ingredients in the atmosphere, although only rarely to the extent shown in Figure 2.23.

Worldwide, particles originating from human activity such as industrial and transportation effluents and the burning of slash (logging debris) are roughly equal in amount to those from natural sources such as volcanoes, sea salt, and the weathering of surface rock and soil. In urban locations, however, the amounts from human sources may far outweigh those from natural ones.

Figure 2.23 The August 1992 eruption of a volcano caused this ashfall of particle pollution in Anchorage, Alaska.

## The Indoor Atmosphere

Figure 2.24 is a reminder that in any closed space, there is a risk that the air's composition will differ significantly from that in the "free air." We conclude this section with a look at the atmosphere that most of us breathe most of the time: indoor air.

### *Oxygen Deficits*

Picture yourself sitting alone in a small room. (Perhaps you are at this moment.) With each breath you take, you alter the composition of the air in the room. If the air is fresh, you inhale a mixture that is 21% $O_2$ and 0.03% $CO_2$. You exhale air that is approximately 16% $O_2$ and 4% $CO_2$. Thus, you have removed about 25% of the $O_2$ in your breath and enriched the air in $CO_2$ by more than 100-fold. Soon the air will develop an oxygen deficit if fresh air does not circulate through. A small room should receive several changes of air per hour, to keep its composition healthy. A lack of oxygen causes sleepiness and fatigue. (And you thought it was studying? Perhaps now would be a good time to open your window.)

### *Radon*

Radon is an element that is emitted to the air as a gas as atoms of uranium and thorium decay within certain types of rock. The air in a house built on such radon-emitting rock may contain elevated levels of radon. Recently, people have come to realize that radon's presence in indoor air is a significant health threat.

The threat, it turns out, is not with the radon itself but with its by-products. Radon is chemically inert, that is, it is unlikely to combine with other atoms. However, its large nucleus is radioactive: it decays spontaneously to form atoms of the elements polonium, lead, and bismuth. These elements, when breathed, may lodge in the lungs and may lead to cancer. The Environmental Protection Agency estimates that as many as 20,000 lung cancer deaths annually may be associated with radon decay. Most states provide instruments with which residents can measure radon levels in their homes. The solution, as in the case of oxygen deficits, is to circulate fresh air frequently.

Figure 2.24 The air we breathe indoors often differs substantially from that in the free atmosphere.

### *Smoking*

Smoking constitutes an obvious and massive alteration of the air's composition. Even secondhand smoke, the smoke a nonsmoker inhales in the company of smokers, constitutes a serious threat to air quality and hence to human health. The Environmental Protection Agency has estimated that secondhand smoke is responsible for 3,800 deaths in the United States annually.

Recent studies have shown that a person who smokes is 15 times more likely to contract radon-related cancer than a nonsmoker. This sobering statistic exemplifies the difficulty in defining the term *pollutant:* the same radon level that is no threat to a nonsmoker may be lethal to a smoker.

## Checkpoint

***Review***

Figure 2.25 represents many of the compositional change processes discussed in this section. For review, follow each of the paths shown and explain in words what is happening.

***Questions***

1. What is the source of the observed increase in atmospheric carbon dioxide?
2. What role do plants play in the carbon cycle? What role do animals play?
3. Where does the sulfur in sulfur dioxide originate? Where does the nitrogen in $NO_2$ originate?
4. Explain in words the meaning of this expression:

$$NO_2 + \text{sunlight} \rightarrow NO + O$$

5. Radon pollution is cited often as an important indoor health danger. However, radon itself is harmless. Explain the danger.

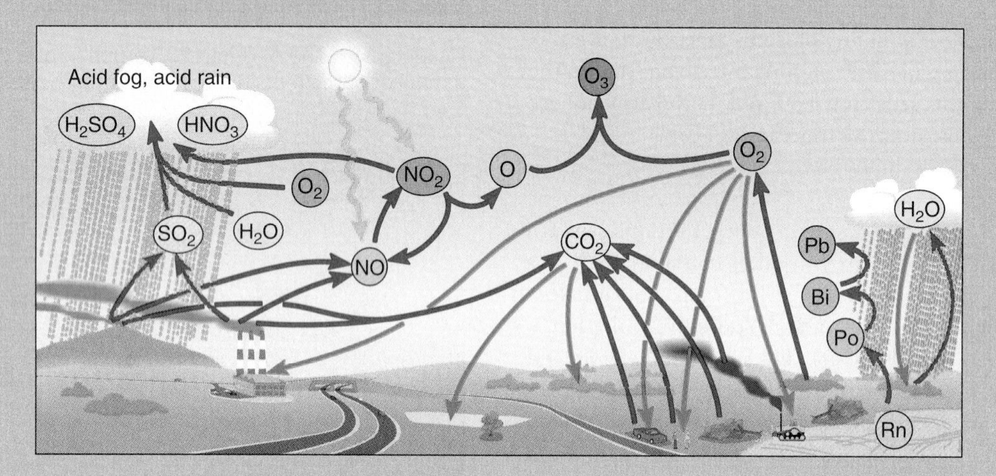


Figure 2.25 A summary of a few ways in which the atmosphere's composition can change.

# Composition Changes in the Upper Atmosphere

Examine closely the map and the graph in Figure 2.26, which depicts conditions 20 kilometers above Antarctica, a location as remote from humanity as any on earth. Note in particular that August-through-November ozone concentrations have decreased drastically over a recent 10-year period. This change symbolizes one of the most important issues you will encounter in this course. You are about to see the "ozone as resource" side of this Janus-like substance. To understand this face of ozone, you first need to become acquainted with the atmosphere's vertical structure, a subject to which we turn now.

## A Vertical Profile of the Atmosphere

In the next few pages, we ask you to imagine making a balloon ascent through the atmosphere. Throughout the ascent, you will be observing atmospheric conditions, and we will provide hypothetical, but typical, atmospheric data at various intervals. We encourage you to record the data by plotting them onto the graphs in Figure 2.27.

Manned balloon ascents for the purpose of meteorological data collection are no fantasy: balloons were used for over 150 years, from about 1750 until the early 1900s. In 1862, a French scientist named James Glaisher ascended by balloon to an altitude of 11.2 kilometers (1800 meters higher than Mount Everest). Hundreds of manned balloon ascents were made for scientific purposes around the turn of the 20th century, until the development of the airplane rendered the manned balloon ascent obsolete. Research balloons have reached altitudes 50 kilometers (30 miles), which is higher than 99.9% of the matter in the atmosphere.

## The Troposphere

At liftoff, the sun is shining brightly, scattered cumulus clouds billow upward above you, the temperature is a cool but pleasant 60°F (15°C), winds are light, and the air pressure is 1013 millibars. You plot these data at point 1 on Figure 2.27A.

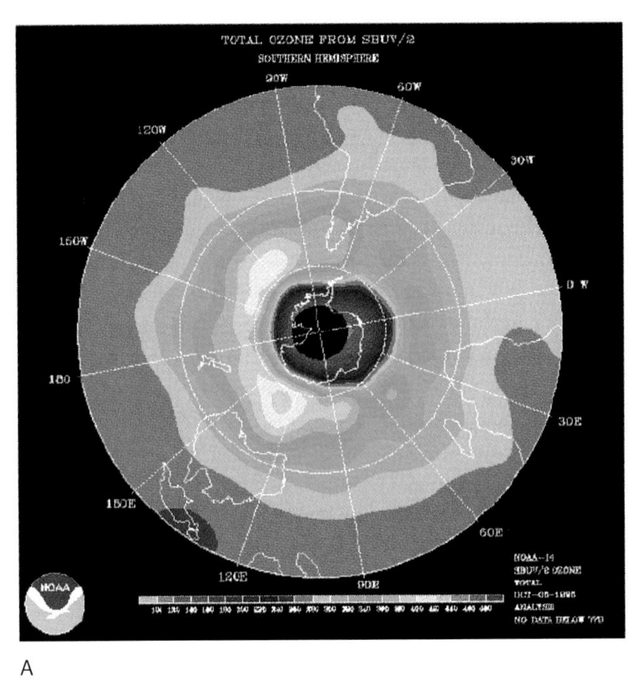

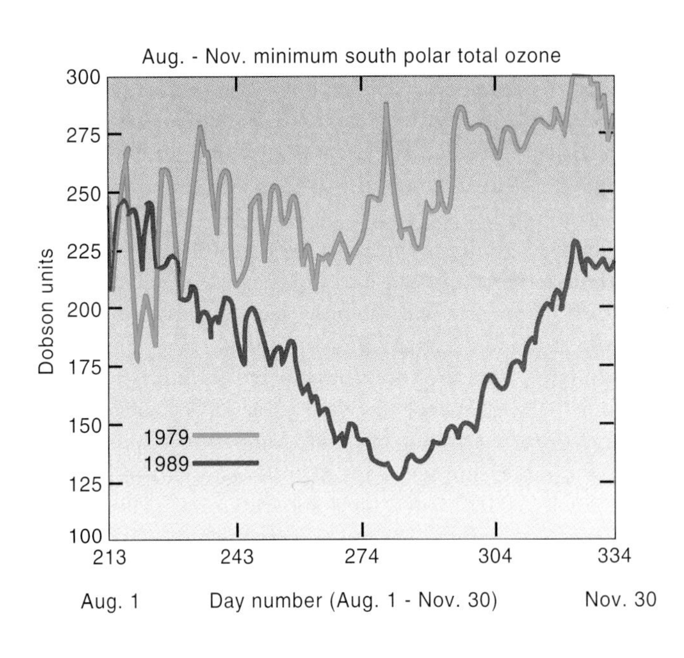


A B

Figure 2.26 (*A*) Antarctic Ozone Depletion. The violet region in the center has suffered substantial ozone loss. (*B*) Decline in ozone at the 20 kilometer level over a 10-year period. Dobson units are "milliatmosphere centimeters," a measure of the ozone's total thickness if all the ozone above the locale being measured were collected and brought to earth's surface at standard pressure and temperature (1013.25 mb, 0°C). Thus, 250 Dobson units corresponds to a total ozone thickness of 250 milliatmosphere centimeters, or 0.25 cm of pure ozone. Values in the range of 250 to 350 Dobson units (0.25 to 0.35 cm) are common worldwide. Source: NASA

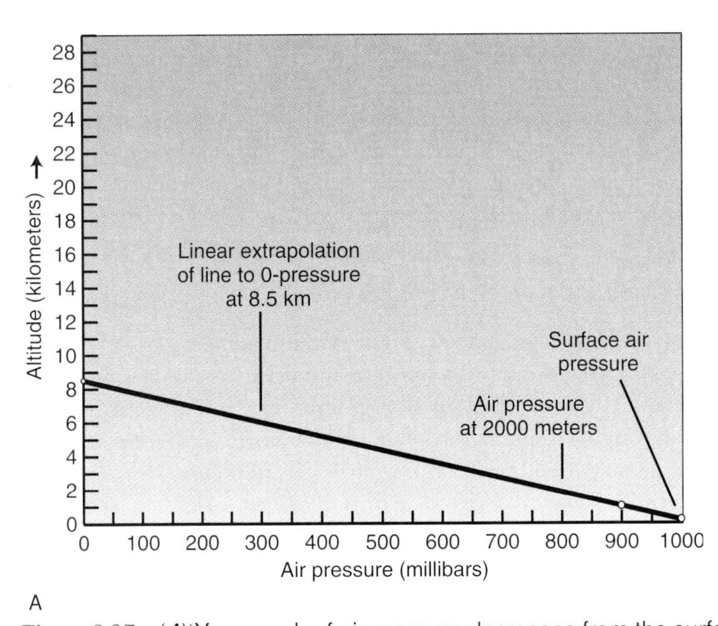


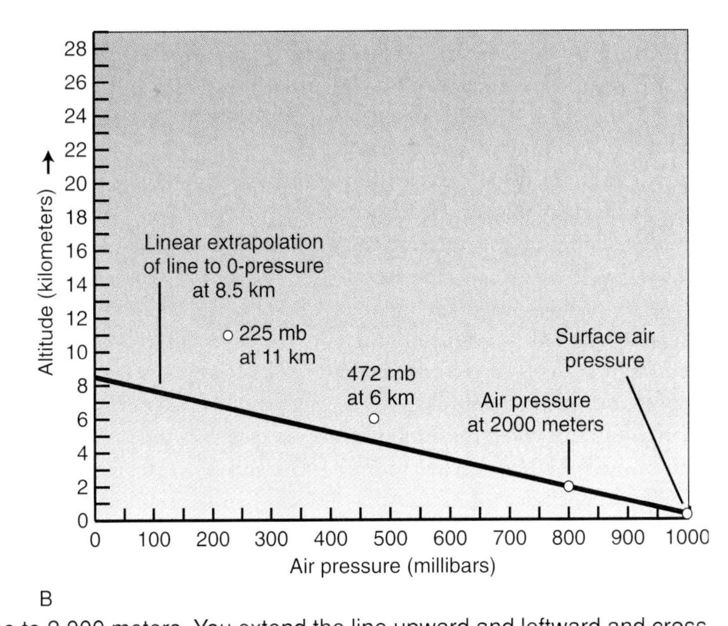


A B

Figure 2.27 (*A*) Your graph of air pressure decreases from the surface to 2,000 meters. You extend the line upward and leftward and cross the zero-millibar axis at 8.5 kilometers. Will the pressure continue to decrease as your line suggests? (*B*) Your pressure-altitude graph, with additional points plotted. Clearly, the air pressure did not decrease along your extrapolated line, but more slowly. Thus, at 6 kilometers, the observed pressure had fallen only to 472 millibars, compared to a value of 300 millibars indicated by the plotted line at that altitude.

A while after liftoff, you begin to feel a chill in the air. Checking the instruments as you pass among cumulus clouds at the 2000-meter (6600-foot) level, you note the temperature has fallen to 2°C (36°F). Again, you plot these data in Figure 2.27A, at point 2. The drop in temperature comes as no surprise; you reflect that mountain tops tend to be cold, often snow-capped, when lower elevations are much warmer. You *are* surprised, however, when you glance at the barometer: its 2000-meter value is just 795 millibars, a decline of over 20% from the launch value of 1013 millibars at sea level.

Why the great decrease in air pressure? You have ascended above a great number of air molecules whose weight pressed down on you (and on the barometer) at sea level. Based on this reasoning, the reading of 795 millibars indicates you have ascended already above 20% of the earth's atmosphere in the first 2000 meters! Extrapolating your plot in a straight line, you predict you will reach the top of the atmosphere (pressure of 0 millibars) at an altitude of about 8.5 kilometers (5.3 miles). You check your oxygen equipment to prepare for this eventuality.

Although the temperature and pressure have fallen radically since liftoff, your instruments show that the 2000-meter air has essentially the same composition as the surface air. And the wind? Judging from the speed of your balloon's progress across the land below, you realize the wind is increasing as you ascend.

As you rise, the trends of falling temperature, pressure and density and increasing wind speed continue unabated. At 6000 meters (about 20,000 feet), the temperature is a chilly −25°C (−13°F), and you are higher than most of the clouds; only a few high cirrus clouds are higher. You check the air's composition: although less dense, the proportions of its constituent gases remain the same, except for water vapor, which is nearly absent in the cold temperatures. You are using an oxygen mask because the air is too thin for easy breathing: you record the pressure at just 472 millibars.

Plotting the 6000-meter value on your pressure graph (Figure 2.27B), you see that the pressure is not falling as fast as you had predicted. In fact, over the last 4000 meters, the pressure fell only as much as over the first 2000 meters. It seems that the less the pressure is, the more slowly it changes with height. The air near the ground is denser, compressed by more pressure, than the air aloft. This suggests that the atmosphere may not have a precise top, as the ocean does, but may become progressively thinner and thinner as you ascend.

All of these observed weather conditions, including the decrease of temperature with altitude, are typical of those in the **troposphere,** the lowest layer of the atmosphere. Earth's surface, heated by the sun, warms the air in the lower troposphere, thereby causing convection currents of warmer air, like those associated with the cumulus clouds you are observing. Colder air from higher levels in the troposphere sinks to replace the rising surface air, with the result that the troposphere is well mixed: throughout this layer, the air's composition remains the same, and clouds and precipitation are found. The vigorous vertical mixing taking place in the troposphere is typical of an **unstable layer** of the atmosphere. An unstable layer is one in which warm, buoyant air bubbles "float" and rise (much as your own balloon is rising and carrying you with it), while cooler, denser air masses sink.

Continuing your ascent, you observe that the temperature continues to fall until, at 11,000 meters, it levels off at a frigid −57°C (−70°F). You have reached the **tropopause,** the boundary separating the troposphere and the next higher layer, called the **stratosphere**.

## The Stratosphere

As you pass through the tropopause and enter the stratosphere, you notice you are now above the tops of the highest cirrus clouds. Above you, the sky is a clear, dark blue. Air pressure has fallen to 225 millibars; you update your graph in Figure 2.27B and see that once again the pressure has decreased less rapidly with height than over the previous span. You also observe that the winds are the strongest of the flight and that air composition is still essentially unchanged from the earth's surface—except for being less dense and drier.

Rising through the stratosphere, you observe that the temperature remains constant for a while, then increases. This temperature increase with height, termed an **inversion,** is the reverse of conditions in the troposphere and causes the stratosphere to be stable: the colder air at lower levels is denser than the lighter air aloft, resulting in very little vertical mixing of the air. By the time you reach the **stratopause,** the upper boundary of the stratosphere at an altitude of 50 kilometers, the temperature has warmed all the way to a relatively balmy −2°C (+29°F). However, you are in no position to enjoy the mild temperature. The combination of intense radiation from the sun and very low air pressure (now just 0.8 millibars) has forced you to remain fully encased in a protective space suit. You record the atmospheric conditions and begin your descent to a more hospitable environment and a happy landing. (We wish you better luck than at a similar landing in rural France in 1860, when the balloon so terrorized local farmers that they attacked it with pitchforks and tore it to shreds.)

Back on earth, you report that as you passed through the stratosphere, you recorded a dramatic increase in the concentration of ozone. At an altitude of 25 kilometers (15 miles), readings reached as much as 10 parts per million, a concentration 500 times greater than at the ground. Where does all this ozone come from? Also, why is it so much warmer in the stratosphere, and why is there intense radiation at that level? Are the warmth and the radiation related to the high ozone concentrations? To answer these questions, we now complete the ozone discussion begun in the previous section.

## Ozone in the Stratosphere

High in the stratosphere, your radiation meters indicated increasing amounts of **ultraviolet radiation**. Ultraviolet radiation is more energetic than visible sunlight: it is so energetic, in fact, that it is able to break the chemical bonds of $O_2$, thereby altering the chemical composition of matter. In humans, such chemical alterations may take the form of skin cancer, cataracts, and other health problems. Thus, incoming ultraviolet (UV) radiation constitutes a serious health threat. However, events occurring in the stratosphere protect us from most UV radiation, as follows:

Ultraviolet radiation entering the stratosphere occasionally dissociates $O_2$ molecules into monatomic oxygen:

$$O_2 + \text{UV radiation} \rightarrow O + O$$

As you learned in the last section, monatomic oxygen quickly recombines, most likely with an existing $O_2$ molecule, to make $O_3$:

$$O_2 + O \rightarrow O_3$$

In this way, ozone is formed naturally in the stratosphere in small amounts. Because of the presence of ultraviolet radiation, ozone is much more common here than in the troposphere.

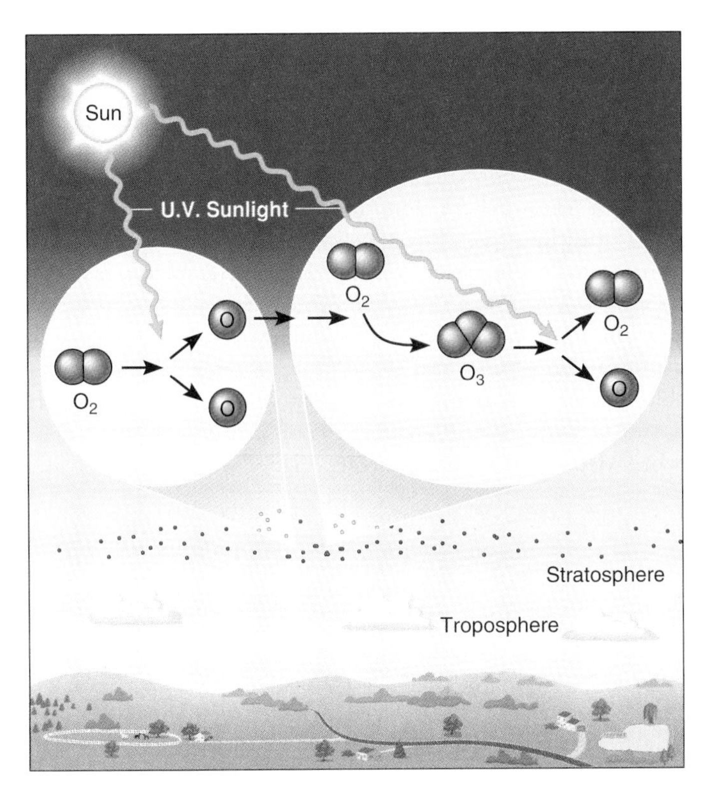


Figure 2.28 Interactions between ozone and UV radiation in the stratosphere. Notice that UV radiation is absorbed in both the formation and destruction processes.

The ozone itself is an excellent absorber of ultraviolet radiation through the following process:

$$O_3 + \text{UV radiation} \rightarrow O_2 + O$$

Thus, UV sunlight operates on both sides of a cycle of ozone production and destruction in the stratosphere, as shown in Figure 2.28. In both steps above, the UV radiation is intercepted and absorbed, warming the stratosphere in the process (just as your body is warmed by absorbing sunlight). By absorbing UV radiation in this fashion, the ozone layer shields life at the earth's surface from this dangerous energy form. Clearly, ozone in the stratosphere is a crucial resource, in contrast to its polluting presence at the ground.

### *Thinning of the Ozone Layer*

In 1985, a team of British scientists shocked the scientific world by presenting the evidence like that you saw back in Figure 2.26. The data revealed precipitous declines in stratospheric ozone concentrations over Antarctica each year around early October. The popular news media has dubbed this phenomenon "the ozone hole," but the name is misleading: it suggests a complete absence of ozone. We will refer to the phenomenon as a thinning or depletion of the ozone layer. Since its discovery, the thinning has been observed each year at the onset of spring in the Antarctic (September-October). In November, the layer begins to recover toward normal values once more.

Discovery of the thinning of the ozone layer was important for at least three reasons: first, its great magnitude, which was generally unexpected, served as a reminder of the depth of our ignorance about many of the complexities of our natural environment. Second, the loss of ozone has serious implications: it portends an environmental health threat of unprecedented proportions, were it to occur in a populated region. And third, it appears now that pollutants of industrial origin play a key role in the ozone layer's depletion.

What could cause such a decline in ozone concentrations? There are a variety of ways in which $O_3$ is destroyed naturally (in addition to UV absorption). These events, however, would not explain the seasonal nature of the huge declines found over the Antarctic.

### *CFCs*

Scientists now believe that chemicals known as **chlorofluorocarbons (CFCs)** play a central role in the thinning of the ozone layer. As their name suggests, CFCs are compounds of chlorine, fluorine, and carbon, and they are produced in great quantity for refrigeration: the coolants in your air conditioner and your refrigerator are CFCs. Worldwide CFC production has averaged about 1 million tons per year in recent years. (Bromine compounds, used in fire extinguishers and elsewhere, are thought to cause problems similar to CFCs. For the sake of brevity, we will consider only the CFC problem.)

If CFCs remained contained forever within the devices for which they were made, they would not be a problem. Unfortunately, however, cooling systems are subject to failure, like any other device. Eventually a leak is inevitable, resulting in a release of CFCs into the atmosphere. (Perhaps you can recall a refrigerator or air conditioner of yours suffering this fate.) CFC concentrations in the atmosphere have been rising steadily at the rate of about 5% per year.

Once in the atmosphere, CFCs affect conditions in several ways. In the troposphere, they warm the lower atmosphere but do not readily react chemically with other substances. Eventually, some CFCs make their way to the stratosphere, carried there perhaps by the updrafts in strong thunderstorms. In the stratosphere, ultraviolet radiation removes the chlorine (Cl) atom from its parent CFC molecule. Now the stage is set for the following two reactions:

$$Cl + O_3 \rightarrow ClO + O_2$$

and

$$ClO + O \rightarrow Cl + O_2$$

Note the remarkable role played by the chlorine atom in this sequence. First, it combines with an ozone molecule, resulting in chlorine monoxide (ClO) and plain old diatomic oxygen, $O_2$. In other words, the ozone molecule has been destroyed. Next, the ClO combines with an oxygen atom (O) to create $O_2$ and to liberate the Cl atom, allowing it to repeat the first step again. Thus, a single Cl atom may cycle through these two steps indefinitely, destroying an ozone molecule each time. Atmospheric scientists have estimated that a given chlorine atom could destroy as many as 100,000 ozone molecules through this process before finally becoming bound in some other stable molecule and thereby rendered harmless.

## *CFCs and Ozone*

If CFCs are responsible for the loss of stratospheric ozone, then why does ozone depletion occur so intensely over distant Antarctica and not more widely over the industrialized countries where the CFCs are produced? This question is not yet answered to everyone's satisfaction. Below, we present an explanation that seems reasonable to many scientists; we suggest you refer to Figure 2.29 as you follow the discussion.

CFCs are emitted continuously into the atmosphere from many locations worldwide. The atmosphere's wind systems distribute the CFCs more or less uniformly across the globe, including Antarctica. Over Antarctica, however, a series of events occurs in the upper atmosphere that seems to bring on the ozone depletion. During the Antarctic winter (June-August), the air becomes intensely cold in both the troposphere and the stratosphere (see Figure 2.29B). This deep cold air mass causes a "polar vortex" of fast-moving stratospheric air that encircles the continent, effectively isolating it for a time from the rest of the atmosphere. The cold becomes so intense that thin clouds form in the stratosphere, a region where cloud formation normally is uncommon.

These cloud particles serve as tiny platforms on which the chlorine molecules react, changing to less stable forms. When the sun returns to the Antarctic skies in September and October after the south polar winter, the sunlight breaks up these unstable molecules, shown in Figure 2.29C, freeing large amounts of chlorine, which then attacks the ozone. Because the polar vortex is still in place, the atmosphere cannot dissipate the resultant ozone depletion by transporting higher ozone concentrations

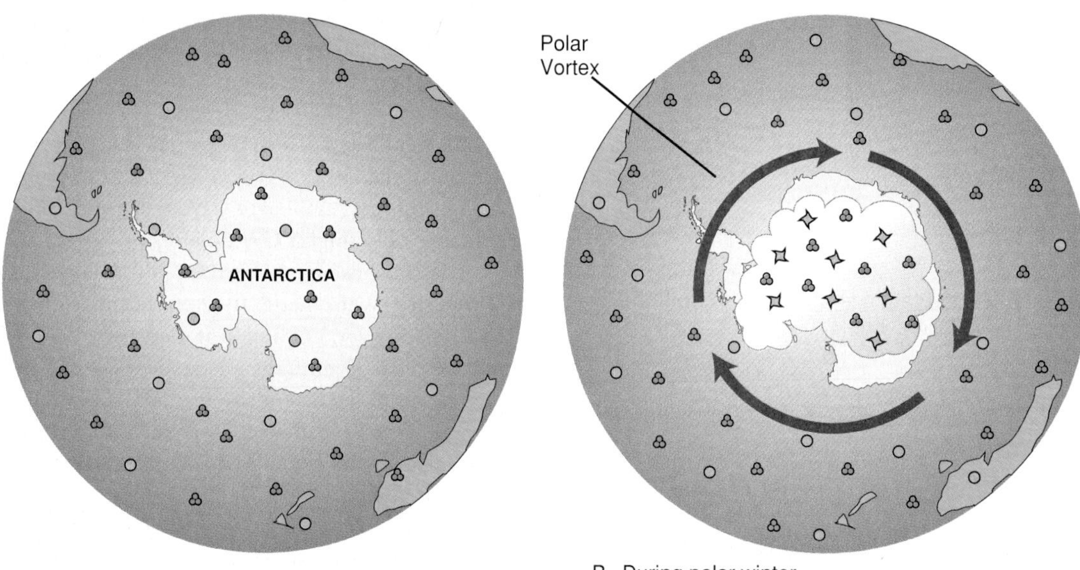

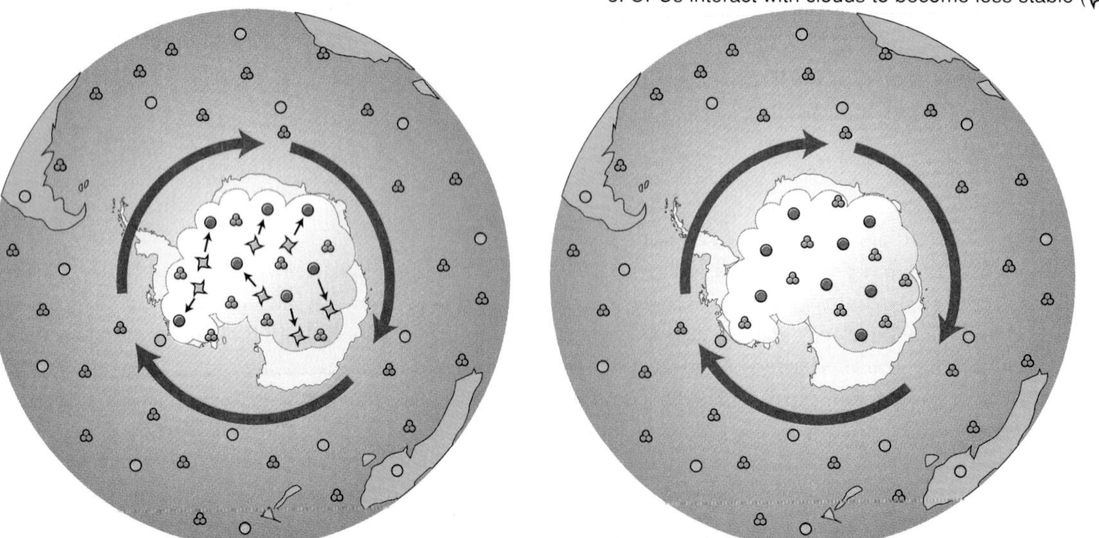


**Figure 2.29** A possible sequence leading to depletion of ozone over polar regions.

into the region. Thus, the hole deepens and persists until the polar vortex dissipates in November. Then the ozone-poor Antarctic air mixes with air of more normal $O_3$ concentration from more northern latitudes, and Antarctic $O_3$ counts rise. (When this occurs, locations in Australia and New Zealand observe concurrent losses of stratospheric ozone, as the $O_3$-poor Antarctic air reaches these latitudes.)

What about the north polar regions? Is a similar sequence of events likely to occur in April, at the outset of spring in the Northern Hemisphere? Since the discovery of the Antarctic ozone layer's thinning, researchers have tested this hypothesis through extensive measurements. One major difference between the two poles is that the north polar winter is not as cold as winter at the south pole. As a result, the north polar vortex does not develop so strongly, and north polar stratospheric clouds are less common. Therefore, the north polar ozone layer is not likely to experience so drastic a thinning as that observed over the Antarctic. However, recent measurements have revealed an increase in ClO concentrations and some ozone depletion in the Arctic stratosphere during the Northern Hemisphere's spring. This discovery is particularly important considering the number of people who live in high northern latitudes and who would be seriously affected if the ozone shield were destroyed.

The sequence of events is complex, and parts of this explanation may be modified substantially as we learn more about the processes. It seems clear, however, that the destructive role of chlorine atoms originating in CFCs is indisputable. As a result, some important steps have been taken. Fifty-nine countries (as of this writing), including the greatest CFC producers and consumers, have pledged to cease production of CFCs by the year 2000 and to reduce production of other ozone-damaging chemicals. Several U.S. chemical companies, responding to both environmental and economic arguments, have announced their self-imposed cessation of CFC production.

While these are encouraging steps, the ozone saga will not end quickly. Injections of CFC will continue long after 2000 as

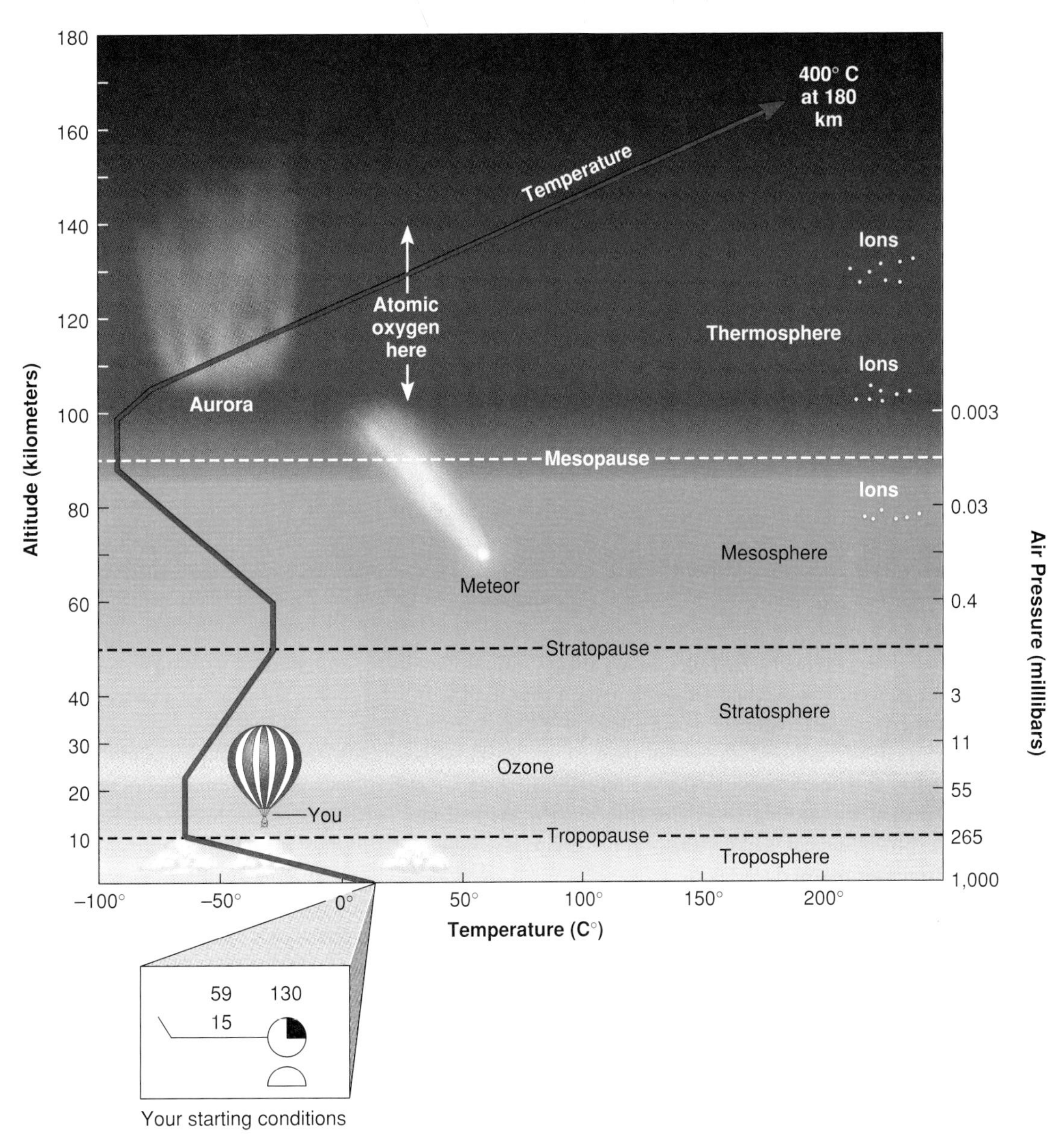


Figure 2.30 Profile of the atmosphere in the vertical.

A

Figure 2.31 (*A*) Layers of ionized particles in the ionosphere reflect radio waves, allowing the signals to reach locations normally shielded by the earth's curvature. (*B*) Sightings of the aurora are more common in polar locales.

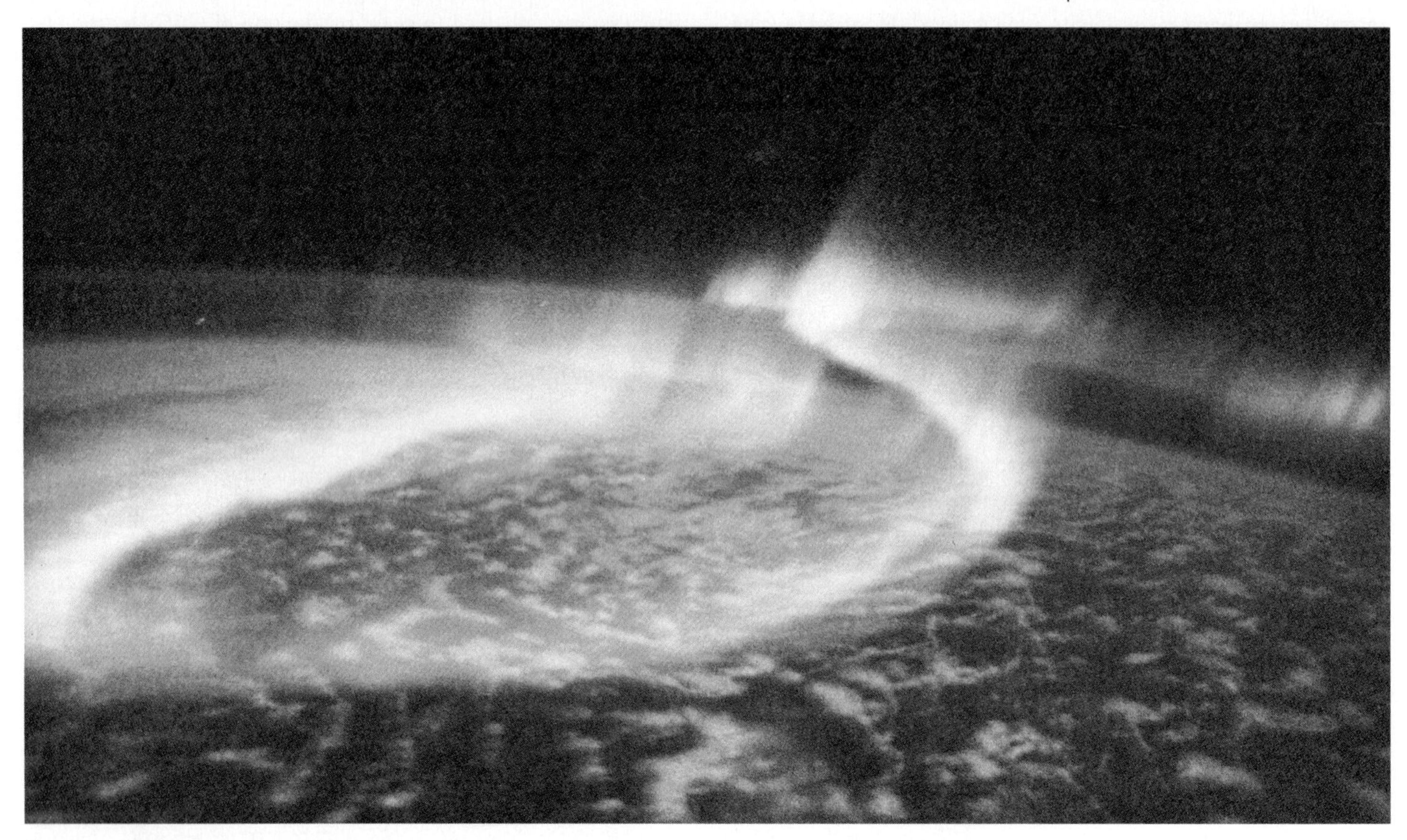

B

refrigeration systems worldwide fail, leaking their CFC coolants into the atmosphere. Researchers have estimated that if all CFC production were to end instantly, atmospheric CFC concentrations would continue to increase for another 10 to 20 years due to leakage from existing cooling systems and the long lifetime of chlorine atoms in the atmosphere. Considering the critical nature of the problem, it seems likely that ozone research will be an important topic for many decades.

## The Mesosphere

Above the stratosphere, the air temperature decreases once more with height, as illustrated in Figure 2.30. Why is the temperature falling again? Consider that it is ozone's absorption of ultraviolet radiation that causes the stratospheric warming and that ozone concentrations decrease at altitudes above 25 kilometers. With no radiation-absorbing molecules to warm it, the **mesosphere** becomes colder the farther you rise from the stratosphere's relative warmth.

The mesosphere is the realm of so-called "shooting stars." Pieces of rock, ice, and dust from outer space enter the earth's atmosphere at speeds of up to 100 kilometers per second. Traveling at such high speeds, they are heated to the point that they glow for a few seconds as they contact the increasingly dense gas of the mesosphere.

At the **mesopause,** which marks the top of the mesosphere, the temperature typically lies in the range of −90 to −100°C (−130 to −150°F), the coldest of any level in the atmosphere. The air pressure here is just 0.01 millibars, very low but not quite zero. Notice again that the air pressure decreases more and more slowly with altitude. Like repeatedly dividing a number in half, it seems the air pressure will never reach exactly zero, no matter how high you go.

## The Thermosphere

Above the mesopause lies the **thermosphere,** the fourth and highest of the atmosphere's four basic layers. In the thermosphere's lowest few kilometers, the temperature remains constant, near mesopause values; with altitude, it begins to increase again. What might account for the warming? Recall that stratospheric temperatures show a similar warming trend with height; there, the warming is brought on by absorption of UV sunlight by oxygen. Might there be a similar explanation for the thermosphere?

This line of conjecture proves fruitful: the proportion of monatomic oxygen (O) increases with altitude in the thermosphere, with a corresponding decrease in $O_2$. This increased O concentration results from $O_2$ absorbing very high-energy radiation (mostly from the sun), thereby becoming photochemical dissociated and warming the air:

$$O_2 + \text{high-energy UV radiation} \rightarrow O + O$$

At lower levels, the O atoms quickly collide and recombine; here, however, the density is so thin that long periods pass before recombination is likely. Thus, the higher you climb in the thermosphere, the greater the proportion of atomic oxygen.

The absorption of high-energy radiation by $O_2$ in this fashion drives the thermosphere's temperature to very high values: at 150 kilometers, the temperature reaches an amazing 180°C (roughly 350°F); at 500 kilometers, it reaches 600°C (1100°F). Although this sounds like an extraordinarily hot environment, actual conditions there are somewhat different. It is true that the atmospheric molecules and atoms have tremendous speeds commonly associated with high temperatures, but there are very, very few of these particles. The density in the thermosphere is so low that an exposed object would receive very few molecular impacts; thus, very little heat would be transferred.

Above 500 kilometers, the atmosphere's density is so low that colliding atoms deflected upward may not experience further collisions to redirect them downward again and hence may escape from the atmosphere into space. This level (500 kilometers, or 300 miles) marks the top of the thermosphere. For most meteorological purposes, you can think of this level as "the top of the atmosphere," even though some atmospheric gases in very low concentrations extend well beyond this point.

## The Ionosphere

In the upper mesosphere and the thermosphere, incoming UV radiation causes some molecules to become ionized, that is, to lose one or more electrons. The separate particles are known as **ions,** and the region in which numbers of ions are found is called, appropriately, the **ionosphere.** (Note that the ionosphere is not separate from the mesosphere but lies within it and the thermosphere.) At times, layers of ionized particles in the ionosphere may reflect radio waves, extending their range far beyond normal limits (see Figure 2.31A).

Another ionospheric event is the **aurora,** the emission of light caused by interactions of fast-moving, charged particles, mostly from the sun, with the ionospheric gas. Figure 2.31B shows a typical aurora viewed from space.

***Review***

Figures 2.28, 2.29, and 2.30 provide summaries of much of the material in this section. In Figure 2.30, notice how the slope of the temperature line changes from layer to layer. The changing slope is a useful reminder of key properties of each layer: the instability of the troposphere, the stability and heating by ozone in the stratosphere, the great cold of the upper mesosphere, and the extraordinary high temperatures in the upper thermosphere. The ionosphere is another name for the ionized regions of the upper atmosphere.

***Questions***

1. Name the four major atmospheric layers, going upward from the earth's surface.
2. What causes the temperature to increase upward through the stratosphere? In the thermosphere?
3. What are CFCs? What is meant by the phrase "thinning of the ozone layer"? How are the two connected?
4. How and why does air pressure change as you rise through the atmosphere?

# Composition Changes: A Broader Perspective

The composition changes you have read about in the previous sections may seem rather minor in magnitude; generally, we spoke of changes measured in parts per million or parts per billion over the past 100 years or so. We conclude this chapter by considering some truly huge atmospheric changes that have occurred over very long periods of time. And for an even broader view, you will also compare earth's atmosphere with those of other planets in the solar system.

## Past Atmospheres on Earth

Has the atmosphere always contained 78% nitrogen, 21% oxygen, and so on? Has there always been an atmosphere? For that matter, has there always been an earth?

A wide variety of scientific evidence suggests that the earth, sun, and other objects in the solar system formed about 4.6 billion years ago from a huge, cold, slowly rotating cloud of gas

Figure 2.32 An artist's conception of the earth's atmosphere at an early stage of its evolution.

Figure 2.33 The red color in the rocks in this Grand Canyon view is the result of iron in the rock rusting, a process that requires atmospheric oxygen. The date of the oldest red rock layers is 1.8 billion years.

(almost entirely hydrogen and helium) and dust in space. Over the span of millions of years, the cloud condensed under the force of its own gravity, forming the sun and its family of planets. The formation process left the young planets very hot: initially, surface temperatures on earth were several hundred degrees C.

Little is known about earth's atmosphere in these very first years. Scientists speculate it may have been a hydrogen-helium mixture typical of the parent gas cloud's composition. However, these hot, lightweight atoms moved so fast they soon escaped earth's gravitational embrace or perhaps were stripped from earth by fast-moving particles from the young sun. In any case, our planet lost its first atmosphere almost as soon as it acquired it, a catastrophic event that luckily no one was present to suffer.

Earth's second atmosphere came from within. Nitrogen compounds, water vapor, sulfates, argon, and carbon dioxide continually escaped from the surface through volcanoes and other fissures, a process known as outgassing (see Figure 2.32). The outgassed molecules have considerably greater atomic masses than hydrogen and helium and therefore were retained by the gravity of the slowly cooling earth. For over 2 billion years, earth's atmosphere was composed almost entirely of these substances, which gradually built up in volume.

The date 1.8 billion years ago marked the onset of another major compositional change in earth's atmosphere. At that time, plant life evolved that was capable of performing photosynthesis, which as you know uses carbon dioxide and water to make plant tissue. We repeat the chemical reaction:

$$CO_2 + H_2O \text{ — sunlight} \rightarrow C_6H_{12}O_6 + O_2$$

Note the presence of $O_2$ on the right side of the reaction, indicating that through photosynthesis, plants began removing $CO_2$ from the atmosphere and replacing it with their waste product, diatomic oxygen. Thus oxygen, which was almost totally absent from the atmosphere up to this time, began to accumulate. The 21% proportion of oxygen we enjoy breathing is a direct result of a change in our atmosphere's composition brought about through plant photosynthesis. Figure 2.33 shows evidence indicating the timing of oxygen's appearance in the atmosphere.

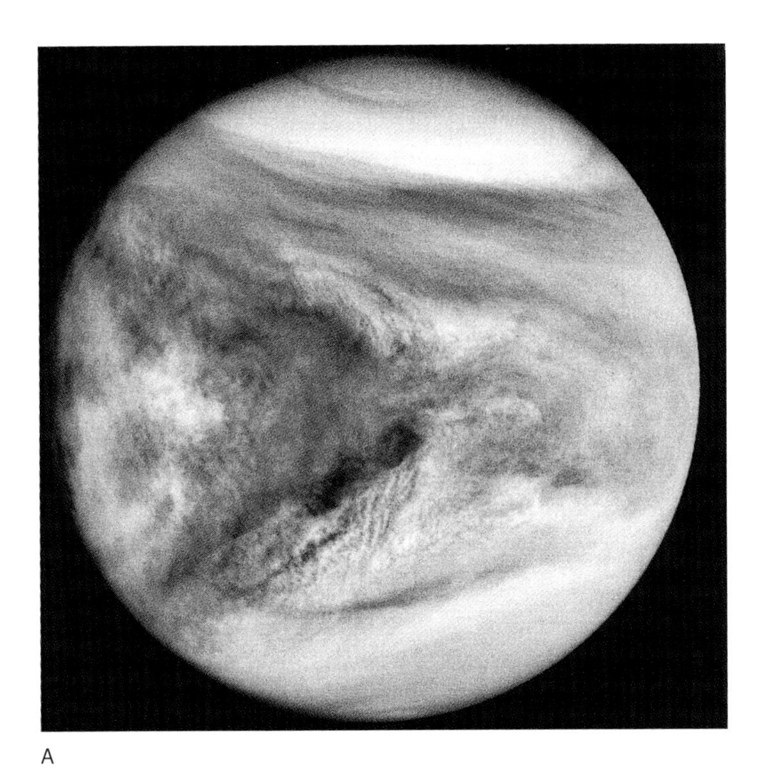

A

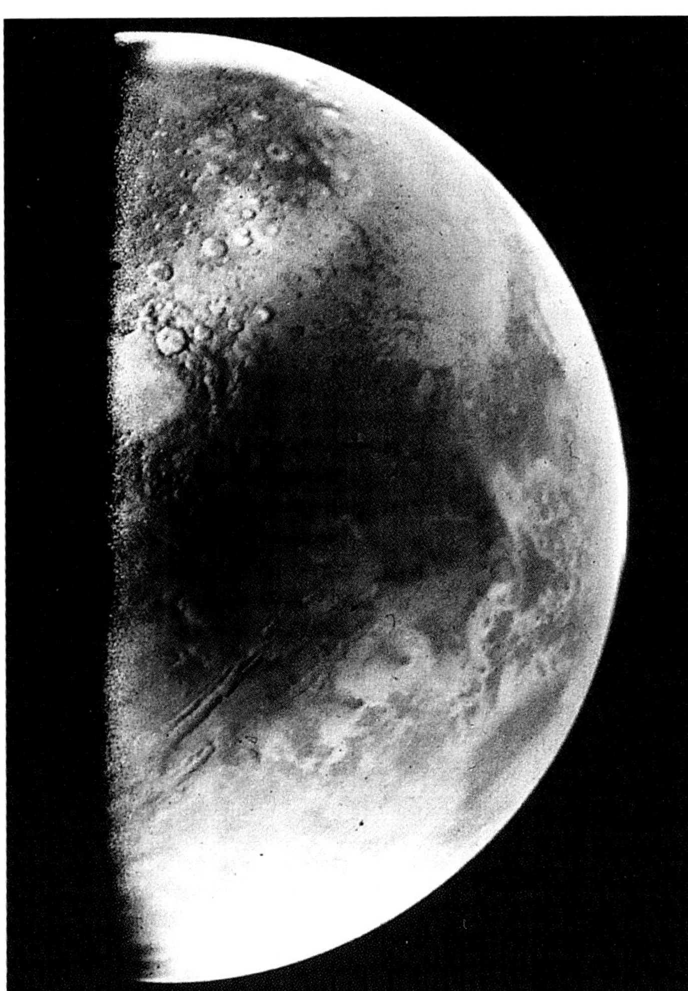

B

Figure 2.34 (*A*) Venus. (*B*) Mars.

## The Atmospheres of Other Planets

Earth's nearest planetary neighbors are Venus and Mars (Figure 2.34). Examine the data on their atmospheres in Table 2.6. What similarities and differences do you see between earth's atmosphere and those of Venus and Mars? The most striking fact about these planets' atmospheres is that they are so different from earth's. Note that the Venusian and Martian atmospheres are composed mainly of carbon dioxide, in contrast to earth's nitrogen-oxygen atmosphere. (In later chapters, you, will deal further with carbon dioxide's influence on a planet's climate and the lessons that Venus' $CO_2$-rich atmosphere may have for earthlings.) Notice also that Venus' atmosphere is about 100 times more massive, and Mars' is more than 100 times less massive, than that of earth. Table 2.6 shows that other conditions are radically different from those on earth. These three planets, celestial neighbors who shared a common birth, have atmospheres that evolved along very different paths.

## Reflection

We conclude this chapter by asking you to think about the extraordinary variety amongst the atmospheres of Venus, Earth, and Mars and on how very different earth's past atmospheres were. What lesson is there to draw from these differences? Drawing lessons is a personal affair, so we leave you to draw your own. To us, the lesson is that nothing is guaranteed about our current atmosphere, except that it can change. It has changed dramatically in the past and presumably can do so again. We must be careful about how we treat it. It is the only one we have.

*Table 2.6* **Venus, Earth, and Mars: The Planets and Their Atmospheres**

| Property | Venus | Planet Earth | Mars |
|---|---|---|---|
| Diameter | 12,100 km | 12,800 km | 6800 km |
| Distance from sun | 108 million km | 150 million km | 228 million km |
| Atmosphere: | | | |
| $N_2$ | 3.4% | 78.1% | 2.7% |
| $O_2$ | <0.01% | 20.9% | 0.1% |
| A | <0.01% | 0.9% | 1.6% |
| $H_2O$ | 0.1% | 0–3% | 0.03% |
| $CO_2$ | 96.4% | 0.03% | 95.3% |
| Average surface temperature | 470°C (880°F) | 15°C (59°F) | −60°C (−75°F) |
| Average surface air pressure | 90,000 mb | 1000 mb | 7 mb |

# Study Point

## Chapter Summary

This chapter has continued the process begun in Chapter 1 of equipping you with a basic inventory of tools for investigating the atmosphere. The two major tools introduced in this chapter are more abstract than the weather instruments of Chapter 1. They are the atomic model of matter and the kinetic theory of gasses. The first states that protons, neutrons, and electrons combine to make atoms, which themselves combine to make molecules of various compounds. The atmosphere, you saw, is a mixture of various atoms, molecules, and larger clumps of molecules called aerosols and particles.

A basic result of kinetic theory is that the temperature of a gas is a measure of the speed at which its molecules are moving. Kinetic theory provides a connection between events at the molecular level and larger-scale properties we can measure, such as temperature and pressure.

Thus equipped, you began in earnest to explore the chapter question: Is the atmosphere's composition changing? The short answer clearly is yes; it always has and always will. The long answer is very long indeed: the atmosphere contains a great number of different substances that can have significant impacts on life and/or the environment. The relative proportions of oxygen and nitrogen, which comprise well over 90% of the atmosphere, are changing little from year to year. However, concentrations of many problem substances, such as ozone, carbon dioxide, and sulfur dioxide, are subject to change, in many cases due to human influences.

You took a tour of the atmosphere's structure in the vertical and studied the depletion of the stratospheric ozone layer. The chapter concluded with a consideration of the very large composition changes our atmosphere has undergone in the geologic past and the highly divergent paths along which the atmospheres of our neighboring planets, Venus and Mars, have evolved.

## Key Words

mass
density
proton
electron
atomic number
isotope
radioactive decay
molecule
diatomic
triatomic
volume
nucleus
neutron
atom
element
atomic mass number
chemical bond
compound
ozone
monatomic
mixture
evaporation
deposition
variables of state
air pollutant
sources and sinks
storage (in budgets)
acid rain and fog
photochemical dissociation
hydrocarbon
unstable layer
stratosphere
stratopause
chlorofluorocarbons (CFC)
mesopause
ion
aurora
kinetic theory of gases
sublimation
condensation
the gas law
hydrologic cycle
flux
photosynthesis
aerosol particles
photochemical smog
troposphere
tropopause
inversion
ultraviolet radiation
mesosphere
thermosphere
ionosphere

## Review Exercises

1. How do monatomic, diatomic, and triatomic molecules differ?
2. How many protons does an atom of oxygen have? How many neutrons? Electrons? What would the answers be for a molecule of ozone ($O_3$)?
3. Hydrogen occurs most commonly with a nucleus of one proton and no neutrons. However, some hydrogen atoms have two nuclear particles; a few others have three. What must these additional particles be? What are these different forms of an element called?
4. Suppose the nucleus of a neon atom underwent fission by ejecting a cluster consisting of three protons and three neutrons. What kind of element would the ejected cluster be? What would the remaining, former neon nucleus be? (Refer to Appendix E.)
5. A bicycle pump forces air molecules into a tire, thereby increasing the air density in the tire. Suppose you double the air density without changing the tire's volume or its temperature. By what amount will the pressure inside the tire change?
6. Explain what happens when you blow up a balloon until it bursts. (Consider the previous exercise.)
7. Why doesn't the densest molecule in the atmosphere settle out on the ground and the lightest float to the top of the atmosphere?
8. What is wrong with this chemical reaction:

$$O_2 + CO \rightarrow NO_2 + CO_2$$

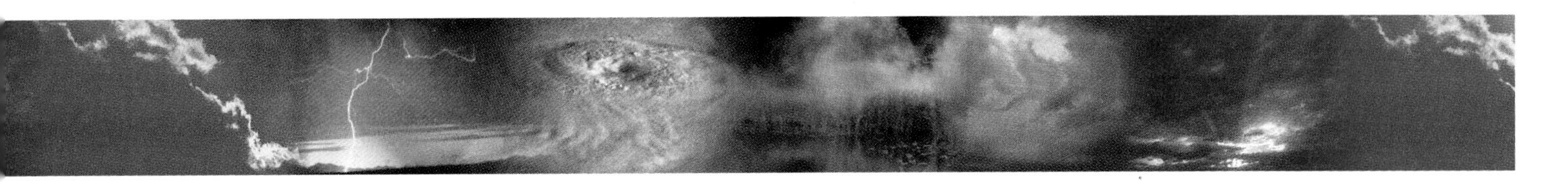

9. Outline the steps by which thinning of the Antarctic ozone layer occurs each year.
10. Explain how it is that a single chlorine atom can destroy many thousands of ozone molecules.
11. Which of these is not a health hazard? Ozone, carbon monoxide, argon, sulfur dioxide.
12. What causes ions to occur in the upper atmosphere?
13. What is the major difference in the composition of Venus' and Mars' atmospheres, compared to that of earth?

## Problems

1. A dorm room has dimensions of 3 meters by 4 meters by 2 meters. What is its volume?
2. Assume the air in the room in problem 1 has a density of 1.2 kilograms per cubic meter (a typical value for air near sea level). What is the mass of the air in the room? How does this compare to the mass of a 65-kilogram student?
3. In Figure 2.14, the atmosphere's storage of 13 units of water is considerably less than the annual incoming flux of 423 units. Explain how the amount in storage can be less than that coming in.
4. Table 2.3 shows a count of 207,500 $O_2$ molecules in a sample of 1 million. Express this number in parts per thousand; in parts per hundred (%); in parts per billion.
5. Make a "budget" drawing of a bank account, labeling fluxes, sources, sinks, and storage.
6. Comment on this statement: "The air pollution 'problem' is over-rated: so-called pollutants are present in such tiny amounts, measured in parts per million or parts per billion, that they have no effect on anything."
7. "One man's meat is another man's poison": Relate this statement to carbon dioxide in the atmosphere.
8. Suppose you were transported back 3 billion years in time. What would be the biggest threat to your survival?
9. State the following event in the form of a chemical reaction formula: A molecule of nitrous oxide combines with one of ozone to make molecules of nitrogen dioxide and diatomic oxygen.
10. By how many times has methane in the atmosphere increased since the year 1700, according to Figure 2.21?

## Explorations

1. Are there relations between air quality and other atmospheric variables, such as cloud cover, wind speed, high or low pressure positions, and so forth? Collect the air quality information for a number of days from local daily weather reports or from Internet sources (some suggested sources are given below). Compare the readings with prevailing weather conditions. For example, does air quality vary according to wind direction?
2. What is the particulate count of the air in your vicinity? Cover a microscope slide with a coating of petroleum jelly or a layer of double-stick cellophane tape. Place the slide outside for 24 hours. Then examine under a microscope, recording your observations. Repeating this procedure for a number of days, correlate your findings with other meteorological data as in the previous exploration.
3. What are the atmospheres of the other planets like? In the last section of the chapter, we looked at atmospheric conditions on only two planets (besides earth). Extend Table 2.6 to include data for other planets and their satellites.

## Resource Links

The following is a sampler of print and electronic references for topics discussed in this chapter. We have included a few words of description for items whose relevance might not be clear from their titles. For a more extensive and continually updated list, see our World Wide Web home page at www.mhhe.com/sciencemath/geology/danielson.

Firor, J. 1990. *The Changing Atmosphere.* New Haven: Yale University Press. 145 pp.

Hedin, L.O., and G. E. Likens. 1996. Atmospheric dust and acid rain. *Scientific American,* December 1996: 88–92.

Otton, J.K. 1992. *The Geology of Radon.* Denver: U.S. Geological Survey. 28 pp. A nice survey of the radon problem for the nonspecialist.

Somerville, R. C. J. 1996. *The Forgiving Air: Understanding Environmental Change.* Berkeley: University of California Press. 216 pp. An informative discussion of ozone chemistry.

Tillery, B. W. 1993. *Physical Science,* (2d ed.). Dubuque, IA: Wm. C. Brown. 618 pp. An introductory physics and chemistry textbook that might serve as a useful reference on topics such as atomic structure.

NOAA's Geophysical Fluid Dynamics Laboratory's report on stratospheric ozone is available at http://www.gfdl.gov/brochure/3Strat_Ozon.doc.html

For readers with access to AccuData, AccuWeathers real time database, AIR provides air pollution data; <send> SWVUSF displays water vapor images. UPD gives height, pressure, temperature and wind data for upper air observations. SKW provides a plot of atmospheric soundings.

Users with access to AccuWeather's AccuData can obtain current air quality data at STAGD and the air quality index at AQI.

# Is the Atmosphere's Temperature Changing?

Are air temperatures on an upward trend from year to year? And if they are, what are the causes? Questions about "global warming" are the focus of periodic attention from the news media and congressional committees. They are the ongoing focus of attention of a number of atmospheric scientists. They are also the focus of this unit.

Global warming is an important topic because a sustained and substantial warming could have catastrophic environmental effects. It is a particularly troubling topic because evidence exists that certain human activities likely stimulate global warming and it is a topic that generates controversy. Scientists are far from unanimous in their interpretations of the data surrounding global warming.

The sun is the atmosphere's ultimate source of warming. In Chapter 3, you will see how and where sunlight is converted into heat, how that heat is exchanged between the earth and the atmosphere, and how the energy eventually escapes the earth. You will develop an "energy budget" to follow these processes. In Chapter 4, you will use these energy budget concepts to study global warming.

# Chapter 3

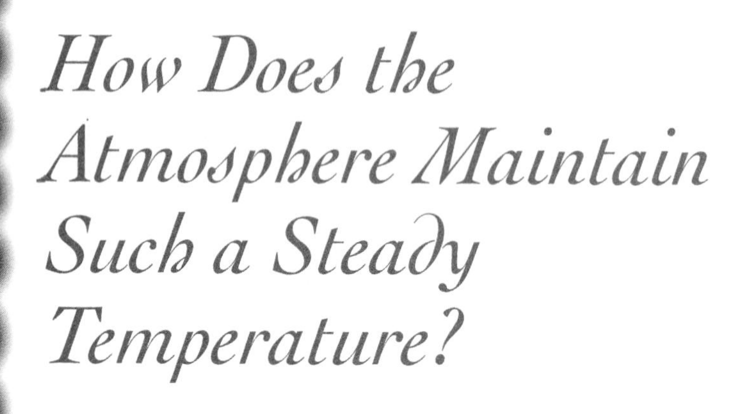

# How Does the Atmosphere Maintain Such a Steady Temperature?

# Chapter Goals and Overview

Every day, the sun floods our planet with the equivalent of more than 4000 trillion kilowatt-hours ($4 \times 10^{15}$ kwh) of energy. This is more energy than all of humanity has used over the past 100 years. The earth and its atmosphere reflect about 30% of this energy back to space. The remaining 70%, nearly 3000 trillion kilowatt-hours of energy are absorbed in the atmosphere and at the earth's surface, warming the planet. With the earth absorbing this extraordinary amount of energy each day, you might well wonder why temperatures on the planet aren't rising day after day.

How do you go about answering such a question? Our strategy will be to make an inventory of the major factors influencing the atmosphere's temperature. Then we will assemble these factors in the form of an atmospheric **energy budget,** which is a diagrammatic tally of all fluxes of energy into and out of the atmosphere. In the process of constructing the atmosphere's energy budget, you will see how the earth manages to keep a relatively even temperature in the face of all that solar energy. You will see how solar energy is transformed and moved throughout the atmosphere and eventually returns to space.

You will observe that the largest fluxes in the earth's energy budget are in the form of radiation, one example of which is sunlight. A major portion of this chapter is devoted to developing a rather extensive model of radiation's properties and behavior. You will be introduced to two additional processes of energy transfer: conduction and convection.

The energy budget you develop will not only help to answer this chapter's main question; you will also find it to be a basic tool for exploring the issue of global warming, which you will do in the next chapter.

## Chapter Goals

Upon completing this chapter you should be able to:

- compare and contrast the concepts energy, heat and temperature;
- define the terms *calorie* and solar *constant*;
- explain why the amount of incoming solar radiation varies with latitude, time of day, and time of year;
- explain how an object's reflectivity and specific heat affect the rate at which it may change temperature;
- relate the concept of the energy budget to the law of conservation of energy;
- calculate an energy budget, given values for various energy fluxes; and
- explain how radiation is emitted and absorbed by objects of different densities, temperatures, and chemical makeup.

You will also improve your skill at interpreting graphs and quantitative diagrams, such as energy budget diagrams.

# Sunlight: The Atmosphere's Ultimate Energy Source

## Sunlight, Heat, and Energy: A Thought Experiment

We begin our exploration of the atmosphere's various energy fluxes with sunlight, the earth's most obvious source of energy. Anyone who has parked a car with its windows closed in the sun knows that sunlight can be transformed to heat. Just how much energy does sunlight contain? We offer the thought experiment in Figure 3.1 to reach a quantitative answer. Study the "experimental setup" carefully. It is designed so that a given amount (exactly 1 gram) of a known substance (liquid water, at 15°C) is

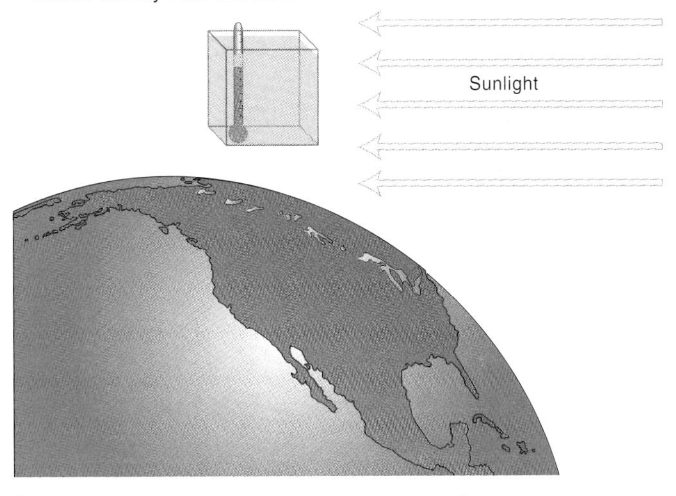


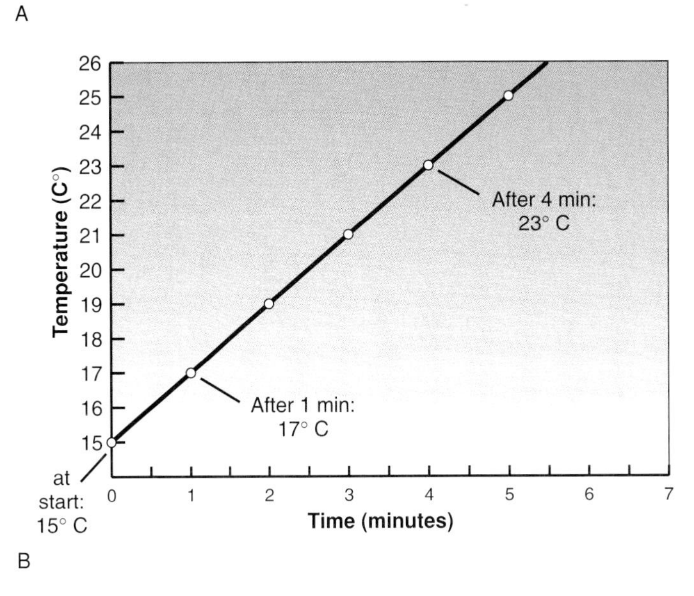


Figure 3.1 (*A*) Experimental setup for the thought experiment. (*B*) Data from your thought experiment. Note the water's steady temperature rise at the rate of 2°C per minute.

subject to just one energy flux, that of incoming sunlight. No other fluxes of energy into or out of the container are permitted. The container is cube-shaped and measures 1 centimeter on a side. All incident sunlight (that is, sunlight striking some surface; here, the side of the cube) is absorbed at the surface and goes into heating the water. Note the container's location: near the earth but above the atmosphere and oriented so that one face is perpendicular to the sun's rays; only this face receives solar energy.

Having placed the experiment in location, you begin recording the change in the water's temperature. Figure 3.1B shows your data. The water's temperature rises nearly 2°C (1.96°C, to be more precise) each minute the water is exposed to the sunlight. This result could be summarized as follows: At earth's average distance from the sun, the solar energy striking 1 square centimeter oriented perpendicular to the sunlight will raise the temperature of 1 gram of water by 1.96°C per minute.

This hypothetical experiment radiates with important concepts. For example, note that solar energy is being transformed into heat. Implied in this transformation is our conviction that energy does not disappear or appear spontaneously. It may be transformed from one form to another, but it can always be accounted for. This concept is embodied in one of the most fundamental laws of science, the **law of conservation of energy.** In everyday language, the law states that energy may change form but is neither created nor destroyed. Figure 3.2 gives some additional examples of the law. The assumption of conservation of energy is basic to the energy budget you will build later in the chapter.

The thought experiment offers a way to define a unit of energy known as the **calorie.** A calorie is the amount of energy required to raise 1 gram of water by 1°C, technically, from 15.5°C to 16.5°C. (The calorie thus defined is sometimes called the small calorie. It is one one-thousandth of the "large Calorie" used in describing the energy content of foods.)

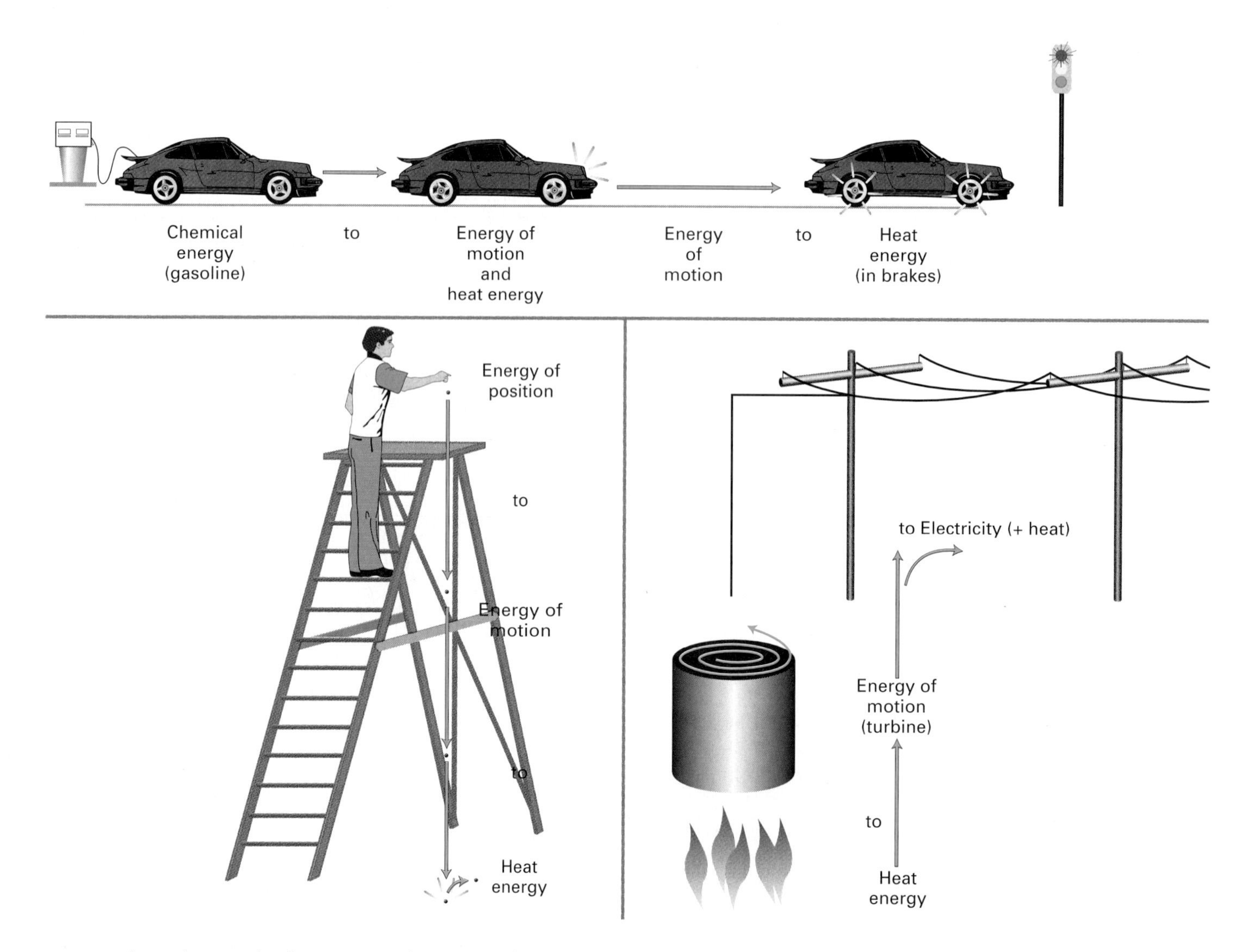


Figure 3.2 In each example above, energy is conserved as it is transformed. Note that energy is often transformed into heat, either deliberately or inadvertently, as in heat loss from electric lines. (An everyday example is a warm appliance cord.)

Look back for a moment at Figure 3.1. How many calories of energy did the water absorb? Since 1 calorie is required to warm 1 gram of water by 1 degree, each degree of warming required 1 calorie. Thus, the water, warming at 2°C per minute, was absorbing 2 calories per minute. The incoming solar beam, 1 square centimeter in cross-sectional area, therefore transported 2 calories of energy to the water per minute.

Measurements over the past 100 years or so indicate that the sun's energy output is nearly constant. Therefore, any time you were to repeat the thought experiment, the result should be virtually the same: close to 2 calories of sunlight per minute will impinge on 1 square centimeter-sized area at earth's mean distance from the sun. This quantity, the amount of solar radiation incident per minute on a 1 square-centimeter surface oriented perpendicular to the beam and positioned at the top of the atmosphere at earth's mean distance from the sun, is known as the **solar constant**. The current value of the solar constant is 1.96 calories per square centimeter per minute. The term provides a convenient way of referring to the flux of energy radiated by the sun.

You will find it useful to know two additional energy terms. The first is **insolation,** which refers to the sunlight incident onto a horizontal surface. The word *insolation* is constructed from the words *incoming solar radiation*. Be careful not to confuse *insolation* with *insulation*, a substance that retards the transfer of heat.

The second term is used in discussions of the *rate* of energy flow, which is known as **power**. The unit of power generally used in atmospheric studies is the **watt**. One watt is an energy flux of 14.33 calories per minute. Expressed in watts, the solar constant equals 0.1367 watts per square centimeter, or 1367 watts per square meter. (Notice that we did not say "watts per minute per square centimeter"; units of time are built into the definition of the watt.) As you might suspect, these watts are related to the units in which people purchase electric energy: a kilowatt-hour is a 1000-watt energy flow lasting for one hour.

Consider for a moment the considerable energy contained in the solar beam. If you ignore for the moment losses from the solar beam as it passes through the atmosphere, you would find that a 1-square meter area oriented like the one in Figure 3.1A receives 1.367 kilowatt-hours of solar energy each hour. Imagine the amount of energy striking a roof exposed to the sun over the course of a day. We invite you to investigate this problem in the Explorations at the end of the chapter.

## The Distribution of Insolation across the Earth

The sun's energy output is constant, or nearly so; however, the amount striking a unit horizontal area at the top of the earth's atmosphere varies greatly with latitude, time of day, and season of the year. This section addresses these variations.

### *Distribution by Time of Day and Latitude*

Figure 3.3 shows the sun's apparent path across the sky on September 23. We call it the apparent path because, as you know, it is the earth's rotation on its axis once per day, and not any actual motion of the sun, that causes the solar paths shown. As earth rotates, different regions are brought into the sunlight and out of it again. On this date (and on March 21), viewed from anywhere on the earth, the sun appears to rise on the eastern side of the sky, travel westward across the sky, and set 12 hours later (on the average) in the western sky. As the figure

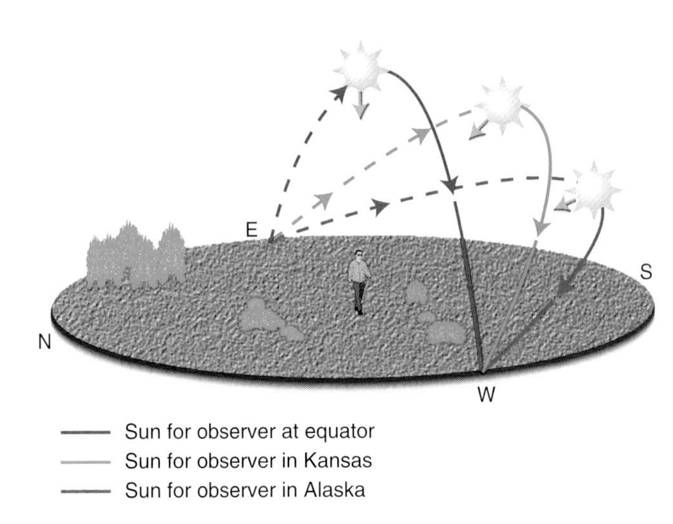


Figure 3.3 The sun's path across the sky on September 23. All three observers see the sun rise due east, reach its highest point in the southern sky, and set due west. The farther north the observer, however, the lower the sun's path.

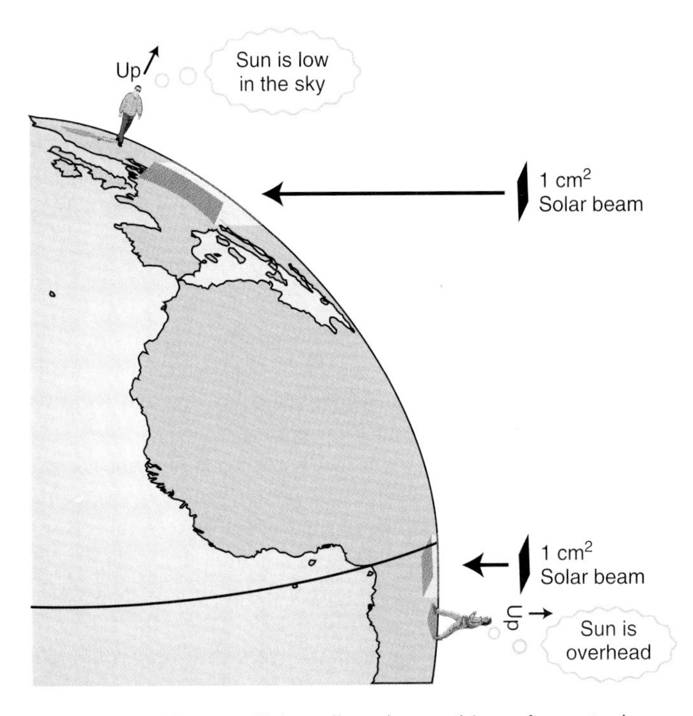


Figure 3.4 When sunlight strikes the earth's surface at a low angle, its energy is diluted over a larger area. The resultant heating is less than in locations where the sun is closer to overhead.

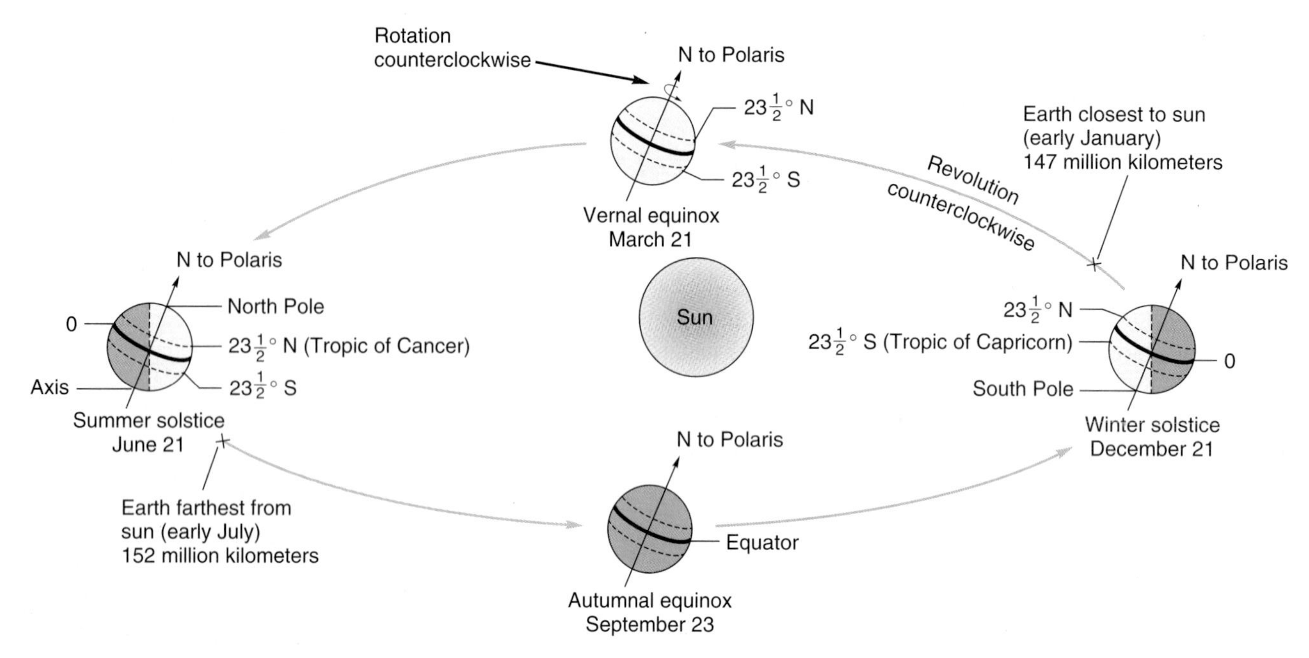


Figure 3.5 The earth revolves annually around the sun in an elliptical orbit. Note that the axis of rotation is tilted and remains pointing in the same direction as it rotates.

indicates, the actual path varies considerably according to your latitude.

How does the sun's daily path affect the amount of insolation available? Unlike conditions in the thought experiment, only rarely, if ever, is the solar beam actually perpendicular to the earth's surface at any given location. And note in Figure 3.4 that the "less perpendicular" the sunlight, that is, the smaller the angle between the incoming beam and the earth's surface, the larger the area over which its energy is spread and therefore the less solar energy per area is received. Thus, in late September, considerably more solar energy is incident per unit area at the top of the atmosphere over equatorial regions than elsewhere. However, even an observer located directly on the equator will see the sun shining directly downward for only a brief time around noon.

## *Distribution by Season of the Year and Latitude*

The sun's daily path does not remain constant throughout the year; instead, it shifts slightly from day to day, which leads to important changes in insolation. In this section, we will consider these variations with the aid of Figure 3.5.

Notice that the earth's orbit in Figure 3.5 is not circular but elliptical. Much of its elliptical shape is the result of perspective in the drawing: a circle looks elliptical when viewed on a slant. However, the earth's orbit around the sun actually *is* slightly elliptical; as the data in the figure indicate, Earth is closest to the sun in January and farthest in July.

Does the earth's varying distance from the sun explain the seasons? Examine Figure 3.6 for evidence. Note that different locations experience summertime (warmest temperatures) in different months. If the earth-sun distance controlled the seasons, then all the temperature curves would run in phase with each other, with warmest temperatures in January, when the earth is closest to the sun.

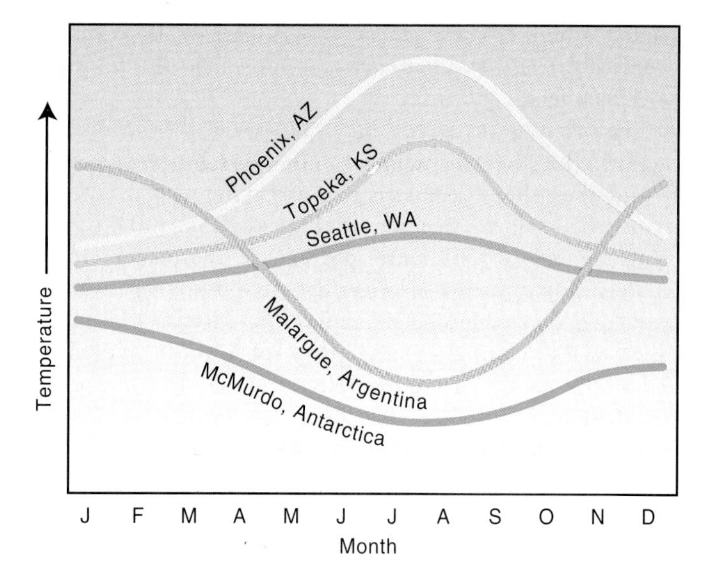


Figure 3.6 Mean monthly temperatures for five locations. Note that two stations experience summer in December through February and winter from June through August.

The explanation for seasonal variations, then, lies not in the earth-sun distance. Instead, solar angle is the crucial factor, as it is for the variation with time of day and latitude. The earth's axis is tilted relative to its plane of revolution around the sun, as you can see by referring back to Figure 3.5. Also notice in that figure that the axis remains pointed in the same direction in space throughout the year. As a result, as the earth orbits the sun, the north end of its axis (North Pole), and thus the Northern Hemisphere, alternately tilt toward and away from the sun. The North Pole's tilt toward the sun is greatest on or about June 21, a moment called the **summer solstice,** which marks the official beginning of summer in the Northern Hemisphere. On this date, the sun shines directly down not on the equator but on latitude 23.5°N, which is known as the Tropic of Cancer. The sun's daily path is highest in the sky and the length of daylight is the longest of the year (for people living north of the Tropic of Cancer). Notice that the situation in the Southern Hemisphere is just the reverse: that hemisphere is tilted farthest from the sun on June 21, and therefore experiences the least amount of sunlight and the onset of winter.

After the summer solstice, the earth's orbital motion causes its axis to point less directly toward the sun each day. As a result, the sun appears to move south in the sky each day. By September 23 (Figure 3.5), the earth's axis points neither toward nor away from the sun; the sun now shines directly down on the equator. This moment is the **autumnal equinox** and is the onset of fall in the Northern Hemisphere (and of spring in the Southern Hemisphere). On this date, each latitude experiences 12 hours of daylight and 12 of night.

Three months later, on December 21, the earth has moved so that the North Pole points farthest from the sun. The sun is directly overhead at latitude 23.5°S, called the **Tropic of Capricorn.** This moment is known as the **winter solstice,** the onset of winter in the Northern Hemisphere. On the winter solstice, the sun is "up" the fewest hours and its daily path is lowest in the sky. Meanwhile, Southern Hemisphere residents are enjoying the beginning of summer, with long hours of sun high in the sky.

By March 21, the earth has reached the **spring equinox.** The sun shines directly down on the equator once again, and all latitudes experience equal amounts of day and night.

Figure 3.7 indicates the sun's daily path for observers at different latitudes at different times of year. You can see that in polar regions, the sun's summertime daily path never dips below the horizon; thus, these regions experience continuous sunlight (24 hours of sunlight each day) during summer months, giving them the highest amounts of insolation (at the top of the atmosphere, at least) at that time of year. Given these high values, you might wonder why summer temperatures in the polar regions aren't higher than anywhere else on earth. A reasonable question, which leads us to the next section.

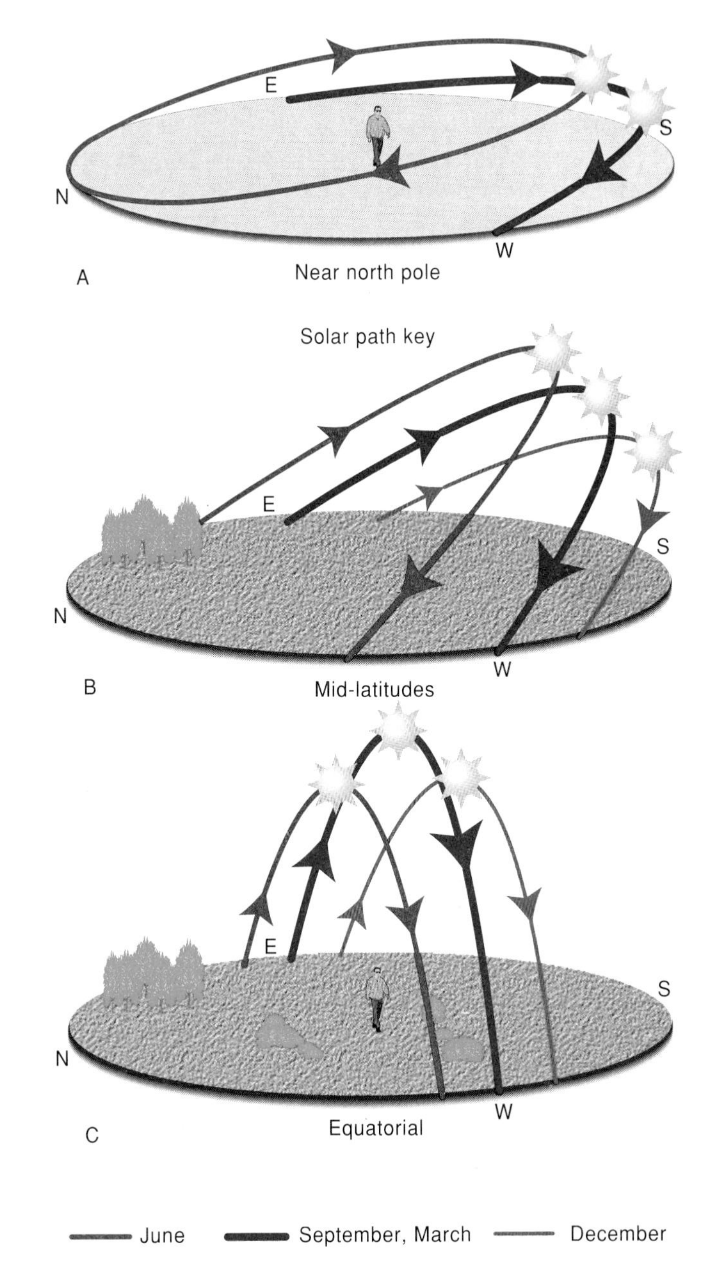


Figure 3.7 Daily solar paths for observers at different latitudes, for different times of year. Notice how the sun's apparent height in the sky changes with the season and that the locations of sunrise and sunset move along the horizon. Note also that the near-polar observer sees no December sun whatsoever, but experiences 24 hours of summer sun.

## Checkpoint

***Review***

The concept map in Figure 3.8 relates the major ideas presented in this section. Note that although the sun's output (as indicated by the solar constant) is nearly constant, the amount incoming per unit area at any given place on the earth varies widely due to three factors mentioned in the map.

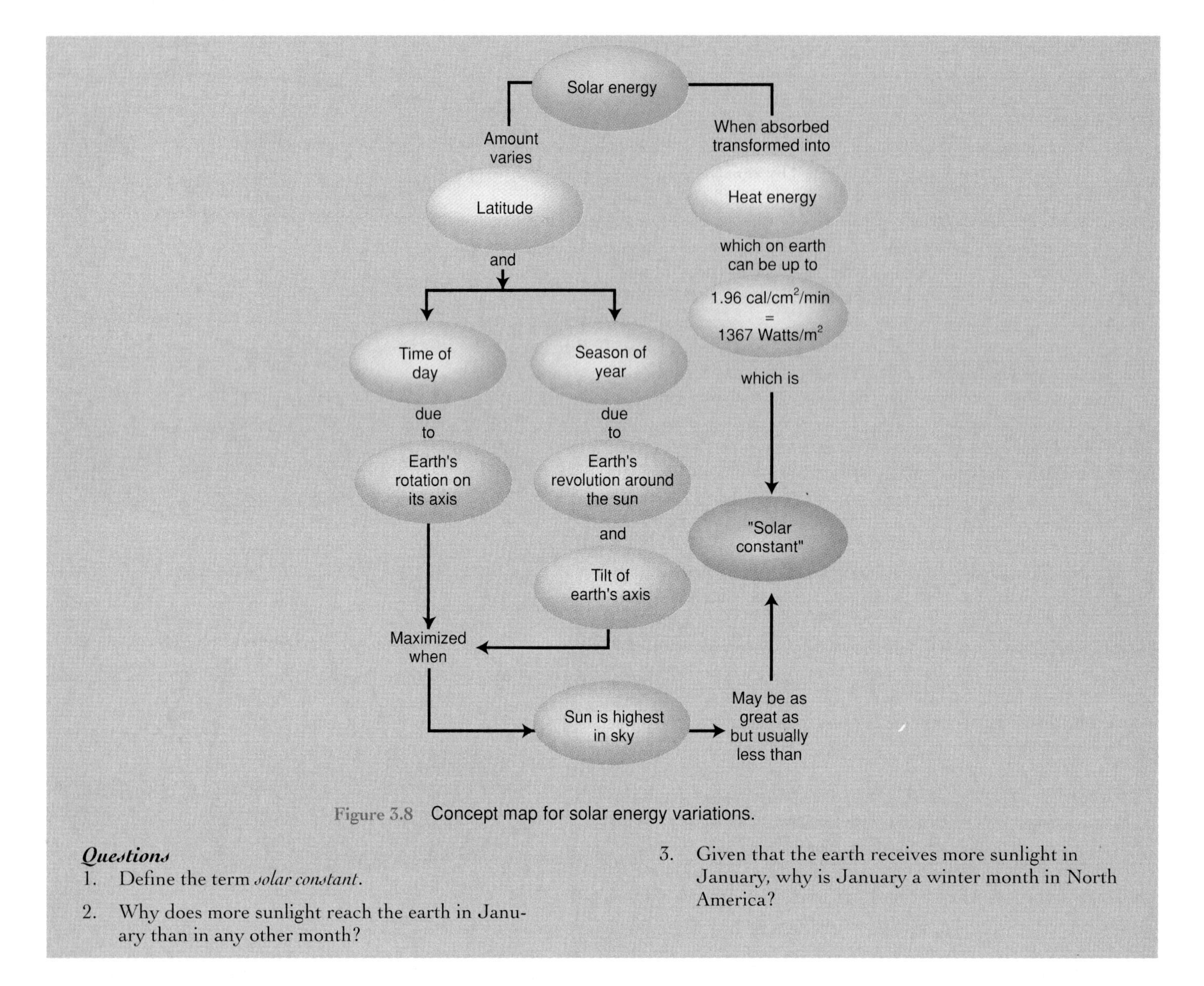


Figure 3.8 Concept map for solar energy variations.

*Questions*

1. Define the term *solar constant*.
2. Why does more sunlight reach the earth in January than in any other month?
3. Given that the earth receives more sunlight in January, why is January a winter month in North America?

# Solar Fluxes in the Atmosphere and at the Earth's Surface

In the last section, we explored the distribution of insolation at the top of the atmosphere at different times and locations. Now we will follow a beam of solar energy and study its interactions first with the atmosphere and then with the earth's surface. We assume the beam contains 100 units of energy as it enters the atmosphere.

## Insolation and the Atmosphere

How much does the incoming sunlight interact with the atmosphere? Your own past experience gives some guidance on this subject. If the air is free of clouds and pollution, you know that sunshine reaches the ground with little apparent interference from the atmosphere. Under cloudy or dirty skies, however, the sun's disk may become completely obscured.

These everyday observations correctly convey the fact that clear air is rather transparent to sunlight. On the average, the atmosphere absorbs (that is, removes from the solar beam and converts to heat) only about 16% of the incoming beam: that is, about 16 of our sample of 100 units (see Figure 3.9). Some of this absorption occurs in the upper atmosphere and is due to ozone and monatomic oxygen, as you learned in Chapter 2. Air molecules and other small particles **scatter,** or deflect, some of the incoming solar beam in all directions. Under clear skies, six of the original 100 units are scattered back to space. This leaves 78 units (100-16-6) to reach the earth's surface as direct (beam) or diffuse (scattered) sunlight. Thus, under clear skies, 78% of the insolation reaches the earth's surface.

Clouds change the picture considerably. Clouds deplete the incoming beam by both absorbing and reflecting insolation. Some clouds reflect as much as 80% of the sunlight striking them. On a worldwide average, clouds reflect back to space 20

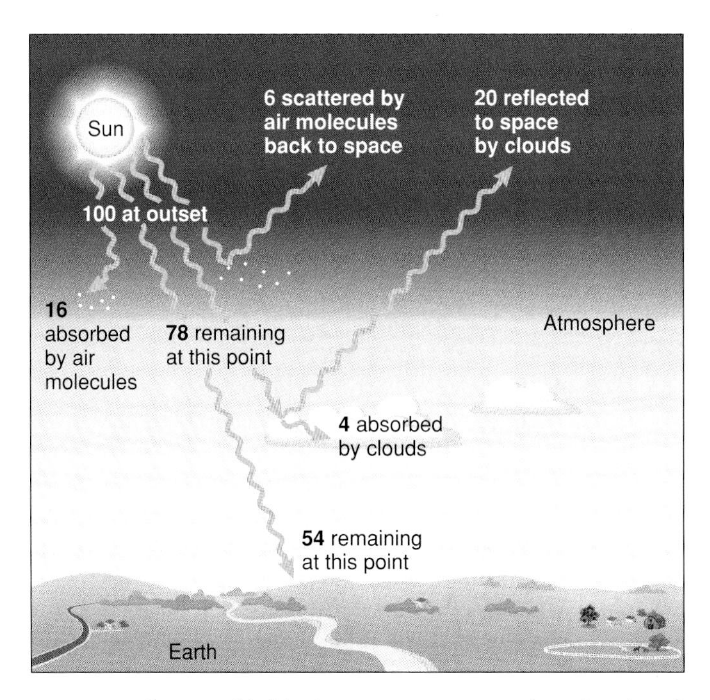


Figure 3.9 On a worldwide, long-term average, air molecules absorb 16% of the incoming sunlight and reflect another 6%. Clouds remove another 24%.

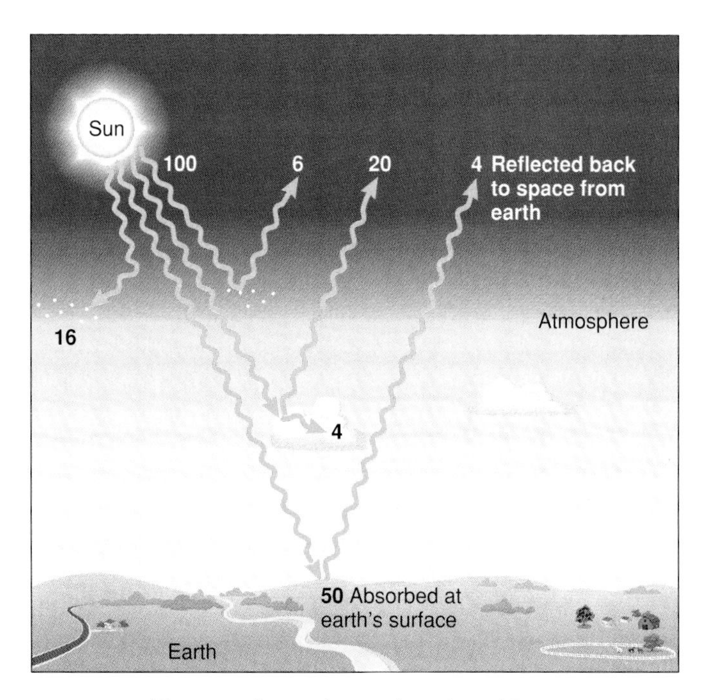


Figure 3.10 The complete solar portion of earth's average energy budget. What percent of the incoming 100 units actually warms the earth and atmosphere?

of the original 100 units of sunlight (i.e., 20% of the original beam) and absorb an additional 4 units. This means that, as you can see from Figure 3.9, on average, 54 of the original 100 units (100 incoming − 16 absorbed by the air − 6 scattered by the air − 4 absorbed by clouds − 20 reflected by clouds) reach the earth's surface.

What happens to the sunlight absorbed by the atmosphere? The law of conservation of energy reminds us that the energy cannot disappear. Most of it is applied to increasing the kinetic energy of the molecules comprising the atmosphere, an increase we observe as a warming of the air. As you will see, however, this source of atmospheric heating is minor compared to others you will learn about later.

We should emphasize that the data in Figure 3.9 are long-term, worldwide averages and that locally, the values may differ radically. For example, in high latitudes, where the sun's angle is low, the solar beam passes through a much longer atmospheric path than at low latitudes. As a result, the beam is more likely to be scattered, or intercepted by scattered clouds, than in places where the beam is more vertical. This loss of energy is partly responsible for the lack of warmth at the North Pole in summer, despite high daily average insolation values at the top of the atmosphere. But there are other, even more important factors, which we turn to next.

## Insolation at the Earth's Surface

### *Albedo*

You have just seen that on a global, long-term average, 54% of solar energy entering the top of the atmosphere reaches the earth's surface. Now, a portion of this energy is absorbed at the surface; but the remainder is reflected. The portion reflected compared to that incoming is known as the **albedo.**

$$\text{albedo} = \text{sunlight reflected}/\text{sunlight incoming}$$

Measurements from satellites, among other sources, indicate that the average albedo of the earth's surface is 0.08. Albedo is usually expressed as a percentage. Thus, 8% of the insolation reaching the surface is reflected, on the average.

If you multiply both sides of the above expression by the incoming term, then the expression becomes

$$\text{reflected} = \text{albedo} \times \text{incoming}$$

Using this form of the expression, we find that

$$\begin{aligned}\text{reflected} &= 0.08 \times 54 \text{ units}\\ &= 4.32 \text{ (equals about 4 ) units reflected}\end{aligned}$$

If 4 of the incoming 54 units are reflected, then the difference, 50, must be absorbed by the earth's surface. Figure 3.10 shows the energy budget updated to include albedo and reflected sunlight.

It is instructive to apply the concept of albedo to the entire earth-atmosphere system. Figure 3.10 indicates that of 100 units of sunlight at the top of the atmosphere, a total of 30 units (20 from clouds, 6 from atmosphere, and 4 from the earth's surface) are reflected back to space. The albedo for the system, then, is

$$\text{albedo} = (20 + 6 + 4)/100 = 0.30 = 30\%$$

Thus, 70% of the insolation reaching the earth is absorbed and transformed into heat; 30% is reflected back to space and has no net effect on the planet's energy budget.

### *Albedo Variations*

In Table 3.1, you can see how greatly different substances may vary in albedo. Note that fresh snow reflects up to 95% of the insolation reaching it, while an ocean surface may reflect only 3%, absorbing the other 97%.

Once again, you already have some direct experience with this topic. Consider that in winter people tend to wear dark (low albedo) clothes, which the limited sunlight more effectively than light-colored ones; in summer, high albedo clothing (especially white) is more common. Also, people feel the greatest need for eye protection such as sunglasses at the beach or when skiing because of the large amounts of reflected sunlight from the high albedo sand and snow surfaces.

**Table 3.1 Albedoes of Various Surfaces**

| | Percents |
|---|---|
| Snow, fresh | 75–95 |
| Snow, old and dirty | 40–60 |
| Ice | 70 |
| Water, high sun | 03–05 |
| Water, low sun | 10–50 |
| Bare ground | 15 |
| Sand | 18–28 |
| Forest | 05–10 |
| Green crops | 15–25 |
| Cities | 14–18 |
| Clouds: | |
| Opaque | 50–85 |
| Thin | 05–50 |

Source: Data from Smithsonian Meteorological Tables, 1966, and J. Peixoto and A. Oort, *Physics of Climate*, 1992. American Institute of Physics, New York, NY.

Let's consider the effects of albedo in a more quantitative way. Imagine sunlight shining down onto equatorial regions, which are mostly ocean-covered. Suppose 54 units of insolation strike the ocean surface at a nearly vertical angle. Taking an albedo of 4% from Table 3.1, the albedo formula becomes

$$4\% = \text{reflected/incoming} = \text{reflected}/54 \text{ units}$$

or,

reflected = 4% × 54 units = 2 units (rounding off to the nearest integer).

Thus, the ocean surface absorbs 52 of the 54 units of insolation.

Now repeat the calculation for the same quantity of insolation, but this time for a high latitude where the surface is covered with fresh snow. Choose an average albedo for these conditions, and do the calculation yourself.

You should have found that the snow surface reflects approximately 46 of the incoming 54 units (85% × 54 units incoming = 46 units reflected), leaving just 8 units absorbed. From these calculations, it should be evident that albedo variations can cause great differences in the amount of sunlight actually absorbed from place to place and from time to time. It also helps to further explain the lack of warmth at the North Pole in summer despite its 24-hour daylight.

Figure 3.11 shows a computer-generated map of satellite measurements of reflected sunlight. Note the locations of highest values: over high albedo surfaces such as ice, snow and sand, and in regions of high insolation.

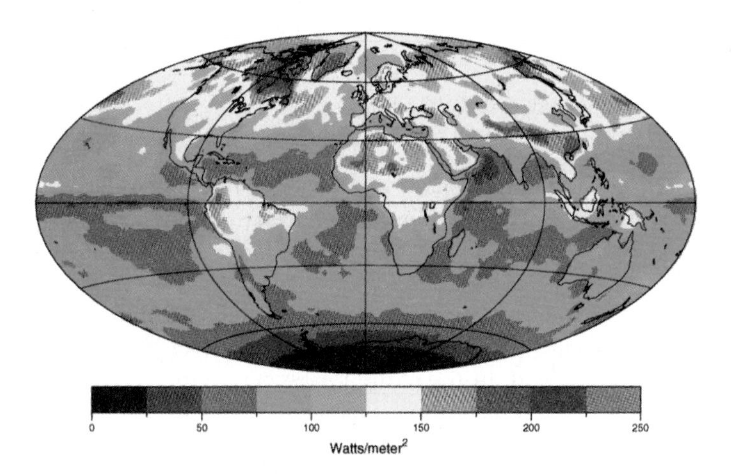


Figure 3.11 Reflected solar radiation, measured by satellite, for the month of April, 1985.

## Temperature Changes Due to Absorbed Sunlight

### *Specific Heat*

How do you translate absorbed solar energy into a change in temperature? Recall from the discussion of the solar constant that 1 calorie is the energy required to raise 1 gram of water by 1°C. But suppose that the incoming sunlight is absorbed by pavement, or snow, or air. Will a calorie warm a gram of these substances by 1°C also?

The answer, in short, is no. To elaborate, Table 3.2 lists the **specific heat** of various substances. Specific heat is the amount of energy required to raise the temperature of 1 gram of a substance by 1°C. Notice that water has the highest specific heat listed; hence, more energy is required to raise the temperature of a gram of water than a gram of any other material on the list. Note that the specific heat of concrete, for example, is only 0.2 calories per gram per °C. Thus, only one-fifth as much energy is required to raise the temperature of a gram of concrete as to cause the same temperature rise in a gram of water.

**Table 3.2 Specific Heat of Various Substances**

| | |
|---|---|
| Pure water | 1.00 calories $g^{-1}$ $°C^{-1}$ |
| Sea water | 0.93 |
| Dry air | 0.24 |
| Dry sand | 0.19 |
| Wet mud | 0.60 |
| Brick | 0.20 |
| Concrete | 0.20 |
| Granite | 0.19 |
| Limestone | 0.22 |

Source: Data from Smithsonian Meteorological Tables, 1966.

The relations between energy input, temperature change, the mass of an object, and its specific heat can be represented in the following formula:

$$\text{temperature change (°C)} = \frac{\text{energy input}}{\text{specific heat} \times \text{object's mass}}$$

To apply this formula, consider the following problem. Suppose 20 calories of sunlight are absorbed by a 10-gram mass of water. By how much will the temperature change?

$$\text{temperature change} = \frac{20 \text{ calories}}{(\text{specific heat of } H_2O) \times (\text{mass of } 10 \text{ g})} = \frac{20 \text{ calories}}{(1.0 \text{ calorie g}^{-1}\,°C^{-1}) \times (10 \text{ g})} = 2°C$$

If the same energy were absorbed by a 10-gram mass of concrete, the temperature change would be

$$\text{temperature change} = \frac{20 \text{ calories}}{(\text{specific heat of concrete}) \times (\text{mass of } 10 \text{ g})} = \frac{20 \text{ calories}}{(0.20 \text{ cal g}^{-1}\,°C^{-1}) \times (10 \text{ g}) = 20/2.0} = 10°C$$

As another example, apply the formula to a sample of air. The specific heat of dry air, near sea level, is 0.24. Verify from the formula that if 10 grams of air were to absorb the 20 calories considered above, the resulting temperature rise would be 8.3°C.

From these examples, you can see the influence of specific heat differences is considerable: the same amount of energy that warms a given mass of ocean water by 1°C warms most other substances considerably more.

## *Mixing*

You have probably noticed that the sand on a beach at mid-day may warm to the point where it is too hot to walk on, while the water nearby remains almost unaffected by the day's heating.

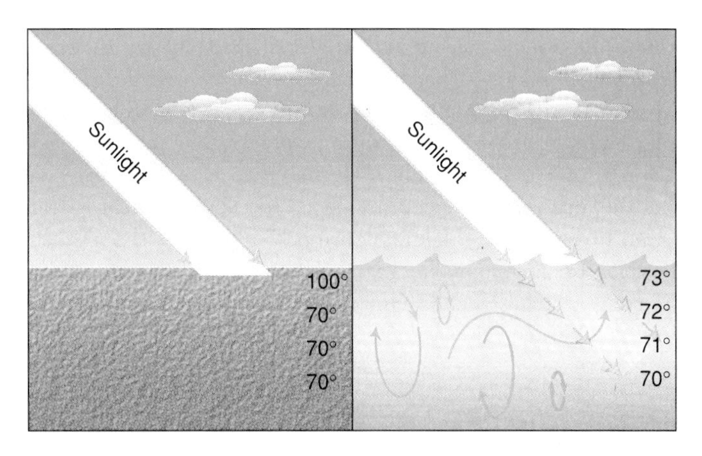


Figure 3.12 Water's transparency to sunlight and vertical motions in the water cause solar energy to penetrate and be mixed to much greater depths over water than over dry land.

Specific heat differences between water and sand explain part of the difference. Even more important in this case, however, is the difference in the volume of matter through which the absorbed solar energy is mixed.

As you can see in Figure 3.12, sunlight striking an object such as sand or rock is absorbed at the very surface. The absorbed energy is concentrated within a small mass of material, resulting in a large temperature change, as you can verify from the specific heat formula. Some of this surface energy gain is transferred into the air, warming it, and some is transferred downward a shallow depth into the ground. (More about both of these processes later.) Sunlight striking the water, however, penetrates to considerable depths before being completely absorbed. Furthermore, the

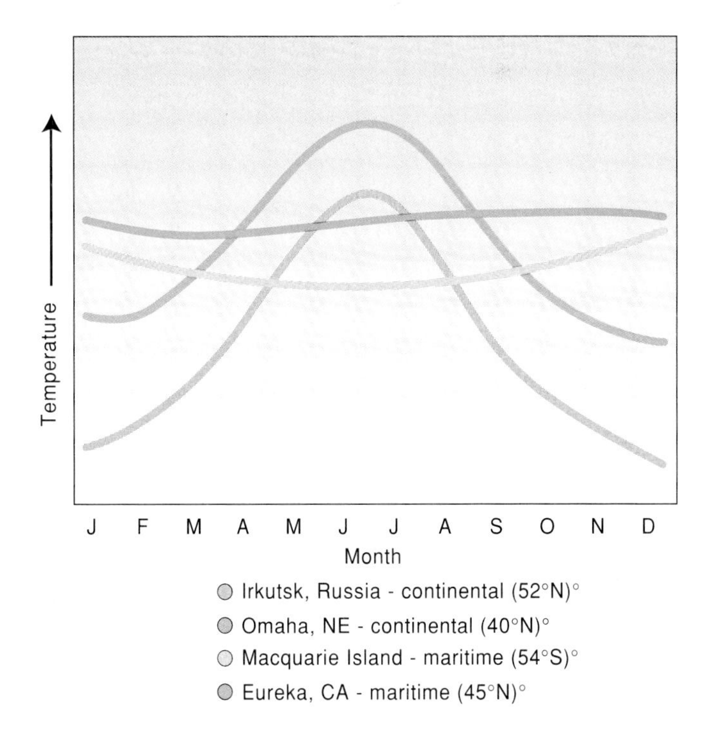


Figure 3.13 Monthly mean temperatures for two maritime and two continental stations. The moderating effect of water is evident in the flatness of the maritime curves.

water's motions serve to mix the absorbed energy to greater depths. The result is that the energy input is spread downward through a far greater mass of water. Finally, some of the absorbed solar energy goes into evaporating water, rather than increasing its temperature. As a result of all these processes, the water surface remains relatively cool by day, and thus, less heat is transferred upward into the air. Figure 3.13 illustrates the impact of these factors in seasonal temperatures at different locations. Note the comparatively small temperature variations at maritime stations compared to those at continental sites.

***Review***

Figures 3.9 and 3.10 serve as a sort of concept map, illustrating the course of events as the solar beam proceeds through the atmosphere to the earth's surface.

The albedo formula with Table 3.1 and the specific heat formula with Table 3.2 capture two other key concepts of this section. They express the facts that some surfaces absorb sunlight more effectively than others and that different materials warm by different amounts in response to the same energy inputs.

***Questions***

1. Suppose a beam of 200 calories strikes a paved parking lot, which absorbs 180 of the calories. What is the albedo of the surface?
2. As a Halloween prank, students paint the parking lot in problem 1 white with black lines, raising its albedo to 0.6. How many of the 200 calories will it absorb now?
3. Define specific heat. Why is specific heat important in meteorology?

# Radiation: A Model

From Figure 3.10 in the last section, you can determine that our planet absorbs 70% of the sunlight reaching it. What happens to this energy? The law of conservation of energy tells us we must be able to account for the energy gain. In energy budget terminology, the law may be expressed

energy input − energy output = change in energy stored

In one sense, this expression contains nothing new: it simply states that if input exceeds output, the storage of energy will increase. In the present case, increased storage appears as a rise in temperature. Using the analogy of a bank account, if your income ("energy input") and your expenditures ("energy output") are not equal, then your balance ("energy stored") changes.

Now, let's assume for the moment that the earth's temperature remains more or less constant from year to year. The constancy of temperature means that the storage change in energy is approximately zero. Thus, for the earth-atmosphere system, over a period of many years,

energy input − energy output = 0 storage change

The implication is that for the earth's temperature to remain constant or nearly constant over the years, its energy output must be very nearly equal to its solar input! Somehow, the system must be releasing roughly as much energy back to space as it receives from the sun. Only one mechanism of energy transfer can export this energy away from earth: the process known as radiation. You need to understand radiation in some detail in order to proceed through this chapter and the next. This section is devoted to that purpose.

## Radiation as Waves and Particles

Most of us have marveled at the way sunlight interacts with a glass crystal such as a prism (see Figure 3.14). Similar experiments inspired Isaac Newton to realize that a sunbeam is actually a composite of beams of many different colors that comprise what is termed the visible spectrum. Newton's discovery was among the first in a series that have spanned the past three centuries: our understanding of the nature of light has come very slowly. Now, however, there exists a fairly complete model of the nature of light, or (speaking more generally) of radiation.

### *Light as Waves*

Extensive experiments performed on light reveal that in some respects it behaves like trains of waves of different **wavelengths.** Wavelength is the distance from a given point on one wave to the same point on the next wave in the same train (for example, from crest to crest), as shown in Figure 3.15. Light's dispersion by a prism is explainable by thinking of light as a train of waves.

Experiments also indicate that light's color varies with its wavelength, in the manner shown in Figure 3.15. Note that red light has the longest wavelengths, and violet the shortest in the visible spectrum. The wavelengths of visible light are extremely short by human standards, as you can see by the values given in Figure 3.15. Scientists use a suitably small unit of length for wavelength: the **micrometer.** A micrometer is one-millionth of a meter or one-thousandth of a millimeter.

Prisms are not the only devices that break sunlight into its spectral colors. The atmosphere itself scatters light differentially, according to its wavelength. Shorter wavelength blue and violet light are scattered more readily than longer wavelength colors. This differential scattering is what causes the clear sky to be blue. (The human eye is not very sensitive to violet wavelengths, which is one reason the sky appears blue, not violet.)

### *Light as Particles*

Still other experiments with a beam of sunlight indicate that under some circumstances, the energy behaves not like a train

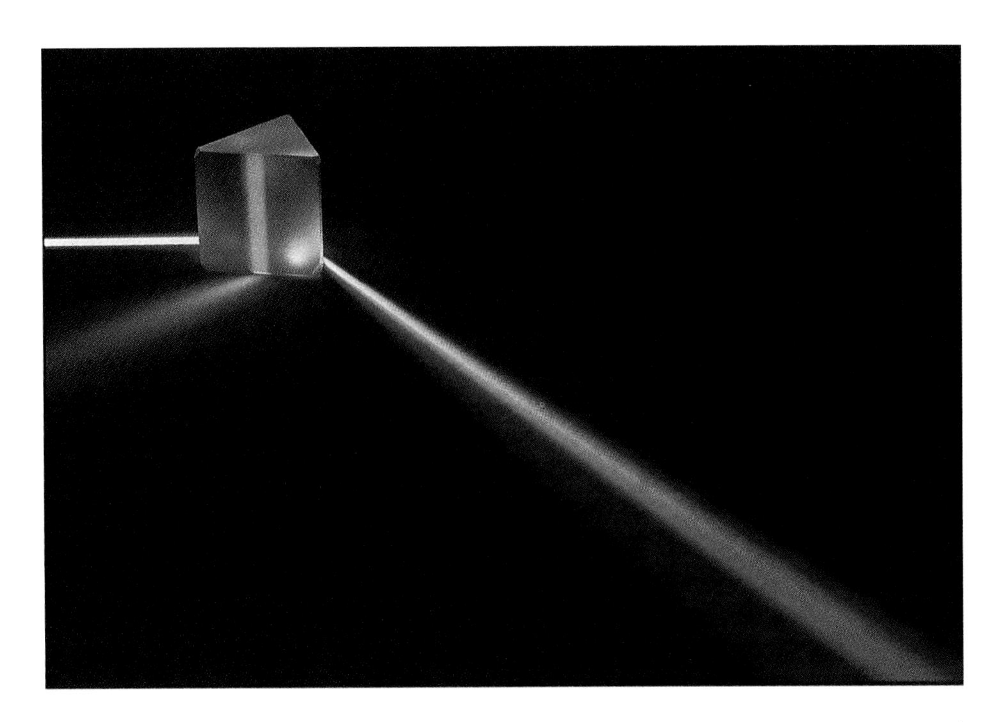

Figure 3.14 White light such as sunlight contains the entire spectrum of colors, which are made individually visible by passing the beam through a prism.

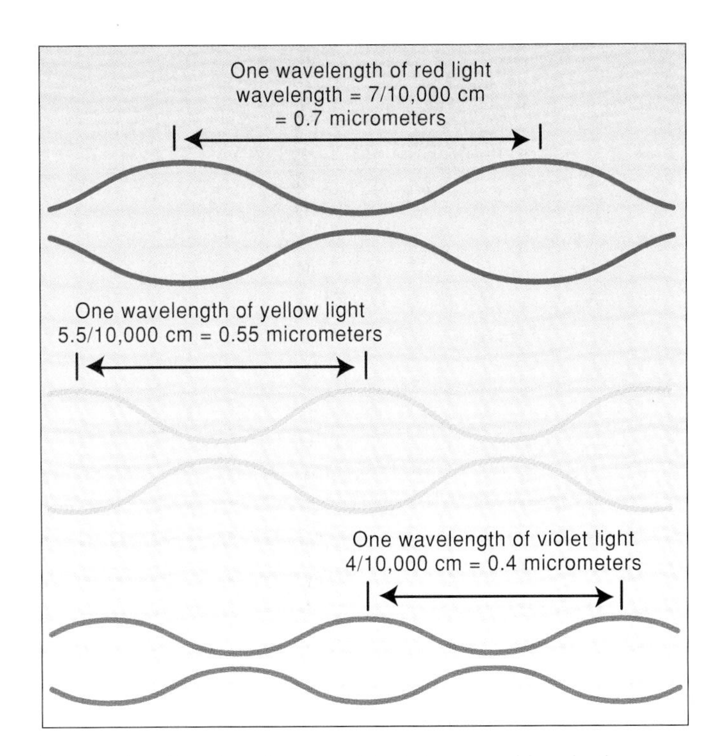


Figure 3.15 Wavelengths of visible light vary with color but are shorter than 1/1000 centimeter.

of waves but instead like a stream of particles. For example, the fact that sunlight passes through empty space is difficult to understand using a wave model. Waves require a medium in which to travel or propagate their energy: ocean waves require water molecules, and it is difficult to understand how they could exist without a medium of water molecules through which to propagate. Similarly, it is difficult to understand how light waves can propagate without a medium. For this and other reasons, scientists often find it convenient to visualize light as a stream of particles of radiant energy known as **photons**. As bundles of pure energy, photons differ from particles of matter in several important ways: they have no mass, they occupy no space, and they travel at the speed of light (because they *are* light).

The energy carried by an individual photon is extremely small and varies inversely with wavelength: the shorter the light's wavelength, the more energetic its photons. Thus, photons of violet light are more energetic than those of red light.

The heating coil in Figure 3.16 provides a good example of the relation between photon energy and color (and wavelength). As the coil heats up, the first visible light it emits is a deep red light of relatively low energy and long wavelengths. At higher temperatures, photons of greater energy and therefore shorter wavelengths (corresponding to orange and yellow) join the red photons, causing the coil to appear orange, then yellow.

You may wonder how light can be described sometimes as a train of waves and other times as a beam of photon particles. Which is it really? There is no simple answer to this question; light is both, and it is neither. Each model describes certain aspects of behavior that the other cannot. This wave-particle duality, as it is called, is one of the most interesting topics in modern physics. It also reminds us that our models of nature are only models and not nature itself.

## Emission and Absorption of Radiation

### *Emission*

Photons differ from particles of matter in another important way: unlike protons, electrons, atoms and molecules, which are relatively permanent objects, countless numbers of photons are

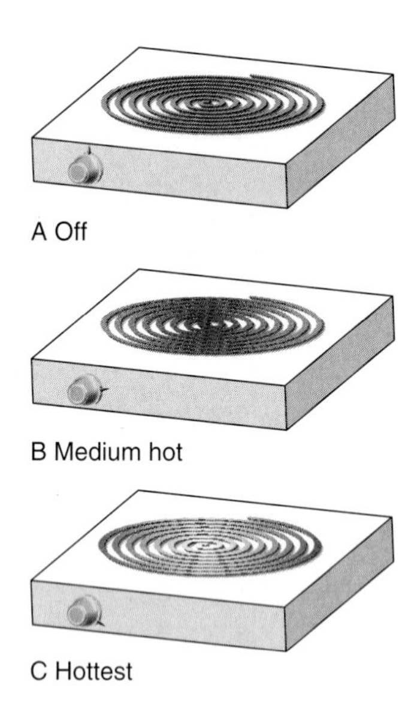


Figure 3.16 As the coil becomes hotter, it emits light first at the red and of the visible spectrum. At higher temperatures, shorter wavelengths (orange, yellow) predominate.

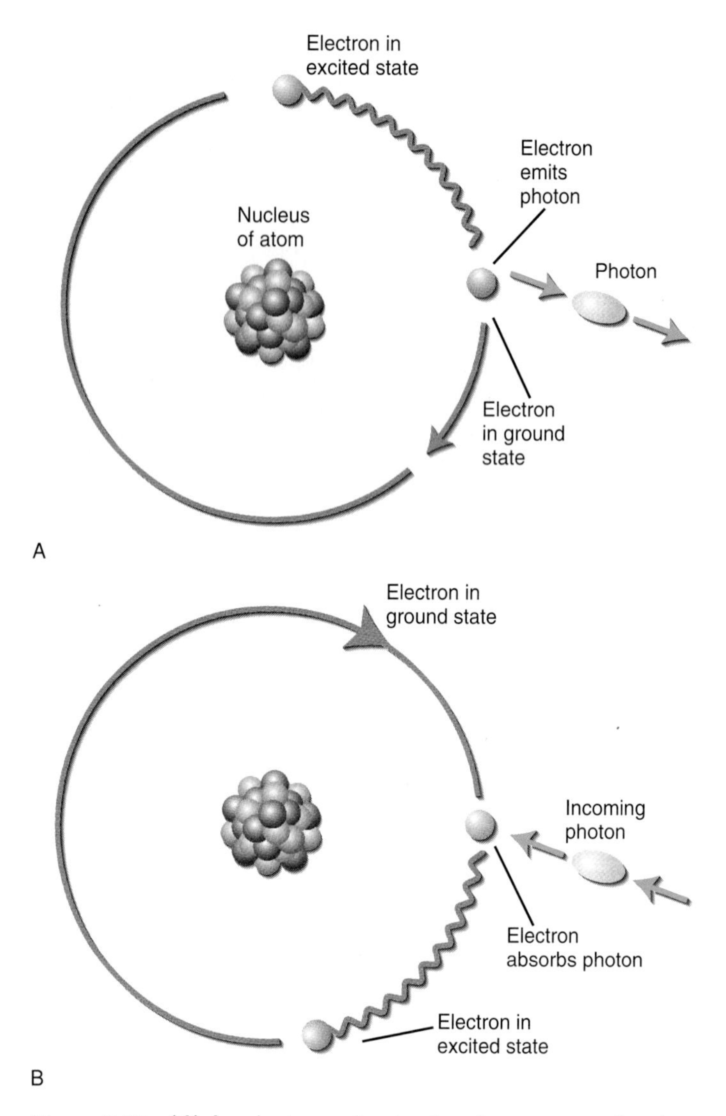


Figure 3.17 (*A*) An electron returning to a lower energy level emits a photon of radiant energy. The photon carries off the atom's excess energy. (*B*) By absorbing an incoming photon, an atom's electron moves to a higher energy level.

created and annihilated every instant. These appearances and disappearances do not violate the law of conservation of energy; photon creation and destruction, called emission and absorption, respectively, simply transform energy from one form to another. In this sense, they are similar to the transformations illustrated back in Figure 3.2.

Where do photons originate? To answer this question, we return to the example of the heating coil. Electricity is flowing in the coil; on the molecular level, this means that energetic free electrons, unassociated with particular atoms, are moving along the wire. Interactions between these free electrons and atoms of the wire coil transfer electrical energy to molecules of the wire. A portion of this transferred energy takes the form of increased molecular vibrational and rotational motion, which you know of as heat. The remainder is absorbed by the molecules' electrons, moving the electrons to higher energy levels. The energized molecules or electrons are said to be in **excited states**.

A molecule or electron remains in an excited state for a very short time before shedding its excess energy and returning to a lower energy level. The particle may drop all the way to its **ground state,** the lowest energy level it can occupy, or it may shed only a portion of its energy and fall to only a slightly lower level than before. The electron sheds its excess energy by releasing a photon, as shown in Figure 3.17A. The amount of energy carried off by the photon and therefore its color depend on the amount of energy absorbed in the first place. By turning the heating coil's switch to a higher setting, you force more electrical energy into the heating coil, causing more energetic interactions and resulting in photons with higher energies and shorter wavelengths.

Electricity is not the only energy source leading to photon production. Any form of energy, such as frictional heating, combustion, or nuclear decay can provide the needed excitation energy.

From this discussion of electricity and wire coils, you should take one major concept: a photon is a bundle of energy emitted by an excited molecule, atom, or electron as it returns to a lower energy level. It is interesting and instructive to consider this connection between the model of matter and that of energy: in the process of gaining and losing energy, the basic particles of matter (that is, electrons, atoms, and molecules) create and destroy photons, which are the basic particles of radiant energy.

### *Absorption*

Absorption of photons is in many ways the exact reverse of the photon emission process. As shown in Figure 3.17B, a photon is absorbed when it interacts with an atom and passes its energy to the atom or one of its electrons. This interaction annihilates the photon and leaves the atom in an excited state. The excited atom may subsequently re-emit one or more photons, thus continuing the exchange from radiant energy to atomic excitation and back again.

## The Electromagnetic Spectrum

Do the visible colors represent the entire span of radiant energy, or might red and violet light simply mark the limits of the human eye's sensitivity to radiation? In other words, might there be radiation with wavelengths longer or shorter than those in the visible spectrum that, though invisible to us, is nonetheless real? In Figure 3.18 you can see that indeed, the visible spectrum is just a narrow segment in a much broader range of radiant energy known as the **electromagnetic spectrum**. The spectrum, its radiation, and its component waves are called electromagnetic because they have both electric and magnetic properties. For brevity, often we will refer to electromagnetic radiation simply as "radiation," or, in the case of visible radiation, as "light."

The properties of visible light presented in the previous section apply to radiation throughout the electromagnetic spectrum. Thus, moving to the left in Figure 3.18 means moving toward radiation of longer wavelengths and less energetic photons. Moving to the right, you encounter progressively shorter wavelength and more energetic photons.

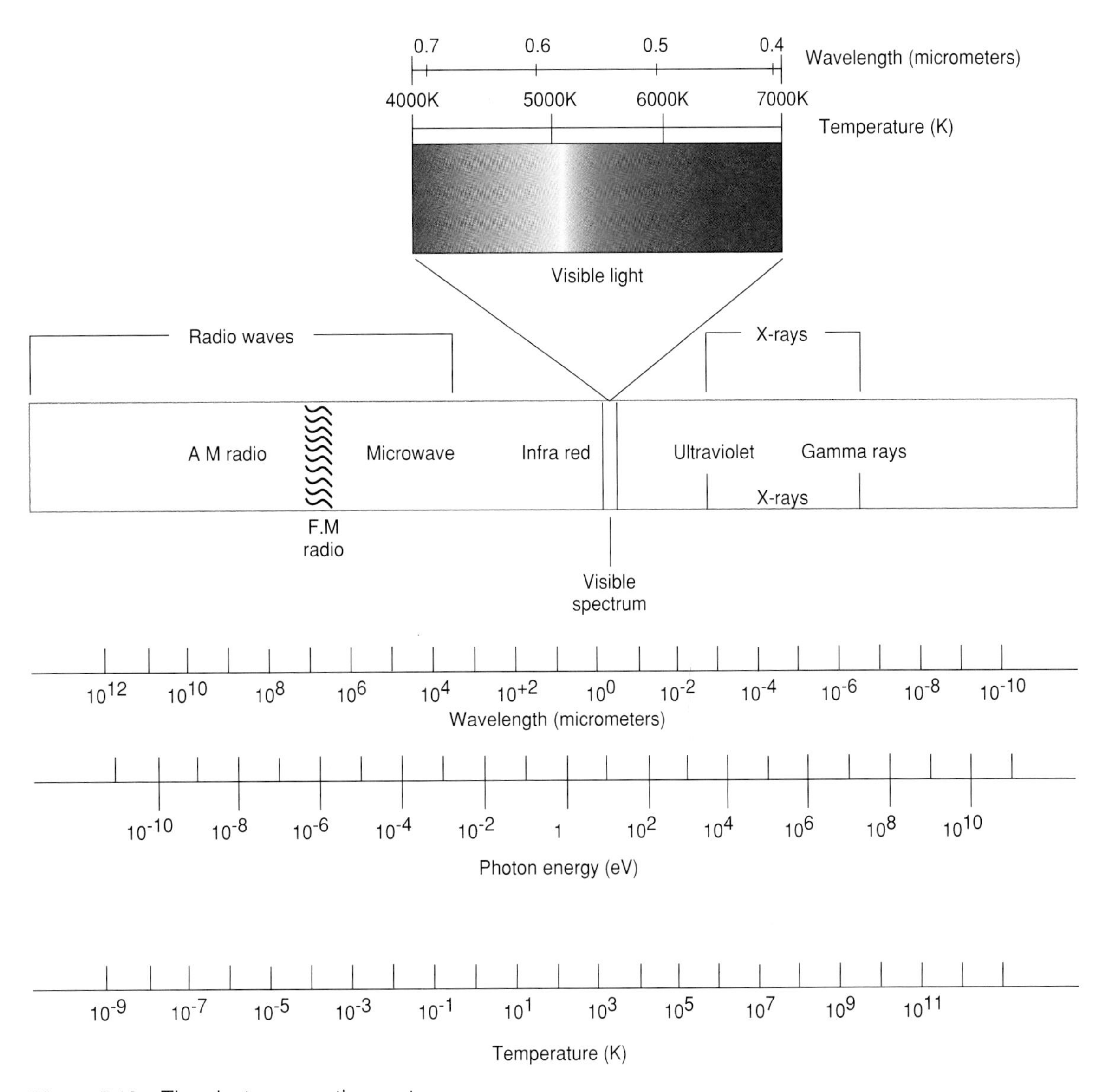


Figure 3.18 The electromagnetic spectrum.

Observe in Figure 3.18 the location of ultraviolet radiation in the electromagnetic spectrum. Recall from Chapter 2 the role that the ozone layer plays in shielding the earth's surface from these photons. The reason that ozone's shielding is important is that the photons of ultraviolet radiation are highly energetic; from the figure you see that they possess higher energy than photons of visible light.

Now examine the expanded version of the visible spectrum in the figure. Note that red light is associated with a temperature of approximately 4000 Kelvin (K). This means that objects with surface temperatures of 4000 K emit a greater share of their energy at wavelengths corresponding to red than at any other color. As you move to the violet end of the visible spectrum, notice that the temperature rises; objects emitting primarily violet light are at temperatures of around 7000 K.

Now look at the spectrum outside of the visible range. Moving from violet to even shorter wavelengths, you see very high temperatures are required. Moving from red to longer wavelengths, you encounter lower temperatures. In fact, Figure 3.18 is a spectrum of radiation across all emission temperatures, as much as across all wavelengths or photon energies. Pick a temperature, any temperature, and Figure 3.18 tells you the properties of the dominant radiation an object at that temperature will emit. In particular, notice that in the infrared region, you find temperatures corresponding to typical everyday values, such at that of the human body (37°C, or roughly 300 K). Does this mean that your body is emitting radiation? The answer is, "Yes, continuously, at the rate of billions of photons per second."

You have just encountered one of the most counterintuitive but important concepts in atmospheric science. Put bluntly and only slightly too inclusively, it is that every object in the universe radiates energy. Look around you; every object you can see (plus the air itself) participates with every other object in sight in an ongoing exchange of countless, invisible photons. This textbook is radiating energy at you, and you are radiating energy back at the textbook. The walls of the room, other people, chairs, pencils, all emit and absorb many millions of photons per second. And the earth and the atmosphere radiate, which is why the topic is important to us here.

One difficult aspect of the concept that all objects radiate is that we are speaking not of radiation *reflected* but of radiation *emitted*. The visible light reaching you from this page is not emitted, or radiated, by the book; the page is merely reflecting the light from its source (a lightbulb?) to your eyes. If you turn out that light and remove other sources of visible light, the page will be invisible to you, but still it emits its own radiation. If you refer once again to Figure 3.18, you will see that an object as cool as this book radiates energy predominantly at wavelengths far longer than anything the eye can perceive.

Similarly, a tree looks green not because it is *emitting* photons of green light; if it did so, it would burst into flames because it would require a temperature of many thousands of degrees to radiate at predominantly green wavelengths. No, the tree is green because visible light is striking it, and it happens to *reflect* the green portion, *absorbing* other colors. Meanwhile, it emits **infrared** radiation, which is radiation of wavelengths longer than that of red light and therefore is invisible to us. At night, in the absence of sunlight, the tree no longer appears green. It is invisible, because we cannot see the infrared radiation it emits without special instruments.

Meteorologists often refer to infrared radiation emitted and absorbed by the earth and the atmosphere as **longwave radiation,** to distinguish it from the much shorter wavelength solar radiation. Longwave radiation tends to result from changes in molecules' vibrational and rotational energies; these are relatively low-energy transitions. Shortwave radiation, such as visible light, generally comes from transitions in electrons' energy levels, which involve greater energy changes.

## Continuous Spectra

We return to the heating coil one more time. Suppose you make careful measurements of the amount of energy emitted at each wavelength as the coil's temperature rises. Because the coil radiates energy at every wavelength, its spectrum is known as a **continuous spectrum.**

Each curve in Figure 3.19 describes the amount of **blackbody** radiation emitted at different wavelengths by an object at a given temperature. *Blackbody* is the term used to describe a perfectly efficient radiator (and does not refer to its actual color). Thus, the blackbody curves represent the maximum radiation an object at a given temperature can emit. Most solids, liquids, and dense gases radiate close to the blackbody rate.

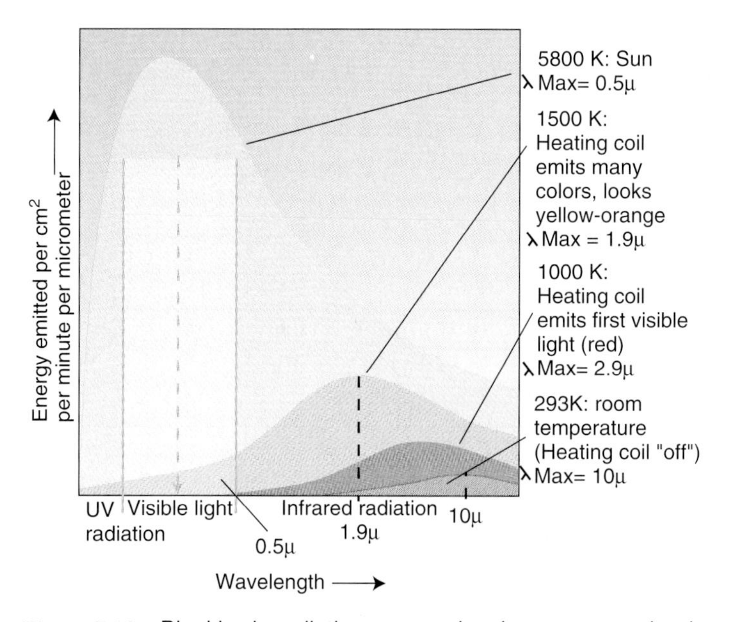


Figure 3.19 Blackbody radiation curves showing energy emitted at each wavelength for objects at various temperatures. Note that the vertical coordinate is not labeled with specific values; the curves do not show exact values, but general relations. (To represent actual values, the sun's curve would have to be 3 million times higher than the peak for the room temperature curve.)

To understand Figure 3.19, consider the curve labeled "sun." Notice that at very short wavelengths, the sun emits very little radiation. However, moving along the curve toward the visible portion of the spectrum, the emission increases sharply, reaching a maximum near the middle of the visible range. At greater wavelengths, the emission is less. Note also that the other blackbody curves, for objects at cooler temperatures, have the same basic shape as that for the sun. However, these curves are lower (indicating less emission) and are centered to the right of the solar curve (at longer wavelengths).

Now examine Figure 3.19 to verify the following properties of blackbody radiation: (a) every object emits photons of all energies, regardless of its temperature; (b) a hotter object emits more photons at every wavelength than a cool object; and (c) a hotter object emits a greater proportion of its energy at shorter wavelengths than a cooler object.

The last property may be stated more specifically in a form known as Wien's law (pronounced "Veen's law"): the wavelength at which a blackbody emits its greatest amount of radiation is inversely proportional to its Kelvin temperature. If wavelength is specified in micrometers, then Wien's law is

$$\lambda_{max} = 2890/T$$

Here $\lambda_{max}$ (pronounced "lambda max"; $\lambda$ is the Greek letter *lambda*) is the wavelength at which the object emits the greatest intensity of radiation and T is the Kelvin temperature.

You should be able to see by inspecting the formula that substituting larger and larger values of T makes the resultant $\lambda_{max}$ smaller and smaller. Thus, hotter blackbody curves peak farther to the right (shorter $\lambda$) in Figure 3.19.

To take a quantitative example, calculate the maximum wavelength for the sun's radiation. Assume the sun's surface temperature is 5800 K. Then,

$$\lambda_{max} = 2890/5800 = 0.498 \text{ micrometers}$$

This is in the center of the visible spectrum, according to Figure 3.19. On the other hand, calculate $\lambda_{max}$ for the earth's surface. Assume a surface temperature of 15°C.

First, convert Celsius degrees to Kelvins:

$$15 + 273.2 = 288 \text{ K}$$

Then,

$$\lambda_{max} = 2890/288 = 10.0 \text{ micrometers}$$

This is a wavelength in the infrared, far beyond the visible spectrum.

How much radiation does a given object emit over all wavelengths combined? To find out, you would have to add up the energy contributions of the photons at all wavelengths. This total energy is represented by the area between the object's blackbody curve and the x-axis, as illustrated in Figure 3.20A. That area, and therefore the energy, is given by the **Stefan-Boltzmann formula:**

$$E = \sigma \times T^4$$

Here, E is the energy in calories emitted by a square centimeter of the object's surface per minute, and T is the Kelvin temperature of the surface; $\sigma$ (the Greek letter *sigma*) is a conversion constant, whose value is $8.132 \times 10^{-11}$ in the units we are using.

As an example of the formula's utility, we will calculate the radiation emitted per minute by a 1-square centimeter area of your body (roughly the size of your little fingernail). Assume your body surface is temperature of 32°C (90°F) and that it ra-

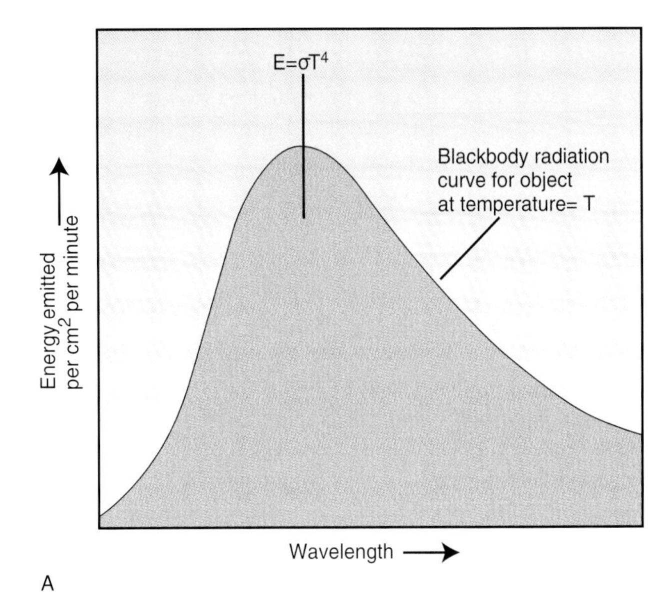


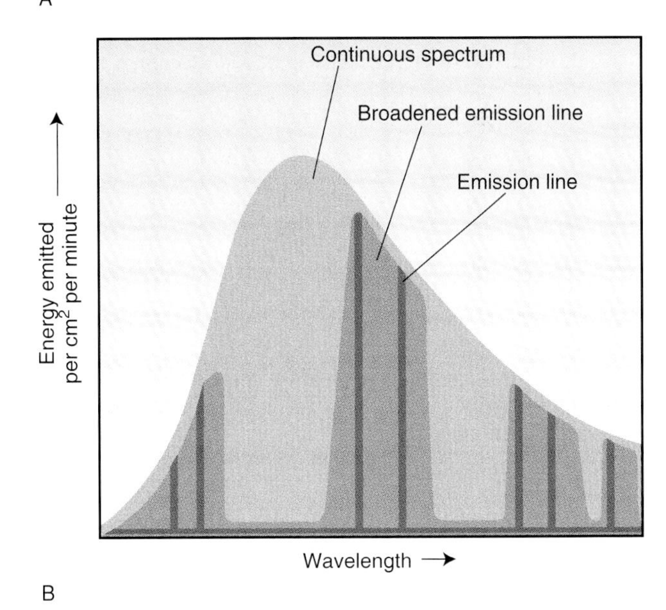


Figure 3.20 (*A*) You can determine the total amount of radiation an object emits at all wavelengths by finding the area under the blackbody curve. The Stefan-Boltzmann formula allows you to calculate the value directly from the temperature. (*B*) A thin gas emits and absorbs radiation at only discrete wavelengths, as shown by the black emission lines. If the gas's density is increased, the emission lines broaden, as shown by the gray area. Further density increase leads to the continuous spectrum.

diates at the blackbody rate. To apply the formula, you must convert 32°C to the Kelvin scale: simply add 273.2. Then,

$$\begin{aligned} E &= \sigma \times (305.2)^4 \\ &= 8.132 \times 10^{-11} \times (305.2 \times 305.2 \times 305.2 \times 305.2) \\ &= 8.132 \times (8.6 \times 10^9) \times 10^{-11} \\ &= 0.70 \text{ calories per minute} \end{aligned}$$

This value is roughly one-third of the solar constant. This does not mean your fingernail is radiating one-third as much energy as a square centimeter of the sun's surface, of course! Remember that the solar constant is the amount of solar radiation *received* at the earth's distance from the sun, not the amount *emitted* by the sun's surface, which is diminished as it spreads through space before reaching earth. To find the radiation rate of the sun's surface, enter its temperature, 5800 K, into the Stefan-Boltzmann formula. You should arrive at an answer of $(8.132 \times 10^{-11}) \times (5800)^4 = 92{,}000$ calories $cm^{-2}$ $min^{-1}$.

## Line Spectra

We are nearly finished constructing our radiation model; just one part remains, which concerns the emission and absorption of photons in thin gases such as the atmosphere. Air molecules are widely separated from each other and therefore interact with each other far less than do molecules of denser substances, such as the heating coil. In this roomier environment, each air molecule vibrates, rotates, and undergoes electronic transitions more or less undisturbed by surrounding molecules. And because each kind of atom and molecule has a distinct mass and structure, each also has certain distinct levels of vibrational, rotational, and electronic energy at which it can exist. In scientific terminology, the atoms' energy levels are said to be quantized, that is, the atoms exist at only certain energy levels. In denser substances, on the other hand, interactions between different molecules tend to blur these quantized energy levels into a continuum of values. Figure 3.20B illustrates schematically the spreading of quantized energy levels into a continuous spectrum as an object's density increases. Substances that interact with photons of only certain energies, illustrated by the black and gray curves in Figure 3.20b, are called **selective absorbers** and **emitters** of radiation. Selective emitters are not blackbodies, because at many wavelengths along the theoretical blackbody curve, they radiate little or no energy at all. Therefore, the Stefan-Boltzmann formula, which gives the total amount of energy beneath the blackbody curve, also does not apply to a selective emitter such as the atmosphere; the formula overestimates the amount of energy emitted.

The spectrum emitted by the thin gas (the black line) in Figure 3.20B is an example of a **line spectrum.** A line spectrum, in contrast to a continuous spectrum discussed earlier, is one in which only certain discrete photon energies and wavelengths are represented. Thus, line spectra are the products of highly selective absorbers and emitters. Figure 3.21 shows line spectra for several important constituents of the atmosphere.

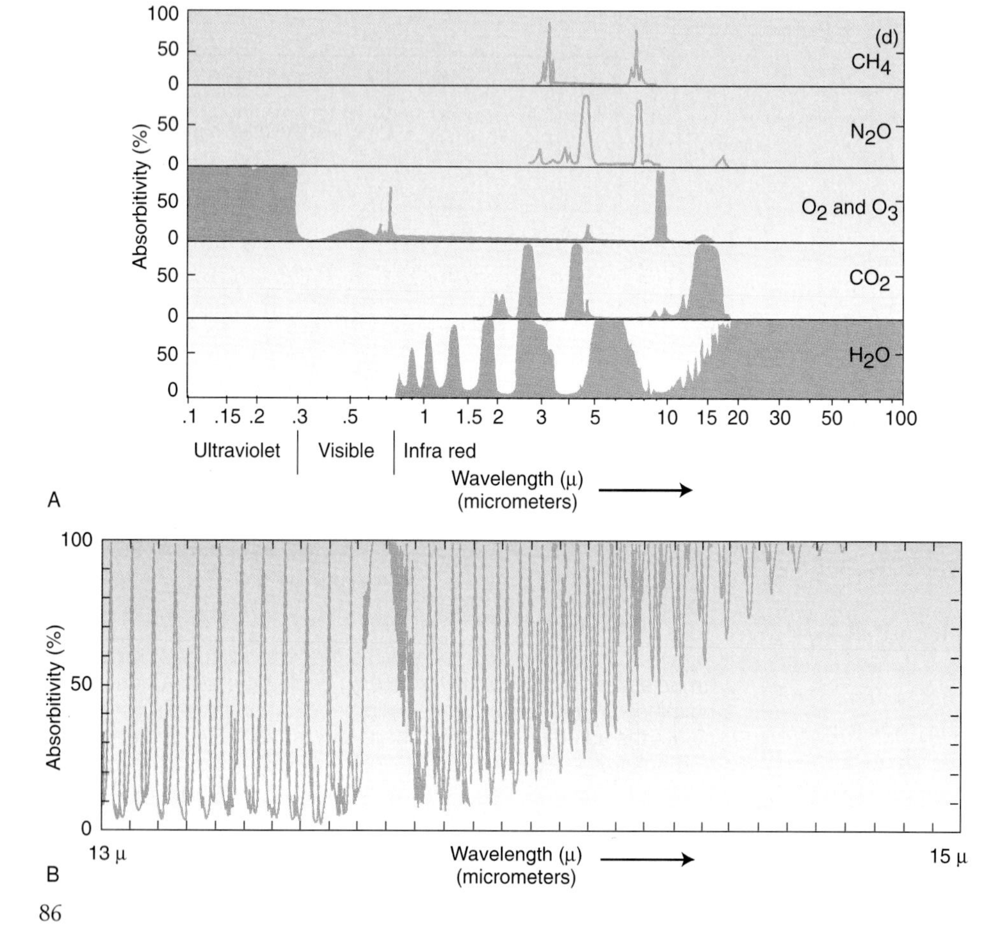


Figure 3.21 Spectral lines for a few atmospheric gases. The presence of a line indicates that the gas will absorb and emit photons at that particular wavelength. (*A*) Large-scale view. (From Peixoto and Oort, 1992.) Source (A): Data from J. Peixoto and A. Oort, *Physics of Climate*, 1992, American Institute of Physics, New York, NY.
(*B*) The lines shown actually have great variation if examined in closer detail, as this view of $CO_2$ absorption between 13 and 15 microns indicates. (Based on Houghton, 1986). Source (B): Data from R. A. McClatchey and J. E. A. Selby, "Atmospheric Transmittache 7–30 Micrometers: Attenuation of $CO_2$ Laser Radiation" in Environ. Res. Paper No. 419, AFCRL-72-0611.

To recapitulate, gases emit and absorb radiation at selective wavelengths, whereas liquids and solids emit and absorb continuous spectra of radiation. These different behaviors arise primarily from density differences: atoms in gases are so dispersed that they rarely interact with neighboring particles; therefore, they occupy only those quantized energy levels defined by their own internal structure. In denser objects such as rocks or soil, the quantized energy levels are blurred by constant interactions with neighboring particles; here, the line spectrum blurs into the continuous spectrum.

## Checkpoint

***Review***

The concept map in Figure 3.22 represents just one way of linking radiation concepts presented in this section. We encourage you to construct your own.

The most important concepts are: the relation between photon energies, wavelengths, and temperatures; the differences in radiation emission and absorption between thin gases and other objects; and the properties

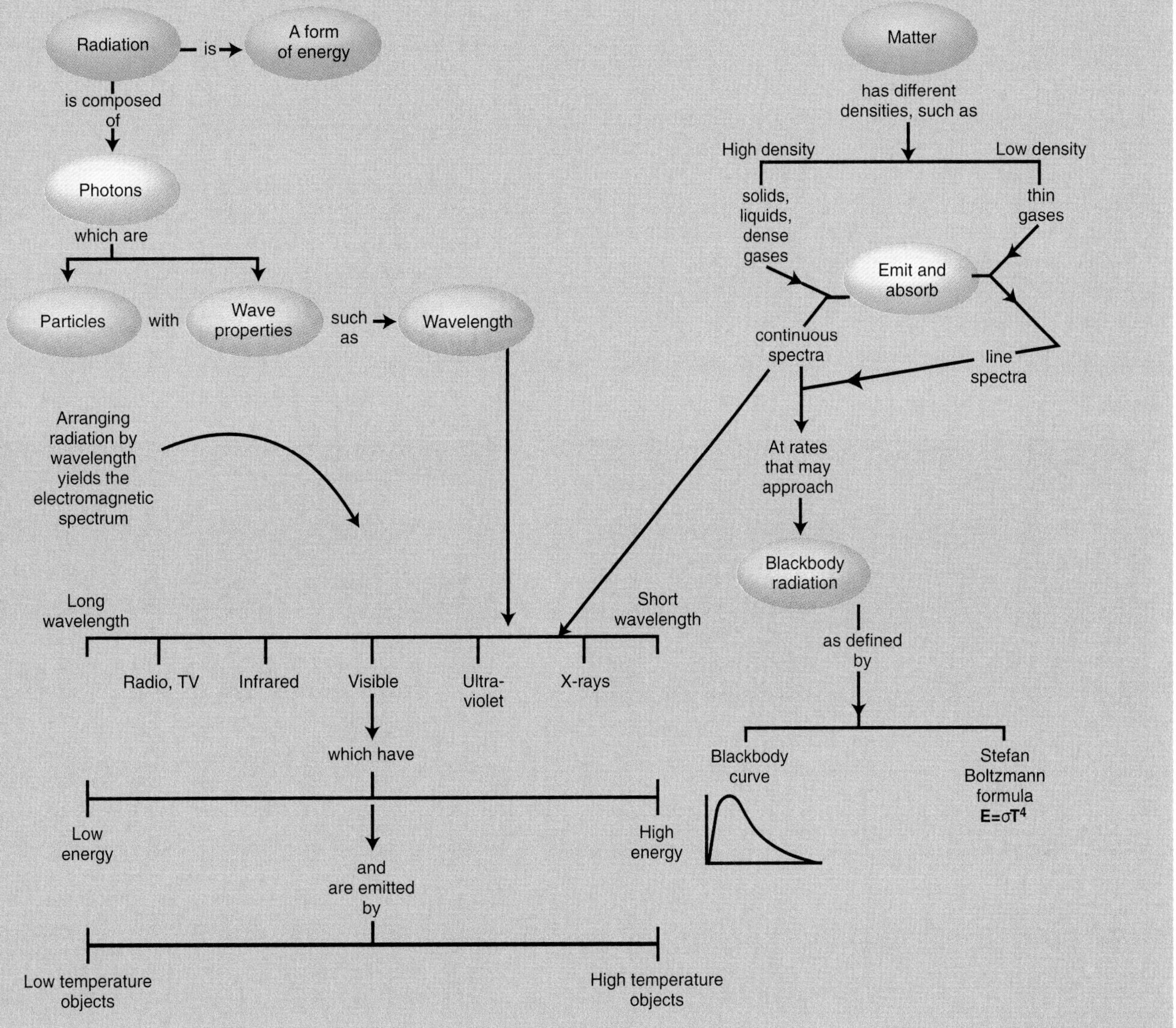

Radiation – is → A form of energy
is composed of
Photons
which are
Particles with Wave properties such as → Wavelength
Arranging radiation by wavelength yields the electromagnetic spectrum
Long wavelength
Short wavelength
Radio, TV
Infrared
Visible
Ultra-violet
X-rays
which have
Low energy
High energy
and are emitted by
Low temperature objects
High temperature objects
Matter
has different densities, such as
High density
Low density
solids, liquids, dense gases
thin gases
Emit and absorb
continuous spectra
line spectra
At rates that may approach
Blackbody radiation
as defined by
Blackbody curve
Stefan Boltzmann formula
$E=\sigma T^4$

Figure 3.22 Concept map: Radiation model.

of blackbody radiation given by the blackbody curves and the Stefan-Boltzmann formula.

***Questions***

1. Distinguish between a proton and a photon.
2. If a glowing object is raised to a higher temperature, how will its color change? How will the wavelengths of its emitted light change?
3. How does the spectrum of a glowing rod of iron differ from that of a thin gas heated to the same temperature?
4. What is the temperature of an object that radiates most intensely at a wavelength of $10^{-5}$ micrometers? What is the name of the wavelength band in which this radiation is located? (See Figure 3.18.)

# Infrared Radiation

At last we are ready to consider the fluxes of infrared radiation on the earth's surface and the atmosphere.

## Radiation Emitted by the Earth's Surface

The substances that make up the earth's surface, being solids and liquids, radiate infrared energy nearly at the blackbody rate. Therefore, to determine the rate at which the surface is emitting infrared radiation, you need to know only the surface temperature. Applying the temperature to the Stefan-Boltzmann formula gives the emitted energy.

Surface data and satellite measurements indicate that the average temperature at the earth's surface is approximately 15°C, or 288 K. Therefore,

$$E = \sigma \times (288)^4$$
$$= 0.56 \text{ calories per square centimeter per minute}$$

Thus, the earth's surface radiates energy away at the average rate of about half a calorie per square centimeter per minute. This average longwave radiative flux upward from the earth's surface is slightly greater than the average solar flux entering the top of the atmosphere. In fact, if we convert the 0.56 calorie per square centimeter per minute surface flux into the units on which our energy budget discussion has been based, we find that the earth's surface radiates 105 units. This flux is shown in Figure 3.23.

## Longwave Radiation Absorbed by the Atmosphere

What happens to this large radiation flux after it is emitted from the earth's surface? Applying the radiation model, you can predict some of the possibilities. For example, you know that because atmospheric gases are only selective absorbers, you would expect some of the earth's radiation to pass straight through the atmosphere and escape to space. This is correct, as you can see in Figure 3.23.

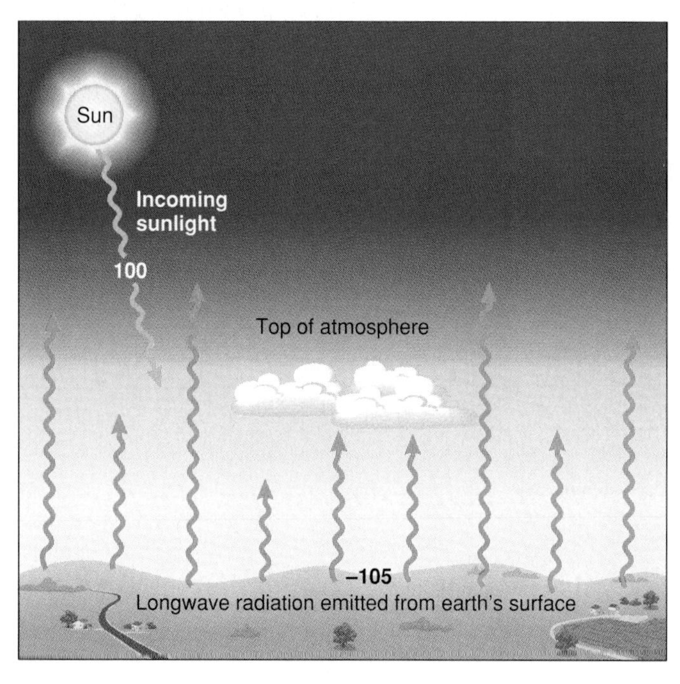


Figure 3.23 Upward longwave radiation from earth's surface is largely absorbed by clouds and certain atmospheric gases. Some, however, escapes directly into space.

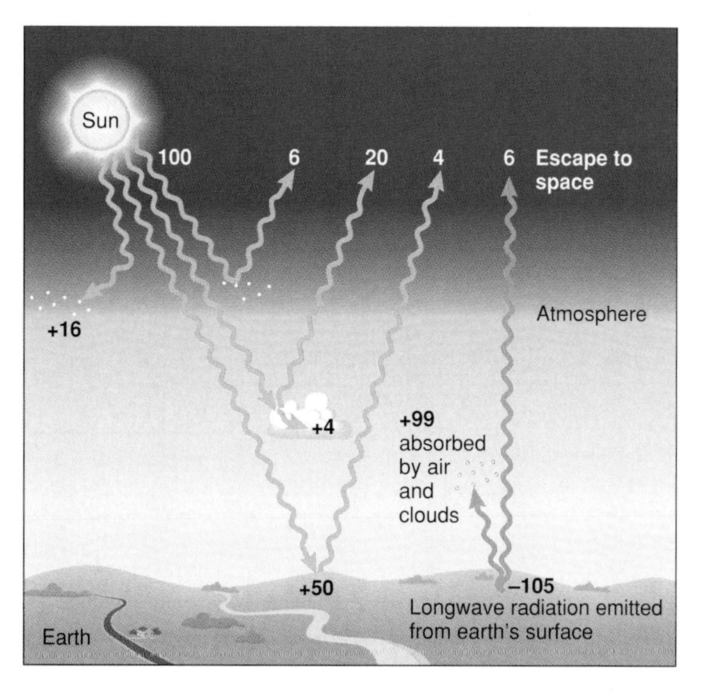


Figure 3.24 The earth's energy budget, here including all solar fluxes plus upward longwave radiation from the earth's surface.

You also know from the previous section (and Figure 3.21) that certain atoms or molecules strongly absorb selected wavelengths of the earth's radiation, while other molecules intercept little or no radiation at all. This, too, describes the situation in the earth's atmosphere: water vapor, carbon dioxide, methane, and CFCs, although comprising only about 1% of the gases in the atmosphere, absorb far more than the other 99%. Finally, you might hypothesize from the radiation model that clouds, because they are composed of larger (solid or liquid) particles, might interact with earth radiation like blackbodies. This is correct: clouds are excellent absorbers of the earth's radiation.

Figure 3.24 presents the energy budget updated to include the 105 units of longwave radiation emitted by the earth's surface and its disposition in the atmosphere. Notice that the net flux at the earth's surface now is negative, as a result of the infrared radiation it emits. The atmosphere, on the other hand, shows a large surplus. In the next section, you will see that radiation emitted by the atmosphere brings things more closely into balance.

## Radiation Emitted by the Atmosphere

The processes by which the atmosphere absorbs the earth's radiation give clear hints of its own emission processes as well. An atom whose energy levels do not allow it to absorb a passing photon will not be capable of emitting a photon of that energy. Therefore, the atmosphere's most effective absorbers, namely water vapor, carbon dioxide, methane, and CFCs, along with clouds, do the lion's share of the atmosphere's emission as well.

The amount of energy the atmosphere radiates depends on its temperature as well as its composition; but as we pointed out earlier, as selective emitters the atmospheric gases do not radiate at the blackbody rate. Therefore, we cannot use the Stefan-Boltzmann formula to calculate atmospheric emission rates; instead, more involved methods are required. Direct measurements from the earth's surface and from satellites also play an important role in determining atmospheric radiation rates. Figure 3.25 gives a beautiful example of the detailed longwave radiation data satellites can provide.

## The Greenhouse Effect

One aspect of atmospheric radiation processes deserves special emphasis. It concerns the behavior of a group of six or so **greenhouse gases:** water vapor, carbon dioxide, methane, tropospheric ozone, CFCs, and nitrous oxide. These gases, like oxygen and nitrogen, are quite transparent to sunlight. Therefore,

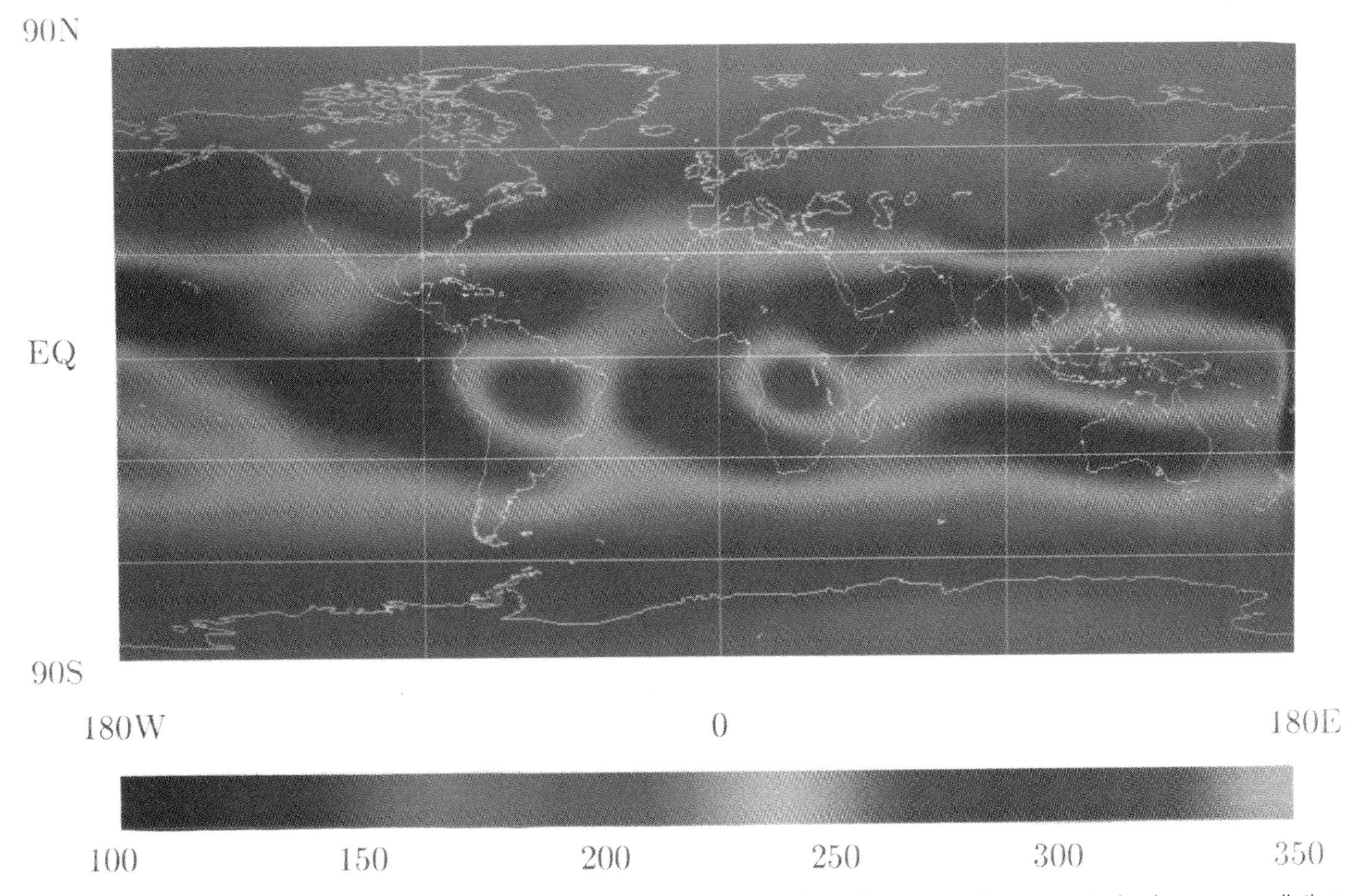


Figure 3.25 This computer-generated map, based on satellite measurements, charts the average January outgoing longwave radiation from the earth's surface and atmosphere to space. Units are watts per square meter.

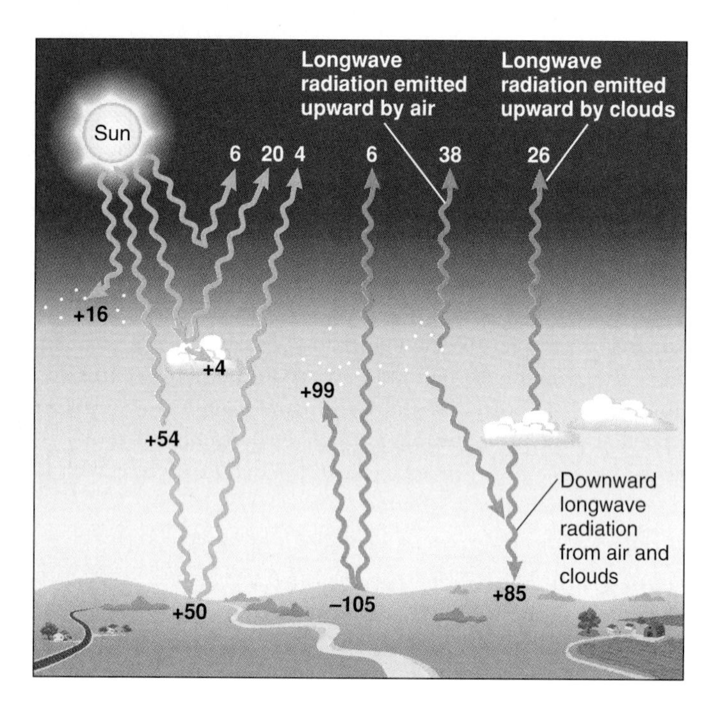


Figure 3.26 The earth's radiation budget. Note that the budget is balanced across the "top of the atmosphere" but is not at the earth's surface.

solar radiation passes relatively undiminished through clear air to reach and warm the earth's surface. However, the greenhouse gases, unlike oxygen and nitrogen, absorb and reradiate longwave radiation back to the surface. The greenhouse gases act like a sort of "radiation valve," allowing the flow of sunlight into the earth and atmosphere but restricting the return flow of longwave radiation back to space. This behavior is commonly referred to as the **greenhouse effect,** because greenhouse glass allows sunlight to enter but inhibits the outward flux of longwave radiation, causing the temperature inside the greenhouse to rise. (Actually, scientists believe now that a greenhouse's warmth occurs less as a result of longwave fluxes and more because the glass simply keeps the heated air inside, preventing it from mixing with the outside atmosphere. As a result, some scientists use the term *atmosphere effect,* rather than *greenhouse effect,* to refer to longwave atmospheric warming.)

How significant is the earth's greenhouse effect? Without the absorption and re-emission of longwave energy by the earth's atmosphere, the earth's average surface temperature would be some 35°C (63°F) colder than at present. To get a feel for such a possibility, imagine that summertime temperatures in Texas suddenly were to average −7°C (20°F). Or check the current outside temperature, then subtract 63° from it (if in Fahrenheit). Clearly, the greenhouse effect plays a crucial role in maintaining a habitable environment on earth. Far more extreme is the impact of greenhouse warming on the planet Venus. With an atmosphere mostly $CO_2$, Venus experiences a 523°C (940°F) *warming* due to the greenhouse effect.

## The Radiation Budget

What, then, is the atmosphere's contribution to the energy budget via radiation? As you can see in Figure 3.26, the atmosphere radiates energy both upward and downward. Direct measurements of longwave radiation with instruments called radiometers, as well as theoretical calculations, indicate that on average, the atmosphere radiates downward to the earth's surface at the rate of 85 units, thus nearly balancing the surface's upward flux of 105 units. The atmosphere also radiates a total of 64 units upward to space; on average, 26 originate in clouds, and the other 38 are radiated by air molecules.

Figure 3.26 shows all radiation fluxes, both long- and shortwave, affecting the earth-atmosphere energy budget. This collection of all radiation terms is often called the **radiation budget.** Notice that the budget is balanced for the earth-atmosphere system: the amount of infrared radiation escaping from the earth exactly equals the input of solar radiation. Note further, however, that the atmosphere's and surface's individual budgets are not in balance: the earth's surface shows an energy surplus, the atmosphere a deficit. The next section will address this imbalance.

### Checkpoint

***Review***

Figure 3.26 serves as a good review of the key concepts in this section. Note especially the large magnitudes of the longwave radiation fluxes. As you can see, the earth's surface overall receives more radiant energy from the atmosphere than it does from sunlight.

***Questions***

1. The "longwave balance" is the difference between incoming and outgoing infrared radiations at the earth's surface. From Figure 3.26, what is the average longwave balance?
2. How much longwave energy is radiated from the earth to space? How does this compare to the amount of sunlight absorbed by the earth-atmosphere system?

# Conductive and Convective Fluxes

You saw in the last section that the earth's surface absorbs more radiation than it emits, on the average, while for the atmosphere, the opposite is true. These imbalances give rise to two additional energy fluxes between the earth and the atmosphere. The earth's surface warms the atmosphere "by contact," and water evaporates from the surface, condensing later in the air. These fluxes are the topic of this section.

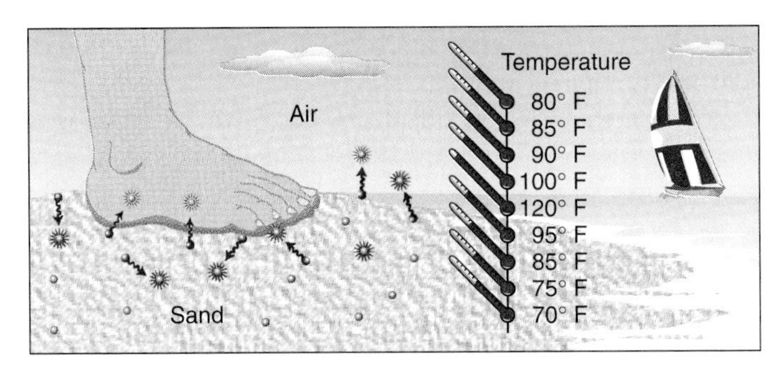


Figure 3.27 Heat is conducted when energetic (hot) molecules strike less energetic (cooler) ones, transferring their heat energy by collision. The rapid decrease in temperature with distance above and below the hot sand surface is a result of the inefficiency of conduction over distances.

## Sensible Heat Flux

Plunge a cold hand into a pan of warm water, and you experience an energy flux—a heat flow from a warmer to a cooler object. This energy flux is carried primarily not by radiation but by molecules in direct contact with each other. Perhaps because such fluxes are driven by differences in temperature, which can be felt or "sensed," atmospheric scientists often refer to them as "sensible heat fluxes."

### *Conduction*

To understand sensible heat fluxes, let's return to the sandy beach we discussed in the section on mixing. Notice in Figure 3.27 that the solar income is concentrated in a very shallow layer of sand, which heats rapidly. When you walk on the hot sand, your feet feel hot because molecules of sand, in great vibrational motion due to their heat, collide with your "foot molecules," passing some of their energy to your feet in the collision. This process of heat transfer by molecular collision is known as **conduction.**

In most substances, conduction is not an efficient mode of transferring heat over great distances. You can verify this by digging down a few inches into the sand, where you find the temperature is much cooler than at the surface; daily cycles of surface heating and cooling penetrate only a small distance downward.

Conduction also transfers the sand surface's heat upward, into the air in contact with it. Again, the process is inefficient when it comes to distributing this heat gain to higher air layers: air molecules in actual contact with the sand warm rapidly, but the heat is conducted very slowly to higher levels. In fact, air's poor conductive qualities are exploited in the manufacture of home insulation: like a fishnet, which is said to be just a lot of holes tied together, insulation is simply air isolated in tiny pockets or cells so that its only significant mode of heat transfer is by the slow process of conduction.

### *Convection*

Once conduction has conveyed heat from the earth's surface into the lowest layer of the atmosphere, another, far more efficient energy transfer process becomes important. This new process, called convection, is illustrated in Figure 3.28. Notice the shallow bubble of air at the earth's surface, which has been heated by conduction. This heating occurs at nearly constant pressure, because the bubble is not contained by walls. From the gas law—pressure = constant × density × temperature—you can predict what happens next: the bubble's temperature increase must be matched by a decrease in its density, in order that the product (density × temperature) remains constant.

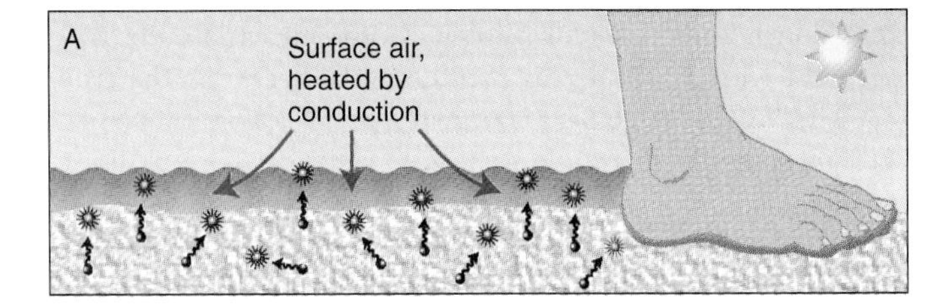


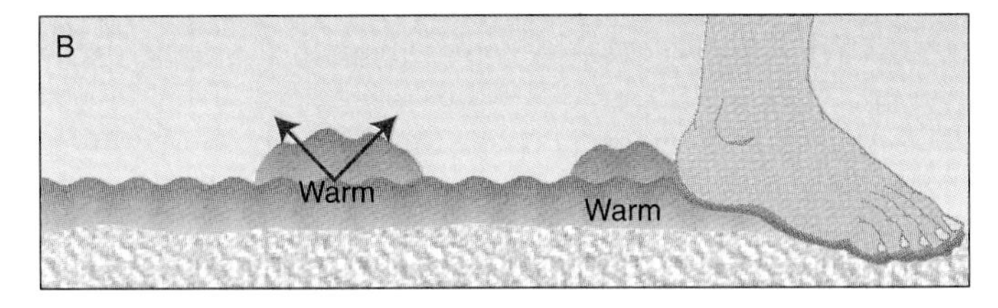


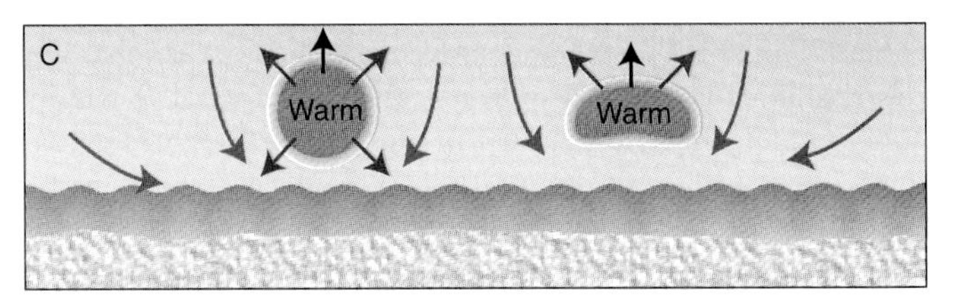


Figure 3.28 Heat transfer by convection. (*A*) Some points inevitably are heated more than others. The air in contact with these spots becomes warmer also by conduction. (*B*) The warmer air expands more than elsewhere. (*C*) As it expands, the air's density decreases. It rises through the denser air. Cells of rising warm air may combine into larger cells.

Its density thus reduced compared to that of the surrounding air, the bubble becomes buoyant and rises. It floats above the denser air, which settles downward onto the hot surface, to be heated in turn. Joined by other buoyant cells, the warmed air parcel continues to rise, for thousands of meters in the case of large, buoyant parcels, until it reaches air of similar density. In this way, the surface heat is transported quickly to much higher and cooler levels in the atmosphere.

This example shows convection as a process in which heat is transferred by the motion of matter from one location to another in the atmosphere. (Recall that convective cells were discussed in the circulation model of Chapter 1.) More generally, **convection** is motion within a fluid that causes transport or mixing of the fluid's properties.

Note that conduction and convection typically operate together. Although conduction is an inefficient method of transferring heat great distances, it plays an essential role conveying heat across the interface between the earth's surface and the atmosphere. *Within* the atmosphere, however, the convective flux is far more significant. Furthermore, by removing heated air from the earth's surface, convection allows cold air to settle onto the surface to be warmed in turn by conduction of heat from the ground.

The rate at which heat is transferred by conduction and convection depends in part on the difference in temperature between the earth's surface and the overlying air. If there is no temperature difference, then, of course, no net transfer occurs. A warmer earth causes an upward heat flux; a colder earth causes a downward flow. Downward fluxes, however, proceed less efficiently. When air is cooled by contact with the colder underlying ground, it becomes denser. If winds are light or calm, this cold, dense surface air remains in contact with the ground, forming an inversion and preventing warmer, overlying air from reaching the earth's surface.

Note from this example that winds can affect the rate of heat transfer by conduction and convection. As a general rule, wind enhances the conductive and convective fluxes. A familiar application of this rule for people in cold climates is the windchill index, which is often included in wintertime weather forecasts. For a given difference between air temperature and your body temperature, the stronger the winds, the faster heat is transferred from your body into the air, and the colder it feels. (Appendix I contains a windchill table.)

Figure 3.29 shows the earth's energy budget updated once more, this time to include heat flow by conduction and convection. These two fluxes are sometimes combined in the single term **sensible heat flux.** Sensible heat flux is heat transfer through the processes of conduction and convection.

## Latent Heat Flux

Meanwhile, back at the beach, the sand and the air are so warm you go for a swim. The water's temperature has risen little during the day for reasons discussed above (see Figure 3.12, for example), so your swim is refreshing and cooling. Leaving the water, however, you suddenly feel more than refreshed: you feel cold and run for the towel to dry off.

Your instinct to dry off to prevent further cooling is well founded. The evaporating water accompanies a flux of energy from the evaporating surface (your skin) into the air.

Why does evaporation cause cooling? For a water molecule to evaporate, it must free itself from forces holding it to its neighboring molecules and escape through the water's surface. To escape these forces requires energy; therefore, the molecules with the greatest energy, those moving most rapidly, are the most likely to break away and evaporate. By losing its most energetic molecules, the water's average molecular motion decreases; therefore, the temperature, which you saw in Chapter 2 is a measure of the amount of molecular motion, is less. Thus, you feel cooler because energetic water molecules are escaping from the surface of your body, thereby lowering your temperature. (Note that this explanation is consistent with Figure 2.9 in Chapter 2, which indicates the absorption or release of heat by objects undergoing phase changes.)

The water molecules thus liberated into the atmosphere eventually condense or freeze as cloud particles, dew, or frost. When they do, they release to the atmosphere the same amount of energy (on the average) they removed from the surface from which they evaporated. Thus, the heat gain to the atmosphere occurs later than, and usually in a different location from, the evaporation. Because this energy transfer proceeds via phase changes (evaporation, etc.) and not via sensible heat transfer, it has come to be called **latent heat flux.** (*Latent* means present, but not evident; hidden.)

The flux of water (and thus latent heat) from earth to atmosphere occurs from open water, ice, and snow surfaces, from moist soil, and from vegetation. Plants, especially their leaves, are major contributors to latent heat flux. This transfer of water from plant surfaces to the atmosphere is known as **transpiration.**

Latent heat flux due to evaporation and transpiration is not a mere scientific curiosity: it is a major element in the energy budget. When 1 gram of water evaporates, approximately 600 calories of sensible heat are converted to latent heat (causing cooling of the source). Similarly, when that gram of water condenses, 600 calories of latent heat are released into the surroundings, warming them.

Latent heat flux involves the physical transport of energy-bearing particles (water vapor molecules) from a place of greater to one of lesser concentration. In that sense, latent heat flux acts like convective flux. Latent heat flux is greatest when large differences in moisture exist from surface to air and when winds are strong. Because of differences in available surface water, atmospheric humidity, temperature, and wind speed, latent heat flux varies greatly in place and time.

Figure 3.30 shows examples of latent heat flux rates compared to other energy fluxes across the earth-atmosphere interface under different conditions. Solar income is shown as a downward bar, because it represents a gain to the earth. The other fluxes are from earth to atmosphere, as shown by their upward-directed bars. The "longwave balance" bar represents the difference between the longwave radiation emitted upward by

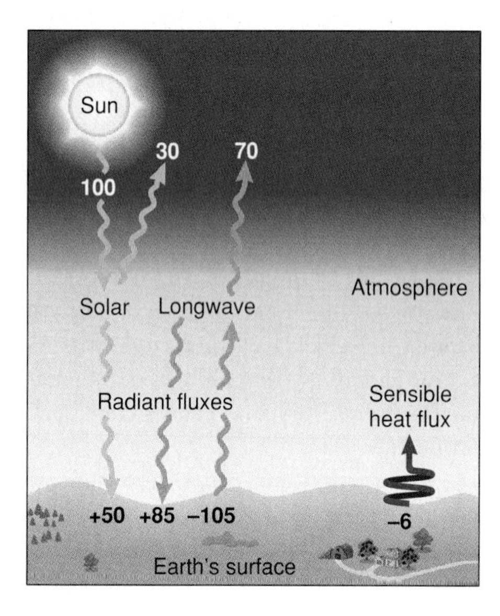


Figure 3.29 The earth's energy budget amended to include sensible heat flux. The gain of 6 energy units for the atmosphere is a loss of 6 units for the earth's surface.

the earth's surface and the longwave radiation arriving at the surface from the atmosphere and clouds.

Notice that the budgets balance as shown: the three upward fluxes equal the downward, solar flux. While this is very nearly true on a worldwide, long-term average, it generally does not hold at any specific place and time. Two other terms must be introduced before the local budget balances. One is **advection,** the horizontal transport of some property (in this case, heat) by the winds or ocean currents. The other term is *storage change,* which is simply the gain of heat due to warming or cooling. Daily and seasonal warming and cooling are examples of storage change. The next (and final) section deals with these concepts.

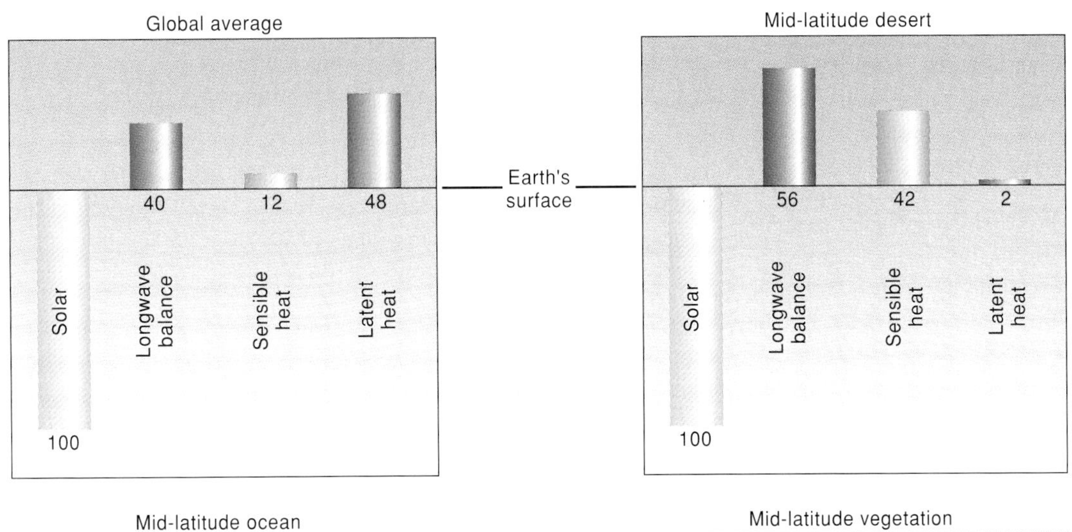


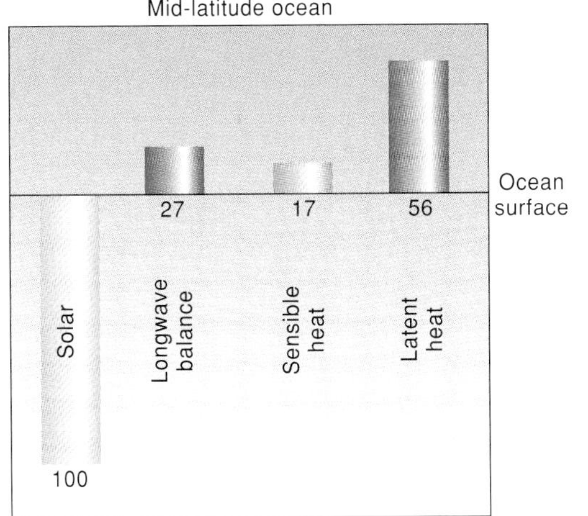


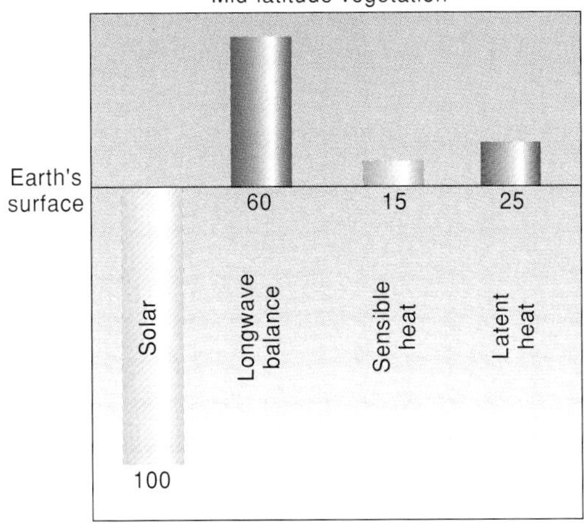


Figure 3.30 Surface energy fluxes for various mid-latitude locations. Units are based on a worldwide average of 100 units of insolation absorbed at earth's surface.

## Checkpoint

***Review***

Sensible and latent heat fluxes occur in response to differences in temperature and moisture between the earth's surface and the overlying air. Overall, these fluxes carry heat from the earth's surface to the atmosphere, thus warming it. On a worldwide average, the warming due to latent heat flux is four times that of sensible heat.

Sensible heat is conveyed by conduction and convection, while latent heat is transferred when water undergoes a change of phase (evaporation, etc.) between the earth's surface and the atmosphere.

***Questions***

1. State briefly the differences in heat transfer by radiation, conduction, and convection.
2. What is "latent heat transfer"?
3. In the first frame of Figure 3.30, how much energy does the earth's surface gain? What is the sum of its losses?

# Earth's Energy Budget and Radiative Forcing

## Balancing the Budget

Figure 3.31 shows all terms in the mean global energy budget and serves as a concept map for this section. Study the budget for a moment. Is it balanced? By a balanced budget, we mean one in which the storage is neither increasing nor decreasing: that is,

$$\text{energy input} - \text{energy output} = 0$$

Stated another way, for balance to exist, the inputs must be equal to the outputs.

Looking at the fluxes across the earth-space boundary in Figure 3. 31, you see the sum of the inputs does indeed equal the sum of the outputs; the budget balances, and the entire system is neither warming nor cooling, according to these statistics. However, it is possible that changes could be occurring within the system; for example, the atmosphere may be warming but the earth itself cooling by the same amount, which would lead to an overall balance of zero.

To check this conjecture, tally up the atmosphere's gains and losses:

(16 sun absorbed by air
+ 4 sun absorbed by clouds )
+ ( 99 longwave from ground
− 85 longwave emitted downward
− 38 longwave emitted upward by air
− 26 emitted upward by clouds)
+ (6 sensible + 24 latent ) = 0

Similarly, for earth's surface,

(54 incoming sun − 4 reflected sun )
− 105 longwave emitted + 85 longwave absorbed
− (6 sensible + 24 latent ) = 0

You see that the earth's surface and the atmosphere are able to maintain a rather constant temperature because they have the means to release energy as well as to absorb it and because their losses are the same magnitude as their gains.

## Radiative Forcing

Among the energy budget terms, a clear hierarchy exists. Some fluxes tend to occur more or less independently of the atmosphere's condition and act as causative agents, while others are reactions to other fluxes and energy stored. Recognizing this hier-

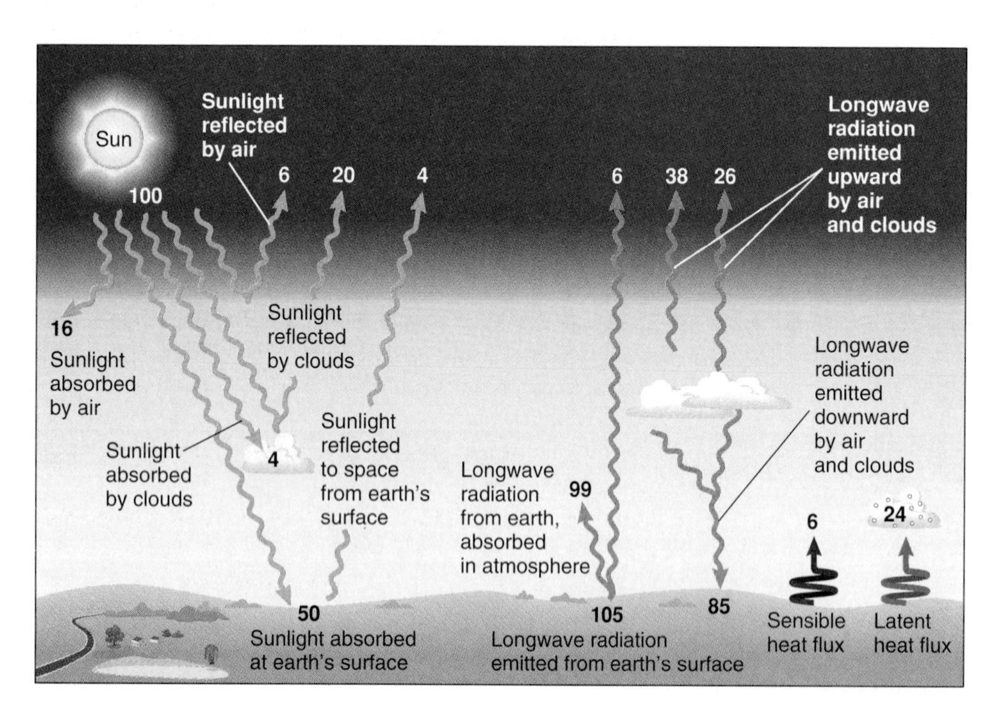


**Figure 3.31** Concept map: Mean global energy budget.

archy of cause and effect helps you understand the energy budget's significance in atmospheric science.

The most obvious element of the hierarchy of energy budget fluxes is the preeminence of solar radiation. The sun's energy is responsible for the warmth of the earth and its atmosphere. It provides the regular deposits of income to the earth-atmosphere energy bank account.

Next in the hierarchy are the longwave terms. Most of these fluxes are large; each always flows in the same direction and changes only modestly in response to other atmospheric conditions. Longwave terms perhaps are the energy budget equivalents of bank account outlays for rent, utilities, and taxes: they are always present but are adjustable in response to changes in income.

At the "effect" end of the cause-effect hierarchy are the conductive and convective fluxes, the sensible and latent heat terms. These fluxes may flow upward, downward, or not at all. They depend on temperature and moisture differences between the earth's surface and the atmosphere. A parallel in the bank account analogy might be discretionary spending: if the bank balance (energy level) is high enough, a flux of money might go into a new TV set; if the balance isn't great enough, the expenditure will have to wait.

It is useful to develop this hierarchy a bit further. To a celestial accountant monitoring earth's energy income and outgo from outside the earth, our planet's energy budget would consist of just two significant events: the absorption of a large quantity of solar radiation by earth and atmosphere, and the return of a roughly equal amount of longwave radiation to space. These two terms determine the overall rate at which energy flows through the earth-atmosphere system. Taken together, the absorbed solar radiation and outgoing longwave radiation comprise what scientists call the system's **radiative forcing.** Radiative forcing is the stimulus provided to the earth-atmosphere system by the absorbed solar and outgoing longwave radiative fluxes. It is called forcing because these fluxes "force" the atmosphere to respond in certain ways, as you will see presently.

Is radiative forcing everywhere the same? If the energy accountant zoomed in for a closer look, she would measure the solar and longwave fluxes as shown in Figure 3.32. Notice that tropical latitudes experience a surplus of radiation; in these regions, solar income exceeds the outgoing longwave flux. In polar regions, the opposite is true.

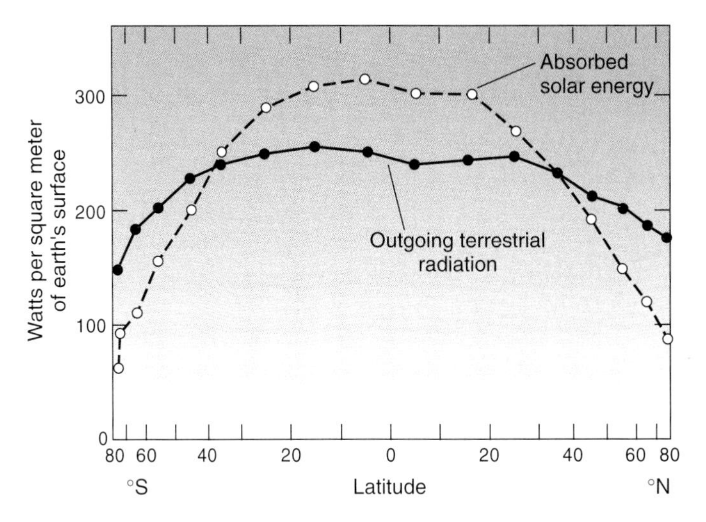


Figure 3.32 The distribution of absorbed solar radiation and outgoing longwave radiation at different latitudes. Equatorial regions experience a surplus, polar regions a deficit. What processes act to achieve a balance?

The accountant's findings might puzzle you. You might be thinking, "Shouldn't the imbalance in Figure 3.32 cause the tropical regions to be getting warmer and warmer, while polar regions cool continuously?" A reasonable question; if it were not for advection of heat by the wind and by ocean currents, the answer would be yes. However, circulations in oceans and air advect excess tropical heat to polar regions, thereby maintaining a zero balance at each latitude. We can state the situation even more strongly, as follows: Differences in radiative forcing, generally between tropical and polar regions, cause atmospheric and oceanic circulations to develop that act to transport the excess heat to regions of deficit.

Thus, radiative forcing ultimately is responsible for virtually the full range of atmospheric and oceanic behaviors, from synoptic scale weather systems to gusts of wind, from huge ocean currents to individual waves, from convection currents to longwave radiation emitted throughout the system. In other words, the atmosphere and oceans are forced to act the way they do because of the amount of radiative forcing they experience at different locations and times.

As long as the earth's overall radiative forcing remains constant, the system will exhibit "normal" behavior, on the average, with the variety of atmospheric and oceanic events we have come to think of as typical. However, a change in radiative forcing could cause changes in weather and oceanic events, which is the topic of the next chapter.

Study Point

## Chapter Summary

In this chapter, we have considered the earth-atmosphere system as a sort of energy-processing device. Overall, the system receives a steady input of radiation from the sun, although its distribution varies considerably with time, season, and location. The system returns a roughly equal amount of energy to space in the form of longwave radiation. Within the system, energy flows in various forms: as radiation and as sensible and latent heat transfer through the processes of conduction and convection. These various energy fluxes are intimately connected to the various atmospheric events we recognize as weather; in fact, weather events are a response to radiative forcing of the atmosphere.

To understand specifics of energy transfer, we developed a rather detailed model of the physics of radiation. The concept of the photon as a bundle of radiant energy, the Stefan-Boltzmann formula, and Wien's law are important components of this model.

## Key Words

energy budget
calorie
insolation
watt
Tropic of Capricorn
Tropic of Cancer
spring equinox
scattering
albedo
wavelength
photon
ground state
infrared radiation
continuous spectrum
Stefan-Boltzmann formula
greenhouse effect
conduction
sensible heat flux
transpiration
radiative forcing
advection
law of conservation of energy
solar constant
power
summer solstice
autumnal equinox
winter solstice
specific heat
micrometer
excited state
electromagnetic spectrum
longwave radiation
blackbody radiation
selective absorbers and emitters
line spectrum
greenhouse gases
radiation
budget
convection
latent heat flux

## Review Exercises

1. State the law of conservation of energy. Then give several examples illustrating the law.
2. What is the solar constant? What are its units?
3. Explain the difference between temperature and heat.
4. How many calories does it take to warm 1 gram of water 1°C? 10°C? To warm 5 grams of water 1°C? To warm 5 grams of water 10°C?
5. Describe the daily path of the sun across the sky in your location in September, December, and June.
6. January is the coldest month for most locations in the Northern Hemisphere, but it is also the month when the earth is closest to the sun. Explain.
7. How large is one micrometer expressed in millimeters? In centimeters?
8. Arrange these forms of radiation from shortest to longest wavelength: yellow light, radio waves, red light, ultraviolet light, longwave radiation, x-rays, blue light.
9. Arrange the radiation forms in the last question according to the energy of their photons.
10. Explain why each of these objects appears blue to us: blue jeans, the blue sky, a bluish star.
11. The atmosphere is said to be a selective absorber. Explain what that means.
12. What substances are effective absorbers and emitters of longwave radiation in the atmosphere?
13. Translate the Stefan-Boltzmann relation and Wien's law from formulas into words.
14. What is a greenhouse gas? Name at least four examples.
15. Compare and contrast energy transfer by conduction, convection, and radiation.
16. What does it mean that the energy budgets for the earth's surface, for the atmosphere, and for the earth-atmosphere system are balanced?

## Problems

1. Figure 3.10 shows the solar portion for the earth's mean energy budget. Suppose the mean amount of cloudiness increased. Do you think the change would increase or decrease the amount of sunlight reflected from clouds back to space? The amount absorbed in clouds? The amount reaching the earth's surface? Explain your answers.
2. Which surface absorbs a greater percentage of incoming sunlight: forest, bare ground, ice, or old snow?

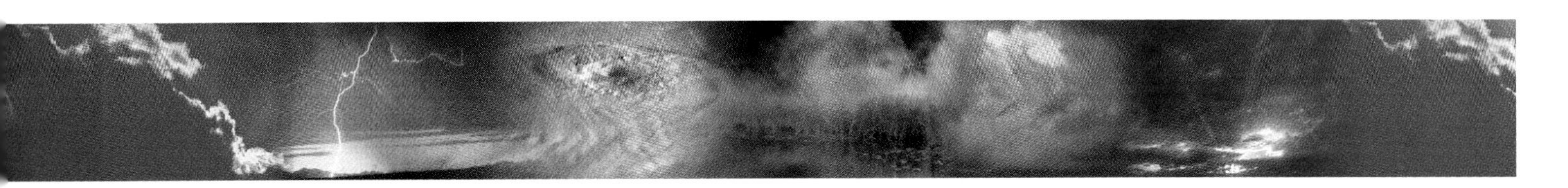

3. Suppose you have 1-gram samples of pure water, dry sand, dry air, and concrete. Which will warm up the most in response to a 1-calorie gain of heat?
4. A metalsmith heats two identical pieces of iron, one to 1000 K, and the other to 2000 K. Which piece emits more radiation? How much more? Compare the wavelengths of the radiation emitted.
5. The constellation Orion, visible during winter and spring, contains two very bright stars: reddish Betelgeuse and blue-white Rigel. Which star has the hotter radiating surface? Explain your answer.
6. A photon of ultraviolet sunlight, streaking toward the earth's surface, is absorbed by an ozone molecule in the stratosphere. The photon disappears; what happens to its energy?
7. The moon has no atmosphere. What fluxes would make up its surface energy budget?
8. The earth absorbs huge quantities of solar energy every day. Why isn't it warming noticeably from one day to the next?
9. Suppose that somehow the mean temperature at the earth's surface were suddenly to become several degrees cooler. Which of the energy budget terms in Figure 3.31 would be immediately affected and in what sense (increase, decrease)?
10. Energy passes from the earth's surface to the atmosphere via longwave radiation, sensible heat flux, and latent heat flux. Explain how these processes differ from each other.
11. Figure 3.31 shows the mean global energy budget. Of course, local values constantly vary from these mean figures. Which fluxes might be zero at a given place and time? Which cannot be, and why?
12. A given input of solar energy heats dry land more than the ocean surface. Discuss the roles of each of these factors in causing the differences: albedo, specific heat, mixing, and wetness of the surface.

## Explorations

1. Investigate the amount of solar energy striking a roof on a sunny day. To begin, choose some typical roof dimensions (7 meters by 10 meters, for example), and find the maximum value possible: assume the sun is shining perpendicularly onto the roof surface at the solar constant rate (i.e., with no loss by the atmosphere) of 1.367 kilowatt hours per square meter per hour, and assume that the roof's albedo is 0%. How many kilowatt hours of energy does the roof absorb per hour? At 10 cents per kilowatt hour, what is the monetary value of one hour of absorbed sunlight? Of one month's worth? Now repeat the calculation, reducing the total because of atmospheric losses (make an estimate based on Figure 3.9), nonperpendicularity of the beam to the roof (reduces daily total by 50% to 90%, depending on season and latitude, and albedo of the roof (10% to 70%). Suggestion: spreadsheet programs are ideal tools for calculating answers for a range of conditions like those in this example.
2. If you have access to AccuWeather's AccuData database, use the CSUN product, which gives sunrise and sunset times, to make a study of how the length of daylight varies with latitude and season.

## Resource Links

The following is a sampler of print and electronic references for some of the topics discussed in this chapter. We have included a few words of description for items whose relevance is not clear from their titles. For a more extensive and continually updated list, see our World Wide Web home page at www.mhhe.com/sciencemath/geology/danielson.

Tillery, B. W. 1993. *Physical Science* (2d ed.). Dubuque, IA: Wm. C. Brown. 618 pp. An introductory physics and chemistry textbook that might serve as a useful reference on topics such as energy, radiation, and so on.

Gruber, A., and R. Ellingson, P. Ardanuy, M. Weiss, S.K. Yang, and S. N. Oh. 1994. A comparison of ERBE and AVHRR longwave flux estimates. *American Meteorological Society Bulletin,* 75 (11): 2115–30. Two different sets of satellite measurements of upward longwave radiation are compared.

Monastersky, R. 1994. The sunny side of weather: How can minute changes in solar rays influence conditions on the ground? *Science News* 146 (23): 380–81.

AccuWeathers home page at http://www.AccuWeather.com provides visible and infrared satellite images, and maps of ultraviolet radiation. For readers with access to AccuData, CSUN enables users to calculate sunrise and sunset times for any location on any day. <send> RAYBAN provides forecasts of ultraviolet radiation. SATMAP gives worldwide satellite images.

Lyndon State College's weather web site at http://apollo.lsc.vsc.edu/weather/weather.html is one of a number that provide current satellite images at both visible and infrared wavelengths.

Chapter 4

# *Is the Atmosphere Warming?*

# Chapter Goals and Overview

Is the atmosphere's temperature changing? Will it rise and melt the polar ice caps, threatening our coastal cities with floods? Will it fall and return us to another glacial age? Will it change in less dramatic ways, or will it remain unchanged altogether? To what extent is human activity responsible for any change? As you can see from the items on the facing page, there is a great deal of interest and difference of opinion about the topic popularly referred to as "global warming." Because of its potentially far-reaching consequences, global warming currently is one of the most actively investigated scientific issues.

This chapter contains less "background science" and fewer new terms (note the brevity of the key words list at the end of the chapter). Instead, you will apply many of the concepts developed in earlier chapters and critically examine scientific data and arguments to determine the answer to the question, "Is the atmosphere warming?" You will begin by considering what sorts of phenomena are capable of triggering global temperature change through their influence on the energy budget. Looking at the research of atmospheric scientists, you will see ways in which warming might lead to changes in other atmospheric variables, such as cloudiness, which in turn could affect the temperature further. You will see how computer simulations can contribute to understanding these complex interrelations. Then, you will examine a wide variety of observational evidence, following the research of a number of different scientists on global temperature records. Finally, you will consider some ways in which humans might respond to the threat of global warming.

## Chapter Goals

Upon completing this chapter, you will be able to:

- use energy budget relations to trace the potential of greenhouse gases and other influences for bringing about global temperature change;
- give examples of how an enhanced greenhouse effect can lead to changes in other atmospheric variables, such as cloudiness, which in turn lead to further temperature change;
- cite observational evidence in support of and in opposition to the hypothesis of global warming and point out limitations or uncertainties in such evidence;
- discuss potential impacts of global warming on the earth's geography and on human activities; and
- present choices facing humanity on the global warming issue.

# Radiative Forcing: Key to Global Warming

In the last chapter, we presented an expression that describes an energy budget for the earth-atmosphere system. The expression was

$$\text{energy input} - \text{energy output} = \text{storage change}$$

Recall that a change in stored energy means a change in temperature within the system.

Just what are the energy inputs for the earth and its atmosphere? Take a moment to examine Figure 4.1, which is a reproduction of last chapter's Figure 3.31, the complete energy budget diagram. Note that there are three energy inputs to the system: 16 units of solar energy absorbed by atmospheric gas molecules, 4 units by clouds, and 50 at the earth's surface, for a total of 70 units of energy income. All other energy gains in Figure 4.1 are the result of fluxes from one part of the system to another, so a gain in one place is matched exactly by a loss somewhere else. (Check Figure 4.1 to verify this statement.) Thus, absorbed sunlight, in its three different branches, is the only significant energy input.

You can also see from Figure 4.1 that energy loss from the earth and the atmosphere occurs in three fluxes: 6 units of longwave radiation that are emitted by the earth's surface and escape to space, 38 units of longwave radiation that are emitted upward from air molecules and particles, and 26 emitted upward from clouds. We shall combine these three fluxes into the single term **outgoing longwave radiation,** whose sum is 6 + 38 + 26, or 70 units.

Rewriting the radiation balance formula with these terms (and abbreviating a bit), we have

$$\text{absorbed solar} - \text{outgoing longwave radiation} = \text{storage change} \quad [1]$$

These two terms comprise the radiative forcing on the atmosphere, as you saw last chapter. Entering the values we just computed for each term, we have

$$70 \text{ (absorbed solar)} - 70 \text{ (outgoing longwave)} = \text{storage change} = 0$$

for the balanced energy budget in Figure 4.1. The fact that storage change is zero means that the temperature change, under a balanced budget, is also zero. Overall, a system neither warms nor cools when energy income is equal to the energy outflow.

What happens if the terms do not balance exactly, causing the storage change to differ from zero? Looking at the formula, you can see that if the incoming solar term is greater than the outgoing longwave term, it means that more energy is entering the system than leaving it. The difference between the terms is positive, so the change in stored energy is positive, which means a rise in temperature. If the outgoing flux predominates, then the change in stored energy is negative. The system is losing more energy than it is gaining, and stored heat energy is used (which

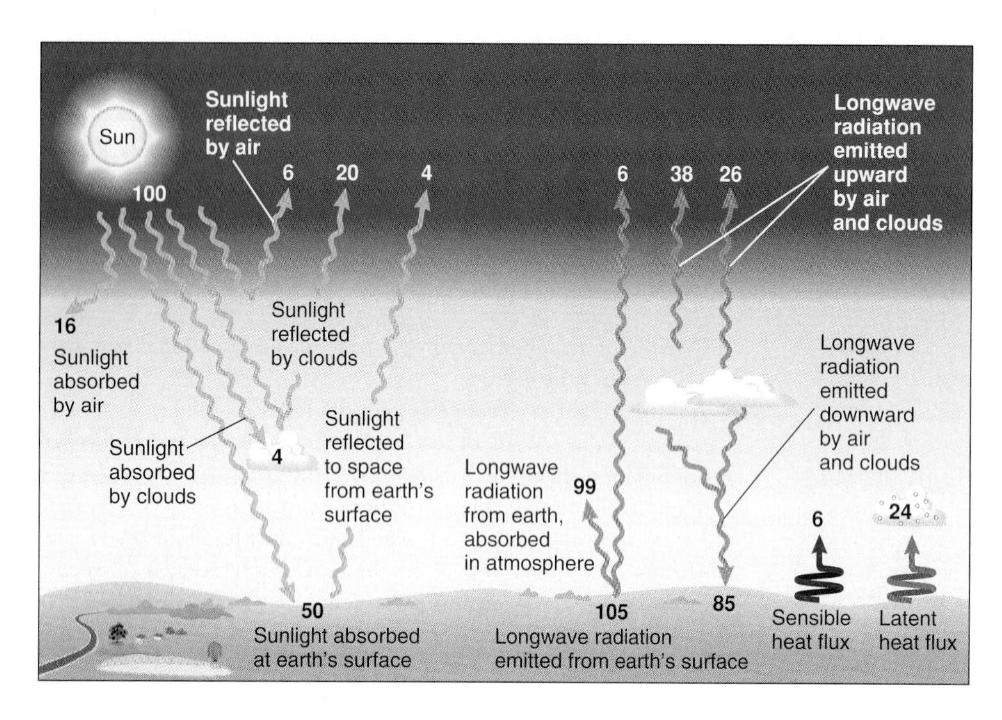


Figure 4.1 The mean global energy budget.

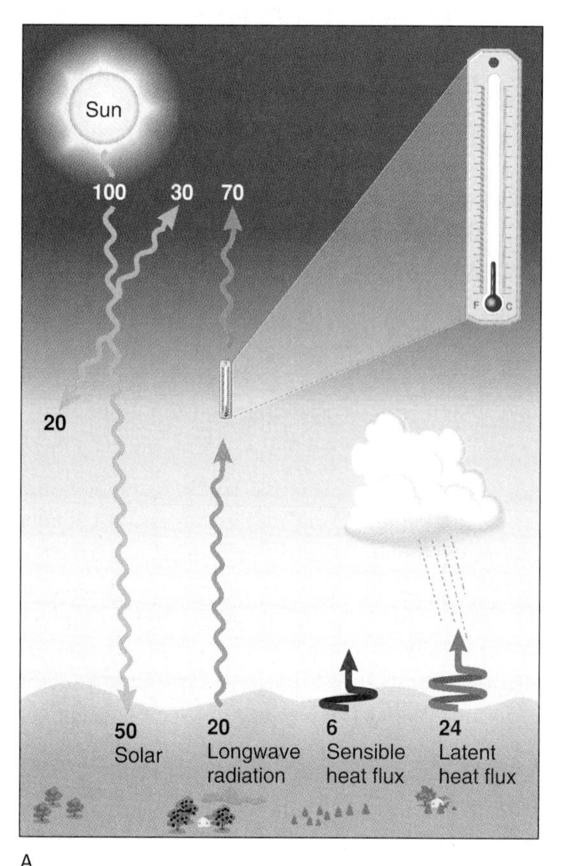


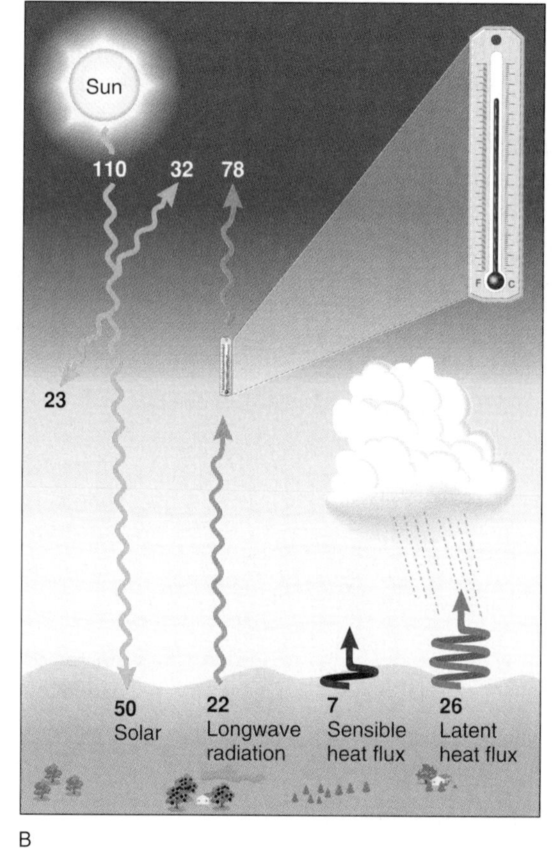


Figure 4.2 Two energy budgets for the earth-atmosphere system. (Notice that some terms from Figure 4.1 have been combined here for brevity: thus, reflected sunlight is shown by a single arrow, and the surface longwave fluxes are shown as a single arrow representing the difference between upward and downward fluxes. Both budgets show a balance. However, in (*B*) more energy cycles through the system, causing higher temperatures and, most likely, other differences: perhaps more evaporation and therefore more precipitation, as suggested in (*B*).

means a decrease in temperature) to fuel the outgoing flux. In the bank account analogy of last chapter, if income and expenses are equal, then the account's balance does not change. The balance grows if either the income increases or the expenditures decrease.

We will refer to formula [1] on p. 99 as the **radiative forcing formula.** Notice that the formula can be interpreted as an expression of the law of conservation of energy: the law requires that any difference between incoming and outgoing energy amount must be accounted for and appears as a change in energy stored. The expression also contains three important facts about the earth's energy budget and global change. The first is that if the two terms on the left are equal, then there is no change in radiative forcing, and the system's total stored energy—its mean temperature—is not changing either. The second fact is that a change in radiative forcing, due to a change in either or both of the radiation terms, leads to a change in stored energy and therefore to a change in temperature. Thus, the formula tells about conditions of temperature stability and of those leading to temperature change.

The third fact relates to the amount of energy flowing through the terms in the formula. The slightly simplified energy budget diagrams in Figure 4.2 illustrate the situation. Suppose the radiative forcing is unchanging, with each flux at 70 units. Then, the formula says

$$70 \text{ input} - 70 \text{ output} = 0$$

Because the storage change is zero, the earth-atmosphere system's temperature is unchanging (on the average) as well.

Suppose now that the solar income somehow increases to 80 units, and the system readjusts to emit 80 units back to space. Then,

$$80 \text{ input} - 80 \text{ output} = 0$$

Again, the fluxes are equal, so the storage change equals zero and the temperature is stable also, as before.

Are the two situations equivalent? No, they are not. In the second case, the atmosphere is "processing" 14% more energy. That means the atmosphere's processes are running at a faster rate, at higher temperatures, perhaps with more vigorous circulation patterns and a host of other attendant changes. The point is that the magnitude of each radiation term in the formula is influential individually, even when each term is equal and therefore cancels each other out. The magnitudes of the terms determine the rate at which the atmospheric engine runs; the difference in their magnitudes determines the rate at which the system is changing temperature.

Finally, we want to point out the generality of the radiative forcing formula. It embraces every possible way in which the earth and the atmosphere may change temperature. Only a change in radiative forcing, that is, an imbalance between incoming solar and outgoing longwave radiations, can lead to overall global temperature change. Thus, the rather simple formula [1] lies at the center of all discussions of global warming.

## Changes in Radiative Forcing

In this section, you will examine three different factors that could lead to changes in radiative forcing and therefore to global warming. By far the greatest scientific interest is focused on the first of the three, an enhanced greenhouse effect.

### An Enhanced Greenhouse Effect

You learned in Chapter 2 that atmospheric carbon dioxide and methane concentrations are increasing annually. From the last chapter, you know that these gases and others are active greenhouse gases, allowing solar radiation through the atmosphere but absorbing much of the return longwave flux. (We suggest you glance back at Figure 3.21 to refresh your memory of these substances' radiation-absorbing properties.)

Just how fast are $CO_2$ and other greenhouse gases accumulating in the atmosphere? In Figure 4.3, you can see some trends. The data for the past 50 or so years are based on direct measurements of the atmosphere's composition. For earlier years, the data come from analyses scientists have made of air trapped in glacial ice. Notice the graphs show that not only are concentrations of greenhouse gases increasing, they are increasing at ever faster rates.

Is there any observational evidence that links changes in greenhouse gas concentrations to global temperature change? Researchers studying air trapped thousands of years ago in Antarctic ice cores have found a distinct correlation between variations in air temperature and those in greenhouse gas concentrations. Their research is shown in Figure 4.4 for $CO_2$ concentrations. Note that the temperature and the gas curves are in phase: warmer temperatures occur simultaneously with higher $CO_2$ concentrations. The patterns' similarity does not prove that the greenhouse gases caused the temperature rise; in fact, some recent evidence suggests the temperature rises may have occurred slightly before the greenhouse gas increases. Clearly, however, the patterns of change in greenhouse gas concentrations and air temperature seem to move in step.

To recap, concern about global warming due to an enhanced greenhouse effect is based on the following chain of evidence:

1. Earth's greenhouse effect is known to keep our planet's surface environment warmer, by as much as 35°C. (Recall the discussion on this topic in Chapter 3.)
2. Atmospheric concentrations of the greenhouse gases responsible for this warming are known to be increasing steadily.
3. Past temperature changes in glacial climates are correlated with greenhouse gas levels.

Does all this mean that an enhanced greenhouse effect, that is, greater warming, is inevitable?

Although the conclusion may seem obvious, the problem is not so simple. As you will see in the following pages, a change in radiation fluxes may lead to other changes in the energy budget, which could nullify or even reverse an enhanced greenhouse effect. Other factors altogether, such as volcanic eruptions, influence the budget in a variety of ways. And even if the greenhouse effect does lead to higher temperatures, the rate of increase may be so low that it is not significant.

On the other hand, it is possible that an increase in temperature could promote even greater greenhouse gas concentrations, which would lead to even higher temperatures, with results that could be catastrophic. Clearly, the question of global warming needs answering.

### Variations in the Solar Constant

It should come as no surprise that a change in the sun's energy output rate would lead to a change in radiative forcing. Such a

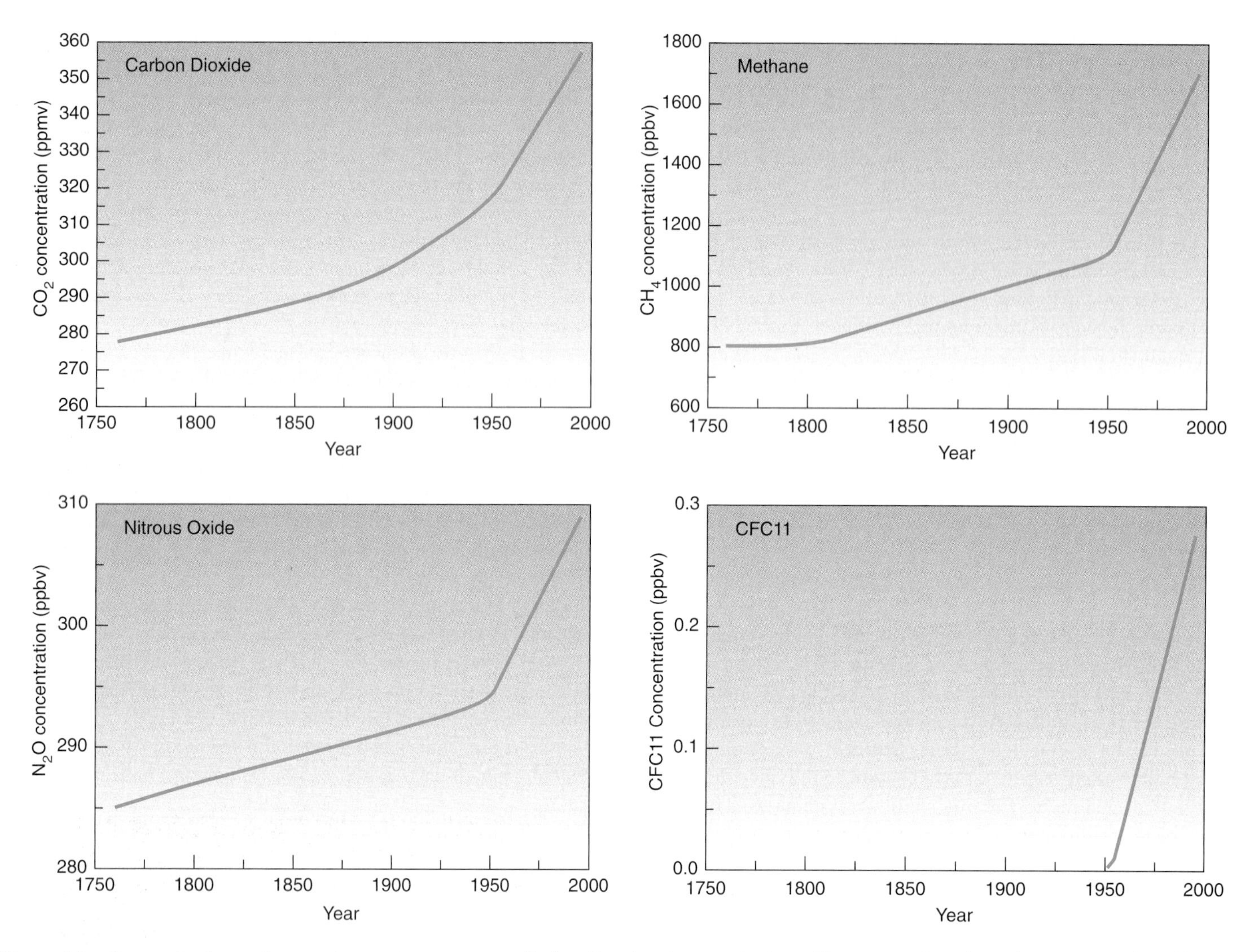


Figure 4.3 Concentrations of several greenhouse gases. Notice the steady recent rates of increase of each gas's concentration. Also note that gases with a longer history of measurement show accelerating rates of buildup. (Note: *Y*-axis scales are not the same.)

change could cause either warming or cooling, depending on whether the change was positive or negative.

Is the sun's output changing or constant? Scientists P. Foukal and J. Lean have developed an estimate of the average solar intensity at the top of the earth's atmosphere, based on satellite and other measurements. Figure 4.5A shows their findings. As you can see, their data exhibit a rhythmic variation in solar output over a period that averages about 11 years. This cycle corresponds to the sunspot cycle. Sunspots are magnetic disturbances that appear, migrate, and disappear from the sun's surface, in loose analogy to migrating weather systems on earth (see Figure 4.5B).

Are there other trends in the solar output data? Examine Figure 4.5A for signs of a single, long-term trend. Notice that radiation values at the top of each cycle seem to show an increase over the past 100 years; perhaps you detect a slight upward trend in average values as well. But note also the values on the graph's vertical axis: the entire graph spans only the radiation range from 1367 to 1368 watts per square meter, or 1.955 to 1.956 calories per square centimeter per minute. Thus, the variations in solar intensity fall within a very narrow range of values. Our tentative conclusion is that changes in the solar constant are unlikely to be causing significant changes in radiative forcing, although not all scientists share our view.

## Changes in Atmospheric Particle Concentrations

Figure 4.6 shows the impact that concentrations of small particles can have on the energy budget. Heavy smoke such as that in the figure can reduce daytime insolation at the earth's surface to nearly zero. Imagine the surface insolation of 54 units in Figure 4.1 changed to zero. Taken by itself, this energy loss would lead to a significant decrease in surface temperature.

Of course, you cannot take such an energy loss by itself. Fifty-four units of energy cannot simply disappear from the energy budget diagram. How does the energy budget adjust to such a loss?

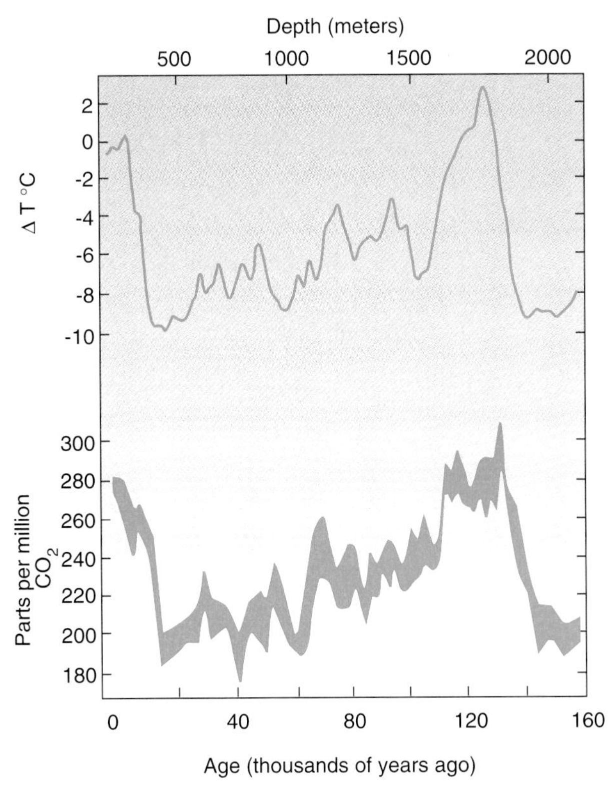


**Figure 4.4** The concentration of atmospheric $CO_2$ is strongly correlated with air temperature as derived from polar ice measurements.

The overall impact of particles on the energy budget is a highly complex one. Particles such as the soot in Figure 4.6 have low albedoes, which allows them to absorb considerable solar energy and thereby warm the atmosphere. The particles also stimulate cloud formation, which alters the atmosphere's ability to absorb and reflect insolation. Thus, the insolation lost to the surface is distributed among absorption and reflection fluxes in the atmosphere.

Particles also affect the longwave part of earth's radiative forcing. As you learned in Chapter 3, solids and liquids interact differently with radiation than gas molecules do. Generally, particles behave much more like blackbodies than atmospheric gases do; therefore, the particles tend to absorb and emit longwave radiation at greater rates than particle-free air. Current scientific opinion is that the net effect of particulate pollution is to increase the amount of insolation reflected more than it enhances greenhouse warming, thus lowering the input term in the radiative forcing formula more than the output term, which leads to cooling. To give a hypothetical example, suppose particles reduce the total absorbed solar term of 70 by 4 units by reflecting sunlight back to space: $70 - 4 = 66$ units. Suppose also that the particles' impact on longwave radiation is to decrease the longwave outflow from 70 by 2 units: $70 - 2 = 68$. In this case, the radiative forcing formula becomes:

$$\underset{\text{(absorbed solar)}}{66} - \underset{\text{(outgoing longwave)}}{68} = \underset{\text{(storage change)}}{-2}$$

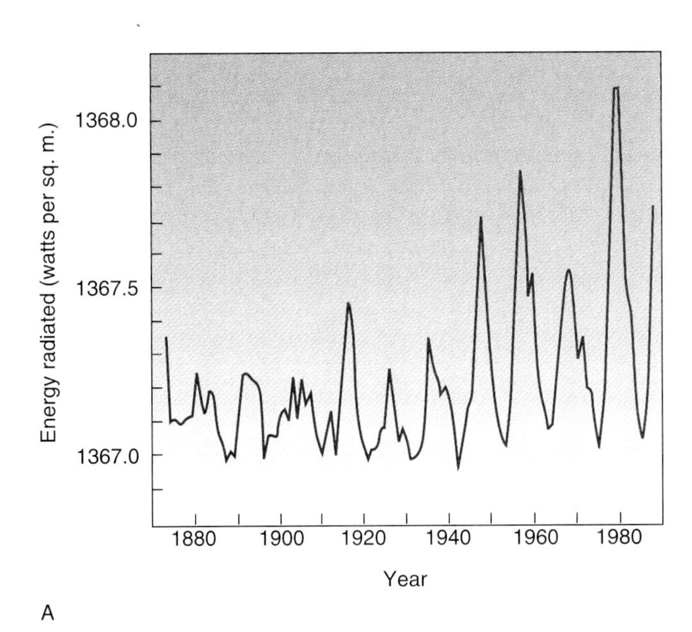


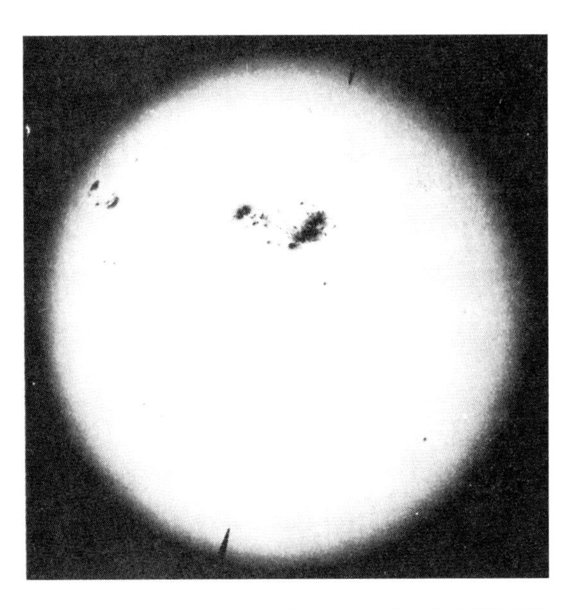

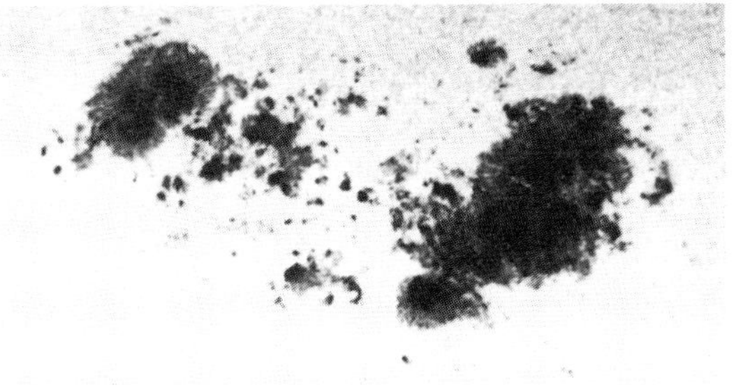


**Figure 4.5** (*A*) Variations in the sun's energy output occur over an 11-year cycle, according to these data compiled by P. Foukal and J. Lean. Is there an overall trend? If so, what is its magnitude? (*B*) A typical sunspot. Sunspots are dark because they are several hundreds of degrees cooler than the average solar surface temperature.

Figure 4.6 This photograph was taken during daylight hours in Kuwait City during the Iraq war in 1991. Clouds of particles from oil well fires have reduced the amount of incoming sunlight to nearly zero.

Here, solar energy loss outweighs the longwave energy saving, with the result that the system cools (negative energy storage change). Notice, however, that the storage change (−2 units ) is a small difference between two large quantities. A small change in the particles' impact on either the solar or the longwave term could change the result dramatically.

Recently, sulfur-based particles have received special attention for their ability to affect the energy budget. Scientists such as Robert Charlson at the University of Washington have found that sulfuric acid particles, both liquid and solid, are particularly effective reflectors of solar energy. The particles also affect longwave fluxes and cloud cover in various ways, thereby altering both income and outflow terms in the radiation budget formula. Charlson and others believe that the particles' net effect is to reduce the solar income, causing cooling. However, this subject is one of active study and debate at present.

Sulfur emissions, as you read in Chapter 2, derive from natural processes and human activity. At present, humans contribute more sulfur, mainly through burning fossil fuels, than do the natural sources such as the oceans and volcanic eruptions. In urban areas such as the eastern United States, the contribution from human sources is far greater than from natural ones. Some researchers have estimated that **anthropogenic** (generated by human activity) sulfur emissions have led to a 7.5% depletion of incoming solar energy in the eastern states.

The sulfur emissions from even a single volcano can be so great that they may affect energy fluxes worldwide. Scientists have estimated that the 1991 eruption of Mount Pinatubo in the Philippines caused the release of 20 million tons of sulfuric acid into the atmosphere. NASA scientist James Hansen believes this single eruption may have caused a cooling of the earth's temperatures by as much as 0.6°C (1.1°F) for two or three years.

***Review***

The concept map (Figure 4.7) relates a number of the major ideas of the first two sections of this chapter. Trace the different paths to reinforce the relations among these ideas.

***Questions***

1. Suppose a certain type of particle pollution caused the earth's albedo to increase. How would that affect the radiative forcing formula? What effect would the change have on the earth's temperature?

2. By what percent has the concentration of atmospheric methane increased from 1750 to 1975, according to Figure 4.3?

3. How do the shapes of the curves in Figure 4.3 indicate that concentrations of greenhouse gases are increasing more rapidly now than in earlier years?

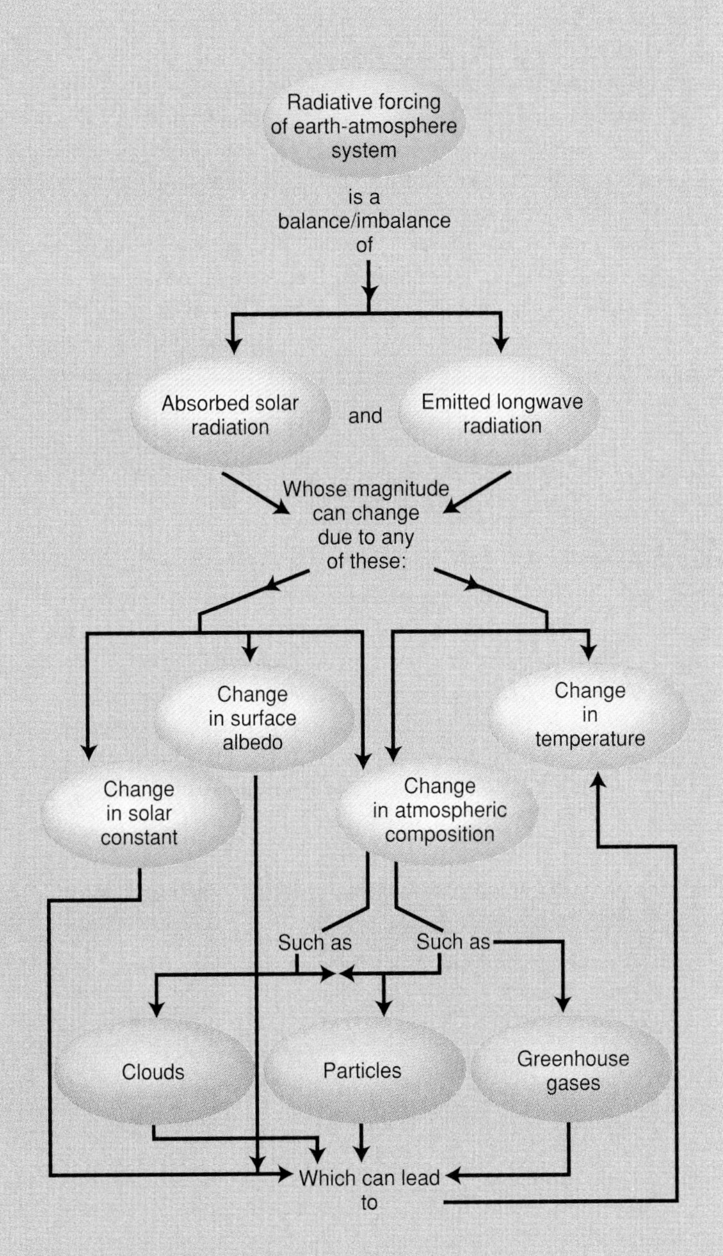


Figure 4.7 Concept map: Radiative forcing.

# Feedbacks to Radiative Forcing

## The Nature of Feedback

What will be the nature of the atmosphere's response to radiative forcing of the sorts just presented? The answer is important: if the atmosphere's response somehow causes the forcing itself to increase, for example, then the atmosphere could conceivably evolve ever more rapidly into some extreme condition. Such a response is an example of **positive feedback**. Positive feedback occurs when a system (the atmosphere, in this case) responds to some forcing in a way that amplifies the original force. A familiar, everyday example of positive feedback is a microphone that picks up sound from the speakers and amplifies the sound further, leading to ear-shattering shrieks. A more pleasant example is that of a parent praising her child in response to a good report card: the child's good grades evoke words of praise from the parent, which in turn motivate the child to do even better work.

Positive feedback cannot continue indefinitely. Eventually, the system "maxes out" in some way (the sound system reaches peak volume, or destroys its own speakers; the student gets all A's and can't improve further). However, positive feedback can be a very important, often destabilizing factor in the shorter term.

Many forms of feedback in nature are **negative feedback**. Negative feedback, as you might guess, is a response to a perturbing signal that tends to diminish the perturbation itself. Examples of negative feedback are all around you. Eating in response to hunger is an example: the more you eat at a meal, the less you want to eat.

## Feedback through Changes in Longwave Radiation

Longwave radiation provides important feedback to many energy perturbations of the earth-atmosphere system, as illustrated in Figure 4.8. The radiation fluxes shown in Figure 4.8A are the ones for a balanced budget developed in the last chapter.

Now suppose that an enhanced greenhouse effect occurs. What happens to the energy budget as a result? You know that a stronger greenhouse effect is caused by greater concentrations of $CO_2$ and the other gases that are efficient absorbers and emitters of longwave radiation. Thus, the atmosphere is more opaque to longwave radiation. As a result, more of the longwave energy emitted by the earth's surface is absorbed by the atmosphere and less escapes directly to space. The atmosphere's greater opacity also causes radiation emitted downward from the atmosphere to be greater and that emitted upward to be less than before, as shown in Figure 4.8A.

Notice that the immediate effect of the greenhouse gas buildup is that the flux of outgoing longwave radiation is reduced and thus no longer equals the incoming flux of solar energy; also, neither the earth's surface nor the atmosphere's budget is balanced any longer. The radiative forcing formula describing this situation now reads

$$70 - 66 = 4$$

How will the earth-atmosphere system react to this change in radiative forcing? The change in radiative forcing causes energy to accumulate in the system (i.e., the temperature rises), both at the earth's surface (land and oceans) and in the lower atmosphere. As long as outgoing radiation is less than the incoming energy (sunlight), the earth's temperature will continue to increase. How long will this extra forcing persist?

Recall from the last chapter that an object's rate of radiation depends strongly on its temperature; the warmer it is, the more energy the object radiates. Each of the longwave fluxes in Figure 4.8B, including those comprising the outgoing radiation, will in-

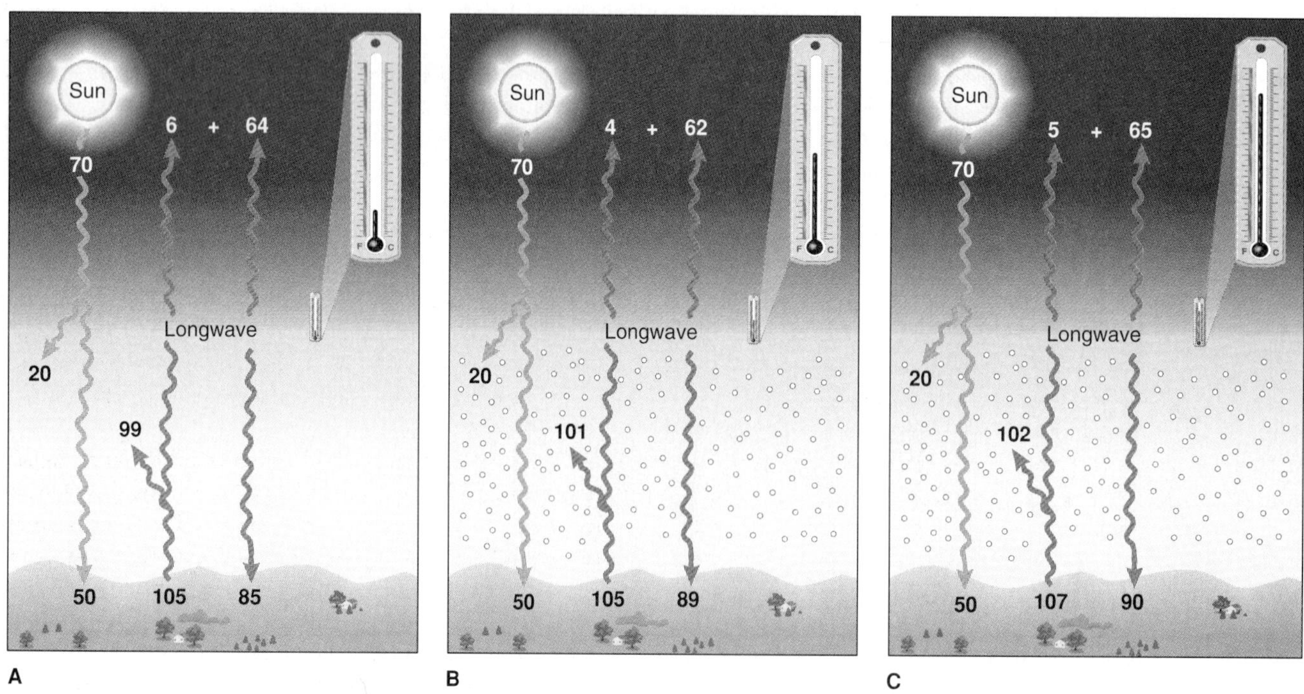


**Figure 4.8** The impact of greenhouse gas buildup on the radiation budget. (*A*) Current values of radiative fluxes, from earlier figures. Notice that longwave radiation upward from atmosphere and clouds is combined into a single flux and some solar fluxes are combined or not included. (*B*) A sudden increase of greenhouse gases would have no effect on solar terms. However, less of the longwave flux from earth's surface would escape to space, and the atmosphere would radiate more energy downward and less to space, leading to imbalances in the radiation budget and warming of the atmosphere and earth. (*C*) Warming would cause all longwave fluxes to increase until outgoing once more balances incoming solar. (Sensible and latent heat fluxes in (*C*) would also be different from those in (*A*).)

crease as the temperature rises. Thus, the warming of the atmosphere creates a negative feedback loop, eventually bringing the atmosphere's longwave output back up to a level in balance with the incoming solar flux. These new fluxes are represented by the values in Figure 4.8C. At this point, the energy budget is balanced once more; the radiative forcing terms are equal and opposite, no further excess heat energy will be stored within the system, and therefore no further warming will occur. Thus, in the simple situation considered here, increased radiative forcing leads to negative feedback, causing a new balance at a warmer temperature, and the extra forcing is eliminated in the process.

## Other Feedbacks to Radiative Forcing

### *Changes in Albedo*

The snow cover depictions in Figure 4.9 reveal a transformation of more than 100,000 square kilometers from bare ground to snow cover, brought about in a single day by a passing low pressure system. The air temperature was just cold enough to cause a snowstorm rather than rain. As a result, the surface albedo is much greater, and the amount of absorbed insolation is much less, than if the storm had given rain. Less absorbed radiation means a reduced energy income, which means lower temperatures, thus tending to maintain the presence of the snow cover, an example of positive feedback. On the other hand, a warmer atmosphere would lower the earth's surface albedo as ice and snow melt, allowing the surface to absorb more solar radiation.

Changes in albedo could also occur in response to changes in vegetation patterns brought about by temperature changes. For example, Gordon Bonan and his colleagues at the National Center for Atmospheric Research in Boulder, Colorado, have suggested that global warming would allow northern spruce forests to advance into colder regions presently covered with tundra. Spruce forests have a lower albedo than tundra and thus would absorb more solar radiation in a positive feedback loop.

Figure 4.9 Changes in snow cover (and thus albedo) from March 31 to April 1, 1997. A single storm changed surface conditions in a way that stimulated various forms of feedback on the energy budget and on future weather.

### *Changes in Atmospheric Water Vapor and Cloud Cover*

Warmer global temperatures would likely lead to greater evaporation from the oceans and thus to increases in concentrations of atmospheric water vapor. Recall that water vapor is an effective greenhouse gas. Thus, greater atmospheric water vapor levels would promote an enhanced greenhouse effect.

Changes in atmospheric water vapor would likely alter cloud patterns, which would affect both solar and longwave radiation fluxes. An increase in cloudiness due to global warming would result in more sunlight being reflected to space and therefore would likely act as negative feedback to the original warming. But clouds are effective emitters and absorbers of longwave radiation; thus, greater cloudiness would lead to a magnified greenhouse effect, which would give positive feedback to the warming.

Current evidence indicates that the expected increase in reflected sunlight generally outweighs the predicted increase in greenhouse warming; thus the net effect of an increase in cloud cover is believed to be a cooling one. However, the effect of clouds on global temperatures is an extremely complex one and remains perhaps the greatest uncertainty in our attempts to understand feedback related to global warming.

### *Changes in $CO_2$ Storage in Oceans*

Earth's oceans contain vast amounts of $CO_2$, more than 50 times the amount in the atmosphere. A continuous flux of $CO_2$ persists between the ocean surface and the atmosphere, similar in some ways to the constant exchange of water vapor between the ocean and the atmosphere. The net effect of this flux, scientists believe, is to remove nearly half the anthropogenic $CO_2$ added to the atmosphere each year.

A change in global temperatures could change this situation significantly. The amount of $CO_2$ that can be dissolved in water decreases with increasing temperature. Thus, a rise in temperature would cause a release of $CO_2$ from the ocean into the atmosphere. Also, warmer temperatures would likely alter ocean currents, possibly bringing $CO_2$-rich water to the surface. This would reduce the amount of $CO_2$ flux from the atmosphere to the ocean, causing a greater buildup in the air, leading to a greater greenhouse warming, leading in turn to less $CO_2$ take-up by the oceans, in a positive feedback loop. Changes in the chemistry of sea water and in the amount and type of biological activity would also occur in a warmer world, leading to additional feedback to atmospheric $CO_2$ levels.

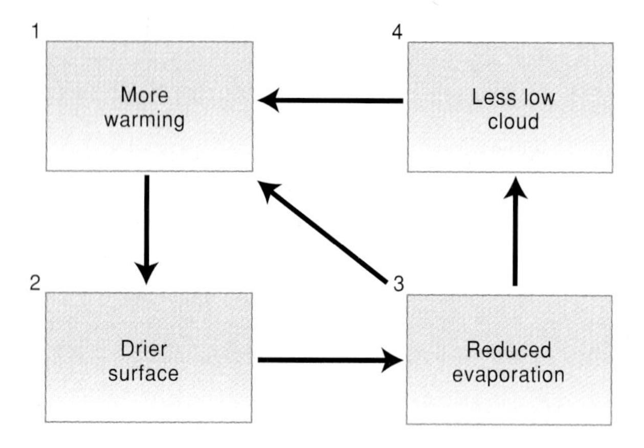


Figure 4.10 A feedback loop between soil moisture and warming. In what two ways, according to the diagram, does a drier surface lead to greater atmospheric warming? Why can this feedback loop not run endlessly?

### *Changes in Soil Moisture*

Look back for a moment at the energy budget values shown in Figure 4.8C. Notice that as a result of the change in radiative forcing, greater amounts of energy must be transferred into the atmosphere by sensible and latent heat fluxes. Over the oceans, heat will move from surface to atmosphere through each of these fluxes. Over land, however, the surface has only limited moisture. The warmer temperatures may cause this moisture to evaporate, leaving the soil drier than before. This results in the feedback loop shown in Figure 4.10. As you see, positive feedback occurs for two reasons: less evaporation means the surface warms more and it leads to less low cloud cover, allowing more warming.

## Checkpoint

***Review***

In this section, we considered the types of feedback the atmosphere and the earth might offer in responses to changes in radiative forcing. Longwave radiative fluxes are important feedback mechanisms, as are changes in albedo, cloud cover, soil moisture, and $CO_2$ storage in the oceans. This variety of feedback mechanisms contributes to the complexity of the global warming question.

***Questions***

1. Give examples of positive and negative feedback.
2. How can snow cover create a positive feedback loop?
3. What role do the oceans play in the greenhouse effect?

# Computer Models of Global Warming

In the previous section, you saw the complexity inherent in any attempt to determine the atmosphere's response to the stimulus of an enhanced greenhouse effect. This complexity is due in large part to the many forms of feedback between radiative forcing and the atmosphere's response. These feedback loops make it a gigantic and tedious chore to calculate the long-term effect of changes in radiative forcing. As you might expect, many atmospheric scientists rely on computers to make such calculations. The computer has become a central tool in global warming research.

## Computer Models

How does the computer proceed? Typically, computations follow a program called a computer model. Many current atmospheric computer models, known as **general circulation models,** or **GCMs,** are mathematical descriptions of atmospheric conditions and the natural laws that dictate those conditions. The "conditions" are simply weather data such as wind, temperature, air pressure, and humidity specified for a number of points across the globe at various altitudes in the atmosphere. The "natural laws" are expressions such as the gas law, radiation laws such as the Wien law and the Stefan-Boltzmann formula, given in terms of variables such as pressure, temperature, and wind.

The goal of GCMs used in global warming studies is not to replicate or forecast actual weather conditions (although weather forecasters use GCMs for this purpose). Instead, climate researchers hope their models will exhibit overall behavior typical of the atmosphere in general or average terms. For example, a good model should produce sequences of air temperatures and pressures, and weather systems that are typical of what is observed; but unlike the weather forecasting models, their output need not correspond to any actually observed individual weather situation. In other words, the models make simulated weather that looks statistically like the real thing.

A typical (highly simplified) climate GCM might proceed as follows:

1. Starting weather conditions are specified for each geographical point at each atmospheric level in the model. The model's running time, t, is set to $t = 0$.
2. The program moves time forward by a period of perhaps 10 minutes, known as one **"time step."** A time step is a fixed interval by which a program advances time in a model: new time = old time + 10 minutes.
3. The program "visits" each geographical point in the model and calculates the 10-minute energy fluxes over a one time-step period (10 minutes, here) at each point; it also calculates the 10-minute changes in variables such as temperature, air pressure, and winds at each point.
4. The program takes these new, calculated conditions and jumps back to step 2, advancing time by another 10 minutes and then onto 3, recalculating conditions at each point.

Then once again, it returns to step 2 and so on and so on. The model continues looping through steps 2 and 3, advancing in simulated time, until it reaches the end time specified in the program. Typically, model termination time is 30 to 100 years after t = 0.

"One hundred years?!" you must be asking. "Can we afford to wait so long?" Remember that the program is a model, and therefore, all conditions are simulated, including time. The powerful supercomputers on which GCMs are run can make all the computations involved for a simulated 10-minute time step in just a fraction of a second. Thus, the computer model may take "only" a few hours of continuous running to reach its 100-year conclusion.

## What the Models Say about Global Warming

Researchers at more than a dozen facilities in the United States, Canada, Germany, Russia, Great Britain, and other countries are using computer GCMs to explore global warming due to the greenhouse effect. A common approach is for researchers to run a simulation as described above, using current values of $CO_2$ and other greenhouse gases. Then, they run the program a second time from the same starting conditions *except with double the current atmospheric $CO_2$ content*. By changing only the $CO_2$ amount between runs, the researchers can see the impact of increased $CO_2$ on the computer-simulated atmosphere.

What do such experiments tell us about global warming? Figure 4.11 shows one of the Canadian Climate Centre's predictions of the atmosphere's response to a doubling of $CO_2$. The model predicts warming of the mean surface air temperature at all locations across the earth in both winter and summer. The average warming predicted by this model comes out to 3.5°C (6.3°F). Notice, however, the great variations in the amount of warming predicted. The polar regions suffer the greatest increase during winter months: that temperature increase may be as much as 12°C (22°F) or more.

How do these results compare with those of other researchers? Although the different models do not give identical results, they present many points of similarity. For example, all models predict that a doubling of $CO_2$ will lead to a warming of the troposphere and lower stratosphere, with surface

December – February

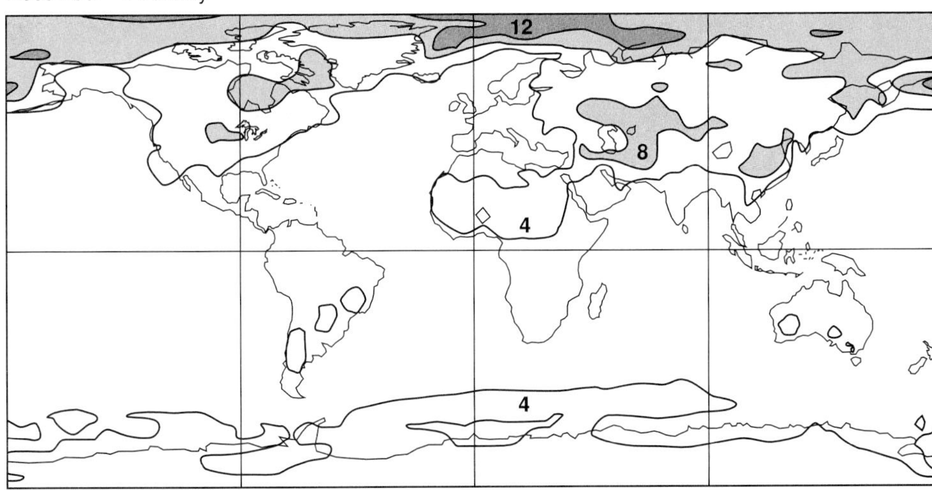

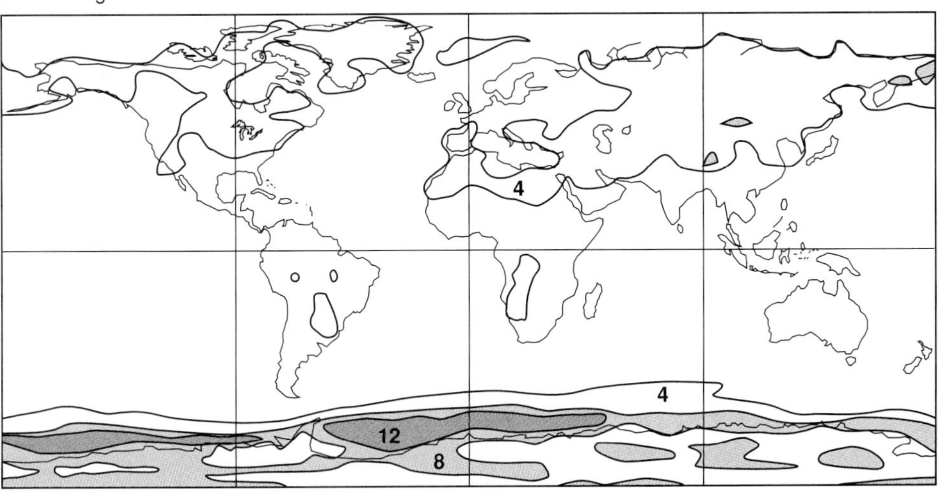


Key

| Color | Temperature Increase |
|---|---|
| White | 0-4°C |
| Yellow | 4-8°C |
| Orange | 8-12°C |
| Red | Over 12°C |

Figure 4.11 Predictions of global warming due to a doubling of atmospheric $Co_2$, according to the Canadian Climate Centre's general circulation model. Where and when are the greatest changes predicted to occur?

temperature increases in the range of 1.5°C to 5.0°C (2.7°F to 9.0°F). The models tend to show the greatest warming occurring in polar regions during winter months and the least in the tropics. The models also generally predict that cooling will occur in the upper stratosphere.

## How Trustworthy Are the Models?

Should we trust these computer simulations? Just how reliable are they? These are important questions; you should maintain a level of skepticism about all scientific evidence, and computer-simulated calculations certainly are no exception. To the unwary, computer output can give a false sense of security. While the computer's calculations may be error-free, the formulas it evaluates generally are idealizations or approximations of real-world processes and therefore do not describe the actual world with perfect accuracy or completeness.

What approximations do the computer modelers employ? The list is long. Some examples found in one or more of the current GCMs include simplifying the earth's geography (see Figure 4.12); representing the entire atmosphere by an assembly of only a few thousand data points; not accurately representing seasonal changes in sunlight or feedback due to changes in albedo, clouds, ocean temperature, vegetation, or surface moisture; and so on. As you can imagine, a model that fails to represent these and other influences is bound to fall short of perfection, and we should regard its pronouncements with a healthy skepticism.

On the other hand, such models give researchers a way of performing experiments, such as $CO_2$ doubling, that could be dangerous, even catastrophic, to try in the actual atmosphere. Furthermore, the atmosphere is simply too large and responds too slowly for climate experimentation. In addition, because you cannot return the actual atmosphere to the same condition when an experiment has ended, you can't run the same experiment on it repeatedly, making systematic changes and observing the changes that result. Thus, despite their limitations, computer models play an important role in advancing our understanding of the atmosphere's behavior.

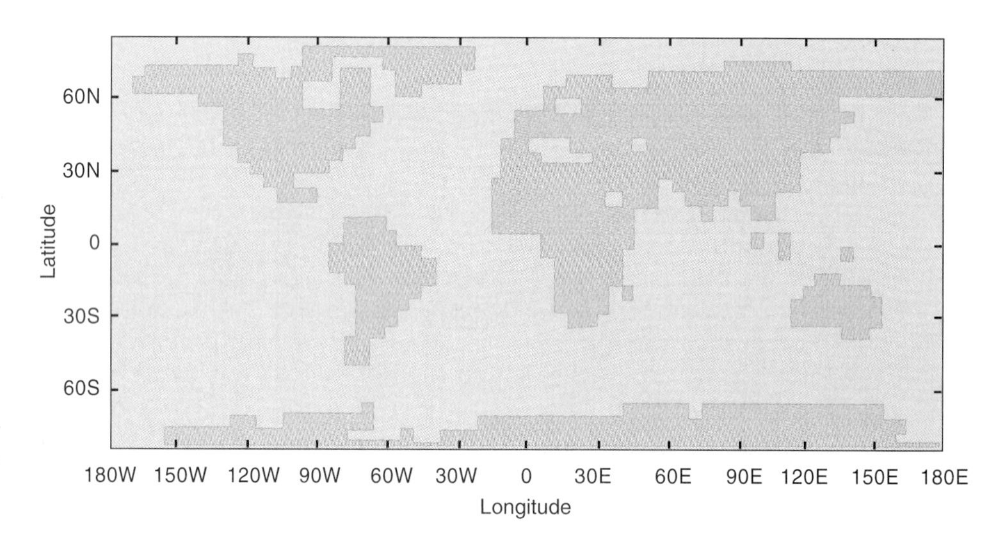


Figure 4.12 General circulation models operate on highly simplified geographical data, as indicated by this typical GCM "map" of the earth's land and oceans.

## Checkpoint

***Review***

General circulation models (GCMs) are computer-based simulations of weather conditions. In global warming research, they are used to simulate typical day-to-day weather conditions that might occur for a period of many years. By running the models several times under different starting conditions, researchers can see the effects of certain variables (such as atmospheric $CO_2$ concentration) on the model's projected weather conditions.

***Questions***

1. Name two advantages of using a computer model for atmospheric experimentation.
2. Name two sources of inaccuracy or error in GCM predictions.
3. What do GCMs predict for the atmosphere's response to a doubling of atmospheric $CO_2$?

# The Temperature Record: What Does It Say?

## Global Mean Temperatures

### *The Data*

How could you measure the thickness of one page of this textbook using an ordinary ruler? Since one page has a thickness of roughly 0.01 centimeters (less than one one-hundredth of an inch), the task would be a daunting one. However, if you measure the combined thickness of 100 pages, say, and then divide your answer by 100, you can find the average thickness of a single page.

What does this have to do with global warming? Scientists expect that in yearly terms, any rate of temperature change is likely to be very small: less than 0.01°C per year. To measure such a small rate, it is helpful to look for changes over long periods of time, such as 100 years or more. In this section, you will be looking for trends over dozens or hundreds of years, as the record permits, to determine what may be happening to the earth's temperature.

Which stations' temperature records would you choose to examine? You might choose a single station that boasts a long history of observations. But the temperature record at a single station can be very different from that at another and not representative of overall global conditions. It would be more reasonable to build a temperature data set using data from every station. However, observing stations are not evenly distributed across the earth's surface. If you count each station's data equally, regions with a high density of stations, such as North America and Europe, figure far too strongly in the average, while the temperature over earth's oceans, which cover 70% of the planet, is barely represented. Another problem derives from the fact that the quality of the meteorological observing record varies with location and time.

Thus, calculating a mean global temperature is more complicated than simply adding a lot of numbers and then dividing. Researchers attempting to determine the mean global temperature must deal with these and other problems.

A number of investigators have confronted these problems and have constructed tables and graphs of the long-term global temperature record. The results of three such groups of researchers are combined into the graphs in Figure 4.13.

The temperatures in Figure 4.13 are based on "near surface" readings on land, as measured typically in instrument shelters such as described in Chapter 1. Ocean values are sea surface temperatures. Observers on many oceangoing vessels regularly measure the temperature in the top meter or so of the sea. It is these readings, and not ocean air temperatures, that you see in Figure 4.13. Altogether, the graphs represent several tens of millions of individual temperature measurements.

### *Interpreting the Data*

Notice that the vertical axes in Figure 4.13 are not temperature but temperature anomaly. This is the departure from a "baseline"

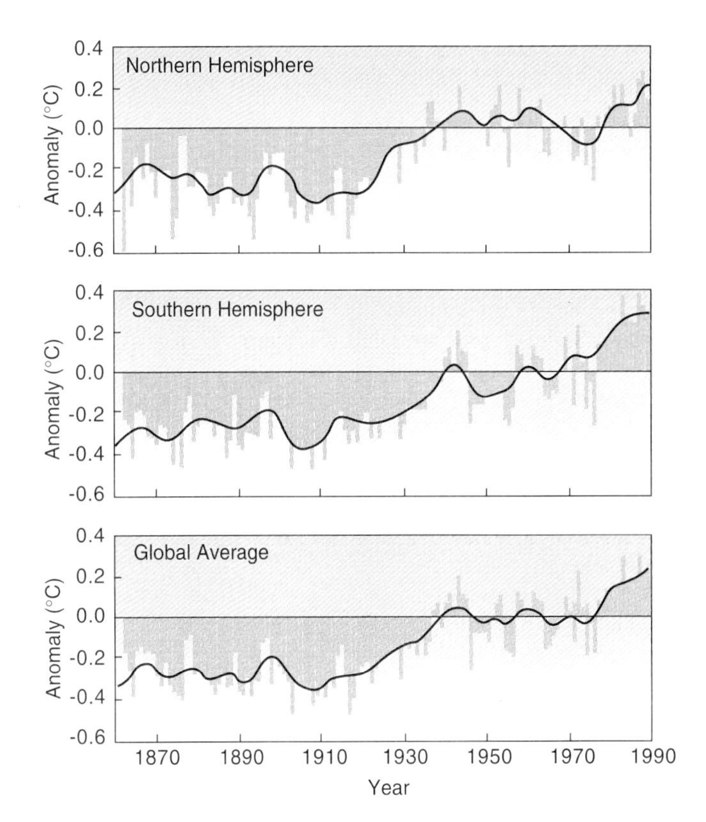


Figure 4.13 Mean global surface temperatures since 1860. In each graph, the "zero" temperature is the mean for the years 1951 to 1980.

temperature, which was chosen as the average temperature for the period 1951–1980. Thus, the graph shows annual global temperatures ranging from 0.6°C (1.1°F) colder to 0.4°C (0.7°F) warmer than the baseline. The vertical bars show each year's departure from the baseline. Note also that the data record in Figure 4.13 begins in the year 1861; the researchers felt that global temperature data before that time were too sparse and too unreliable to be useful.

The curved lines in Figure 4.13 are constructed to eliminate some of the year-to-year variability, making it easier to see longer-term trends. The curve of a type called a **running mean,** is constructed by averaging the values for several years on each side of each year on the graph. Thus, the 1980 position of the curve is determined by averaging the values for 1978 through 1982; the 1981 value represents the averages of 1979 through 1983, and so forth.

What do you see in the graphs of Figure 4.13? Certainly, recent temperatures are warmer than those in the 19th century. However, the pattern is anything but straightforward. From 1861 until about 1920, there is no evident trend at all. Then the mean temperature seems to increase rather steadily to a mean departure of about 0.0°C, which persists from 1930 until 1970 or so. The late 1970s and the 1980s mark another period of rather sudden warming.

Note also that the curves in Figure 4.13 reveal a series of ripples that peak roughly every 11 years. It is believed that these ripples reflect solar cycles discussed earlier in this chapter.

Finally, if you examine the bars representing each year's temperature, you will notice a great deal of variation from one year to the next in the global temperature record. Scientists often use the terms **signal** and **noise** when discussing data like those in Figure 4.13. The signal is the message, the information contained in the data. In this case, since you are looking for signs of global warming, the signal would be a clear upward trend in the data with time. The noise is the natural variability of the data, which tends to mask any signal that might be present. An example of noise would be the year-to-year variation in the temperature values shown by the bars. Then the question becomes, "Is there a clear signal in the data of Figure 4.13, or are the variations you see only noise?"

Statistical methods can help resolve the signal-noise puzzle. Statistical calculations on the Figure 4.13 data reveal a warming trend from 1881 to 1989 of 0.45°C (0.8°F), a signal at least twice as large as the year-to-year noise; therefore, the likelihood that the trend is real, and not the result of random noise, is greater than 98%. The statistics also give a "probable error" of ± 0.15°C in the 0.45°C value. This means that although the warming likely is not exactly 0.45°C, the value probably lies between 0.30°C and 0.60°C.

A temperature rise of 0.45°C over 100 years may not sound like an alarming proposition. More alarming, however, is the prospect that this warming trend might continue or even increase in rate, a prospect we will examine shortly.

## Sources of Error in the Temperature Data

Statistical methods have given us 98% confidence that the rise in global temperatures is real. However, this confidence rests on the assumption that the original temperature data are trustworthy. Although measuring air or sea surface temperature appears to be a straightforward task, many complications arise when you attempt to follow a record of observations over the course of a century or more.

For example, the weather observing sites in many cities have changed location, often several times, over the period of record. Is the new site equivalent to the old one? In many locations, the environment surrounding the observing site has changed due to urbanization. How does the replacement of nearby woodland with pavement and buildings affect the data? Many weather stations are located at airports. Is airport weather typical of conditions elsewhere?

Furthermore, instrumentation, observers, and observing procedures change over the years. Generally, the intent of changes in instruments and procedure is to achieve greater accuracy in the data record. The implication is that earlier methods gave different, and probably less accurate, results. Table 4.1 gives examples of the concerns that arise when atmospheric scientists begin to scrutinize the long-term weather record at a station. Although considerable effort is made to adjust for these effects in published temperature data sets, some inaccuracy no doubt remains.

*Table 4.1* **Some Problems in Comparing Temperature Records over Many Years at a Single Site**

Has the same thermometer been used throughout the period? If not, how do the different instruments compare?
Has the thermometer's mounting and shelter changed?
Has the thermometer's location changed?
Has the nature of the surrounding environment changed during the period?
Has the time of observation (e.g., 6:00 A.M. or noon) changed?
Has observing procedure changed?
Has the observer changed?

So what's a person to do? Do you ignore the concerns about the data's trustworthiness and accept the statistical assurance that the 0.45°C trend exists and is real? Or do you disbelieve the trend because the data record may be flawed? This is the sort of question scientists confront frequently. In scientific journals and at meetings, debates rage over such questions. After all, research means exploring the unknown, and your knowledge of the unknown necessarily will be incomplete. Sometimes decisions or commitments can be postponed until the evidence is absolutely clear, but other times decisions cannot wait. Therefore, scientists sometimes have to act or commit themselves in the face of incomplete or conflicting evidence. In such circumstances, no sequence of steps exists to guide you to the correct choice. The decision ultimately is a personal one: you carefully study all the evidence, seeking the option that has the ring of truth. Then, your communications on the topic to other scientists and to the nonscientific community should indicate your level of uncertainty in your position.

## Other Temperature Data

One way of resolving doubt about the 0.45°C global warming trend is to look at other data sources. For example, do regional and local temperature records by and large reflect the trend? Figure 4.14 gives two examples.

Figure 4.14A shows mean January temperatures measured at Blue Hill Observatory, Massachusetts. Blue Hill lies several miles southwest of Boston, at an elevation of several hundred feet above the surrounding land. Weather observers have maintained a continuous record at the site since 1845. The region has urbanized considerably over the past 150 years, although most urbanization has occurred to the northeast, in Boston. The station's location atop a rugged hill and its altitude above the surrounding terrain probably protect it to some extent from urbanization effects. Urbanization, as you will see presently, often causes local warming of air temperatures. The data in Figure 4.14A show a 10-year running mean, which removes year-to-year variations in favor of longer trends.

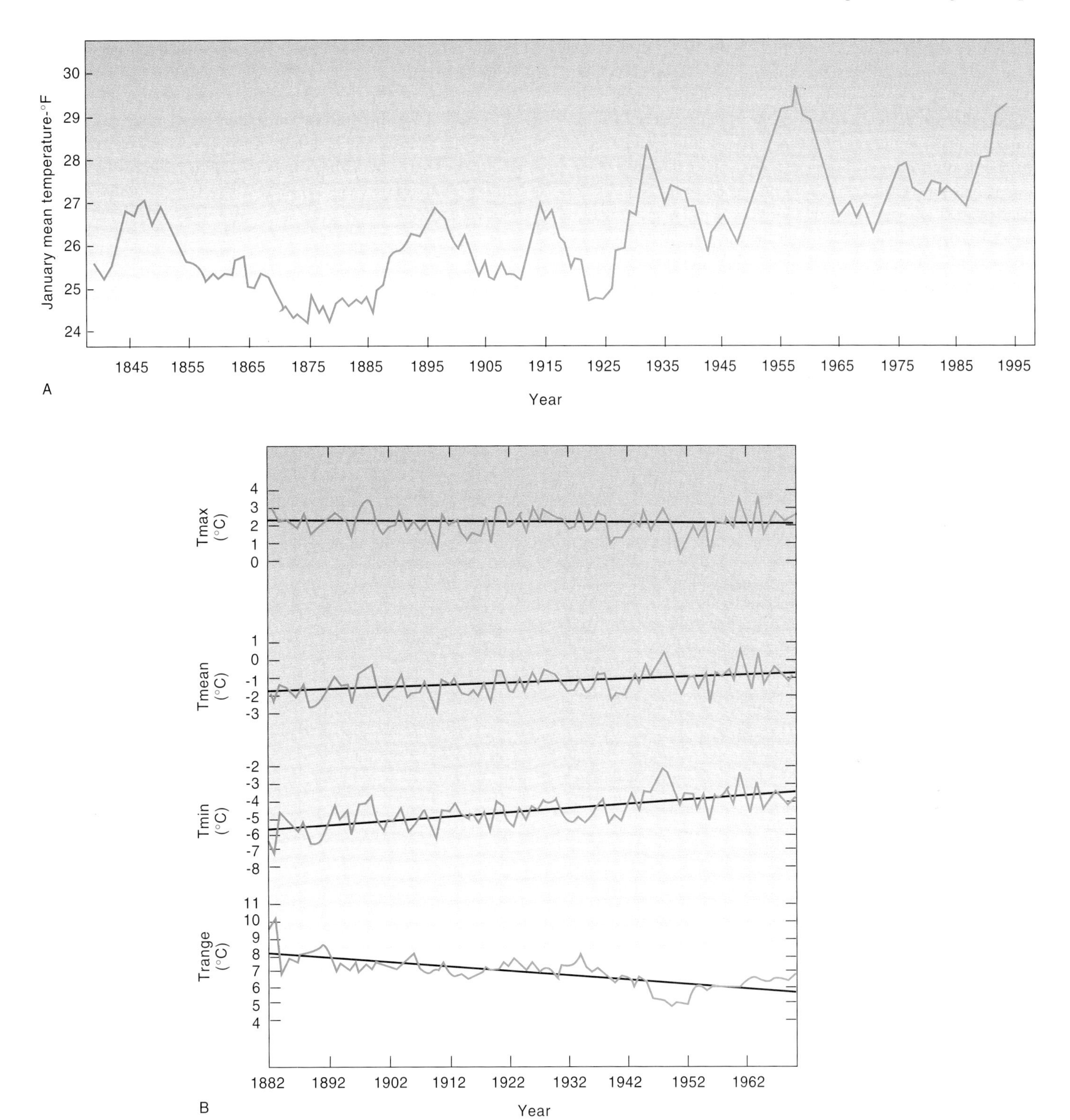


Figure 4.14 Long-term temperature records for two sites: (*A*) Blue Hill, Massachusetts, 185 meters above sea level, and (*B*) Pic du Midi, 2863 meters above sea level. What trends do you observe at the two sites? Are they mutually consistent?

In Figure 4.14B are data from the Pic du Midi Observatory high in the French Pyrenees mountains, as analyzed by A. Bucher and J. Dessens of the observatory staff. At an altitude of 2862 meters (9387 feet) above sea level in a remote location, the observatory's temperature record should be quite free of local urbanization effects. Statistical analysis of the mean temperature gives an increase through the period of 0.83°C (1.5°F). Notice that the daily high and low temperatures have changed in opposite directions. The mean daily maximum has actually decreased slightly over the period, while the daily minimum has risen by more that 2°C (3.6°F). These changes, the researchers suggest, may be caused by an increase of cloudiness (of 15%, which was observed over the period). The increased cloud cover would reduce the amount of incoming solar energy, causing lower maximum temperatures. At night, however, the clouds would tend to trap longwave radiation, causing higher minimum temperatures.

Of course, the two examples we have presented in this section do not by themselves constitute proof of global warming. However, they are typical of many local records from all parts of the globe that show similar trends. Taken collectively, these data do lend support to the validity of the 0.45°C global trend discussed earlier.

## Proxy Data of Global Warming

Figure 4.15A suggests that you can find clues to past temperatures on earth in a number of places besides meteorological observations. Many natural processes and features, such as length of the glacier pictured in the figure, are affected by temperature

A

B

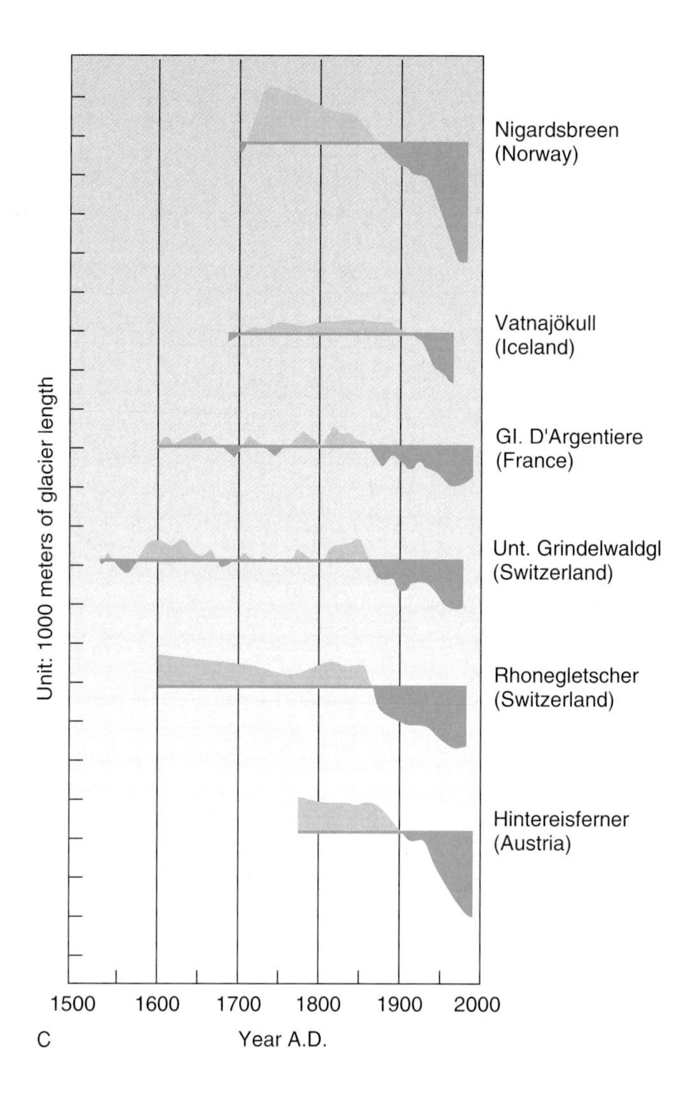


Figure 4.15 (*A*) & (*B*) These photos of the Alpine Glacier de la Brenva clearly show its retreat over the period from 1767 to 1966. Such evidence is proxy data of global temperature trends. (*C*) Variation in the length of six European glaciers. European glaciers are shown because they have been observed for longer periods of time than glaciers in other locales. The horizontal line running through the graph for each glacier represents the average length of that glacier. The glacier's actual length is represented by the wavy line, shaded in dark for periods when the length was shorter than average.

Figure 4.16 Masses of ice such as the one pictured here break loose from the Antarctic ice sheet and often meander equatorward.

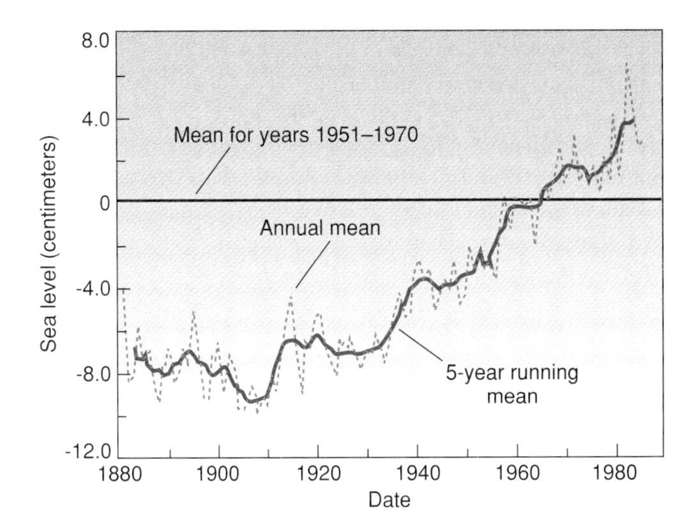


Figure 4.17 Global mean sea level variations since 1880, according to an analysis by T. P. Barnett. The "0-level" is the mean for the years 1951–1970. The dashed line is the annual mean, and the solid line is a 5-year running mean.

and are examples of **proxy data.** Proxy data are data from climate-sensitive phenomena such as glacier lengths, pollen deposits, and tree ring spacings that can be used to reconstruct estimates of past climate conditions. Proxy data provide less accurate and less broad-based data than a global observing network, of course; on the other hand, they can greatly extend our knowledge of earth's thermal history to times and places where no instrumental record exists.

### *Glacier Length Changes*

In Figure 4.15C, you can see the change in length of six European glaciers over the past 200 to 450 years. Notice that all six glaciers have shrunk in length over the period of observation.

What causes a glacier to become shorter? The answer, like so many in the environmental sciences, is complex. The main factor seems to be summer air temperatures, as you might expect: warmer temperatures cause accelerated melting. However, greater accumulations of snowfall during winter months can cancel or reverse summer melting. Thus, like radiative forcing, which is the difference between solar income and longwave outflow, "glacial forcing" is the difference between winter growth and summer melting. Most researchers agree, however, that the decrease in glacier length shown in Figure 4.15C is due to warmer summer temperatures. Scientists studying glaciers in other parts of the world find changes similar to those in Figures 4.15A and B.

Notice that the major changes in length in Figure 4.15C occurred in the mid-20th century and that for the past decade or so, several glaciers have remained essentially unchanged in length or even have begun to grow once more.

### *Sea-Level Changes*

Consider the huge floating ice sheet in Figure 4.16. If all that ice melted, what do you think would happen to the height of sea level?

The answer is, "Nothing." Because the ice is floating in the ocean, it is already contributing its mass and volume to sea level. (Check this assertion yourself with a glass of water and ice.) Thus, melting of floating ice in the Arctic or Antarctic Oceans would have no impact on global sea-level values. However, most of the world's ice rests not on the ocean surface but on land, particularly on Antarctica and Greenland. If warmer temperatures were to cause some or all of this ice to melt and run off to the oceans, sea level would rise in response. Furthermore, a warming of sea water would cause it to expand, thereby occupying greater volume and making an additional contribution to the rise in sea level.

These lines of reasoning have led scientists to treat records of sea-level depth as proxy data for past temperatures. Sea-level depth records are collected by the Permanent Service for Mean Sea Level in the United Kingdom. Like worldwide temperature data, the sea-level data are subject to various errors and incompletenesses, and great judgment must be exercised in deciding just which data to use and what corrections to make. Figure 4.17 shows you the results of T. P. Barnett's research on global sea-level change. Based on this figure, how would you describe the record of sea-level change since 1880?

### *Warming of Lakes*

Several researchers have reported recently on changes in lake temperatures. Andrew Solow has collected data showing the number of times Lake Constanz in Europe has frozen over each century for the past 1000 years—an extraordinarily long data record! Meanwhile, Howard Hanson, Claire Hanson, and Brenda Yoo have analyzed the date of the ice's disappearance each spring for the past 35 years at a number of locations on the shores of the Great Lakes. Both the Lake Constanz and the Great Lakes data are consistent with a trend toward milder temperatures over the past few decades.

## Is the Greenhouse Effect to Blame?

Data such as those you have examined in the last few pages have convinced many (but not all) scientists that earth's climate indeed has experienced a measurable warming over the past century or so.

### *Correlation and Causality*

The question "Is the greenhouse effect responsible for the warming?" takes you into a topic on which there is considerable discussion and disagreement. Not every scientist will agree with what we say here, and you may not agree with us, either.

Having said that, our answer to the question of greenhouse gases' role in global warming is, "We cannot conclude beyond a reasonable doubt that the observed buildup of greenhouse gases has caused the observed global warming." By themselves, the statements "The atmosphere has warmed over the past century" and "Greenhouse gas concentrations in the atmosphere have increased over the past century" do not constitute proof that one observed condition caused the other. This issue of causation is an important one and bears closer scrutiny. Therefore, we offer another example: If you conducted a study of elementary students' heights, you would likely find that a student's grade in elementary school and his or her height exhibit a relationship like that shown in Figure 4.18: that is, older students tend to be taller. Statisticians would say that grade and height are **correlated**; that is, a trend exists that relates values of one variable (school grade) and values of the other (height). Although the correlation is not exact—for any given grade, a variety of student heights is observed—the trend is evident. You can even use the relation in Figure 4.18 to predict a child's approximate height if you know his or her grade, or vice versa.

Although school grade and height are correlated, they are not **linked causally**: that is, they are not connected by a cause and effect relationship. (Notice the word is *causal,* not *casual.*) School grade and student height are each determined by a number of other factors. The situation is the same with graphs of global warming and greenhouse gas buildup: the fact that both trends occur simultaneously is no proof they are linked causally. The important point here is that correlation does not imply a causal link.

You may object that our comparison is flawed; whereas there is no reason to expect a causal link between a student's height and his or her school grade, you know that greenhouse gas buildup *does* change radiative forcing of the atmosphere, leading to warming of the troposphere. Doesn't awareness of this causal link between greenhouse gas increase and an enhanced greenhouse effect allow us to state that the greenhouse buildup has caused the observed 0.45°C global warming? We still argue that it does not. We cannot rule out the hypothesis that greenhouse gas buildup is causing the warming, but we cannot yet accept that hypothesis beyond a reasonable doubt, either.

### *Long-Period Temperature Cycles*

Our focus in this chapter has been on relatively current temperature trends, generally those over the past 100 years or so. (In a later chapter, you will explore in more depth the topic of climate change over earth's history.) However, a little deeper historical perspective from earth's past climates may be useful at this point.

Figure 4.19 shows a reconstruction of global temperatures over the past 1000 years, based on a variety of proxy data. Notice that temperatures have varied over a range of nearly 1.5°C (2.7°F) during the period and were at least as warm around 1200 A.D. as they are now. The temperature increases that occurred from 1000 to 1200 A.D. and from 1650 to 1700 A.D. were as great or greater than the present episode, and they occurred without benefit of an enhanced greenhouse effect. These data suggest that the current observed warming could be the expression of some natural, long-term cycle, rather than the greenhouse gas buildup.

### *Looking for a Greenhouse Fingerprint*

Most computer climate models agree on several features of the atmosphere's response to a buildup of greenhouse gases. Like identifying a fingerprint at the scene of the crime, many scientists feel that detecting these features in the observational record would support the contention that greenhouse warming is present.

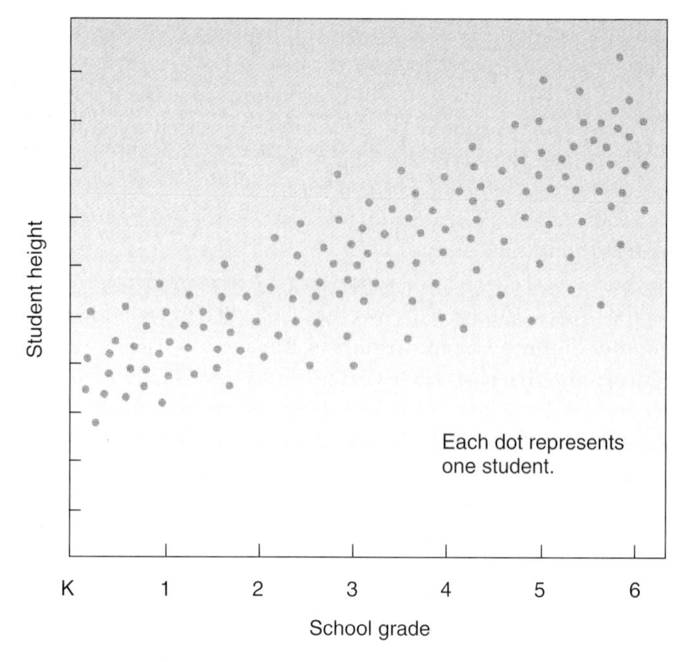


**Figure 4.18** A student's height and her school grade are correlated, as you can see from this scatter diagram. However, school grade does not cause a student to be a certain height, nor does a student's height cause her to be in a certain grade. The data are correlated but not causally linked.

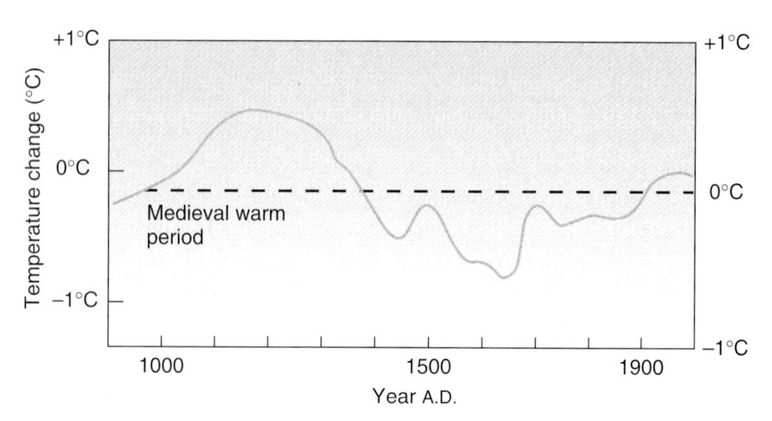


**Figure 4.19** An estimate of global temperatures since A.D. 900. Because nearly all of this record is based on proxy data, you should regard it as an indication, not an exact record. Nonetheless, note the significant changes that presumably have occurred in worldwide temperature in past centuries.

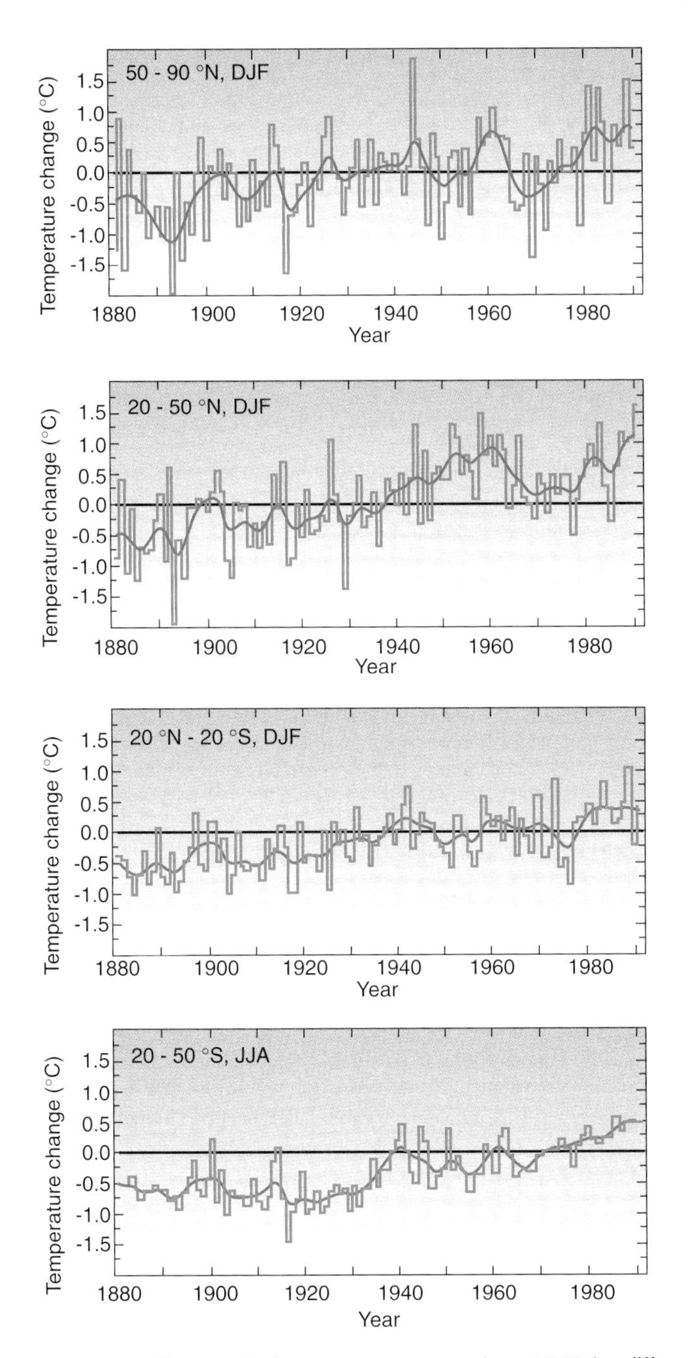


**Figure 4.20** Observed winter temperatures since 1880 for different latitude belts. Which zone exhibits the greatest change?

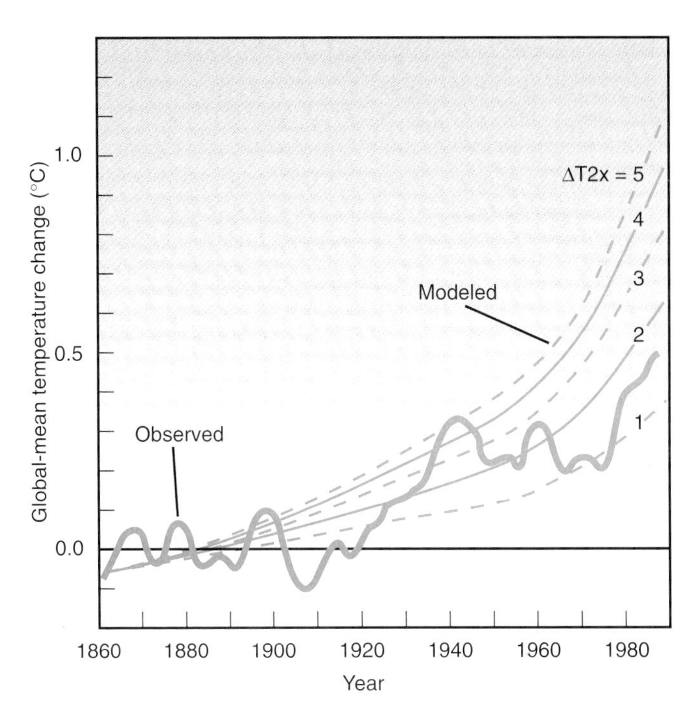


**Figure 4.21** Observed global temperature change and computer model predictions for temperature changes due to a doubling of $CO_2$. The "1" curve shows the predicted rate of temperature rise if a $CO_2$ doubling causes a 1°C temperature increase; the "2" curve shows the predicted rate if doubling causes a 2°C rise, and so on. The observed data seem to be in the 1–2°C range, but there is considerable noise in the data.

One such feature is the fact that wintertime polar temperatures are expected to show the greatest amount of warming in response to a greenhouse gas buildup. Figure 4.20 depicts observed winter temperature data organized according to latitude. As you can see, the region from 50°N to 90°N shows the most warming, in agreement with the models' predictions. Another feature predicted by most GCMs is a cooling of the upper stratosphere due to changes in the flux of longwave radiation. Observations in this region, while sparse, do seem to suggest a cooling trend, in agreement with greenhouse warming.

Finally, computer predictions indicate that greenhouse warming should increase with time and should respond to a $CO_2$ doubling by increasing in the range of 1 to 5°C (1.8 to 9°F). Figure 4.21 compares observed global temperatures with these computer predictions. The smooth curves show different rates of warming according to the atmosphere's sensitivity to greenhouse gas buildup. Clearly, the 1°C and 2°C curves follow the observed data best. Notice, however, that the bulk of the warming actually occurred from roughly 1910 to 1940, whereas the models predict change to be greatest in the most recent years.

Thus, matching GCM predictions with observed temperature patterns gives mixed results. Some features are present as predicted, others are not.

### *Our Opinion*

At the time of writing, our opinion about global warming is as follows. We accept the reality of the increase of greenhouse gases over the past century, and we agree with the assertion by a group of scientists comprising the International Panel on Climate Change (IPCC) that the global temperature record indicates a warming of roughly 0.45°C during that time. We believe it is likely, but so far unproved, that part or all of that increase has been caused by the buildup of greenhouse gases in the atmosphere. We also recognize, however, that the atmosphere's natural, long-term variability is capable of producing changes on the scale observed. We also note some discrepancies between predictions of computer models and conditions actually observed.

Finally, we have a healthy respect for the complexities of feedback in the global warming problem. For all these reasons, we are unable to state that an enhanced greenhouse effect is the cause of the warming seen in data such as those back in Figure 4.13.

This is not to say, however, that we are indifferent to environmental problems. The fact that we feel the link between global warming to date and greenhouse gas buildup has yet to be proved does not mean we think we can ignore the problem. On the contrary, we regard the cause of global warming as one of the most crucial questions facing us today. Our inability even to recognize and quantify the extent of global warming in the data record acts to heighten, not diminish, our concern. Therefore, we conclude this chapter with a discussion of possible impacts of global warming and ways humanity can respond to the threat.

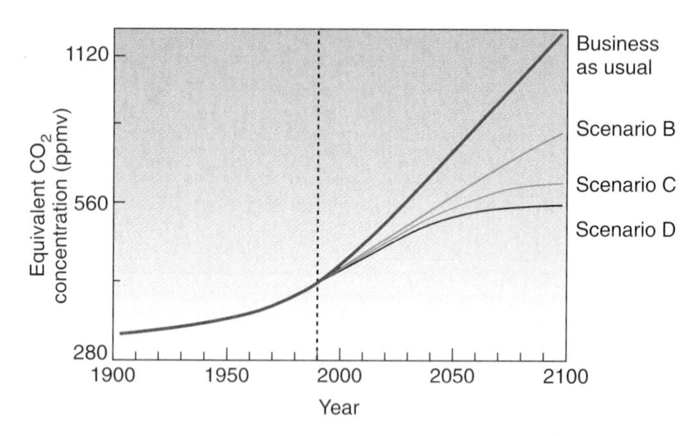


Figure 4.22 Predictions of changes in greenhouse gas concentrations according to four different scenarios. All greenhouse gases are included in the diagram; their impact is translated into "equivalent $CO_2$ concentration." Briefly, the scenarios, as defined by IPCC scientists, are: (*A*) Business as usual: "The energy supply is coal-intensive and on the demand side only modest efficiency increases are achieved. Carbon monoxide controls are modest, deforestation continues until the tropical forests are depleted and agricultural emissions of methane and nitrous oxide are uncontrolled." (*B*) "The energy supply mix shifts towards lower carbon fuels, notably natural gas. Large efficiency increases are achieved. CO controls are stringent; deforestation is reversed. . . ." (*C*) "A shift towards renewable energy sources, such as solar and nuclear energy takes place in the second half of the 21st century. CFCs are now phased out and agricultural emissions limited." (*D*) "A shift to renewables and nuclear in the first half of the 21st century reduces the emissions of $CO_2$ . . . $CO_2$ levels are reduced to 50% of 1985 levels by 2050."

***Review***

Over the past century or so, temperatures at individual stations as well as global mean temperatures have shown a small but significant increase. Proxy data, in the form of records such as lengths of glaciers and sea-level changes, also indicate a warming trend. Atmospheric greenhouse gas concentrations have increased during this same period. This correlation between increases in temperatures and greenhouse gas buildups does not mean necessarily that the greenhouse gas buildup has caused the temperature rise, however. Over past centuries, the earth's temperatures have varied considerably; thus it is possible that the present warming is due, in part at least, to some long-term natural cycle.

***Questions***

1. Has earth's temperature record (Figure 4.19) shown a steady trend, or has it changed irregularly?
2. What is meant by the term *proxy data?*
3. Give an example of two variables that are correlated but not causally linked. Then name two that are both correlated and causally linked.
4. What is your opinion about the relation between global warming and an enhanced greenhouse effect?

# Global Warming and the Future

## How Much Warming, If Any, Can We Expect?

The amount of warming we experience in future years depends on how nature and humans behave. As you saw in Figure 4.19, global temperatures have undergone major changes in the past due entirely to natural causes. At present, we simply cannot tell how natural variability will influence future temperatures. At least some of the natural causes, such as volcanic eruptions and meteor impacts, are quite unpredictable, nor is there much we can do to influence their course. Furthermore, global warming may be subject to other significant but as yet undiscovered natural factors. In view of these facts, we will focus on the human factors, which—in theory, at least—are controllable, if not predictable.

What is the most important anthropogenic agent of future temperature change? By now, you should be able to answer that question in your sleep: the buildup of greenhouse gases, particularly $CO_2$. How much, if any, will the temperature rise in response? This answer depends on whose computer climate model you believe and on how much greenhouse gas we add to the atmosphere.

You read earlier that different models predict anywhere from a 1.5°C to a 5.0°C (2.7 to 9.0°F) increase in global temperature in response to a doubling of atmospheric $CO_2$. Assuming the model builders are on the right track, the correct value lies somewhere within this range, and the uncertainty will diminish with time as the models improve. How long will it take for $CO_2$ doubling to occur? And how much will other greenhouse gases, such as methane and CFCs, add to $CO_2$'s contribution? Answers to these questions depend, of course, on how humans respond to the global warming threat. Figure 4.22 shows several possibilities: option A is the "business as usual" scenario, which means a continued, largely uncontrolled increase in greenhouse gas emis-

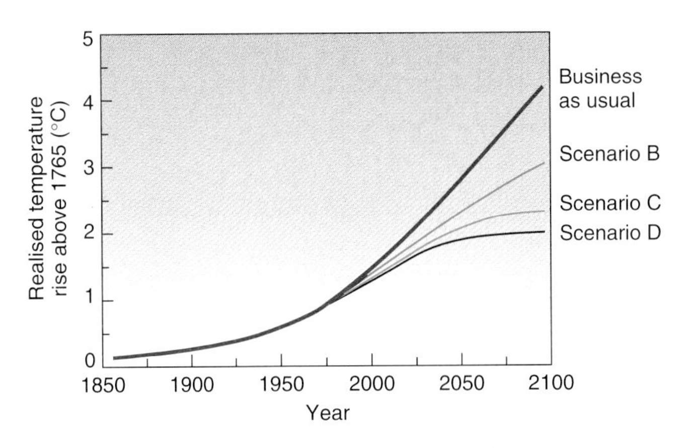


Figure 4.23 The projected global temperature increase as a result of following the four scenarios outlined in Figure 4.22. Note that all four scenarios project some warming.

Figure 4.24 The Antarctic ice sheet rests on the Antarctic landmass. If it were to melt or slide into the ocean, a large and potentially disastrous rise in sea level would follow.

sions; scenarios B, C, and D represent increasingly stringent controls on greenhouse gas emissions, as defined by the IPCC and described further in Figure 4.22. Adopting options B, C, or D requires increasingly drastic steps, such as an almost complete switch from coal and oil to natural gas, nuclear, or solar energy sources, in addition to controlling $CO_2$ emissions and reversing deforestation. Notice that Figure 4.22 shows the combined effect of all greenhouse gases (except water vapor) in terms of the equivalent $CO_2$ concentration.

You can see from Figure 4.22 that greenhouse gas concentrations are predicted to quadruple in the next century under the "business as usual" scenario. In fact, all four scenarios, even those requiring substantial reductions in greenhouse gas emissions, project a significant increase. In only scenario D, requiring a 50% reduction in $CO_2$ emissions by the year 2050, are greenhouse gas emissions truly controlled.

How does this increase translate into greater global warming? Figure 4.23 portrays the expected warming for the same four scenarios. Note that the "business as usual" scenario is expected to yield an additional warming of 3°C (5.4°F) by the end of the next century. The other scenarios suggest a warming of 1 to 2°C (1.8 to 3.6°F) from present values. To eliminate any and all further warming due to greenhouse gas buildup would require such drastic reductions and changes in energy use that implementation is very unlikely. A best estimate, therefore, is that over the next century, we are destined to experience at least a doubling of greenhouse gas concentrations and a warming of 1 to 4°C (1.8 to 7.2°F). Remember, however, that this statement is based on imperfect and incomplete information, and doubtlessly will be revised as we learn more through research.

## Impacts of Global Warming

A warmer earth can mean a different earth in many ways. Weather, geographical features, agriculture, and plant and animal distributions will all show the effect of a pronounced warming. In this section, you will consider a few examples of the impact of global warming.

### *Changes in the Weather*

When it comes to detailing specific regional weather changes in response to global warming, again we find we have a great deal still to learn. Reasoning suggests, and computer models tend to confirm, that greater warming in the polar regions than in the tropics could lead to a weakening of the synoptic scale highs and lows that transport tropical heat and water vapor poleward. Also, a warmer tropics suggests a change in the frequency of tropical storms such as hurricanes: some studies predict more such storms, others predict fewer. Presumably, the tracks followed by frontal systems and tropical storms in a warmer world would differ from present tracks, thereby affecting precipitation and wind patterns; however, the specifics of such changes are not known. As you can see, much important work awaits the future atmospheric scientist.

### *Changes in Sea Level*

You read in the last section that sea level is thought to be rising, possibly due to global warming. IPCC scientists' projections through the next century indicate a rise of from 30 to 65 centimeters (1 to 2 feet) above present levels, depending on the scenario followed. In some locations such as Bangladesh, where large numbers of people live essentially at sea level, even this apparently small increase could have disastrous effects, particularly at times of storminess or flooding. The IPCC scientists state, "The prospect of . . . an increase in the rate of sea level rise should be of major concern to many low lying coasts subject to permanent and temporary inundation, salt intrusion, cliff and beach erosion, and other deleterious effects."[1]

The Antarctic ice sheet (Figure 4.24) is a source of some uncertainty in sea-level projections. It is not expected that global warming will affect the ice sheet during the next century. If that prediction proves incorrect, however, and the ice sheet were to collapse and slide into the ocean, sea level could rise by as much as 6 meters. Such an event would flood hundreds of thousands of square miles of coastal land, including many major cities.

1. Houghton, J. T., et al., eds. *Climate Change: The IPCC Scientific Assessment.* (New York: Cambridge University Press, 1990), p. 279.

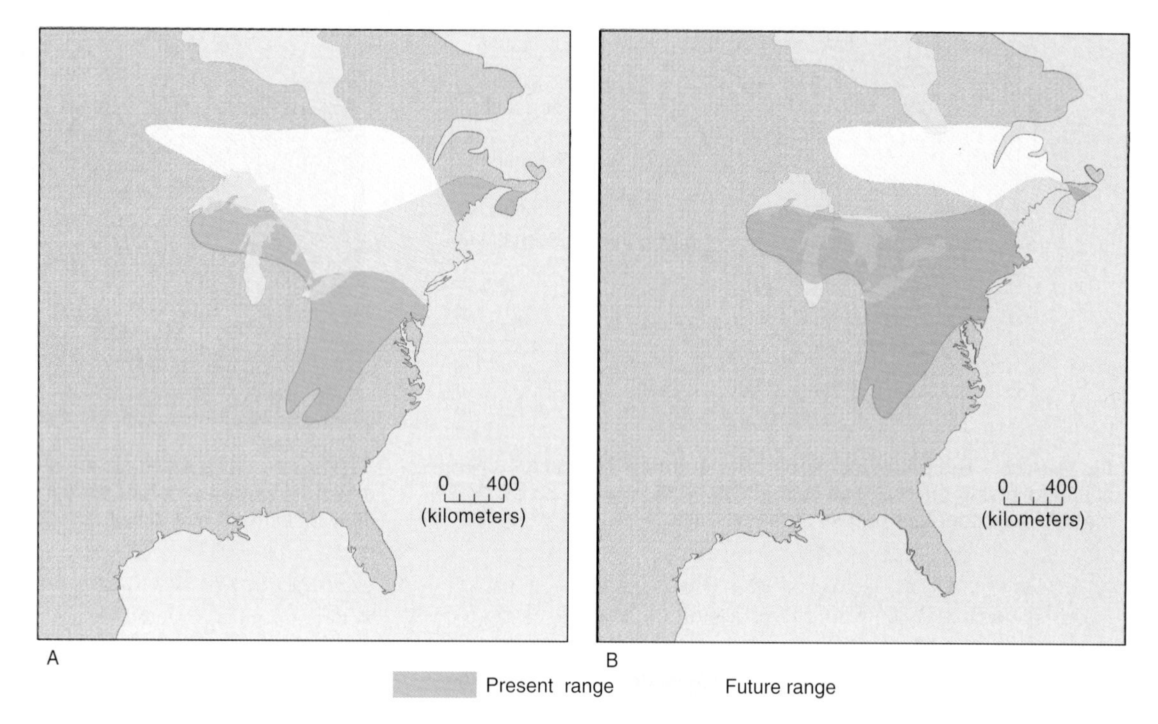


Figure 4.25 Present and future ranges of the eastern hemlock tree, according to two different GCM projections. The future ranges are in response to $CO_2$ doubling. Note that both models project a northward migration and shrinkage of the hemlock's range.

### *Changes in Plant and Animal Communities*

Global warming may affect plant and animal life in a number of ways. Several GCMs predict that higher temperatures will stimulate evaporation, which will diminish soil moisture. Such changes may combine to force the extinction of some species or cause them to migrate from one region to another. As an example, Figure 4.25 shows two different computer models' estimates of the change in range predicted for the eastern hemlock tree over the next 100 years if $CO_2$ were to double during that time. Note the northward migration and shrinkage of the range of this important forest tree.

Changes in ocean life may be particularly significant. Alteration of temperature, evaporation, and precipitation patterns could lead to changes in oceanic circulations that transport nutrients to marine organisms. As discussed earlier in this chapter, the resultant changes in marine populations could lead to a reduced ability of the oceans to act as a sink for atmospheric $CO_2$, thus establishing a positive feedback loop.

Global warming may have profound effects on our ability to feed ourselves. Agriculture is perhaps the single most important human activity to be influenced by weather and climate. Cynthia Rosenzweig, an agronomist at Columbia University in New York, has led a team of investigators who explored the question of crop yields in a warmer world. Figure 4.26, which shows one of the results of their work, indicates the change in crop yields in the year 2060 for a greenhouse-warmed climate, compared to yields in the same year with no climate change. Notice that in the great majority of countries, crop yields are expected to be reduced due to global warming. The results also suggest that the poorest countries, which can least afford to invest in irrigation and other means of climate control, are the ones facing the greatest potential losses in food production due to global warming.

## How Can We Respond to the Threat?

One possible response to the threat of global warming is, "We don't know if global warming is a real threat; until we know better, we should do nothing. It costs money to change, and we should not make the expense if there is no need to do so." Elements of this argument make some sense. It would be foolish to act in panic, on the assumption that global warming is certain and that its impact will be catastrophic. And, as stated earlier, our present understanding of the specific effects of global warming is sketchy at best; much research remains to be done.

On the other hand, to do nothing now may be courting disaster later. We can act in a variety of ways to lessen the probability or the severity of the problem. Two basic strategies are to reduce emissions of greenhouse gases into the atmosphere and to manage an enhanced greenhouse effect through technological countermeasures.

### *Reduction of Greenhouse Gas Emissions*

In 1992, representatives from 150 nations met in Rio de Janeiro, Brazil, to discuss global warming. The Rio conference attracted more heads of state than any event in human history, a sign of humanity's recognition of the problem's significance. Although in some ways the conference was a disappointment because it produced no specific commitments, nonetheless 150 nations signed a treaty agreeing to reduce greenhouse gas emissions.

How do you reduce greenhouse emissions? The most obvious answer—Stop using fossil fuels—is simple in theory but enormously difficult to implement for a variety of reasons. Developed and developing countries are accustomed to consuming more energy each year, not less. Furthermore, developing and undeveloped countries, in attempting to catch up economically with the developed world, expect to increase their fuel consump-

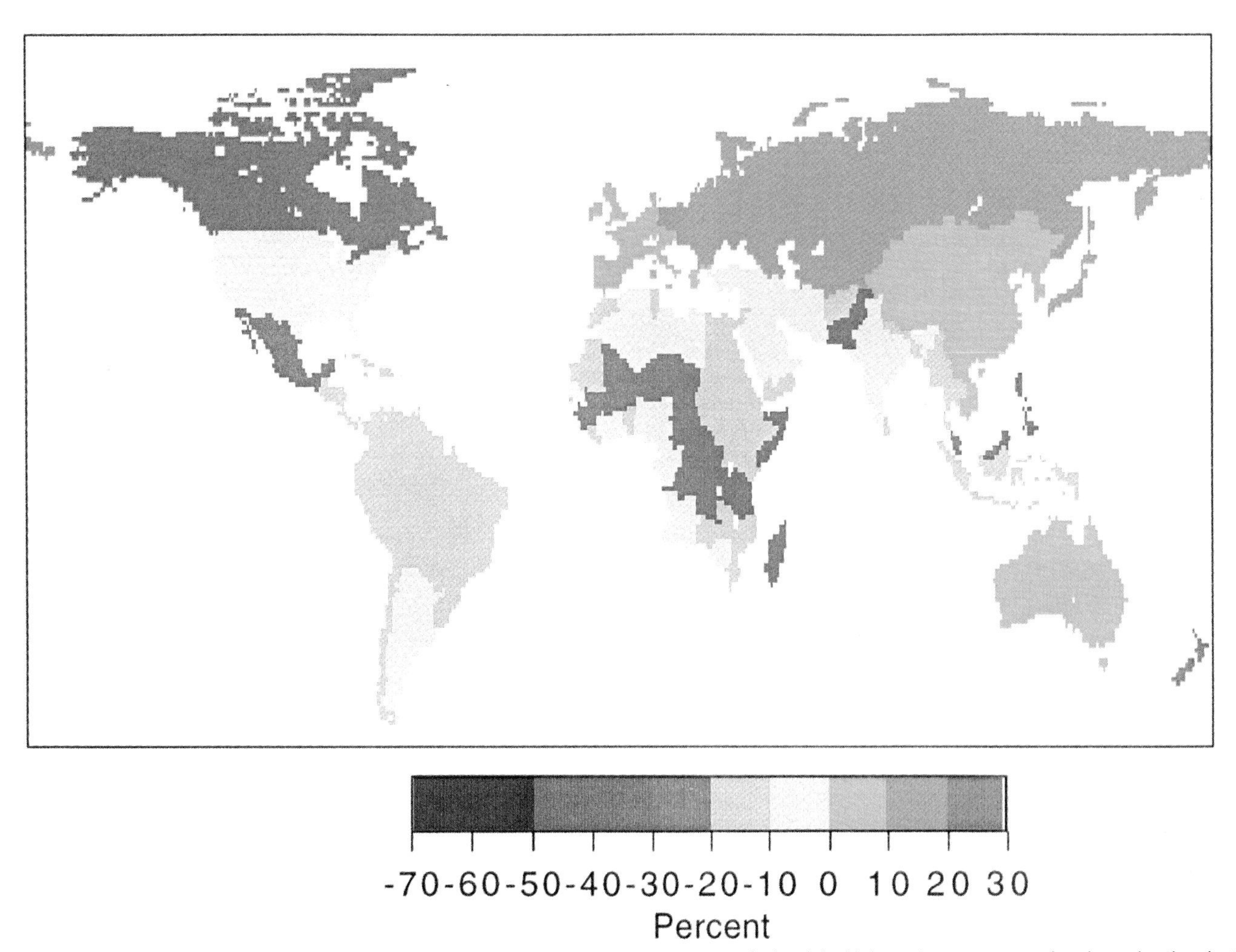


Figure 4.26 This computer plot generated by Rosenzweig and her colleagues at Columbia University compares food production in the year 2060 under two different scenarios: one with no additional greenhouse warming, the other with greenhouse warming. Note that most countries, particularly tropical nations with large populations and weak economies, are projected to experience the greatest losses due to greenhouse warming.

tion dramatically. Under these circumstances, there is a great and general disinclination simply to halt the use of fossil fuels, which are among the least expensive in the short term.

Nonetheless, a number of options exist. Conserving energy by being aware of the energy we use and using it more efficiently can make a substantial difference. Changing from oil, coal, and wood as energy sources to natural gas, which is free of $CO_2$, would be a significant advantage.

Changing from all fossil fuels to solar energy would have an even greater impact. Solar energy is nonpolluting, abundant, and (except for the collecting equipment) free; but sudden, large-scale conversion to a solar energy base would be a costly transition. Conversion to solar energy has a basic logic and appeal beyond its abundance and cleanliness. Recall that ultimately all of earth's energy (except for nuclear energy) comes from the sun. Fossil fuels are fossilized sunlight in a sense, so burning them amounts to consuming reserves of ancient sunlight. In a few hundred years, we will have consumed millions of years worth of fossilized sunlight. This is akin to spending a bank balance you inherited, without depositing or earning anything yourself. Eventually, the account's balance will fall to zero. On the other hand, living on the present flux of solar energy is like living on a steady, reliable stream of income and is in tune with other natural processes on earth. Figure 4.27 suggests several ways in which solar energy is becoming an everyday reality.

### *Countermeasures*

A more controversial strategy for managing global warming is by altering environmental conditions to cancel the warming. One example, which you have read about earlier, is based on the cooling effects of particle pollution, particularly sulfates, on the energy budget. Although at present we are not deliberately adding particulate pollution to the atmosphere as a global warming countermeasure, the fact is that combustion of fossil fuels and biomass burning appear to reduce the enhanced greenhouse effect. It is ironic that the air pollution/acid rain problem may tend to mitigate the global warming problem.

Other countermeasures are conjecture at present. Clearly, it would help to cease deforestation practices and replace them with reforestation programs. Ecologist George Woodwell has estimated that planting 2 million square kilometers of trees annually for the next 50 years would significantly impact the rate of $CO_2$ accumulation in the atmosphere. Unfortunately, 2 million square kilometers is an enormous area—roughly three times the

A

area of Texas. Scientists at the Electric Power Research Institute have calculated that a similar area (2.6 million kilometers) of seaweed or salt-tolerant desert brush would also trap significant amounts of $CO_2$.

A number of other suggestions have been offered. For example, some scientists have suggested fertilizing the southern oceans with iron. In theory, iron fertilization should stimulate growth of tiny marine organisms called phytoplankton, which would engage in increased photosynthesis, thereby removing large amounts of $CO_2$ from the atmosphere. Experimental tests of the hypothesis did record dramatic phytoplankton growth but found little impact on $CO_2$ levels.

Figure 4.27 Capturing the sun's energy with solar collectors or wind-driven generators results in no $CO_2$ emissions because photosynthesis is not involved.

B

C

The unfortunate fact is that attempts to solve environmental problems through technological countermeasures often fail, or they produce unforeseen new problems, or both. Our poor success record in this area is a firm reminder of the great complexity of natural systems such as the atmosphere and of the incompleteness of our present knowledge. Thus, it is crucial—particularly now, when we have other options—that we not put blind faith in technological countermeasures to save the day.

## Urban Heat Islands: Climates of the Future?

In many ways, cities provide us glimpses of future trends. It is the cities, and not the surrounding suburbs or rural places, that tend to offer the latest in fashion, the avant garde in the arts, and forward looking social and cultural trends, for better or for worse. Cities may be setting the pace in global warming as well.

In many cities in the United States and worldwide, a substantial warming in the temperature record is an undisputed fact. The phenomenon is so pervasive it has a name of its own: the **urban heat island.** An urban heat island is a region of warmer air temperatures compared with those in the surrounding countryside. Figure 4.28 gives an example of Washington, D.C.'s heat island.

Why has it become warmer in the cities? Concentrations of greenhouse gases tend to be higher there, of course. This leads to a locally enhanced longwave warming compared to outlying areas. But greenhouse warming is only a small part of the story. A city alters virtually every term in the energy budget formula, often in ways that result in an increase in surface temperatures. Figure 4.29 illustrates some of these mechanisms. Note, for

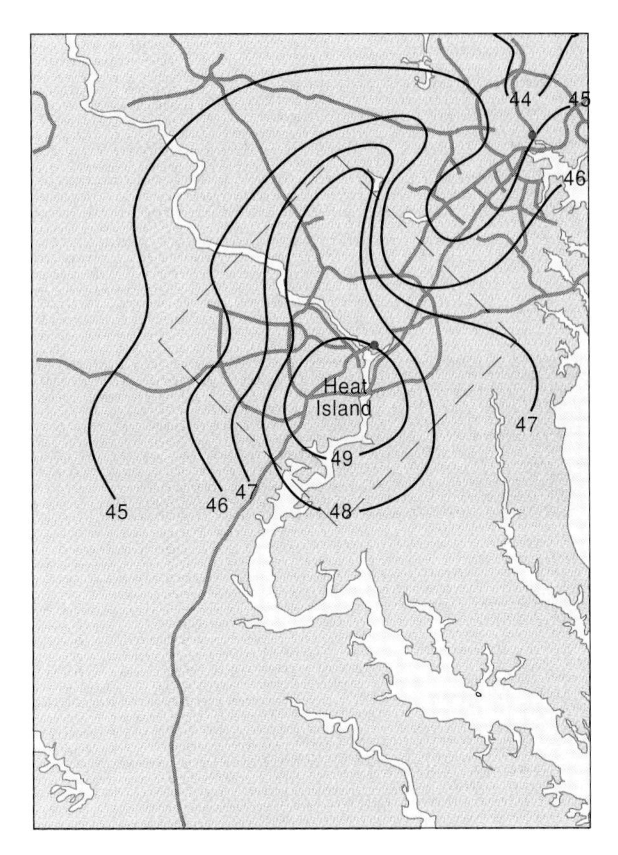


Figure 4.28 The Urban Heat Island over Washington, D.C.

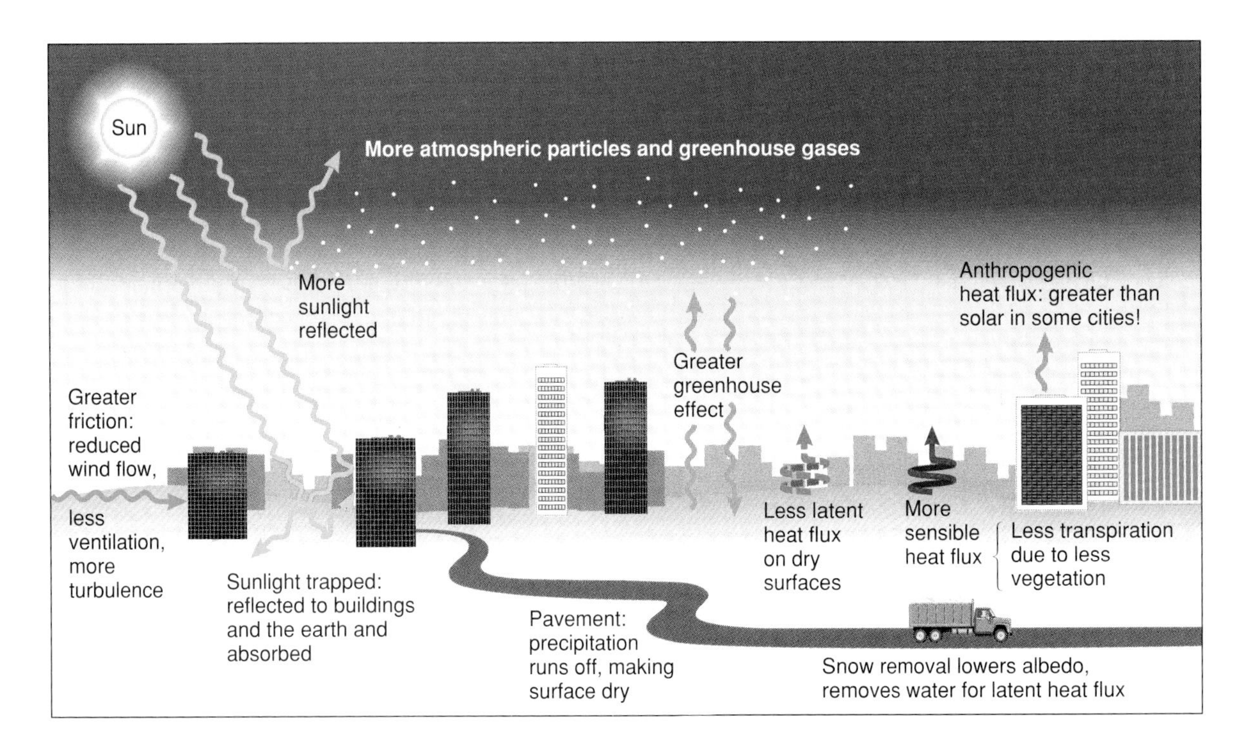


Figure 4.29 Some of the ways in which an urban environment affects the energy budget and hence the local climate.

example, that city pavement causes precipitation to run off, thereby reducing the potential for latent heat exchange with the atmosphere. The scarcity of vegetation in cities also means less latent heat flux from leafy surfaces to the atmosphere. With less latent heat transfer, surfaces become warmer, which leads in turn to an increase in the flux of sensible heat. City buildings affect the energy budget in several ways. They offer greater resistance to low-level wind flow, thus retarding the mixing of city air with that in the outlying regions. Buildings also absorb and retain heat effectively. Particulate air pollution, which is often 10 or more times greater in the city than in the country, affects nearly every term in the radiation budget, both directly through the radiation it absorbs, emits, and reflects, and indirectly through enhanced cloudiness.

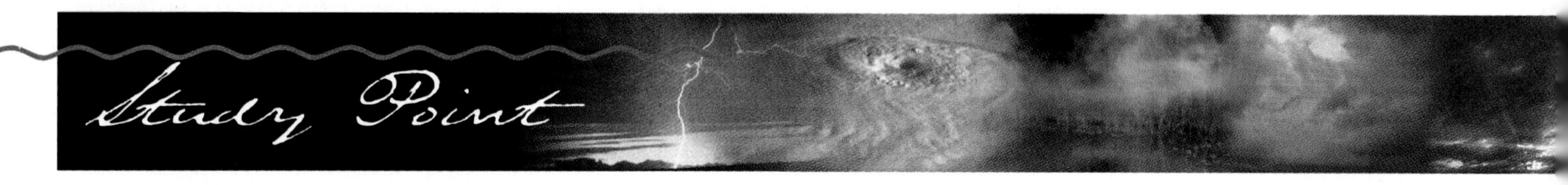


## Chapter Summary

In this chapter, you have considered a single big question and several related ones. The big question was, "Is the atmosphere warming?" Related questions include "What reasons do we have to suspect that the atmosphere is warming" and "What might cause the warming?"

You saw that theory and GCM computations point to an increase in air temperatures due to an enhanced greenhouse effect. The observational data also indicate a modest warming over the past 100 or so years. However, issues such as feedback and an imperfect data record make the problem a particularly difficult one to solve. As a result, it is not evident at this time that an enhanced greenhouse effect has caused the warming.

You have seen scientists' projections of the impact of global warming on the natural environments and on human activities such as agriculture, and you learned of options we have to respond to the threat.

## Key Words

outgoing longwave radiation
radiative forcing formula
positive and negative feedback
general circulation model (GCM)
running mean
anthropogenic
time step
signal and noise
correlation
urban heat island
proxy data
causal link

## Review Exercises

1. What is "radiative forcing"? State the radiative forcing formula.
2. Which of these are not greenhouse gases: nitrogen, oxygen, argon, water vapor, carbon dioxide, methane, ozone.
3. Explain step by step the sequence of events in the feedback loop relating soil moisture and warming, as shown in Figure 4.10.
4. What is responsible for the 11-year rise and fall of energy radiated by the sun, evident in Figure 4.4? How large is the variation, compared to the mean value radiated?
5. From Figure 4.13, determine the number of years prior to 1930 that the global mean temperature was warmer than the 1951–1980 average. In how many years after 1930 was it below the 1951–1980 average? Comment on your results.
6. According to GCM predictions, in what regions will the lower atmosphere warm the most due to greenhouse gas buildup? How great do the models predict the warming will be? How are stratospheric temperatures predicted to change?
7. A few weather stations have continuous or nearly continuous data records for a period of 150 years or more, making them potentially useful sources of information about long-term temperature change. What problems are liable to crop up in long-period records of this sort?
8. How is sea-level change linked to global warming?
9. Give an example of a correlation that does not imply cause and effect.
10. Why do most cities experience warmer temperatures than the surrounding countryside?

## Problems

1. Why are latent heat and sensible heat terms absent from the radiative forcing formula?
2. Suppose that 100 units of solar energy are absorbed by the earth (atmosphere and surface) over the same period it radiates 103 units back to space. Discuss the implications of these values.
3. If you had global temperature data (in Figure 4.13) only up to 1975, would you feel the data contained a warming trend? Why or why not?

Figure 4.29 contains one flux you have not encountered before. It is the direct release of heat into the atmosphere from anthropogenic activity: furnaces, motor vehicles, air conditioners, and industrial activity all alter the city's air directly by pumping heat into it. Researchers have calculated that in New York City, the magnitude of the anthropogenic flux is greater than that of absorbed solar energy.

Is the weather in cities, then, the precursor of global weather in the future? Although other scientists may disagree, we think not: too many factors specific to cities are at work in Figure 4.29 to make a generalization from present urban weather to future weather across the whole earth. On the other hand, urban heat islands do offer an example of humanity's potential to significantly alter our environment in unintended ways.

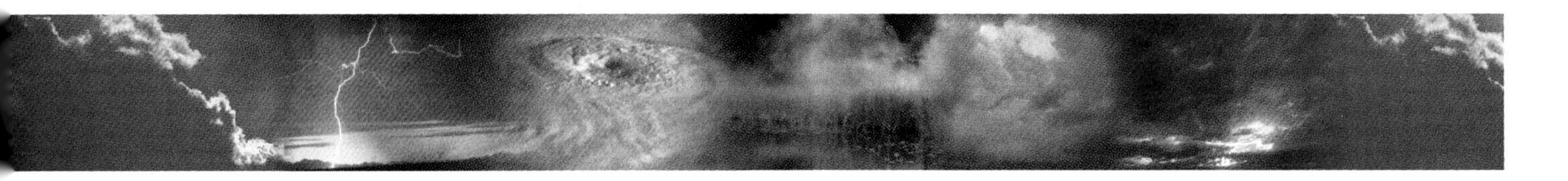

4. Greater particle emissions from combustion may reduce the impact of greenhouse gas buildups. Why might this be called "fighting fire with fire"?
5. Discuss the pros and cons of using computer models to study global warming.
6. Explain why the statements "Greenhouse gases have increased in the atmosphere over the past century" and "The atmosphere has warmed over the past century" don't necessarily imply that the greenhouse gas buildup is responsible for the warming.
7. In what way(s) might increased air pollution act to warm the lower atmosphere? How might it act to cool it?
8. Give two examples of proxy data for temperature. What might serve as proxy data for precipitation during the growing season? For mean wind direction?
9. Explain why planting trees over vast regions would help to slow the increase in atmospheric carbon dioxide. Then explain why this would be only a short-term solution.
10. What are some long-term measures humanity could take to reduce atmospheric $CO_2$ increases?

## Exploration

Review articles written over the past two years on global warming and the greenhouse effect. Determine if and how current thinking is changing about global warming. (If possible, use your library's computer-based search facilities to locate and select relevant articles.)

## Resource Links

The following is a sampler of print and electronic references for topics discussed in this chapter. We have included a few words of description for items whose relevance is not clear from their titles. More challenging readings are marked with a C at the end of the listing. For a more extensive and continually updated list, see our World Wide Web home page at www.mhhe.com/sciencemath/geology/danielson.


Houghton, J. T., et al., eds. 1990. *Climate Change: The IPCC Scientific Assessment*. New York: Cambridge University Press. 364 pp. A basic reference from which several charts in this chapter were taken.

Houghton, J. 1994. *Global Warming: The Complete Briefing*. Elgin, Il: Lion Publishing. 192 pp. A highly regarded summary by a world expert.

Karl, T. R., R. W. Knight, D. R. Easterling, and R. G. Quayle. 1996. Indices of climate change for the United States. *American Meteorological Society Bulletin* 77 (2): 279–92. Many interesting tables and graphs. C

Monastersky, R. 1995. Dusting the climate for fingerprints: Has greenhouse warming arrived? Will we ever know? *Science News* 147 (23): 362–63.

National Academy of Sciences. 1990. *One Earth, One Future*. Washington, D.C.: National Academy Press. 196 pp. Clear discussion of global warming and resultant changes.

National Academy of Sciences. 1991. *Policy Implications of Global Warming*. Washington D.C.: National Academy Press.127 pp. A look at the interface between science and public policy.

For readers with access to AccuData, PLDAY allows users to plot average and record temperatures for north American locations. DAY/R/X gives daily record temperature data. RECD lists current record setting weather.

*Clouds, Radiation, and Climate Sensitivity,* a brief report from the National Oceanographic and Atmospheric Administration's Geophysical Fluid Dynamics Laboratory, is located at http://www.gfdl.gov/brochure/7Clouds_Rad. doc.html.

The National Climate Data Center's Global Warming Update, at http://www.ncdc.noaa.gov/gblwrmupd/ global.html, presents graphical data on climate trends through 1993.

# How Does the Atmosphere Produce Rain?

Examine the photos on the facing page, which illustrate the flux of water into the atmosphere (left panel) and out of the atmosphere (right panel). Why do the two photos look so different? In other words, why is the process by which water enters the atmosphere so different from the way in which it exits? Stated more generally, the question is "How does the atmosphere make rain?"

Consider these questions a bit further: Water vapor enters the atmosphere via evaporation. It does so constantly, invisibly, quietly, molecule by molecule. Although you can't see it, the tranquil scene in the *left-hand* photo is a hotbed of evaporation. On the other hand, most water returns to earth in comparatively sporadic, sudden, sometimes violent events, such as the one in the *right-hand* panel, whose occurrence it would be difficult *not* to notice. Further, water exiting the atmosphere via precipitation does so in truly huge units by molecular standards: in contrast to evaporation, which occurs molecule by molecule, just a single small raindrop returns quadrillions of water molecules to earth. What causes the flux of water into the atmosphere to be so very different from the flux out?

Apparently water vapor molecules find ways of clumping together in the atmosphere, thus ensuring their return to earth generally in the form of precipitation and not condensation. Chapter 5 deals with the evaporation of water vapor molecules and their subsequent condensation as cloud particles. In Chapter 6, you investigate the large-scale properties of clouds and the conditions surrounding their formation. Chapter 7 considers processes of transforming cloud particles into precipitation and prospects for precipitation modification.

# Chapter 5

## Humidity

## Chapter Goals and Overview

On an average day, more than a trillion ($10^{12}$) tons of water enter the atmosphere through evaporation and transpiration. In more personal units, each day a mass of water more than 10 times the mass of the human race evaporates into the atmosphere. Furthermore, the impact of this flux is as extraordinary as it is varied, including phenomena such as blizzards, thundershowers, freezing rain storms, rainbows, dew, and frost.

Humidity, the title and focus of this chapter, is a general term for water vapor in the air. The first topic is evaporation, after which you will survey the variables and instruments used to describe and measure atmospheric humidity. Finally, you will study processes by which the atmosphere's water vapor condenses onto the earth's surface and onto minute airborne objects, many of which are destined to become cloud particles.

### Chapter Goals

By chapter's end, you will be able to:

- define equilibrium, a key concept, and explain how the amount of water vapor at equilibrium varies with temperature;
- compare five different methods used to describe the amount of water vapor in the air and identify several instruments or methods by which humidity is determined;
- use the relative humidity formula, psychrometric tables, and the saturation mixing ratio chart to find numeric solutions to a variety of humidity problems;
- explain why indoor humidity is often unhealthfully low in the wintertime;
- identify conditions under which dew and frost occur; and
- explain the roles played by atmospheric particles, dissolved impurities, and supercooling in the formation and growth of cloud particles.

# Evaporation

We begin the study of atmospheric water with the quiet process of evaporation. Evaporation, which proceeds largely unnoticed, frees countless water molecules, which break free of the earth's waters and join the atmosphere, future ammunition perhaps for some great cumulonimbus blunderbuss.

## Evaporation As a Molecular Process

What factors affect the rate at which water evaporates into the atmosphere? In this section, you will conduct a series of simulated experiments to answer this question.

The setup for our simulated experiments is illustrated in Figure 5.1A. Note the starting conditions: the sealed tank contains pure water at 10°C (50°F), an air-tight divider rests at the water surface, and the space above the divider is devoid of all matter: it is a vacuum. What happens at time $t = 0$, when the divider is removed?

Suppose you could observe individual molecules of liquid water and air in the container. Among other things, you would notice that energetic liquid water molecules moving upward toward the surface with enough energy would escape the liquid and evaporate into the space above, as shown in Figure 5.1B. You might expect that because the more energetic molecules escape, the liquid would cool. You prevent this by adding heat as needed to maintain a constant water temperature.

As water molecules evaporate, they move through the space, colliding with the container walls, with each other, and with the water surface below. Occasionally, a downward-moving molecule collides with and rejoins the liquid water below, a process called condensation.

### *Equilibrium*

As time passes, liquid water molecules continue to evaporate, which in turn leads to greater numbers of molecular collisions against the walls of the container and the water surface below—and therefore more condensation. Eventually, an equilibrium is reached (Figure 5.1C): the rate at which vapor molecules rejoin the liquid water equals the rate at which water molecules are evaporating; that is, the flux of water vapor from liquid to gas equals the reverse flux, from gas back to liquid. Once this state is achieved, the amount of water vapor in the space above does not change. However, the individual molecules comprising the vapor continue to change constantly because evaporation and condensation are still proceeding as before. Now, however, they are in balance; the liquid water surface is in equilibrium with the water vapor above. When this equilibrium condition is reached in the presence of a clean, plane water surface, the air is said to be **saturated.** (Note that the "air" in this particular experiment is composed only of water vapor.)

### *Vapor Pressure*

Consider for a moment the collisions made by the water vapor molecules on the sides of the container. Recall from Chapter 2 that such collisions collectively are the cause of "air" pressure. In this case, the gas is composed entirely of water vapor molecules, but the concept is the same: water vapor molecules' collisions exert a pressure. Therefore, you could use a barometer to measure the **vapor pressure,** which is the pressure caused by water vapor molecules in the container.

How much vapor pressure would you measure in the experiment above? Until time t = 0 (Figure 5.1A), the vapor pressure would be zero, since at that time no molecules have evaporated. As the experiment progresses, you would see the vapor pressure rise to about 12.3 millibars at equilibrium (see Figure 5.2A). This quantity is known, logically, as the **equilibrium vapor pressure.** The equilibrium vapor pressure is defined by conditions in Figure 5.1C, in which evaporation from a plane water surface equals the rate of condensation.

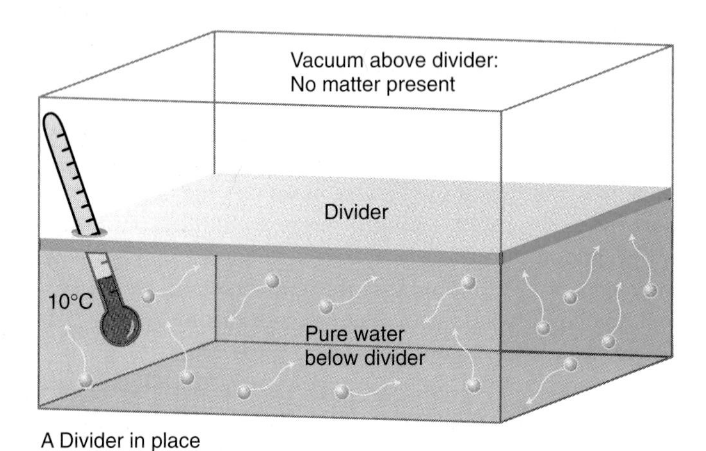


A Divider in place

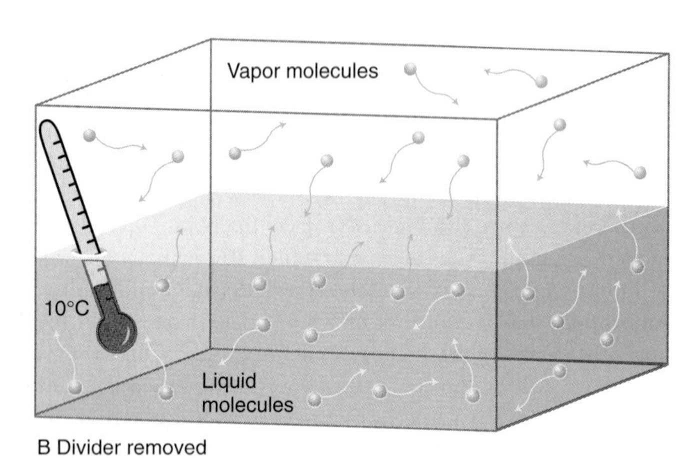


B Divider removed

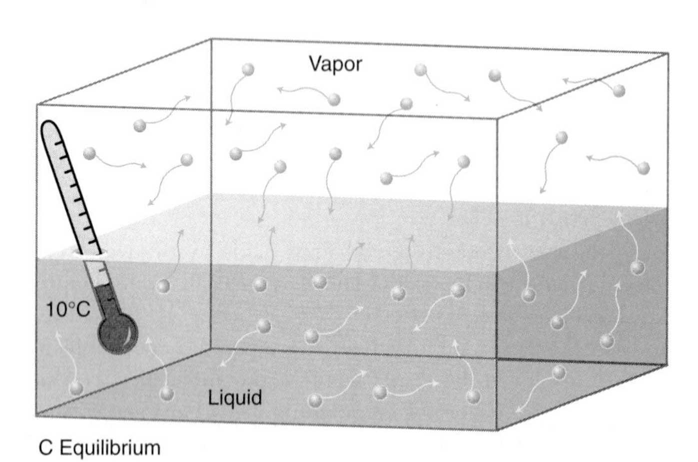


C Equilibrium

Figure 5.1 When the divider is removed (*B*), evaporation commences into the vacuum above the water. Eventually, equilibrium (*C*) is reached; evaporation is matched by the flux of condensing vapor.

## Variations in Equilibrium Vapor Pressure

### *Variation with Temperature*

Suppose you run the previous section's experiment again. You keep everything the same except the temperature, which you set

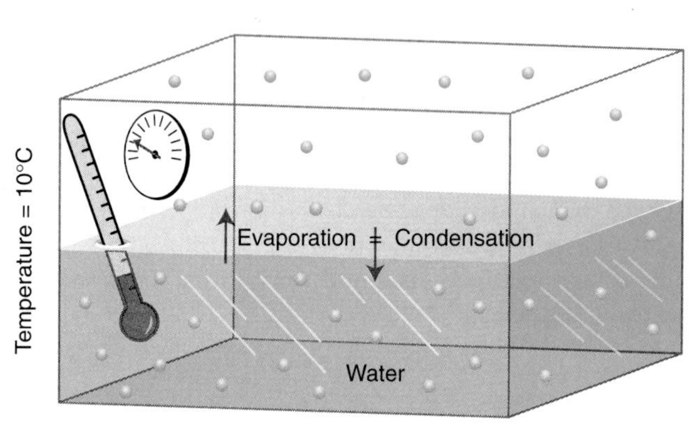


A Vapor pressure = 12.3 millibars at equilibrium

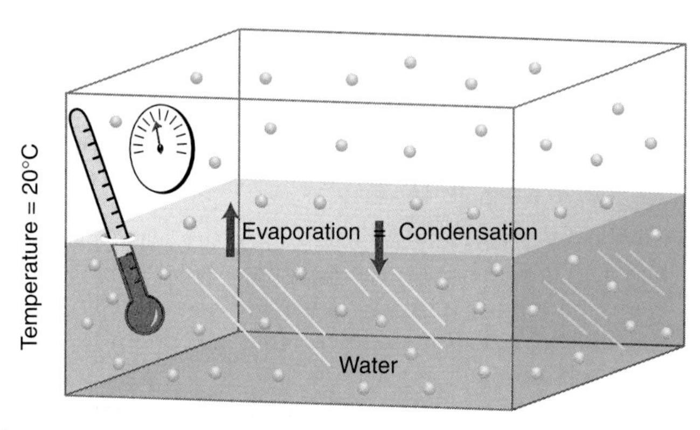


B Vapor pressure = 23.4 millibars at equilibrium

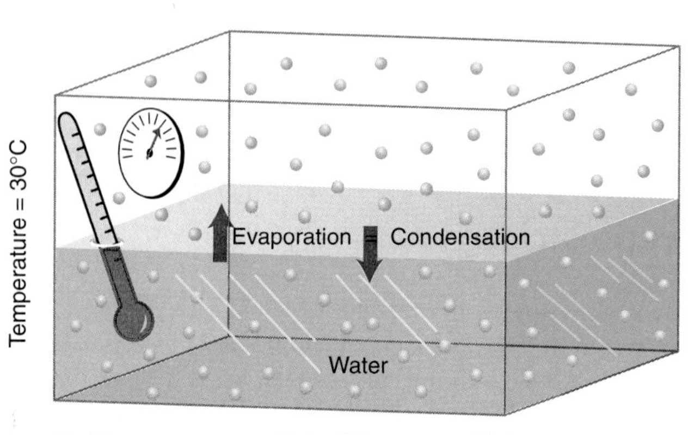


C Vapor pressure = 42.4 millibars at equilibrium

Figure 5.2 Equilibrium conditions for evaporation experiments conducted at three different temperatures. What information in the drawings indicates that each experiment is at equilibrium?

to 20°C (68°F). What do you expect will happen? Will the results change?

The results of this trial appear in Figure 5.2B. Notice that equilibrium is not reached until the vapor pressure reaches 23.4 millibars.

Why are these results so different? You know that for a molecule to evaporate, it must possess sufficient thermal energy. Thus, we might speculate that at a higher temperature, a larger proportion of the water molecules possess the energy required to evaporate, and equilibrium is reached at a higher vapor pressure.

Suppose you repeat the experiment once again, this time at a temperature of 30°C (86°F). What would you predict the equilibrium vapor pressure to be? You can read the result in Figure 5.2C: at 30°C, the equilibrium vapor pressure is 42.4 millibars.

These are dramatic results! They indicate that a 10°C (18°F) increase in temperature causes nearly a doubling of the equilibrium vapor pressure. Does this doubling occur every time you increase the temperature by 10°C? Does it happen across any range of temperature? If you repeated the experiment a number of times and plotted a graph of equilibrium vapor pressure at different temperatures, your results would be as shown in Figure 5.3.

What information does the graph show? Notice that the trend of the curve from lower left to upper right expresses the fact that vapor pressure increases as temperature increases. The steepening of the curve's slope to the right indicates that at higher temperatures, the equilibrium vapor pressure changes more rapidly with temperature than it does at lower temperatures. (Thus, from 20°C to 30°C, the change was 19.0 millibars, whereas from 10°C to 20°C, the change was only 11.1 millibars.) The relation between equilibrium vapor pressure and temperature is extremely important and plays a key role in the formation of clouds and precipitation.

Note also that the curve divides the graph space into two regions. Below the line, conditions are **"unsaturated"**; in experiments like the one above, evaporation will exceed condensation until the equilibrium vapor pressure (saturation) is reached. Above the curve is the **supersaturated** region; in experiments like the one above, condensation will exceed evaporation, causing the vapor pressure to decrease until saturation is reached.

Finally, observe that in Figure 5.3 there are two equilibrium curves in the region of temperatures colder than 0°C. If the surface in the experiments is solid, that is, an ice surface rather than to a liquid one, the curve labeled "ice" indicates equilibrium with respect to the ice surface. However, as you will see later in more detail, water in the free atmosphere may remain liquid below 0°C. This condition will be referred to as "supercooled." The "water" curve below 0°C in Figure 5.3 represents that circumstance.

### *Variation with Air Pressure*

You might be wondering why the experiments above were conducted with no atmosphere above the liquid water surface. Certainly, such a situation is not representative of conditions on earth, where the atmosphere exerts a pressure of as much as 1000 millibars or more on evaporating water surfaces. Why not fill the space above the water in Figures 5.1 and 5.2 with ordinary air and determine the values of equilibrium vapor pressure in that more typical case?

The reason we imagined the experiments running in a vacuum was to make the point that evaporation, condensation, and equilibrium all occur whether an atmosphere is present or not. Strange as it may seem at first, the atmosphere plays almost no role in these experiments. If you performed the experiments once more with normal air at typical atmospheric pressure in the space above the water in Figures 5.1 and 5.2, you would end up with essentially the same graph in Figure 5.3. In short, the equilibrium vapor pressure, although highly sensitive to temperature, is practically independent of air pressure.

Water is not unique in having an equilibrium vapor pressure. Every gas comprising the atmosphere contributes its own partial pressure to the total air pressure, and each has its own equilibrium value, dependent on temperature, and analogous to water's vapor pressure. Only water vapor, however, is present in sufficient quantities and at the appropriate temperatures to reach equilibrium in earth's atmosphere. In the atmospheres of other planets, gases such as methane and ammonia, as well as water vapor, may reach equilibrium values. On the other hand, the most common gases in earth's atmosphere, oxygen and nitrogen, have equilibrium vapor pressures at ordinary temperatures that are far greater than atmospheric pressure. Therefore, these gases are not found as liquids on earth's surface.

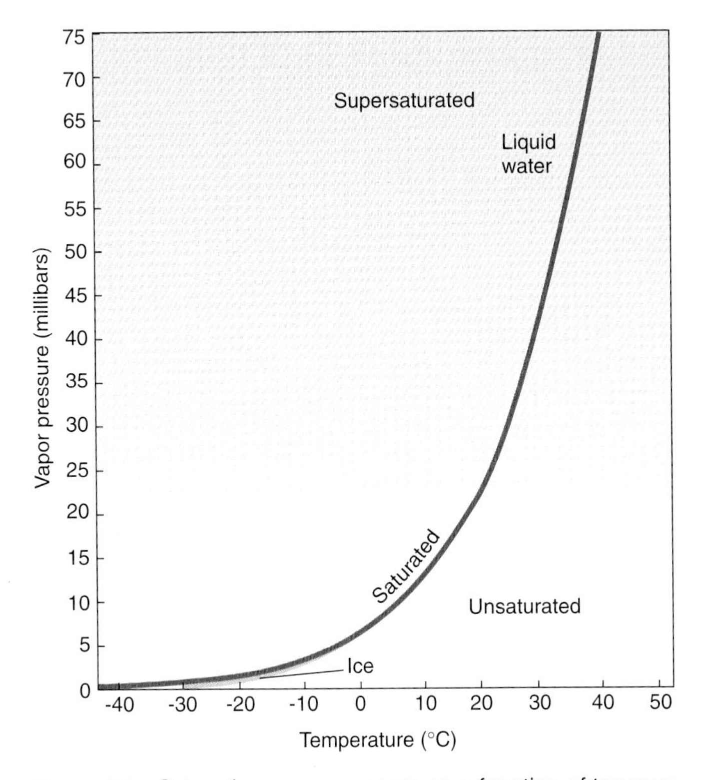


Figure 5.3 Saturation vapor pressure as a function of temperature. Is air at 15°C with a vapor pressure of 15 millibars unsaturated, saturated, or supersaturated?

## Checkpoint

***Review***

Perhaps the single most important concept in this section is that equilibrium or saturation values of water vapor pressure depend strongly on temperature: equilibrium vapor pressure values are higher at higher temperatures.

In contrast to this strong temperature dependence, the equilibrium vapor pressure is nearly independent of total air pressure.

Remember also that equilibrium means not that evaporation has ceased but only that evaporation is balanced by an equal amount of condensation. Under unsaturated conditions, evaporation exceeds condensation.

***Questions***

1. Describe equilibrium conditions at a liquid surface.
2. Suppose you perform this section's experiment at a temperature of 0°C. What would be the equilibrium vapor pressure? (Refer to Figure 5.3.)
3. Imagine conducting an experiment like the one at the beginning of this section, in which you maintained a constant temperature of 20°C but varied the air pressure. Draw a graph showing the relation between equilibrium vapor pressure (on the vertical axis) and air pressure (on the horizontal axis) you would obtain for such an experiment.

# Humidity Variables and Measuring Instruments

Vapor pressure is just one of a number of different terms that describe the quantity of water vapor in an air parcel. (The term *air parcel* is often used when referring to a small sample of air under study.) You already know something of two other humidity variables: relative humidity and dew-point temperature. In this section, you will learn several more and see how they are related.

## Mixing Ratio

Suppose that you perform last section's experiment one more time (see Figure 5.4), this time with exactly 1 kilogram of dry air in the space above the liquid water. By "dry" in this case, we mean air in which no water vapor molecules are present. You measure the air pressure—1000 millibars—and the temperature—10°C. (A 1-kilogram parcel with these properties would occupy a volume of about four-fifths of a cubic meter, or roughly a cubic yard.)

From earlier experiments, you know what will happen: water will evaporate, faster than it recondenses at first, until the vapor pressure reaches 12.3 millibars. Then equilibrium, or saturation, is reached, and conditions remain constant.

Another way of describing the humidity of the air parcel in this experiment is by the **mixing ratio,** labeled "r," which is the number of grams of water vapor present in the kilogram of initially dry air. Initially (at time $t = 0$), no water vapor is present in the dry air, so the mixing ratio is equal to 0.0 grams of water vapor per kilogram of dry air (written "$r = 0.0$ g/kg"). As evaporation proceeds, the mixing ratio increases. By the time the system has reached equilibrium, the value of r has risen to *7.76* g/kg; that is, *7.76* grams of water vapor have evaporated into the kilogram of initially dry air. In meteorology, this equilibrium or saturation value is called the **saturation mixing ratio,** we use "$r_S$" for its label.

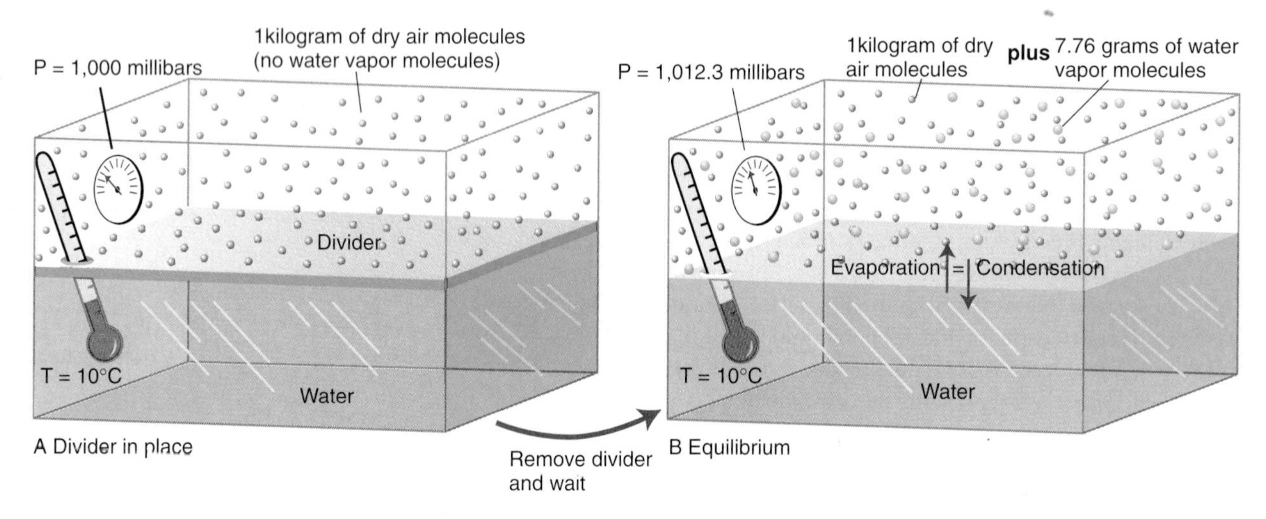


Figure 5.4 The presence of air in the chamber above the water does not affect the value of the saturation vapor pressure. Notice the total pressure increases by 12.3 millibars, the contribution of the saturation water vapor pressure. The mixing ratio changes from 0 to 7.76 grams of water vapor per kilogram of dry air.

Does equilibrium mixing ratio vary with temperature, like equilibrium vapor pressure? Figure 5.5 shows the relationship. Notice its strong resemblance to the equilibrium vapor pressure graph of Figure 5.3. Unlike vapor pressure, however, equilibrium mixing ratio varies considerably with air pressure. In the examples that follow, air pressure is assumed to be 1000 millibars, unless otherwise specified.

You will find the saturation mixing ratio curve to be a useful tool for visualizing and solving humidity problems. As an example, consider the point A in Figure 5.5. This is the initial condition of the air above the water surface in the experiment shown in Figure 5.4: temperature = 10°C, mixing ratio = 0 g/kg. As the experiment progressed, the air parcel's temperature remained at 10°C, while its mixing ratio increased. You can visualize this by moving vertically on the 10°C isotherm (line of constant temperature). Point B represents the parcel some time after the divider was removed.

What was the mixing ratio at the time the parcel reached point B? Moving horizontally to the left axis, you see that B lies on the 4 g/kg line. Thus, 4 grams of water had evaporated into the kilogram of initially dry air. Has equilibrium been reached at this point? No, you can see that it is only about halfway from point A to the equilibrium (saturation) value at point C, which, for a temperature of 10°C, is slightly under 8 g/kg. (Recall from above that a more accurate value is *7.76* g/kg.)

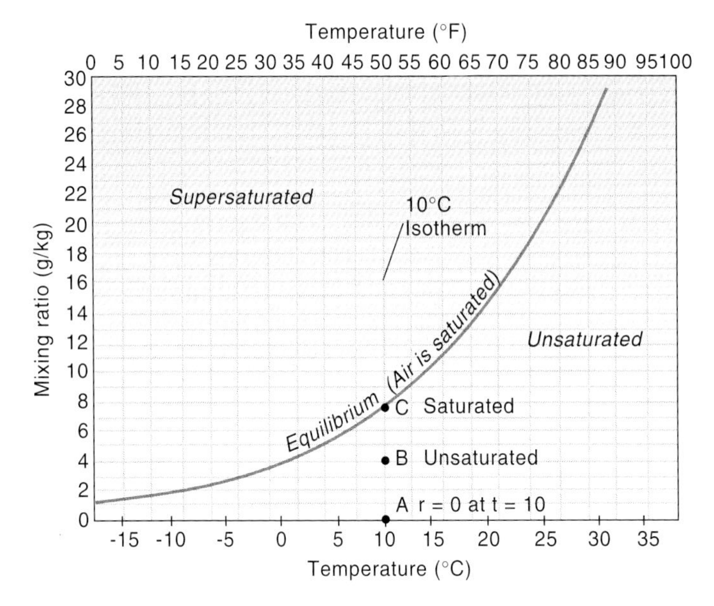


Figure 5.5 Saturation mixing ratio as a function of temperature. The experiment is represented by moving from A to C along the 10°C isotherm.

## Relative Humidity

Continuing the last example, suppose you compare the air's "actual" mixing ratio (r) at point B with its saturation mixing ratio ($r_S$) for that temperature. Expressed as a percent,

$$\frac{\text{mixing ratio (r)}}{\text{saturation mixing ratio } (r_S)} \times 100\% = \frac{4.0 \text{ g/kg}}{7.76 \text{ g/kg}} \times 100\%$$
$$= 52\%$$

This quantity is defined as the **relative humidity** (RH). Defined here more precisely than in Chapter 1,

$$\text{relative humidity} = \frac{\text{(the air's mixing ratio)}}{\text{(its saturation mixing ratio)}} \times 100\%$$

or, in symbols,

$$RH = \frac{r}{r_S} \times 100\%$$

Figure 5.6 represents relative humidity as a comparison of the length of the bar AB with the bar AC.

As another example, imagine an air parcel whose temperature is 25°C and whose mixing ratio is 6 g/kg. What is the parcel's relative humidity?

To solve this problem, locate the parcel in Figure 5.6. Point X marks the spot: T = 25°C, r = 6.0 g/kg. Notice that X is only a small fraction of the way from the chart's bottom to the saturation line: thus, the relative humidity should be low. To make the calculation, you need to divide the mixing ratio (which you are told is 6.0 g/kg) by the saturation value, whose value is not given

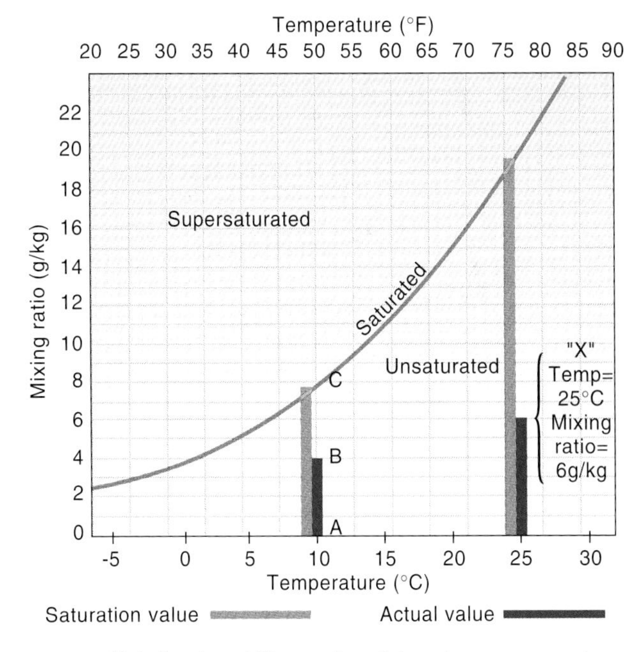


Figure 5.6 Relative humidity can be pictured as a comparison of the length of the "actual" mixing ratio to the value required for saturation at the same temperature.

directly. However, Figure 5.6 shows that for a temperature of 25°C, the saturation mixing ratio is roughly 20 g/kg. Therefore,

$$RH = \frac{r}{r_S} = \frac{6.0 \text{ g/kg}}{20 \text{ g/kg}} \times 100\% = 30\%$$

Note that relative humidity is a measure of a parcel's actual moisture content compared to the saturation value; since the saturation value depends strongly on temperature, the relative humidity itself depends strongly on temperature. Thus, if the air

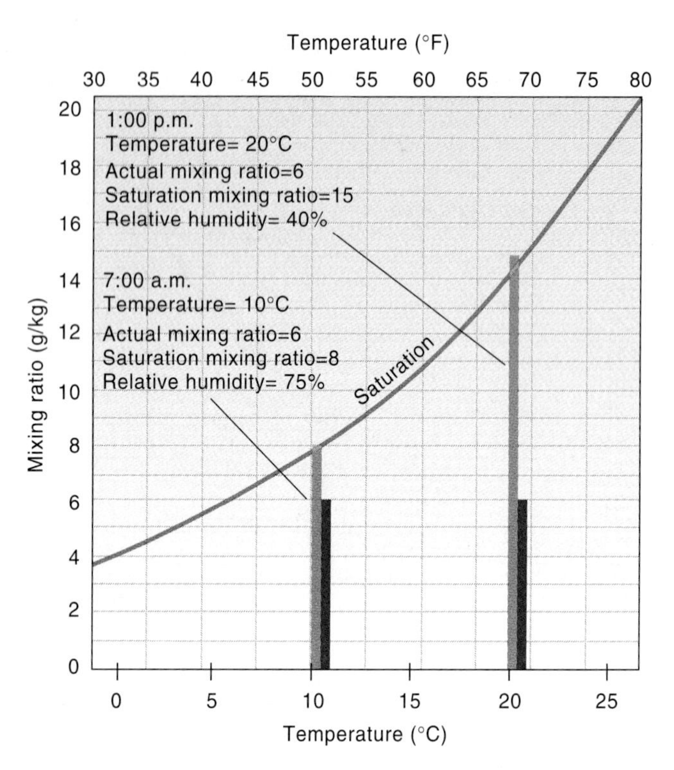


Figure 5.7 The air's actual mixing ratio at this site remained unchanged (dark purple bars) from 7:00 A.M. to 1:00 P.M. However, the temperature warmed from 10°C to 20°C, which increased the air's saturation mixing ratio from 8 to 15 g/kg. As a result, the relative humidity fell from 75% to 40%.

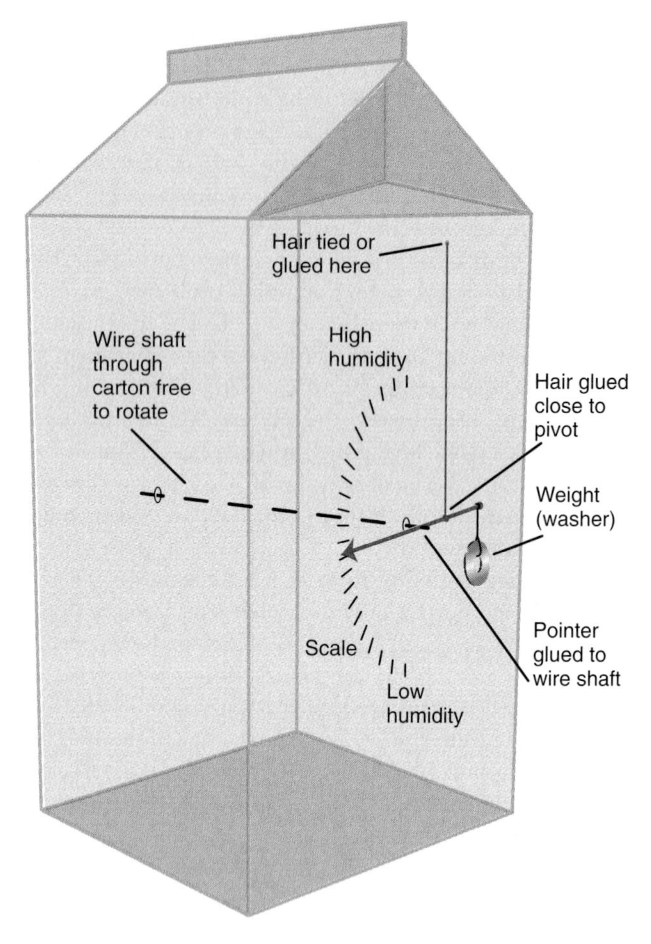


Figure 5.8 A hair hygrometer. As relative humidity increases, the hair becomes longer, allowing the weight to pull the pointer clockwise.

temperature changes, the relative humidity changes, even if there is no change in the actual amount of water vapor in the air. Figure 5.7 illustrates this situation.

The relationship between temperature and relative humidity is so important we want to emphasize it by stating it another way. Assuming no other changes in the air, if the temperature rises, the relative humidity falls; if the temperature falls, the relative humidity rises. This is an example of an "inverse relationship." In an inverse relationship, an increase in one variable is matched by a decrease in another variable.

Sometimes it is useful to employ the relative humidity formula in reverse. As an example, consider conditions on a recent September afternoon in Houston: the air pressure was near 1000 millibars, the air temperature was 30°C (86°F), and the relative humidity was 50%. Suppose you want to know how much water vapor was actually in the air at the time. (Meteorologists use this sort of information when predicting precipitation amounts.) The relative humidity formula applied to this situation states

$$\mathrm{RH} = \frac{\mathrm{r}}{\mathrm{r_S}} = 50\%$$

We seek the mixing ratio. First, isolate the desired quantity, mixing ratio, on the left side of the expression by multiplying both sides of the formula by the saturation mixing ratio (whose value we will find presently):

$$\mathrm{r} = 50\% \times \mathrm{r_S}\ (\text{at } 30°\mathrm{C})$$

Looking back at the saturation mixing ratio graph in Figure 5.5, you can see that the saturation mixing ratio at 30°C is approximately 28 g/kg. Thus,

$$\mathrm{r} = 50\% \times 28\ \mathrm{g/kg} = 14\ \mathrm{g/kg}$$

Therefore, mixed with each kilogram of dry air in Houston at the time of observation were approximately 14 grams of water vapor.

The partial pressure and mixing ratio of water vapor are not easy to measure directly. On the other hand, you can measure relative humidity in any of several ways. One method is related to the fact that human hair is less "manageable" on humid days. This is due in part to the fact that human hair changes length as relative humidity changes. A strand of hair may lengthen by as much as 2.5% as relative humidity changes from 0 to 100 percent. Instrument makers have capitalized on this fact in constructing **hygrometers,** instruments that measure relative humidity. Figure 5.8 gives an example of such an instrument; we encourage you to construct, calibrate, and test it for various temperatures and humidities.

Another everyday observation leads to a second method of measuring relative humidity. You have probably noticed that in humid summer weather, salt is reluctant to flow from its shaker. The salt has absorbed water vapor from the air, which has caused it to become sticky. The water's presence on the salt also makes the salt a better conductor of electricity. Thus, if you pass an electric current through salt, you will observe a greater electric flow in humid weather. This relation is used in electronic hygrometers. The "salt" used is lithium chloride, rather than table salt; but the principle is the same. Such devices are used widely in instruments such as radiosonde sensors.

## Dew-Point Temperature

We are not yet finished in Houston. Figure 5.9 shows the Houston air plotted on the now-familiar mixing ratio-temperature graph. Notice that the parcel lies halfway between the zero mixing ratio line and the saturation curve (at point T), consistent with its relative humidity of 50%. With the help of Figure 5.9, we will examine some ways in which the relative humidity might change; specifically, we will consider how the air might become saturated.

One way to saturate Houston's air would be to add water vapor at constant temperature, in effect moving the parcel up the 30°C isotherm until it reaches the saturation curve. This process is illustrated by the red line.

Another way would be to leave the air's mixing ratio unchanged but to cool the air. Cooling is equivalent to moving the parcel to the left on Figure 5.9. Cooled sufficiently, the parcel eventually meets the saturation curve. This process is illustrated in blue. The temperature at point D, reached by cooling an air parcel to saturation without changing its mixing ratio, is the **dew-point temperature.** In the case of the Houston air, you can verify that the dew-point temperature was 19°C.

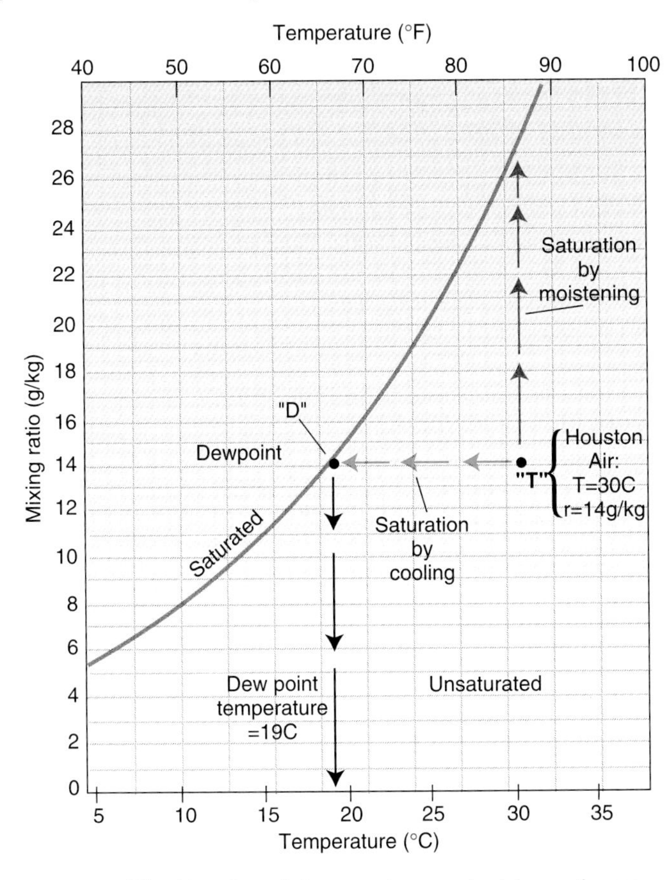


Figure 5.9 The Houston air temperature and mixing ratio are plotted at T, on the right. Cooling this air without changing its mixing ratio takes it to saturation at D; this point is known as the dew point. The dew-point temperature is 19°C.

What is the relation between the dew-point temperature and the air's mixing ratio? Figure 5.9 relates these two variables in the same way it relates air temperature and saturation mixing ratio. Thus, knowing the dew-point temperature, you can use Figure 5.9 to find the mixing ratio, and vice versa. Further, all of these variables are related through the relative humidity formula. In summary, we can write schematically

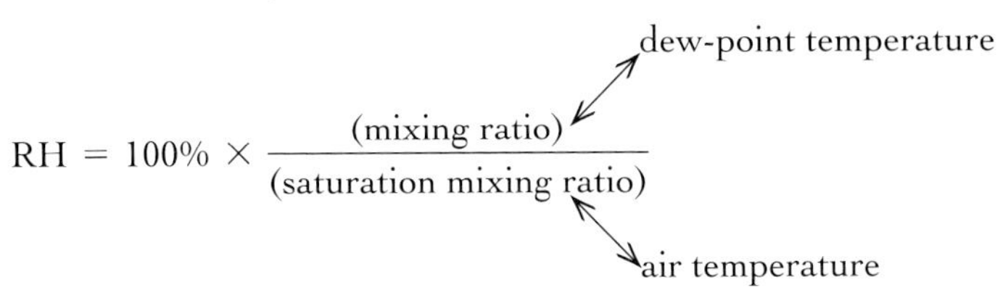


where the ⟷ symbols mean "If you know one of these quantities, you can find the other through Figure 5.9." Note, however, that the actual calculation of relative humidity involves only the mixing ratios, not dew-point or temperature.

Let's use the relative humidity formula above to solve the following problem. A football game played in Phoenix between the Phoenix Cardinals and the Chicago Bears on August 15, 1992, is known as "the hottest football game ever played." At game time, the temperature on the playing field was 43°C (109°F), and the dew-point temperature was 24°C (75°F). Air pressure was close to 1000 millibars. What was the relative humidity? (Assume the saturation mixing ratio at 43°C = 60g/kg.)

Before reading further, try to solve this problem on your own. If you need help, remember that the air temperature determines the saturation mixing ratio, and the dew-point temperature determines the air's mixing ratio.

From Figure 5.10, you can see the mixing ratio was approximately 19 g/kg, while the saturation mixing ratio was roughly 60 g/kg. Therefore,

$$\text{RH} = \frac{19 \text{ g/kg}}{60 \text{ g/kg}} \times 100\% = 32\%$$

A relative humidity of 32% may not seem like a very high value. However, paired with such a high temperature, it created conditions that were extremely dangerous for players and spectators alike.

Like relative humidity, dew-point temperatures can be measured directly. In Figure 1.10 of Chapter 1, you saw how a pitcher of ice water can serve as such an instrument. Figure 5.11 shows a more precise dew-point apparatus. Notice, however, that these instruments are based on condensation onto a solid, initially dry surface (the pitcher's outside wall, or a mirror), and not a liquid surface as in the experiments that define equilibrium. Thus, you might question whether condensation onto a nonliquid surface occurs at the same temperature as over a pure, plane water surface.

As it turns out, dew-point instruments that are based on condensation onto glass or polished metal surfaces are capable of

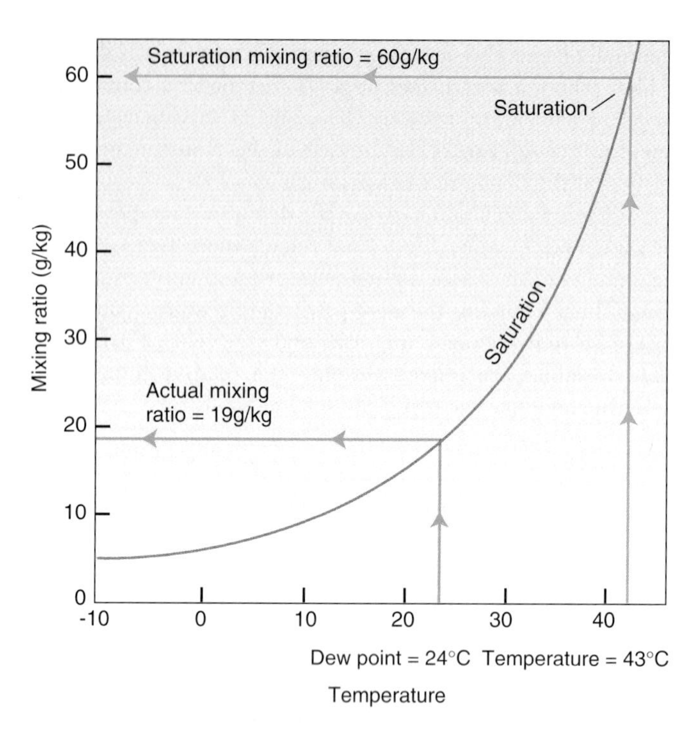


Figure 5.10 Finding actual and saturation mixing ratios from dew-point and air temperatures, respectively. Notice you could apply this process in reverse: if somehow you knew the actual and saturation mixing ratios, you could move along the lines in the opposite directions and determine dew-point and air temperatures.

Figure 5.11 A Dew-point hygrometer. An air sample is drawn into a chamber within the instrument and cooled. Sensors measure the temperature at which moisture condenses onto a mirror in the chamber.

giving quite accurate measurements. However, on other substances (salt, for example), condensation may commence at a temperature considerably higher than the dew-point value over a pure water surface—an important fact whose implications you will study in later sections.

## Wet-Bulb Temperature

A simple but effective device for measuring atmospheric humidity is the **psychrometer,** illustrated in Figure 5.12. As you see, the instrument consists of two thermometers mounted on a swivel. A piece of absorbent cloth is wrapped around the bulb of one thermometer and thoroughly moistened; this thermometer is known as the "wet-bulb thermometer." The other thermometer is called the "dry-bulb thermometer."

To use the psychrometer, you swing it by its handle. Water evaporates from the wick of the wet-bulb thermometer (assuming the relative humidity is less than 100%). This evaporation cools the wet-bulb thermometer by absorbing the latent heat required from the wet-bulb thermometer. This thermometer, therefore, becomes colder than the air temperature measured on the dry-bulb thermometer.

How far does the wet-bulb thermometer cool? The drier the air, the greater the evaporation and hence the greater the cooling. However, it never cools as far as the dew point (unless the relative humidity is 100%), because a negative feedback mechanism comes into play. The temperature difference between the wet-bulb thermometer and the surrounding air temperature causes a flow of sensible heat from the warmer air to the cooler wet-bulb. The greater the wet bulb cools due to evaporation, the greater this compensating heat flow of sensible heat. Eventually, the wet-bulb's reading stabilizes when its heat loss due to latent heat flux is just balanced by the sensible heat gain from the surrounding air. This equilibrium temperature is known as the **wet-bulb temperature** (see Figure 5.12C).

Although the explanation of the wet bulb's behavior may seem somewhat complicated, the measurement itself is simple. Furthermore, once you have determined wet- and dry-bulb temperatures, you can use a psychrometric table like that in Appendix J to determine other humidity variables, such as dew-point temperature, relative humidity, mixing ratio, and vapor pressure. Thus, the psychrometer offers a convenient and practical way of finding atmospheric water vapor content.

Let's consider a typical psychrometric problem. Suppose you receive an e-mail message from a friend at the University of Colorado telling you he just measured an air temperature there of 16°C along with a wet-bulb reading of 10°C. Knowing you're studying meteorology, he wonders if you can tell him the relative humidity.

You can, indeed. You turn to Appendix J. Along the bottom of the chart you find the temperature, 16°C. The vertical scale is labeled "wet-bulb depression," which is the difference between dry- and wet-bulb readings. In your friend's case, the depression is 16 − 10 = 6°C. So you enter the chart at the 6°C depression line, move horizontally until you reach the 16°C dry-bulb column, and read the relative humidity. You e-mail your answer back to your friend: "Relative humidity = 40%."

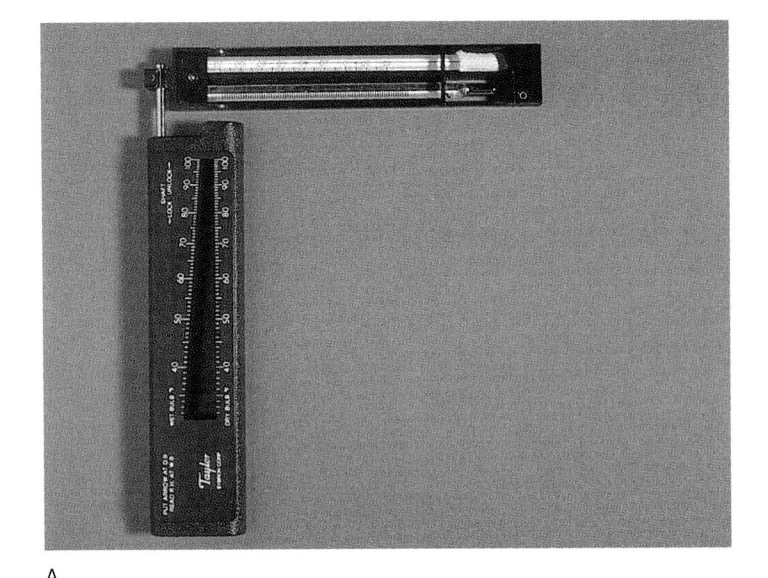

A

B

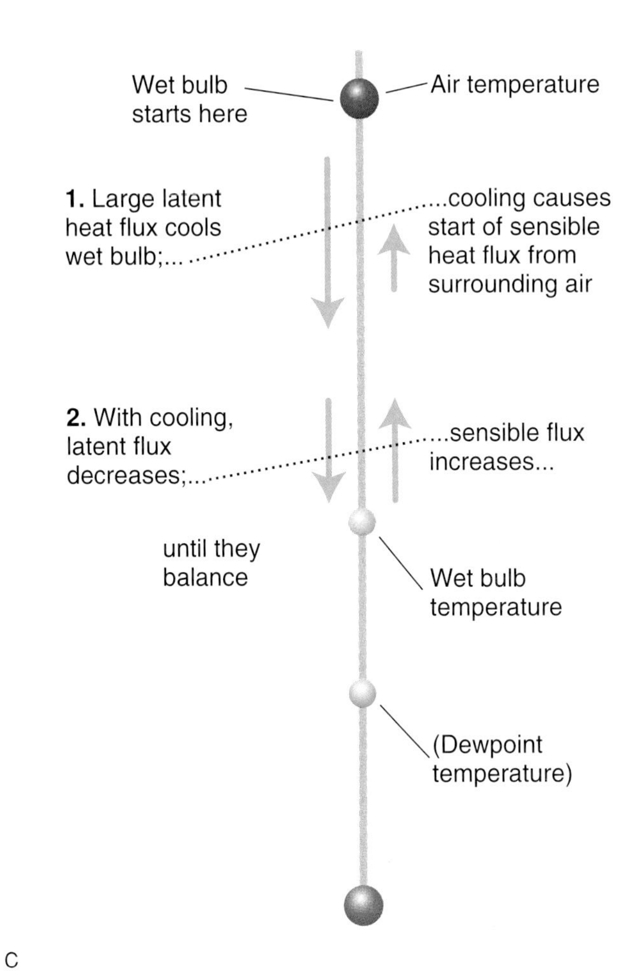


C

Figure 5.12 (*A*) The sling psychrometer and (*B*) its use. (*C*) As the wet bulb is slung, it cools until evaporative heat loss is balanced by sensible heat gain from the atmosphere.

## Checkpoint

***Review***

In this section, you learned about several humidity variables and three tools for determining their values: the saturation mixing ratio chart, the relative humidity formula, and psychrometric tables. The diagrams in Figure 5.13 summarize the various ways to move among these quantities. At the top, you see that the saturation mixing ratio diagram allows you to find the saturation mixing ratio if you know the temperature, and vice versa; it works similarly for the dew point and the mixing ratio. In the lower diagram, if you know the value of any two variables connected to a square, you can employ the tool in that square to find the third variable. Thus, if you know the temperature and the wet-bulb temperature, you can use the psychrometric table in Appendix J to find the relative humidity (RH). If you know the dew point and the relative humidity, you can employ the RH formula to find the temperature (and also the saturation mixing ratio, if you wish).

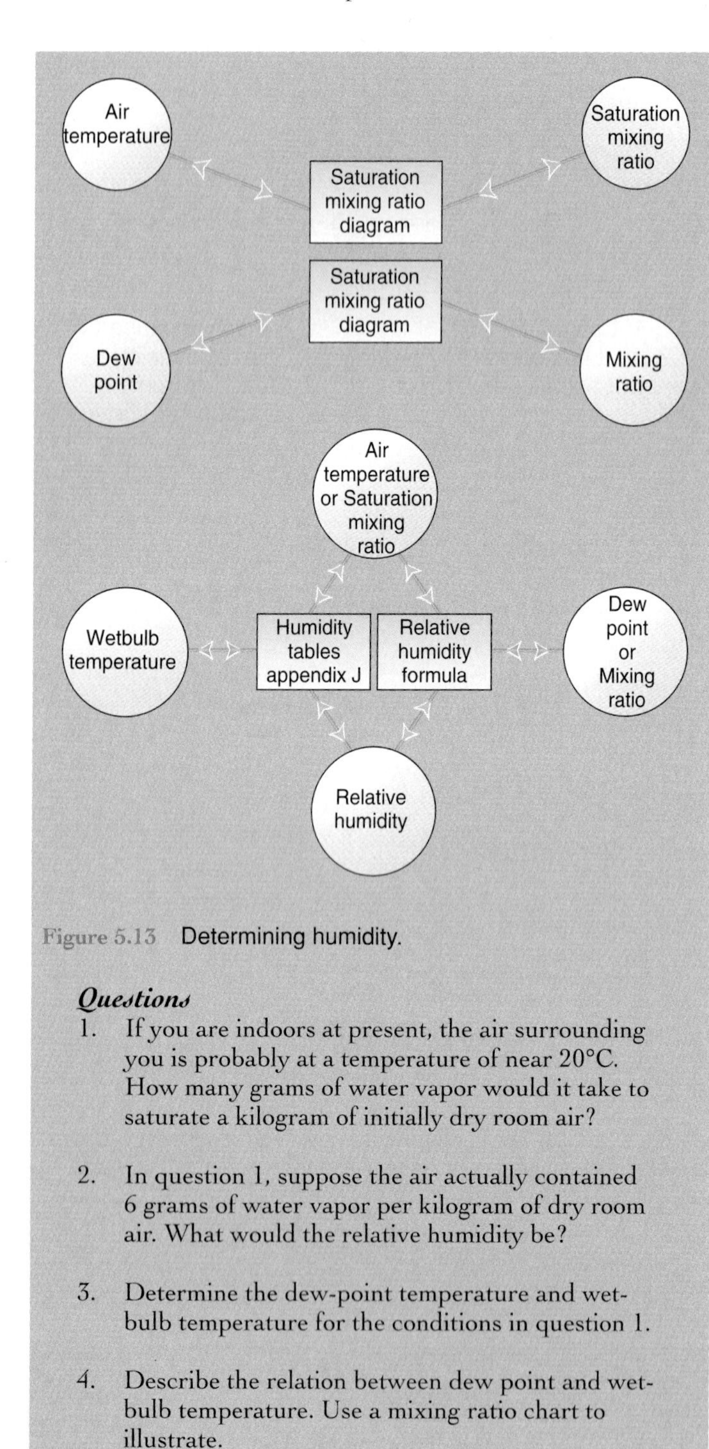


Figure 5.13 Determining humidity.

*Questions*

1. If you are indoors at present, the air surrounding you is probably at a temperature of near 20°C. How many grams of water vapor would it take to saturate a kilogram of initially dry room air?

2. In question 1, suppose the air actually contained 6 grams of water vapor per kilogram of dry room air. What would the relative humidity be?

3. Determine the dew-point temperature and wet-bulb temperature for the conditions in question 1.

4. Describe the relation between dew point and wet-bulb temperature. Use a mixing ratio chart to illustrate.

# Humidity Variations and Their Effect on Human Comfort

Now you possess some powerful tools with which to investigate atmospheric humidity. In this section, you will explore its variation with time of day, season of the year, and location.

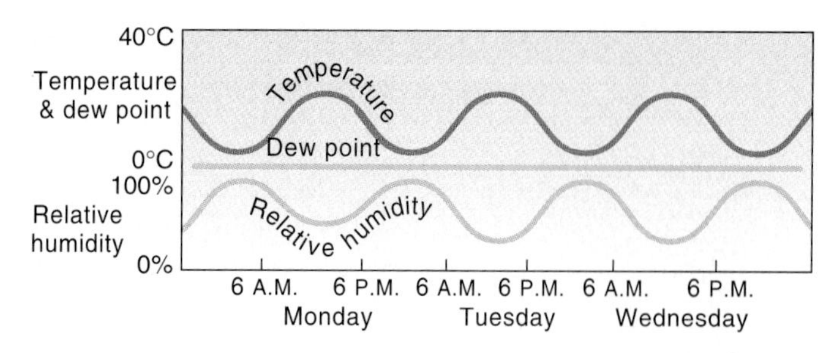


Figure 5.14 Compare the temperature and relative humidity traces on this graph. Note the inverse relation: a decrease in relative humidity accompanies an increase in temperature. What causes this correlation? Does the dew-point graph exhibit a diurnal trend?

## Humidity Variations with Time of Day

How does atmospheric humidity vary during the day? In the absence of frontal passages or sudden wind shifts, Figure 5.14 tells it all. Even though the data represent a single location for just a few days, the patterns they reveal are typical of most places and times.

And just what do the data reveal? First, you can see the strong diurnal (daily) air temperature cycle, driven by solar forcing. Second, notice that in contrast to the temperature pattern, the dew-point temperature shows little diurnal trend; the actual amount of water vapor in the atmosphere does not vary greatly with time of day. Finally, the relative humidity exhibits a strong diurnal cycle, like the temperature curve, but is of opposite phase from the temperature. Thus, low relative humidity occurs with high temperature, and vice versa. You should recall that the reason for this inverse relation between temperature and relative humidity is the dependence of saturation mixing ratio on temperature. Refer back to Figure 5.7 as a reminder.

## Seasonal and Regional Variations in Humidity

According to a favorite expression during the dog days of summer, "It's not the heat, it's the humidity." This expression is so overused, it has been widely parodied. Just where do extremes of humidity occur?

### *Summertime Humidity Conditions*

We might hypothesize that humidity would be highest near large sources of water, such as oceans and the Great Lakes, and lowest far inland. Figure 5.15 shows the mean July distributions of two humidity variables, relative humidity and dew-point temperature. Evidently our conjecture is valid, but only partially. The lowest values of both relative humidity and dew point occur over the desert regions of Nevada, Arizona, California, and Utah. The air in these regions is dry both in relative and in absolute terms. Notice also the eastward bulge of lower relative humidity through the Midwestern states. This region's relative warmth and distance from large water sources results in lower noontime relative humidity values than in neighboring regions.

"But why noontime?" you may be asking. You saw in Figure 5.14 that relative humidity varies with time of day; thus, it is necessary to base the relative humidity map on one specific time. On the other hand, dew-point temperatures vary little with time of day, given no changes of air due to frontal passages, and so forth.

Where are summer humidities the highest? In contrast to the low humidity situation, the answer depends on which humidity variable you consider. Compare the two maps, and notice that highest relative humidities occur along the northwest and northeast coasts, where cold surface water chills moist air. These locations, however, do not report the highest values of dew point. This honor goes to southeastern coastal locations, which receive their humidity from the warm waters of the Gulf of Mexico. Why the discrepancy? Recall once again that relative humidity is a comparison of the air's mixing ratio to its saturation value, and the latter is controlled by the temperature. Even though dew points (and mixing ratios) are highest in the southeast, the high air temperatures there make for high saturation mixing ratio values, hence lower relative humidity than along the northern coasts.

Why is it that a combination of high temperature and high humidity makes people so uncomfortable? The reason is that your body strives to maintain a temperature of approximately 37°C (98.6°F). If your body temperature rises above this value, you perspire. As the perspiration evaporates, its latent heat flux removes heat from your body, lowering its temperature to a more comfortable level. In hot, dry weather, perspiration evaporates readily because the air is unsaturated, and you are able to maintain proper body temperature (as long as you drink enough water). In hot, humid weather, however, the perspiration evaporates more slowly; thus, its cooling effect is greatly diminished. Prolonged exposure to such conditions can lead to fatigue, heat stress, heatstroke, or even death.

So, how much of a strain on the body are, say, conditions of 38°C (100°F) and 30% humidity? Is this worse than a temperature of 30°C (86°F) and humidity of 80%? To answer such questions, meteorologists have devised a variable called the **heat index** (Figure 5.16) to indicate the effects of both heat and humidity on human comfort. Knowing air temperature and relative

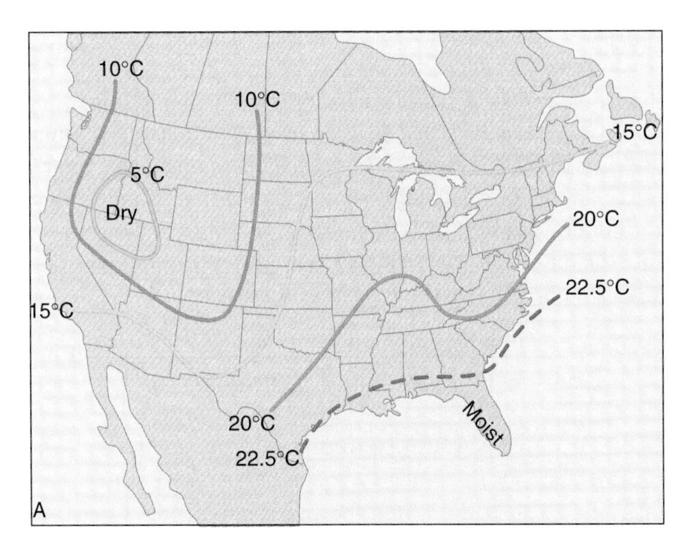


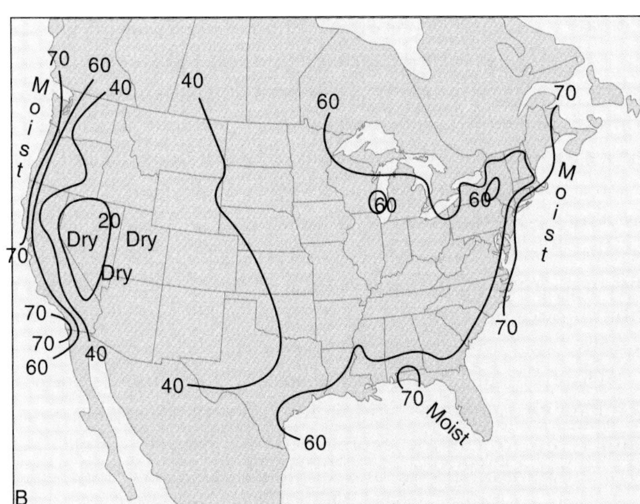


Figure 5.15 Mean July dew-point temperatures (*A*) and noontime July relative humidities (*B*). Notice that lowest values of both variables occur in the desert regions of the West. The location of highest values depends on which variable is considered.

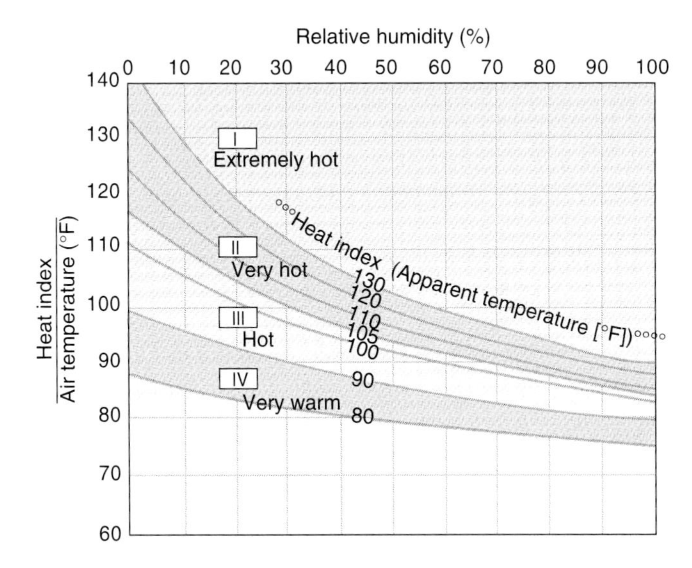


| Category | Apparent temperature (°F) | Heat symptoms |
|---|---|---|
| I | 130° or higher | Heatstroke or sunstroke |
| II | 105° - 130° | Sunstroke, heat cramps, or heat exhaustion likely. Heatstroke possible with prolonged exposure and physical activity. |
| III | 90° - 105° | Sunstroke, heat cramps, and heat exhaustion possible with prolonged exposure and physical activity. |
| IV | 80° - 90° | Fatigue possible with prolonged exposure and physical activity. |

Figure 5.16 The heat index. What level of risk does a person take working in a temperature of 38°C (100°F) and relative humidity of 50%? Note: Heat index temperatures are calculated for weather instrument conditions and therefore underestimate the discomfort experienced by a person in the sun.

humidity, you can determine the stress level on your body. Use the heat index chart to compare the conditions mentioned at the beginning of this paragraph. What stress level does each present to your body?

### *Indoor Humidity Conditions in Summer*

Air conditioning is widely used where there is oppressive heat and humidity in the United States. Air conditioning generally modifies the indoor weather in two ways: by lowering the air temperature, and by removing water vapor from the air. When hot, humid air moves across the air conditioner's cold coils, it is cooled below its dew point. Water vapor condenses onto the coils and is thus removed from the air. (The puddle of water you might have noticed under an air conditioned car was part of the atmosphere, in the form of water vapor, a few minutes earlier.) The amount of water vapor removed by an air conditioner can be substantial: it can amount to several gallons per day in a single home.

### *Wintertime Humidity Conditions*

Mean wintertime surface temperatures across much of the United States are roughly 25°C (45°F) colder than in summer. Does this seasonal cooling affect atmospheric humidity?

Recall that the saturation mixing ratio curve (Figure 5.5) indicates that the air's saturation mixing ratio is lower at lower temperatures. Therefore, the maximum amount of water vapor in the air is limited by the temperature. Stated another way, because the dew-point temperature cannot exceed the air temperature, wintertime dew points and therefore mixing ratios—are lower on the average than in summer.

In extremely cold air, the saturation mixing ratio is very low, indeed: at −25°C (−14°F), only 0.4 grams of water vapor suffice to saturate a kilogram of dry air; at –40°C (–40°F), the saturation value is just 0.08 g/kg. Perhaps recognition of these low saturation values has led to the popular myth that at such temperatures it is too cold to snow. This is a fallacy. Snow may occur at any temperature below freezing, although the amount of snow produced is diminished because of low saturation mixing ratio values at very cold temperatures.

### *Indoor Humidity Conditions in Winter*

Most people are aware that indoor air tends to be very dry in winter. Skin creams and, oils, humidifiers, and vaporizers are among the devices employed to combat the low humidity. Why is indoor dryness in winter such a problem?

To see how this occurs, consider the situation presented in Figure 5.17. The outdoor temperature of −4°C and dew point of −8°C represent mean January values for Detroit, Michigan.

What is the relative humidity of the outside air in Detroit? Entering a saturation mixing ratio diagram with the Detroit temperature and dew point values and using the curve with respect to water, you find that

$$\text{RH} = \frac{\text{mixing ratio at dew point of } -8°\text{C}}{\text{saturation mixing ratio at temperature of } -4°\text{C}} \times 100\%$$

$$= \frac{2.1 \text{ g/kg}}{2.8 \text{ g/kg}} \times 100\% = 75\%$$

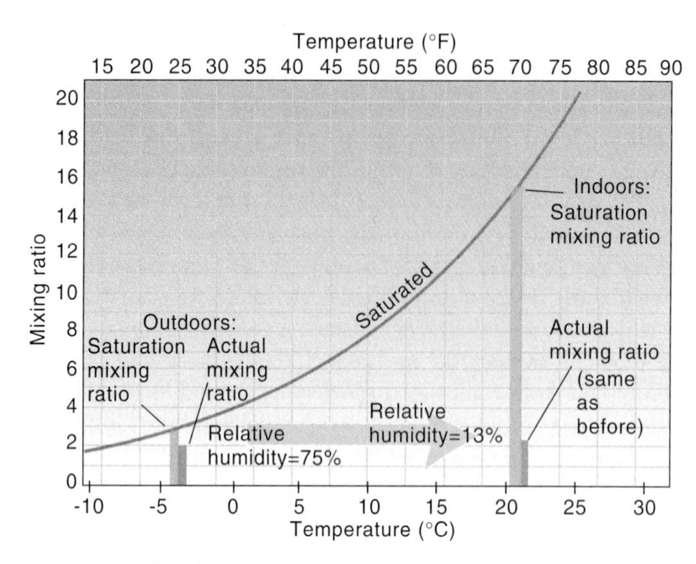


Figure 5.17 As air comes indoors and is heated, its actual mixing ratio remains relatively unchanged. Note, however, the increase in saturation mixing ratio and the corresponding decrease in relative humidity.

Thus, the outdoor relative humidity is fairly high. However, consider what happens when this air travels indoors. The temperature rises, to 21°C (70°F), let us say. The mixing ratio, and therefore the dew point, do not change, however, because moving the air indoors neither adds nor removes water vapor molecules from the air. Thus, the inside air is as shown on the right portion of the graph. What is the relative humidity now?

$$\text{RH} = \frac{\text{mixing ratio at dew point of } -8°\text{C}}{\text{saturation mixing ratio at temperature of } 21°\text{C}} \times 100\%$$

$$= \frac{2.1 \text{ g/kg}}{16 \text{ g/kg}} \times 100\% = 13\%$$

The remarkable result is that the air's relative humidity has fallen from 75% to 13% in the process of coming indoors and being warmed to "room temperature." Similar situations occur all winter long in most locations throughout all but the southernmost states. As a result, indoor air in winter frequently is dry to the point of discomfort. Moisture constantly evaporates from a person's skin, leaving it dry and scaly. Sales of body creams and oils reach a peak during winter months. The oils restore moisture to the skin and retard evaporation, thereby protecting it from excessive dryness. Nasal passages also become dry in the dry air and may crack, allowing infection to enter the body, thus contributing to the frequency of head colds during the winter months.

How do you combat the low indoor humidity of wintertime? You have three choices: add water vapor to the air, lower the air temperature, or employ a combination of these two steps. Vaporizers and humidifiers can add significant amounts of water vapor to the air, as anyone who has had to keep them filled with water can testify. Note also how reducing the indoor temperature helps to raise the relative humidity. Suppose you keep the indoor temperature at 17°C (63°F) instead of 21°C. At 17°C, saturation is only 12.3 g/kg, compared to 16 g/kg at 21°C. Thus, at 17°C,

$$RH = \frac{2.1 \text{ g/kg}}{12.3 \text{ g/kg}} \times 100\% = 17\%$$

Thus, lower wintertime temperatures indoors are more moderate environments in terms of humidity.

***Review***

The main message of this section is that major variations occur, in both time and location, in the amount of water vapor in the atmosphere and its relative humidity. The sensitivity of relative humidity to air temperature is a particularly important concept; often, changes in relative humidity are caused entirely by changes in air temperature, with no variation in the amount of water vapor actually in the air.

***Questions***

1. Use Figure 5.15 to determine the mean July values of dew point and relative humidity in your locality.
2. Explain how one location can have a higher mixing ratio (or dew point) but a lower relative humidity than another.
3. Name some ways in which low humidity is evident indoors in winter.

## Formation of Dew and Frost

Up to this point, the discussion has concentrated on how water vapor *enters* the atmosphere and on how to determine the quantity of water vapor present. For the remainder of this next chapter and the others in this unit, we will deal primarily with mechanisms that *remove* water vapor from the atmosphere. We begin by considering dew and frost formation, which in some ways are "evaporation running in reverse." In Chapters 6 and 7, you will see how clouds and precipitation remove water vapor in more complex ways.

### Dew

It's a pretty safe bet that the hikers in Figure 5.18 are going to have wet feet by the time they finish their walk. Dew formation is a regular summer weather feature in many places. In some locations with sparse rainfall, the nightly occurrence of dew represents an important source of water for small plants and animals.

How does dew form? To answer this question, it is helpful to turn once again to the saturation mixing ratio chart, reproduced in Figure 5.19. Suppose it is a sunny mid-afternoon, and the air's temperature and dew point are 25°C and 15°C, respectively. (Verify for yourself that point A on the chart represents these conditions.)

Now imagine the sun setting. In the absence of incoming sunlight, surface objects cool. Why? Recall from Chapter 3 that under clear skies, the earth's surface emits more longwave radiation than it receives from the atmosphere. The air well above the surface cools much more slowly, because it does not emit radiation so effectively as the surface below; thus, it loses heat energy more slowly. However, the air in contact with the surface is also cooled by conduction. This cooling of the air in contact with the surface is shown by the blue horizontal line in Figure 5.19.

What does it mean when the air temperature reaches its dew point (point B on the chart)? It means that the air's mixing ratio equals its equilibrium, or saturation, value. Thus, water vapor molecules condense onto the surface as rapidly as they evaporate (assuming the surface behaves like the plane water surface in the experiment opening this chapter). A slight addi-

Figure 5.18 In many locations, dew formation is a regular occurrence on clear summer late afternoons and evenings.

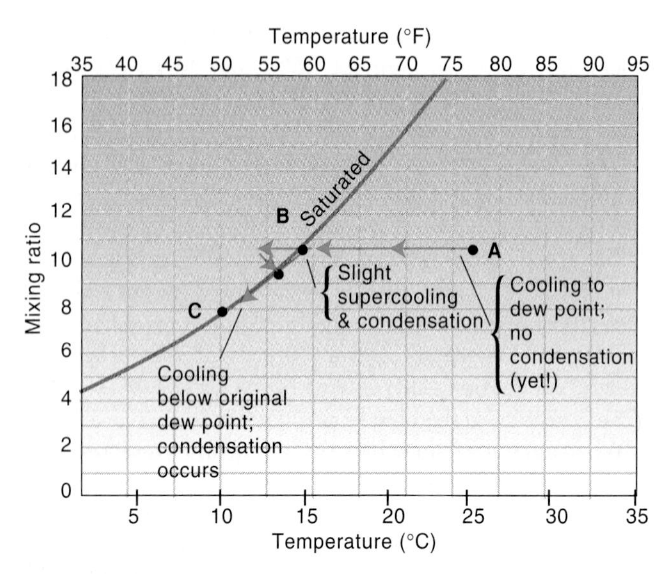


Figure 5.19 Cooling of surface air late in the day and evening (*A* to *B*) may cause saturation and condensation in the form of dew (*B* to *C*). What is the dew point as the temperature falls from *A* to *B*? From *B* to *C*?

tional cooling moves the parcel into the supersaturated region of the chart. At this point, condensation exceeds evaporation, and tiny droplets of liquid water begin to accumulate on the surface, condensed from the air in contact with it. The more the cooling persists, the more water vapor is removed from the air and deposited onto the surface. The process is shown by the line from B to C in Figure 5.19. Note that in this example, the temperature and dew point decrease together from 15°C to 10°C, and the air's mixing ratio decreases from 11 to 8 g/kg. The difference, 3 g/kg, has been removed from the atmosphere and deposited onto the surface as dew.

The following morning, a bright sun will warm the surface and the air in contact with it, providing the energy for the dew to evaporate once more. Some of this water may remain behind, however, absorbed by surface plants or the earth itself.

## Frost

When the dew-point temperature is colder than 0°C, it may be called the **frost-point** temperature. Cooling air to its frost point results in the formation of frost, much as dew forms. However, ice is a crystalline solid, and as frost deposits grow, they often exhibit ice's crystalline structure in spectacular and beautiful ways, as you can see in Figure 5.20.

What happens if dew forms at temperatures just above freezing, and then the dewdrops' temperature falls below freezing? Do the dewdrops turn into frost?

In this situation, the dewdrops do not freeze into frost patterns; rather, each dewdrop simply becomes a tiny sphere of ice sometimes called "frozen dew."

You probably do not think of frost or dew as lethal substances, but under certain circumstances they can be. On clear winter nights when the relative humidity is high, frost or frozen dew may form on roads and bridges; the process may be augmented by water vapor from vehicles' exhaust systems. Dubbed "black ice" in some locales, such deposits are nearly invisible to the driver, and, if thick enough, can make driving extraordinarily hazardous. Black ice is a contributor to many vehicular accidents each winter.

Figure 5.20 These graceful, fern-shaped frost patterns formed as atmospheric water vapor was deposited as ice crystals onto the glass surface.

***Review***

Dew and frost represent a direct return flux of water vapor to the earth's surface. Much of this return is only temporary, as evaporation the next morning returns it to the atmosphere once more. Most dew and frost formation occurs due to surface radiational cooling; therefore, clear, calm nights offer the most favorable conditions for it to occur.

*Questions*

1. It's a clear, still summer evening. The air temperature in the country is 24°C; in the nearby city, it's 27°C. In both locations, the dew point is 19°C. Where is dew formation more likely? Why?
2. Why is dew or frost formation more likely on a clear than a cloudy night?
3. What is the difference between frost and frozen dew?

## How Does Water Vapor Nucleate in the Free Air?

Now, we turn to the processes by which water vapor changes phase from gas to liquid or solid in the free air; by "free air," we mean air not in direct contact with the earth's surface. **Nucleation** is the condensation, freezing, or deposition of water vapor in the free air. When nucleation occurs on microscopic particles (called "nuclei") in the air, the process is known as **heterogeneous nucleation.** (An object is heterogeneous if it is composed of dissimilar elements or parts. For example, a class composed of both women and men is heterogeneous in terms of gender. Similarly, nucleation is heterogeneous if water condenses or freezes onto particles composed of substances other than water, such as salt particles.) Nucleation is an important step toward the formation of clouds and precipitation.

### Condensation

#### *Condensation Nuclei*

Imagine for a moment a parcel of air in the free atmosphere, far from liquid water or ice surfaces. Suppose the parcel's temperature falls. What happens to the parcel's water vapor as the temperature reaches the dew point? Specifically, will condensation occur in the free atmosphere in the absence of a nearby surface such as a blade of grass or liquid water surface on which to condense?

A closer look at the air parcel gives an important clue. A typical cubic centimeter sample of air contains, in addition to its component gases (including water vapor), anywhere from a hundred or so to thousands of minute solid or liquid particles. Many of these particles provide appropriate sites at which condensation can occur and are therefore called **condensation nuclei** (singular: *nucleus*). Thus, as the temperature in the parcel approaches its dew point, condensation commences at many different points within the parcel.

Table 5.1 shows the distribution of condensation nuclei according to size. The units used are micrometers.

*Table 5.1* **Condensation Nuclei Concentrations**

| Nucleus Name | Diameter in Micrometers | Typical Concentration per cc |
|---|---|---|
| Giant | 1 to 10 | none to a few |
| Large | 0.1 to 1.0 | hundreds to thousands |
| Small ("Aitken") | 0.01 to 0.1 | thousands to millions |

Note: Aitken nuclei are named for John Aitken, a Scottish cloud physicist of the late 19th century.

Note the inverse relationship between particle diameter and concentration; larger nuclei are less common. (Reminder: 1 micrometer = $10^{-6}$ m, or 0.001 mm.) In fact, small particles are vastly more common than large ones. However, all the particles listed in Table 5.1 are microscopic.

What is the origin of these particles? The world's winds pick up innumerable microscopic particles of dust, rock, sand, and soil from the earth's surface. At the ocean surface, breaking waves and air bubbles release great numbers of salt particles into the atmosphere. Forest fires and volcanoes contribute significant numbers of condensation nuclei. In addition, anthropogenic sources, particularly combustion in its many forms, add important numbers of nuclei to the atmosphere.

#### *Terminal Velocity*

Why do condensation nuclei remain airborne? Why don't they fall out of the atmosphere, like larger objects made of the same substances?

The answer is that the particles in fact do fall, but at rates so slow that they remain airborne for long periods of time. Furthermore, wind currents may keep very small particles aloft almost indefinitely.

Perhaps you are wondering why smaller particles should fall more slowly than large ones. After all, didn't Galileo prove at the Tower of Pisa that objects of different mass fall at the same rate?

Galileo's experiment (which is a favorite part of the Galileo legend but which he probably never actually performed) is valid for falling objects only if air resistance is of little or no significance. In such a case, the force of gravity accelerates all objects equally, causing them to fall at the same rate as the legend indicates. Cannonballs dropped from a height of a few tens of meters satisfy this condition. However, smaller objects, less dense objects, and objects falling greater distances are affected by a second force, that of air friction, which tends to limit the object's speed of fall.

Figure 5.21 shows how the forces of gravity and friction combine to act on a falling object. Gravity's force on an object is essentially constant; however, frictional forces increase with the object's speed and its physical characteristics. As a result, every

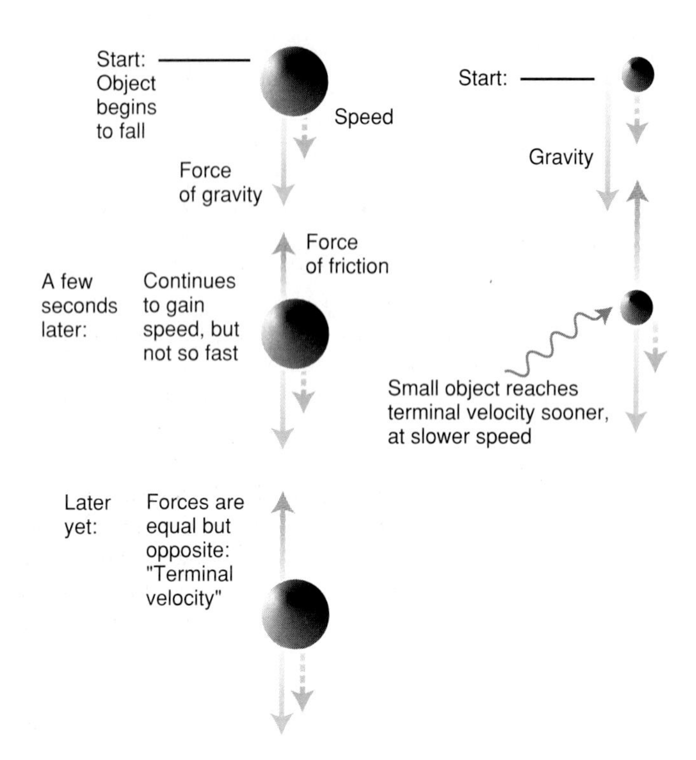


Figure 5.21 As a falling object gains speed, air friction increases. When it equals gravity's force, the object gains no additional speed; it continues to fall at a constant terminal velocity. An object with more of its molecules at its surface experiences a greater frictional force and reaches a slower terminal velocity.

falling object accelerates at first under the greater influence of gravity but eventually reaches **terminal velocity,** the speed at which the motion-resisting effects of friction just match gravity's downward force.

Although every falling object would experience the same acceleration due to gravity alone, the frictional forces would not be equal. Consider a very small object, like the one on the right in Figure 5.21. Nearly all its mass lies close to its surface and is therefore influenced by both gravity and air friction. Friction, directly applied to nearly all the mass, acts as a check on gravity's accelerating influence, limiting the object's terminal velocity to a low value. In a large object, however, like the one on the left, most of the mass lies not at the surface but in the interior, where friction has no effect. Therefore, friction does not directly affect most of this object's mass, and it reaches a faster terminal velocity. Table 5.2 lists terminal velocities for particles of different sizes falling through air.

In summary, the air contains countless minute particles, of both anthropogenic and natural origin, which serve as platforms on which condensation may occur in the free air. Because of their tiny size, these particles have very low terminal velocities and therefore remain airborne for long periods of time. In the free air, they serve as sites for condensation.

### *Water-Attracting and Water-Repelling Nuclei*

Can we assume that water vapor condenses onto these tiny airborne nuclei just as it does onto a plane water surface? In other words, do our findings in the simplified environments presented in the first part of this chapter apply to actual atmospheric conditions? These are important questions; the answer is that there are some important differences, which we need to explore in this section.

Consider the air parcel at point A in Figure 5.22. Assume for a moment that it contains ample condensation nuclei and that condensation onto these nuclei occurs exactly as it does for a plane water surface. We will refer to such particles as "ideal nuclei." If the air parcel is cooled, following the blue line in the figure, you know its relative humidity becomes 100% as it reaches point B. If the parcel is cooled further, it becomes supersaturated; then condensation onto the ideal nuclei in the parcel exceeds evaporation from these nuclei, and the droplets grow.

If the air is warmed, represented by a shift back toward the saturation curve at B, droplet growth slows; at B, evaporation equals condensation (which, as you recall, is the definition of saturation), and the droplets are in equilibrium, neither growing nor shrinking. If the air is warmed further, moving toward A in

**Table 5.2 Terminal Velocities of Atmospheric Particles**

| Particle Diameter | | Terminal Velocity | |
|---|---|---|---|
| Micrometers | Millimeters | Millimeters per second | Typical object |
| 5000 | 5.0 | 9000 (20 mph) | large raindrop |
| 1000 | 1.0 | 4000 | small raindrop |
| 100 | 0.1 | 300 (0.7 mph) | fine drizzle |
| 10 | 0.01 | 3 | cloud droplet |
| 1.0 | 0.001 | 0.04 (5 in/hr) | large nucleus |
| 0.1 | 0.0001 | 0.0004 | Aitken nucleus |

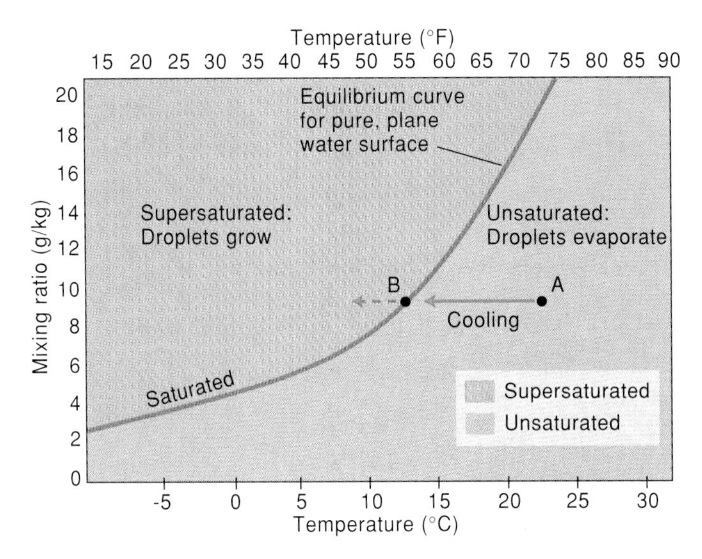


Figure 5.22 Cloud droplet growth on ideal condensation nuclei. The air's location with respect to the saturation curve determines whether ideal droplets will grow, evaporate, or remain unchanged.

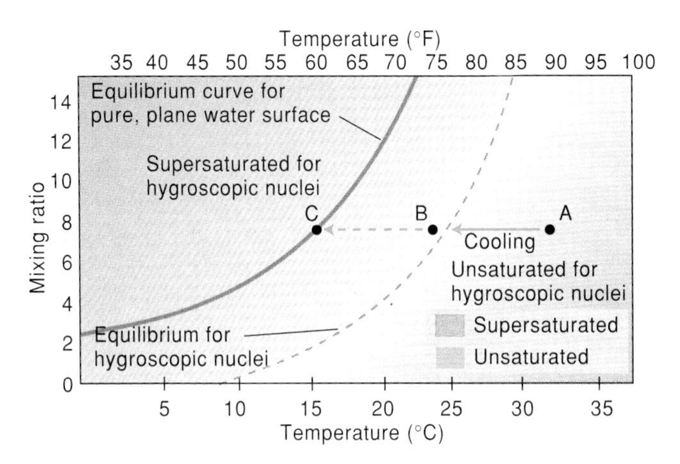


Figure 5.23 Condensation may commence on hygroscopic nuclei well before a cooling air parcel reaches the saturation curve for pure water.

the figure, the air is unsaturated, evaporation exceeds condensation, and the droplets evaporate.

In summary, for droplets to grow on ideal condensation nuclei, the air must be supersaturated (relative humidity > 100%); for the droplets to be in equilibrium, the air must be exactly at saturation (relative humidity = 100%). Droplets tend to evaporate if the air is unsaturated.

With this background, now let's add some realism. The salt grains in a shaker clogged by humid summer weather offer a vivid example of a class of nonideal nuclei called **hygroscopic** substances. A hygroscopic substance is one that attracts water vapor and on which condensation occurs at relative humidity of less than 100% compared to a plane water surface. Examples of hygroscopic substances in the free air include many combustion compounds, such as sulfuric and nitric acid aerosols, in addition to salt particles.

To see how hygroscopic nuclei affect condensation, consider the saturation mixing ratio chart in Figure 5.23. Note that the air parcel represented at point A is far from saturation. Suppose the parcel cools, as indicated by the path in the figure. You know that in the presence of a plane water surface or ideal condensation nuclei, cooling would continue to point C before condensation occurred. If salt nuclei are present in the air parcel, however, condensation may commence as early as point B, where the relative humidity (with respect to a pure water surface) is well below 100%. Thus, hygroscopic nuclei initiate condensation in air that is unsaturated with respect to pure water. Summertime haze in humid weather (Figure 5.24) is often made more visible by condensation onto hygroscopic nuclei at relative humidities well below 100%. The demonstration suggested in the explorations at the end of this chapter provides another striking example of the effective role of hygroscopic nuclei in condensation.

Not all condensation nuclei are hygroscopic. Some, such as particles of oil or fat, are **hydrophobic,** that is, water vapor-repelling. The ambient relative humidity must exceed 100% before hydrophobic nuclei are activated as condensation nuclei.

### *Spontaneous Nucleation*

Imagine a sample of free air has been filtered to remove all condensation nuclei. What happens to its water vapor if the parcel is cooled to its dew-point temperature? In such a case, condensation does not occur at the dew-point temperature. With continued cooling, the parcel becomes supersaturated. In laboratory experiments, parcels have been cooled to achieve high degrees of supersaturation—corresponding to relative humidities of several hundred percent. At greater and greater supersaturations, however, the probability rises that small numbers of water vapor molecules will randomly clump together and remain together long enough to provide a condensation site without benefit of a foreign nucleus. This process of changing to a denser phase without benefit of a seed nucleus is known as **homogeneous nucleation.** The term *homogeneous* is apt, as no substances other than water (i.e., no foreign nuclei) are required for condensation to occur.

Homogeneous nucleation of water vapor into liquid droplets is rare in the free atmosphere. Generally, there is an abundance of suitable nuclei on which condensation proceeds at humidities near 100%, thus preventing the extreme supersaturation required for homogeneous nucleation. The situation is different, however, in the case of ice crystal formation, as you will see shortly.

### *Curvature and Solute Effects*

We have now explored nearly all the steps that lead to condensation and therefore cloud droplet formation in the free atmosphere. In short, as long as some water vapor is present, you can achieve saturation by cooling, and generally abundant condensation nuclei exist on which nucleation can occur. Just one

Figure 5.24 Haze. The relative humidity of the air pictured here is considerably less than 100%.

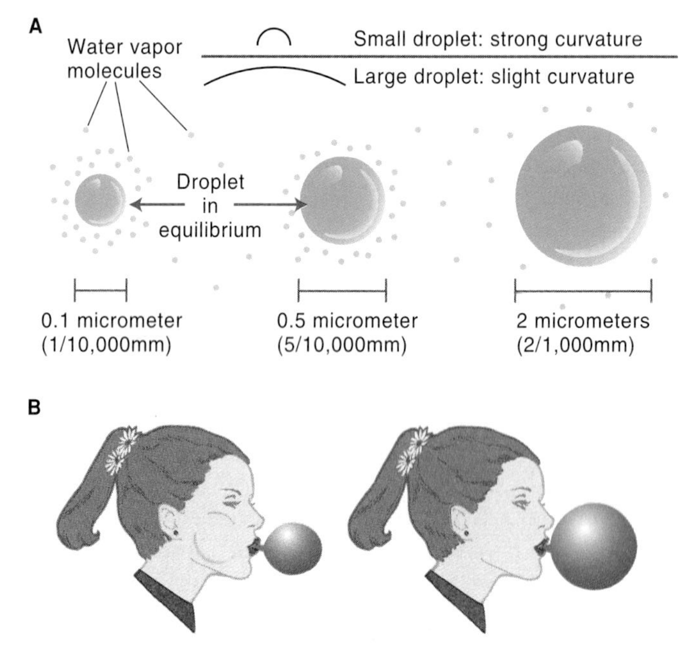


Figure 5.25 The curvature effect. (*A*) Equilibrium vapor pressure is higher over the strongly curved surfaces of small droplets, which prevents them from growing unless the surrounding air is substantially supersaturated. Larger droplets require less supersaturation. (*B*) The inflating balloon analogy. Initially, it takes greater pressure to cause the balloon to expand. Once past that point, inflation is easier.

problem remains, which is illustrated in Figure 5.25 and is known as the **curvature effect:** the smaller the droplet, the greater its surface curvature and the greater its equilibrium vapor pressure. As a result, the droplet is at equilibrium with its surroundings not at a relative humidity of 100% but at a considerably higher value. The smaller the droplet, in fact, the greater the equilibrium vapor pressure of its surface. For the droplet to form in the first place and then to begin to grow, the surrounding air would have to be supersaturated beyond this equilibrium level, a condition rarely observed. The problem is analogous to inflating a balloon, shown in Figure 5.25B. How, then, are cloud droplets actually able to form?

Two properties of certain condensation nuclei extract us from this dilemma. One is nucleus size. Water condensing on a large condensation nucleus has much less curvature than on a smaller particle. In fact, droplets greater than about 2 micrometers in diameter show no significant curvature effect; thus, a condensation nucleus of this size provides an escape from the curvature problem. However, you saw in Table 5.1 that such "giant nuclei" are rather rare in the atmosphere.

Hygroscopic nuclei such as salt that dissolve upon condensation offer another way out of the dilemma. The dissolved substance (the "solute") causes a decrease in equilibrium vapor pressure. Therefore, water vapor molecules may condense onto the droplet at humidities less than 100%, a circumstance called the **solute effect.**

In Figure 5.26, you can see conditions under which the solute effect is important. Notice that once the nucleus dissolves, the solution becomes progressively weaker as the droplet grows; therefore, the solute effect, like the curvature effect, is most important when droplet diameter (and volume) is small.

The curvature effect and the solute effect are both important at small droplet diameters. How do they combine to affect an actual growing droplet? The curves in Figure 5.27 show the relative humidity required for equilibrium for droplets of several nucleus diameters. Notice that the horizontal axis is labeled "droplet diameter," with values increasing from left to right. Thus, a growing droplet will trace some trajectory from left to right in Figure 5.27.

Of course, not all droplets grow. For a droplet to grow, the relative humidity of the air surrounding the droplet must remain greater than the droplet's equilibrium value.

Let's turn to an example. Consider the fate of a droplet that forms on a condensation nucleus 1 micrometer in diameter. The equilibrium curve for this droplet is shown in green in Figure 5.27. Will this droplet grow? It depends on the relative humidity of the surrounding air. Suppose the relative humidity is measured to be 100.5% and remains at this value throughout the period we consider. Notice that the droplet's equilibrium curve lies entirely below the 100.5% level; therefore, the droplet will continue to grow throughout the range of diameters shown in the figure. On the other hand, if the relative humidity is just 100%, the same droplet will not grow beyond point X in the figure: on reaching X, the droplet is in equilibrium with the air at the current relative humidity (100%), and no further growth will occur.

Observe in Figure 5.27 that the blue curve, representing the equilibrium curve for a 0.1 micrometer nucleus droplet, lies entirely above the green (1 micrometer nucleus) curve. Thus, a droplet forming on a smaller condensation nucleus requires higher values of relative humidity at each step in the growth process than does a larger droplet.

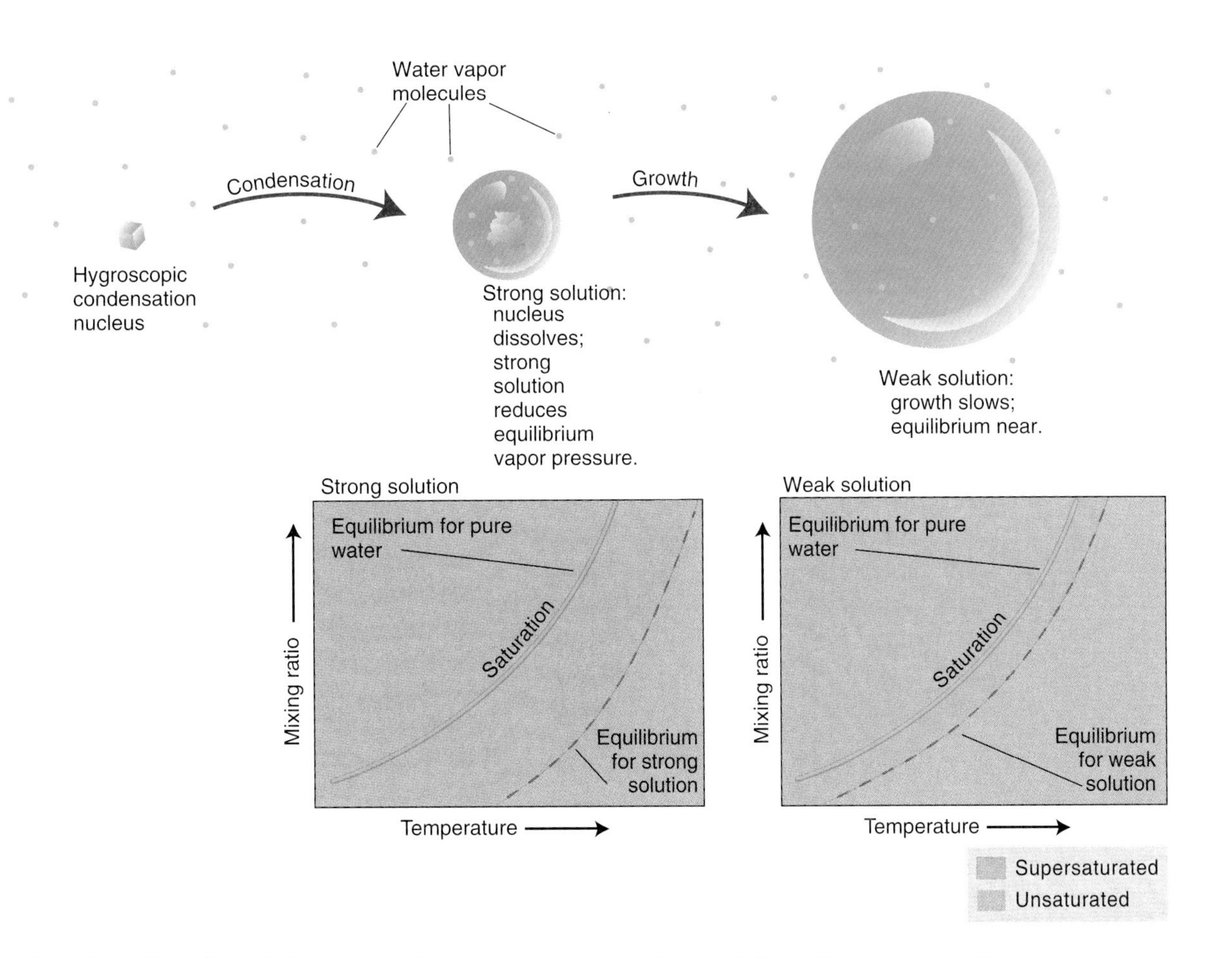


Figure 5.26 The solute effect. A small droplet with dissolved nucleus is a strong solution with reduced equilibrium vapor pressure. The droplet will grow under a wide range of relative humidities (green line). As it grows, the additional water dilutes the solution, and the equilibrium vapor pressure increases.

At smallest droplet diameters, Figure 5.27 indicates that equilibrium occurs at humidity values less than 100%, evidence that the solute effect prevails at first. The equilibrium curves indicate, however, that humidity in excess of 100% is required for growth to larger diameters, a consequence of the curvature effect. Once the droplets grow larger than a few micrometers, both effects become insignificant; the equilibrium lines converge on 100%, the value for plane water surfaces.

Figure 5.27, despite its detail, does not capture all the intricacies of cloud droplet growth. Other effects, such as electrical forces, may affect droplet formation and growth. These are important areas of current research in meteorology.

## Nucleation at Temperatures below 0°C

Still other complexities arise when nucleation occurs at temperatures below 0°C. Then, as you would expect, nucleation often is in the form of ice crystals.

But note that we said "often," not "always." Frequently, nucleation at temperatures colder than 0°C results in liquid droplets, not ice. This section deals with this surprising and important circumstance.

### *Supercooling*

Fill a plastic container with water, insert a thermometer, and place this apparatus into a very cold freezer. What happens? You know that the water will freeze eventually. How will the temperature change throughout this process? The blue line in Figure 5.28 shows a typical plot.

Notice from the graph that the water temperature shows a rather steady decline until it reaches 0°C. Soon after that time, ice begins forming around the edges of the water surface. For some time, the temperature remains at 0°C, as ice formation proceeds. Once all the water is frozen, the temperature resumes its decline until it reaches the same temperature as the surrounding air.

Now, suppose you repeated this experiment but substituted cloud droplets in the free atmosphere in place of the container of water. Your results for three different droplets are plotted in red in Figure 5.28. What do these plots indicate? Observe that the droplets did not freeze at 0°C but were **supercooled.** Supercooled water is water that remains liquid at temperatures colder than 0°C. Notice also that when the supercooled droplets finally did freeze, they did so at a variety of different temperatures. The photo in Figure 5.29 shows supercooled droplets freezing under laboratory conditions (the freezing compartment of this author's refrigerator).

Unlike supersaturation, which occurs to only a very slight extent in the free air, considerable supercooling of liquid cloud droplets is the norm. Furthermore, water vapor colder than 0°C often condenses as supercooled droplets rather than undergoing deposition to ice. Why don't water vapor and droplets in the free air freeze at "freezing"?

### *Ice Nuclei*

The "problem" turns out to be not with the water but with the nuclei onto which it would freeze. Ice, a crystalline substance, requires an appropriately shaped nucleus on which to form. An existing piece of ice works just fine, but how does *that* piece form? To initiate the process of ice crystal formation, an **ice nu-**

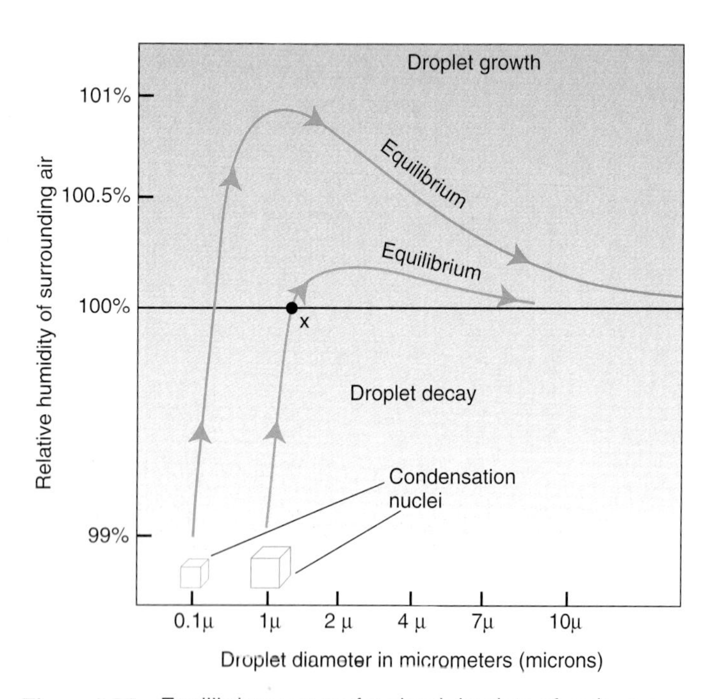


Figure 5.27 Equilibrium curves for cloud droplets of various nucleus diameters. The curves are shaped by the solute and curvature effects.

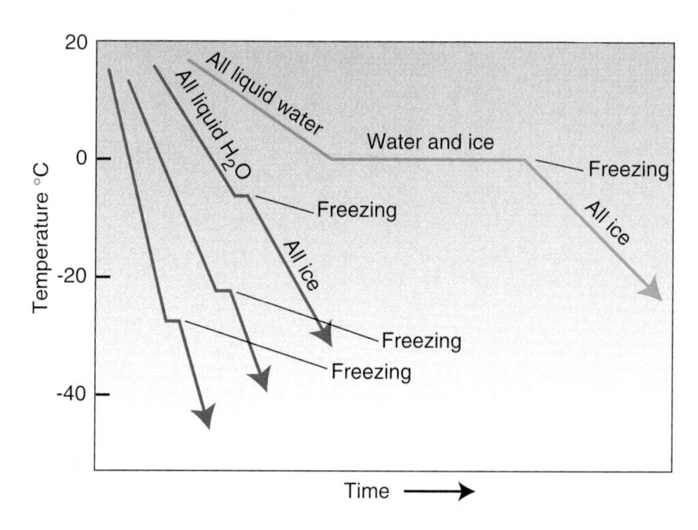


Figure 5.28 Unlike a container of water that freezes at 0°C (blue curve), droplets may be supercooled below 0°C (red curves) before freezing. A droplet that happens to contain relatively effective freezing nuclei freezes at a temperature only slightly colder than 0°C, as shown in the right-hand red curve. Droplets more deficient in freezing-nuclei droplets require considerable supercooling before freezing, as shown by the other red curves.

A

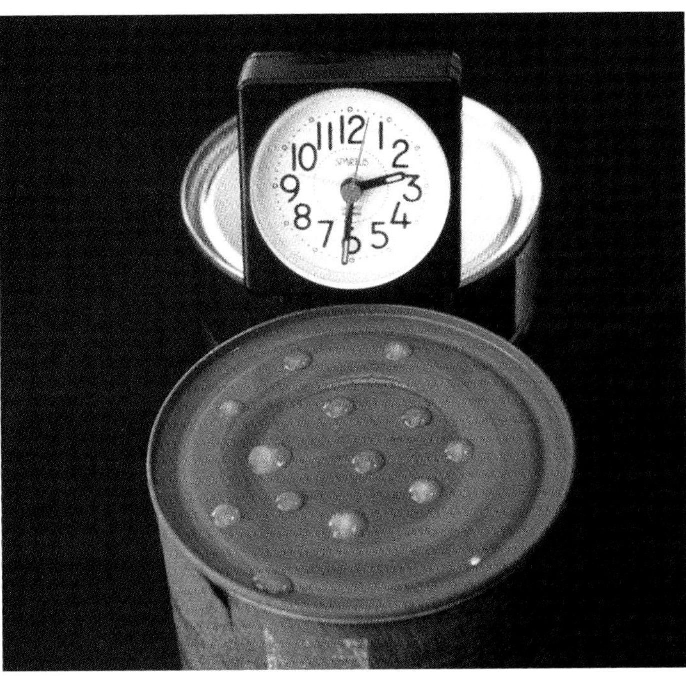

B

Figure 5.29 (*A*) Water drops on a tin can were placed in a freezer at 2:20 P.M. Temperature in the freezer was −15°C (+5°F). (*B*) After ten minutes in the freezer, some drops remained liquid, while others had frozen entirely. (Thanks to C. Bohren for the inspiration.)

**cleus** is required, that is, a particle whose molecular structure is similar to the crystal structure of ice. Such shapes are relatively uncommon in nature (except in ice crystals themselves); therefore, water vapor and water droplets in the free air frequently lack a suitable site onto which to freeze, and freezing simply does not occur at 0°C.

Figures 5.28 and 5.29 suggest one source of ice nuclei. The figures indicate that a droplet *does* freeze eventually if its temperature falls far enough below 0°C. How far is "far enough"? The answer depends on the shape of the nucleus and the size of the droplet, as well as on the temperature. Figure 5.30 shows events within typical air parcels as they approach saturation at temperatures near or below 0°C. Note that in a cloud just a few degrees colder than freezing virtually every particle is in the liquid phase. Typical clouds at −10°C to −20°C consist of a mixture of ice and water droplets. However, at temperatures colder than −39°C, known as the temperature of **spontaneous nucleation,** all cloud particles are ice crystals, regardless of the shape of their nuclei. Spontaneous nucleation means that liquid droplets freeze whether they contain ice nuclei or not.

Another mechanism of ice nuclei production in clouds is known as **contact nucleation.** Contact nucleation is the freezing of supercooled droplets as a result of collision with other particles. The colliding particles need not be ice nuclei themselves; the act of collision appears sufficient to initiate freezing in the supercooled droplet.

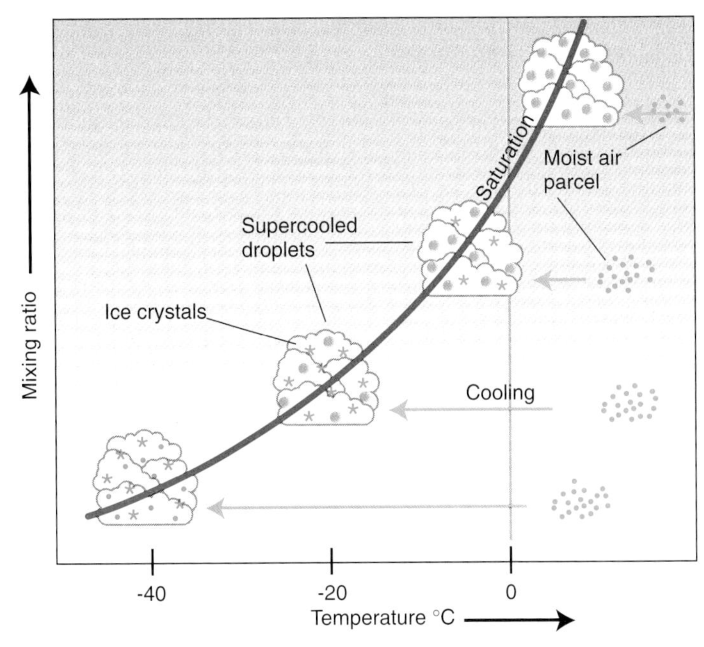


Figure 5.30 Droplet and ice crystal populations within clouds colder than 0°C. The farther below 0°C, the smaller the proportion of liquid droplets. Below −39°C, all droplets have frozen.

A

B

Figure 5.31 (*A*) Deposits of rime ice and frozen precipitation must be removed by aircraft deicing operations, shown here. (*B*) A pine tree in the California mountains is a victim of heavy rime ice deposits, caused by the freezing on contact of supercooled cloud droplets.

## *Evidence of Supercooling*

The concept of supercooled water may seem so foreign that you may be wondering about its significance outside of laboratory experiments. Examples abound, such as when supercooled cloud (or fog) particles impinge on some object such as a tree branch or an airplane wing and freeze on contact. This freezing of supercooled droplets on an object leads to an ice coating known as **rime ice.** Figure 5.31 gives evidence of the reality of supercooling. Rime ice is a serious hazard for aircraft flying through supercooled clouds. The rime buildup on the wing's leading surface alters the air flow, which in turn diminishes the amount of lift the wing can provide.

# Checkpoint

***Review***

The concept map in Figure 5.32 summarizes nucleation processes.

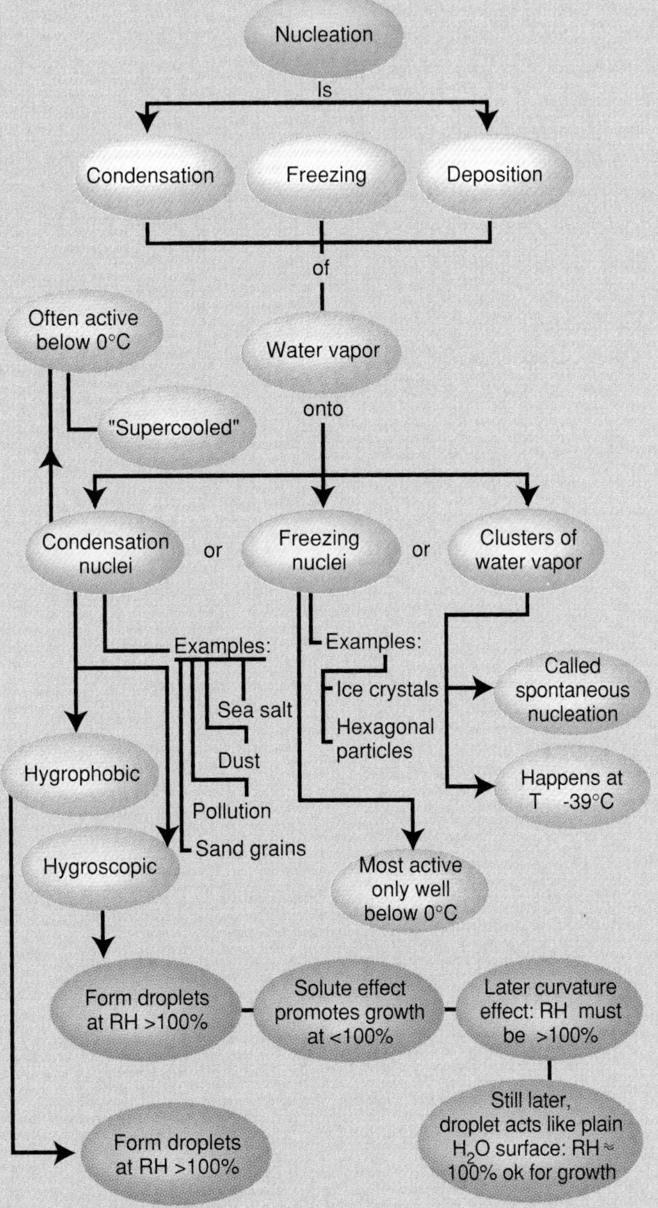


Figure 5.32 Concept Map: Nucleation.

***Questions***

1. What is a hygroscopic condensation nucleus? How do hygroscopic nuclei affect the onset of condensation in a cooling air parcel?
2. How does curvature affect the growth of very small droplets?
3. What is the temperature of spontaneous nucleation? What happens at that temperature?
4. Compare the condensation of water vapor above and below 0°C.

# Study Point

## Chapter Summary

In this chapter, you studied the evaporation of water from the earth's surface (primarily its oceans) into the atmosphere and then its condensation or deposition within the atmosphere as cloud particles—many of which have nuclei of sea salt. In the process, you learned how to solve a variety of problems relating temperature, relative humidity, dew point, wet-bulb temperature, and mixing ratio. You saw that relative humidity is a measure of how much water exists in vapor form compared to the equilibrium or saturation value. You saw that relative humidity is highly sensitive to temperature changes. You also considered the geographic and temporal variations of humidity, both absolute and relative, and some ways humidity affects the human body.

The chapter continued with a discussion of condensation as dew and frost and (more important for our purposes) in the free atmosphere, onto condensation and freezing nuclei. Such condensation and deposition are the genesis of cloud particles. You learned that a cloud droplet's diameter and the amount of impurity within its water mass can influence the efficacy with which the water vapor condenses onto the drop, hence its rate of growth. Finally, you explored the formation of frozen cloud particles and the process of supercooling, a common atmospheric phenomenon.

In the next chapter, you will study the specific circumstances leading to formation of different cloud types. How does the air become saturated? How are these conditions sustained? What cloud types result? This leads to the discussion of precipitation mechanisms and attempts at precipitation modification in Chapter 7.

## Key Words

| | |
|---|---|
| saturated | vapor pressure |
| equilibrium vapor pressure | unsaturated |
| supersaturated | mixing ratio |
| saturation mixing ratio | relative humidity |
| hygrometer | dew-point temperature |
| psychrometer | wet-bulb temperature |
| heat index | frost point |
| nucleation | heterogeneous nucleation |
| condensation nucleus | terminal velocity |
| hygroscopic | hygrophobic |
| homogeneous nucleation | curvature effect |
| solute effect | supercooled water |
| spontaneous nucleation | ice nucleus |
| contact nucleation | rime ice |

## Review Exercises

1. Distinguish between vapor pressure and mixing ratio.
2. What is the equilibrium water vapor pressure at a temperature of 15°C? What is the saturation mixing ratio at 15°C and air pressure of 1000 millibars?
3. As a rough rule of thumb, you can figure that the saturation mixing ratio doubles for each 10°C increase in temperature. Suppose outside air at −10°C in winter is warmed indoors to +20°C. By how many times has the air's saturation mixing ratio increased? What would be the trend of its relative humidity in such a warming?
4. Perhaps the highest dew-point temperature ever recorded was a value of 34°C (93°F), in Sharjah, Saudi Arabia. What was the vapor pressure at the time? If the relative humidity at the time was 70%, what was the temperature? What was the heat index value for this air?
5. For air whose temperature and dew point both are 17°C, what is the relative humidity? What is the wet-bulb temperature? Where would such an air parcel be plotted in a mixing ratio chart?
6. Explain how clothes dryers and hair dryers enhance evaporation.
7. If you were to hang your laundry to dry out of doors, what would be the best time of day or night to do so? Why?
8. Why are the two sets of lines on the map in Figure 5.33 so similar?
9. Dew tends not to form on a night when it is windy. Why?
10. Why does dew form more readily on objects like blades of grass than on sidewalks or streets?
11. Compared to a 1-micrometer droplet, how many times faster does a 10-micrometer droplet fall? A 100-micrometer droplet? A 1-millimeter raindrop?

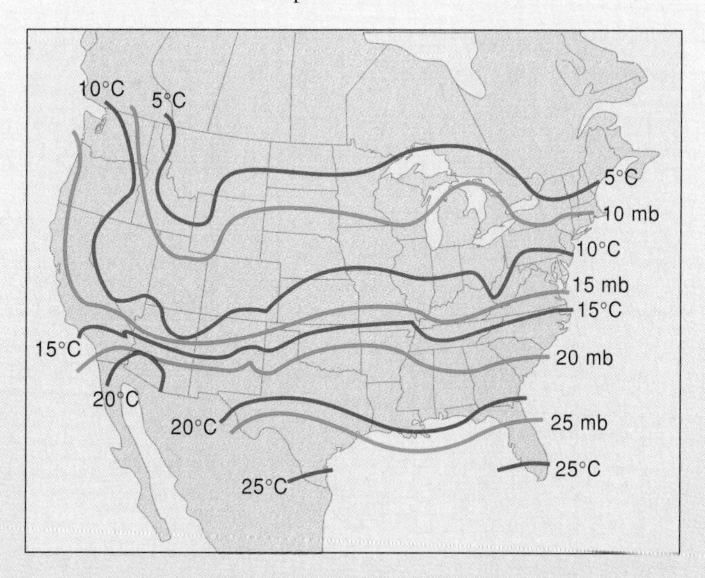


Figure 5.33 Mean April temperature (red) and mean April saturation vapor pressure (green).

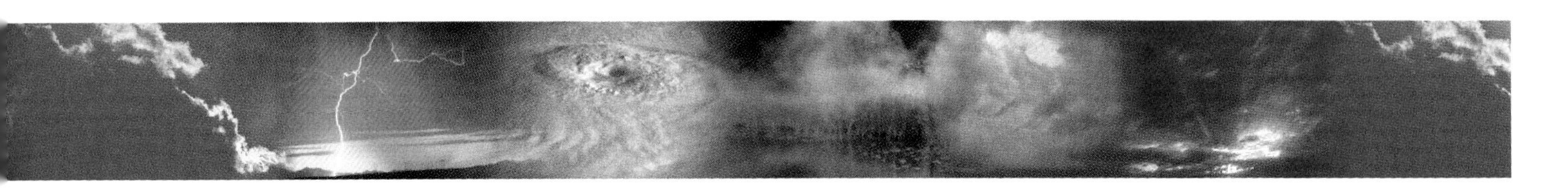

## Problems

1. A common but incorrect expression is that "the air holds water vapor." The expression implies that the air somehow supports the water molecules. How do the experiments at the outset of this chapter show that this notion is incorrect?
2. Imagine your hair is 25 centimeters (one foot) long. By how many centimeters does it change length as you move from a living room whose relative humidity is 25% (typical indoor value in winter) to a shower room in which the humidity is 100%?
3. A condensation nucleus 1 micrometer in diameter floats along at an altitude of 1 kilometer above the earth's surface. What is the particle's terminal velocity? At that rate, how long will it take the nucleus to reach the ground? Assume there are no vertical winds.
4. You are curious about the humidity in your room, so you make a simple psychrometer and record the following measurements: temperature = 21°C, wet-bulb temperature = 9°C. From a TV weather program you learn the air pressure currently is 1000 millibars. (a) What is the air's relative humidity? (b) Its dew point? (c) Its mixing ratio? (d) Its saturation mixing ratio?
5. Suppose your room in the previous problem measures 3 meters by 4 meters by 2.5 meters high. Assume that the air's density is 0.8 kg per cubic meter. (a) What is the mass of air in the room? (Assume the entire volume is filled with air.) (b) How many grams of water vapor are in your room?
6. A parcel of air has a temperature of 25°C and a mixing ratio of 10 g/kg. It contains hygroscopic nuclei that become active at relative humidity = 85%. At what temperature will the water vapor in the parcel begin to condense?
7. How long would it take a cloud droplet 0.01 millimeters in diameter to fall to the ground from a cloud 1000 meters in altitude? How long would it take for an Aitken nucleus?

## Explorations

1. Construct a human hair hygrometer; refer to Figure 5.8 for suggestions. A key feature is some mechanism for magnifying the hair's change of length. (One possibility is shown in Figure 5.8.) Once constructed, calibrate it. (A steamy shower room can provide a 100% humidity environment. A second instrument is useful, however, for a complete calibration.) Then record the instrument's readings, compared to those of another instrument.
2. Investigate the fall velocities of different objects in different media. For example, drop pebbles of various diameters into a tall, water-filled tube; measure the time taken to fall as a function of particle diameter or mass. Experiment with different fluids: water, oils of different thicknesses, and so on.
3. Bring some water to a boil in a teakettle. Observe the tiny cloud that forms at the spout. Now light a match, extinguish it, and wave the match's smoke past the spout. Observe the changes in the cloud.
4. Which humidity variable typically undergoes the greater change over a 24-hour period—dewpoint or relative humidity? Use real data to support your answer as follows: log onto the Internet to collect hourly temperature and dew-point data for one station near you for a 24-hour period, and then calculate relative humidity for each hour in the period. (See this chapter's Resource Links for suggested Internet addresses.) Plot the data on a graph by hand or using a computer spreadsheet program.

## Resource Links

The following is a sampler of print and electronic references for topics discussed in this chapter. We have included a few words of description for items whose relevance is not clear from their titles. For a more extensive and continually updated list, see our World Wide Web home page at www.mhhe.com/sciencemath/geology/danielson.

Bohren, C. F. 1987. *Clouds in a Glass of Beer.* New York: John Wiley and Sons. 195 pp. Lively writing on many topics including wet-bulb temperature and condensation.

Bohren, C. F. 1991. *What Light through Yonder Window Breaks?* New York: John Wiley and Sons. 190 pp. Much of interest, especially a chapter on dew.

Hourly dew-point data for stations throughout North America are available at Pennsylvania State University's weather page, at http://www.ems.psu.edu/cgibin/wx/uswxstats.cgi , and at AccuWeather. AccuWeather also provides relative humidity values.

U.S. maps of total atmospheric water content ("precipitable water") are available from Purdue University's weather server at http://wxp.atms.purdue.edu/maps/upper_air/ua_con_prec.gif .

AccuWeather's home page at http://www.AccuWeather.com provides visible and infrared satellite images, and Doppler radar images.

For readers with access to AccuData, <send> SWVUSF displays water vapor images, SURA allows the user to plot and analyze worldwide data including dewpoint, mixing ratio and relative humidity. SFC gives hourly weather observations.

# Chapter 6

## How Do Clouds Form?

## Chapter Goals and Overview

Imagine you are a weather forecaster working the graveyard shift at the local airport. Throughout the night, you have watched the temperature gradually fall; now, at 4:00 A.M., it is just 3°C (5°F) above the dew point. Several inbound aircraft are due to land near sunrise, two hours from now. Their crews need to know if fog will close the airport; if so, they must change course now for another city. What is your forecast? The problem is not an unrealistic one; it is typical of the problems thousands of meteorologists across the globe face each day.

This chapter deals with the formation of clouds, specifically with the meteorological and geographic circumstances that produce different cloud types. You will find this information important as you investigate this unit's central question on causes of precipitation and prospects for precipitation modification. However, the subject of cloud formation is important in its own right, as the example above suggests.

As you know from Chapter 5, cooling an air parcel can raise its relative humidity toward 100%, which in the free air can lead to cloud formation. In fact, cooling plays a role in the formation of virtually every cloud. Despite this single unifying factor, a surprisingly wide variety of cloud forms occur. First, we'll identify and explore the cloud types, then examine how atmospheric and geographic conditions cause the differences in cloud forms. Since the cooling that takes place when air rises is so significant, we need to explore this subject in some depth.

### Chapter Goals

Upon completing this chapter, you will be able to:

- identify representative clouds from each of the major cloud genera;
- for a given cloud, give a plausible explanation of how it formed;
- distinguish between stable and unstable clouds;
- interpret atmospheric soundings to identify stable layers, unstable layers, inversions, clouds; and
- use thermodynamic diagrams to solve a variety of problems in atmospheric thermodynamics.

## Basic Cloud Genera

Meteorologists classify clouds according to three properties: the cloud's basic shape, its height above the ground, and whether or not it generates precipitation. Figure 6.1 maps the basic cloud genera according to these properties.

Unfortunately, each of the three classification properties has a degree of imprecision about it. For example, you know from personal observation that cloud shapes are not always distinct; a given cloud may change shape radically or blend with another cloud. Similarly, classification by height has its difficulties: in the tropics and in warm weather, middle and high clouds tend to form at higher altitudes than under colder conditions. These problems serve as a reminder that clouds do not have distinct identities in the way an elephant, say, is distinct from an eel; instead, a single cloud may evolve through several different genera during its life span. And as it evolves, there will be times when it does not fit neatly into one of the 10 categories shown in the figure. Once you recognize these limitations, however, we expect you will find the cloud classification scheme of Figure 6.1 has considerable value. With this system as a reference, you will now take a close look at each of the 10 genera. Study the photos closely; spend more time on them than on the accompanying text. A cloud photo is worth a thousand words.

### Low Clouds

**Low clouds** are those whose bases typically are within 2000 meters (6500 feet) of the earth's surface. As you see from the examples in Figure 6.2, low clouds include stratus, stratocumulus, and some nimbostratus clouds. Low clouds generally are composed of water droplets or, if below freezing, supercooled droplets but generally not of ice crystals.

#### *Stratus*

**Stratus** clouds (Figure 6.2C) may be so low, their bases actually touch the earth's surface; a surface-based cloud is called **fog.**

Identifying stratus generally is straightforward: its closeness to the ground often is apparent. Stratus skies generally look dull and featureless. Fog and stratus clouds are far more common in early morning than at other times of day. As you will see, the underlying surface often exerts a strong influence on stratus formation.

#### *Stratocumulus*

As its name suggests, **stratocumulus** is a hybrid type, a stratiform cloud possessing roll or cellular structure typical of cumulus clouds. The photo of stratocumulus in Figure 6.2D may look familiar to you; stratocumulus is one of the most common cloud genera.

#### *Nimbostratus*

Stratiform clouds yielding precipitation are called **nimbostratus.** (Recall from Chapter 1 that *nimbus* means rain.) Typical nimbostratus skies remain dull, featureless, and continuously rainy for hours at a time.

### Middle Clouds

You can recognize **middle cloud** names by the prefix *alto-* (an unfortunate choice, perhaps, since *alto* in Latin means "high." But then, the alto voices in a choir are not the high ones either.). There are two middle-cloud genera, as you can see in Figure 6.2: **altostratus** and **altocumulus.**

Middle-cloud bases generally occur between 2000 and 7000 meters (6500–23,000 feet) above the ground. Cloud tempera-

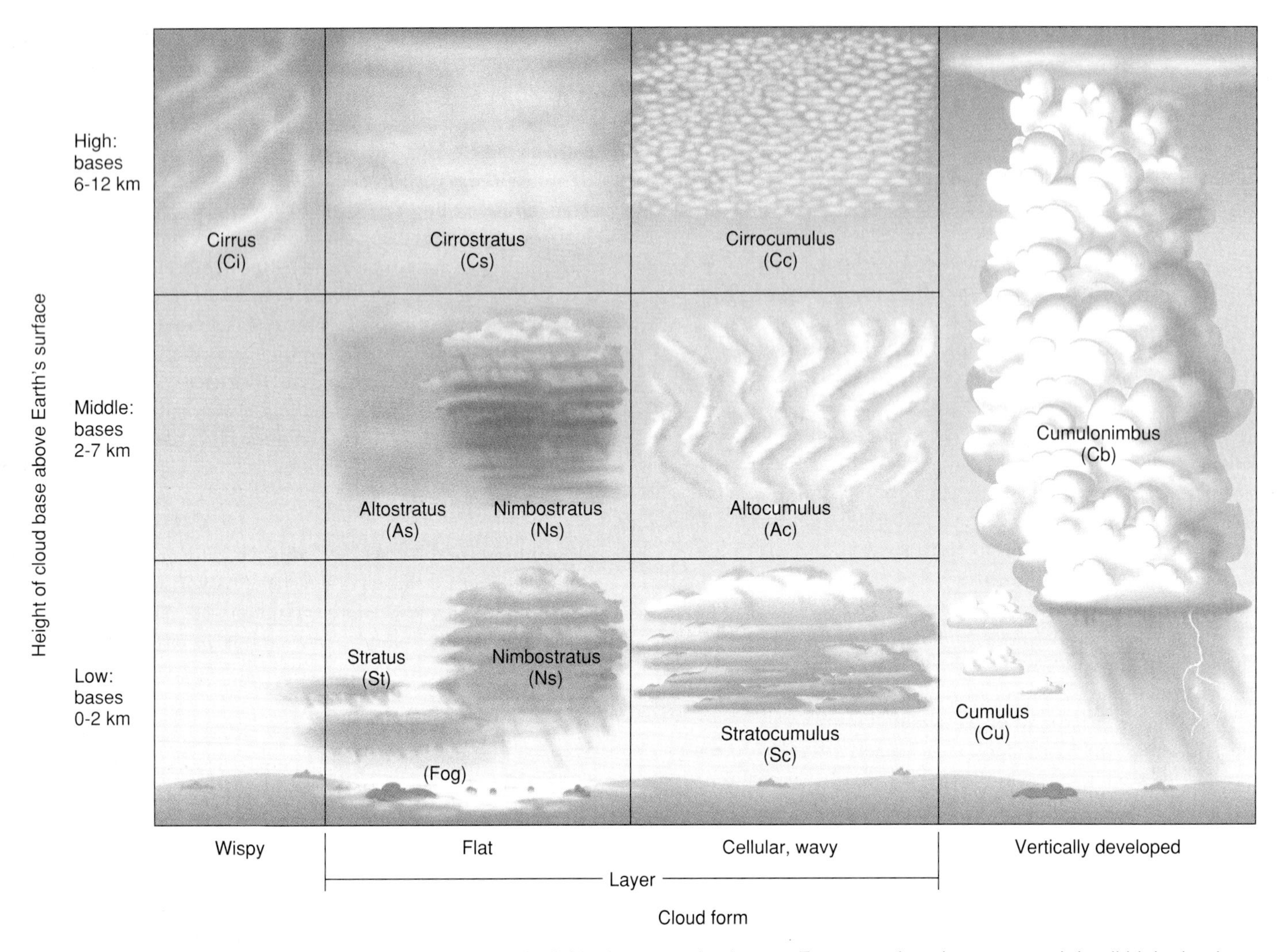


Figure 6.1 The 10 cloud genera. Note the similarities in individual rows and columns. For example, what root word do all high clouds have in common? All flat clouds?

tures at the higher end of this altitude range are frequently well below 0°C. As a result, middle clouds often are composed of mixtures of supercooled water droplets and ice crystals.

While most low cloud formation is strongly influenced by conditions at the earth's surface, such as surface temperature and topography, different factors are responsible for most middle and high clouds. Large-scale circulation systems, fronts, and thunderstorms are the primary sources of middle and high clouds.

### *Altostratus*

Altostratus clouds are flat, relatively featureless middle clouds. They commonly form in advance of warm fronts.

Note from Figure 6.1 that precipitation sometimes falls from altostratus clouds. When this occurs, the cloud is classified as a middle-level nimbostratus. Thus, nimbostratus clouds may be found at either low- or middle-cloud altitudes. Typically, precipitation from middle-level nimbostratus is lighter in intensity than from its low-level counterpart.

The water droplets in altostratus clouds cause a bending of sunlight or moonlight called diffraction, which results in a **corona** of light surrounding the illuminating object.

### *Altocumulus*

As the name implies, altocumulus clouds are middle clouds exhibiting cumuliform structure. In contrast to the rather featureless nature of altostratus, altocumulus comprise some of the most picturesque of all clouds. The beauty of altocumulus derives in part from their tendency to occur in rolls, waves, or semiregular patterns of cells. Reflection of red and orange light near sunrise and sunset sometimes enhances their beauty in spectacular displays. Perhaps for these reasons, altocumulus clouds long have been favorite subjects of artists.

All clouds in the low, middle, and high categories exhibit some stratiform tendencies. Thus, altocumulus clouds generally appear not as individual cumulus buildups but as layers of cells, a middle-cloud version of stratocumulus.

## High Clouds

**High clouds** typically form between 6000 and 12,000 meters (20,000–40,000 feet) above the ground, depending on the season and the latitude (higher where warmer). In any case, high-cloud tops generally are limited by the height of the tropopause (recall from Chapter 2 that the tropopause is the boundary between troposphere and stratosphere). At high altitudes, the air temperature tends to be far below freezing; as a result, high clouds are composed almost entirely of ice crystals, even in summer.

The names of all three high-cloud genera contain the word *cirrus* in some form. Notice that the cirrus form applies only to high clouds; there is no middle- or low-cloud equivalent of the wispy cirrus form.

### *Cirrus*

**Cirrus** clouds are unmistakable. Their characteristic fibrous texture (in Latin, *cirrus* means "curl or lock of hair") is evident in Figure 6.2H. A popular name for cirrus is "mares' tails," because of their resemblance to the flying tails of running horses. Pure white in midday, cirrus may turn beautiful shades of yellow, orange and red in the scattered light of the rising or setting sun.

### *Cirrostratus*

You may recall days when the sky started out clear blue but later changed color to a milky blue-white hue. A thin deck of **cirrostratus** clouds can cause such changes. As the layer becomes thicker, it becomes more opaque, and the color changes to white or gray-white.

Because stratiform clouds often are featureless, it can be difficult to estimate their height: "Is that gray-white veil at the middle- or the high-cloud level?" One way of distinguishing is by looking for optical effects. You read that altostratus clouds sometimes cause a corona to appear. High clouds composed almost entirely of ice crystals refract light in the manner of a prism, resulting in a **halo**. The differences are dramatic: the corona, caused by an optical process called diffraction, lies close to the moon or sun; it seems almost in contact, in fact. The halo, a result of refraction, lies much farther from sun or moon, a great ring encompassing a large part of the sky at a span of roughly two fists at arm's length (22 degrees in radius).

### *Cirrocumulus*

**Cirrocumulus,** the least common of the 10 cloud genera, consists of cirrus clouds in cellular form. The cells appear small in size and generally are arranged in rows or waves (note the examples in Figure 6.2). As you can also see from the figure, they often appear in combination with cirrus or cirrostratus.

Cirrocumulus clouds may contain significant numbers of highly supercooled liquid water droplets as well as ice crystals. Reflection of sunlight off the cloud droplets and refraction by the ice crystals can give cirrocumulus beautiful colors. Their resemblance to fish scales has earned cirrocumulus and small-celled altocumulus clouds the nickname "mackerel sky."

How do you distinguish cirrocumulus from altocumulus and from stratocumulus? Comparing images in Figure 6.2, you see that apparent cell size offers one clue. The higher the cloud base, generally the smaller the individual cells. Also, higher forms of cumulus tend to be arranged more tightly and obviously into patterns. Cloud color provides another clue: lower clouds, being warmer, contain more water, which absorbs and scatters sunlight more than higher clouds; therefore, lower clouds tend to be darker gray in color than high clouds. Cirrocumulus show very little, if any, gray shading. This grayness criterion also applies in distinguishing altostratus from cirrostratus: cirrostratus is whiter and altostratus is grayer.

## Clouds of Vertical Development

Examine the cumuliform clouds in Figure 6.2A and B. To which height category do they belong? With bases below 2000 meters (6600 feet), they fall in the low-cloud group. However, their tops may reach as high as the tropopause; therefore, it seems inappropriate to classify them as low clouds. Cumulus and cumulonimbus clouds, which are capable of extending to great heights, are categorized not as low, middle, or high clouds but in a separate category as "clouds with vertical development."

### *Cumulus*

**Cumulus** clouds, as you recall from Chapter 1, are puffy, heap-shaped clouds that form in rising columns of air. Typical cumulus cloud bases are in the low cloud range.

Notice in the photos in Figure 6.2 that growing cumulus clouds assume a range of shapes, from the modest "cumulus humilis" to towering cumulonimbus. Not every cumulus reaches great altitudes, for reasons you will study in a later section. Those that do, however, may become cumulonimbus clouds.

### *Cumulonimbus*

A cumulus cloud that generates precipitation is called a **cumulonimbus.** Cumulonimbus clouds can be awesome sights: swelling masses of cloud may rise as much as 16 kilometers (10 miles) above the earth's surface. Atop these cloud pillars may be a flat cirriform cap known as an anvil cloud. Many of these features are apparent in Figure 6.2B, as well as further on, in Chapter 12. Thunder and lightning often are produced in large cumulonimbus clouds, along with rain, hail, and strong, gusty winds.

From beneath, you might see little or nothing of the cumulonimbus cloud structure, except perhaps an extremely dark (even in daytime), menacing base. The cloud structure may also be obscured by surrounding clouds. However, the occurrence of showers, as opposed to the steady precipitation of stratiform clouds, is indicative of a cumulonimbus cloud overhead. Hail, lightning and thunder, and gusty winds are confirming evidence that the cloud type is cumulonimbus, regardless of whether or not you can see the cloud's cumulonimbus structure.

Because they are such powerful and important weather phenomena, you will study cumulonimbus clouds, thunderstorms, and their related weather at a number of points in future chapters. They are the main focus of Chapter 12.

Figure 6.2 The 10 basic cloud types. *A:* Cumulus; *B:* Cumulonimbus; *C:* Stratus; *D:* Stratocumulus; *E:* Nimbostratus; *F:* Altocumulus; *G:* Altostratus and altocumulus; *H:* Cirrus; *I:* Cirrostratus with halo; *J:* Cirrocumulus; *K:* Cirrus and cirrostratus at sunset.

H

J

K

I

## Checkpoint

***Review***

Use Figure 6.1 as a framework for organizing the information in this section. In the figure, you are reminded that clouds are classified by height, basic form, and whether they yield precipitation. For practice, match the clouds in each photo in Figure 6.2 to their corresponding category in Figure 6.1.

***Questions***

1. What property do all high clouds have in common? All *nimbo-* or *-nimbus* clouds?

2. Suppose you observe a cloud deck composed of rows of gray-white cells. To which two genera might they belong? What properties would determine the cloud's correct genus?

3. Why are cumulus and cumulonimbus clouds not classified by height in Figure 6.1?

## Mechanisms of Cloud Formation: Overview

In the last chapter, you studied the molecular-scale processes by which water vapor condenses to form cloud droplets or ice crystals. Those processes, however, do not seem to explain the great variety of cloud forms you viewed in the last section. What determines whether condensing droplets make up a billowing cumulus cloud, a sheet of stratus, or a layer of fog?

In the remainder of this chapter, you will explore the atmospheric and geographic circumstances that bring water vapor in the air to its saturation point and determine the properties (and the genus) of the resultant cloud. The concept map in Figure 6.3 outlines this subject. As you see, the map indicates three distinct ways of achieving saturation: by cooling the air, by adding moisture, or by mixing air parcels of different properties. (Although these three mechanisms are distinct, more than one may be active in the same cloud.) Note also, from the width of the lines in the map that by far the most important mechanism is the first:

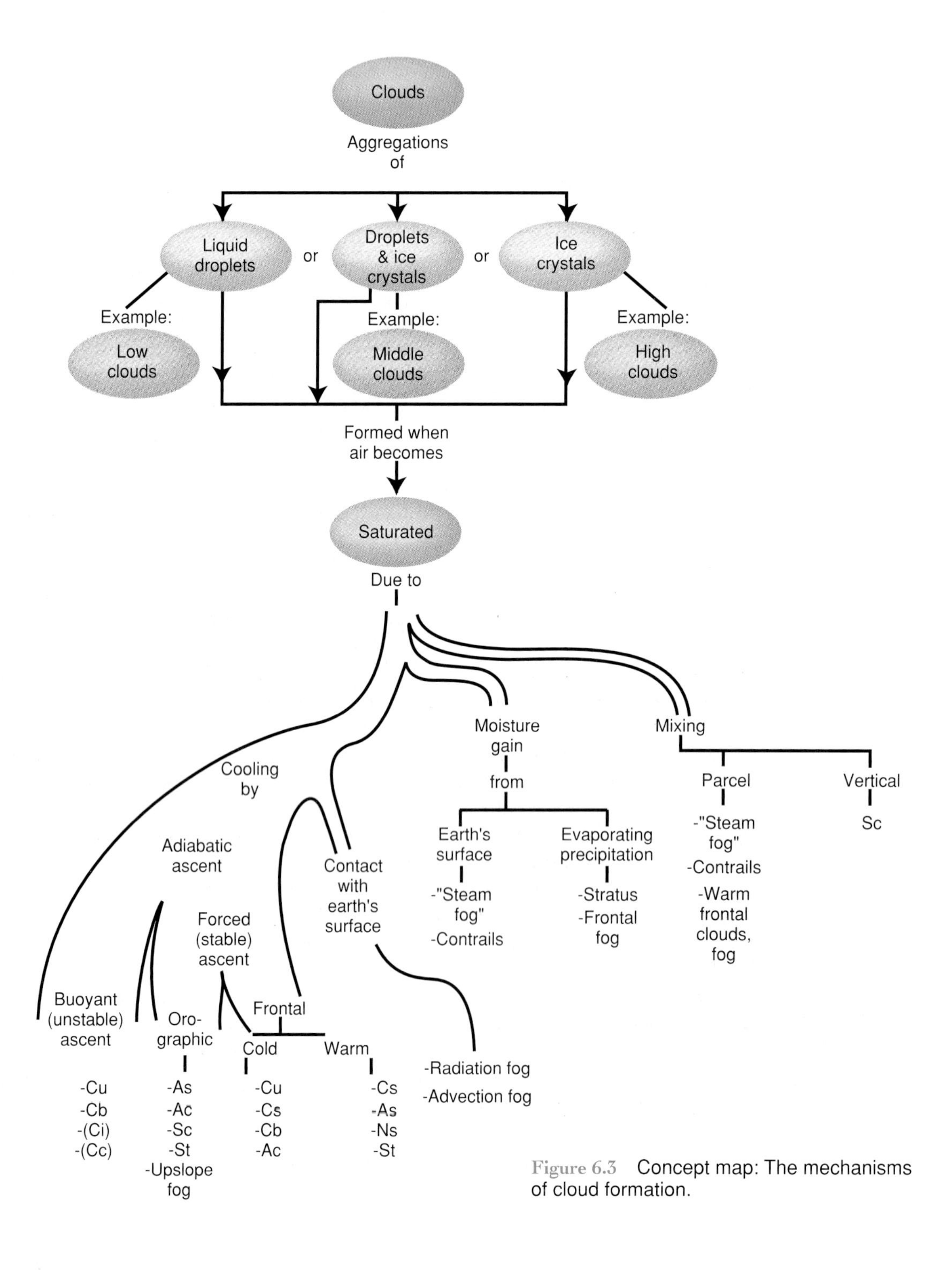


Figure 6.3 Concept map: The mechanisms of cloud formation.

the great majority of clouds form as air is cooled to its dew point. Further, notice that of the several mechanisms for cooling, by far the most significant is upward air motion ("adiabatic ascent"). Because it's so important, we'll spend the most time on this cause of cooling.

# Upward Motion As a Cooling Mechanism

You may have noticed that in most places, cumulus clouds are most common in warm weather—in summer rather than in winter and in the afternoon more than in the morning. Watching them grow on a summer afternoon, you might have speculated that these upward-billowing masses of air rise because they are buoyant. That speculation happens to be correct; a typical cumulus cloud forms in a parcel of air that rises because it is warmer, and therefore less dense, than the surrounding air.

These observations seem to lead to a paradox: on the one hand, cumulus clouds are known to form at the *warmest* time of day, in currents of warm, rising air. On the other hand, you saw in the last chapter that it is by *cooling* an air sample, not heating it, that the air's relative humidity increases to values necessary for water vapor to condense into cloud droplets. You will resolve this paradox in the next few pages as you analyze the changes that occur within rising air currents.

## Adiabatic Changes in a Rising Air Parcel

Consider the air in Figure 6.4, whose surface pressure, temperature, and density are designated P, T, and $\rho$ (the Greek letter *rho*), respectively. Take a small portion of this air and call it a parcel. Since the parcel is just a portion of surface air, of course its T, P, and $\rho$ are the same as that of its surroundings.

Imagine now that the parcel is displaced upward, as shown in Figure 6.4B. Assume that during this displacement, no exchange of heat energy occurs between the parcel and its surroundings. Such a process is called an **adiabatic** process. For most clouds that form due to vertical air motion, the adiabatic assumption is approximately true: rising air parcels exchange relatively little energy with their surroundings.

Now consider what will happen to the parcel's variables of state (e.g., its temperature, pressure, and density) as it rises. Air pressure, as you know, decreases upward; therefore, as the parcel rises, the pressure in the air surrounding the parcel becomes steadily less than the parcel's initial pressure. This pressure imbalance means that there is a net outward pressure force at the parcel's boundary, which causes the parcel to expand. The expansion leads to other changes as well. We consider each in turn.

### *Changes in Density*

Recall that

$$\text{density } (\rho) = \frac{\text{mass}}{\text{volume}}$$

When the parcel expands, its density decreases because its mass occupies an ever-greater volume. Therefore, spreading the same mass over a greater volume results in a decrease in density, as the formula indicates.

### *Changes in Temperature*

Does the parcel's expansion affect its temperature? Think of the experience of inflating a bicycle tire with a hand pump (Figure 6.5). To push air from the pump into the tire, you have to exert energy on the pump. The energy required in this case comes from an external source: you pushing down on the pump handle.

The air parcel's expansion into the surrounding air is similar in some ways to the process of inflating a bicycle tire: air is being pushed from the original volume of the parcel into space occupied by the surrounding air. This process also requires energy. Since we have assumed the process is adiabatic, we are saying in effect that there is no external source of energy (like you pushing on the pump) to drive the expansion. From where, then, will the parcel derive the necessary energy? It must come from within the parcel itself. There is only one available internal energy source: the thermal motions of the parcel's component molecules. It is this energy source that fuels the expansion. As a result, the parcel's temperature decreases as it expands adiabatically. The temperature change is considerable: in this example, the parcel will cool at the rate of 9.8°C for every kilometer (or 5.4°F for every 1000 feet) it rises.

This rate of temperature change is of fundamental importance in meteorology. It is known as the **dry adiabatic lapse rate** and is defined as the rate at which an unsaturated air parcel changes temperature due to an adiabatic pressure change. *Dry* means not that the air is devoid of all moisture but only that condensation does not occur during the process. *Adiabatic* means the parcel neither gains nor receives heat from its surroundings, and *lapse* means the temperature falls (lapses) as the parcel rises. Think of "lapse rate" as meaning "change rate." Again, the numerical value of the dry adiabatic lapse rate is 9.8°C per kilometer, or 5.4°F per 1000 feet. You will use this quantity often in this and later chapters.

### *Changes in Pressure*

We have now established that in an air parcel rising adiabatically, both density and temperature decrease. We can now employ the gas law,

$$\text{pressure} = \text{constant} \times \text{density} \times \text{temperature}$$

or symbolically,

$$P = C \rho T$$

to determine what will happen to the parcel's pressure. Since both density and temperature decrease, the entire right side of the gas law formula becomes smaller in magnitude, therefore, the left side—air pressure—must decrease an equal amount for the expression to remain in balance. Thus, the gas law indicates the parcel's air pressure must decrease, as expected. In fact, it will decrease until the pressure of the gas in the parcel equals the

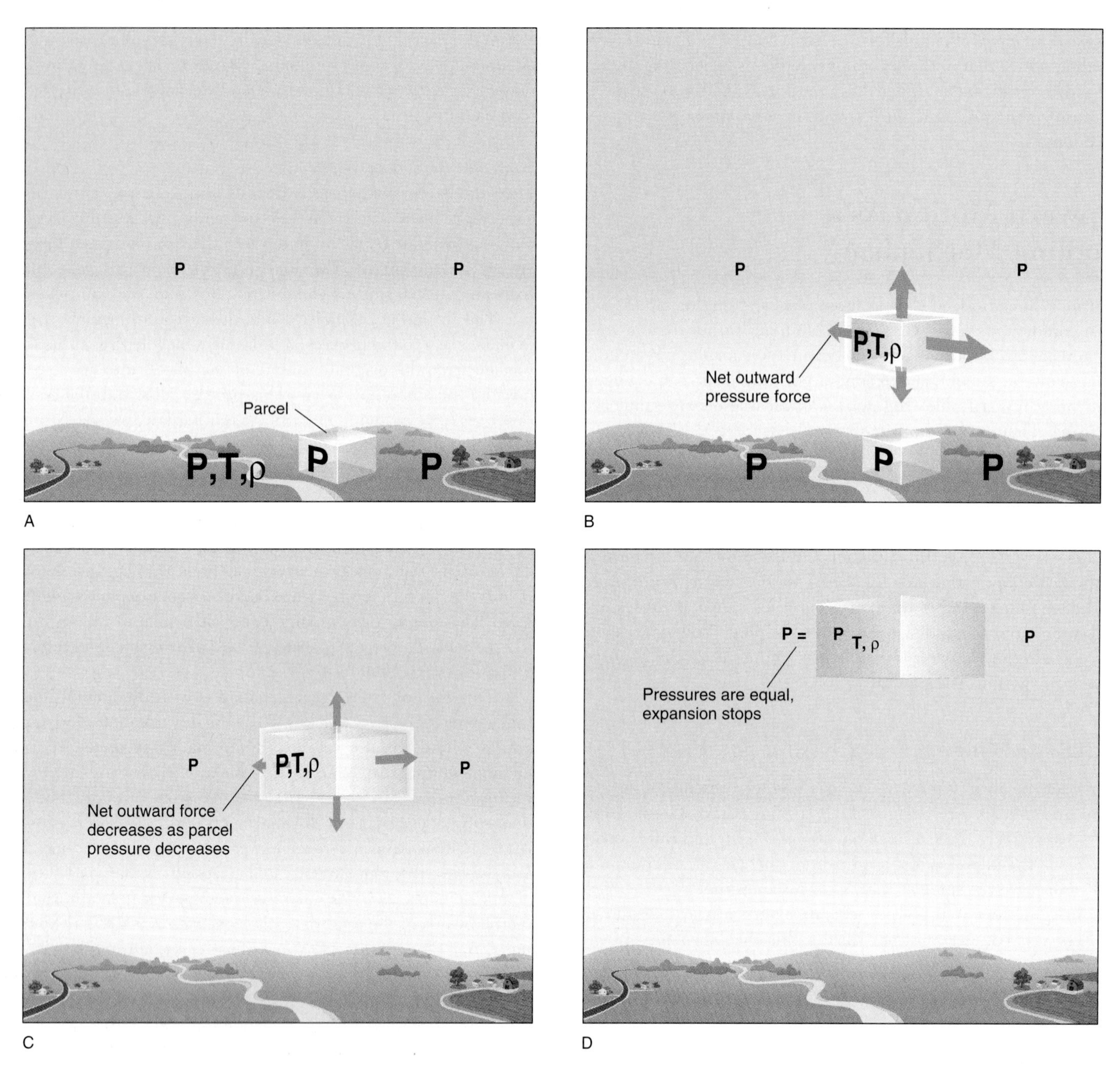


Figure 6.4 Adiabatic expansion. (*A*) Parcel of surface air with same P, T and ρ as the surrounding environment. (*B*) The parcel rises to where environmental P is less than initial parcel's P. Result: a net outward pressure. The parcel expands. (*C*) Expansion causes decrease in parcel's density and temperature and therefore pressure. (*D*) Expansion ceases when parcel's pressure has decreased to equal that of the surrounding air. Parcel pressure, temperature, and density all are less than they were initially.

pressure of the surrounding environment; otherwise, a pressure gradient will exist, and the parcel will continue to expand.

In summary, an air parcel rising dry adiabatically (i.e., without gaining or losing heat from the surrounding air) experiences a decrease in temperature, density, and pressure. The parcel's pressure decreases until it equals that of the surrounding air.

### *Why Clouds Form in Rising Air*

Now, part of the dilemma of clouds forming in rising warm air currents is resolved, at least qualitatively: adiabatic expansion in ascending air causes cooling. Although we haven't yet considered changes in the air's moisture variables (such as relative humidity or dew point), at least we see the temperature is moving in the right direction. Therefore, you might surmise that cumulus clouds form due to cooling caused by upward vertical motion.

A remaining part of the dilemma is why the cumulus clouds form most frequently in warm, afternoon air. Why not in morning air, for example, which is colder and therefore closer to its dew point to begin with? You may well have an explanation for this; we will address it after exploring adiabatic changes in more detail.

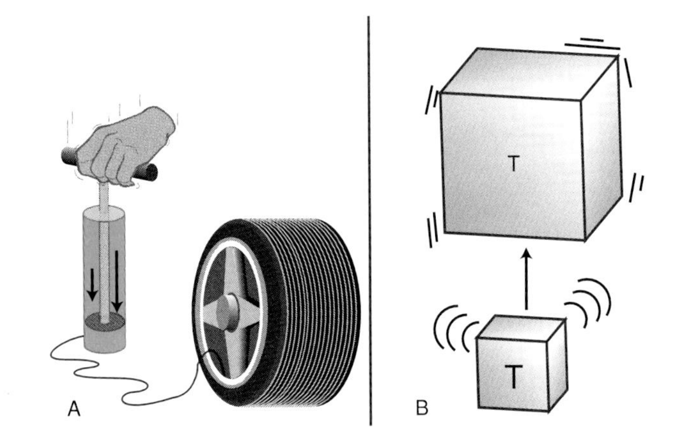


Figure 6.5 It requires energy to push air from the pump and into the tire and to push surrounding air back from an expanding parcel. For the adiabatically expanding parcel, the only energy source is its internal thermal energy; thus, the parcel's temperature decreases.

## Graphing Dry Adiabatic Processes

Adiabatic cooling is such a common occurrence in the atmosphere and the resultant temperature changes are so large that they play a dominant role in cloud formation. Therefore, in this section, you will learn how to solve a variety of problems dealing with adiabatic processes by using a chart called the thermodynamic diagram.

### *Variables of State*

We begin with Figure 6.6, a grid for plotting air parcels with various values of pressure and temperature. Notice that, unlike most graphs, values of the vertically plotted variable—air pressure—decrease upward. The advantage of this orientation is that "up" on the graph corresponds to "up" in the atmosphere.

Carefully measure the spacing on the grid between the 1000-millibar and the 800-millibar levels; compare it to the distance from 600 to 400 millibars. You will find that isobars are not spaced evenly. Instead, at lower pressures (i.e., higher altitudes), the isobars are more widely spaced on the grid, again in correspondence with the way pressure changes vertically in the actual atmosphere. (This particular pressure scale is called logarithmic; the logarithm of the pressure variable is evenly spaced in the vertical.) The altitudes listed along the right margin are only average values, because the height of a particular pressure level depends on the atmosphere's temperature and surface pressure.

On Figure 6.6, locate an air parcel whose temperature is 10°C and whose pressure is 1000 millibars. You should arrive at point *A* on the diagram. Verify that air with the same temperature but a pressure of 500 millibars would be at *B* and that air at −10°C and 500 millibars would be at *C*. In fact, for any given set of T and P values, there is one and only one point on the chart corresponding to that set of conditions. Moreover, given the parcel's temperature and pressure, the air's density is also specified: the gas law tells you the density, $\rho = P/CT$. Thus, each of the parcels plotted in Figure 6.6 has all three of its variables of state (pressure, temperature, and density) determined.

### *Dry Adiabatic Temperature Changes*

Suppose the parcel at *A* in Figure 6.6, whose temperature is 10°C and pressure is 1000 millibars, rises adiabatically to an altitude where the air pressure is 800 millibars. From the earlier discussion, you know that the parcel's temperature will decrease as the parcel ascends, but by how much? Using Figure 6.7, you can determine the answer graphically. As you can see, the diagram has the same grid of pressure and temperatures as in the previous figure, plus an additional set of lines called **dry adiabats,** plotted in blue. Dry adiabats indicate the dry adiabatic lapse rate; that is, they slope upward and toward colder temperatures at the rate of 9.8°C per kilometer. Thus, an air parcel rising dry adiabatically will follow a path along (or parallel to) the dry adiabats in Figure 6.7.

You can use the dry adiabats in Figure 6.7 to determine the changes the parcel initially at A will experience as it rises dry adiabatically to the 800-millibar level. Beginning at A, follow the parcel's temperature change by "moving" the parcel upward along the dry adiabat blue line) that passes through A. What is the parcel's temperature at 800 millibars? From the figure, you should arrive at 800 millibars at point B on the chart, at a value of −8°C. Thus, in rising from 1000 to 800 millibars, the parcel has cooled from 10°C to −8°C, a change of $10 - (-8) = 10 + 8 = 18$°C, over an altitude change of only about 1800 meters!

Suppose now that the parcel sinks adiabatically to its original location at 1000 mb. What will be its temperature upon return? To find out, again take the parcel along the dry adiabat back to the 1000-millibar level. Since you return along the same adiabat, you find that the temperature has warmed to its original value of 10°C.

### *Dew-Point Changes*

Reflecting on the last chapter's discussion of atmospheric humidity, you might wonder what happens to variables such as dew-point temperature and mixing ratio during the parcel's ascent. By including one more set of lines—saturation mixing ratio lines (shown in green in Figure 6.8 and labeled near the top of the chart)—you can account for humidity changes as well.

To illustrate the use of the mixing ratio lines, suppose the air's dew-point temperature in the previous example was −8°C. Plot it, along with the temperature, on the 1000-millibar line in Figure 6.8, at the points labeled DP and T.

Notice that the dew-point temperature, DP, happens to lie directly on a saturation mixing ratio line; reading from the labels at the top of the chart, you see that the line is the "2 g/kg" line. Recall from Chapter 5 that the saturation mixing ratio value corresponding to the dew point gives the actual amount of water vapor present in the air. Therefore, the air sample plotted at T and DP is composed of 2 grams of water vapor per kilogram of dry air.

Now if the parcel rises, how does the dew-point temperature change? As it ascends, it maintains the same molecules within its boundaries: that is what defines it as a parcel. Therefore, those 2 grams of water vapor remain within each kilogram of dry parcel air; thus, its mixing ratio remains constant at 2 g/kg.

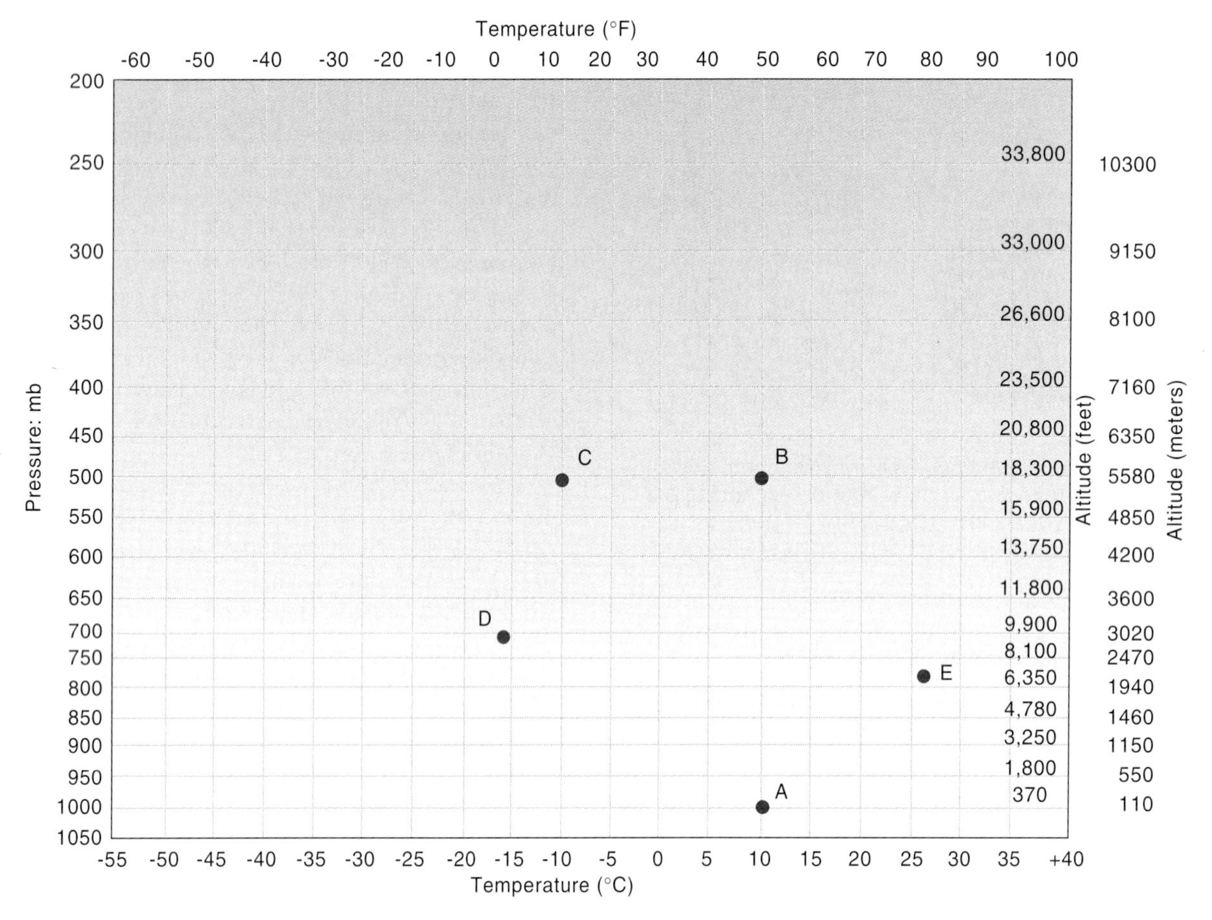


Figure 6.6 A temperature-pressure plotting chart. Which two parcels have the same temperature? Which have the same pressure? What are the approximate temperature and pressure of parcel D? (Interpolate between −15°C and −20°C.) Of parcel E?

Figure 6.8 shows the process. The parcel's temperature decreases along the dry adiabat, that is, along the red lines, as before. The dew-point temperature follows a line of constant mixing ratio, as just explained and as shown in the figure. Notice that the dew-point temperature decreases much more slowly with height than the air temperature (which decreases at the dry adiabatic rate), so eventually the two lines intersect at point C in the figure. You can see from the chart that although the mixing ratio remained constant, the dew point in the rising air parcel decreased slightly (from −8°C to −11°C) as the parcel ascended to point C. We will have more to say about the significance of point C in a moment.

## *Determining Relative Humidity*

From Chapter 5, you know that given a parcel's temperature and dew point, and a saturation mixing ratio chart, you can determine the parcel's relative humidity. Since Figure 6.8 contains mixing ratio lines, it can serve as a mixing ratio chart. You can determine the relative humidity by using the mixing ratio values on the chart that correspond to the temperature and the dew point.

For example, you have already determined the parcel's mixing ratio is 2 g/kg. Looking at the point representing temperature (at T in Figure 6.9) to find the saturation value, you notice that no mixing ratio line happens to pass directly through T. You must interpolate between the 5 g and 10 g mixing ratio (green) lines lying to the left and right, respectively, of point T. Doing so, you should arrive at a value of about 8 g/kg for the saturation mixing ratio at T. Knowing both the actual and the saturation mixing ratios, you can find the parcel's relative humidity as you did in Chapter 5:

$$\text{RH} = \frac{\text{mixing ratio}}{\text{saturation mixing ratio}} \times 100\%$$

$$= \frac{2\text{g /kg}}{8\text{g/kg}} \times 100\% = 25\%.$$

## *Determining the Condensation Level*

What happens at point CL in Figure 6.9, where the temperature and the dew point are equal? The relative humidity is 100%, and water vapor begins to condense into cloud droplets. This point is known as the **condensation level,** or **CL**. In brief, the CL is the level at which a rising air parcel's temperature and dew point become equal; that is, where they meet on a thermodynamic diagram. In practical terms, the CL commonly marks the level of cumulus cloud bases.

Let's review the procedure you just followed: knowing the temperature, dew point, and pressure at the earth's surface, you

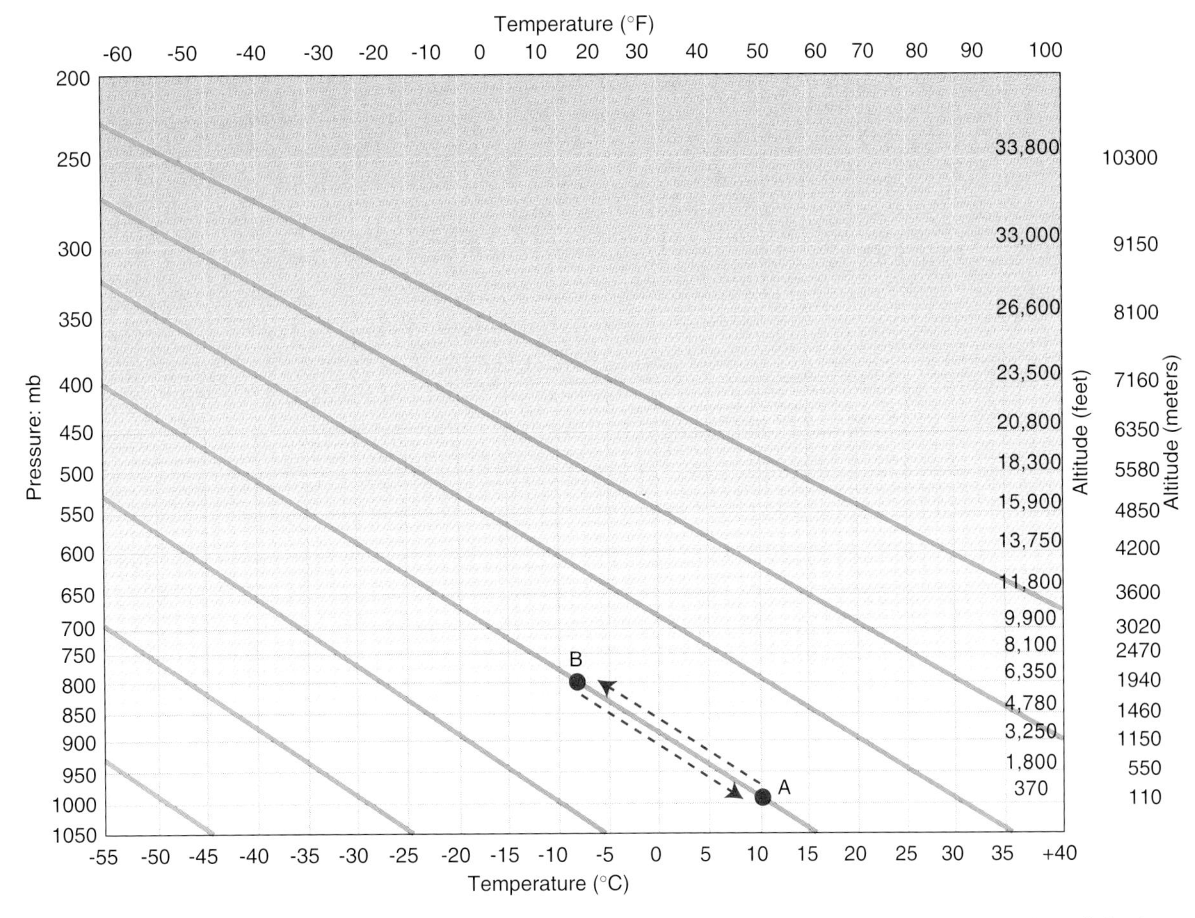


Figure 6.7 A portion of the temperature-pressure chart with dry adiabats added. Parcel A, rising dry adiabatically to 800 millibars, follows the path indicated to point B. If the parcel returns to 1000 millibars, its temperature also returns to its original value of 10°C.

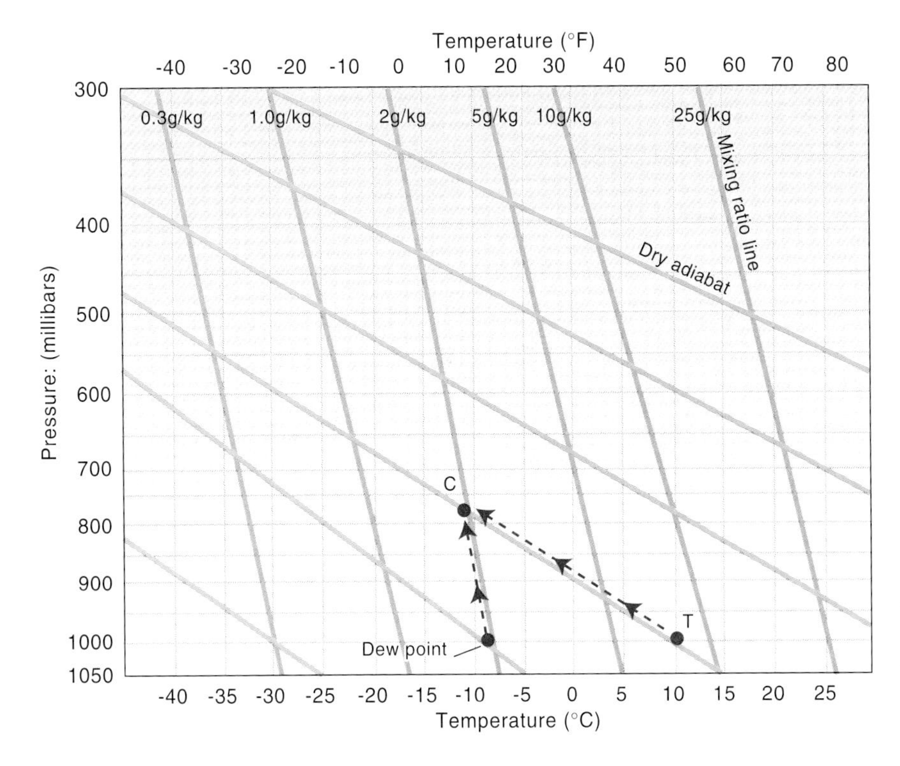


Figure 6.8 The addition of saturation mixing ratio lines (in green) allows you to follow dew-point changes in rising and sinking parcels. Note the dew-point temperature cooled by about 3°C in the air's ascent to point C.

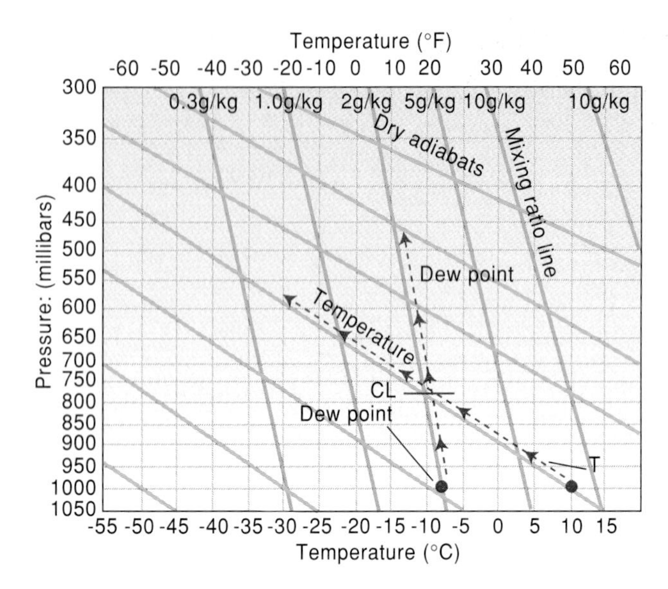


Figure 6.9 If temperature and dew point continued to follow dry adiabats and mixing ratio lines above the condensation level (CL), the air temperature would be colder than the dew point, causing the air to be greatly supersaturated.

lifted the air adiabatically, using the diagram in Figure 6.8, and found the height of the cloud base. The fact that this method regularly gives accurate results suggests (but doesn't prove) that cumulus clouds do indeed form in air parcels rising from the earth's surface. It also suggests that if surface weather conditions are fairly uniform within a locale, then all cumulus clouds will form at nearly the same altitude. If you have ascended or descended by airplane through cumulus clouds, you may remember noticing that their flat bases lie at nearly the same level, an observation captured in one of the photos in Figure 6.2.

## Moist Adiabatic Processes

We are not yet finished with our overworked air parcel. Reaching the CL is like "cloudbirth"; in a sense, we have just reached the beginning. Picture, therefore, in Figure 6.9, the air parcel rising through the CL, its water vapor beginning to condense, the first wisps of cloud beginning to form.

How do the temperature and the dew point change as the parcel moves upward above the CL? Can they continue changing along the same lines as below the CL? Figure 6.9 illustrates why this is impossible. If the dew point were many degrees warmer than the air temperature, as indicated by the values at 800 millibars, for example, the air would be highly supersaturated, an occurrence not observed. Instead, relative humidity in the cloud typically is never far from 100%; this means that the temperature and dew point must remain essentially equal to each other as the saturated air continues to ascend. Thus, on a thermodynamic diagram, they should follow the same line. But which line?

As water vapor condenses to droplets, it changes conditions in the parcel in two important ways. First, as water molecules

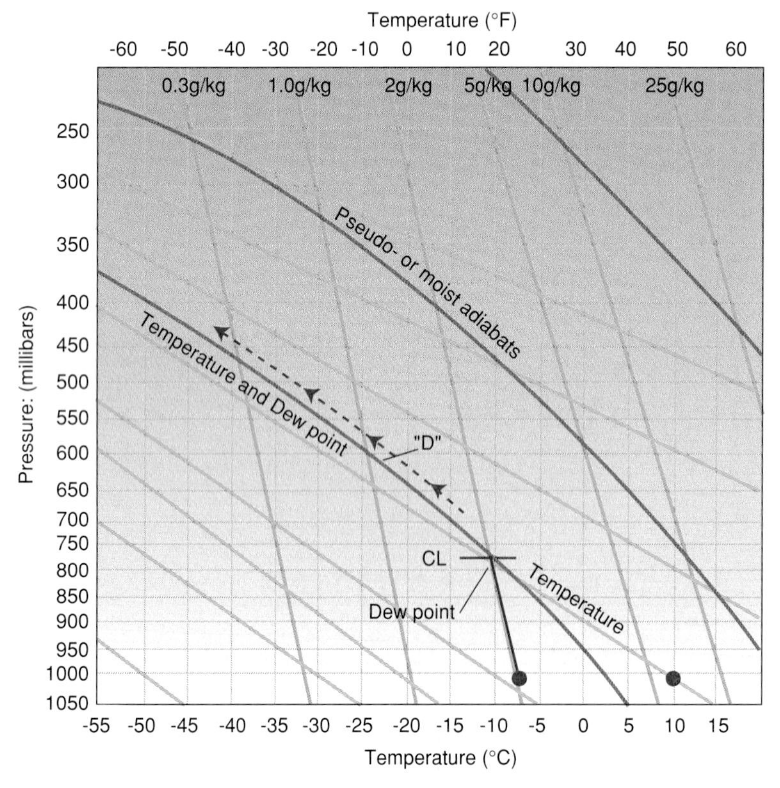


Figure 6.10 When a saturated air parcel rises, its temperature and dew point change at rates shown by the moist adiabats, shown in violet. At what temperature is the parcel as it crosses the 700-millibar isobar?

condense from gas to liquid, they cease to contribute to the mixing ratio. The mixing ratio consists only of water *vapor* molecules. Therefore, the parcel's mixing ratio decreases. Second, recall that water releases considerable latent heat as it condenses from gas to liquid, thus warming the parcel. This warming partially offsets the cooling due to adiabatic expansion, typically by several degrees C per kilometer. The net result is that temperature and dew point in a rising, saturated parcel follow a path above the CL indicated by the violet lines in Figure 6.10; these lines are known as **moist adiabats.**

Now we are prepared to determine temperature and dew-point changes as the saturated parcel continues its ascent. Both the parcel's temperature and dew point changes are described by the moist adiabat (violet line) indicated in Figure 6.10. Thus, at 600 millibars, for example, the cloud's temperature and dew point both will be −24°C, represented at point D in the figure. What will the relative humidity be? Since the air temperature equals the dew-point temperature, it will be 100%, as it has been at each level above the CL. What is the mixing ratio of the air at 600 millibars? Interpolating between values of the (green) mixing ratio lines at point D gives a value of 1.1 g/kg. Compare this value with that of 2.0 g/kg at the CL: in the ascent from the CL to 600 millibars, the parcel has lost 2.0 − 1.1 = 0.9 grams of water vapor from each kilogram of dry air. Where has this moisture gone? It has changed phase, forming cloud droplets or ice crystals. The moisture may still be in the cloud, or it may have fallen out as precipitation; in either case, it is no longer water vapor, and therefore no longer contributes to the air's mixing ratio.

Notice that at very cold temperatures, the moist adiabats in Figure 6.10 are nearly parallel to the dry adiabats but diverge from them considerably at high temperatures. Why the difference? Remember that at low temperatures, the saturation mixing ratio for water vapor is very low; therefore, little water vapor is available to condense (or deposit), and little latent heat is released. In warm air, the saturation values are far higher, thus more water vapor is available to condense, and correspondingly more latent heat is released. As a result, the **moist adiabatic lapse rate**—the rate of temperature decrease in a rising, saturated parcel—is not a constant like the dry adiabatic rate but varies with air temperature. Typical values range from 4 to 7°C per kilometer (2 to 4°F per 1000 feet), compared with the dry adiabatic rate of 9.8°C per kilometer (5.4°F per 1000 feet).

Figure 6.10 is the final form of the temperature-pressure diagram. In this form, containing dry and moist adiabats and saturation mixing ratio lines, the chart is called a **thermodynamic diagram.** As you have seen, the thermodynamic diagram is a powerful tool for solving a wide variety of problems related to changes in temperature, pressure, humidity, and cloud formation. You will use thermodynamic diagrams often throughout this book.

***Review***

Figure 6.10 illustrates many of the major concepts developed in this section as applied to the cumulus cloud. If you know the air's surface temperature, dew point, and pressure, then with the help of the diagram you can determine the conditions of temperature and humidity at every level from the ground to and within the cumulus cloud.

***Questions***

1. Why does rising air cool more rapidly if it is unsaturated than if it is saturated?
2. Consider a flow of air crossing the crest of California's Sierra Nevada Mountains. Suppose its temperature and pressure at the crest are −10°C and 700 millibars, respectively. If this air sinks dry adiabatically on the lee side of the mountains to a pressure of 1000 millibars, what will its temperature be?
3. What is the CL? Describe in words how you determine its value on a thermodynamic diagram.
4. Suppose an air parcel has a pressure of 1000 millibars, a temperature of 15°C, and a dew point of 10°C. Assume the parcel rises adiabatically, and use Figure 6.10 to determine the parcel's temperature and dew point at 900, 800, and 700 millibars.

# Cloud Formation by Buoyant Lifting

One aspect of the cumulus cloud dilemma posed earlier remains: why do these clouds form when surface temperatures are warmest and therefore the relative humidity is the least? In general terms, the answer is as follows: Heating of the earth's surface may lead to convection, and convection consists in part of rising air currents. Thus, the rising parcel is part of a convective circulation. As the earth's surface becomes warmer, more air is carried aloft convectively, cooling as it rises until its water vapor begins to condense to form cumulus clouds.

This description, though basically correct, is incomplete. For example, it gives us no way of telling how high a rising parcel will go. Will it stop rising before reaching the CL and form no cloud at all? Will it reach the CL and form a cumulus humilis, or will it grow to a giant cumulonimbus? This section deals quantitatively with these issues of cumulus cloud formation and development.

## Stability: Basic Concepts

Consider the marble in the bowl in Figure 6.11A. If you perturb the marble slightly by giving it a gentle poke, it will roll a short distance up the inside of the bowl, then back to its point of origin. (It may overshoot and roll up the opposite side, but eventually, it will settle back in its original location.) In such a circumstance, the marble resting at A is said to exist in a stable configuration with its surroundings.

If you invert the bowl (Figure 6.11B) and balance the marble at its highest point, you will have placed the marble in an

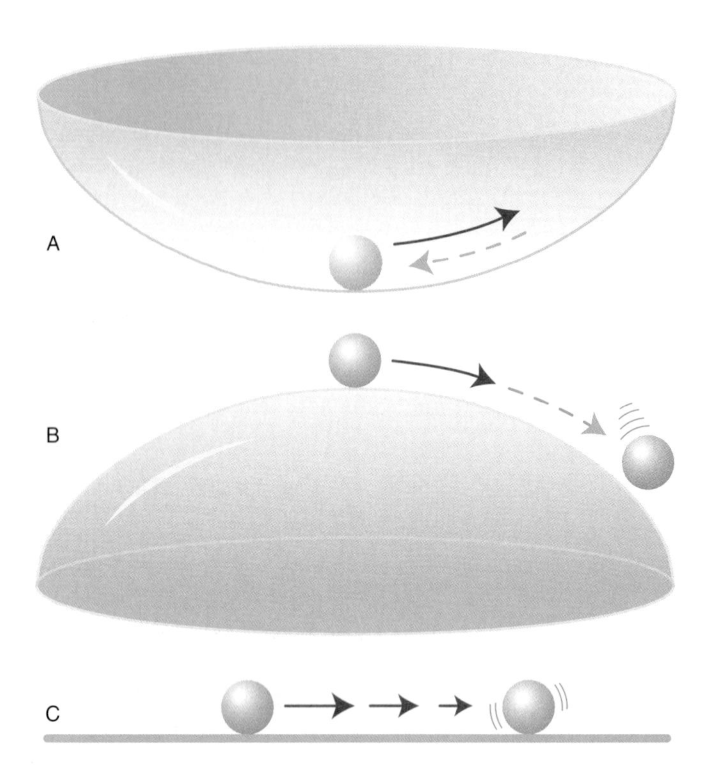


Figure 6.11 Three types of stability. In (*A*), stable, a perturbation generates negative feedback from gravity, restoring it to its initial position. In (*B*), unstable, positive feedback accelerates the marble away from its original position. In (*C*), neutral stability, gravity exerts no feedback.

unstable environment. If you perturb the marble from its position, it will accelerate away from its starting point. Finally, if you remove the bowl altogether and place the marble on a flat surface such as the floor (Figure 6.11C), perturbing the marble will cause it to roll away some distance and stop. In this case, one location on the surface is equivalent to any other; the flat surface is said to offer the marble an environment of neutral stability.

The bowl-and-marble model is closely analogous to an air parcel's **static stability.** Static stability is a measure of the restoring force an air parcel experiences upon undergoing a small vertical displacement. Suppose an air parcel becomes buoyant and begins to rise through the surrounding air, a situation comparable to the initial push and motion of the marble. Like the marble, once in motion the air parcel may act in any of a variety of ways: it may stop moving altogether, it may sink back toward its starting place, or it may accelerate farther upward. Thus, for the air parcel, static stability can assume a spectrum of values from highly stable (when a displaced parcel encounters forces that tend to return toward its point of origin) to neutral (when there is no net restoring force) to highly unstable conditions (when the parcel moves farther away from its point of origin). And also like the marble-bowl example, static stability of an air parcel depends on the relation of the parcel to its environment. This is a crucial concept: the stability of an air parcel, which determines whether a cumulus cloud will form and grow, depends on the parcel's relation to the environment in which it is situated.

## Parcels versus the Environment

Just what do we mean by an air parcel's "environment"? We mean the surrounding atmosphere in which the parcel is embedded and whose properties are measured more or less instantaneously at all levels. For example, consider the incredibly high tower, shown on the right edge of Figure 6.12, which is outfitted with barometers, thermometers, humidity gauges, and so on at a number of different heights. These instruments give a continuous record of the conditions of the atmospheric environment at each height. Suppose you collected one complete set of tower measurements, all made at the same time, and you plotted them onto a thermodynamic diagram. What do you expect the data would look like? Would they be dry adiabatic, moist adiabatic, or none of the above?

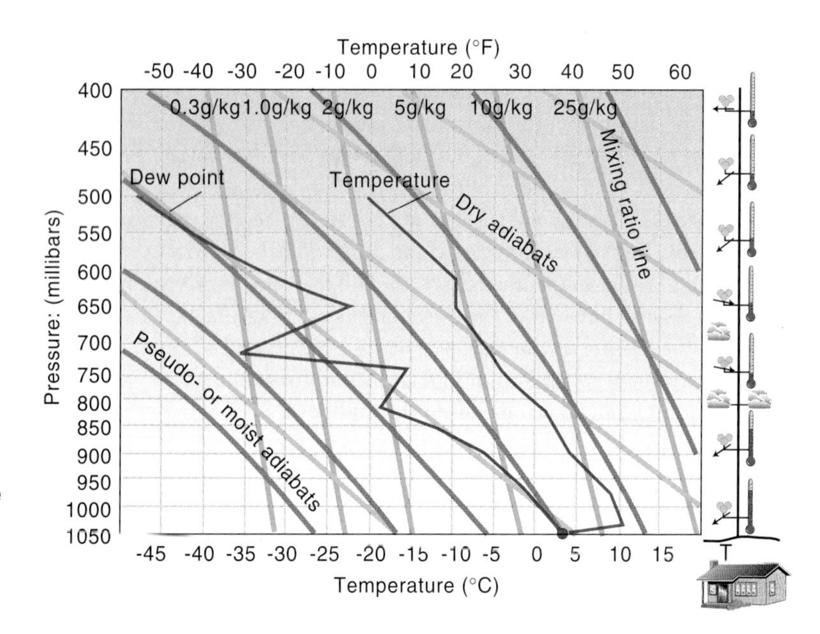


Figure 6.12 An instrumented tower, like that depicted on the extreme right, would give environmental (not rising parcel) data like those plotted for Charleston, South Carolina. (The measurements were actually made by radiosonde, not by tower, of course).

In Figure 6.12, you can see an actual set of such measurements, collected by radiosonde (since no tower of this height exists). Notice that the temperature and dew-point profiles do not follow any of the families of curves, unlike the parcel changes you studied in the previous section (Figure 6.8, for example). Why is this? Because the profiles in Figure 6.12 show the properties of the environmental air at each height; they do not represent the changes a single rising parcel would experience. Recall that a **parcel** is a specific group of gas molecules that does not mix with the surrounding air but changes temperature with altitude at either the dry or moist adiabatic rates. The **environment,** on the other hand, is not a discrete mass of air; it consists of *different* air molecules at each level whose temperature and humidity values may differ erratically from one level to the next.

The distinction between parcel and environment is so important we offer one more example. You can see in Figure 6.13 that the people on different floors of the building maintain their environments at different temperatures. As a result, room temperature at any given level may be higher, lower, or equal to that above or below. On the other hand, the air in the elevator car behaves like a parcel (again, assuming it does not mix with the environmental air at each floor). As the elevator moves up and down, its air temperature changes in a predictable way, due to adiabatic expansion (when rising) or compression (when sinking).

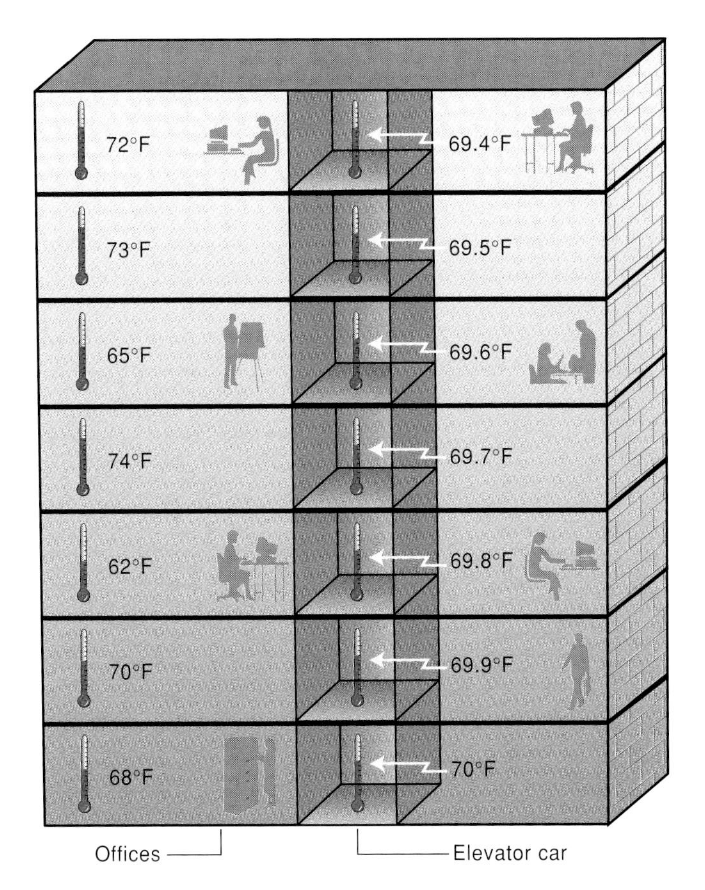


Figure 6.13 Environment temperatures: The air in the offices is not moving from floor to floor. The temperature on each floor results from influences on that floor. Parcel temperatures: The air in the elevator car hebaves like a parcel. As the car rises and falls, its air, rising and sinking within it, warms and cools adiabatically.

## Cloud Formation Due to Static Instability

Under what conditions is the atmospheric environment unstable enough to promote cloud formation? How can it exert positive feedback on a perturbed, rising air parcel and sustain its upward motion to the point of saturation? You will find the answers to these questions as you work through the following example.

### *Stability of Unsaturated Air*

Figure 6.14 is a plot of environmental conditions at Albuquerque, New Mexico, early on a spring morning. Notice the temperature inversion in the lowest layers: from 850 millibars to 810 millibars, the environmental temperature increases with height. (Albuquerque's high altitude, 1600 meters above sea level, causes surface air pressure to be only 850 millibars.) Such surface-based inversions are common at night under clear skies and light winds, due to radiational cooling of the surface.

How stable is the surface air in Figure 6.14? To find out, imagine that a parcel of surface air rises, perhaps due to uphill wind flow that carries it upward through the environment some distance and then releases it—at the 800-millibar level, for example. How will the parcel and the environment interact?

As the parcel rises, it cools adiabatically. Does it cool dry or moist adiabatically? From the plotted data, you can see that the surface air temperature is considerably warmer than its dew point. Therefore, the surface air (and the parcel) is unsaturated, and the rising parcel will cool dry adiabatically. The arrow illustrate the adiabatic temperature change the parcel experiences in its ascent to 800 millibars.

When released at 800 millibars, how will the parcel respond? Compare conditions within the parcel to those of the environment at 800 millibars. The air pressures of environment and parcel are equal, of course, at 800 millibars. Temperatures are not equal, however; you can see that the parcel's 800-millibar temperature is 5°C, compared to the environment's temperature of 19°C. Therefore, the parcel is 14°C colder than the environment. Now we can apply the gas law, $P = C \rho T$, to the parcel and to the environment to see how parcel and environment densities ($\rho$) compare. Since the parcel's P equals the environment's P, the product $C \rho T$ (constant times density times temperature) for the parcel must also equal the environment's $C \rho T$. And since the parcel's T is smaller in value than the environment's, its density must be correspondingly greater to compensate. (Recall we employed this same argument in Chapter 3, discussing convection.) Therefore, being denser, the lifted parcel will sink back toward the surface from which it was perturbed; the air is stable.

By late afternoon of the same day, surface and near-surface air at Albuquerque had warmed under bright sunlight to the values shown in Figure 6.15. How has the stability changed? To find out, again perturb a parcel of surface air by lifting it. Notice that in this case, as the air parcel ascends, it is warmer than its

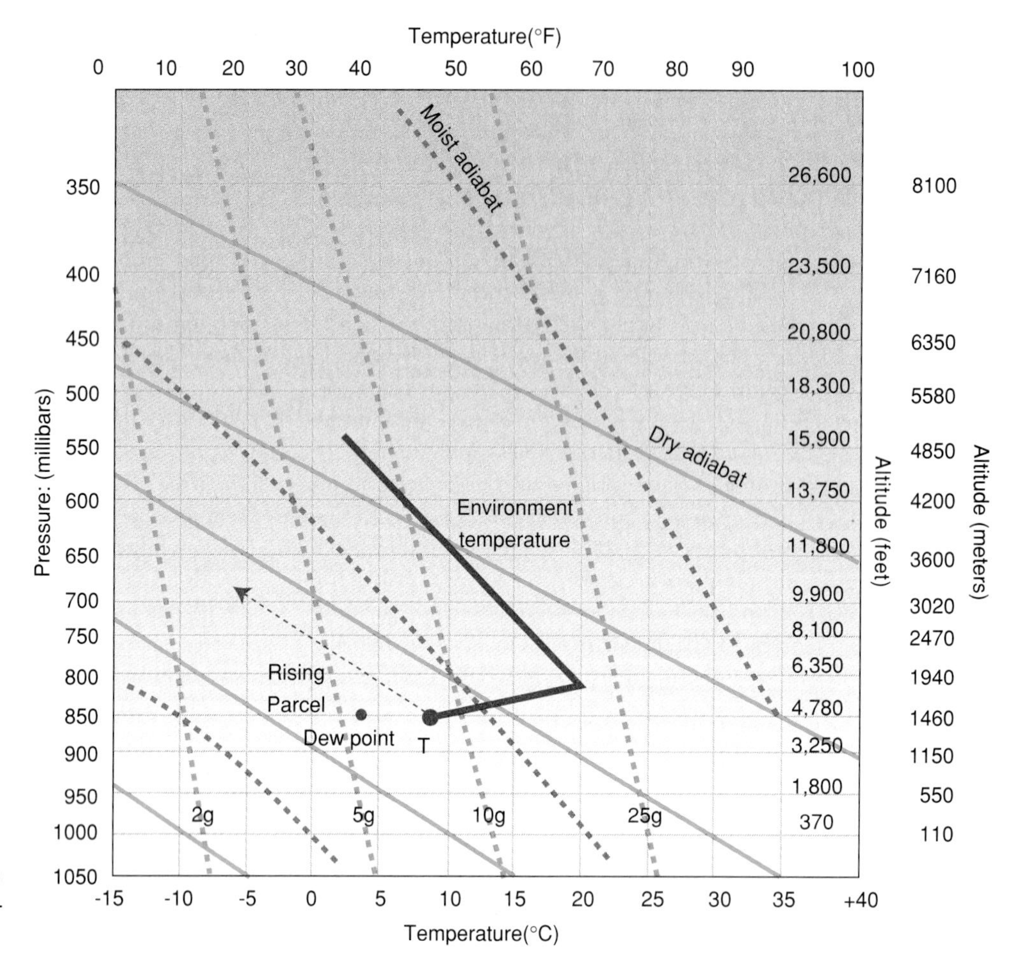


Figure 6.14 The air over Albuquerque in early morning shows a surface-based inversion. A parcel of surface air lifted to 800 millibars remains colder than the surrounding air throughout its ascent.

surroundings. Warmer parcel temperature implies lower parcel density than in the surrounding air; therefore, the parcel is buoyant. The air in this case, therefore, is unstable.

Notice that in both examples above, the parcel's lapse rate was the same: the dry adiabatic rate. It was the change in the environment's lapse rate that brought about the change in stability.

The conditions at Albuquerque illustrate why cumulus clouds occur with greater frequency in the afternoon than in the morning. By late afternoon of a typical day, the earth's surface is at its warmest, the near-surface environment is at its most unstable, and convective currents leading to cumulus formation are therefore most likely to exist. Near sunrise, by contrast, the ground is coldest, causing the near-surface atmosphere to be at its most stable, with a corresponding absence of deep convection and cumulus cloud development.

## *Stability of Saturated Air*

Examine the sounding plotted in Figure 6.16. Suppose an air parcel at 800 millibars experiences an upward perturbation. How will the environment act on the parcel?

This problem is similar to that in the last section. However, notice that the 800-millibar air is saturated. Therefore, a lifted parcel of that air cools not at the dry adiabatic lapse rate but moist adiabatically. In this case, therefore, the air's stability depends on how the environment's lapse rate compares to the *moist adiabatic* rate followed by the parcel. Use the thermodynamic diagram in Figure 6.16 to lift a parcel from 800 to 700 millibars and determine its temperature compared to that of the environment. Is the air stable or unstable? You should find that the parcel is slightly warmer than its environment at 700 millibars; therefore, the air is unstable.

It is interesting to consider what would have resulted if the 800-millibar parcel had been unsaturated. In that case, it would cool at the dry adiabatic rate. You can verify from the thermodynamic diagram that it would have remained slightly colder than the environment; thus it would be stable and would descend toward the 800-millibar level.

## *Stability and Lapse Rate*

With the help of Figure 6.17, we can summarize the static stability relations between parcel and environment. Any environment whose lapse rate is greater than dry adiabatic, such as those shown in green on the chart, is called **absolutely unstable.** In such environments, a lifted air parcel, whether it cools dry or moist adiabatically, will remain warmer than the environment and hence will be buoyant.

Suppose the environment's lapse rate falls between dry and moist adiabatic, such as the examples shown in red in Figure 6.17. Is this air stable or unstable? As you just saw in the last section, the answer depends on the air's humidity. If it is saturated, a parcel will cool moist adiabatically, be warmer than its surroundings, and be unstable. If it is unsaturated, it will cool at the dry rate, remain colder than the surroundings, and therefore be

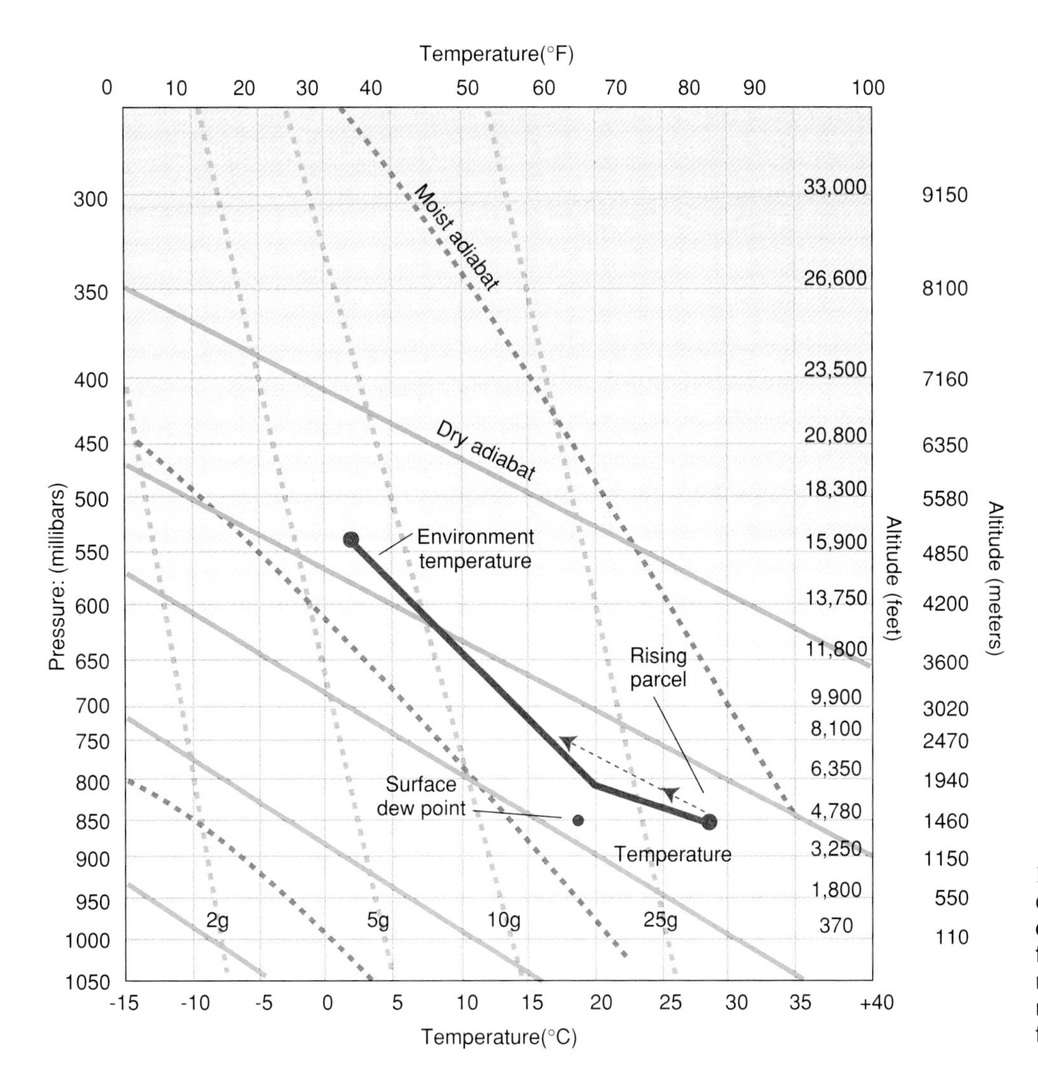


Figure 6.15 Late afternoon temperatures over Albuquerque show the effect of a day's warming at the surface. A rising surface parcel remains warmer than the surrounding environment. How does its 800-millibar temperature compare to that of the environment?

stable. An environment with a lapse rate between dry and moist adiabatic, such as those in red in Figure 6.17, is said to be **conditionally unstable**. Notice from this discussion that moist air has a greater potential for instability than dry air.

Finally, if the environment's lapse rate resembles the examples plotted in black in Figure 6.17, the air is said to be **absolutely stable**. Absolutely stable air has an environmental lapse rate less than the moist adiabatic rate. Stability in these last cases does not depend on the air's humidity: whether the parcel rises dry or moist adiabatically, it remains colder than any of the examples in black.

## *Changes in Stability*

Daytime heating is not the only process that alters the atmosphere's stability and therefore promotes or suppresses cloud formation. If you think of stability in terms of the slope of the environment's temperature profile, it is not difficult to see what other processes might be.

Consider the "original" environment lapse rates plotted in Figure 6.18. What would make the air less stable? Any process that tilts the plot as shown in the lower three diagrams will achieve the affect. That is, any process that warms the lower regions or cools the upper regions will tend to destabilize the layer. Thus, daytime heating is a destabilizer.

Horizontal winds may advect warmer or colder air into the region at various levels and thereby alter the atmosphere's stability. For example, notice in Figure 6.18D that advection of cold air aloft destabilizes the air and can lead to cloud formation.

What processes might stabilize the air? Clearly, just the opposite ones from those above. Low-level cooling, upper-level warming, or both will tilt the environment's lapse rate toward isothermal (vertical on a thermodynamic diagram) or even toward an inversion and therefore are stabilizing influences.

Such changes in stability may take place at any level in the atmosphere, of course; if the air is destabilized, buoyant cloud formation may result at any level as well. The altocumulus and cirrus clouds in Figure 6.19 have formed from buoyant lifting of air at the middle- and high-cloud levels, respectively. Sometimes, several stability-altering processes operate simultaneously at different altitudes over one location, causing clouds to form at several different levels.

Atmospheric stability is such an important determinant of atmospheric behavior that weather forecasters closely monitor its value and changes. Several formulas have been developed

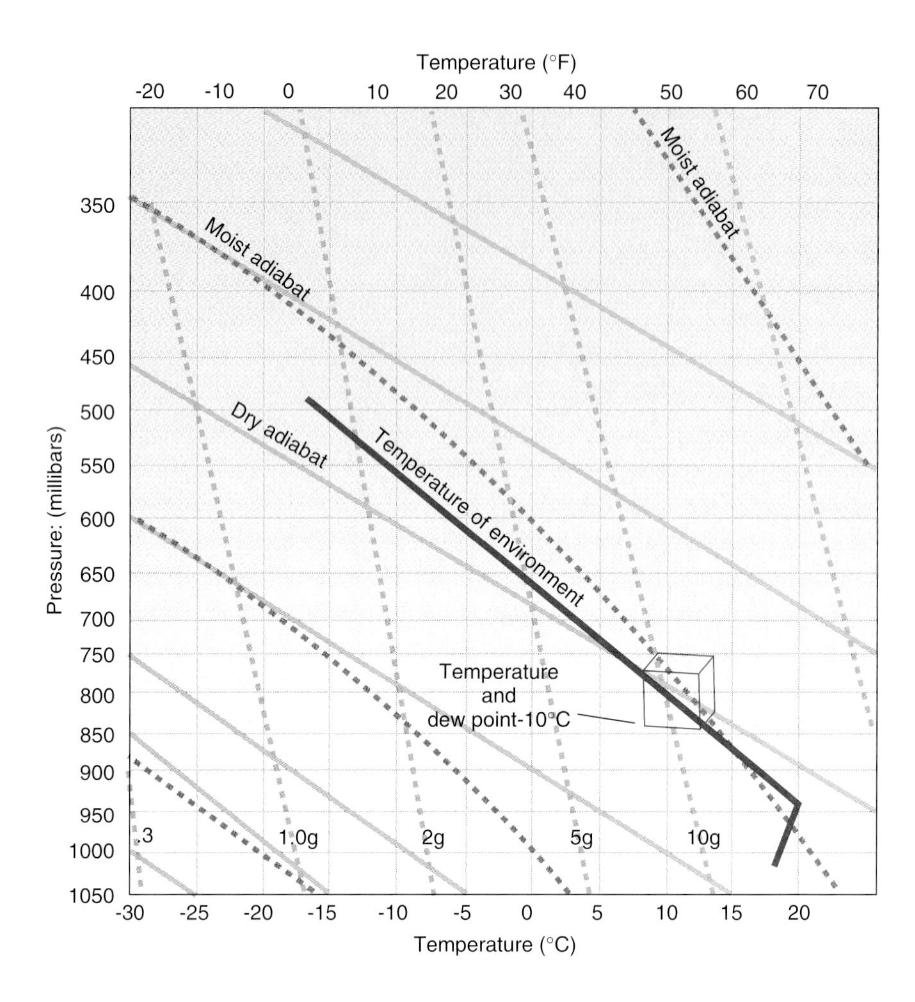


Figure 6.16 The saturated air at 800 millibars will cool moist adiabatically if lifted. What will be its temperature at 600 millibars? Will it be warmer or colder than the environment? Will it be stable or unstable? Answer the same questions for the case of dry adiabatic lifting.

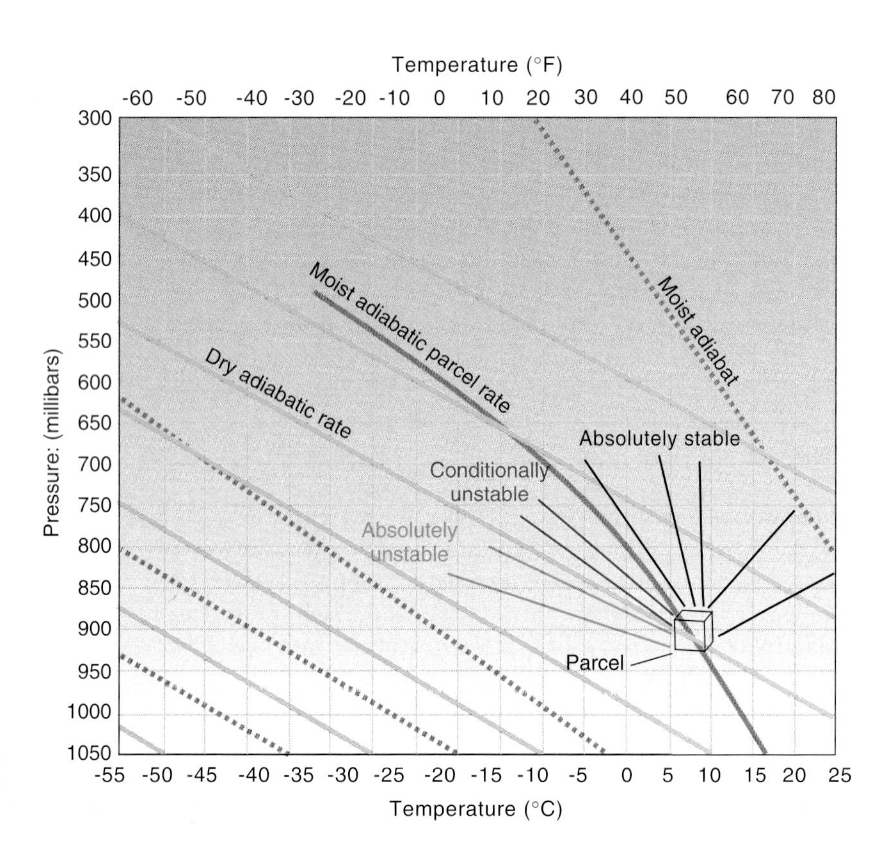


Figure 6.17 Stability ranges. The more the environment lapse rate slopes to the left, the less stable the air. The parcel's lapse rate, by contrast, is always either the dry or the moist adiabatic rate.

What temperature changes increase the air's stability?

Pressure

Temperature

A Warming aloft Or B Cooling at surface Or C Both

What temperature changes destabilize the air?

Pressure

Temperature

D Cooling aloft Or E Warming at surface Or F Both

Original temperature New temperature lapse rate

Figure 6.18 (Top) Any atmospheric process that causes the environment's lapse rate to tilt as shown stabilizes the air. (Bottom) A change in the opposite sense destabilizes the air.

A

B

Figure 6.19 (*A*) Altocumulus castellanus clouds ("castles in the sky") result from instability at the middle cloud level. (*B*) Cirrus clouds form in buoyant updrafts within the high cloud range; light precipitation falls from their bases, trailing behind in the lighter winds at lower altitudes.

that allow calculation of the atmosphere's stability. You will see these "stability indices" in the study of thunderstorms in Chapter 12.

## Cumulus Cloud Growth

You have seen that cumulus cloud formation results from buoyant forces acting on a perturbed air parcel. The parcel, being warmer and less dense than the environment, rises and cools at the dry adiabatic rate. Eventually, condensation occurs; thereafter, cooling proceeds at the moist adiabatic rate as the parcel continues to rise.

When does the cloud stop growing? You know that cumulus clouds do not grow thousands of miles high. What brings the process to a halt?

Once again, the thermodynamic diagram offers some insight. Consider the cumulus cloud that has formed in Figure 6.20. You see the surface air is unstable, resulting in a rising parcel that is warmer than the environment from the surface up to the CL at 900 millibars. As the parcel (i.e., the cloud) rises moist adiabatically above the CL, it remains warmer than the environment through 850 millibars, through 800 millibars, until finally at 735 millibars, the parcel and the environment temperature are equal, at point A. Further lifting causes the parcel to become colder and therefore denser than the environment, thereby losing its buoyancy. Thus, point A represents the top of the cumulus cloud.

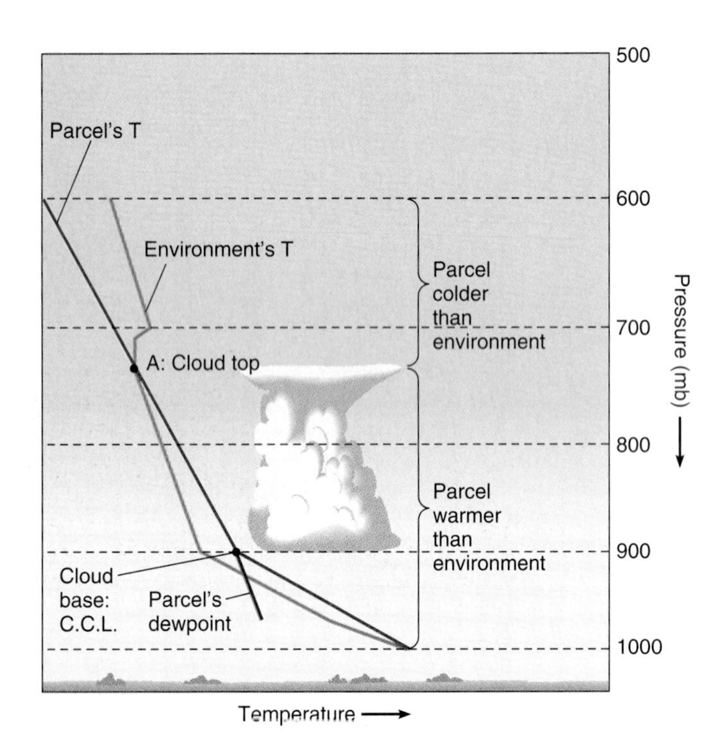


Figure 6.20 The rising parcel of surface air is warmer than the environment up to a pressure level of 735 millibars. Further lifting would cause the parcel to be colder that its environment.

Inversions, which are highly stable features, frequently define the upper limit to cumulus cloud growth. Stratocumulus layers also are often capped by inversions that limit further vertical growth. Cumulonimbus clouds frequently are capped by the inversion at the tropopause; look back at the example in Figure 6.2. The air inside the cloud may rise rapidly through the tropopause and penetrate a short distance into the stratosphere, but soon it loses its buoyancy in the warmer environment of the stratosphere. Unable to rise further, the air spreads horizontally, forming the anvil shape characteristic of many cumulonimbus cloud tops.

Another factor influencing cumulus growth is **entrainment,** the mixing of environmental air with that of the cloud. Although we have assumed a rising parcel does not mix with the surrounding air, in fact some mixing does occur as the parcel pushes upward. Entrainment mixes cooler, drier air with the warm moist air within the cloud and therefore, it is thought, tends to weaken the buoyant forces sustaining the cloud. A poorly understood and difficult phenomenon to observe directly, entrainment is the subject of considerable current interest among cloud physicists.

## Checkpoint

***Review***

The central idea of this section is that heating can make an air parcel unstable, causing it to rise and cool adiabatically and to form a cloud. Another major concept is that air's stability depends on the environment's temperature lapse rate, compared to that of a rising parcel. Table 6.1 summarizes the steps involved in determining the static stability of an air parcel. Compare this diagram with Figure 6.17.

***Questions***

1. Explain the difference between an air parcel and its environment.
2. On a recent October afternoon, the surface air in East Lansing, Michigan, was 15°C and the dew point 6°C. The environment temperature cooled steadily with height, to 10°C at the 800-millibar level. Was the surface air stable or unstable? (Use the thermodynamic diagram in Appendix L.)
3. The following morning, East Lansing skies were clear, the wind calm. Would the stability of the surface air likely be more, less, or unchanged from the previous afternoon? Explain your answer.

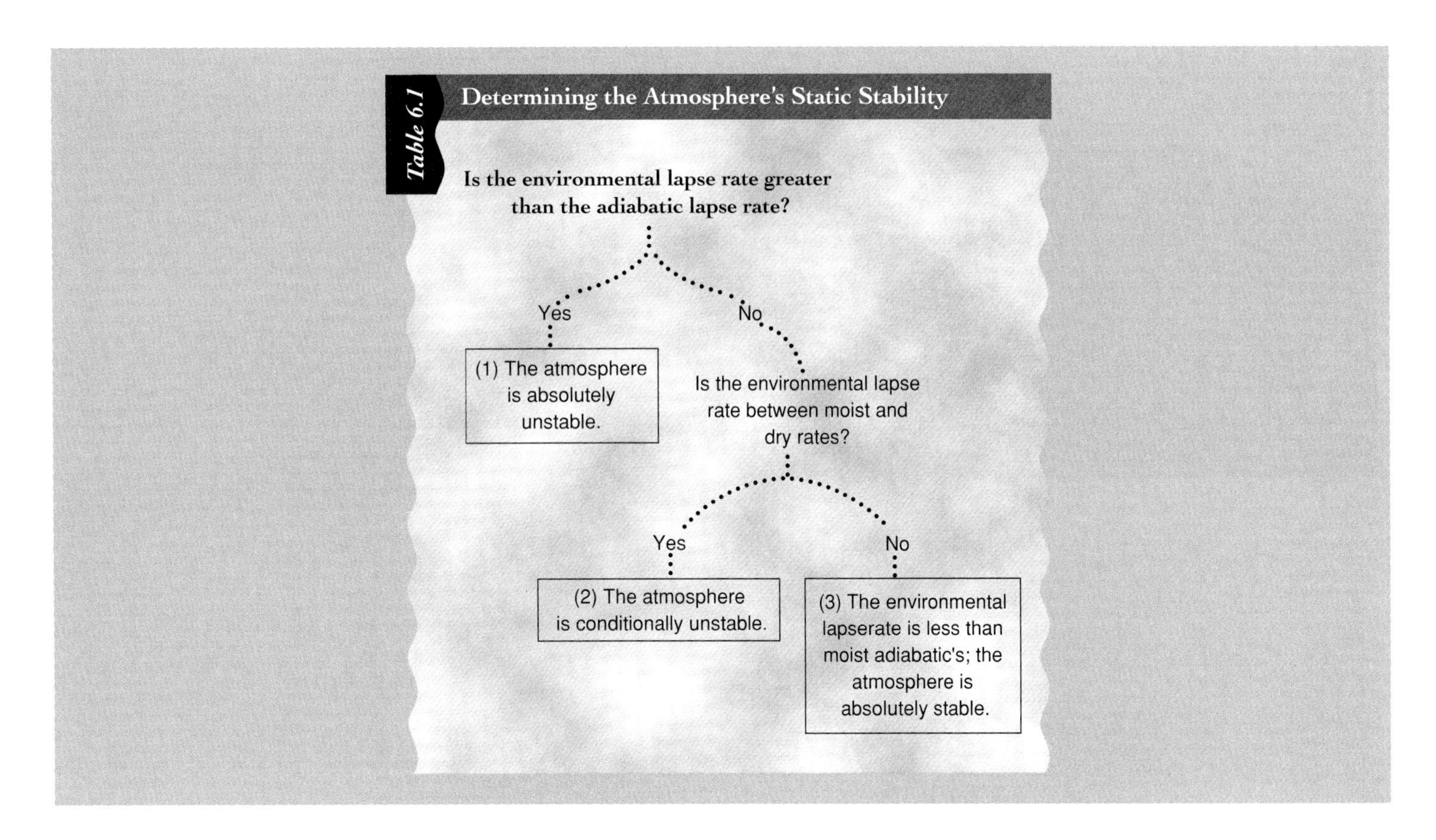


## Cloud Formation by Forced Lifting

You might be surprised to learn that the stratiform clouds in Figure 6.21 were formed by lifting. Why, then, do they bear so little resemblance to cumulus clouds?

This section deals with the formation of layered clouds like those pictured. As you will see, the air comprising these clouds does not rise buoyantly; instead, some external mechanism forces the entire layer to ascend. Evidently the manner in which air ascends has a great deal to do with the resulting cloud form. Or, stated the other way around, the resulting cloud form is a message about the process—or processes—that caused it. Many clouds, it turns out, are a result of several influences at work at once.

### Orographic Lifting

Perhaps the most obvious type of forced lifting is called **orographic** lifting, which is upward air motion caused by flow over sloping terrain. The clouds in Figure 6.21 are orographically induced. Unlike buoyant lifting, in which individual cells of air rise, the entire lowest layer of the atmosphere in the figure has ascended by moving uphill. The result is the uniform sheet of clouds pictured.

The altocumulus clouds formed in Figure 6.22 are also orographically formed. Because of their characteristic lens shape, these clouds are called lenticular altocumulus.

Figure 6.21 Winds blowing onshore from the ocean often cause low clouds due to orographic lifting.

In Figure 6.22, you can follow the process of lenticular cloud formation. The air streaming over the ridge cools and reaches saturation before A and remains cloudy to D; then, the air sinks back down on the lee side of the ridge, warming and its cloud evaporating as it sinks. The lenticular cloud, then, does not represent a parcel but rather a volume of space above the ridge

A

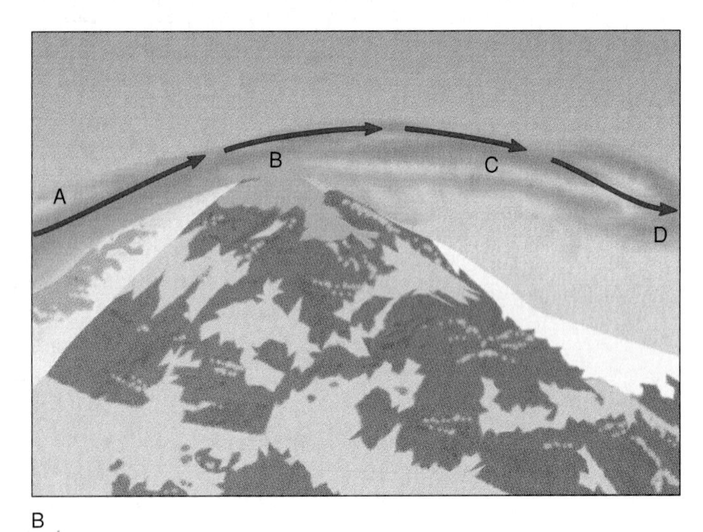


B

Figure 6.22 Forced lifting over the mountains formed these lenticular altocumulus clouds. How would the clouds differ in appearance if the air had been unstable?

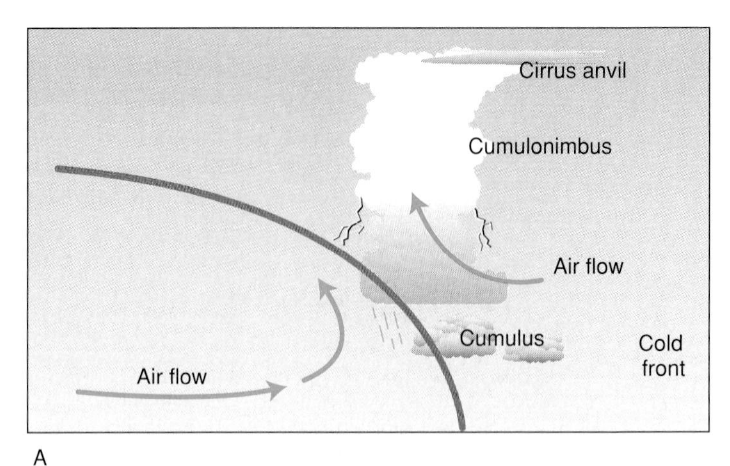


A

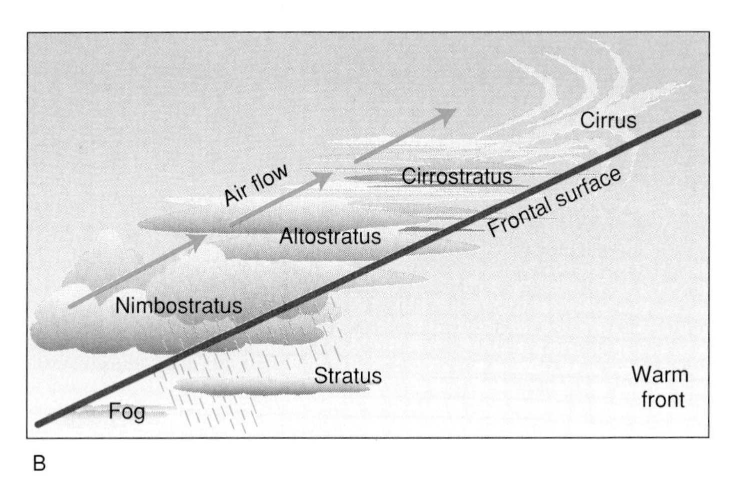


B

Figure 6.23 Cloud types typically associated with warm fronts and cold fronts. Note the prevalence of stratiform clouds in the gentler-sloping warm front.

within which passing water vapor temporarily condenses. Lenticular clouds often remain in the same place for hours at a time, despite the strong wind flow.

Why do the clouds in Figure 6.22 have stratiform or lenticular, and not cumulus, shapes? The answer lies in the air's stability. Orographic lifting of stable air causes stratiform and lenticular clouds. The air, perturbed upward orographically, receives no positive feedback from buoyancy and settles back to lower altitudes once past the perturbing influence of higher terrain.

If the orographically lifted air is unstable, however, then cumulus clouds result from the combined influences of buoyancy and orographic lifting. The high frequency of cumulus clouds and thunderstorms in the mountains is a result of orographic lifting and buoyancy working together.

## Lifting in Fronts and Low Pressure Centers

As you can see from Figure 6.23, a great variety of cloud types occurs in association with fronts and frontal cyclones. Consider, for example, the air gliding upward over the warm frontal surface, typically extending horizontally over a thousand kilometers or more, as the air moves from the surface to many kilometers aloft. This ascent over colder, denser air is similar to orographic lifting and generates sheets of stratiform clouds at all levels: cirrostratus, altostratus, stratus, and nimbostratus if precipitation occurs.

You read in Chapter 1 that cold fronts generally have steeper slopes than warm fronts. In addition, the air above a typical cold front tends to be less stable than above a warm front. Both of these circumstances cause cold fronts to generate more turbulent, convective motion and cumuliform cloud types than do warm fronts.

Because they are associated with low pressure and convergent wind flow and becaus'e of their sloping geometry, fronts are particularly active regions of cloud development. However, as air converges toward any low pressure center, whether fronts are present or not, it experiences cloud-forming influences. For example, the air's pressure decreases to match its surroundings; therefore, it expands and cools. Furthermore, the air rises as it

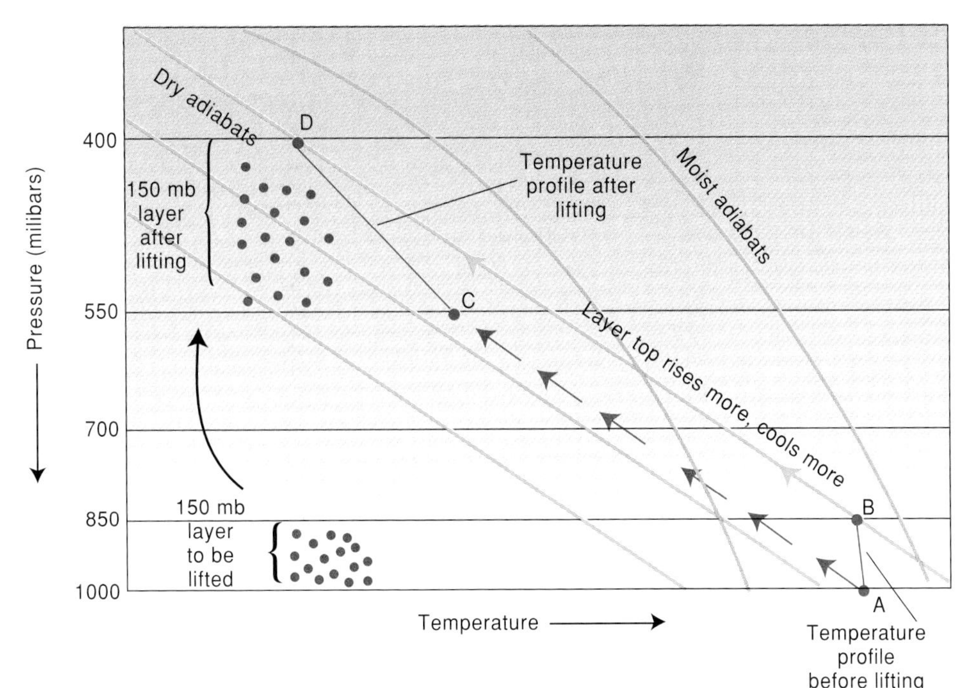


Figure 6.24 The top of the lifted layer rises more meters than does the base; thus, the layer top cools more than the base, destabilizing it. Notice the change in slope of the temperature profile. (Note: The increase in thickness with height has been somewhat exaggerated in the figure to make the effect more evident.)

converges on the low center, causing additional cooling. It is not surprising, then, that frontal cyclones tend to be centers of clouds of every description.

## Lifting Destabilizes an Air Layer

Examine the segment of the environmental sounding labeled "AB" in the lower-right portion of Figure 6.24. How stable is this air? You can see that its slope is more vertical than the moist adiabatic rate. Thus, the layer of air is absolutely stable. (For a reminder about stability values, refer back to Figure 6.17.)

Now suppose the entire layer AB is lifted, due to orographic or frontal lifting. What will the layer's temperature profile be after the lifting? Since the layer is far from saturated, it will rise dry adiabatically. The arrows illustrate the paths taken by the top and the bottom of the layer in rising to its final location, CD. (The change illustrated is a very large one: we chose such an extreme case to emphasize the point.)

Note the change in slope of the layer's temperature as it rises from AB to CD. The process of lifting has destabilized the layer (i.e., made it less stable). Similarly, a layer with starting conditions CD that sinks will become more stable in changing to the trace AB. Thus, layer lifting is an inherently destabilizing process, and sinking is a stabilizing one. Forced orographic or frontal lifting of stable air in this manner sometimes destabilizes the air sufficiently that the lifted air becomes buoyant. The result is a mixture of different cloud types, such as cumulus buildups protruding from the tops of warm frontal stratus clouds.

If the layer is moist at its base but dry aloft, then lifting causes even greater destabilization. The layer top cools at the dry adiabatic rate, 9.8°C per kilometer, while the base cools more slowly due to release of latent heat. The geography of the southern Plains states promotes destabilization of this sort. Warm, moist south winds from the Gulf of Mexico may move inland and upslope below a dry westerly cold flow of air aloft. The resultant lifting and cooling can generate intense cumulonimbus cloud formation, sometimes accompanied by violent weather such as torrential rain, strong winds, and tornadoes.

***Review***

The main point of this section is that an entire air layer (and not just air parcels) may be lifted and cooled, resulting in the formation of layered clouds. Another important concept is that layer lifting is an inherently destabilizing process.

***Questions***

1. What is orographic lifting? How does it differ from buoyant lifting? How are the resulting clouds different?
2. In Figure 6.24, is the air before lifting stable, unstable, or conditionally unstable? After lifting?

## Cloud Formation Due to Cooling by the Earth's Surface

Let's consider the processes at work forming the fog pictured in Figure 6.25. How was this air brought to its dew point? The cloud's base is at the earth's surface, so buoyant lifting cannot be the explanation. Furthermore, the fog rests in a valley, with higher elevations in all directions, so orographic lifting cannot have been the cause. Apparently, the cloud has formed without benefit of the cooling-through-lifting process discussed in previous sections.

This fog layer, and others considered in this section, were formed by cooling air through contact with a colder underlying surface. Because of the cooling from beneath, these clouds, like most fogs, occur in stable air.

Figure 6.25 Fog has formed overnight in this valley under calm winds. What caused its formation?

### Radiation Fog

Sometimes, the clearest of skies can be a factor in forming the thickest of clouds. Recall from Chapter 3 that under clear, wind-

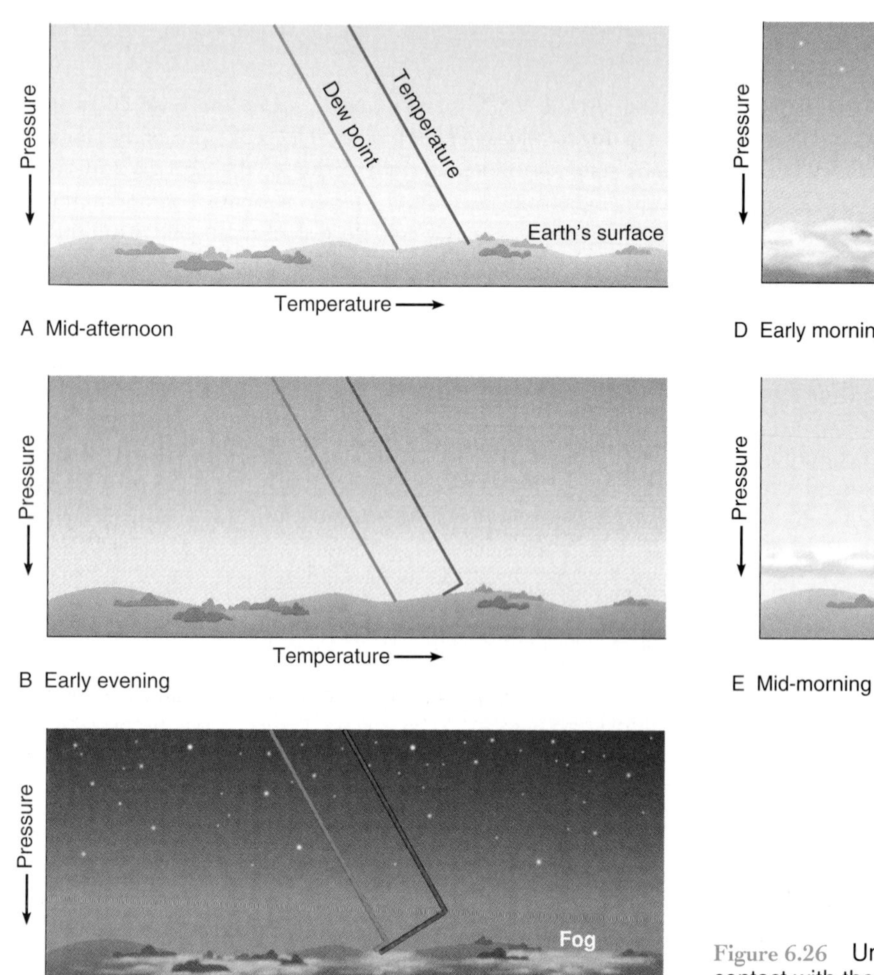


Figure 6.26 Under clear skies, the surface air temperature cools in contact with the earth throughout the night. Radiation fog forms at the surface and builds upward. The fog also dissipates upward from the surface under solar heating the following morning.

less night skies, the earth's surface radiates considerably more longwave radiation upward than it receives from the atmosphere. This net radiative energy loss causes the earth's surface to cool faster than the overlying air, which may result in the near-surface temperatures changes shown in Figure 6.26. Notice that the cooling at the ground increases the air's stability; typically, a surface-based inversion forms, which deepens throughout the night. If the air is rather dry, dew or frost might be the only form of condensation to occur. At higher humidities, however, fog may form as well.

As the temperature cools to the dew point at various levels, cloud formation occurs: first at the surface (C), later at higher elevations (D). Thus, the cloud grows upward from the earth's surface. Such a cloud is known as **radiation fog,** so named because the negative radiative energy flux at the surface is responsible for the cooling leading to condensation. Radiation fog is also known as ground fog.

Why has the fog formed in the valley and not over higher terrain? As cooling of the surface air progresses through the night, the very coldest and therefore densest air slides downhill under the influence of gravity, collecting in the lowest places—the valleys. Thus, it is in valleys that the air is most likely to reach its dew point, followed by fog formation. (The tendency of water to collect in valleys often contributes to the fog formation process.) For these reasons, the radiation fog like that in Figure 6.25 also is called "valley fog."

Once the fog has formed, its upper surface behaves radiatively like the earth's surface before fog formation, radiating more energy upward than it receives from the clear atmosphere above. Thus cooling occurs at the fog's top layer, which leads in turn to further fog formation. Although very shallow at the onset, radiation fog can grow to 100 meters or more in depth in a single night. Its effect on highway and airport traffic can be devastating.

In the morning, solar heating at the earth's surface warms the surface air more than the foggy air above. This surface heating causes the fog to begin to evaporate, first at the ground. The fog then appears to "lift," resulting first in a layer of stratus, then sometimes stratocumulus, as it breaks up and evaporates.

## Advection Fog

If you live in a cold climate, you may have observed dense fog forming as warm air blows across a cold or frozen surface. Figure 6.27 illustrates the process, which is known as **advection fog.** Advection fog occurs when winds carry in air whose dew-point temperature is higher than the temperature of the underlying ground. As you can see from the figure, the surface air is chilled to that of the earth's surface, resulting in saturation of surface air within a well-defined, surface-based inversion. Like ground fog, advection fog is quite variable in density and height and can change greatly over short times or distances, depending on the nature of the underlying ground.

Advection fog does not require a snow-covered surface. Any relatively cold surface exposed to air with higher dew points offers suitable conditions. Such conditions exist along both the west and northeast coasts of North America and are illustrated in Figure 6.28. Advection fog can be so thick and so frequent in some of these regions that it has become a part of the local culture and folklore, like hurricanes in Florida and thunderstorms in the Great Plains.

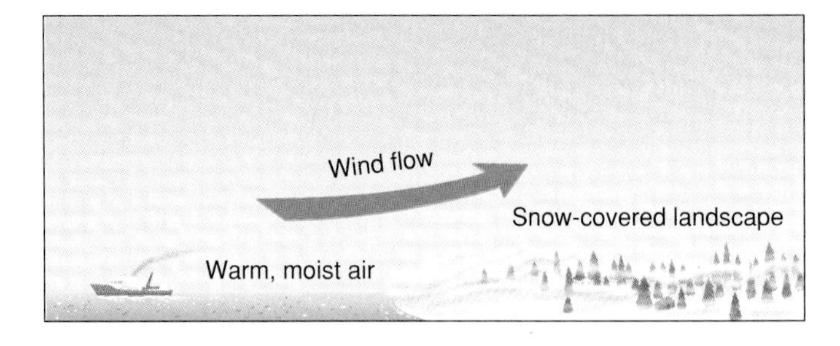


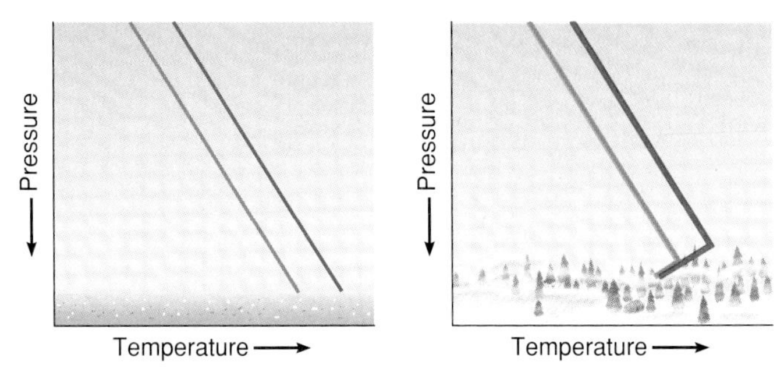


Figure 6.27 Advection fog caused by warm air passing over a snow-covered surface. The air's temperature and dew point, both initially greater than 0°C in contact with the snow.

Figure 6.28 San Francisco's Golden Gate Bridge shrouded in coastal fog.

## Checkpoint

***Review***

This section has presented you with the first examples of clouds in which cooling through lifting is *not* the main mechanism leading to saturation. These clouds, which consist of two kinds of fog, form due to cooling from contact with a cold earth's surface.

***Questions***

1. Why is radiation fog most likely on clear, calm nights?
2. Why would advection fog tend to be more likely (in North America) when surface winds are from the south?

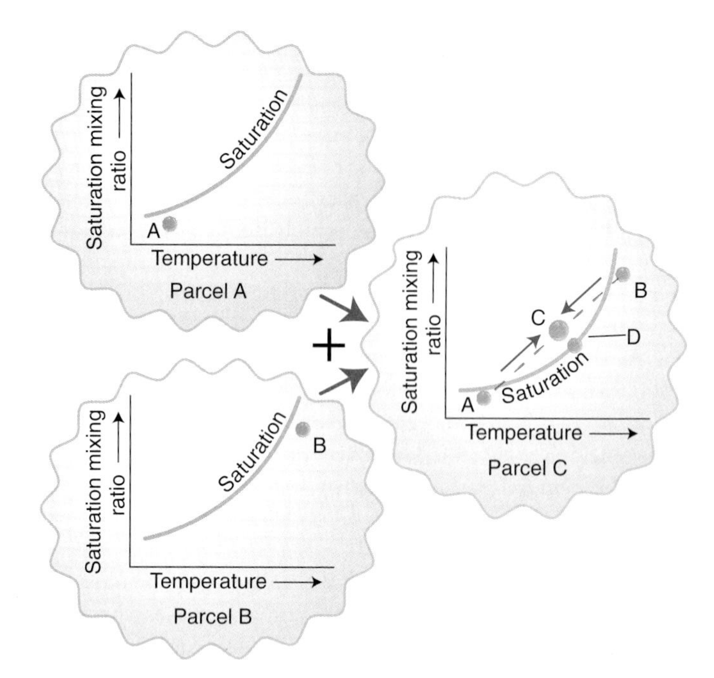


Figure 6.29 Parcels A and B are humid but unsaturated. Combining them into parcel C results in supersaturated air. Would the resulting air be supersaturated if A and B were at the same temperature? Why not?

# Some Other Cloud Formation Mechanisms

Every cloud formation process you have studied to this point has relied on cooling to achieve saturation. In this section, you examine several other mechanisms that may combine with cooling to play a role in forming clouds.

## Cloud Formation by Mixing

At first, it sounds impossible: in a frontal zone, air parcels from two unsaturated, cloud-free air masses of different temperatures come together and mix thoroughly. Due to mixing only, the air becomes saturated and cloudy.

### *Parcel Mixing*

How can two unsaturated air parcels produce a cloud simply by mixing together? In Figure 6.29, notice that the two air parcels A and B, although close to the saturation curve, are in fact unsaturated. They also have markedly different temperatures. Mixing equal amounts of each parcel would create air with properties halfway between A and B, represented by point C in the figure. Notice that point C lies above the saturation mixing ratio curve. Thus, the mixed air is supersaturated, and cloud formation occurs as the excess vapor condenses. From the figure, you can see that mixing can lead to saturation because the saturation mixing ratio curve turns upward sharply at higher temperatures. If the curve were linear, parcel mixing would not be a cause of cloud formation. (Satisfy yourself that this is true: draw a hypothetical straight-line saturation curve and try to achieve saturation by mixing.)

Figure 6.30 shows several examples of cloud formation in which mixing plays a role. In frontal fog (Figure 6.30A), contrasts between the two air masses provide the requisite temperature difference. (Precipitation often plays a role in frontal fog formation, as you will see shortly.)

In Figure 6.30B, the air closest to the ground is warmer and moister than the air a few meters higher. The warmed surface air is also buoyant; it rises and mixes with the air above, briefly achieving condensation. As it continues to mix with more dry air at higher levels, the cloud quickly dissipates, to be replaced by another rising from the surface. Unlike other forms of fog, which are stable clouds, the "steam fog" in (B) is composed of unstable clouds, like miniature cumulus clouds. The steam cloud appearing at the spout of a kettle of boiling water (Figure 6.30C) is another example of this type of mixing cloud.

The aircraft contrails in Figure 6.30D are also examples of mixing clouds. The air coming from the jet engines is far warmer and moister than the air of the upper troposphere or lower stratosphere with which it mixes. At times, contrails are so extensive and persistent, they reduce significantly the amount of insolation that reaches the ground.

### *Vertical Mixing*

Examine the temperature and dew-point profiles for the air layer plotted in Figure 6.31. Is the air well mixed vertically? At first glance, you might think so: in any mixed substance, you would expect to find similar properties throughout, and in this air layer, the temperature and dew point vary only slightly in the vertical. In fact, the air represented in the figure is not well mixed; thorough mixing will alter its temperature and dew-point profiles significantly.

Consider the dew-point profile first. What are values of mixing ratio along this profile? Notice that they range from 8 g/kg at the base to 12 g/kg at the inversion top, with an average value of

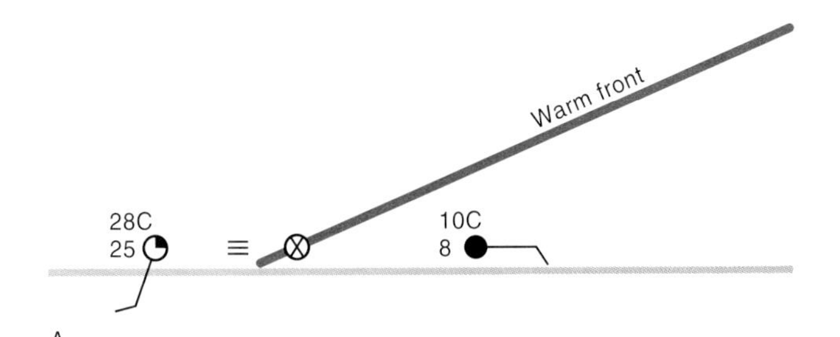


A

B

C

D

Figure 6.30 Each of these clouds forms due to mixing of air of different properties. In (*A*) assume the temperature and mixing ratio of the air in the frontal zone are averages of the values in the two air masses. What will the temperature and dew point be? (*B*) Steam fog or Arctic Sea smoke. (*C*) Tea kettle steam. (*D*) Aircraft contrails.

10 g/kg. Thus, there are more water vapor molecules in the upper levels than below. Thorough mixing will cause the distribution of water vapor molecules to be the same at all levels. The net effect, as you see in Figure 6.31B, is that the lower part of the layer becomes moister, the upper region drier.

Turbulent mixing alters the temperature profile also. If the layer is undergoing thorough mixing, there is no passive "environment" through which isolated parcels rise and sink; instead, the entire layer may be considered to be composed entirely of rising and sinking parcels. Since parcels' lapse rates are always adiabatic, the mixed layer must have an adiabatic lapse rate. This realignment, as you see in Figure 6.31C, causes surface warming, and cooling aloft. The average temperature of the layer remains unchanged.

Figure 6.31D shows the two profiles after thorough mixing. Notice that the temperature and dew-point curves are wider apart in the mixed air at the surface but meet at some point within the mixed layer: at this level, a cloud deck forms. The cloud type is typically stratocumulus—a sheetlike low cloud with convective, cellular shape but only modest vertical extent. Often, an inversion exists from the cold top of the mixed layer to the warmer, unmixed air above; the inversion is a stable cap to the cloud deck below.

Cloud formation by vertical mixing is a frequent occurrence, and stratocumulus clouds formed in this way are one of the most common cloud types.

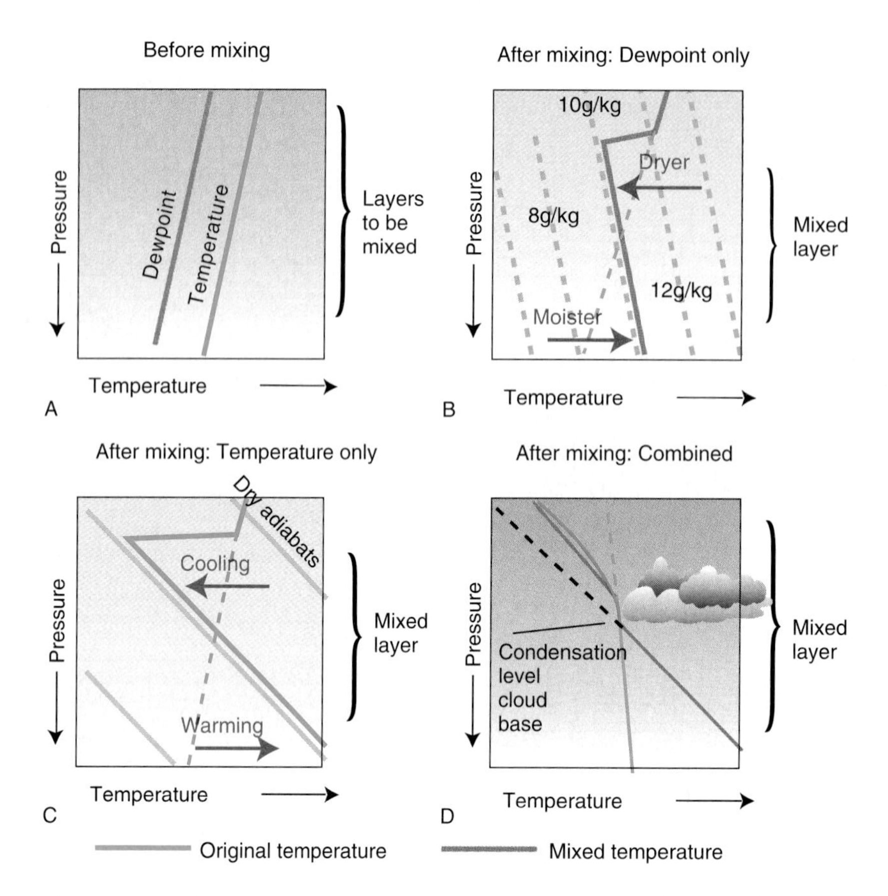


Figure 6.31 How can vertical mixing lead to cloud formation? If a stable, moist layer (*A*) is mixed, two changes occur: (*B*) Mixing ratio through the layer becomes constant; (*C*) Temperature change becomes adiabatic: warmer at surface, colder aloft. Net effect (*D*) is that air may become saturated at some level within the mixed layer.

## Cloud Formation by Water Vapor Uptake

At one time or another, you have probably heard a meteorologist report that rain was falling, but the relative humidity was only 70 or 80%. You might have wondered how the humidity could be so low in the presence of rain.

While it is true that in clouds the relative humidity always is very close to 100%, precipitation from these clouds may fall into much drier air below. Thus, a relative humidity observation well below 100% at the onset of precipitation is not uncommon. However, evaporation from the precipitation raises the humidity and cools the temperature of the air through which the precipitation falls. Then just slight additional cooling of this air, perhaps through mixing with colder air, can result in the formation of additional clouds. The development of low stratus clouds beneath a rainy nimbostratus layer is a typical example of precipitation-stimulated cloud formation. Frontal fog also often forms through this process.

Water vapor increase is also partly responsible for the formation of "steam fog" and aircraft contrails discussed in the last section. In each case, high mixing ratios in the warmer air provide the majority of the moisture needed for the clouds to form due to mixing. Without this injection of moisture, clouds would not form.

## Cloud Metamorphosis

Some clouds owe rather little of their appearance to processes at work when they first formed and much more to subsequent influences. We conclude by considering a few examples of these clouds that have undergone metamorphosis from other cloud genera. An example you encountered earlier is that of radiation fog changing to stratus and thence to stratocumulus as it lifts and dissipates. Figure 6.32 shows others.

The stratocumulus deck in Figure 6.32A has formed from the proliferation and spreading out of cumulus, a common event. Like

A

B

Figure 6.32 (*A*) The stratocumulus layer pictured has formed by the widespread development and spreading out of cumulus. (*B*) Layers of middle and high cloud generated from cumulonimbus development.

individual fair weather cumulus clouds, such cloud decks tend to dissipate late in the day, as the earth's surface begins to cool.

Thunderstorms carry vast amounts of water vapor into the middle and upper atmosphere; as these storms dissipate, residual patches of altocumulus, altostratus, and cirrostratus clouds may persist for hours. Small vertical waves in the wind flow may generate roll or cellular structures observed in some layered clouds.

Finally, some altostratus and cumulus clouds grow to the point where they yield precipitation, thereby becoming nimbostratus and cumulonimbus, respectively. Just how are clouds able to make the transition from nonprecipitators to precipitators? This important question is the focus of the next chapter.

# Study Point

## Chapter Summary

The chapter opened with a survey of the 10 cloud genera. The remainder of the chapter was devoted to a discussion of the different mechanisms leading to cloud formation and the types of clouds that result in each case. (The concept map in Figure 6.3 provides a visual summary of this information.) By far, the most effective mechanism of cloud formation is the adiabatic cooling of upward-moving air. Because of its preeminence, you spent considerable time studying adiabatic processes and atmospheric stability with the aid of thermodynamic diagrams. You saw that buoyant ascent of unstable air may generate clouds of vertical development, while forced lifting of stable air may produce layered clouds.

A variety of other mechanisms play roles at certain times and places: cooling at the earth's surface causes radiation and advection fog, while the mixing of unsaturated air of different properties may contribute to the genesis of certain types of fog and layered clouds. The chapter concluded with a brief look at cloud metamorphosis from one genus to another.

## Key Words

| | |
|---|---|
| low clouds | stratus |
| fog | stratocumulus |
| nimbostratus | middle clouds |
| altostratus | altocumulus |
| corona | high cloud |
| cirrus | cirrostratus |
| halo | cirrocumulus |
| cumulus | cumulonimbus |
| adiabatic | dry adiabatic lapse rate |
| dry adiabatic | condensation level (CL) |
| moist adiabatic | moist adiabatic lapse rate |
| thermodynamic diagram | static stability |
| parcel | environment |
| absolutely unstable | conditionally unstable |
| absolutely stable | entrainment |
| orographic lifting | radiation fog |
| advection fog | |

## Review Exercises

1. A student at Ohio State University observes a corona surrounding the sun; another student, at Penn State, notices the sun is surrounded by a halo. What are the cloud types at each location?
2. Suppose an air parcel with pressure of 650 millibars and temperature of −5°C rose dry adiabatically to 500 millibars. What would be its new temperature?
3. Suppose the parcel in the last example sank adiabatically. What would be its temperature as it passed through the 850-millibar level?
4. What is the major difference in cloud particle structure between high and middle clouds? Between middle and low clouds?
5. What is the CL? How do you determine its value?
6. What causes the air in which radiation fog forms to be stable?
7. Picture an unsaturated air parcel rising through the atmosphere. At what rate does its temperature change? Its dew point? What are the rates if the parcel is saturated? What lines on a thermodynamic diagram represent these rates of change?
8. An air parcel in stable air receives a small upward perturbation. When lifted, will it be warmer or colder than its environment? Denser or less dense? When the perturbation ceases, will the parcel continue to rise, remain where it is, or return to its original level?
9. By what mechanisms can stratocumulus clouds form, according to Figure 6.3? Altocumulus?

## Problems

1. One winter day, the surface air temperature in Austin, Texas, is 5°C; the temperature decreases steadily to the 800-millibar level, where it is −5°C. Is the surface air stable, unstable, or conditionally unstable?
2. By how many degrees does air cool if it rises dry adiabatically from sea level to the highest elevations of the Appalachian Mountains (elevation of approximately 2000 meters)?
3. If cloudy air rises moist adiabatically 2000 meters, how would its temperature change compare to that in the last example? Why would the temperature change at a different rate?

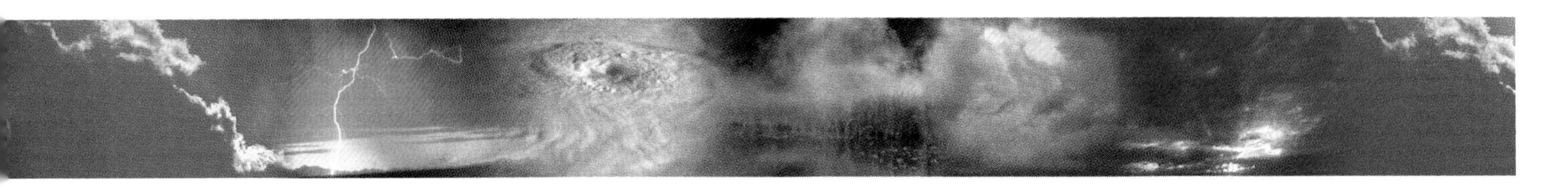

4. Students at Hartford College for Women measure the surface pressure, temperature, and dew point one late summer afternoon to be 1000 millibars, 30°C, and 25°C, respectively. Scattered cumulus clouds dot the Hartford sky. What is the height of the cloud bases?
5. In the previous problem, what was the relative humidity at the surface? At the cloud bases?
6. Picture a deep layer of saturated air at a temperature of 5°C (41°F) blowing inland over California from the Pacific Ocean. The air rises moist adiabatically as it ascends the Sierra Nevada Mountains, reaching an altitude of 3000 meters (700 millibars); it loses much of its moisture via precipitation as it ascends. Then, it descends dry adiabatically down the east slopes of the mountains to an altitude of 500 meters (950 millibars). What is its final temperature?
7. What would cause summer afternoon cumulus clouds to be so much more common over land than over water?
8. A radiosonde measures the Tucson, Arizona, temperatures to be 20°C at 850 millibars and 10°C at 700 millibars, with dew point at 0°C at both levels. Is the air in this layer stable? If the 700-millibar temperature cooled to 0°C, how would that affect the air's stability?
9. A front lies between Tulsa, Oklahoma, (where the air temperature and dew point are 5°C and 3°C, respectively) and Little Rock, Arkansas, (temperature = 20°C, dew point = 18°C). If equal portions of these air masses mix at the front, will fog form?

## Explorations

1. Develop a key to assist a person to identify the genus of a cloud. (If you know computer programming, consider writing the key as a program.) The key should ask the user various questions about a cloud the user had observed and then announce the cloud's genus.
2. With a camera (film or video), you can make a number of interesting studies of clouds. For example, you can prepare a cloud atlas; make time-lapse recordings of growing cumulus, dissipating fog, or cloud motions; record hour-to-hour changes in cloud cover accompanying passage of a front; and so on.
3. If you have access to a weather service such as AccuWeather, explore variations in cloud types and amounts by time and location. Examples include variations in cumulus or fog amounts by time of day; occurrence of fog along coastal regions versus inland; and occurrence of cumulus clouds at Rocky Mountain stations versus elsewhere.
4. Investigate the role of clouds in artists' paintings. What cloud genera appear, and how do they contribute to the composition or mood of the painting?
5. Use camera images from World Wide Web sites, plus observations at nearby weather stations, to make a computer-based atlas of cloud types. (See this chapter's Resource Links for suggested Internet addresses.)

## Resource Links

The following is a sampler of print and electronic references for topics discussed in this chapter. We have included a few words of description for items whose relevance is not clear from their titles. For a more extensive and continually updated list, see our World Wide Web home page at www.mhhe.com/mathscience/geology/danielson.

World Meteorological Organization. 1987. *International Cloud Atlas* Vol. 2. Geneva: World Meteorological Organization. 212 pp. A standard reference for cloud types with hundreds of color photos.

Bohren, C. F. 1991. *What Light through Yonder Window Breaks?* New York: John Wiley and Sons. 190 pp. Excellent chapter on inversions.

Bohren, C. F. 1987. *Clouds in a Glass of Beer.* New York: John Wiley and Sons. 195 pp. Fine discussions of cloud formation.

For readers with access to AccuData, SKW and SKWT plot upper air soundings and provide data on condensation level and various humidity indices.

At over 100 web sites, you can "see" current weather conditions across the United States, thanks to online cameras. These offer an interesting way of studying cloud types. The University of Michigan's Weathernet web site at http://cirrus.sprl.umich.edu/wxnet/ provides a list of "cloud-cam" sites.

A contour map of atmospheric stability is available from Purdue University's weather server at http://wxp.atms.purdue.edu/maps/upper_air/ua_con_lift.gif.

# Chapter 7

# *Precipitation*

## Chapter Goals and Overview

On March 15, 1993, the great "Blizzard of '93" moved northward along the Atlantic coast. Every state east of the Mississippi River felt the influence of this mammoth storm. Coastal regions from Florida to the Canadian Maritime Provinces experienced the storm's full fury in the form of hurricane-force winds, torrential rain, hail, snow up to 10 feet deep in drifts, and damaging surf.

Over a 48-hour period, clouds in this single weather system managed to generate over 10 trillion ($10^{13}$) gallons of rain and snow, the equivalent of a backyard-sized swimming pool of water for every man, woman, and child in the United States.

This chapter deals with processes by which raindrops and snowflakes form in the atmosphere: precipitation on the scale of the '93 blizzard and in less violent forms as gentle tropical showers or the misty drizzle of a foggy morning. You will see how different types of precipitation (rain, snow, sleet, etc.) form, and you will learn of the distribution of precipitation across the earth. Finally, you will consider the unit's main question: Can we significantly modify precipitation patterns for human advantage?

### Chapter Goals

By the end of this chapter, you should be able to:

- outline two processes by which precipitation forms;
- explain why most clouds never generate precipitation;
- explain how different types of precipitation form;
- give approximate values for mean and extreme precipitation amounts over various times and locations;
- explain the theory behind modern cloud seeding; and
- discuss technical and societal issues related to precipitation modification.

# How Is Precipitation Generated?

We are certain you know what precipitation is. Nonetheless, we begin by defining the term precisely: to the meteorologist, **precipitation** is any form of water (raindrops, snowflakes, etc.) that falls from the atmosphere and reaches the ground. Notice the definition specifies that precipitation must fall and that it must reach the ground. Thus, dew, frost, and water vapor are not considered precipitation, because they do not fall. Cloud droplets are so small they do not fall at significant rates and thus are not considered precipitation either.

Three processes occurring within clouds act to collect water molecules into precipitation-sized particles. You will consider each of these processes in the following sections.

## Condensation and Deposition

Recall from Chapter 5 that cloud particles grow as a result of condensation or deposition of water vapor onto appropriate nuclei. As long as the humidity is high enough, these processes exceed evaporation, and the droplet grows. However, you can see from Figure 7.1A that an enormous amount of growth is required for a cloud droplet to reach precipitation size. Can condensation or deposition alone cause so much growth? To answer this question, examine the data on droplet growth rates in Figure 7.1B.

As you can see from the graph, a typical cloud droplet grows rapidly at first. In less than five minutes, the particle graphed has reached a diameter of 5 micrometers. Thereafter, growth proceeds ever more slowly. After two hours, the graph indicates the droplet is only 30 micrometers in diameter, well short of the 200-micrometer (0.2 millimeter) minimum diameter for liquid precipitation. From the graph, it is clear that many hours would be

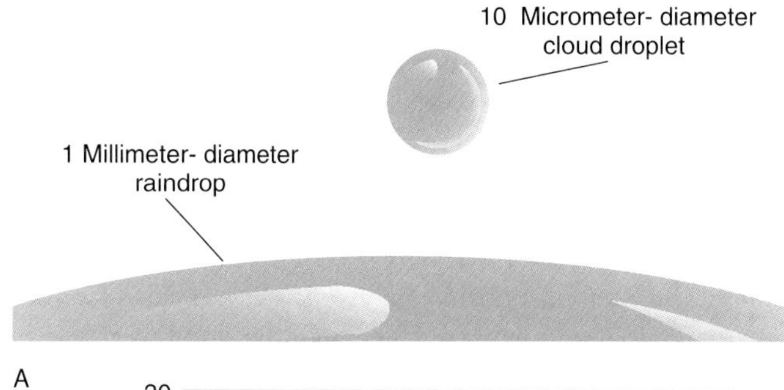


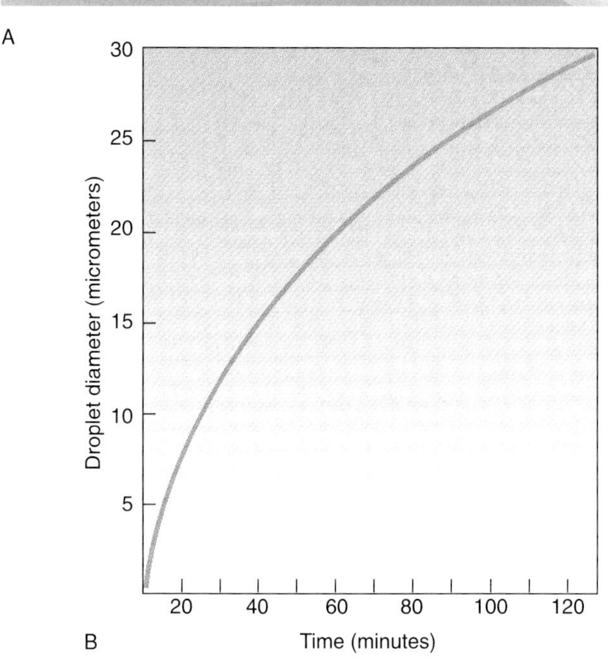


Figures 7.1 (*A*) Typical cloud droplet and raindrop sizes compared. At 100 times the diameter, a raindrop contains 1 million times the water of a cloud droplet. (*B*) Cloud droplet growth due to condensation (under typical atmospheric conditions) onto a salt nucleus. In just a few minutes, the particle reaches normal droplet size (10 micrometers or so). How long does it take for the droplet to reach 25 micrometers in diameter? Source: (*B*) Data from Fleagle and Businger, Introduction to Atmospheric Physics, 1980, Acadamy Press.

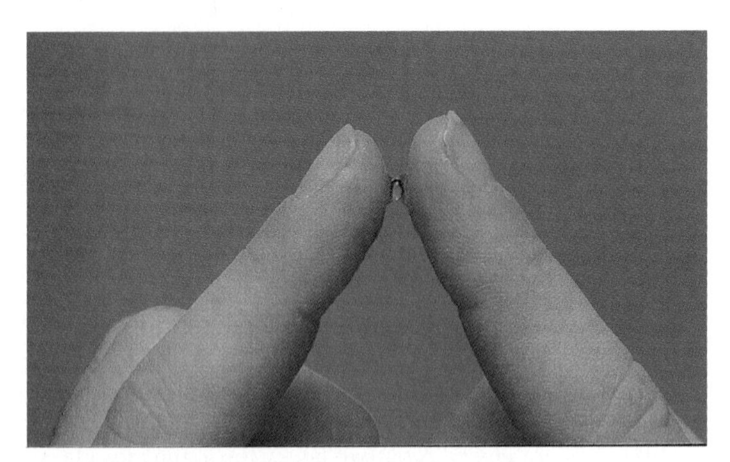

Figure 7.2 Colliding water drops tend to coalesce.

required for precipitation-sized particles to form by condensation alone.

Compare this data with an observation you likely have made yourself: that cumulus clouds can form, grow into cumulonimbus towers, and generate heavy rain, all within an hour. If precipitation can develop that rapidly, evidently other processes besides condensation must be at work in such clouds. In fact, the only precipitation that forms entirely due to condensation or deposition is the fall of fine snow crystals that occurs at extremely cold temperatures. Such crystals are equivalent to cirrus cloud particles that form close to the ground in frigid air. In all other cases, condensation and deposition play only secondary roles in the growth of precipitation particles beyond the cloud particle stage.

## Collision and Coalescence

If you suspend water drops from two wet fingertips and move them close together until they collide (as in Figure 7.2), the drops will join, or **coalesce,** on contact. A similar processes of droplet collision followed by coalescence plays a significant role in precipitation formation.

### *Collision*

Under what circumstances might collision occur? Figure 7.3 suggests some possibilities. From the figure, you can see that if there is a range of droplet sizes in the cloud, the larger droplets will fall faster, overtaking and possibly colliding with smaller, slower-moving ones. On the other hand, if all droplets are the same size, they all fall at nearly the same rates, and little overtaking—and colliding—will occur.

Collision is not simply a matter of position in Figure 7.3A, however. Very small droplets in the path of a falling larger drop may be swept out of the drop's range, as you can see in Figure 7.3C. Electrical forces may repel or attract droplets, altering the number of expected collisions. **Wake capture,** in which a droplet is swept behind a larger particle and collides from the rear, occurs as well. Finally, turbulence within the cloud may cause different-sized particles to travel different trajectories and hence collide, as shown in Figure 7.3D.

Cloud physicists use the term **collision efficiency** to describe collision rates among cloud droplets. The collision efficiency is

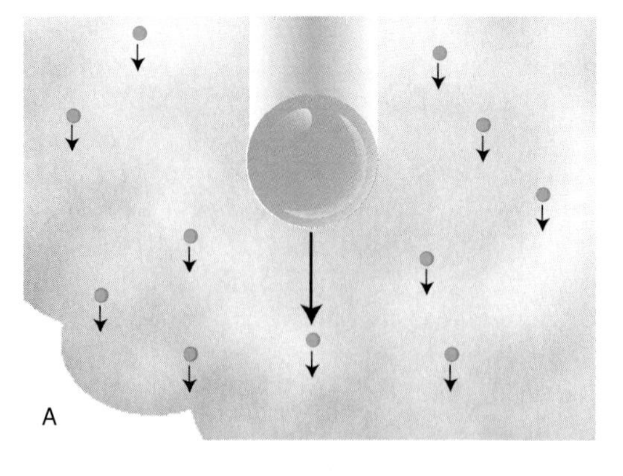


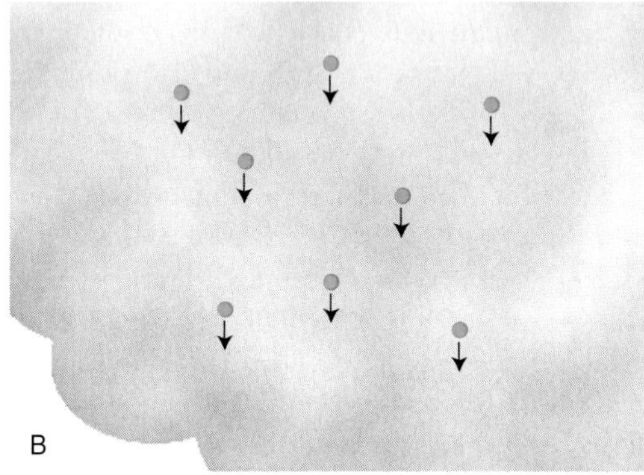


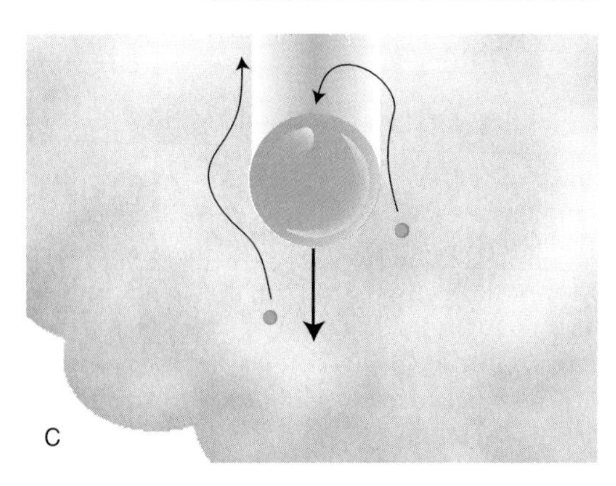


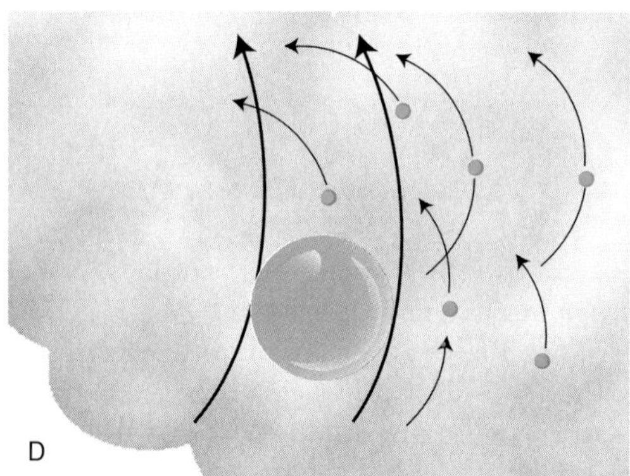


Figure 7.3 Droplet collisions within a cloud. (*A*) A larger droplet falls faster and may collide with smaller, slower droplets in its path. (*B*) In a cloud with uniform sized droplets, little collision occurs. (*C*) Factors affecting collision efficiency. Some droplets in the path are swept aside; others out of the path experience wake capture. (*D*) Turbulent motion within the cloud can increase collision rates. Large droplets turn less sharply, crossing paths with smaller droplets.

the number of droplets actually colliding with a given droplet compared to the number in its path. If, of 100 droplets lying in the path of droplet A, all 100 collide, then

$$\text{collision efficiency} = 100 \text{ colliding}/100 \text{ in the path} = 1.0$$

Suppose, however, that of the 100 in the path, only 80 collide, because 20 are swept aside, and suppose an additional 4 droplets collide due to wake capture. Then

$$\text{collision efficiency} = (80 + 4)/100 = 0.84$$

### *Coalescence*

Unlikely as it may seem, sometimes colliding water droplets do not coalesce but bounce off one another instead. A tension exists on all liquid water surfaces, caused by forces that hold the water molecules together. On small droplets, this surface tension is strong enough to reduce the chance that two colliding droplets will actually coalesce. Electric forces also help to reduce the number of coalesce. In close analogy to the collision efficiency, a droplet's **coalescence efficiency** is the ratio of the number of droplets actually captured compared to the number that collide. As an example, suppose that of the 84 droplets colliding in the last example, 63 actually coalesce while the other 21 bounce off. Then,

$$\text{coalescence efficiency} = \frac{63 \text{ droplets coalesced}}{84 \text{ droplets colliding}} = 0.75$$

The overall efficiency of droplet capture is known as the **collection efficiency** and equals the number of droplets coalescing compared to the number in a droplet's path. Using values from the previous example once more,

$$\text{collection efficiency} = \frac{63 \text{ droplets coalesced}}{100 \text{ droplets in the path}} = 0.63$$

Clearly, the higher the collection efficiency, the more effective is the collision-coalescence process in causing cloud droplets to grow.

### *Raindrop Breakup*

Luckily for all of us, basketball-sized raindrops do not occur. In fact, once a drop has grown to about 5 millimeters in diameter, it breaks into smaller drops due to turbulence and other factors. Each of these smaller drops, still far larger than the cloud droplets on which it feeds, continues to grow by collision and coalescence, dividing again if it reaches critical diameter. In this way, a single large drop can be the parent of numerous raindrops.

### *How Important Is the Collision-Coalescence Process?*

Like condensation, collision and coalescence are ongoing processes in all clouds. Under certain conditions, the process may lead to precipitation, as in the examples in Figure 7.4. The clouds in this figure are called **warm clouds** because they are warmer than 0°C at every level. Thus, they are composed solely of liquid water: no ice is present. Because the collision-coalescence

Figure 7.4 Fog and drizzle settle across one of Lake Motts' rain-flooded shores in Flint, Michigan.

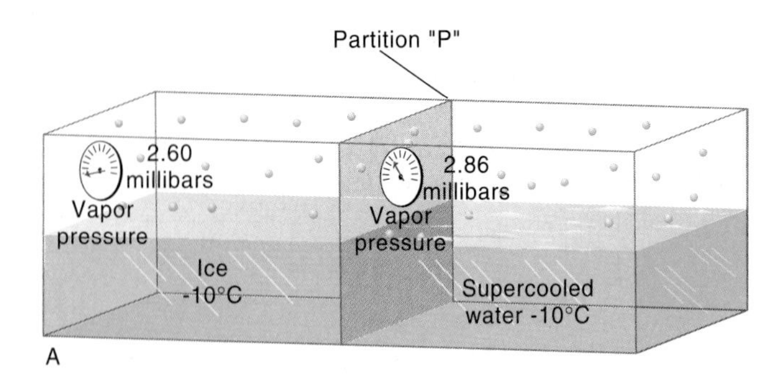


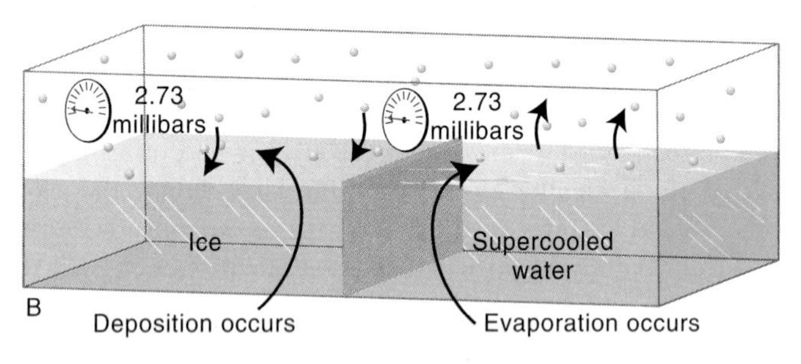


Figure 7.5 (*A*) Equilibrium (saturation) vapor pressures over ice and supercooled water surfaces at −10°C. (*B*) If the partition is removed, water vapor molecules mix, supersaturating the space over ice and unsaturating it over water. Note the effects on evaporation and deposition.

process causes warm cloud precipitation, it is sometimes referred to as the **warm rain process.**

## The Bergeron Process

By themselves and in combination, condensation-deposition and the warm rain process account for only a small fraction of the earth's precipitation, particularly at middle and high latitudes. Where does the remainder—including virtually all snow, sleet, and hail, and most rain from frontal low pressure systems (such as the Blizzard of '93)—come from?

### *Saturation over Ice versus over Water*

To answer this question, it is useful to consider another evaporation box experiment like those done in Chapter 5. In this experiment (illustrated in Figure 7.5A), supercooled water at −10°C and ice at −10°C, separated by a barrier, occupy the bottom of the container. Initially, the spaces above the water and ice surfaces are free of water molecules and are separated by the partition. You allow evaporation to proceed on each side as in earlier examples, and equilibrium is achieved, as shown in the figure. Is this equilibrium the same on each side?

From the figure, you can see it is not. Equilibrium vapor pressure is less over the ice than over liquid water (except at 0°C, where they are equal). This difference in equilibrium pressures can have important consequences, which we will now examine.

Suppose you remove the partition, allowing the air to mix throughout the container. How will the vapor pressure change? The water vapor molecules will tend to spread uniformly throughout the space. At the moment the partition is removed,

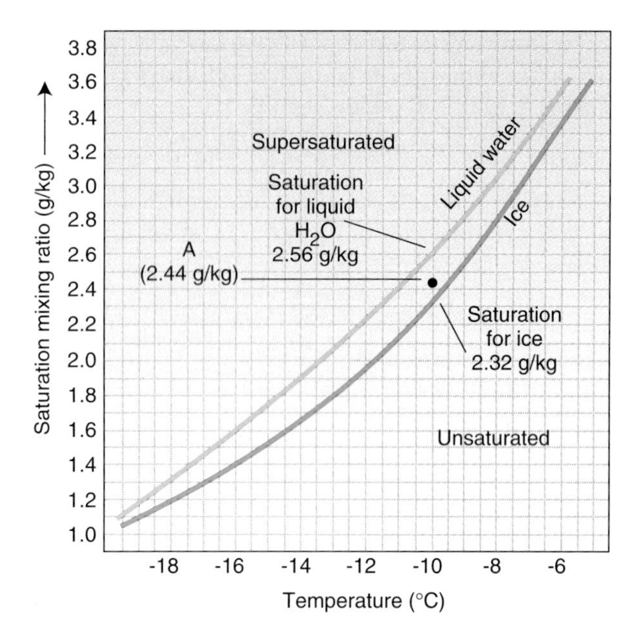


Figure 7.6 Saturation mixing ratio curves for supercooled water and ice. Notice that air unsaturated with respect to liquid water may be supersaturated with respect to ice.

there is more water vapor on the liquid side; thus, mixing will have the effect of moving water vapor from the liquid to the ice side of the container (shown in Figure 7.5B). But the air on the left is already saturated with respect to the ice surface, so introducing additional vapor causes supersaturation, leading to deposition of vapor molecules onto the ice.

Mixing has the opposite effect over the water surface: in this space, some water vapor is lost, causing the air to become unsaturated; therefore, more water molecules proceed to evaporate from the water surface into the space above. Further mixing carries this freshly evaporated water over the ice surface, again raising the vapor pressure past the equilibrium point for ice, causing additional deposition on the ice. In this fashion, water is transported from the supercooled liquid surface, into the air, and thence onto the ice surface. The ice grows at the expense of the liquid water. Equilibrium cannot be achieved at this temperature unless all the liquid is evaporated.

### *Relative Humidity in Supercooled Clouds*

Let's get quantitative for a moment. Suppose you do the evaporation box experiment at a number of subzero temperatures and at an air pressure of 700 millibars (which are typical conditions for supercooled altostratus clouds). You measure the saturation mixing ratio for each temperature and plot your data on the graph in Figure 7.6. Note there are two saturation curves: one for liquid water and one for ice.

Now consider an air sample at a temperature of −10°C, as in the previous example. At saturation, you measure the mixing ratio over water to be 2.56 g/kg, while over ice it is just 2.32 g/kg. After you remove the partition and the air becomes thoroughly mixed, its mixing ratio is 2.44 g/kg—the average of 2.32 and 2.56. Where would air with temperature of −10°C and mixing ratio of 2.44 g/kg appear in Figure 7.6? By entering the diagram with the appropriate values, you can verify that the air parcel is at point A. What is the relative humidity of this air?

To find out, first calculate the value with respect to the water droplets. The air's mixing ratio is 2.44 g/kg, while the diagram shows that at −10°C, the saturation mixing ratio with respect to water is 2.56 g/kg. Thus,

$$\text{relative humidity} = \frac{\text{mixing ratio}}{\text{saturation mixing ratio for liquid water}} \times 100\%$$

$$= \frac{2.44\ \text{g/kg}}{2.56\ \text{g/kg}} \times 100\% = 95\% \text{ for liquid water}$$

Now, repeat the calculation, this time with respect to the ice surface. The mixing ratio remains the same, but the saturation value is now only 2.32 g/kg. Therefore,

$$\text{RH} = \frac{\text{mixing ratio}}{\text{saturation mixing ratio for ice}} \times 100\%$$

$$= \frac{2.44\ \text{g/kg}}{2.32\ \text{g/kg}} = 105\% \text{ for ice}$$

Thus, you arrive at a remarkable and important result: the relative humidity of the air in the chamber is both greater than 100% and less than 100% at the same time, depending on the surface to which it is referred! Under these conditions, ice crystals will tend to grow, while liquid droplets will tend to steadily evaporate.

### *Precipitation Formation in Supercooled Clouds*

What does the difference in saturation over ice and water have to do with precipitation formation? In the 1920s, the Swedish meteorologist Tor Bergeron and German scientists Alfred Wegener and Walter Findeisen conjectured that events within a supercooled cloud might develop much as they do in the evaporation box. They pictured the cloud's water droplets and ice crystals as equivalent to the liquid and ice surfaces in the evaporation box. They realized that the air would be supersaturated with respect to ice, while at the same time, it would be unsaturated with respect to liquid water. Thus, water droplets would tend to evaporate at the same time that ice crystals would grow by deposition.

You might be skeptical that this process could generate precipitation-sized particles. Isn't water simply being moved around from a lot of water droplets to a lot of ice crystals?

If ice crystals were equally numerous as water droplets, then your objection would be valid. However, in many clouds between 0°C and −15°C, ice crystals are relatively rare. They may be outnumbered by water droplets by as much as 1 million to one. Transferring the water of 1 million water droplets to a single ice crystal can easily make a particle large enough to precipitate (recall Figure 7.1).

This process, whereby supercooled droplets evaporate as neighboring ice crystals grow by deposition, is known as the **Bergeron process.** It is also known as the Bergeron-Findeisen process and the three-phase process. Remember, the process depends on the fact that equilibrium over an ice surface occurs at lower water vapor pressures than over supercooled water at the same temperature: the relative humidity is greater than 100% over the cloud's ice crystals at the same time and place it is less than 100% for the water droplets.

As the ice crystals formed in the Bergeron process grow, they fall more rapidly and begin to collide and coalesce with other crystals and droplets. In this way, the collision-coalescence process plays a supporting role in three-phase precipitation.

### *How Important Is the Bergeron Process?*

The Bergeron process is not just a clever theoretical possibility: it is by far the most important of the three precipitation-generating mechanisms. It is responsible for virtually all snow, all frontal precipitation, and most or all thundershower rain. (Think of it! Even in a summer thundershower, the precipitation forms as snow, melting only as it falls through warmer air near the ground.) In fact, for some time after Bergeron proposed it, many atmospheric scientists suspected that the process might account for virtually all precipitation. However, as you saw in Figure 7.4, it is known that warm clouds, in which the Bergeron process cannot operate, also are capable of generating rain.

### *Why Doesn't Every Cloud Generate Precipitation?*

At least two of the precipitation-generating mechanisms discussed above are continually at work in every cloud; in supercooled clouds, all three processes operate. Why, then, does not every cloud generate precipitation?

Any of a number of factors can prevent a cloud from precipitating. The most obvious is the quantity of water available: some clouds simply lack the water density or the height to provide the necessary moisture. In addition, warm clouds must contain some large droplets for collision and coalescence to occur. In supercooled clouds, there must be a proper mix of supercooled droplets and ice crystals: if ice crystals are absent, the Bergeron process cannot commence; if they are too numerous, the water cannot become concentrated sufficiently to generate precipitation-sized crystals.

Finally, some clouds do create precipitation-sized particles that fall toward the earth but evaporate in the drier air between cloud base and the ground; such particles, visible in Figure 7.7, are known as **virga.**

## Using Radar to Observe Precipitation in Clouds

How can you observe the precipitation forming in the interior of clouds? You can't see these regions directly because cloud particles reflect and scatter visible light, hiding the cloud's interior from view. Longer wavelengths of radiation are attenuated far less, however, as indicated in Figure 7.8; if your eyes were sensitive to microwaves, whose wavelengths are 100,000 times longer than those of visible light, you could look directly into cloud interiors.

In essence, **radar** provides just such a long-wavelength "eye." Radar is a device that detects objects by transmitting and then receiving reflections of radiation in microwave wavelengths. A typical weather radar, operating at the 10-centimeter wavelength range instead of the 0.00005-centimeter range of visible light, is able to receive radiation directly from objects within cloud interiors.

Figure 7.7 The wispy streaks extending earthward from the bases of these clouds are called virga. Technically, virga is not considered precipitation, as by definition it evaporates before reaching the ground.

## *How Radar Works*

Radar differs from the human eye in another important way besides wavelength: whereas the eye is a passive device, receiving whatever light impinges on it, Figure 7.8 shows that the radar system sends out its own radiation and then collects reflections (sometimes called echoes) of that radiation. An echo means the radiation has reflected off some target.

What sorts of targets does the radar pick up? Most objects larger than cloud droplets reflect microwaves. Therefore, buildings, hills and mountains, and airborne objects such as airplanes and even large flocks of birds all may appear on radar images. To the meteorologist, however, the interesting targets are precipitation particles. Although cloud droplets are too small to reflect radar waves, snowflakes, raindrops, and other precipitation forms are good reflectors. Therefore, weather radar images show precipitation within clouds but generally not the clouds themselves.

How does a radar system determine a target's location? The antenna scans the sky in various directions, alternately transmitting and receiving microwaves in rapid succession. As each pulse is transmitted, a highly accurate clock records the time until its return echo is received (assuming the pulse has struck a target). The target's direction is easily found: it is the same as the direction the antenna is facing at the moment. The target's distance is computed from the time elapsed between transmission and reception; the longer the time, the farther away the target.

Recent advances in radar technology have made it possible to measure motions as well as positions of radar targets. These methods rely on the **Doppler effect** (see Figure 7.9), the shift in wavelength of radiation emitted or reflected from an object moving toward or away from the observer. You are familiar (but not too familiar, we hope) with an everyday application of the Doppler effect: highway radar guns employed to catch speeders.

A **Doppler radar** utilizes the Doppler effect to measure motions of targets toward or away from the radar instrument, as well as the target's position. Under the U.S. government's NEXRAD program, a network of approximately 160 Doppler radar systems is being phased into operation. Figure 7.10 shows some of the capabilities of this new system.

## *Information Received from Radar*

The most basic radar image is the "plan view". Imagine yourself high above the radar installation and looking down. The radar transmitter and receiver are at the center. As the radar beam sweeps around the sky, targets are "painted" on the screen, often in different colors to represent different echo intensities.

You will see many more radar images in the discussions of various weather systems in later chapters.

Figure 7.8 Short wavelength visible radiation reflects off cloud droplets. Longer wavelength radar waves are not reflected until they strike something larger, such as precipitation particles. Note also that radar transmits the radiation it subsequently senses, unlike the eye.

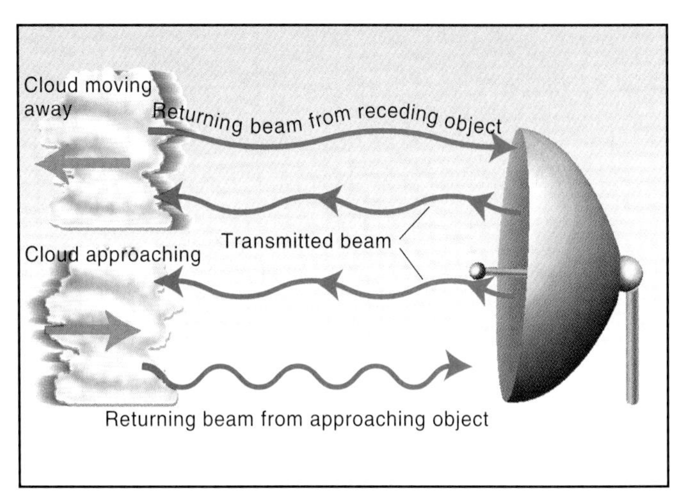


Figure 7.9 The Doppler effect. How is the radar beam's wavelength altered by reflection off an approaching object? A receding one?

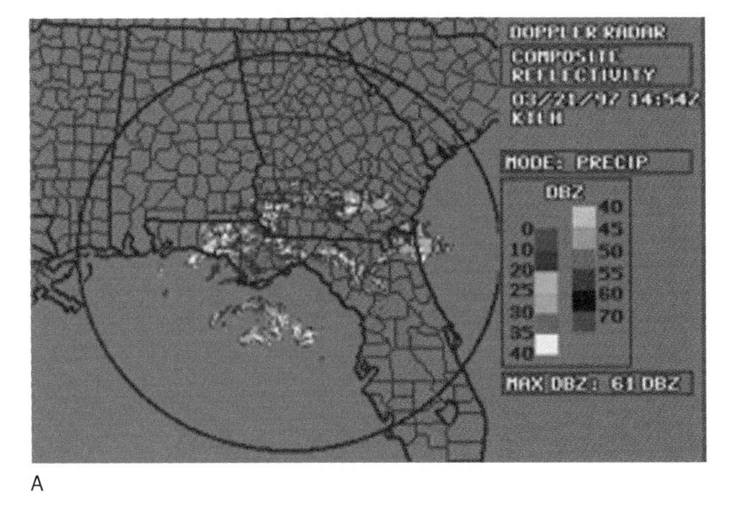


A

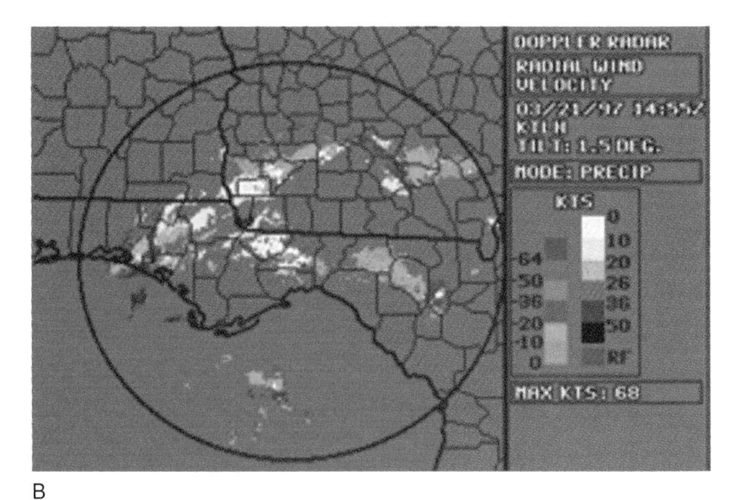


B

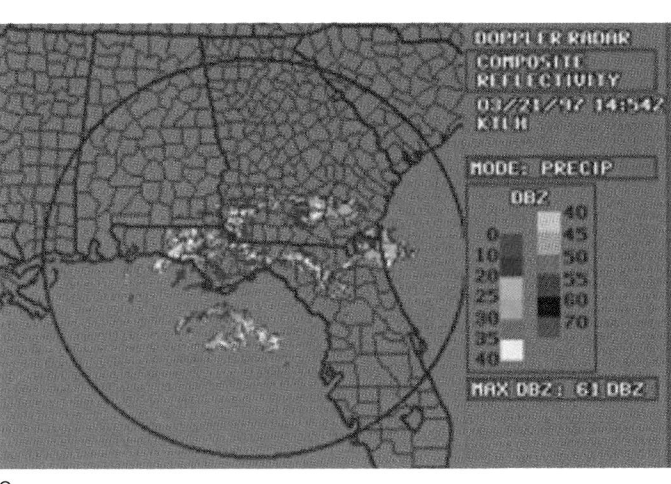


C

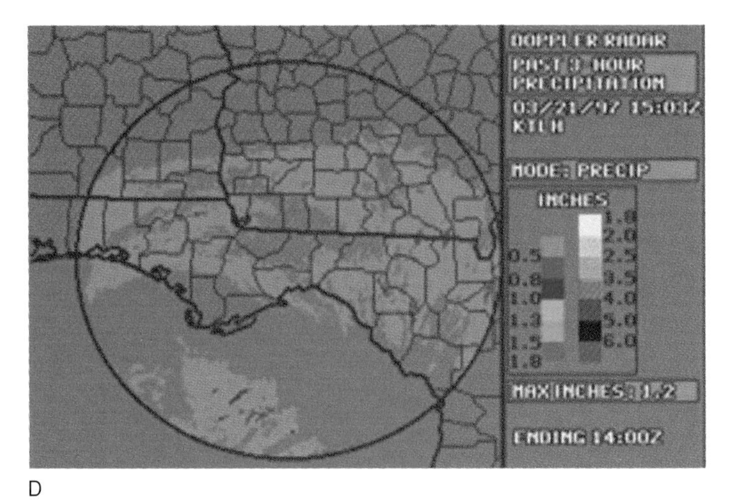


D

Figure 7.10 Doppler radar images of thunderstorm activity over Florida and Georgia. (*A*) Composite reflectivity: The strongest signal measured over each point. Units are in decibels (DBZ), a measure of the reflected signal's intensity. (*B*) Echo tops: The altitude of the highest target at each location, in thousands of feet (KFT). (*C*) Radial wind velocity: The speed at which echoes are moving toward (indicated by negative values) or away from (positive values) the radar instrument. In regions labeled RF, valid data was not obtainable. (*D*) Past three-hour past precipitation: An estimate of the amount of precipitation reaching the ground, based on the signal intensity over the past three hours.

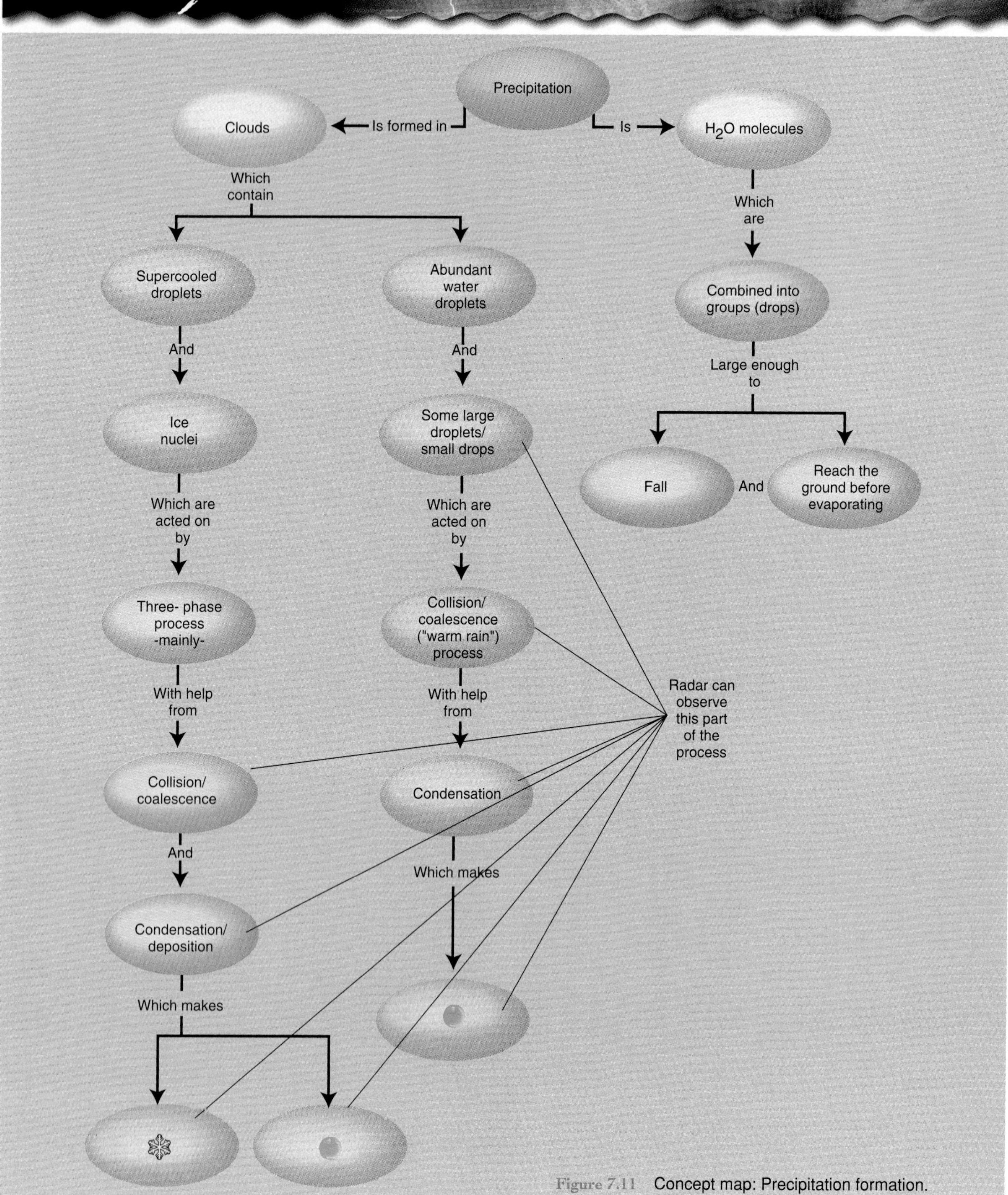


Figure 7.11 Concept map: Precipitation formation.

***Review***

The concept map in Figure 7.11 outlines the main concepts in this section. Notice the different "ingredients" required by the three-phase and warm rain processes and that other processes lend assistance to each of these main mechanisms. The comment about radar reminds you that radar "sees" precipitation-sized particles, not typical cloud droplets.

Not shown in the concept map are details of collision and coalescence and the important difference between saturation mixing ratios over ice and supercooled water at the same temperature. Try including these concepts on the concept map.

***Questions***

1. Compare traffic on a highway to falling droplets in a cloud. When are collisions most likely: when all traffic travels at 55 mph, or at various speeds—some at 55, some at 45, some at 35? Why? How do maximum and minimum speed limits influence the collision efficiency?
2. A falling droplet has 1000 smaller droplets in its path. If it actually collides with 850, what is its collision efficiency? If, of those 850 droplets, it coalesces with 750, what is its coalescence efficiency? What is its collection efficiency?
3. In a paragraph, explain how precipitation forms via the Bergeron process.
4. What are two ways in which a radar system differs from a visual system like the eye?

# Precipitation Types

During the Blizzard of '93, Boston, Massachusetts, reported the weather conditions shown in Table 7.1. Note the variety of precipitation forms that occurred. Could they all have originated in the processes you read about in the last section? If so, then what factors were responsible for shaping the precipitation into the particular forms it assumed when it arrived at the ground? These questions are the subject of this section.

## Snow

After a three-hour period on March 13, 1993, during which the Boston observer noted virga, snow began to reach the ground. As you know, snow crystals are formed in clouds via the Bergeron process. Figure 7.12 illustrates some of the great variety among snow crystals.

In some cases, snow arrives at the ground as individual crystals, just as formed in the cloud. Often, however, crystals collide and stick together, reaching the ground as **snowflakes.** A snowflake is a clump of a few to as many as a hundred or so individual snow crystals.

What conditions must be met for a snowflake to reach the ground as snow? First, the snow crystals must be large enough and the atmosphere below the cloud base moist enough that the precipitation does not evaporate or sublimate. Notice from the data in Table 7.1 that in Boston, it required several hours for the evaporating virga to moisten the subcloud atmosphere sufficiently before the snow began to reach the ground.

*Table 7.1*

**Observed Weather at Boston, Massachusetts March 13–14, 1993**

| | Present Weather | Wind | Temperature (°C) |
|---|---|---|---|
| March 13 | | | |
| 12–14 UTC | Virga | SE/10 | + 1 |
| 15–22 UTC | Snow | E/30 | 0 |
| 23 UTC | Thunder, snow | NE/45 | − 1 |
| March 14 | | | |
| 00 UTC | Ice pellets, snow | NE/40 | 0 |
| 01–04 UTC | Rain | NE/25 | + 2 |
| 05 UTC | Freezing rain | NE/15 | 0 |
| 06–08 UTC | No precipitation | NE/10 | + 2 |
| 08–09 UTC | Rain | S/30 | + 5 |
| 10 UTC | Snow, ice pellets | S/30 | + 3 |
| 11 UTC | Snow | S/30 | + 2 |

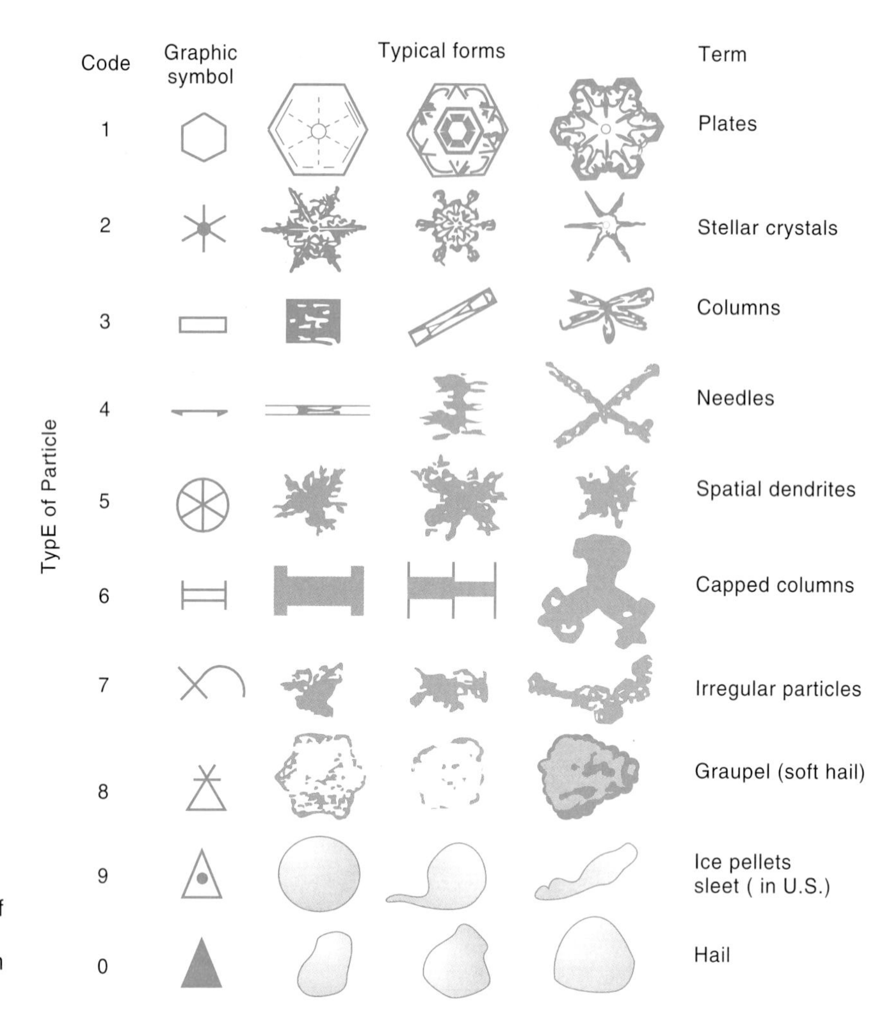


Figure 7.12 The international classification system of snow and ice forms. As you can see, the basic snow crystal shape is hexagonal; however, the specific form taken may vary considerably.

The second condition required for snow is that the environment through which the snow falls must remain cold enough to prevent the crystals from melting. Figure 7.13A shows the temperature conditions aloft over Boston during the time snow fell on March 13. Notice that at no point was the cloud environment warmer than freezing. Therefore, the snow that formed in the Bergeron process also arrived at the ground as snow.

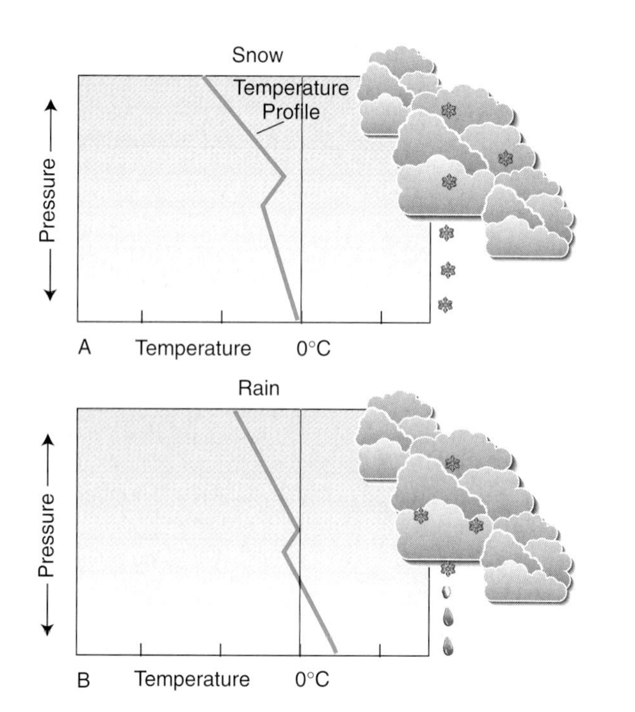


Figure 7.13 Temperature profiles over Boston during snow (*A*) and rain (*B*). Note that the precipitation originated as snow in both cases; in *B*, it melted before reaching the ground.

## Rain

Table 7.1 indicates that after a number of hours of snow, Boston experienced a changeover to rain and then, a while later, a return to snow. During the transition periods, other precipitation forms occurred, with which we shall deal presently. First, however, let's consider the simpler case of rain.

Figure 7.13B shows the temperature profile over Boston during the rain period. Follow a falling snow crystal as it makes its way down the temperature profile in the figure. As you see, the crystal remains in a subfreezing environment only part of the way to the ground. In the last part of its descent, it falls through above-freezing temperatures. If the time it spends in this warmer air is sufficient to melt the crystal, it arrives at the surface as rain. This was the case from 01 UTC to 04 UTC and from 08 to 09 UTC on March 14 in Boston. For the Bergeron process to produce rain, then, a sufficiently deep layer of the lowest atmosphere must be warmer than freezing.

If the above-freezing surface layer is shallow, the precipitation may not have time to melt before reaching the ground. Notice this occurred in Boston at 10 and 11 UTC on March 14; snow was reported at the same time the surface temperature was greater than 0°C.

Before moving on to other forms of frozen precipitation, we will consider briefly the form of liquid precipitation known as **drizzle.** How does drizzle differ from rain? The distinguishing feature is the diameter of the drops. If the drop's diameter is less than 0.5 millimeters, the precipitation is considered drizzle. Because of their smaller size, drizzle drops are able to form in thinner cloud layers than raindrops. A thick stratus or fog layer often is sufficient to initiate drizzle through the warm rain process. Because drizzle drops have such small volumes, precipitation totals from periods of drizzle generally are also small.

## Snow Pellets

Sometimes snow crystals fall through a region of cloud rich in supercooled water droplets. As these droplets collide with the snow crystals, they freeze, forming rimed snowflakes called **snow pellets,** or **graupel.** Several examples are pictured earlier, in Figure 7.12. Snow pellets are somewhat soft and crunchy, and are known by the unscientific but descriptive name "popcorn snow."

## Ice Pellets

Table 7.1 indicates that twice during the March blizzard, Boston experienced **ice pellets,** which are also known in the United States as **sleet.** Ice pellets are simply frozen raindrops.

How do ice pellets form? For precipitation to melt, then refreeze, it must fall into warmer air and then into colder air once more. Figure 7.14, showing the temperature profile over Boston at the time ice pellets were observed, shows a typical ice pellet signature. Notice that the layer of subfreezing temperatures near the ground is rather deep: the precipitation must be in the subfreezing environment long enough to freeze.

You may wonder why the rain doesn't refreeze as snow crystals, rather than as ice pellets. Remember that a snow crystal forms by deposition of water *vapor* molecules onto a *single* freezing nucleus, from which the crystal's hexagonal structure develops. A raindrop, in contrast, is a collection of vast numbers of *liquid* water molecules and nuclei. Freezing within the drop typically progresses from *many* points simultaneously, so that the overall hexagonal shape does not develop.

Another form of ice pellet results from the continued coating of a snow pellet with ice, so that the snow pellet is encased within. This form of ice pellet is sometimes called "small hail," although the formation process is quite different from that of true hail.

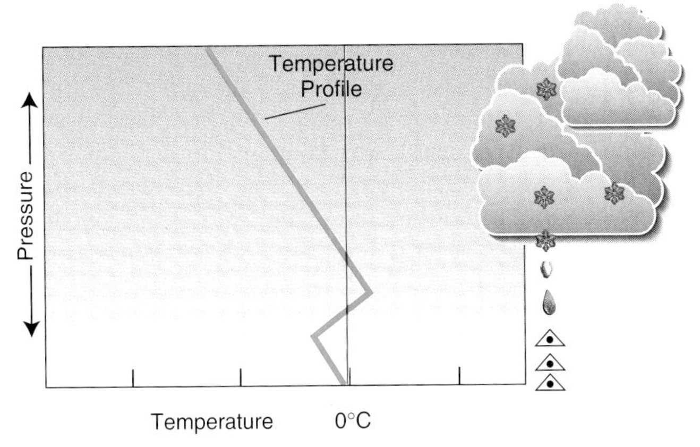


Figure 7.14 A typical temperature profile that accompanies the occurrence of sleet. What is the sequence of forms the precipitation passes through on its way to the ground?

## Freezing Rain

After a period of rain on the night of March 13, the observer in Boston reported **freezing rain** (refer back to Table 7.1). Freezing rain is (liquid) rain that freezes when it strikes the ground, trees, or other structures. Figure 7.15A illustrates that an extended

A

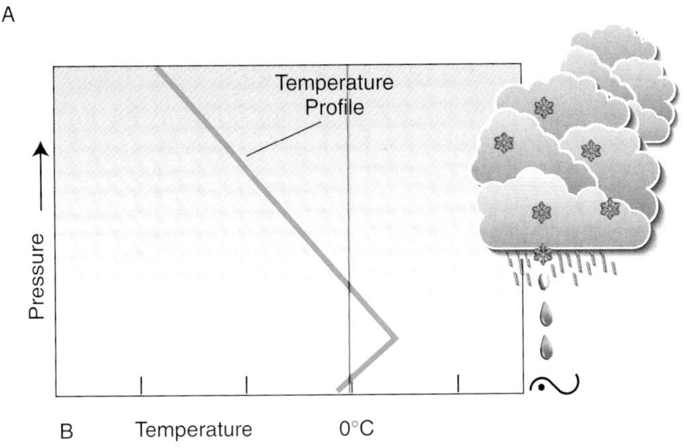


Figure 7.15 (*A*) Freezing rain scene. (*B*) Temperature profile for freezing rain. The rain freezes only on contact with surface objects.

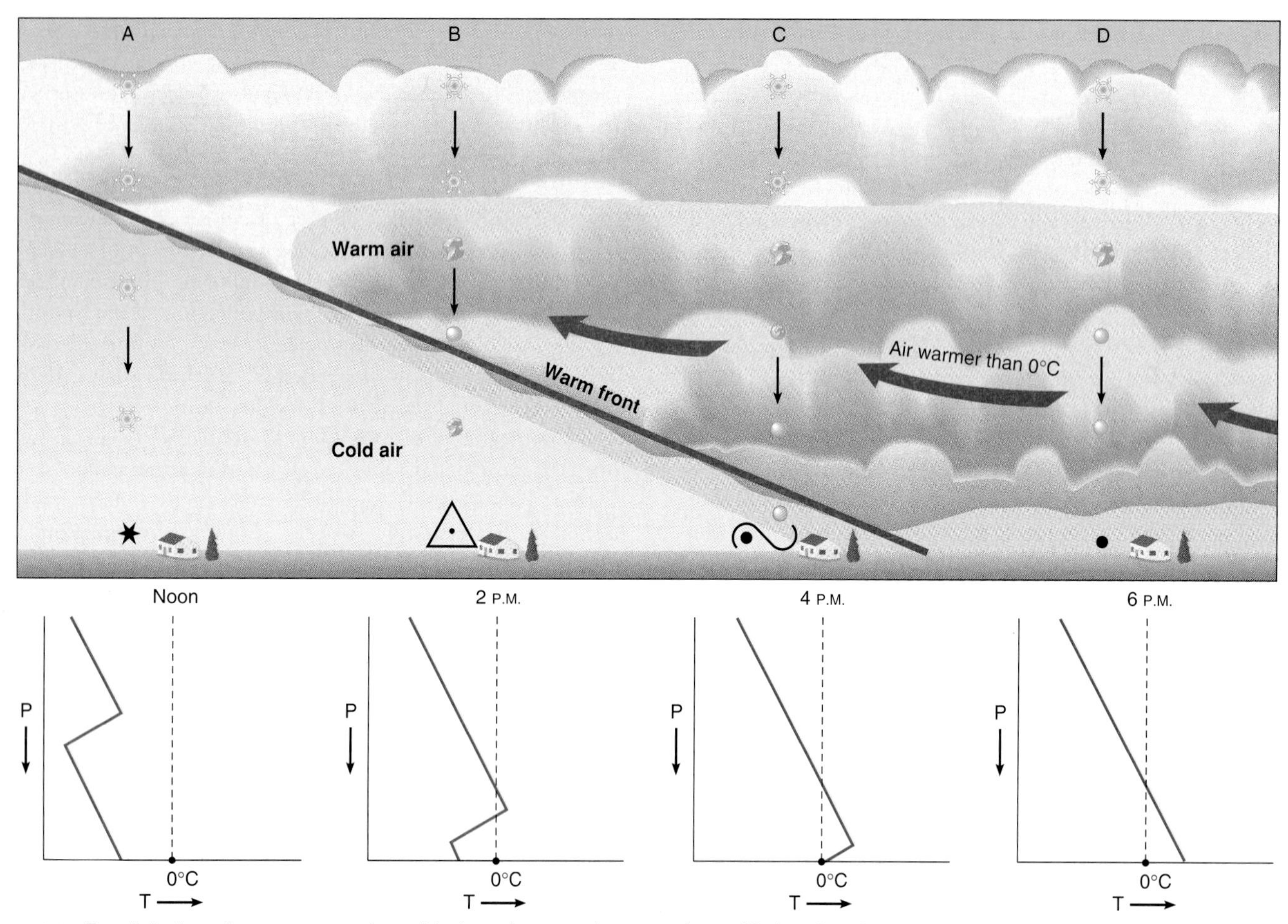


Figure 7.16 Precipitation changes occurring with the advance of a warm front. Notice that the warm air moves in aloft first and only later at the surface. At noon, temperatures above the observer are colder than 0°C at all levels. By 4:00 P.M., a deep layer of warm air has moved in.

period of freezing rain can be calamitous. Glazed surfaces become so slippery that walking may be literally impossible, driving out of the question. Trees, power lines, and even buildings may collapse under the weight of the ice.

What atmospheric conditions lead to the occurrence of freezing rain? Evidently, the original snow crystals must melt to form rain; then, the rain must fall into subfreezing temperatures at the earth's surface. However, the cold surface air layer must be shallow, or else the rain would refreeze in the air and land as ice pellets. Figure 7.15B shows temperature conditions aloft over Boston at the time of freezing rain. (Freezing rain may also occur when the near-surface air is slightly above freezing but the ground itself is colder than freezing.)

The changeover from snow to sleet to freezing rain and thence to plain rain is an all too common and depressing event for snow lovers in the Northeast. One way this sequence develops is illustrated in Figure 7.16. As a warm front approaches, warmer air appears aloft before it does at the surface. The temperature rises above freezing in the layer indicated in sounding (B); this warm layer causes snow to melt before reentering colder temperatures below and refreezing as sleet. As the warm layer thickens and lowers (C), the rain is unable to refreeze until it strikes the ground as freezing rain. When the warm air extends all the way to the ground (D), surface freezing ceases and the precipitation is simply rain.

Notice that the weather in Boston did not follow this sequence exactly. The surface temperature rose above freezing during the ice pellet stage; therefore, no freezing rain was observed before the onset of rain. However, falling surface temperatures caused a brief period of freezing rain at 05 UTC.

By carefully observing the precipitation forms during a winter storm, you can learn a great deal about conditions aloft and impending weather changes. For example, if you notice a changeover from snow to sleet, you know that warmer air has moved in aloft. Even if the surface temperature is steady, it is likely that further warming will occur; therefore, a further changeover to freezing rain and/or rain is a strong possibility. On the other hand, a change in the opposite direction, as experienced in Boston at 10 UTC, means colder air aloft and a trend back toward snow and colder surface temperatures.

## Hail

If you have ever been caught in the open during a severe hailstorm, you know it can be a frightening, even life-threatening, experience. Hailstones as large as those in Figure 7.17A, weighing

A

B

Figure 7.17 (*A*) The world's largest recorded hailstone, which fell at Coffeyville, Kansas, in 1970. This stone had a diameter of 14 centimeters (5.5 inches) and weighed 0.75 kilograms (1.67 pounds). It is estimated to have hit the ground at a speed of 45 meters per second (100 mph). (*B*) A cross-sectional slice of the record Coffeyville hailstone, viewed under polarized light. Note the different layers of growth. What causes these differences?

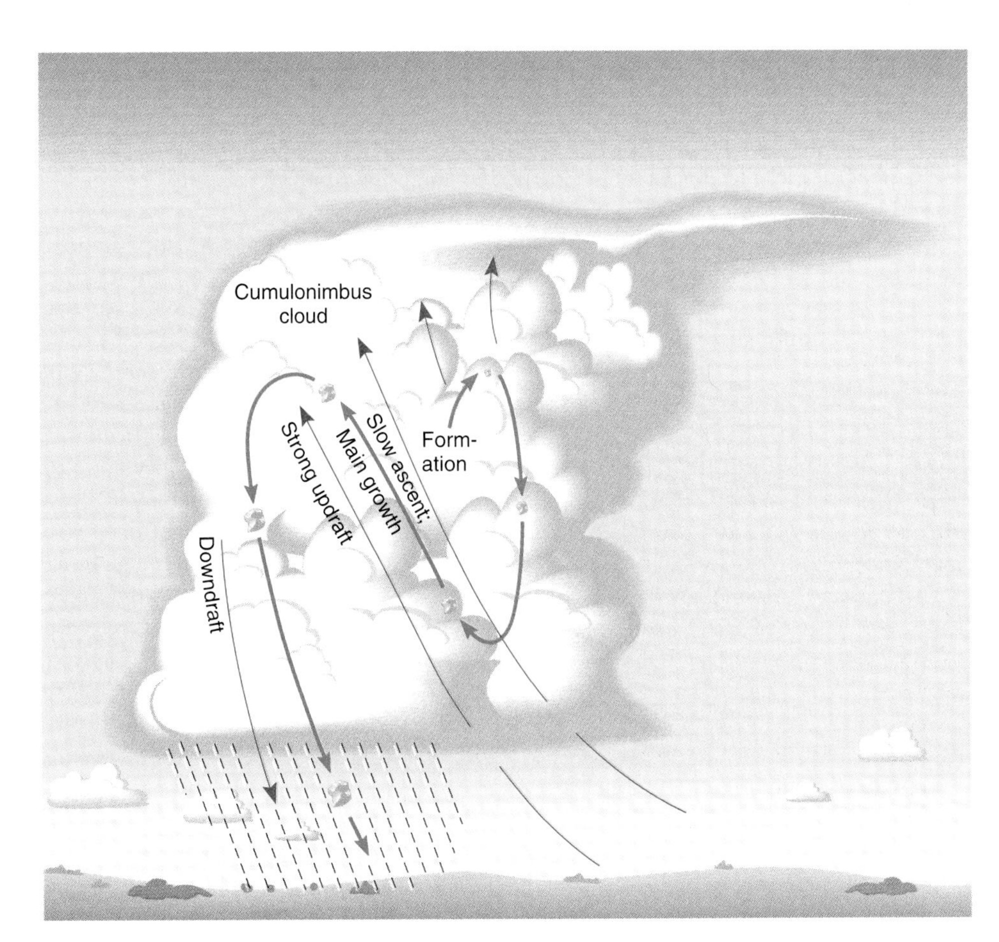


Figure 7.18 Growing hailstones gain most of their mass in strong updraft regions, where they may be suspended for many minutes.

more than a half a kilogram (over 1 pound), can be lethal missiles. A hailstorm in the former Soviet Union is reported to have killed 200 cattle. Hail causes hundreds of millions of dollars in losses from crop and property damage each year.

**Hail** is precipitation in the form of spherical or irregular chunks of ice 5 millimeters or more in diameter. It is formed in cumulonimbus clouds, typically in thunderstorms. Other frozen forms of precipitation such as ice pellets and snow pellets do not occur in diameters greater than about 5 millimeters. How is it that hail particles are able to grow so very much larger?

Figure 7.18 illustrates one current explanation of hail formation. The original particle forms in an area of weak updrafts, grows to a millimeter or so in diameter, and begins to fall. If it falls into a region of stronger updrafts, it will be carried aloft once more. If the speed of this updraft is greater than its terminal fall velocity, the particle will be swept upward yet again. The closer these two speeds match, the longer the particle remains aloft, colliding with supercooled particles that freeze to its surface. It is at this point that the main growth occurs. Eventually, the hailstone may grow so heavy that its terminal velocity is sufficient to carry it downward, or it may be carried out of the updraft and swept earthward in a downdraft.

As it falls into warmer air at lower altitudes, the hailstone begins to melt and evaporate. Depending on its size, as well as on the air temperature and humidity below the cloud base and on the distance between cloud base and the ground, the hailstone may melt or evaporate entirely. Researchers have found that only 10% of the hail falling through the freezing level in Arizona thunderstorms actually reaches the ground as hail. In Alberta, Canada, where the subcloud weather is colder and more humid and the cloud bases are lower, the figure is close to 60%.

If you slice a hailstone in half, you find it consists of a series of layers (see Figure 7.17B). The layers' structure tells something of cloud conditions during the hailstone's growth. When water freezes quickly on the hailstone's surface, tiny air bubbles are trapped in the ice, causing it to appear white, or opaque. Thus, the white ice layers are indicators of colder temperatures and only moderate amounts of supercooled water, which freeze very quickly. Clear ice forms more slowly, allowing the bubbles to escape. These layers form in a warmer and wetter environment, where the greater amounts of available water take longer to freeze than in the white ice layers.

***Review***

Figure 7.16 contains many of the key ideas of this section. Make sure you can explain what is happening in the sequence from A to D.

A useful exercise is to extend the concept map of last section (Figure 7.11) by adding each form of precipitation at the appropriate locations.

***Questions***

1. What kinds of precipitation would each of these temperature profiles produce? (a) Colder than freezing at all levels. (b) Colder than freezing everywhere except for a shallow layer high above the ground.
2. How do snow pellets and ice pellets differ? Ice pellets and hailstones?
3. Which precipitation types may originate in the warm rain process? Which cannot? Explain.

# The Distribution of Precipitation

*Strepsiades: Tell me,*
*Which theory do you side with, that the rain*
*Falls fresh each time, or that the Sun draws back*
*The same old rain, and sends it down again?*
*Amynias: I'm very sure I neither know nor care.*
*—from Aristophanes,* The Clouds *(423 B.C.)*

This passage from a play written over 2400 years ago is rich in meteorological insight. Aristophanes appears to recognize that solar energy drives evaporation and that evaporation provides the water that makes the rain. (Aristophanes knows the correct answer to Strepsiades's question; in a later passage, he argues that if "fresh rain" fell each time, then the oceans should be becoming fuller year by year. Since sea level is more or less constant from year to year, the "same old rain" must be recycled via evaporation.)

The passage also offers a good starting point from which to survey the amount of precipitation that falls across our planet. To a first approximation, at least, the oceans are not becoming deeper or drying up. Therefore, annual precipitation onto the earth must equal the annual evaporation from it. We can state the relation more strongly, in fact: the mean annual worldwide precipitation is determined by the amount of evaporation. Therefore, one way to estimate the amount of precipitation is to begin with the energy budget's latent heat flux, which tells the rate at which energy is being spent on evaporation, and calculate the amount of water this energy represents. The amount of water evaporated should equal the worldwide mean precipitation.

If you do this calculation, you arrive at a value for evaporation (and precipitation) of approximately 97 centimeters—"waist deep," roughly—which is approximately the figure that rain gauge measurements indicate.

Is this correspondence an impressive verification of theory? Not entirely. Energy budget specialists work in the other direction, taking rainfall data and using it to check their estimates for the latent heat term in the energy budget! On the other hand, the fact that the latent heat flux value is consistent with the precipitation measurements and the other energy budget terms does lend plausibility to both estimates.

Although the value for mean global precipitation has a certain significance, it doesn't begin to tell the whole story. Precipitation amounts vary radically with location and time. To determine global patterns and totals of precipitation, enormous numbers of observations must be made and processed. The graphic essay on the following page (Figure 7.19) offers you a slight taste of the quantity of data needed and the variability of precipitation on various scales of area and time. We invite you to explore the charts and Tables 7.2 and 7.3, and to consider the questions posed in the Checkpoint on this page. (Appendix M contains additional precipitation information for specific locations.)

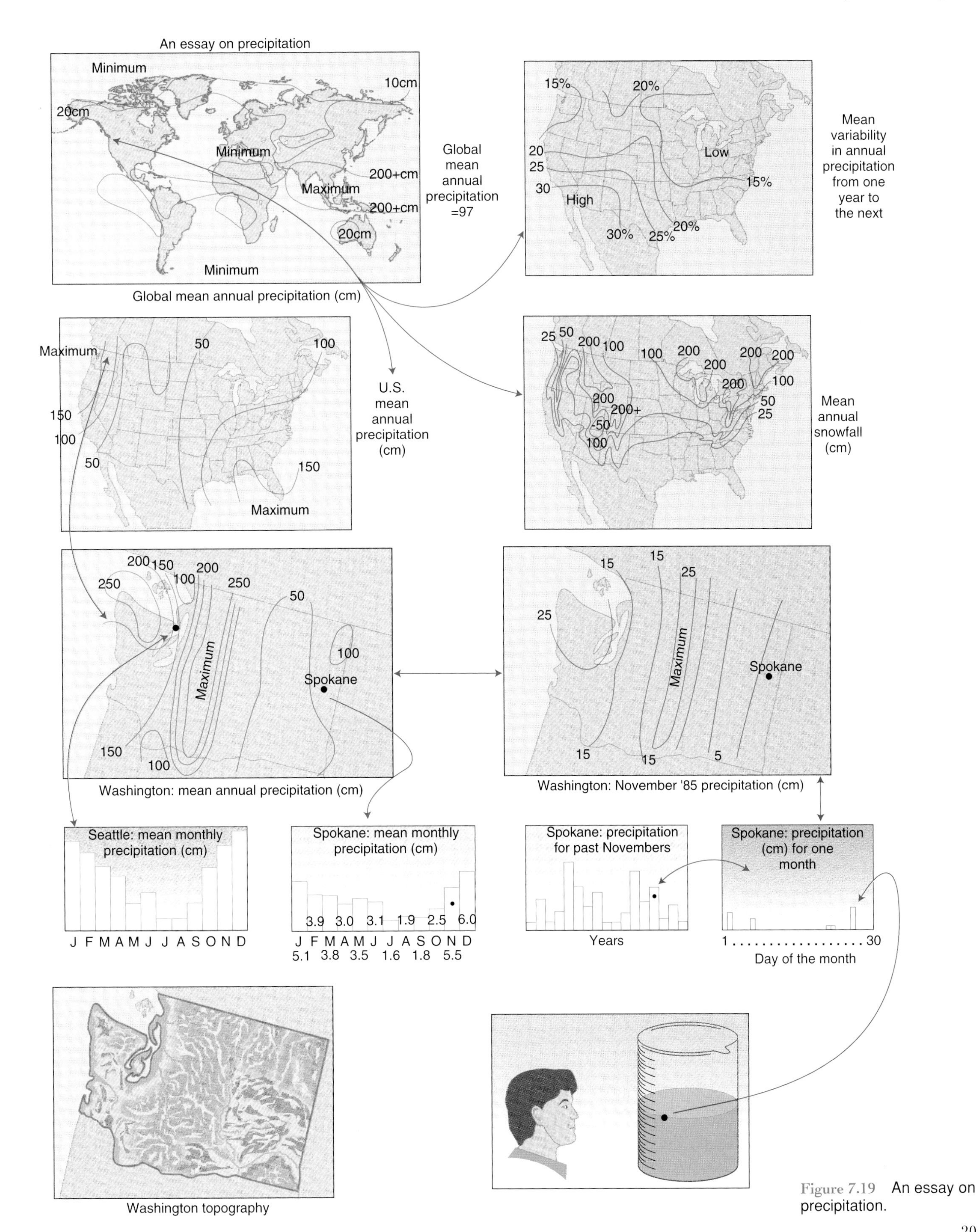


Figure 7.19 An essay on precipitation.

Table 7.2

### Precipitation Extremes

| | World Record | U.S. Record |
|---|---|---|
| PRECIPITATION TOTALS | | |
| Greatest mean annual | 11,684 mm (460 in.) Mount Waialeale, Hawaii | (same) |
| Least mean annual | 0.8 mm (0.03 in.) Arica, Chile | 42.1 mm (1.66 in.) Death Valley, California |
| Greatest 1-year total | 26,470 mm (1042 in.) Cherripunji, India | |
| Greatest 1-month total | 9300 mm (366 in.) Cherripunji | 1803 mm (71 in.) Helen Mine, California |
| Greatest 1-day total | 1168 mm (46 in.) Baguio, Philippines | 1092 cm (43 in.) Alvin, Texas |
| Greatest 1-hour total | | |
| Greatest 1-minute total | 38.1 mm (1.5 in.) Barot, Guadeloupe | 30 mm (1.2 in.) Unionville, Maryland |
| U.S. SNOWFALL | | |
| Greatest 1-year total | | 2860 cm (93.5 ft.) Mount Ranier, Washington |
| Greatest 1-month total | | 991 cm (32.5 ft.) Tamarack, California |
| Greatest 1-day total | | 193 cm (6.3 ft.) Silverlake, Colorado |

Table 7.3

### Precipitation Oddities

— Longest rain-free period (U.S.): 993 days, Bagdad, California
— "Rains" of small fish, frogs, eels, and periwinkles (snails) have been reported a number of times. Presumably, the unfortunate animals were swept up from the ocean by strong winds such as waterspouts (see Chapter 13).
— Bright red, yellow, and green rains have occurred due to dust or pollen mixing with the precipitation.

To begin the graphic essay, you might start with the global mean annual value of 97 centimeters in the top center of Figure 7.19. Moving to the left, you see how that value varies on the largest scale, that is, across the entire globe. From there, you focus on one region of the globe, the United States, for which three maps of many possible are shown: mean annual total precipitation, snowfall, and the year-to-year variability in total precipitation. Notice that in the Southwest, which has the least total precipitation, the year-to-year variability is the largest.

From the U.S. annual map, you can focus on a single state—Washington—and then on just two cities, on smaller and smaller scales of time. Finally, you arrive at the observer, recording the precipitation for one day at one station. Traveling in the opposite direction, from observer back to the global average, you see how the individual observation contributes to a more general understanding of precipitation's distribution over the earth.

***Review***
The graphics and tables in this section illustrate the great variation in precipitation from place to place and over different time periods.

***Questions***

1. What regions of the world receive the most precipitation? The least amounts?
2. What relation(s) do you see between precipitation in the United States and topography?
3. Which is the driest season in Seattle? In Spokane?
4. Why do you suppose Spokane is so much drier than Seattle?
5. On how many days did precipitation occur in Spokane in November 1985?

# Precipitation Modification: The Theory

In this section and the next, you come to grips with the unit's main question: Can humans make it rain or modify precipitation rates in other ways? First, you will reexamine the precipitation-formation processes with modification in mind: how and at what points in the process would it be possible to alter conditions within a cloud to affect its production of precipitation? Then, you will review some of the weather modification experiments recently performed, their results, and other related issues.

## Warm Rain Modification

What strategy might you follow to stimulate the formation of rain by the collision-coalescence method? Figure 7.20 reminds you that in order for the process to proceed, several prerequisites are required. To begin with, there must be a cloud, it must contain sufficient water in the form of droplets, and some of these droplets must be considerably larger than others. Given these conditions, collision and coalescence efficiencies must be great enough and the cloud of sufficient thickness to allow the large droplets to grow and divide repeatedly.

Which of these factors might you be able to influence? In theory, probably every one of them. For example, by igniting huge fires, such as forest fires, it would be possible to create a convective updraft and thus make a cumulus cloud on a day with clear skies. However, we want to restrict ourselves to practical means. We are looking for leverage points in which a relatively small expenditure of effort or expense might lead to a significant increase in precipitation. Therefore, we require that nature provide the cloud of sufficient moisture content and thickness.

Suppose a cloud possesses the appropriate moisture and thickness but lacks the large droplets necessary to initiate collision and coalescence. Could you artificially introduce sufficient large droplets into the cloud to trigger precipitation?

This appears to be the most promising point of leverage on the warm rain process. Researchers have hypothesized that in fact, some warm clouds fail to precipitate because of a lack of large droplets; therefore, they have seeded warm clouds by introducing millions of large droplets or small raindrops into the clouds' upper regions to see if the clouds might be stimulated to grow. A number of experiments have been conducted to test this hypothesis. Results suggest that some warm clouds can be induced to yield more precipitation than they otherwise would have. Because far more precipitation occurs by the Bergeron process, however, particularly in middle and high latitudes, there is more interest in modifying supercooled clouds, which is the subject of the next section.

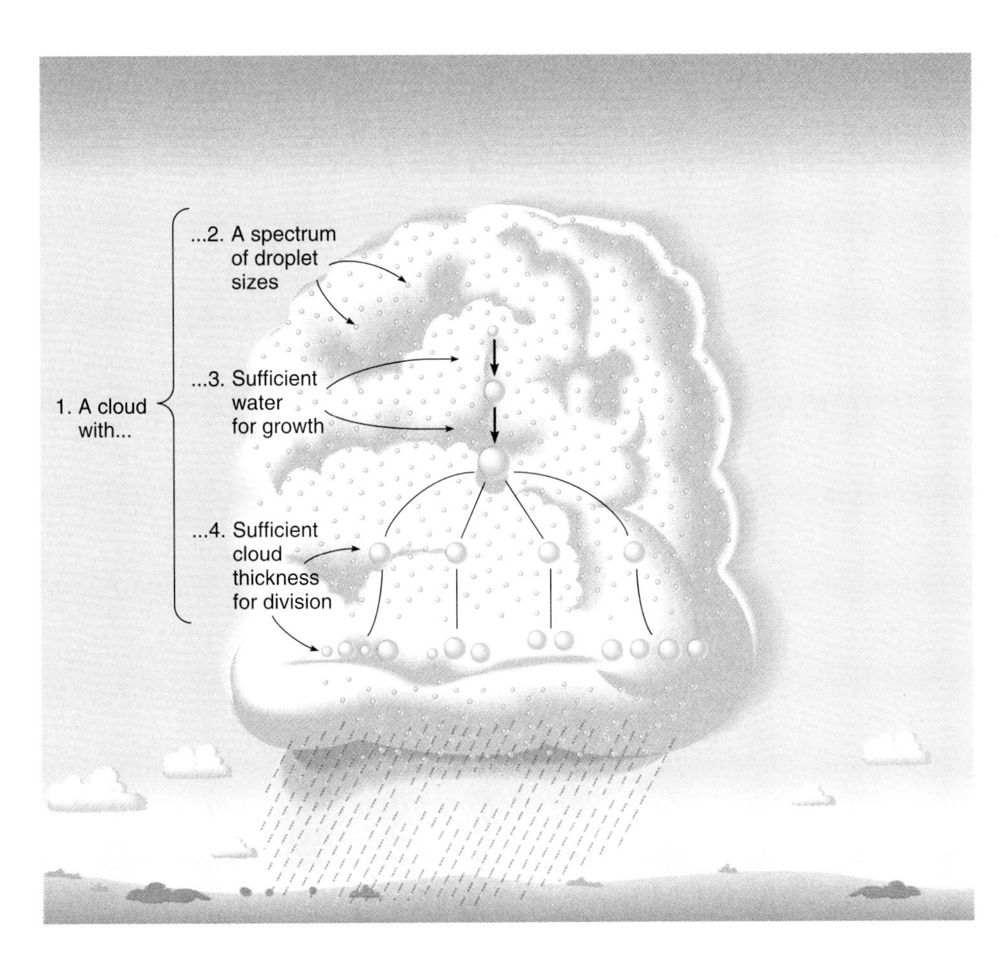


Figure 7.20 Requirements for precipitation to occur through collision and coalescence.

## Three-Phase Rain Modification

Figure 7.21 outlines the main features of the Bergeron process. What leverage points can you identify? What factors, if changed, would significantly influence the amount of precipitation generated?

Notice that the ice nuclei in Figure 7.21 are literally the focal point of the Bergeron process and that they are relatively rare. If they are too rare or entirely absent, the entire process cannot proceed. If they are too numerous, they are no longer focal points and do not concentrate the cloud's water effectively. Scientists have developed two methods of seeding supercooled clouds to increase the numbers of freezing nuclei.

### *Dry Ice Seeding*

**Dry ice,** which is frozen carbon dioxide, has a temperature of −78.5°C (−109°F), far below the temperature of spontaneous nucleation of −39°C for supercooled water. In 1949, Vincent Schaefer, an atmospheric scientist in New York, hypothesized that a piece of dry ice dropped into a supercooled cloud would lower the cloud temperature below −39°C along its path, causing water vapor to nucleate spontaneously. It would also freeze on impact any supercooled cloud droplets it encountered.

It is relatively easy to verify this hypothesis on a small scale, as you can see from Figure 7.22. Make a supercooled cloud by exhaling into an ice chest; then scrape some chips of dry ice into the cloud, and observe the countless tiny ice crystals that appear in the wake of the dry ice. Schaefer has estimated this seeding method can generate as many as 20 billion ice nuclei per cubic centimeter.

Note carefully that the intention is not to freeze water onto the fragments of dry ice. The role of the dry ice is merely to lower the cloud temperature in the wake of each dry ice fragment to a temperature colder than −39°C, so that *spontaneous nucleation* of the cloud's water vapor may occur. In theory, any very cold substance could act as a seeding agent.

Tests of Schaefer's hypothesis on actual clouds such as those in Figure 7.23 have shown conclusively that enriching the cloud with ice crystals in this fashion can significantly stimulate the Bergeron process in some cases. Later in the chapter, we will discuss seeding results in more detail.

### *Silver Iodide Seeding*

In 1947, Bernard Vonnegut, a colleague of Schaefer's, developed a different solution to seeding supercooled clouds. Working from the same supposition, that some clouds do not precipitate because of a lack of freezing nuclei, Vonnegut searched for substances whose crystal structure was similar to that of ice. He reasoned that such crystals, if enough like ice crystals, would serve as ice nuclei when introduced into supercooled clouds.

The substance Vonnegut chose was **silver iodide,** a compound consisting of one atom of silver (Ag) and one of iodine (I) and whose chemical formula, therefore, is **AgI.** Small crystals of silver iodide have a structure and dimensions much like crystals of ice. When introduced into a supercooled cloud, the silver iodide crystals become centers of deposition of water vapor, the focus of the Bergeron process. Experiments have proved that silver iodide seeding can have a significant impact on precipitation generation by the Bergeron process.

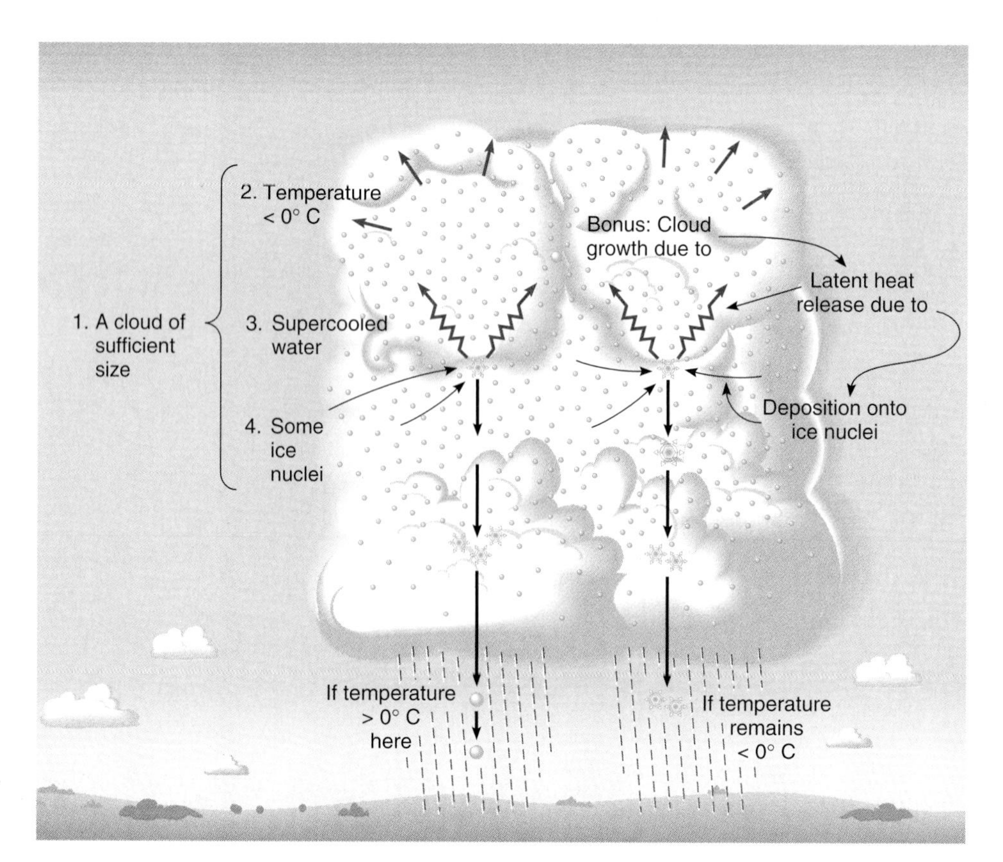


Figure 7.21 Conditions attending the formation of precipitation by the three-phase process.

Since silver iodide is not identical to ice, it is slightly less effective as a freezing nucleus than ice itself. One way to describe a substance's effectiveness as a freezing nucleus is by the temperature at which it "becomes active." Consider the rising air in the cumulus cloud pictured in Figure 7.24. As the air rises and expands, it cools. When will deposition onto freezing nuclei occur?

If you inject ice particles into the cloud at any temperature warmer than 0°C, of course, no deposition of water vapor occurs; instead, the ice simply melts. At cloud temperatures colder than 0°C, however, any ice particles you add will become centers of water vapor deposition. Thus, ice is said to become active at 0°C.

If you observe particles of other substances in the cloud, you notice they become active nuclei not at 0°C but at colder temperatures. Soil particles become active at around −12°C. That is, water vapor does not deposit onto these particles at temperatures warmer than −12°C. Airborne diatoms (microscopic, one-celled organisms) may serve as nuclei, but they are less effective because they do not become active until the temperature falls to −30°C. By comparison, note from Figure 7.24 that silver iodide crystals are highly effective, because they become active at just −4°C. Thus, any cloud region colder than −4°C is a possible target for silver iodide seeding.

Notice that silver iodide operates in a quite different way from dry ice as a seeding agent. Whereas dry ice particles simply cool regions of a cloud so that ice crystals form by spontaneous nucleation, silver iodide actually provides nuclei onto which the water vapor deposits. Thus, ice crystals created by dry ice seeding have centers of ice (not dry ice), whereas those produced by silver iodide seeding have silver iodide centers.

Is it possible to inject enough silver iodide crystals into a cloud to make a difference? Silver iodide cloud seeders can produce vast numbers of crystals: as many as $10^{17}$ crystals per gram of silver iodide. To appreciate the magnitude of this number, suppose ice crystals formed on each of $10^{17}$ nuclei, and they all grew to snowflake size. A snowfall of $10^{17}$ snowflakes would be sufficient to cover an area the size of Connecticut to a depth of 25 centimeters (10 inches).

### *Dynamic Effects*

As dry ice or silver iodide promote deposition of water vapor, they may induce important **dynamic effects** as well. A dynamic effect is one causing motion within the cloud.

How would deposition of water vapor cause dynamic effects? As water vapor deposits onto ice nuclei, it releases its latent heat of deposition, warming the cloud. Heating the air, as you saw in Chapter 6, increases its buoyancy; therefore, the seeded cloud is stimulated to grow vertically. For a cloud on the verge of instability, this extra heat energy may be all it needs to become unstable and grow dramatically, developing stronger updrafts and thereby drawing more water vapor into the cloud, which promotes further growth. Such dynamic effects of cloud seeding may lead to much greater precipitation than simply condensing a cloudful of water vapor and precipitating it onto the earth. The dynamic effect is the "bonus" mentioned in Figure 7.21.

Figure 7.22 Each of the trails in this photo is composed of countless, tiny ice crystals caused by spontaneous nucleation as chips of dry ice fall through the supercooled cloud. The larger dry ice particles are visible at the lower end of each trail.

Figure 7.23 One of the original dry ice cloud seeding experiments conducted by Vincent Schaefer over Pittsfield, Massachusetts, in 1948. The seeding path, from lower left of the photo to distant center, then to lower right edge, is evident. The seeding accelerated the three-phase process, causing the cloud to precipitate itself away, leaving only thin remnants of cirrus.

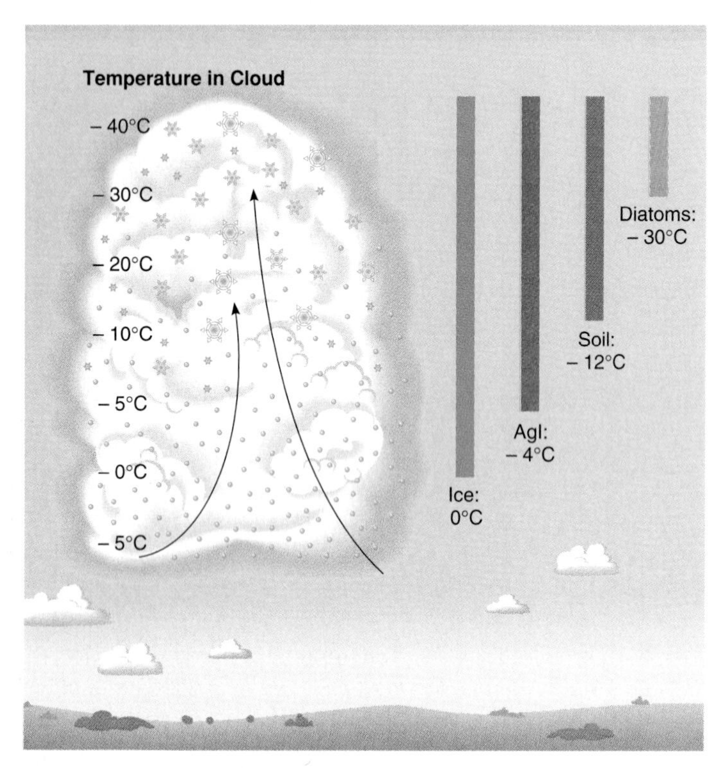


Figure 7.24 Ice crystals (blue) are active as freezing nuclei at 0°C and colder. Silver iodide is active below −4°C; other substances require much colder temperatures before they act as freezing nuclei. Why does the number of supercooled water droplets (blue circles) decrease upward within the cloud?

***Review***

Figures 7.20 and 7.21 should help you visualize the concepts on which cloud seeding specialists base their thinking. Remember the importance of finding a leverage point: building a cloud from scratch, although possible in theory, is infeasible in practice. Also, remember the potential of dynamic effects in dry ice or silver iodide seeding.

***Questions***

1. Why would seeding a warm cloud with silver iodide be a waste of time?
2. Suppose you discovered a substance that became active as a deposition nucleus at −2°C. Explain what this means.
3. What are dynamic effects in cloud seeding?
4. Suppose you seeded a cold radiation fog (−10°C) that was obscuring an airport runway. What might happen? Explain.

# Precipitation Modification in Practice

## Experimental Design

Imagine for a moment that you have seeded a growing cumulus cloud with silver iodide. How would you prove that your seeding had any effect?

"Well," you might say, "I seeded a cloud, observed it for an hour after I seeded it, and saw it grow to twice its original height." Is this proof of the seeding's efficacy? Of course not; perhaps you picked a vigorous cloud that was destined to grow no matter what you did to it. Perhaps it would have grown even larger if you *hadn't* seeded it.

How could you demonstrate the effect of your seeding? Ideally, you would like to perform the experiment on many clouds identical in every respect: seeding some clouds (which are said to be in the "test" group) and not seeding the others (which therefore are in the "control" group), and then observing the differences. Such test-and-control experiments can provide compelling evidence that a given treatment (cloud seeding in this case) causes some change.

But individual clouds, like many meteorological phenomena, cannot be duplicated exactly. As a result, developing solid proof that some seeding procedure "A" lead to a result "B" is very difficult. The task is further complicated by the fact that most cloud seeding influences rainfall rates by relatively small amounts, while natural precipitation amounts can vary greatly over small regions of space or short time periods.

In the face of these challenges, researchers must design their cloud seeding experiments with great care if they are to achieve meaningful results. Typical procedures they follow include (1) studying only those clouds that have similar initial characteristics (height, temperature, water content, etc.); (2) establishing test and control groups of clouds to be seeded or not; (3) assigning each cloud at random to one group or the other; and then (4) studying the development of every cloud, no matter which group it was assigned to.

Once the data are collected, researchers do statistical tests to see if there is a significant difference between the test group and the control group. If there is, then they have evidence their treatment was effective.

## Cloud Seeding Programs: Some Examples

Over the years, hundreds of initiatives in the central and western United States, as well as at locations as varied as France, South Africa, Australia, Israel, and Spain, have aimed at modifying precipitation. Some of these are research studies, designed to test the efficacy of some particular technique. Others are "operational" in nature: they are conducted not to answer research questions but simply to change precipitation rates for some customer or agency (farmers, etc.).

Let's look at a few examples.

### *Precipitation Enhancement*

Wintertime silver iodide seeding of orographic clouds in California and Colorado has been conducted for many years. The goal is to increase the snowpack in the mountains, thereby providing runoff during the warm season for the western states, which are increasingly prone to water shortages. Participants in these programs believe that seeding properly done can lead to water supply increases of 10% to 20%. Calculations suggest that increasing water supplies in this way is at least 10 times more cost-effective than other alternatives, such as desalinating sea water.

An operational program conducted in Israel over many years consists of seeding silver iodide just below the bases of growing cumulus clouds. Updrafts then carry the silver iodide into the clouds. Data from hundreds of days of seeding suggests a precipitation increase from seeded clouds of as much as 15%, at the 2% **significance level.** Significance level is a statistical gauge of the probability that a particular finding (an increase in rainfall in seeded clouds compared to unseeded ones, in this case) might have occurred by chance, rather than due to the seeding.

Notice that because of the way the term is defined, *small values* of significance level, such as 2%, indicate the *greatest likelihood* that any observed differences are real and not the result of random variations in the data. On the other hand, *large values,* such as 15%, suggest a *lower probability* that the differences are real.

How small must the significance level be before you can accept a difference as real? For many purposes, scientists consider the 5% level (or lower) to indicate **statistically significant** differences between groups. Five percent means that there is only one chance in 20 (1/20 = 5%) that a given difference happened by chance. A significance level of 2% or lower (one chance in 50 or less) is called **highly significant.** Measures of statistical significance offer crucial guidance to those attempting to evaluate whether an experiment produced real results.

The fact that the Israeli experiments were significant at the 2% level means that there was only a 2% chance that the observed differences between seeded and unseeded clouds might have occurred by chance. Thus, it is highly likely that the seeding caused the observed rainfall differences.

Results such as these have led most atmospheric scientists to accept precipitation enhancement as a reality. Granted, not every attempt at cloud seeding is successful; the natural variability of clouds is great, and we still have much to learn. Note also that precipitation enhancement is not the same as precipitation control. However, the consensus is that properly designed precipitation modification programs can increase local precipitation totals from orographic or convective clouds by 10% to 15%.

### *Hail Suppression*

Crop damage due to hail in the western Plains states adds up to hundreds of millions of dollars per year. A number of European countries experience hail damage on a similar scale. Can cloud seeding reduce these losses?

In a sense, hail is an example of overefficient precipitation formation: hail occurs when too much water freezes onto too few nuclei, resulting in precipitation particles so large they cause damage. This viewpoint has led scientists to hypothesize that if they injected vast numbers of freezing nuclei into a hail-bearing cloud, the cloud's water might be distributed onto more centers of deposition, resulting in more precipitation particles of smaller diameter. These smaller particles would do less damage or might melt as they fell to earth.

In Serbia, an operational hail suppression program based on this supposition was conducted over a 40-year period. Clouds were seeded by a network of silver iodide rocket launchers arranged across the countryside. Over the years, the network expanded; by 1990, it covered nearly all of Serbia, with an average spacing between launch sites of about 5 kilometers. When observers spotted a threatening cloud, rockets were fired into it at five-minute intervals until the cloud was no longer considered a danger. Serbian scientists reported using as many as 25,000 rockets in a single season.

What effect did this seeding have? The Serbian program was an operational one, not a research effort; thus, no effort was made to assign clouds at random to test or control groups. However, researchers have compared the frequency of hail occurrence at sites before and after the network became established locally. They have also compared the frequency of hail occurrence within Serbia to that in nearby, unseeded regions. Thus, the prenetwork sites and the sites outside the network provide a sort of control group with which to compare the test data.

Comparing the number of days in which hail was reported at the test and control sites, scientists found a reduction of 15% to 20% at the seeded locations. The result is significant at the 1% level. Thus, as long as you feel the lack of a randomized design has not affected the results, you would conclude it is highly probable that the Serbian program reduces the occurrence of hail.

## Inadvertent Precipitation Modification

It is ironic that proof of the success of deliberate weather modification attempts often is elusive, while considerable evidence exists that humans inadvertently alter precipitation patterns every day.

### *Modification Caused by Urban Areas*

How would a city be likely to alter precipitation patterns? In a paper published in *Science* magazine, Stanley Changnon, a scientist at the Illinois State Water Survey, hypothesized that "low level winds moving across an urban area define a plume of urban-altered air (aerosols, heat and moisture) that can affect rainfall over and beyond or downwind of the city."[1]

The aerosols might serve as condensation or freezing nuclei, the heat might stimulate convection, and the moisture would increase the air's water content.

Changnon collected rainfall and wind direction data in the St. Louis area over a five-year period. For each time it rained, he determined the amount of precipitation that fell upwind, downwind, to the left, and to the right of the low-level wind flow. Figure 7.25 illustrates some of his results for the summer months. Notice that downwind, the precipitation was more that 20% greater than the upwind total and that the difference was statistically significant at the 3% level. Recently, he found similar results for St. Louis for the months of September through November, which were significant at the 1% level.

[1] Changnon, S. *Science* 207 (27 July 1979): 402.

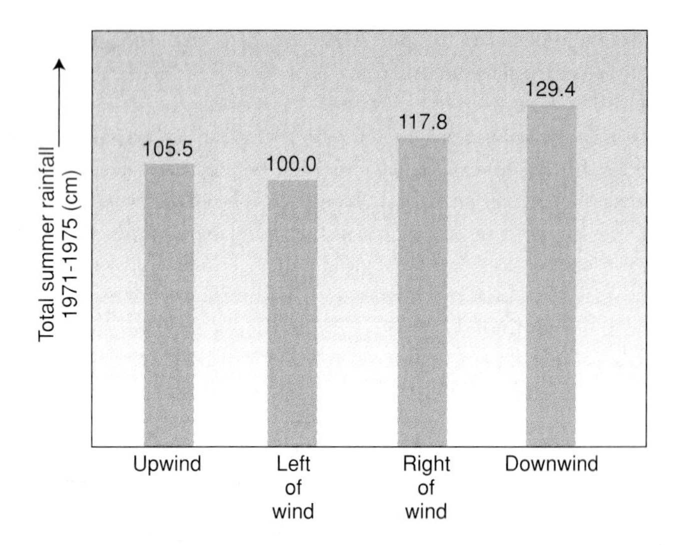


Figure 7.25 Variations in summer (June-August) rainfall totals in the vicinity of St. Louis according to wind direction at the time of precipitation. According to the graphs, does wind flow over the city increase or decrease the amount of precipitation?

### *Other Examples*

Many other types of human activity probably modify natural precipitation patterns, often on a scale too small to observe. For example, farming and ranching can alter the earth's surface in major ways, some of which may lead to changes in precipitation patterns. Irrigation of arid areas increases evaporation in those regions manyfold, affecting the atmosphere's energy budget as well as its water vapor content. On the other hand, overgrazing of rangeland or clear-cutting timber can reduce the amount of natural evaporation and transpiration, influencing energy budget fluxes and water vapor levels.

Modern transportation habits may induce precipitation variations as well. Particulate emissions from internal combustion engines in cities and along major highways may seed clouds. What is the effect of this seeding? Acting as condensation or freezing nuclei, the particles may stimulate precipitation in nuclei-deficient clouds, or they may lessen precipitation by spreading the available moisture over many more sites, thus hindering precipitation-forming processes.

As a final example, consider the effect of the jet aircraft contrails in Figure 7.26. Notice that the ice particles falling from contrails may seed a supercooled middle cloud, thereby stimulating the Bergeron process. This type of seeding may occur naturally as well, as falling cirrus cloud particles seed lower clouds and induce them to precipitate.

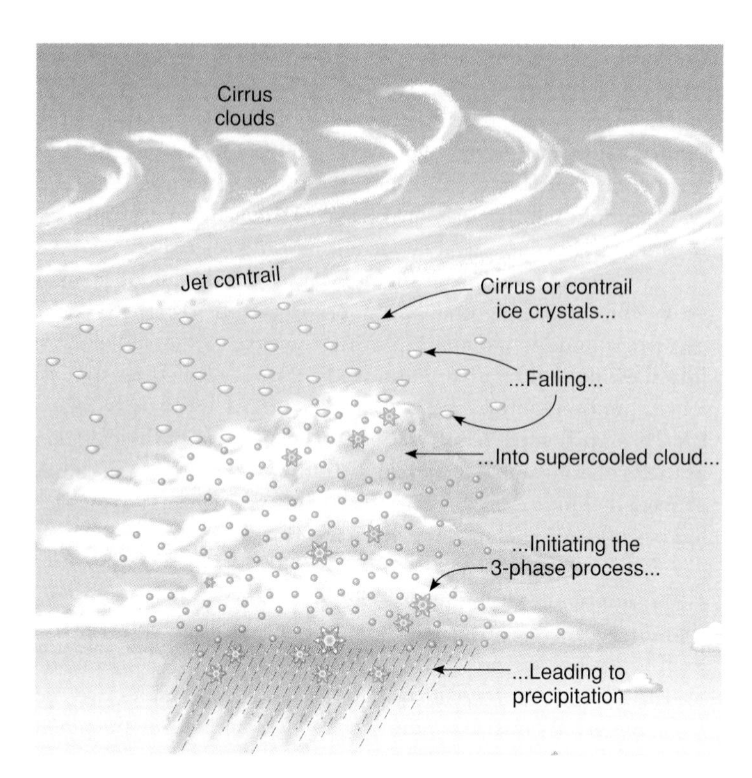


Figure 7.26 Jet contrails can seed lower, supercooled clouds, initiating the three-phase process.

## Social and Legal Concerns

As the human population continues to grow, the demand for earth's resources, particularly fresh water, will grow as well. This expansion of population and demand for water will likely lead to an increase in both deliberate and inadvertent forms of precipitation modification. Several difficult issues lurk at this intersection of science, technology, and society. We conclude this unit by looking at three of them.

### *Does Seeding Reduce Rainfall in Unseeded Areas?*

Suppose you own a ranch in Nebraska. This year, rainfall on your ranch has been sufficient, though not plentiful. But drought conditions develop to the west, and ranchers in those regions hire a cloud seeding company to increase rainfall. The seeding program seems to be successful there: rainfall increases substantially in the drought-stricken area. However, once the seeding begins, no further rain falls on your ranch.

Could the cloud seeding have taken rainfall destined for your ranch? Does cloud seeding actually increase the total amount of precipitation that falls, or does it only redistribute it? These are crucial questions—questions that as yet do not have clear answers.

### *Who Decides What the Weather Will Be?*

You are the manager of an amusement park in Minnesota. Your business depends greatly on the weather: rainy days are disasters. This summer, business has been booming because of a long stretch of sunny days. You learn that local farmers, suffering from a drought, are seeking to engage a cloud seeding operation. Can they do that and perhaps ruin your business? Can you stop them?

As weather modification efforts become more commonplace, who will decide what the weather should be? How will the decision be made in a way that is not only fair to humans but environmentally responsible as well?

### *Who Is Liable?*

The following two examples are *not* hypothetical:

> In a 1961 experiment, research scientists seeded Hurricane Esther well offshore from the U.S. mainland. Subsequent to the seeding, Esther brushed the East Coast from North Carolina to New England, then executed a clockwise loop and struck New England a second time. Figure 7.27 shows Esther's path.
>
> A 1972 cloud seeding project was underway in the vicinity of Rapid City, South Dakota, when torrential rains occurred, causing a devastating flash flood.

In the aftermath of each of these events, project scientists found themselves in the awkward position of having to prove that the seeding program did *not* cause the unusual weather that ensued. Imagine being found personally responsible for the damage caused by a hurricane or a flash flood! As it turned out, virtually no evidence existed that seeding caused either the strange loop in Esther's course or the Rapid City flood, but these two cases have served to highlight a number of legal issues related to this new technology.

At present, most states have laws regulating weather modification programs. The legislation, which differs considerably from state to state, variously requires prior approval for cloud seeding projects, prohibits projects that might affect weather in neighboring states, and requires licensing of those engaged in cloud seeding. These are useful steps. However, we have a great deal still to learn about navigating through these turbulent waters where currents of science, society, and technology converge.

## Perspective

Finally, we ask that you keep the following perspective about weather modification: even the most successful programs affect the weather in only limited ways, in limited regions, for limited periods of time. Weather modification is not "weather control" in the sense that we could cause it to rain or not, be cold or hot, at a given place and time. Weather control of this sort is not even remotely possible at present, even if it were deemed desirable, for at least two reasons. First, the energy fluxes that drive weather systems are far too great for humans to control. And second, gaps in our understanding of the atmosphere's behavior are too great for weather control to be a possibility.

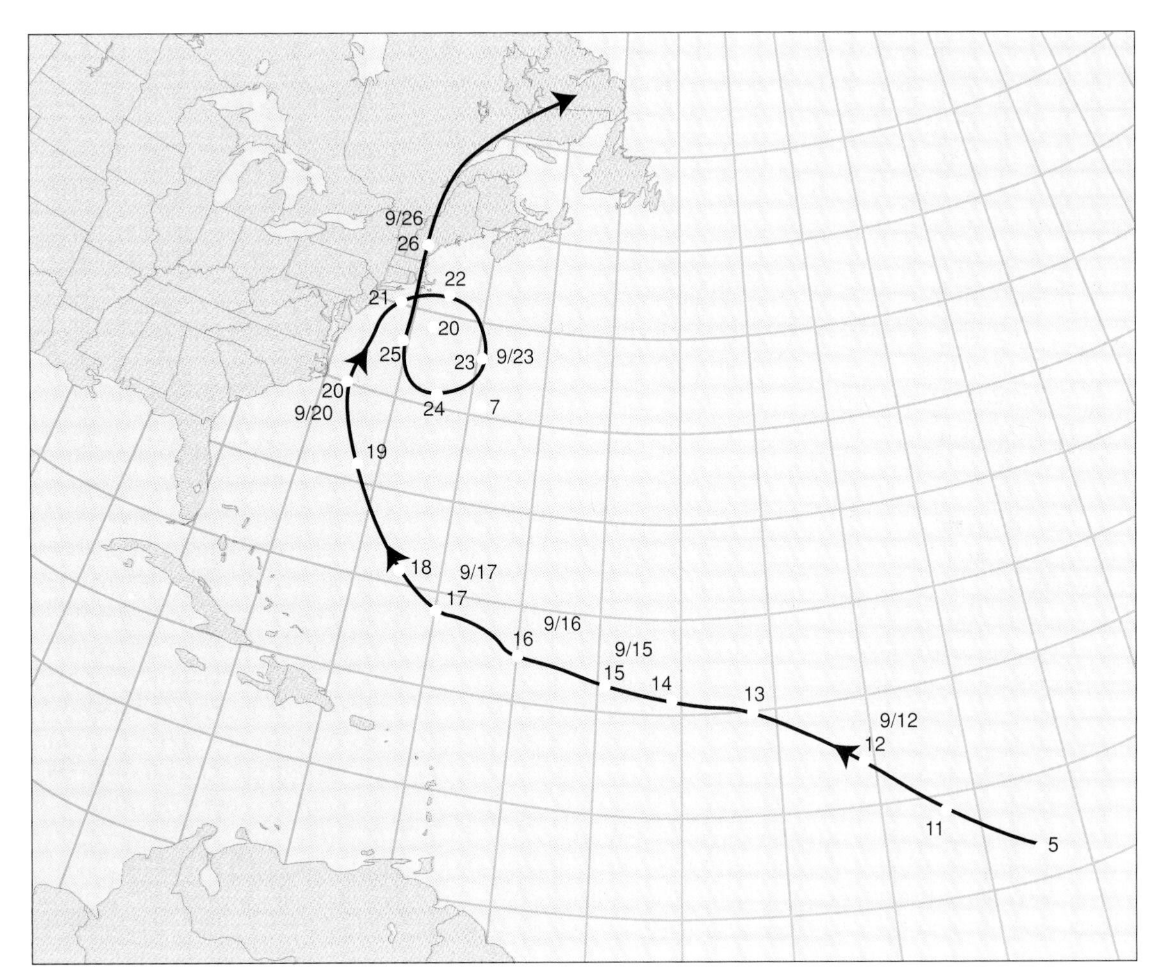


Figure 7.27 The path of Hurricane Esther in 1961. The hurricane was seeded on September 16 and 17.

# Study Point

## Chapter Summary

Precipitation forms within clouds through two main processes: the collision-coalescence process and the Bergeron process. In the former, raindrops grow by colliding and joining with other drops and droplets. The Bergeron process occurs as water droplets evaporate in supercooled clouds and their moisture is transferred to growing ice crystals. Precipitation types such as ice pellets, snow pellets, and freezing rain result mainly from temperature variations in the air through which rain or snow fall.

Worldwide, annual precipitation averages approximately 1 meter per year. This figure masks enormous variations in precipitation from place to place and in time.

Most efforts at precipitation modification are founded on the theory that some clouds are deficient in freezing nuclei. Seeding clouds with dry ice and silver iodide increases the number of freezing nuclei, sometimes induces evident changes in cloud structure, and may increase precipitation under the right circumstances by 10% to 15%. However, many scientific, legal, and ethical questions about this new technology remain unanswered.

## Key Words

precipitation
wake capture
coalescence efficiency
warm clouds
Bergeron process
radar
Doppler radar
drizzle
graupel
sleet
hail
silver iodide
dynamic effects
statistically significant
coalescence
collision efficiency
collection efficiency
warm rain process
virga
Doppler effect
snowflake
snow pellet
ice pellet
freezing rain
dry ice
AgI
significance level
highly significant

## Review Exercises

1. Use Figure 7.1B to determine how long it takes a droplet to grow by condensation from 5 to 10 micrometers, from 10 to 15, from 15 to 20, and from 20 to 25.
2. What is the difference between collision, coalescence, and collection efficiencies?
3. Why is condensation insufficient by itself as a precipitation agent?
4. Explain the role of each phase of water—solid, liquid, and gas—in the Bergeron process.
5. What evidence is there that two different processes of precipitation generation (warm rain and three-phase) actually operate? Cannot one or the other process account for all precipitation?
6. Why doesn't every cloud generate precipitation?
7. What are two major differences between the processes of human seeing and radar "seeing"?
8. Why do police position radar guns as close to the highway as possible? Why not some distance off the highway, where it's safer?
9. Explain how virga can become precipitation.
10. Use Figure 7.19 to estimate the mean annual precipitation and its variability for your location.
11. Why is cloud height an important consideration in selecting a warm cloud for seeding?
12. What is the logic behind seeding a supercooled cloud with a substance like dry ice or silver iodide?
13. What does it mean that silver iodide becomes active as a freezing nucleus at −4°C?
14. Skeptics say that cloud seeding is "robbing Peter to pay Paul." What do they mean?

## Problems

1. Suppose 1000 droplets lie in the path of a larger falling drop. If the drop collides with 750 droplets and actually coalesces with 500, what are the collision, coalescence, and collection efficiencies?
2. A large drop divides in a warm cloud, splitting into four smaller drops. These grow, each dividing again into four more drops. This grow-and-divide process continues through two more cycles before the drops fall from the cloud base. How many drops did the original drop generate?
3. What is the relative humidity with respect to ice in a supercooled cloud having a mixing ratio of 2.0 g/kg and temperature of −12°C? With respect to water? Based on your answers, what will happen in the cloud?
4. Why is the Bergeron process incapable of producing rain from a warm cloud?

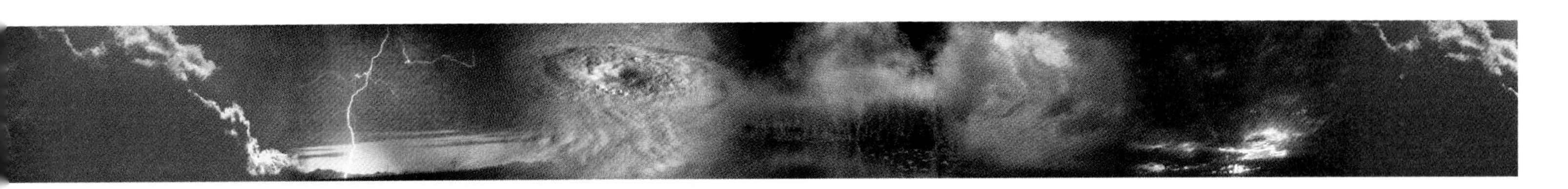

5. Calculate the weight of ice supported by a tree limb carrying 0.5 cm of ice. Choose a reasonable area to represent the limb. (Hint: One cubic centimeter of ice has a mass of approximately 0.9 grams.)
6. What is wrong with this statement: "Each type of precipitation (snow, ice pellets, freezing rain, etc.) is formed by a different process within clouds?"
7. How is a bat's system of echolocation similar to radar? How is it different?
8. You are standing on the sidewalk when a fire engine speeds by, its siren wailing. You notice the siren's pitch drops as the truck passes. Use the Doppler effect to explain the pitch change. (Hint: The shorter the sound wave, the higher its pitch.)
9. Explain how injecting dry ice into a cloud may increase the number of freezing nuclei.
10. Why would anyone (other than ski resort owners, perhaps) spend money on wintertime seeding of clouds in the mountains?
11. What was Changnon's stated hypothesis in the St. Louis precipitation modification study? How large were the differences in rainfall? What was the significance level of these differences? Was the hypothesis proved or disproved?
12. Compose a scenario like those in the last section of this chapter to illustrate some social issue related to inadvertent precipitation modification.

## Explorations

1. A real-time database like Accu-Data or some World Wide Web sites offer countless opportunities to explore precipitation patterns in time and space. You might investigate the sequences of rainy and dry days at your location over the past year, differences between two nearby sites, or the location of low pressure centers on days with precipitation, or you might develop a series of charts like those in Figure 7.19 that focus on your location.
2. What effect do objects like trees and buildings have on small-scale precipitation patterns? Establish a rain gauge network (frozen juice cans serve well) around and under a large tree or close to a building.
3. For a more complete picture of two research studies on precipitation modification, read the paper on the St. Louis study by Changnon (see the reference in Resource Links).
4. Read the novel *Cat's Cradle,* written by Bernard Vonnegut's brother Kurt, and compare Kurt's "ice nine" freezing nuclei with Bernard's silver iodide.

## Resource Links

The following is a sampler of print and electronic references for some of the topics discussed in this chapter. We have included a few words of description for items whose relevance is not clear from the titles. For a more extensive and continually updated list, see our World Wide Web home page at www.mhhe.com/mathscience/geology/danielson.

Middleton, W. E. K. 1966. *A History of the Theories of Rain.* New York: Franklin Watts. 223 pp.

Bohren, C. F. 1991. *What Light through Yonder Window Breaks?* New York: John Wiley and Sons. 190 pp. Clear exposition of Doppler effect.

Changnon, S. 1979. Rainfall changes in summer caused by St. Louis. *Science,* 207 (27 July 1979): 402.

Various authors, 1995. The 12–14 March 1993 superstorm. *American Meteorological Society Bulletin* 76(2):165–212. Three articles, each with multiple authors, about the big coastal storm described in this chapter.

Hughes, P., and R. Wood. 1993. Hail: The white plague. *Weatherwise* 46(2):16–21.

AccuWeathers's home page at http://www.AccuWeather.com provides visible and infrared images, and Doppler radar images.

For readers with access to AccuData, NEXRAD provides many Doppler radar images and data including reflectivity, rainfall accumulations, cloudtops, and radical wind velocity. LIGHT and LGTA plot the location of lighting seconds after they occur.

Real-time Doppler radar images are available at the University of Michigan's Weathernet at http://cirrus.sprl.umich.edu/wxnet/ and at AccuWeather, to name just two of many sites. You can choose presentations that show echo intensity, velocity toward or away from the radar site, and maximum cloud tops, among others.

Many web sites provide charts of precipitation totals. Two examples are the U.S. Government Climate Page at http://www.cdc.noaa.gov/Usclimate and Pennsylvania StateUniversity's hourly U.S. Weather Statistics at http://www.ems.psu.edu/cgi-bin/wx/uswxstats.cgi.

0
50
100
150
200
250
Watts/meter²

# How Are Weather Forecasts Made?

The central question of this three-chapter unit deals with the most important product that meteorology has to offer on a day-to-day basis: the weather forecast. The question is so central that it will occupy us not just in Chapters 8 through 10 but through most of the remaining chapters as well.

In this unit, we concentrate on prediction of weather conditions from a few hours to a few days in advance. Weather changes occurring within this time frame are generally due to the influence of synoptic scale weather systems. These are the highs, lows, and fronts that you studied in Chapter 1 and that are the focus of much of the discussion in television, radio, and newspaper weather presentations. Therefore, we will begin by reacquainting you with these systems; then, we will explore their structure and behavior in greater detail than in Chapter 1. We will examine some of the factors that lead to cyclone formation, intensification, and decay. Finally, in Chapter 10, we will survey a number of weather forecasting techniques and follow a meteorologist through the process of preparing a forecast.

As you will see, there is no single sequence of steps to follow in making a weather forecast. Forecasting is a creative act requiring the forecaster to blend scientific knowledge and logical reasoning with personal experience and intuition. It is a process in which the meteorologist absorbs a vast amount of data, then applies a variety of forecasting methods (both "manual" and computer-based), and finally makes a personal judgment of what the weather will be. Making a weather forecast is a great intellectual adventure, an adventure taken daily by meteorologists the world over. We hope you sense the same excitement we feel about predicting the weather.

# Chapter 8

# Structure of Large Mid-Latitude Weather Systems

# Chapter Goals and Overview

In this chapter you will study synoptic scale weather systems from many different points of view—through weather maps and diagrams depicting surface weather conditions, upper air patterns, and cross-sectional slices through the atmosphere, as well as through the eyes of Doppler radar. You will observe patterns and similarities that persist throughout these various depictions. To a large extent these patterns will match the circulation model developed in Chapter 1. However, we will extend and refine that model here.

You will also see that many aspects of the atmosphere's behavior can be understood in terms of basic natural laws, such as Newton's laws of motion. With this understanding, you will be prepared to explore questions such as how weather systems develop and how weather forecasts are made—the central question of the next three chapters.

## Chapter Goals

By the end of this chapter, you will be able to:

- state Newton's three laws of motion and give examples illustrating each;
- name the important forces affecting air's behavior;
- explain the concept of hydrostatic balance;
- define geostrophic balance and draw a force diagram to illustrate it;
- explain how temperature patterns determine pressure patterns aloft;
- explain how fronts and jet streams are related; and
- identify warm and cold air advection in a layer through vertical changes in horizontal wind direction.

# How Pressure Differences Arise

Transport your mind to northern Canada and Alaska in winter (see Figure 8.1). It's the land where the tender tundra is obliterated by the first snows of September, not to be seen again until the following June. It's a land where one must go prepared or not go at all—a land where intense cold is your constant companion. It's a place where the sun doesn't shine for weeks at a time as earth's annual journey keeps the polar region tilted away from the sun's warming radiation. The land and the Arctic Ocean grow ever colder.

To get an idea of how quickly it cools over the interior of Alaska and northwest Canada, examine the temperature data for Fairbanks, Alaska, in autumn, as shown in Table 8.1. Temperatures listed in the "Actual" columns are for one specific year, 1993; they are followed by "normal" values for the date. Normal temperatures are mean values for a 30-year period (1961 to 1990, in this case). Note that the dates in Table 8.1 are spaced 10 days apart, except between the horizontal bars, from November 11 through 20, where values are listed for each day.

Figure 8.1 Winter in interior Alaska can be both harsh and beautiful.

The column labeled "Dept." is the day's departure from the mean; hence, it is equal to the "Actual (1993) Ave." reading minus the "Normal Ave." value. Thus, on October 1, 1993, when the average temperature was 44 (compared to a normal value of 37 for the date), the departure from normal (Dept.) was +7: it was 7 degrees warmer than the 30-year normal. HDD refers to heating degree days, computed by subtracting the average for the date from 65. The final two columns show the all-time highest and lowest temperatures for each date and the year in which the record was set.

Note the abrupt change from November 11–20, when the temperature went from just 2 degrees short of the record high for the date to tie the record low at the end of the period—a decline of 72 degrees in a week and a half! This change occurred when an unusually mild air mass was replaced by an Arctic blast. In most years, the transition is more gradual.

How did such an intensely cold air mass form? And once formed, what caused it to move over the Fairbanks region? Consider the first question first. Imagine a mass of air sitting over the frigid, sunless surface north of Fairbanks, north of the Arctic

Table 8.1 Autumn Temperature Data for Fairbanks, Alaska, 1993*

| Date | Actual Hi | Actual Lo | Actual Ave. | Normal Hi | Normal Lo | Normal Ave. | Dept. | HDD | Record High | Year | Record Low | Year |
|---|---|---|---|---|---|---|---|---|---|---|---|---|
| 11 Oct. | 41 | 25 | 33 | 36 | 22 | 29 | +4 | 32 | 62 | 1969 | −1 | 1939 |
| 21 | 25 | 7 | 16 | 28 | 15 | 21 | −5 | 49 | 60 | 1938 | −17 | 1958 |
| 31 | 28 | 16 | 22 | 20 | 6 | 13 | +9 | 43 | 50 | 1954 | −27 | 1992 |
| 1 Nov. | 21 | 13 | 17 | 19 | 5 | 12 | +5 | 48 | 46 | 1970 | −30 | 1907 |
| 11 | 36 | 23 | 30 | 13 | −4 | 4 | +26 | 35 | 38 | 1981 | −34 | 1990 |
| 12 | 33 | 21 | 27 | 12 | −4 | 4 | +23 | 38 | 41 | 1976 | −36 | 1989 |
| 13 | 32 | 24 | 28 | 11 | −5 | 3 | +25 | 37 | 45 | 1976 | −37 | 1956 |
| 14 | 30 | 19 | 25 | 11 | −6 | 3 | +22 | 40 | 40 | 1908 | −41 | 1956 |
| 15 | 25 | 16 | 21 | 10 | −6 | 2 | +19 | 44 | 43 | 1916 | −33 | 1969 |
| 16 | 21 | 15 | 18 | 10 | −7 | 2 | +16 | 47 | 44 | 1916 | −41 | 1969 |
| 17 | 19 | −13 | 3 | 10 | −8 | 1 | +2 | 62 | 41 | 1916 | −39 | 1969 |
| 18 | 1 | −12 | −6 | 9 | −8 | 1 | −7 | 71 | 41 | 1916 | −33 | 1969 |
| 19 | −9 | −32 | −21 | 9 | −9 | 0 | −21 | 86 | 47 | 1949 | −33 | 1969 |
| 20 | −15 | −36 | −26 | 8 | −9 | 0 | −26 | 91 | 40 | 1943 | −36 | 1993 |
| 21 | 4 | −15 | −6 | 8 | −9 | −1 | −5 | 71 | 41 | 1947 | −35 | 1904 |
| 1 Dec. | 8 | −8 | 0 | 5 | −12 | −4 | +4 | 65 | 37 | 1976 | −47 | 1990 |
| 11 | 1 | −15 | −7 | 3 | −14 | −6 | −1 | 72 | 44 | 1940 | −54 | 1964 |
| 20 | 14 | −6 | 4 | 1 | −16 | −7 | +11 | 61 | 40 | 1909 | −48 | 1933 |

*All temperatures are in degrees Fahrenheit.

Circle, for many days. The ground emits longwave radiation upward but receives little radiation back from the cold, dry air above. Therefore, the surface temperature falls, causing the air in contact with it to cool also. Radiosonde records in this situation indicate the presence of an intense surface-based cold, and the resulting low-level inversion. Little mixing occurs between this cold, stable surface air and the somewhat less cold air aloft. Gradually, however, the depth of the cold air mass increases upward. It is through such processes that pools of frigid Arctic air form.

These changes in low-level air temperatures lead to other changes. Recall from earlier chapters that when an air parcel cools, its density increases. Thus, as the low-level air becomes progressively colder, it contracts, occupying less and less volume. The air's reduced volume leaves room for other molecules higher in the atmosphere to migrate into the region above the cold surface air, causing the total mass over the cold air to increase. For now, we won't go into how this "new" air is brought in. But clearly, air with increased mass weighs more than it did before the mass was added. Since we express atmospheric pressure as weight of the air above a given area, we conclude that pressure increased in this example.

We now have a cold high pressure area in northern Canada. Since we have not mentioned anything that would change the pressure in any other region, it follows that the pressure outside a high pressure area is lower than in the high. (Actually, since there is only so much air around the globe, the pressures elsewhere must have dropped a little in order for the pressure to rise in northern Canada. That change spread over the whole earth would be imperceptible.)

The fact that there is cold high pressure in the Arctic does not by itself explain why the cold air moves south. However, recall from Chapter 2 that air pressure is a force and that a force is a push or a pull. Indeed, air pressure differences deliver the force that leads to air motions. To understand the relation between forces and atmospheric motions, we need to make a brief detour into the physics of forces and motions, which is the subject of the next section. Then, we will return to apply these concepts to the atmosphere.

## Forces and Motions

In 1687, Sir Isaac Newton, considered by many to be history's most brilliant and influential scientist, published a treatise titled *Mathematical Principles of Natural Philosophy*. The *Principia*, as it is often called, contains work that Newton did many years earlier, as a young man in his twenties. Perhaps the most important of the *Principia's* many scientific contributions are Newton's three laws of motion and the law of universal gravitation. To this day, these laws remain basic tools for scientists and engineers engaged in a wide variety of investigations and problems. In the following paragraphs, we present a brief introduction to Newton's laws.

### Newton's First Law of Motion

Picture rolling a marble across a thick carpet. The marble rapidly loses speed and soon stops, due to the retarding force of friction. Repeat the experiment on hard-packed ground; the marble travels more or less in a straight line, maintaining its speed more than in the first example, but eventually it also slows and stops. But now imagine the marble rolling along an incredibly smooth, hard floor that offers virtually no frictional resistance to the marble's motion. The marble moves in a straight line at nearly constant speed, losing speed only very slowly.

With friction eliminated, the third case exemplifies **Newton's first law of motion,** which states that an object at rest remains at rest and an object in motion moves in a straight line at constant speed unless acted on by an unbalanced external force.

What is meant by an unbalanced force? An unbalanced (or "net") force is simply one that is not offset by other forces acting on the same object with equal magnitude but in the opposite direction. In our first two examples, friction represented an unbalanced force; hence, the "unless" clause in Newton's first law applied, and the motion was not at constant speed in a straight line.

The word *external* refers to the fact that the force must come from or be applied beyond the object itself. Thus, if you sit in a rowboat and push on the seat in front of you, you are creating an internal force and the boat will not accelerate. However, if you push on the water with a hand or an oar, the force is applied externally, and the boat will accelerate.

Notice that Newton's first law refers to two aspects of an object's motion: its direction ("in a straight line") and its speed. Such a quantity, which is specified by both a magnitude and a direction, is known as a **vector**. Wind is an example of a vector: both direction and speed must be included to specify the wind completely. Consider the difference between the wind, a vector quantity, and a variable such as temperature, which is specified by only a magnitude ("23°C") and is called a **scalar** quantity.

The vector that describes an object's motion is termed **velocity**. Thus, the wind velocity might be from the northwest at 20 knots; an airplane might maintain a velocity toward the north at 500 knots. Many quantities in this chapter are vectors.

Using the concept of a velocity vector and recognizing that a stationary object is one whose speed is zero, we could rephrase Newton's first law more succinctly as follows: An object's velocity remains constant, unless the object is acted on by an outside force. Here, the term *velocity* covers both speed and direction in the earlier statement of the law.

## Newton's Second Law of Motion

Newton's first law described the motion of objects affected by balanced forces (or no forces). The law's final words, "unless acted on by an unbalanced force," foreshadow the second law, which states that when an unbalanced force *is* applied to an object, it causes the object to change motion, that is, to **accelerate**.

Acceleration is a change in an object's velocity vector; thus, it is a change in the object's speed, its direction of motion, or both. Note the word *change*. Thus, an object at rest or moving in a straight line at constant speed is not accelerating, because neither its speed nor its direction of motion is changing. However, a car pulling away from a stoplight or slowing down at an intersection is accelerating because its speed is changing. So is a car rounding a curve, even if it maintains a constant speed, because its direction changes.

**Newton's second law of motion** states that an object subject to an unbalanced external force accelerates in the same direction as that in which the force is applied and that the magnitude of the acceleration is inversely proportional to the object's mass. Thus, the same force applied to a small mass will cause a larger acceleration than if applied to a larger one. As a simple example, if you push equally hard against a child's wagon and a car, you expect that the wagon, having much less mass, will accelerate much more rapidly than the car.

Newton's second law may be expressed algebraically as

$$F = ma$$

where F is the force, m the object's mass, and a the acceleration produced. Both force and acceleration are vector quantities; each consists of a magnitude and a direction.

We will illustrate Newton's second law further with the help of Figure 8.2A. In this idealized situation, friction between the mass m and the table is taken to be nil; the only force acting on the mass in the horizontal is the continual tug it receives from the person pulling from the left side of the figure. The spring balance is included to show that a constant force is applied to the mass for the duration of the experiment. If you pull for several seconds with constant force (i.e., the scale's pointer maintains a constant, nonzero value), the mass accelerates steadily to the left, as indicated by the data listed in the figure.

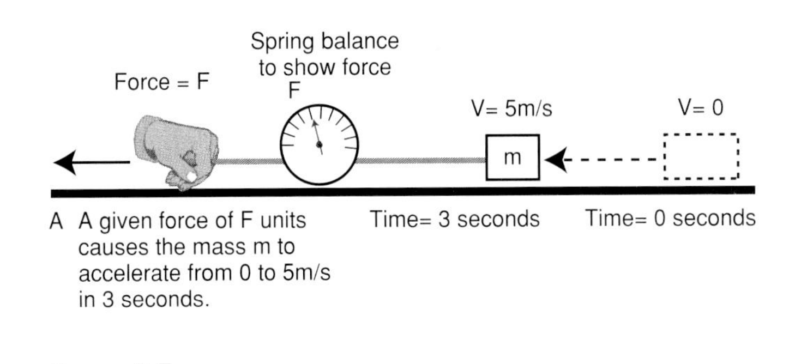


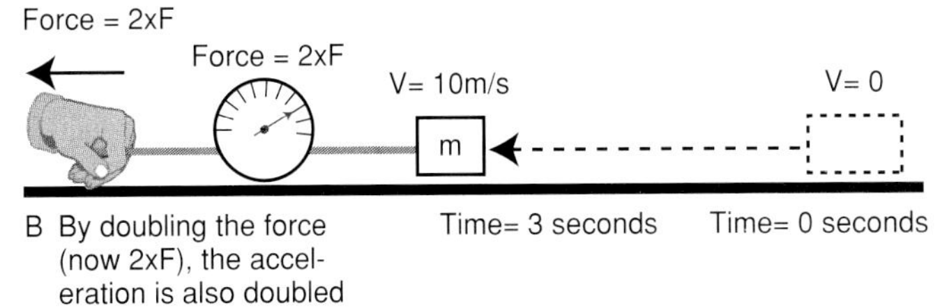


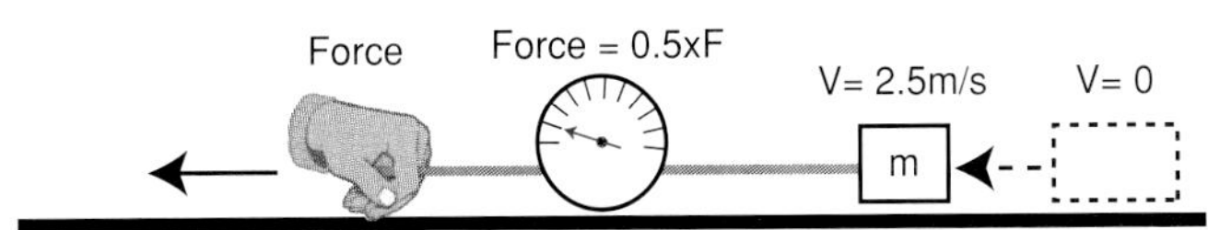


**Figure 8.2** Newton's second law of motion.

Figures 8.2B and 8.2C illustrate what happens when you change the amount of force or the object's mass. Notice that doubling the force doubles the acceleration, while doubling the object's mass reduces the magnitude of its acceleration by half. In general, applying more force causes greater acceleration, whereas increasing the mass of the object to be moved causes less acceleration.

## Newton's Third Law of Motion

**Newton's third law of motion** states that every force acting on an object is accompanied by an equal and opposite force exerted by the object. (Newton called these forces "action" and "reaction.") Figure 8.3 offers an example. Note that the push of your foot against the ground is the action force. That force is met by an equal, oppositely directed force from the ground to your foot, the reaction force, which you feel as the ground's resistance to your pushing against it. The action and reaction force vectors are equal in magnitude but opposite in direction.

A first encounter with Newton's third law can be confusing. You might be thinking along these lines: "If every force is met with an equal and opposite force, aren't we saying that every force is balanced? But Newton's first law says that if there are no unbalanced forces, then objects don't accelerate. So, how can anything ever accelerate?"

Indeed, if every force on an object were matched by an equal, opposite force on the same object, then nothing would accelerate. The way out of this puzzle is to realize that the "equal and opposite" forces of the third law are directed on *different* objects, as you can see in Figure 8.3.

Let's consider what happens when the forces are applied in this example. It is the reaction force from the ground that propels you forward. The greater the force with which you push, the greater the reaction force from the ground, and thus (from Newton's second law), the faster you accelerate. The second law also indicates that the greater your mass, the less you will accelerate, for a given force.

Now consider the action force, which you applied against the earth. It also causes an acceleration: the earth is accelerated

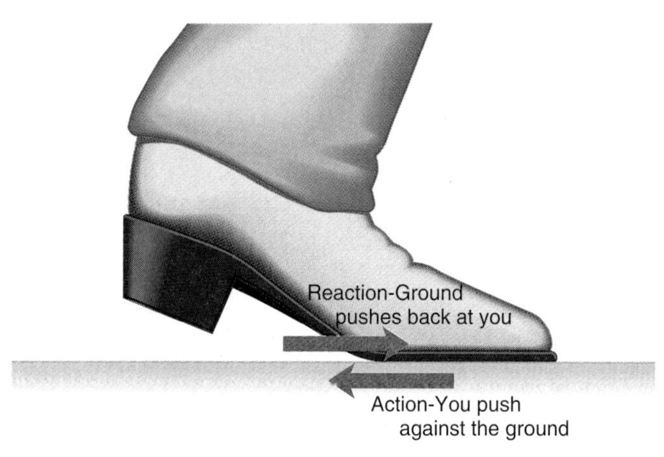


Figure 8.3 Newton's third law of motion. The action force (the push of your foot) is met by an equal, opposite force (the push of the ground on your foot).

a tiny amount in the direction of your push. How much is this acceleration? Since the action and reaction forces are equal, the product ma for the earth's mass and acceleration must equal the product ma for your mass and acceleration. And since the earth's mass is approximately $10^{23}$ times greater than your mass, it accelerates only $1/(10^{23})$ (i.e., 1/100,000,000,000,000,000,000,000) as much as you do.

Newton's laws of motion provide a basic framework for understanding how motions arise. With this framework in place, we can now move forward to deal with the motion of our Arctic air mass.

## Checkpoint

***Review***

At the beginning of this section, you were presented with a description of how a cold high pressure air mass develops in northern Alaska, above the Arctic Circle. Next you were presented with a problem: how does this air mass move south? To answer this, it was necessary to introduce the physics of motion by describing and providing examples of Newton's three laws of motion.

Newton's first law of motion states that the motion of a body will be in a straight line with constant speed as long as no unbalanced force acts on it.

Newton's second law of motion states that a body will accelerate if acted on by an unbalanced force. Furthermore, the acceleration will be in the same direction as the unbalanced force, and the magnitude of the acceleration will be directly proportional to the force and inversely proportional to the mass of the body. This law may be summarized by the formula $F = ma$, where F is the unbalanced force, m the object's mass, and a its acceleration.

Newton's third law of motion states that when one body exerts a force on second body, the second exerts an equal and opposite force on the first.

These laws of motion will be used shortly to answer the question posed previously: What causes the cold high pressure area in Alaska to move south?

***Questions***

1. In this section, it was explained that the air cools as it comes in contact with the cold land surface below. Explain how this cooling results in an increase in atmospheric pressure in this air mass.
2. Explain the difference between a vector and a scalar quantity. Give a few examples of vectors and a few examples of scalars.
3. Give an example of each of Newton's three laws of motion, and explain how each example illustrates the law.

# Forces Acting on the Atmosphere

We return now to consider our Arctic high pressure cell. What forces are acting on this air? In general, two kinds of forces affect the atmosphere: (1) forces that exist whether or not the air is moving, and (2) forces that act only on air in motion. The first class of forces can initiate motion; examples are gravity and pressure. The forces that act once motion is underway include friction, or drag, and Coriolis forces. We will examine each of these forces in turn.

## The Force of Gravity

Gravity's attraction is in the vertical direction, of course. Thus, it would seem that if we are looking for an explanation for how horizontal winds develop, we should look elsewhere. However, vertical and horizontal motions in the atmosphere often are linked, in ways you will see later in this chapter.

We owe much of our understanding of gravity to Newton also. In the *Principia,* he presented his law of universal gravitation, which states that every object in the universe exerts an attractive force on every other object. The magnitude of this "force of gravity" between any two objects depends on the objects' masses (more massive objects exert more gravitational force) and the distance separating them (the greater the distance, the weaker the force). Thus, an object falls to earth because the earth exerts an attractive force on the object. The falling object exerts an equal and opposite attractive force on the earth, according to Newton's third law; but because the earth is so much more massive, the earth's acceleration is insignificantly small.

It is the force of gravity that binds earth's atmosphere to the planet. Thus, every particle and parcel of air is continually subject to gravity's downward force. Now recall from Chapter 6 that if a parcel of air is denser than neighboring air at the same elevation, it will sink. Thus, we can say that, due to its greater density, our Arctic air experiences a stronger gravitational attraction to earth than does the surrounding, less frigid air; hence, the denser parcel tends to sink. When the sinking air reaches the ground, it tends to be deflected horizontally, and this is one way in which horizontal wind is created (see Figure 8.4).

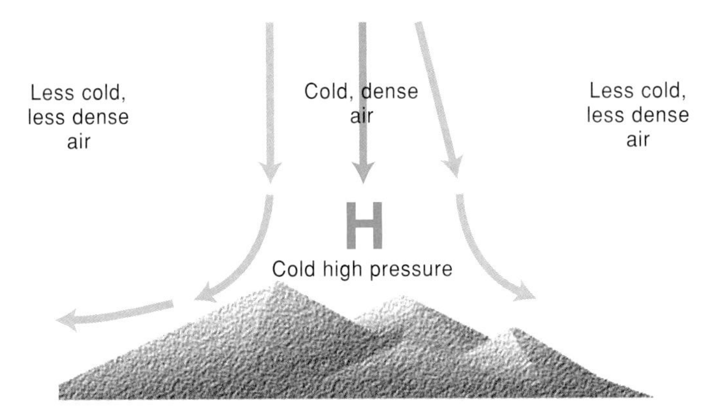


Figure 8.4 Cold, dense air sinks due to the force of gravity. When it reaches earth's surface, it spreads horizontally.

Because vertical motion leads to changes in stability and relative humidity, it is also important for its effect on local weather conditions. Recapping briefly from Chapter 6, sinking air becomes warmer and its relative humidity decreases. The result is often clear skies and fine weather. Rising air cools, eventually to the condensation point. Clouds and precipitation result.

You may be wondering why the gravitational attraction between earth and atmosphere doesn't pull the entire atmosphere right down onto earth's surface, as it does to other unsuspended objects. Some other force must be acting to balance that of gravity. We consider that force next.

## Pressure Gradient Force

Air pressure is the second of the atmosphere's fundamental forces. You know from studying weather charts and radiosonde traces that the pressure varies, sometimes substantially, from place to place. It is this change in air pressure from place to place that constitutes an unbalanced force that contributes importantly to atmospheric motion.

### *Pressure Gradients*

Consider the air pressure in the vicinity of Santa Cruz in Figure 8.5. Notice that the isobars in the region trend from north to south; thus, moving north or south, you tend to remain at the same air pressure. On the other hand, the greatest horizontal pressure change lies at right angles to the isobars, in the east-west direction. If you measure the east-west pressure change, you find a change of 12 millibars over a distance of 300 kilometers. Expressed in millibars of change per kilometer of distance, the change is 12 mb/300 km, or 0.04 mb/km, from west toward east.

This quantity is known as the **pressure gradient** (or, more specifically, the horizontal pressure gradient). The pressure gradient at any point is the rate of pressure change, measured at right angles to the isobars. Notice the gradient involves both a magnitude and a direction; thus, it is a vector quantity.

In symbols, the magnitude of the horizontal pressure gradient is often written $\Delta P/\Delta s$, where the symbol $\Delta P$ ("delta P") means a change in pressure, and $\Delta s$ ("delta s") means a change in horizontal distance. In the example above, $\Delta P$ was (1016 − 1004 = 12 millibars), while $\Delta s$ was 300 kilometers.

The pressure gradients at various other points in Figure 8.5 are represented by arrows. The arrows' lengths are proportional to the magnitude of the pressure gradient. Notice that the gradient always lies at right angles to the isobars; the pressure gradient, like all gradients, by definition points in the direction of greatest pressure increase. Note, however, that we have reversed the arrows' directions in Figure 8.5 so they point toward *lowest* pressure. Why have we done that? So that they point in the direction of the **pressure gradient force,** which is the net pressure force on an air parcel due to an existing pressure gradient. In other words, a pressure gradient causes a net force on air toward lower pressure. All this might sound

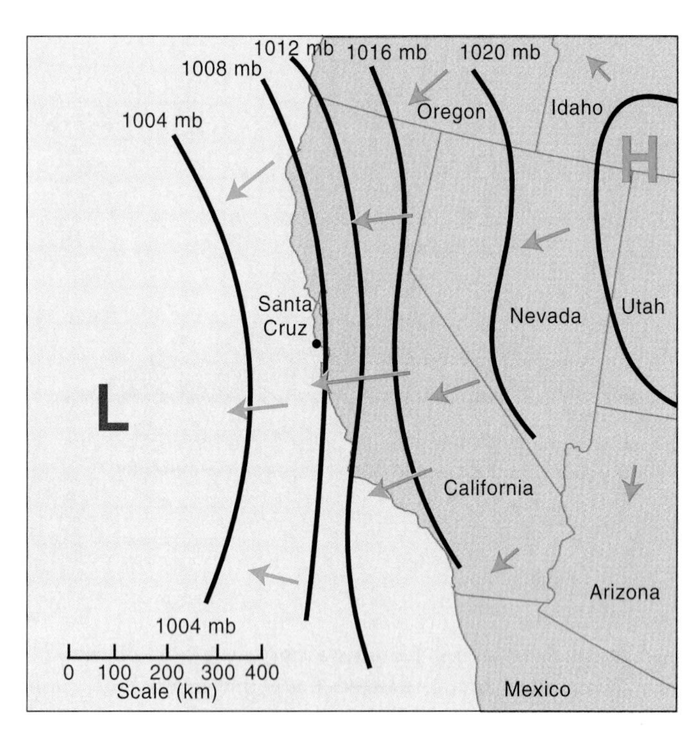


Figure 8.5 Pressure gradients. How do the arrows' directions compare to the direction of the isobars? What do the arrows' lengths represent?

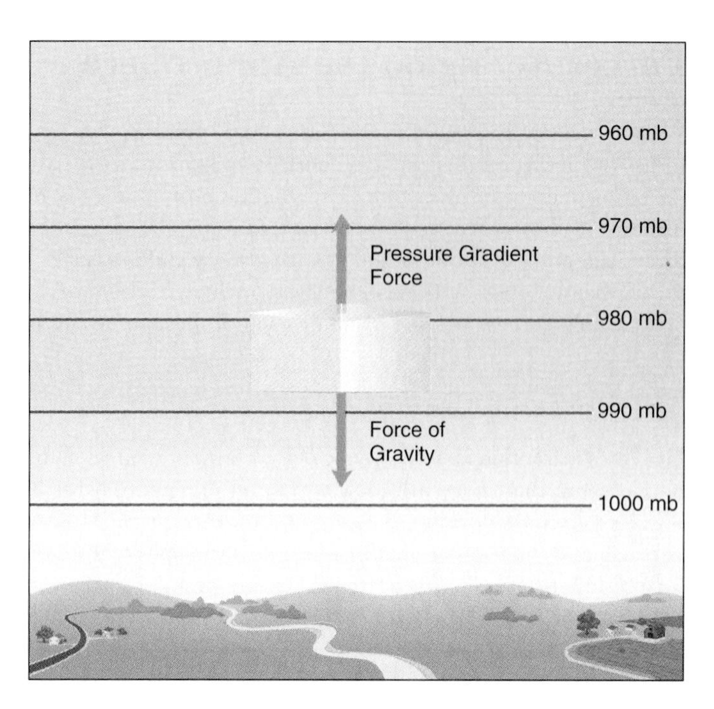


Figure 8.6 When the upward component of the pressure gradient force is balanced by the downward force of gravity, an air parcel is said to be in hydrostatic balance.

mysterious, but truly it is not. Just as the gradient of land (which by definition points uphill) causes water to flow downhill, so the air pressure gradient, which is defined as pointing toward higher pressure, leads to a net force on the air accelerating it toward lower pressure.

### *Hydrostatic Balance*

Let's consider the vertical component of the pressure gradient force in Figure 8.6. Because air pressure decreases upward, lower atmospheric levels exert a greater force upward than do higher levels downward. This vertical pressure gradient constitutes a net upward force on the air. If unbalanced by other forces, the atmosphere would accelerate upward and escape from earth! Gravity, however, provides the counterbalancing force. The balance of the upward pressure gradient force and the downward force of gravity is called **hydrostatic balance**. A parcel in hydrostatic balance experiences no net force in the vertical, and hence, from Newton's first law, it does not accelerate vertically.

The atmosphere is nearly always in or close to hydrostatic balance. However, exceptions exist. A warm air parcel in a growing cumulus cloud has reduced density, experiences less gravitational force downward than pressure force upward, and hence accelerates upward until it reaches a balance between pressure and gravity. Similarly, in our cooling, dense Arctic air mass, the attraction of gravity exceeds the upward pressure gradient force, and the chilled air accelerates downward until a new balance is achieved.

### *Motion Due to Horizontal Pressure Gradient Force*

Now, we will follow a parcel of air as it is influenced by the pressure gradient force. Consider the weather pattern shown in the map and cross section in Figures 8.7A and 8.7B. Imagine for a moment that you were looking at water instead of air. The high pressure area would be represented by a big pile of water, the low by a depression in the water surface. What would happen? The pile would quickly collapse as the water spread out in all directions, toward regions of lower pressure. Eventually, the water level, and hence the water pressure, would be equal everywhere. The atmosphere, as a fluid, behaves in much the same way. In response to the horizontal pressure gradient, air begins moving (is accelerated) from higher toward lower pressure. Note that if this process continues long enough, the high pressure area will disappear as the pressure becomes equal everywhere. The fact that we see large, persistent areas of high and low pressure on daily weather maps suggests there must be other forces at work.

## Coriolis Force

Once the pressure gradient force launches air into motion, other forces arise in response to that motion: the Coriolis force and friction. The Coriolis force is named after the French mathematician G. G. deCoriolis who published work on the effects of earth's rotation in 1835. You can picture the Coriolis force by imagining a ball rolling across a counterclockwise rotating turntable (see Figure 8.8). To a person standing on the ground, the ball appears to be obeying Newton's first law, traveling in a straight line

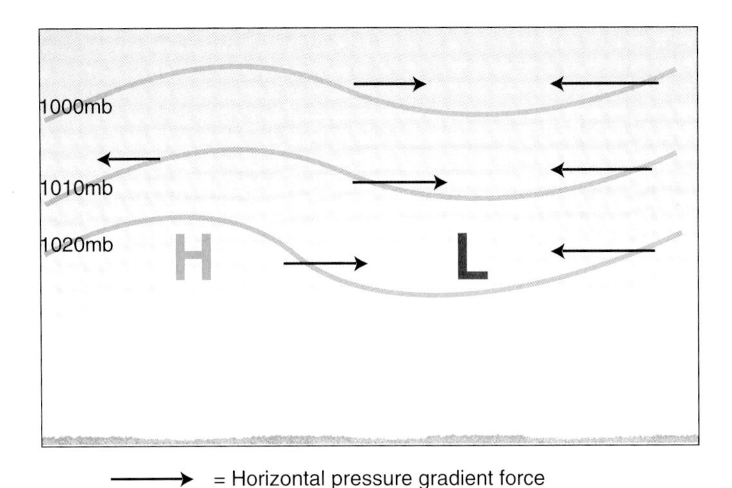


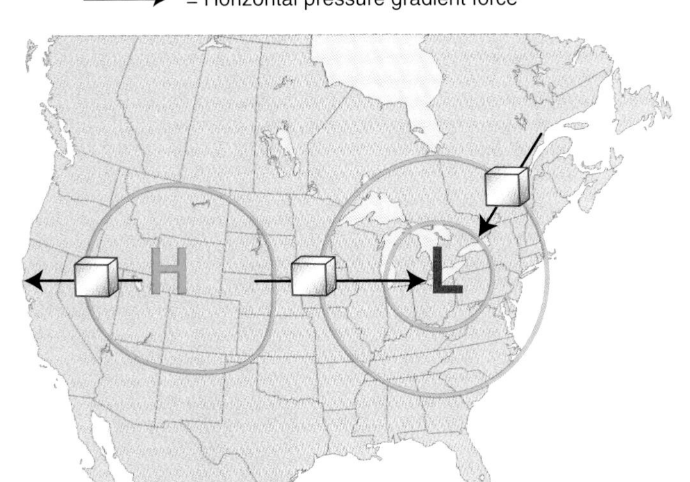


Figure 8.7 The horizontal pressure gradient force acts to "flatten" pressure patterns by moving air out of highs and into lows.

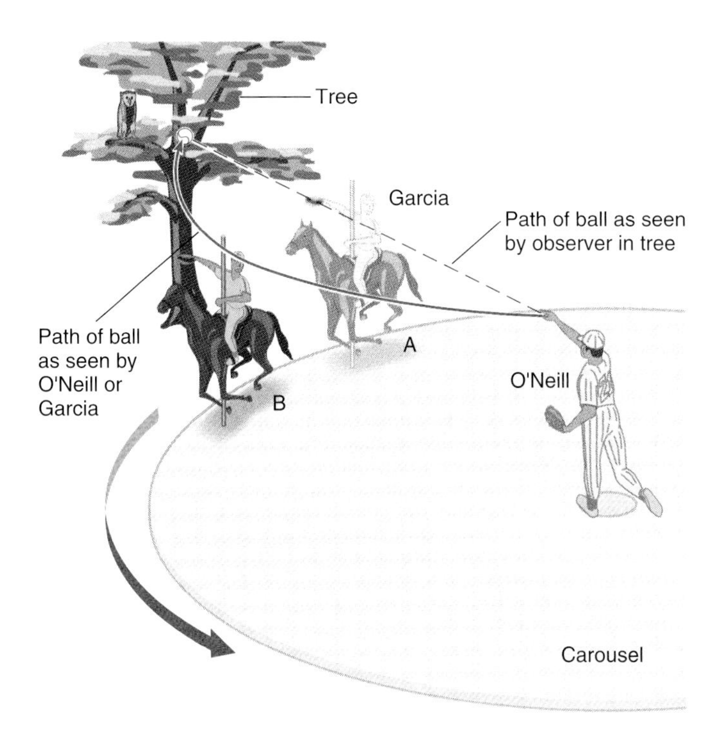


Figure 8.8 Illustration of the Coriolis effect. O'Neill throws a ball to Garcia when Garcia is at A, between O'Neill and a distant tree. The ball travels straight and strikes the tree. Meanwhile, the carousel rotates carrying Garcia to B. To Garcia and O'Neill, the ball follows a curved path, shown in red.

at constant speed. To an observer on the turntable, however, the ball is following a curved path, constantly turning to the right. This apparent deflection of an object with respect to a rotating reference system is called the Coriolis effect, and the equivalent force required to cause the observed turning is called the **Coriolis force**. Coriolis force is not a true force but is the result of rotation and your system of reference. However, the effect is real: objects moving across the surface of the earth, such as winds and ocean currents, experience a deflecting Coriolis "force."

The three most important properties of the Coriolis force are (1) that it acts at right angles to the right (in the Northern Hemisphere) of the object in motion, (2) its strength is proportional to the object's speed, and (3) it is stronger at higher latitudes than near the equator.

## Geostrophic Balance

Let's see what happens to an air parcel under the combined influence of pressure gradient and Coriolis forces. Assume that initially, the air parcel is held stationary, in the pressure pattern shown in Figure 8.9A. Notice that the isobars in this example are parallel, equally spaced and oriented in an east-west direction. Thus, the pressure gradient force (labeled "pg" in the diagram), which always is aligned perpendicular to the isobars and pointing toward lower pressure, points northward. Furthermore, since the isobars are straight and equally spaced throughout the map, the pressure gradient force is constant everywhere.

Now imagine releasing the air parcel. At the moment of release, pressure gradient force is the only force to affect the parcel in the horizontal. (Coriolis force is zero for the moment, because Coriolis force is proportional to the parcel's speed, and the parcel is not yet moving.) Newton's second law tells us that the parcel will accelerate northward, toward lower pressure.

Once the parcel begins to move, Coriolis force begins to act. Its deflecting action, at right angles to the right of motion, is represented in Figure 8.9B by the arrow labeled "Coriolis." Thus, the parcel now is subject to two forces: the pressure gradient force toward the north and Coriolis force (weak at first, because the parcel has not gained much speed as yet) toward the east. The parcel moves in response to both forces, traveling mostly northward but deflected slightly eastward by Coriolis force, as shown in Figure 8.9B.

Now that the parcel is moving northeastward at greater speed, how have the forces acting on it changed? The pressure gradient force is everywhere constant (a consequence of the fact that the isobars were said to be equally spaced), so there is no

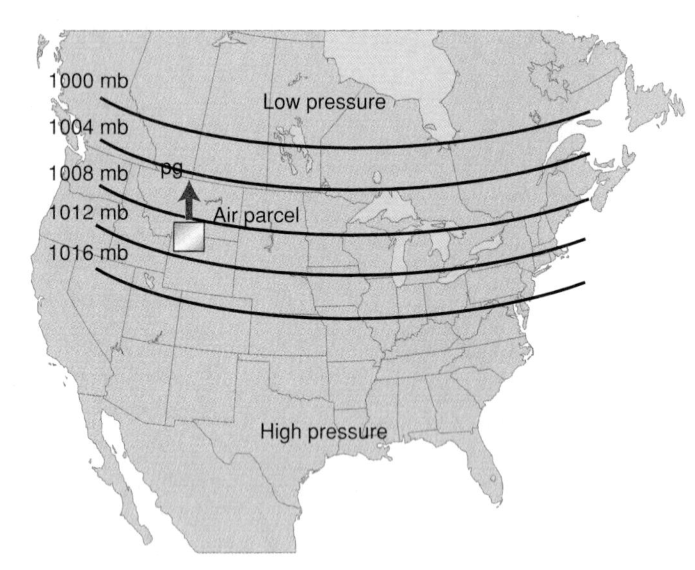


A Parcel held stationary. Only natural force acting is pressure gradient (pg).

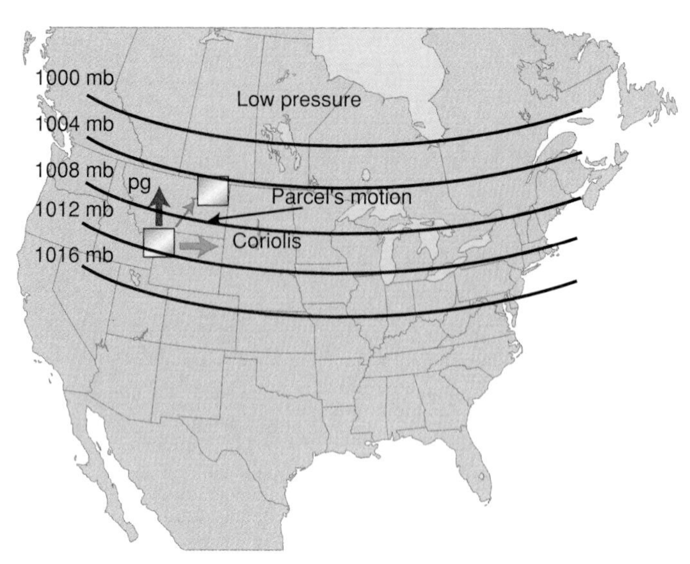


B Parcel is allowed to move, accelerates northward due to pg. Once motion begins, Coriolis acts to right of motion. Parcel deflected slightly to northeast.

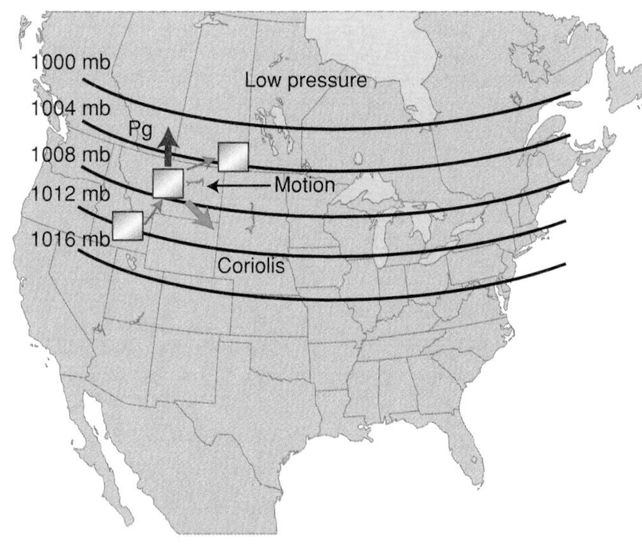


C Pressure gradient force remains constant in direction and magnitude, but Coriolis, acting to the right of parcel's motion, points southeast.

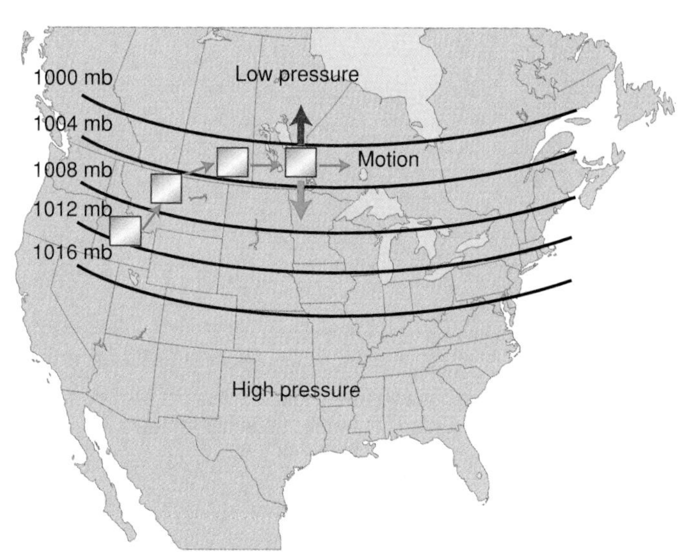


D Eventually, balance is reached with pg force acting 90° to left of motion, Coriolis equal in strength and 90° to right of motion. This balance is called geostrophic balance.

**Figure 8.9** Motions of an air parcel subject only to pressure gradient and Coriolis forces. Wide arrows represent forces. Thin arrows represent the motion of the parcel.

change in its direction or magnitude. However, the Coriolis force is changing both magnitude and direction: in magnitude because the accelerating parcel is now traveling at a greater speed and in direction because now the parcel is moving northeastward. Remember that the Coriolis force acts to the right *of the parcel's motion;* therefore, the Coriolis force is now directed southeastward, as shown in Figure 8.9C.

As you can see, the continued influences of pressure gradient and Coriolis forces cause the parcel to move in a rightward-turning trajectory. Eventually, the Coriolis force deflects the parcel's direction to the situation shown in Figure 8.9D. At this point, there is a balance between pressure gradient and Coriolis forces. Now, the air moves at a 90-degree angle to the pressure gradient force and to the Coriolis force, with lower pressure to the left of the flow and higher pressure to the right. This balance of forces is called **geostrophic balance,** and the wind thus produced is known as the **geostrophic wind**. To summarize, geostrophic balance is the balance between pressure gradient and Coriolis forces acting on a parcel, such that the forces are equal in magnitude but in opposite directions. Geostrophic wind is the motion of air under geostrophic balance.

Remember that the geostrophic wind is based on a major simplifying assumption: that no forces other than pressure gradient and Coriolis are considered. You might rightly ask how valid this assumption is and whether observed winds in the atmosphere are geostrophic. In fact, the real wind generally does behave approximately geostrophically at levels above the planetary boundary layer (i.e., where friction is not significant), where the isobars are straight and where the flow pattern is large and well established. The latter point is important because the Coriolis force does not have time to be a significant player in short-lived circulations, such as air moving around a hill or a whirlwind of dust and leaves on a blustery spring day.

The geostrophic wind approximation is a very useful concept. It allows us to estimate wind direction and speed knowing only the pressure pattern and the location's latitude. At altitudes above a kilometer or so, where the influence of the earth's surface is generally not significant, the wind direction is parallel to the isobars with lower pressure to the left. The wind speed is proportional to the pressure gradient, which is easily estimated from the spacing of the isobars and, for a given spacing, is stronger at higher latitudes. Although the geostrophic approximation is far from perfect, it generally gives a good estimate of actual winds, a remarkable fact considering we are ignoring all but two forces—pressure gradient and Coriolis.

Sometimes it is useful to apply the geostrophic relation in the opposite sense: if you know the wind speed and direction at some locale, you can infer things about the pressure pattern in the vicinity. For example, if you observe the wind direction to be from the south, you can conclude that isobars are oriented in a north-south direction, with lower pressure to the west. If the wind is strong, you know the isobars must be closely spaced; light winds imply a weak pressure gradient. This kind of reasoning is the basis for Buys Ballot's law, which you read about in Chapter 1. Recall Buys Ballot's law states that if you stand with your back to the wind, low pressure is to your left and high pressure is to your right.

You can check the validity of the geostrophic approximation by comparing the plotted wind speeds and directions on a weather map with the isobars or pressure contours. Try this for yourself, using the 500-millibar chart in Figure 8.25B. Notice, however, that the 500-millibar map represents not the pressure pattern at a particular altitude (as with sea-level pressure "surface" maps) but the height of a particular pressure surface (the 500-millibar surface in this case). The change is made for "upper-level maps" because the mathematics describing atmospheric motion are easier to handle if we use constant pressure surfaces rather than constant height surfaces. Thus, at all levels above the surface, weather charts portray variations in height of constant pressure surfaces.

It might be reassuring to see that the two kinds of representation give similar results. Look back at Figure 8.7. Looking at the cross-sectional view (top), trace the 1010-millibar isobar, for example, from left to right. Notice that it rises in the high pressure region and falls in the low pressure region. Thus, height changes in the 1010-millibar isobar—or any isobar—correspond to pressure changes on a constant height surface—such as the sea-level pressure map in Figure 8.7 (bottom).

As an additional check, compare the two maps in Figure 8.10. The first map, which we now call a **constant height map,** shows variations in air pressure at one particular altitude in the atmosphere (sea level, in this case). The lower map is an example of a **constant pressure map,** which shows height variations of one particular pressure surface (1000-millibar surface, in this case). Note that although individual isobars and contour lines do not match, the overall patterns are the same. Thus, both maps show a high pressure (high height) area over New England and eastern Canada, a low pressure (low height) area north of Minnesota and another high pressure area (high height area) over western Canada.

Let's apply the geostrophic approximation to the charts in Figure 8.11. Estimate the wind direction and relative speed at points A, B, and C in Figure 8.11A. At which point is the wind speed greatest? Where is it lowest? What would you predict the wind directions to be at each point? To see how well you did, consult the map in Figure 8.11B. It has the same basic contours, but the original data with wind speeds and directions are plotted.

To recap, you have seen that on the constant pressure (500-millibar) chart, the geostrophic approximation is generally valid:

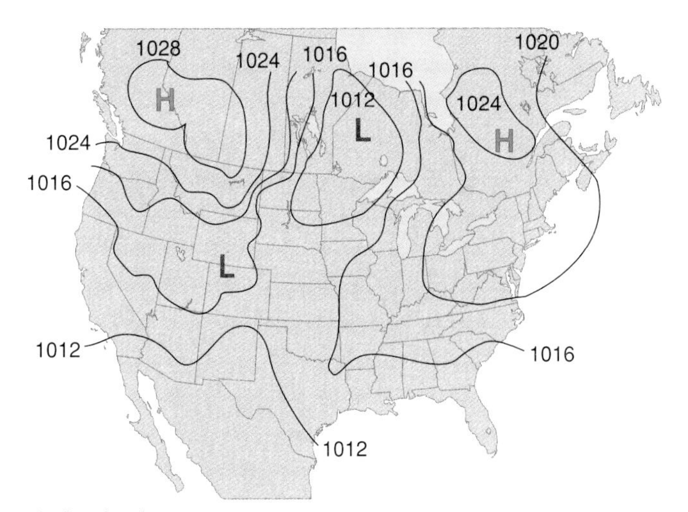

A Sea-level pressure map

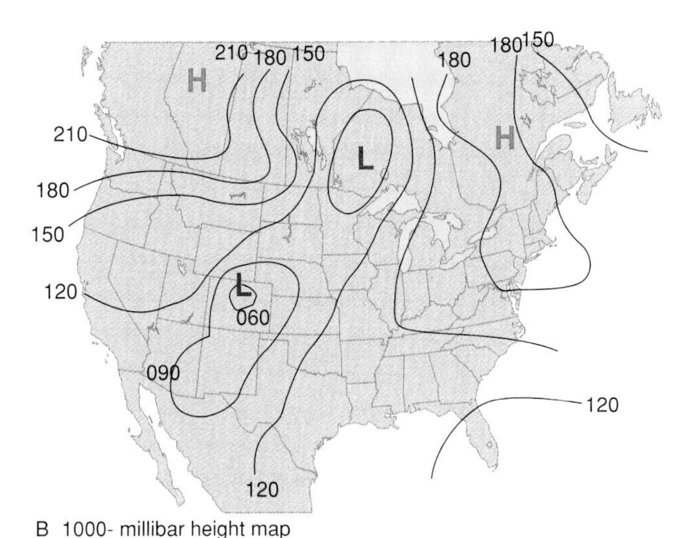

B 1000- millibar height map

Figure 8.10 Comparing constant height and constant pressure maps.

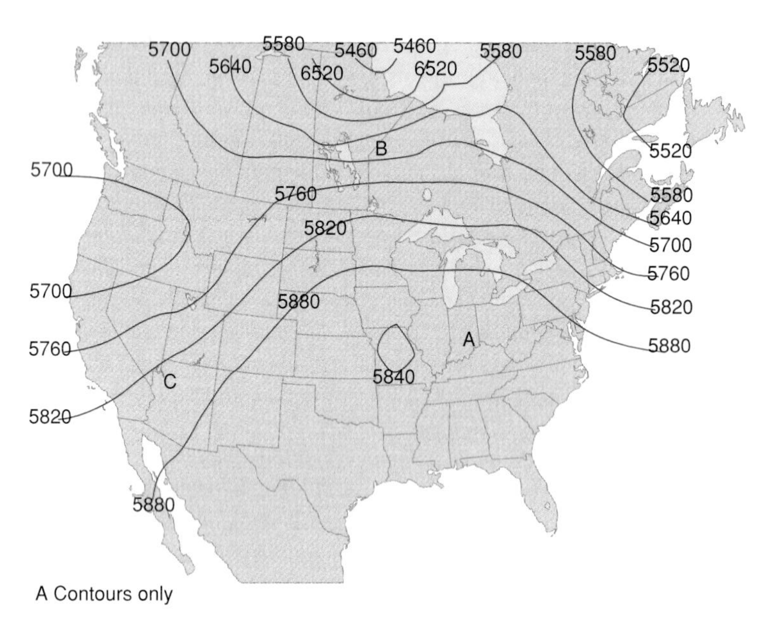


A Contours only

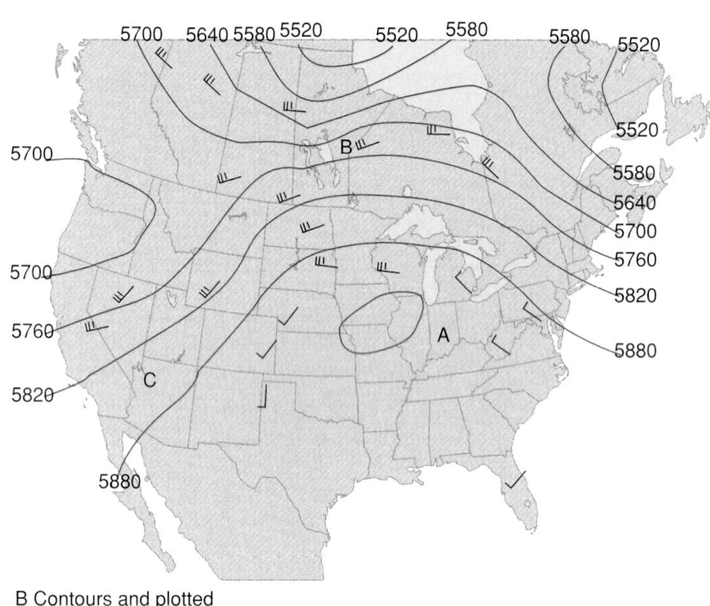


B Contours and plotted

**Figure 8.11** Checking the geostrophic approximation. Estimate winds at A, B, and C in the upper map, then check your estimates below.

winds blow parallel to the contour lines of the pressure surface, and the closer the contour spacing, the faster the winds. It is the pressure gradient force that initiates the wind. In straight line flow, we assume, the wind adheres to geostrophic principles in large-scale systems at altitudes above the layer of air affected by friction.

## Gradient Flow

You just saw that the flow aloft tends to obey the geostrophic approximation when the isobars or contours are straight. However, a glance at any weather map reveals that pressure patterns rarely are so simple and that curved isobars are more typical. Is the geostrophic approximation valid in "curved flow"?

Consider the wind flow at the 500-millibar level in Figure 8.12. In this diagram, we already know that where the flow is straight—at location 'X', for example—the Coriolis force and the pressure gradient force are at least approximately in balance and the actual wind is geostrophic or nearly so. However, if the pressure gradient and Coriolis forces were in balance everywhere in Figure 8.12, we would have a problem. The air under balanced forces would move in a straight line at constant speed, according to Newton's first law. But you can see that the airflow in Figure 8.12 curves in most locations. Thus, the pressure gradient and Coriolis forces must be unbalanced in those places. This lack of balance is said to be due to a third force, called the centripetal force. The word *centripetal* comes from a combination of *centri* (meaning "center") and *petal* (meaning "push"); hence, a force pushing toward the center around which the object is turning. The size of the centripetal force depends on the wind velocity (V) and the radius of the wind's path (r), as given by the formula

$$\text{centripetal force} = \frac{V^2}{r}$$

Think about the implications of this equation for a moment. As the isobars become straighter, the radius of curvature increases. Approaching the straight-line stage, r becomes infinite; thus, the denominator of this expression grows so large it dwarfs the numerator, and the centripetal force approaches a value of zero. Thus, centripetal force is zero for straight-line flow. This reduces to the geostrophic situation we discussed before, in which just two forces, the pressure gradient and the Coriolis, are in balance. When a balance exits between the pressure gradient force, Coriolis force, and a (nonzero) centripetal force, the flow is called **gradient flow** (or **gradient wind**). In a sense, the geostrophic wind is a special case of gradient wind: the case when the curvature is zero.

Before we leave gradient flow, let us look at two interesting implications it has about wind speed. Consider Figure 8.12 once more. Since the isobar spacings around the high and low are identical, the magnitude of the pressure gradient force, directed from high to low pressure, must be the same also. In the low pressure area, the pressure gradient force and centripetal acceleration must be strong enough to keep the air flowing in a circle. This can happen only if the Coriolis force is weaker than the pressure gradient force. In turn, this means the wind speed must be less than geostrophic. (Recall that Coriolis force is proportional to wind speed; thus, a weaker wind speed creates a weaker Coriolis force.) Studies have shown that winds in curved flow around some strong low pressure areas are more than 20% weaker than geostrophic.

In the high pressure area, notice that the Coriolis force must be stronger than the pressure gradient force, in order for the wind to be deflected around the high center. (If the forces were equal, the wind would not continue around the circle.) What is required for a greater-than-geostrophic Coriolis force to exist? The wind speed must be stronger than geostrophic. This conclusion may seem contrary at first to the commonly observed fact that winds in low pressure areas ("storms") generally are stronger than those around fair weather, high pressure areas. The explanation lies in the fact that pressure gradients around lows tend to be much stronger than those around highs. In fact, it can be shown mathematically that there is an upper limit to the pressure gradient force for high pressure areas. This in turn means there is a theoretical upper limit for high pressure area winds. No such limit exists for wind speeds in low pressure areas.

## Friction

Examine the pressure pattern and the wind arrows in the surface weather map in Figure 8.13. Does the wind appear to be acting geostrophically? In other words, are they blowing parallel to the isobars, stronger where the gradient is tighter? You probably noticed that many of the wind arrows cross the isobars, heading toward lower pressure. What causes this clear violation of geostrophic flow? Apparently an additional force must be at work. That force is friction. Near the ground, the air encounters a frictional drag. Which way is friction's effect directed? Friction acts as a brake on any motion, so it must be directed opposite to that motion. If we take the equilibrium (i.e., geostrophic) motion of Figure 8.14A and then apply friction (B), a new equilibrium will become established, as shown in Figure 8.14C. Note that because the frictional force is always applied opposite the direction of motion, its direction changes as the air parcel's direction of motion changes. (These changes are evident if you

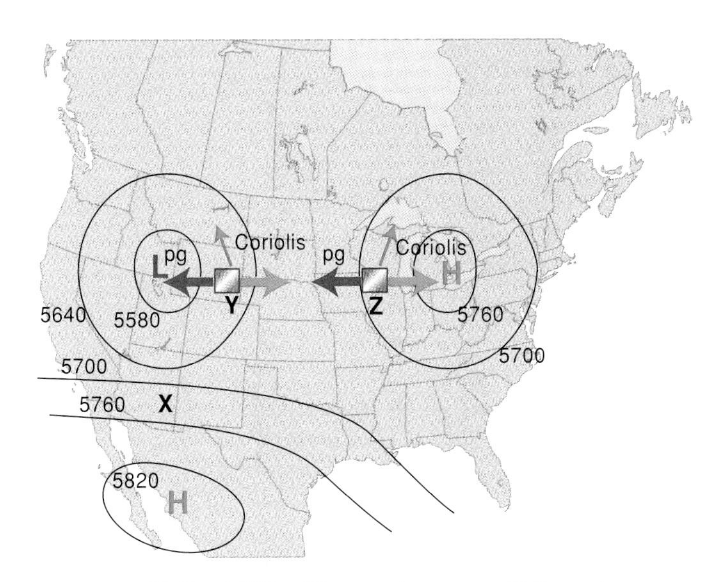


Figure 8.12 Stylized 500 millibar contour map. At X, contours are straight and the flow is geostrophic. At Y and Z, the contours are curved, and flow is gradient flow. Contour values are in meters above sea level.

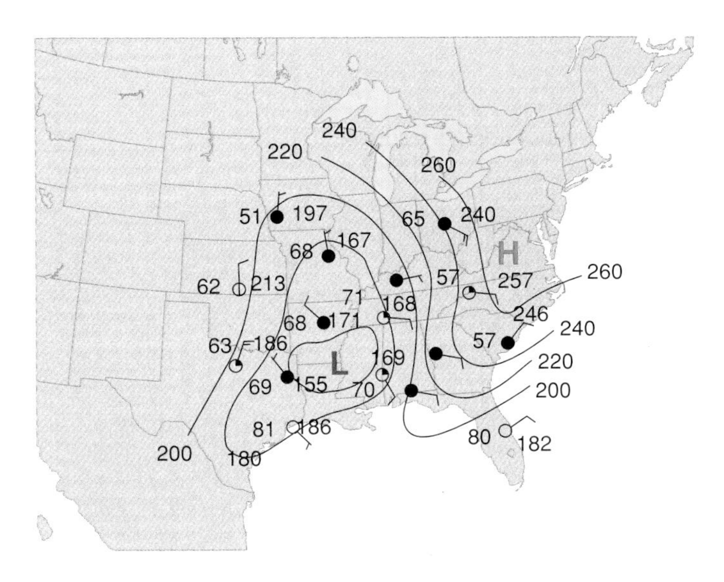


Figure 8.13 The effect of friction on surface wind directions. Note the tendency of the wind arrows to cross isobars flowing toward lower pressure. Are these winds geostrophic? Why or why not?

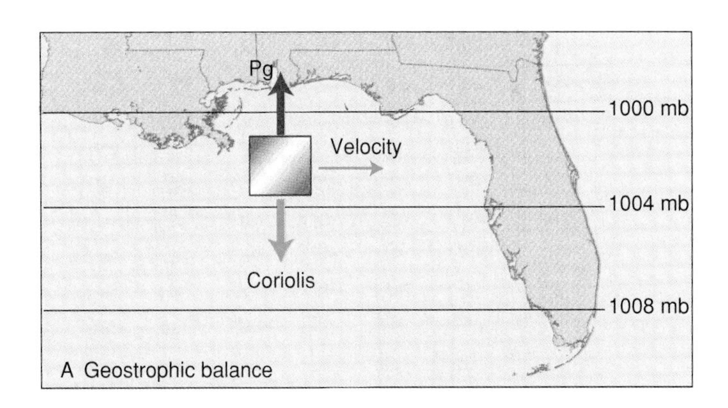


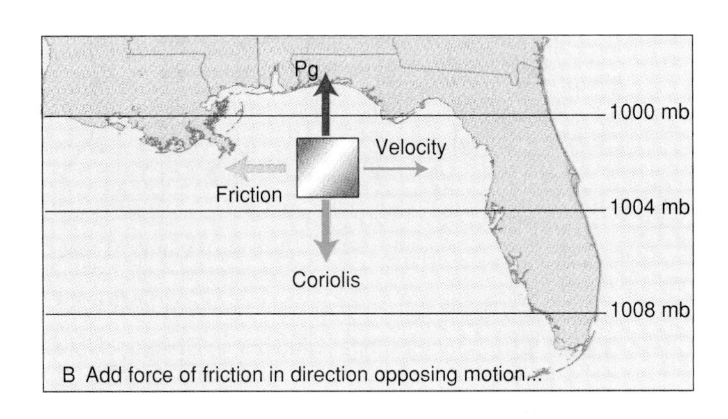


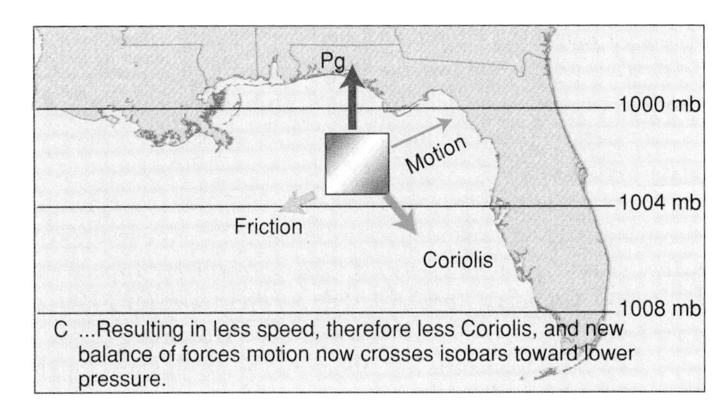


Figure 8.14 The effect of surface friction on geostrophic flow.

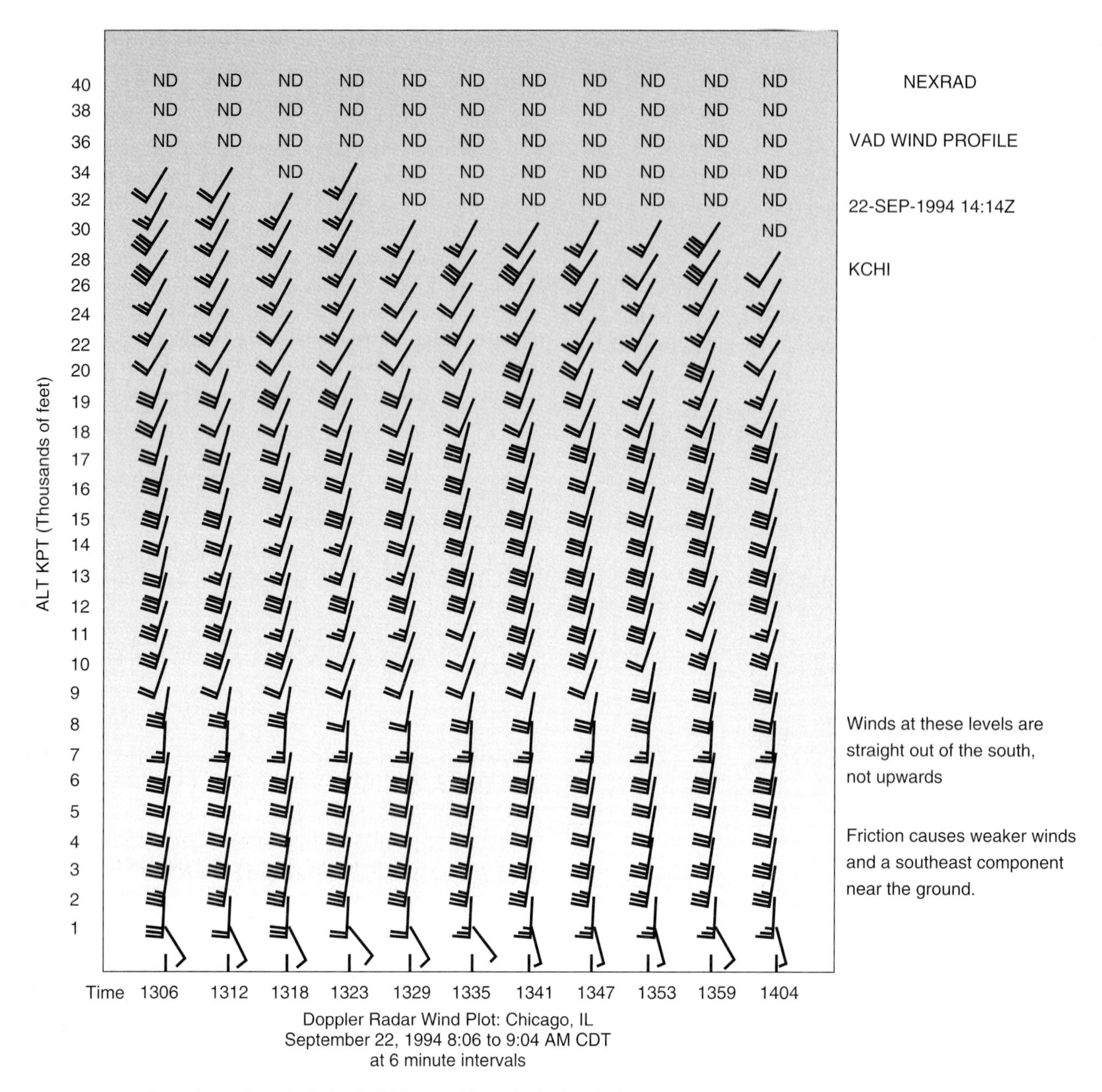


**Figure 8.15** Doppler radar wind plot in Chicago. Note the light wind speeds near the earth's surface.

compare Figures 8.14B and C.) Because it reduces the parcel's speed, friction also reduces the magnitude of the Coriolis force. Thus, the pressure gradient force is slightly stronger than Coriolis, causing the parcel to move somewhat across the isobars toward lower pressure.

The impact of friction on low-level wind flow is evident on any surface weather map, such as Figure 8.13. Notice the consistent across-isobar flow toward lower pressures on these surface maps. Compare them to the parallel-to-isobar flow on upper-level maps. Another vivid example of friction's effect in the lowest atmospheric levels shows up well on the Doppler radar (NEXRAD) wind velocity plot in Figure 8.15. Note the change in direction in the lowest 2000 feet.

***Review***

This section applied Newton's laws of motion to begin to explain the dynamics of atmospheric motion. The forces acting on the atmosphere were divided into those that act on the atmosphere regardless of whether the air is moving and those that act only when the air is in motion. The forces acting on both still and moving air include gravity and pressure gradient force. The forces operating only on moving air include friction and Coriolis force.

These four forces were then used to describe and explain important concepts of atmospheric motion that included pressure gradients, hydrostatic balance, horizontal motion, geostrophic balance, gradient flow, and centripetal force.

Using these concepts, you are now beginning to be able to explain the causes of vertical and horizontal motions, why some air motions are parallel to isobars and others are not, and the circular motion around high and low pressure areas.

***Questions***

1. How does gravity, which acts in the vertical, influence horizontal motion?
2. If the pressure gradient points from low toward high pressure, then why does the pressure gradient force move air from high toward low pressure?
3. You observe cirrus clouds moving in from the southwest. What can you conclude about the pressure pattern aloft?

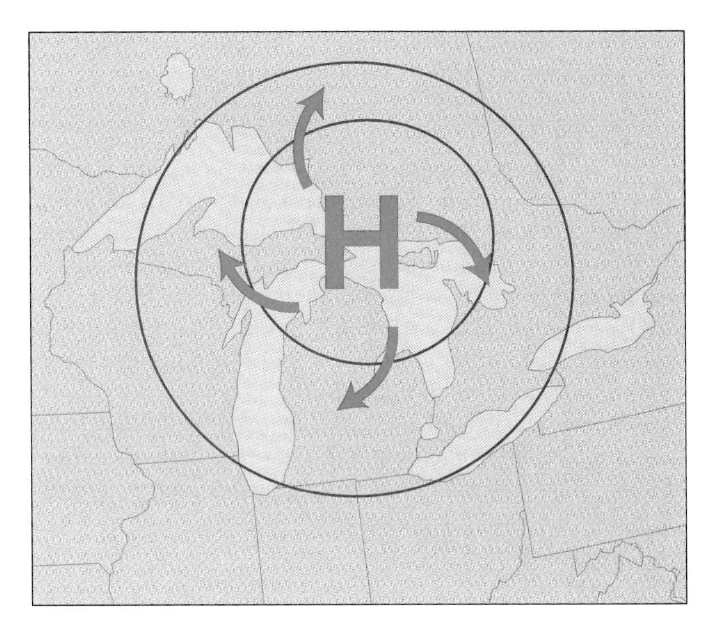


Figure 8.16 Low-level wind flow in a cold high pressure cell is clockwise (anticyclonic) and divergent.

# Vertical Structure of Large-Scale Weather Systems

Most people are familiar with weather maps that show surface positions of systems such as high and low pressure areas and fronts; however, features in the upper part of the troposphere are less familiar and may seem unrelated to surface weather. And yet, if you draw on your own experience, it should be apparent that various layers of the atmosphere are related. While we feel directly only the weather that exists at the ground, we can easily see how it is influenced by what's happening overhead. We see swirls of clouds going in different directions at different elevations; we get rain and snow from clouds that may be many kilometers above us. In this section, we will add considerable detail to our models of high and low pressure systems by considering their vertical structure.

## Wind Flow in Three Dimensions

We begin by applying the concepts of forces and motions to our cold Arctic high pressure area to understand how the winds circulate around and through it.

### *Convergence and Divergence*

Figure 8.16 shows the wind flow pattern near the earth's surface. Notice that the arrows show the air moving around the high in an anticyclonic (clockwise, in the Northern Hemisphere) direction while following a spiraling motion away from the high pressure center. Such spreading out of the air is known as **divergence.** The low-level wind flow in all high pressure systems, including the one we are studying, is divergent. From your previous readings, can you recognize that this pattern is typical of airflow shaped by pressure gradient, Coriolis, and friction forces?

As you look at the wind flow pattern in Figure 8.16, you may sense a problem. If the low-level winds are divergent, causing air to be removed from the high, then won't the high lose its extra pressure and cease to be a "high"? Indeed, that would be the case if this were the whole story. However, divergence at one level in the atmosphere (low-level divergence, in this case) is generally compensated by an inward flowing air pattern, known as **convergence,** at other levels.

Let us assume our high pressure area is maintaining its strength. In this case, air must be converging in the upper atmosphere at the same rate that it diverges at ground level. Thus, the net amount of air in the column reaching from the ground to the top of the atmosphere remains constant, and the pressure likewise remains unchanged. What is the vertical motion of the air like in this high pressure area? By inspecting Figure 8.17, you can see that if air is converging aloft and diverging below, it must be sinking. The circulation shown in Figure 8.17 is typical of that in virtually all high pressure cells.

If convergence aloft exactly balances divergence below, then the surface air pressure will not change. However, if the

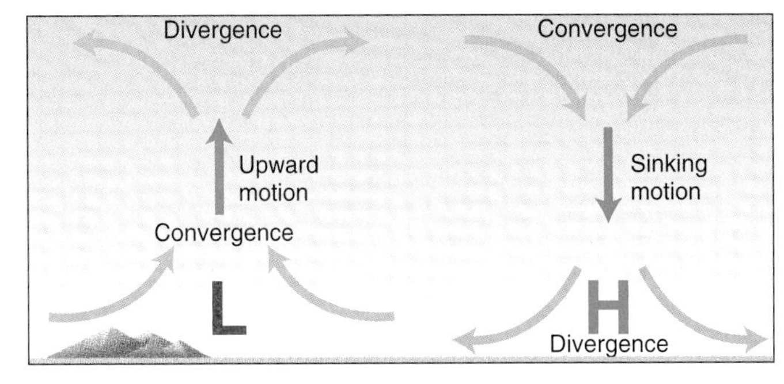


Figure 8.17 A cross-sectional view of air motions in high and low pressure systems. Note the symmetries (surface convergence in lows, divergence in highs, and vice versa aloft). Notice also how the air moves from lows to highs and back again.

convergence exceeds the divergence, then air is entering the column faster than it is being removed, and the surface air pressure must increase as a result. Similarly, if divergence exceeds convergence, the surface pressure must decrease.

Notice the symmetries in Figure 8.17: in low pressure systems, the exact opposite arrangement of flow patterns prevails from that in highs. Low-level wind flow is convergent, vertical motion is upward, and upper-level wind flow is divergent. Notice also that the patterns for lows and highs may link together, with one feeding the other. Such linkages between between highs and lows occur at every scale of atmospheric motion from the very small to the global.

The concepts we have just developed are applications of a basic principle known as continuity. The equation of continuity states, in simple terms, that the change in the amount of matter in a given space is equal to the difference between the amount entering (convergence) and that departing (divergence). Or, even more simply, the net flux of matter must equal the change in the amount of matter present. The continuity equation is itself an application of an even more basic principle, the law of conservation of matter, which states that matter is neither created nor destroyed. In any process, we must be able to account for all matter; it does not spontaneously appear or disappear.

## *Motion of High and Low Pressure Centers*

To this point, our discussion has dealt with winds within a high pressure cell but not with motion of the high pressure cell itself. What does it actually mean when we say a high pressure area is moving? The movement of an entire circulation system is a different concept from the wind flow within the system.

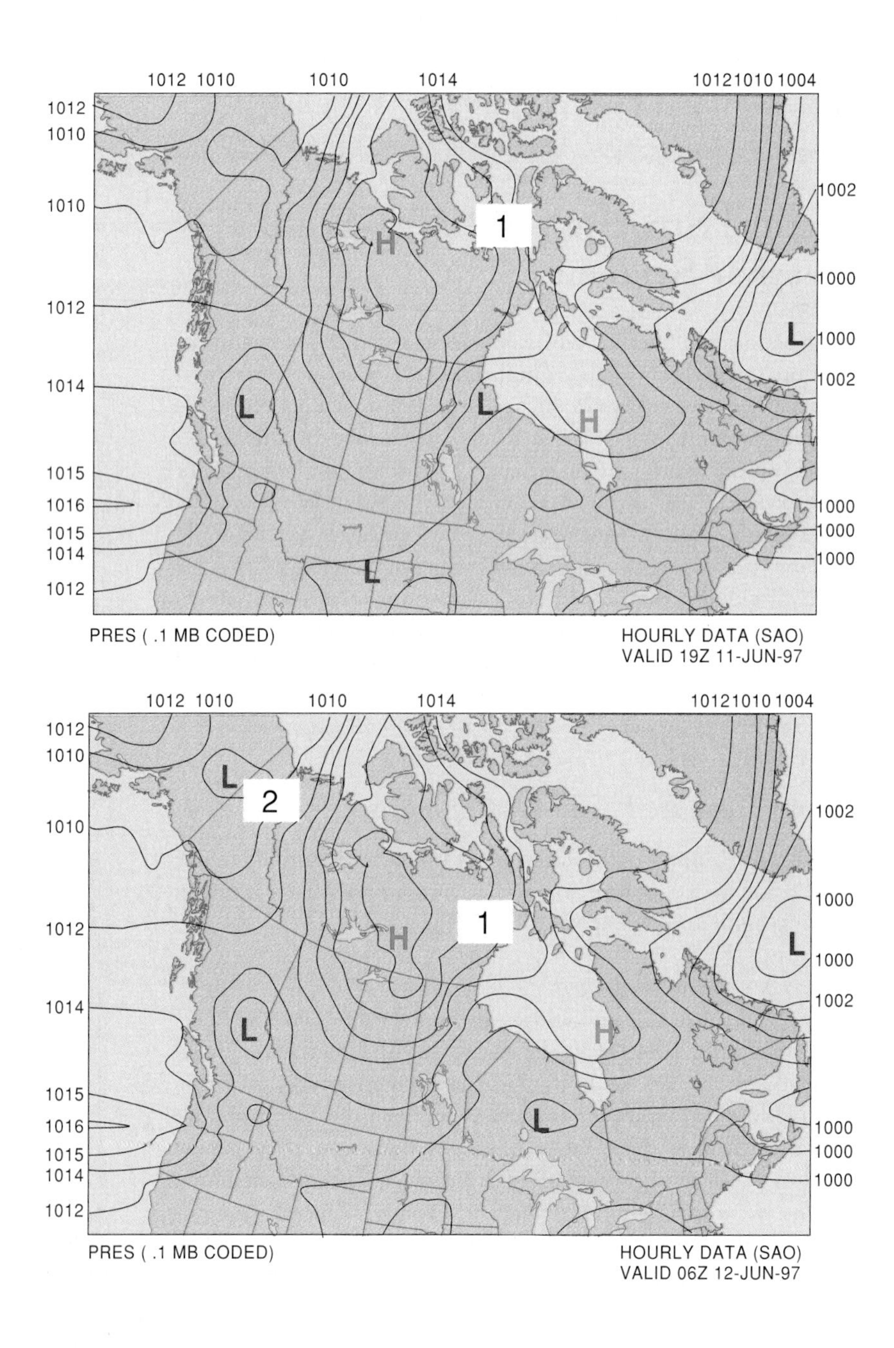


**Figure 8.18** Surface air pressures typically rise in advance of a high pressure cell and fall in its wake.

Our high pressure cell is not an entity like a block of wood or a car, whose motion involves moving all the object's molecules from one place to another. Instead, you might compare the undulating motion of a high pressure area to the "Wave" that spectators make at football games. By raising their arms in sequence, participants cause the wave's crest to ripple around the stadium, while the people causing the wave remain in place. Much as the "Wave" ripples through the spectators, the high pressure center ripples through the atmosphere. Of course, the air does not stay in place, like the spectators; but it is the circulation of air around and through the high, and not the transport of the high pressure cell itself, that moves the bulk of the air from place to place.

With these thoughts in mind, we are ready to give at least a partial answer to the question of what caused our Arctic air mass to move south. Movement is in the direction in which there are the greatest pressure rises, which are determined by maximum amounts of air convergence at upper levels. Thus, the surface high moves under regions of maximum convergence.

The key to observing the motion of pressure systems, then, is not in watching the winds but in watching for changes in pressure. For a high pressure area to move, pressures must rise and fall in new places. If our Arctic high pressure area moves south, the pressure must rise at locations south of the original high pressure center. At the same time, the pressure must fall in the region where the center formerly existed. See Figure 8.18 for an example. Using pressure changes thus to predict a high's movement assumes that the high pressure area is not changing strength. In a rapidly strengthening high pressure cell, the pressure may rise on all sides, even as it moves. However, the pressure will rise the *most* in the direction toward which the center is traveling. Conversely, high pressure areas may weaken, in some cases so quickly that the pressure even falls in areas toward which the system seems to be moving. Remember, the pressure rises or falls at any given place according to whether more air is being added or taken away overhead at that location.

This knowledge of pressure changes associated with movement of high pressure areas can lead us to conclusions about what kind of weather to expect around a high pressure area. If the pressure is rising at your location, it is because more air is being added at higher levels of the atmosphere than is being carried away below, allowing us to conclude that the net vertical movement of air overhead is downward, resulting in stable air and decreasing humidity associated with subsidence. This is why the center of clear weather is often found in advance of an approaching high and not centered on the region of highest pressure. Once the high pressure center passes and the pressure begins to fall, a different process is at work. Now, more air is being removed than is being replaced above any given spot. The vertical air motion here must be predominantly upward, resulting typically in cloud formation and, later, perhaps precipitation. The result is that usually a larger area of clear sky prevails in advance of a high pressure area than behind it (see Figure 8.19). A similar relationship holds for low pressure areas as well. This distribution of vertical motions and its impact on cloudiness and

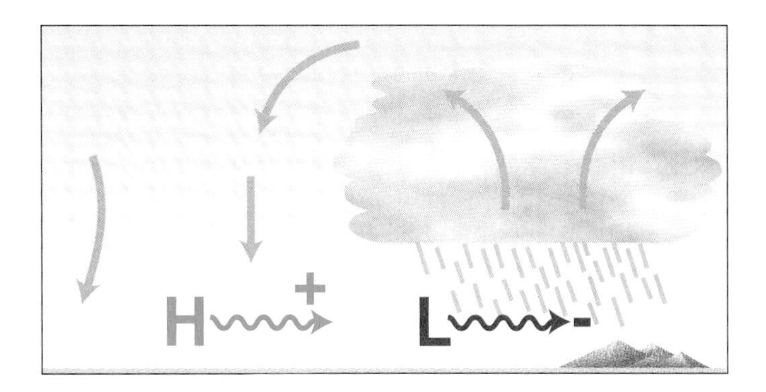


Figure 8.19 Lows move (re-form) toward regions of greatest pressure falls which are areas of upper-level divergence, vertical motions, clouds, and precipitation. Thus, the stormiest weather often occurs in advance of lows. By similar reasoning, explain why fair weather "leads" highs.

precipitation patterns is an important refinement of the frontal cyclone model we developed in Chapter 1.

Now, picture our mass of clockwise-turning, surface-diverging Arctic high pressure moving south, with the highest barometric pressure at the core of the air mass. How does this circulation interact with surrounding air masses? Imagine you are in Iowa, in the path of the high. So far, none of the cold air has reached you. In fact, the last few days have been fairly mild with a patchwork of clouds and clear sky overhead. If you venture north, you will eventually reach a place where the cold air mass is just arriving. From that point on, it grows ever colder as you leave the warmer air to the south and move toward the center of the Arctic high to the north. What should we call the front edge of this change to colder weather? How about cold front? A cold front is defined as the leading edge of a change to colder weather. The air mass on the warm side of the front is usually relatively uniform; all the change to colder conditions occurs after the cold front passes. In Chapter 9, we will consider fronts in more detail.

## Pressure, Temperature, and Thickness Relationships

Examine the two charts in Figure 8.20, both of which represent conditions near the top of the troposphere, at the 300-millibar level. What correlation do you notice between 300-millibar heights (in Figure 8.20A) and temperatures (in Figure 8.20B)?

No doubt you recognized the similarity of the two patterns: the height (or pressure) pattern resembles the temperature pattern, with high temperatures occurring where heights (pressures) are highest and low temperatures in regions of low pressure. This correspondence is not at all peculiar to the specific weather situation shown in Figure 8.20. Rather, it is a regular occurrence on upper-level weather charts, particularly in middle latitudes. Let's see how this correspondence comes about.

### *Layer Thickness, Temperature, and Wind*

Figure 8.21 shows a cross-sectional view of the atmosphere. Note that the horizontal axis represents position along an east-west line running from Los Angeles (LAX) to Savanna, Georgia

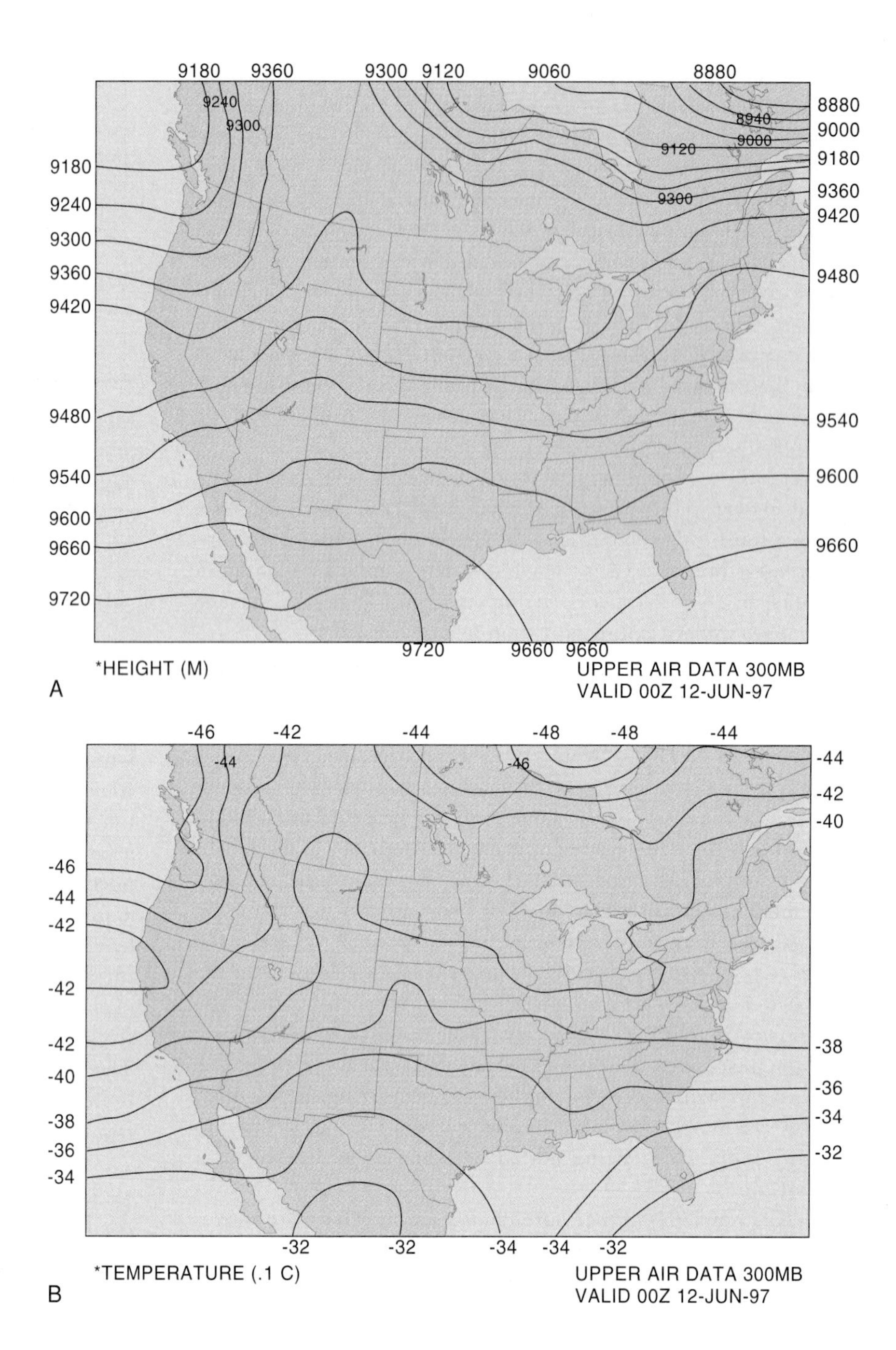


**Figure 8.20** Three-hundred-millibar patterns of heights (*A*) and temperature (*B*). What resemblance do you see?

(SVN). The vertical coordinate is altitude. Notice also that the vertical scale is greatly exaggerated compared to the horizontal.

Observe that the 1000-millibar isobar is lower (i.e., it lies closer to sea level) between Dallas (DAL) and Jackson, Mississippi, (JAN) than elsewhere along the cross section. Therefore, sea-level pressure is lower in the DAL-JAN area than elsewhere. (Why? Recall the discussion relating heights and pressures on p. 223.) The specific pattern is of no concern here; we simply chose a typical one. We also chose a certain (but not typical) temperature gradient: notice that at any given altitude, the temperature increases steadily from west to east. Under these conditions, what will the pressure pattern at upper levels look like?

To answer this question, recall once more that the gas law, P = C $\rho$ T, requires that at constant pressure, a colder air parcel will be denser than a warmer one. Therefore, the surface air in Figure 8.21 is progressively less dense as you travel from west to east. Now, the greater density of the western air causes the pressure to decrease faster in the vertical there than it does in the east, where the air is less dense and molecules of air are spread farther apart from each other. Thus, the thickness of the layer between 1000 millibars and some upper pressure level, 700 millibars, say, is greater in the warm air to the east than in the cold, dense western air. Notice also that the profile of the 700-millibar isobar shows little or no evidence of the surface (i.e., 1000-millibar) pressure pattern; the isobar's altitude varies mainly because of thickness differences in the 1000-700-millibar layer.

This relation between an air layer's thickness and its temperature is not specific to the 1000-700-millibar layer nor to

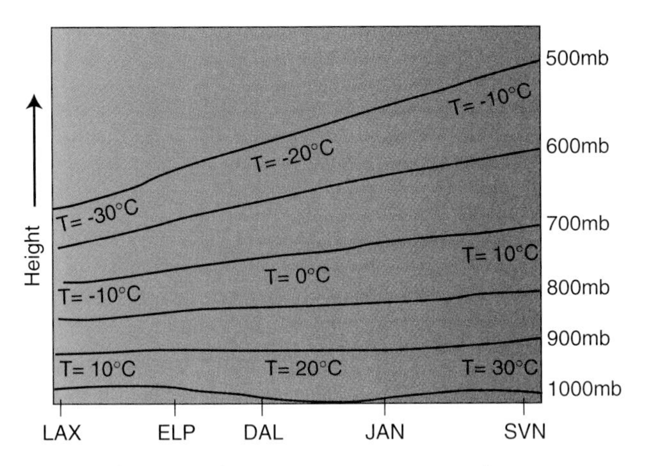


Figure 8.21 Relations between temperature and pressure patterns, in cross section. Where is the 800–780-millibar thickness greatest? Why? Where is the isobaric slope greatest? Why?

near-surface conditions. As long as the air's composition is constant, the relation is completely general and can be stated: The atmosphere's thickness between any two pressure levels varies with the mean temperature of the layer, being greater where the temperature is higher. This "thickness rule," as we will refer to it, is an enormously useful one, one you will employ often.

You can see from the trends in Figure 8.21 that as long as the horizontal temperature patterns remain the same (colder in the west than in the east, in this case), the pressure surfaces will tilt ever more steeply with altitude. It is common in the atmosphere for a given temperature pattern to persist through the entire troposphere. Therefore, the uppermost tropospheric pressure surfaces often have the steepest slopes.

Now we have an explanation for the correlation you saw back in Figure 8.20: higher tropospheric temperatures cause greater atmospheric thickness between lower and upper levels and hence greater upper-level heights. Where it is colder, the height of the upper-level surface is less.

We are not quite finished with Figure 8.21. You can make important inferences from it regarding the wind. Recall that the geostrophic wind is determined locally by the horizontal pressure gradient, that is, by the slope and orientation of the isobars. Applying this rule, you see that the geostrophic wind resulting from the isobars shown in the cross section must be directed into the cross section at nearly all points; it must be a south wind. The only exception is at lowest levels from El Paso (ELP) to somewhat east of Dallas (DAL), where the 1000-millibar isobar's slope is downward toward the east, that is, opposite from everywhere else. Thus, in this region, there is a component of the geostrophic wind directed out of the cross section: a north wind.

We can also reason that, since the strength of the geostrophic wind at any given location is proportional to the horizontal pressure gradient (isobars' slope), the geostrophic wind speed in Figure 8.21 must increase with height, being greatest where the slope is the greatest. This conclusion is consistent with the fact that mid-latitude wind speeds generally are observed to increase upward, reaching greatest speed near the tropopause. Further, we have uncovered the reason prevailing winds blow from west to east in the middle latitudes: it is colder in high latitudes than in the tropics, which causes high latitudes to be regions of low pressure resulting in cyclonic westerly flow aloft.

### *Thickness, Fronts, and Jet Streams*

TV meteorologists often show pictures of the **jet stream,** a narrow zone of strong winds (see Figure 8.22) that meanders across the country at upper tropospheric levels. Wind speeds in a strong jet stream can exceed 200 knots. Notice in the figure that it is warmer on the right side of the jet stream than on the left (as seen from the direction of jet stream travel); note also that a front is present. Is this configuration of jet stream, temperature contrast, and frontal position typical, or is it merely a coincidence? To find out, we will apply what we know about temperatures, thickness, and wind speeds to this problem.

Figure 8.23 illustrates conditions in a vertical cross section from Bismarck, North Dakota (BIS), through North Platte, Nebraska (LBF), and Oklahoma City (OKC) to Brownsville, Texas (BRO). Notice the location of these cities with respect to the position of the front and the jet stream. Thus, Bismarck lies deep within the cold air, north of both the jet stream and the surface frontal position. North Platte lies between the jet and the surface front's position. Oklahoma City and Brownsville lie in warm air south of both the front and the jet.

Observe that the pressure decreases rapidly with height in the cold, dense air north of the front, while in the warmer air to the south, isobars are spread farther apart in the vertical. Notice also that within each air mass, the spacing between any two isobars is nearly constant, an indication that temperatures within one air mass are fairly uniform horizontally. On the other hand, sharp temperature contrasts exist in the frontal mixing zone between the two air masses. The thickness rule dictates that these contrasts must lead to sharp changes in isobar heights and

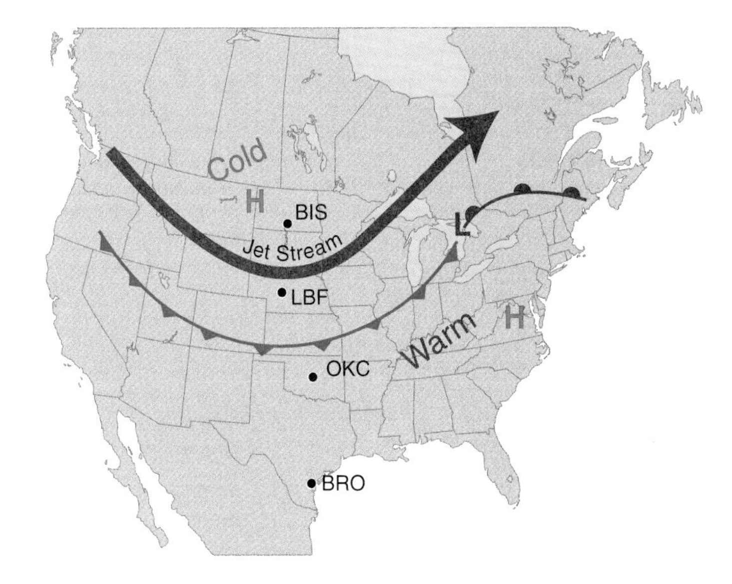


Figure 8.22 A typical configuration of jet stream and surface weather features. Note the position of the jet stream compared to the fronts.

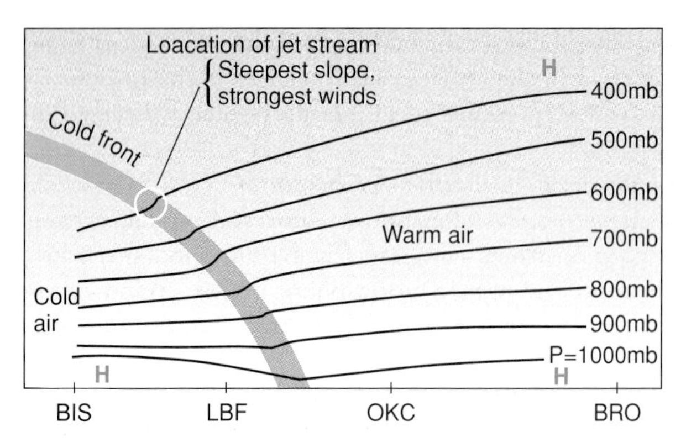


**Figure 8.23** The jet stream is located in the frontal temperature contrast zone, in the upper troposphere, north of the surface front.

thus to strong pressure gradients (steepest isobar slopes). Indeed, if you follow any one isobar horizontally across the chart, you see it undergo a sudden change in slope in the contrast zone from one air mass to the other. This temperature contrast zone is the narrow mixing zone associated with the front.

Thus, because the front is a zone of concentrated horizontal temperature contrasts, it is also a zone of great horizontal thickness change and hence large pressure gradients, which result in the narrow band of jet stream winds. Furthermore, notice that the cumulative effect of the thickness variations in the vertical causes the greatest pressure gradients (and thus the strongest winds) to occur in the upper troposphere, rather than at lower levels. In conclusion, we see that the configuration of features in Figure 8.22 is not coincidental but rather is a natural consequence of frontal temperature contrasts.

The polar front and the jet stream are bound in an interesting feedback loop, as follows. The location and strength of the front determine the strength and location of the jet stream. On the other hand, the jet stream affects the motion and strength of the front by seeming to "steer" it, by transporting warm and cold air into its vicinity and by causing it to intensify or disperse (through processes you will study in Chapters 9). These changes to the front lead to changes in the jet stream, and so on.

Such feedback loops make weather prediction especially difficult. Consider: as the wind continues to transport colder air in behind a cold front, the zone of maximum temperature gradient (change) moves as well. Since the jet stream is located over this zone, the jet stream will shift. Thus, any forecast based on some previous position of the jet stream needs to be continually readjusted as the front and the jet stream both move.

## *Vertical Continuity of Pressure Systems*

Let's return to our Arctic high pressure cell for a moment and see what the thickness rule tells us about the cell's structure aloft. In Figure 8.24A, you see a vertical cross section of the cell, with its cold high pressure bubble at the earth's surface. Now, if we assume that the coldest air aloft is centered over the surface high, the thickness of the 1000-900-millibar layer must be least over the high center and greater on its periphery. As a result, notice the high cell is barely distinguishable at 900 millibars. The weaker pressure gradient causes winds circulating around the high center at this altitude to be weaker also.

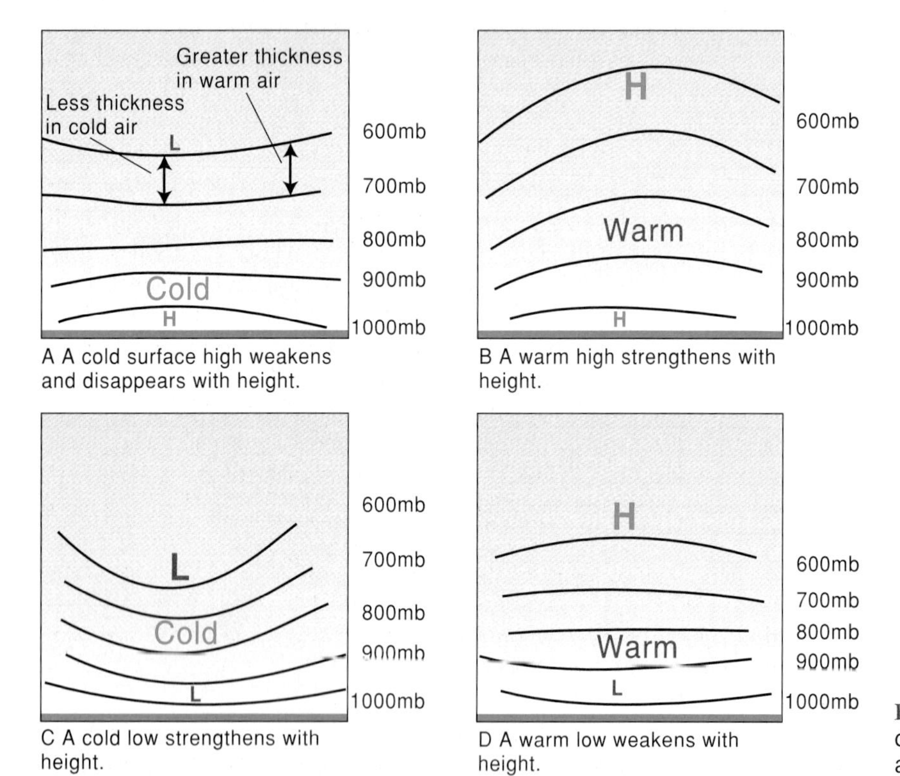


**Figure 8.24** Vertical continuity of warm- and cold-cored highs and lows. Note: Slopes of isobars are exaggerated in these drawings.

Continuing to higher altitudes, you can see the isobars actually curve downward over the surface high cell, indicating lower, not higher, pressures in the cold air, with winds at these levels circulating in a cyclonic, rather than anticyclonic, sense. Thus, cold surface high pressure cells are observed to be shallow weather systems, which weaken or completely disappear at upper levels.

Consider, on the other hand, a surface high pressure cell located in warm air, in Figure 8.24B. The famous "Bermuda High," which controls summer weather over much of the eastern seaboard, is an example of such a system. Warm temperatures cause greater thicknesses in the air layers over the surface high, thereby strengthening the system with altitude. Thus, the Bermuda High, like all warm highs, has great continuity through upper tropospheric levels.

You can apply the same reasoning to surface low pressure systems, illustrated in Figures 8.24C and D. Note that lows with cold air aloft intensify with altitude, while warm-cored lows (such as hurricanes) are strongest at or near the surface and weaken with height.

Often, cold and warm air masses are not centered on surface high and low pressure systems. Thus, the configurations shown in Figure 8.24 are only a few of many possibilities. What is the vertical structure of weather systems when warm and cold air centers lie to the side of the surface pressure centers? We will explore this important question next by studying a typical frontal cyclone.

### *Tilting of Pressure Systems*

In Figure 8.25A, highs are centered in Idaho and along the New England coast, and the center of a frontal cyclone lies over Ontario, Canada. From what you know about fronts, see if you agree with the cold frontal location in the central Plains. Remember that a cold front marks the leading edge of a change to colder weather. Since a front is a boundary between air masses, it makes sense that other air mass properties will change across the boundary as well. Indeed, in most cases, we see changes in temperature, moisture, wind direction and speed, as well as air pressure in the vicinity of fronts.

Now examine the upper air flow (in Figure 8.25B) that existed at the same time as the weather shown on the surface map. Notice that this chart shows conditions at the 500-millibar level. Why have we chosen this particular pressure surface? Since 500-millibars is roughly one-half the pressure at the earth's surface, we can conclude that half of the atmosphere's mass is below this level and half is above. Thus, the 500-millibar level often is representative of mid-atmosphere conditions.

Notice in Figure 8.25 that the lowest pressures aloft are located not directly above the surface low but to its northwest. Thus, the low exhibits **tilting** to the west with height. Here, tilting means the sloping with height of a pressure system whose center at upper levels does not lie directly above the surface low center. Why does the low tilt northwestward? Look at the temperatures on the charts in Figure 8.25B. Notice that at both levels, they are colder to the west of the surface low than to the east. Thus, the system is tilting toward the cold air with height because, as you just saw, this is where layer thickness is least.

The tilting of the system in Figure 8.25 is typical of many frontal cyclones. After all, the presence of fronts in these systems implies that there are significant horizontal temperature contrasts from one side of the low to another. Therefore, frontal cyclones typically are not vertical stacked but exhibit tilting with height toward the coldest air, which commonly lies northwest of the surface low.

### *Backing, Veering, and Temperature Advection*

Now, let's compare conditions above three different sites on the charts in Figure 8.25. Look first at Iowa on the surface map. Notice that the temperature at Des Moines (DSM) is 80°F and the dew point is 66°F. The wind is from the southwest. Isobars are oriented from southwest to northeast. Turning next to the 500-millibar chart, you see the height contours (like isobars) over Iowa are also oriented from southwest to northeast. What is the geostrophic wind at these two levels of the atmosphere? Since it blows parallel to the contours with lower pressures to the left, we see the geostrophic wind is from the southwest at both levels. At the ground, the actual wind may not be geostrophic, however, because geostrophic flow occurs only in the absence of friction—a condition that does not exist at ground level.

Second, consider conditions in western Nebraska (WNB), just "behind" (to the west of) the cold front. ("Behind" or "ahead of" a front describes whether a front has or has not passed a given locale. Iowa is ahead of the cold front in this case because the front has not reached there yet. Western Nebraska is behind the cold front because the front has already gone by.) The geostrophic flow at the ground in Nebraska is from the northwest. Note again the actual wind is not geostrophic. Friction causes the wind to angle from higher toward lower pressures at the ground. At 500 millibars, the flow over western Nebraska is from the west.

Finally, look at the situation in central Ontario (ONT) on Figure 8.25. Here, ahead of the warm front, the surface geostrophic wind is blowing from south-southwest to north-northeast. However, the flow at 500 millibars is more westerly.

From looking at just three locations, we have noted three different arrangements of the wind at the ground and aloft. Is there some pattern here, or does this happen at random? To answer this question, turn to the simplified version of the 1000- and 500-millibar flow patterns illustrated in Figure 8.26.

In Nebraska, where you can see the wind shifts from northwest to southwest as you move upward from the ground, the wind is said to be **backing** with increased height. Whenever the successive wind directions describe a counterclockwise trend, we say the wind is backing. In contrast, the wind over Ontario is **veering** with height. Veering is the opposite of backing. When the wind veers, the change in wind direction proceeds in a clockwise manner. Over Iowa, the flow is unidirectional. Unidirectional flow is motion in one direction, neither backing nor veering.

Veering and backing winds can provide the meteorologist with important clues about changes aloft, which you can

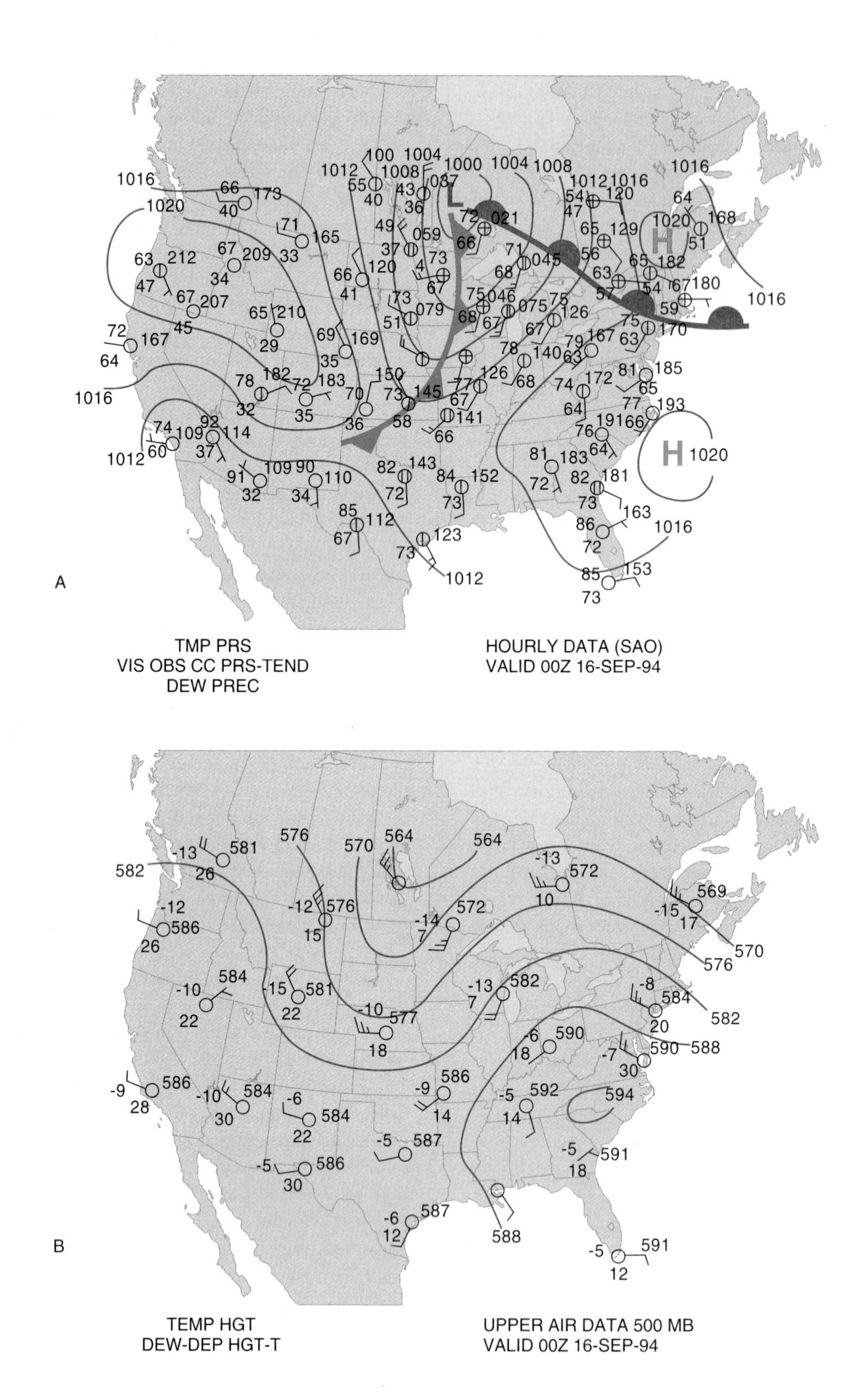


**Figure 8.25** Surface and 500-millibar maps for OZ, September 16, 1994. Note the "tilt" of the pressure systems.

understand with the help of Figure 8.26. The blue lines represent three typical 500-millibar contours from Figure 8.25A, in arbitrary height units. The red lines represent three contours of the 1000-millibar surface, again in arbitrary units. To reduce clutter in Figure 8.26, we show winds as arrowheads on the height contours. The numbers at each intersection show the thickness (in arbitrary units) between the two levels; they are obtained simply by subtracting the value of the lower isobar from the upper one. Thus, at the intersection of the 500 upper contour with the 25 lower contour, the thickness is 500 − 25 = 475.

Notice that the thickness values increase from west to east over Nebraska in Figure 8.26. Therefore, applying the thickness rule, the air must be colder in the west than in the east in the layer up to 500 millibars over Nebraska. Furthermore, since the wind has some component from the west (southwest, west, or northwest) at every level up to 500 millibars over Nebraska, cold air advection must be occurring in this layer.

Notice that we determined the presence of cold air advection in the example above without the benefit of any temperature data! Our result is based solely on the fact that the wind direction backed with height, which implies changes in the pressure patterns, which in turn imply changes in the temperature pattern.

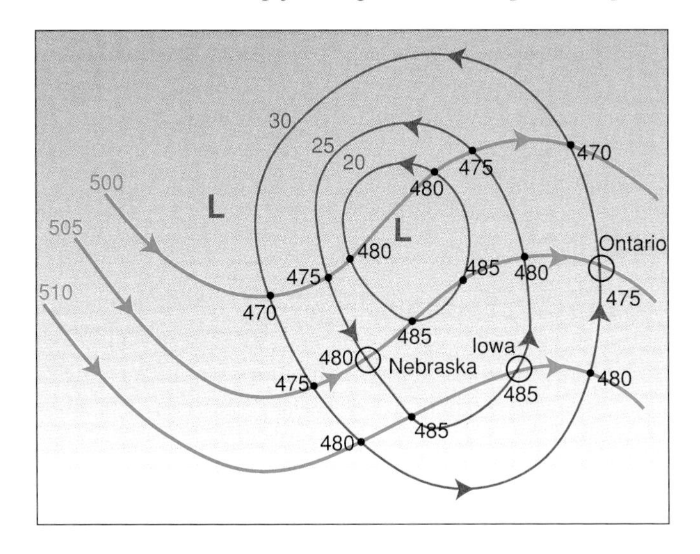


**Figure 8.26** Figure 8.25's weather patterns, combined and simplified. Blue represents 500-millibar contours and red is 1000 contours. Numbers at intersections are 500–1000-millibar thicknesses.

It can be shown that our results are not specific to the example above but can be generalized as follows: In layers through which the wind backs with height, cold air advection is occurring.

Let's apply the same reasoning to the air over Iowa. There, the wind direction is from the southwest at both levels; hence, lower pressure must lie to the northwest at the surface and aloft. Since the wind direction doesn't change at all with elevation, it implies the pressure pattern keeps the same relationship to the left and right at that point. This means the pressure drops off uniformly with elevation to the east and west of Iowa. Isobars at lower and upper levels are parallel, causing thickness values to be parallel to the flow. The result: no advection. Indeed, this is what we expect from conditions in the warm sector of a frontal cyclone. In the warm sector (on the warm side of a cold, stationary, or warm front), the air mass is uniformly warm, according to the frontal cyclone model. Since the wind is bringing in neither colder nor warmer air, we say there is no advection, or the advection is neutral.

Finally, we turn to conditions over Ontario. Notice that here the wind veers with height. You can construct a diagram similar to Figure 8.26 to determine the relation between the veering wind, thickness, and advection in the layer to 500 millibars. As you might surmise, veering, being the opposite of backing, produces opposite results. Thus, in layers through which the wind veers with height, warm air advection is occurring.

We can verify our conclusion by examining the map. A warm front is approaching Ontario, and cold air is retreating to allow warm air to advance. Thus, we see warm advection occurring ahead of a warm front, which is consistent with the frontal model of Chapter 1.

The relations between vertical changes in wind direction and temperature advection are summarized in Table 8.2.

You have seen that changes in wind direction with altitude can give important information about conditions attending highs, lows, and fronts in the free atmosphere. When warming is taking place, the wind veers with height. When colder air is moving in, the flow backs with increasing elevation. And, in the uniform warm sector, the wind direction is basically the same at the ground and aloft. As you examine different weather patterns in the coming chapters, keep these relationships in mind. They help explain a great deal about what the weather is doing at any one place and what might be expected next.

*Table 8.2* **Vertical Changes in Wind Direction and Temperature Advection**

| Situation | Temperature | Wind Change with Height | Advection |
|---|---|---|---|
| Ahead of cold front | Steady | None | Neutral |
| Behind cold front | Falling | Backing | Cold |
| Ahead of warm front | Rising | Veering | Warm |
| Behind warm front | Steady | None | Neutral |

# Checkpoint

*Review*

This section related surface weather features to patterns in the upper atmosphere. According to a simplified circulation model, above the divergent, clockwise winds that circulate around a surface high pressure area are convergent, counterclockwise winds that maintain the high pressure. Conversely, convergent, counterclockwise surface wind flow lies below regions of divergent, clockwise flow, maintaining low surface pressure.

Movement of pressure systems was differentiated from winds that circulate around pressure centers. Surface high and low pressure centers move in response to changes in the location of upper-level divergence and convergence.

# Study Point

## Chapter Summary

We began this chapter, the first of three dealing with how forecasts are made, by considering a cold, high pressure cell in the far north and raising the question, "When and where will this high pressure cell move?" Since forces initiate motion, we briefly explored the nature of forces and Newton's three laws of motion. We next considered a variety of forces that act on the atmosphere, including pressure gradient and Coriolis forces, friction, and gravity. Pressure gradient and Coriolis forces were seen to interact to produce geostrophic balance, a useful approximation in the free atmosphere.

Returning to the motion of the high pressure cell itself, we developed the concept that pressure systems move through the atmosphere not because they are blown around but in response to patterns of divergent and convergent wind flow aloft. Highs move toward regions of upper-level convergence, lows toward upper-level divergence.

We concluded the chapter with an exploration of relations between temperature, pressure, layer thickness and wind flow. Greater atmospheric thickness and high pressure aloft were seen to occur in warm air layers, with opposite conditions prevailing (lesser thickness, low pressure aloft) in cold air layers. This relation explains the tilting of mid-latitude weather systems, as well as the backing and veering of winds with height in regions of cold and warm temperature advection.

With this introduction to atmospheric motions and structure, we are now prepared to deal with air masses, fronts, and cyclones. Thorough familiarity with these is crucial for those wishing to understand how weather forecasts are made, which is the central question of this unit.

## Key Terms

| | |
|---|---|
| Newton's three laws of motion | vector |
| | scalar |
| velocity | acceleration |
| pressure gradient | pressure gradient force |
| hydrostatic balance | Coriolis force |
| geostrophic balance | constant height map |
| geostrophic wind | constant pressure map |
| gradient flow or wind | divergence |
| convergence | jet stream |
| tilting | backing wind |
| veering wind | |

## Review Exercises

1. Which of these are scalar quantities and which are vectors? (a) temperature = 12°C; (b) wind is NW at 6 kts; (c) a skydiver falling at 50 mph; (d) an air pressure of 1020.3 millibars; (e) a pressure gradient of 2 millibars per 100 kilometers toward the northwest.
2. Give three examples of objects accelerating.
3. A racing car streaks due eastward at 120 mph for one minute on a straightaway. During that period, what is the car's speed? Its velocity? Its acceleration?
4. If gravity acts to attract objects toward the earth's surface, explain why all of the earth's atmosphere is not found at the earth's surface.
5. Suppose isobars run in a north-south direction across the state of Missouri, with lower values to the east. Imagine an air parcel in the midst of this pattern. (a) Sketch this situation. (b) Assume the parcel is stationary for the moment. Sketch and label arrows to indicate forces acting on the parcel.
6. What force causes winds to flow not parallel to isobars but across them toward lower pressure?
7. You hear a meteorologist report that winds are northwest at present but that they will be backing overnight. What does this mean?

Using the gas laws, we related the thickness of atmospheric layers, and hence upper-level pressure patterns, to temperature variations. Higher pressures occur aloft above regions of warm air; low pressure aloft is found above cold air zones. It was shown that low pressure areas "tilt" with height toward colder air, while surface high pressure cells tilt with height toward warmer air.

Backing winds are those that turn counterclockwise; veering winds turn clockwise. We concluded this section by showing that winds that back with height are associated with cold air advection, while those that veer are associated with warm air advection.

***Questions***

1. Suppose the air above a surface high pressure cell was not converging but diverging instead. What would happen to the surface air pressure, and why?
2. A low pressure center over Iowa has converging air aloft to the south, and diverging air aloft to the north. In which direction is the surface low likely to move, and why?
3. Why are large low pressure regions found aloft near the poles, with high pressure aloft in the tropics?

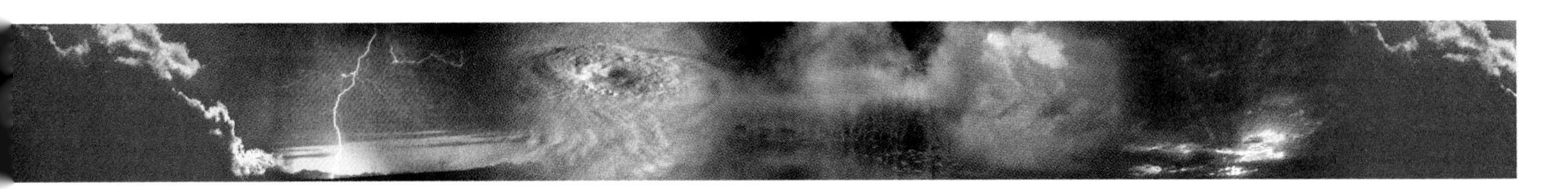

## Problems

1. A car accelerating from a stoplight gains speed at the rate of 5 mph each second for a period of 10 seconds. What will the car's speed be after 10 seconds?
2. Three triplets (same mass, same strength) are taking turns riding on and pushing a swing. First, two kids ride and are pushed by the third. Then, one rides and is pushed by the other two. In which case will the acceleration be greater? How many times greater? Assume the mass of the swing is negligible.
3. You're driving along with a stack of meteorology books in the back seat. As you make a right turn, the stack of books tips over and slides across the seat to the left. Explain why.
4. Explain why, at middle and upper tropospheric levels (above 700 millibars, say), pressures are lower at high latitudes than at the equator. Then, explain why winds in this region are generally from the west.
5. Why do many North American frontal cyclones tilt vertically toward the northwest?

## Explorations

Access a weather service provider such as Accu-Weather or other Internet web site (see suggestions in Resource Links) and try any of the following:

1. Download images of maps for a given time. Study the effects of friction at the surface, 850-, 700-, and 500-millibar levels by comparing plotted wind arrows' directions with the directions of isobars or contour lines.
2. Find examples of backing and veering winds with height in the 850-500-millibar range. Then check 700-millibar temperature changes over a 12-hour period to determine if the regions of backing winds did experience cold air advection, and so on.
3. Examine surface, 850-, 700-, 500-, and 300-millibar maps and follow the continuity of surface low and high pressure systems. Which lows are present at all levels? Which highs? Can you find evidence of tilting?
4. On the Internet, access the latest 300-millibar map for the United States. To what extent does it support the claim that high temperatures aloft occur with high pressure aloft? Cite specific data in your answer. (See Resource Links below for suggested Internet addresses.)

## Resource Links

For a more extensive and continually updated list, see our World Wide Web home page at www.mhhe.com/sciencemath/ geology/danielson .

AccuWeather's home page at http://AccuWeather.com is an extensive source of current weather conditions. It also provides visible and infrared satellite images, and Doppler radar images, and current and forecast maps.

For readers with access to AccuData, HODO plots changes in wind speed and direction with height. UIP plots and analyzes weather data for nine different atmospheric levels. UPR gives an array of upper air data for worldwide locations.

Upper-level charts (500 millibars, 300 millibars, etc.) are available at the University of Michigan's Weathernet at http://cirrus.sprl.umich.edu/wxnet/ , for one, and at AccuWeather.

Surface maps with isobars and surface weather plots are offered at Ohio State's weather page, at http://asp1.sbs.ohio-state.edu/

You can construct a great variety of customized maps at The University of Illinois' Daily Planet website, at http://covis1.atmos.uiuc.edu/covis/visualizer/

Chapter 9

# Air Masses, Fronts, and Frontal Cyclones

# Chapter Goals and Overview

". . . For these six years there is no record. It is during this interval that I am inclined to place a snowless term of years referred to in family reminiscences as the time when snowstorms were supposed to have permanently gone out of fashion and people talked about selling their sleighs."

". . . On all sides of us, as far as the eye can penetrate, there is a desolating expanse of virgin snow. There is not a mountain around us that does not exhibit its towering masses of snow, shining like precious stones in the morning sun, and moulded into fantastic shapes that the wildest fancy can invent. Every road is embargoed. . . . In truth, the entire land from Albany to Buffalo is an ocean of snow. The railroads have lost their occupation."

The opening quotes illustrate people's endless fascination with weather events. The first, on the topic of mild winters, might seem to be recent: it looks like evidence of global warming. Alas, the talk of sleighs betrays the era: the piece was written by Ellen Larned, historian of Windham County, Connecticut, in the 1820s. The second quote matched almost exactly the scene witnessed by residents of Pennsylvania and New York following the "Storm of the Century" in mid-March 1993. However, the account was written about a storm in 1843.

Weather is certainly a topic of conversation. Yet as Charles Dudley Warner observed in the *Hartford Courant* in 1897, "Everybody talks about the weather, but nobody does anything about it." And though our daily bombardment with fresh stories makes it seem like our weather has gone haywire, historic accounts like those above suggest it has always been this way.

In this chapter, you will study the air masses, fronts, and storms of the daily weather map, those weather systems that bring the "oceans of snow" or the "snowless periods" referred to above. Perhaps following these wanderers of the westerlies was what piqued your interest in meteorology in the first place.

## Chapter Goals

The study of air masses, storms, and fronts falls in the discipline known as **synoptic meteorology.** Synoptic meteorology includes the study of weather systems that are on the order of 1500 kilometers across. Thus, they are much smaller than the global circulations that govern world climate but larger than thunderstorms, snow flurries, and individual clouds.

By the end of this chapter, you will be able to:

identify different air mass source regions;
classify air masses on a weather map;
locate and label fronts according to their type on a weather map;
describe typical weather conditions that accompany different types of air masses and fronts;
sketch and label stages in the Norwegian cyclone model;
identify examples of divergence and vorticity in wind flow patterns;
describe how cyclone development is related to patterns of divergence and vorticity, and to other factors; and
find examples of the conditions listed above by looking at actual weather maps.

# Air Masses

When air lingers over a region for a number of days, it acquires characteristics associated with the underlying surface. Thus, if the surface is warm and moist, for example, the air in the region becomes warm and moist also. If the air then moves to a different region, it takes these properties along, gradually changing in response to the new area of residence or passage. We begin this chapter by exploring the nature of the large air masses found around the world, air masses on the order of 1500 kilometers in diameter.

## Air Mass Source Regions

When you consider only synoptic scale air masses, classification is simple. Air masses are designated as cold or warm and as moist or dry. With the help of Figure 9.1, we will identify air mass source regions and the types of air masses these regions produce.

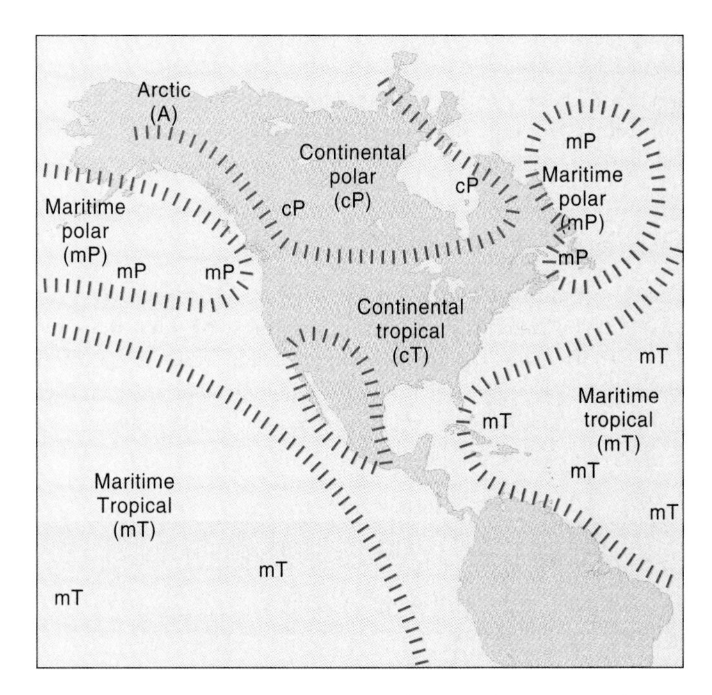


Figure 9.1 Air mass source regions. Boundaries are not sharp but vary with the season and changes in surface conditions.

Cold air masses originate over the polar and near-polar regions, whereas warm air masses form (you guessed it) in the tropics and subtropics. Air masses acquire moisture through evaporation, which you learned in Chapter 5 depends on the temperature of the evaporating surface, as well as on available moisture. Thus, air masses originating over tropical oceans are the moistest, whereas air masses that form over cold, dry land are the driest.

Cold air masses are called **Polar** air masses (abbreviated "P" on some weather charts). Some classification schemes also include **Arctic** abbreviated "A" to represent air masses from the very coldest part of the polar region. North American meteorologists sometimes refer to an inversion of Arctic air as the "Siberian Express," because Siberia is the source region for the Northern Hemisphere's coldest Arctic air masses.

Warm air masses originate in low latitudes and are called **Tropical** air masses (abbreviated "T").

An air mass that originates over a landmass is called **continental** (abbreviated "c"); if it originates over water, it is called **maritime** ("m"). The names of the major air masses are derived from various combinations of the designations polar, tropical, maritime and continental: thus, a warm moist air mass is **maritime tropical (mT)**, and so on, as shown in Table 9.1.

What is the value of identifying and labeling air masses in this manner? Since each major type of air mass has certain temperature and humidity characteristics, knowing the air mass type provides a clue about the kind of weather you can expect. Furthermore, contrasts in conditions between neighboring air masses can produce a great variety of unsettled weather conditions along air mass boundaries (fronts). Yet the masses themselves can bring some of the most settled and serene days we experience. Thus, we are always experiencing either the relatively tranquil weather associated with a single air mass or the more turbulent weather caused where air masses clash. Now, we turn our attention to the individual air masses and how they are represented on the weather map. Later in the chapter, we will focus on the storminess that develops along air mass boundaries.

## Continental Polar Air Masses

Picture conditions in a far northern land in winter. Very little solar energy reaches the ground, and most of what does is reflected. However, the frozen surface emits longwave radiation upward, at a rate greater than that which it receives from the atmosphere. As a result of this longwave radiation imbalance, the surface experiences a net energy loss. The land becomes progressively colder, and the adjacent air chills through contact. Gradually, the coldness spreads through the lower layers of the atmosphere, but because the coldest air is so dense, it does not extend more than a kilometer or two in the vertical. These are the conditions under which continental polar (cP) and Arctic (A) air masses form. The radiosonde plot for Bismarck, North Dakota, in Figure 9.2 shows typical cP air. Note the strong surface inversion and the low temperatures and dry air throughout the sounding. Recall from last chapter that these intense inversions can trap pollutants close to the ground and cause air pollution problems.

When such an air mass moves equatorward from its source region, topography comes into play. In North America, the vast central plains offer little resistance to the advance of cold air. However, the Rocky Mountains pose a formidable barrier to its westward advance. Only rarely is a cold air mass deep enough or persistent enough to mount the mountains and vault through the valleys.

The Appalachian Mountains in eastern North America also offer some resistance to the advance of cold air masses, but since their elevations are almost entirely below 2000 meters (6600 feet), their influence on cP air mass movement is generally of minor significance. While the Appalachians may retard the air mass's very lowest layers, its upper layers may simply spill over the mountains.

In contrast, a smaller and more shallow variety of polar air mass often enters New England from eastern Canada and proceeds southward east of the Appalachians, sometimes as far south as Georgia. Such air masses do not extend as high as the tops of the mountains and thus are trapped between the Appalachians and the Atlantic Ocean.

**Table 9.1 Air Mass Designations**

| Origin | Polar (P) | Tropical (T) |
|---|---|---|
| **Continental (c)** | cP (cold, dry) | cT (hot, dry) |
| **Maritime (m)** | mP (cool, moist) | mT (hot, humid) |

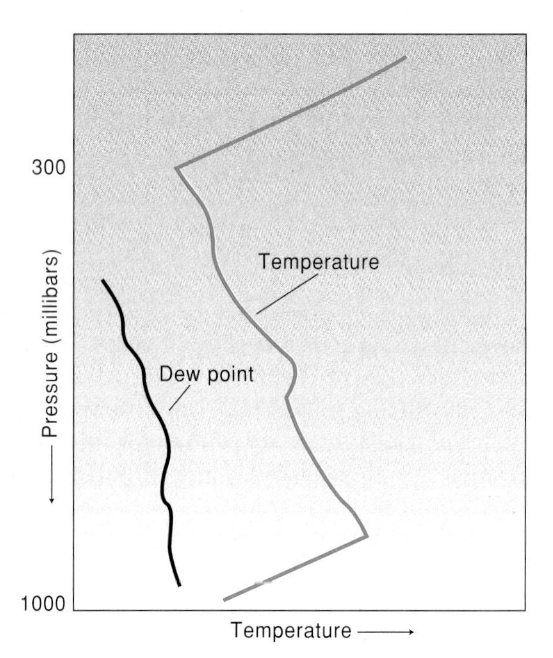


Figure 9.2 Radiosonde data over Bismarck, North Dakota, in cP air. Note the steep surface inversion and dryness of the air.

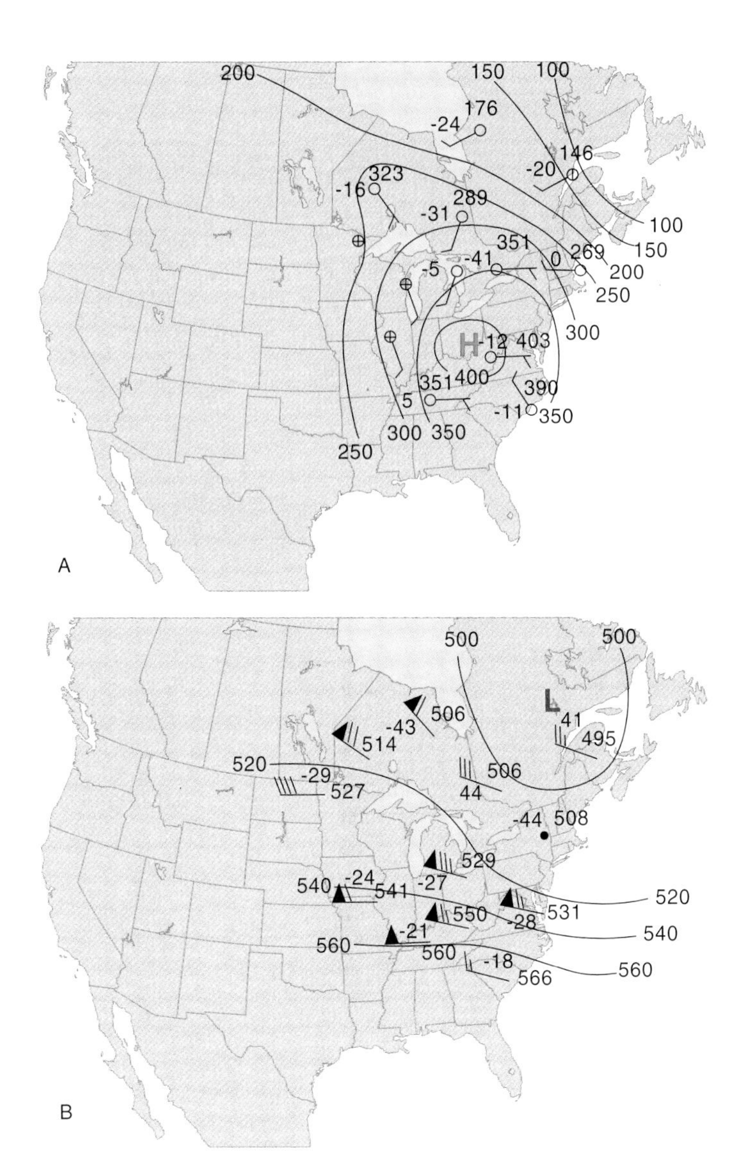


Figure 9.3 (*A*) Continental polar air mass at surface. (*B*) Five-hundred millibar conditions over a cP air Mass. Note the absence of a high center over surface H. Why?

## *Wintertime Conditions*

The surface map in Figure 9.3 shows a typical continental polar (cP) air mass, centered on a cold, dense, low-level high pressure area. Notice, however, that the high is not evident at the 500-millibar level. Since the air is dense, the pressure decreases rapidly with increasing elevation. As a result, the continental polar high is progressively weaker at higher altitudes, eventually disappearing altogether above several kilometers. Since the air in high pressure areas sinks, and sinking produces warming and drying, we would not expect clouds or precipitation in regions under the influence of this kind of air mass.

Many air masses migrate from their source regions and become modified in other locales. For example, many North American cP air masses move southward, where eventually they begin to warm from below as they move to sunnier, less frigid latitudes. Surface heating breaks the temperature inversion, allowing vertical mixing of the air, with the result that pollutants are diluted through a great depth of the atmosphere. What may once have been a smoky-looking cold air mass overlain by pristine, somewhat less frigid air is now a relatively clean and fresh-looking air mass as it continues its journey toward lower latitudes.

Air mass modification is not always such a gradual process. Consider what happens when a cP air mass passes over a source of warmth and moisture, such as the Great Lakes during fall and early winter. The originally cold surface air is heated rapidly from below, and moisture evaporates from the lake surfaces. The stable, dry air mass is rapidly modified into an unstable, moisture-laden snow machine that causes the famous "lake effect" snowstorms in Michigan, Ohio, Pennsylvania, and New York. You will study lake effect snowstorms in more depth in Chapter 11.

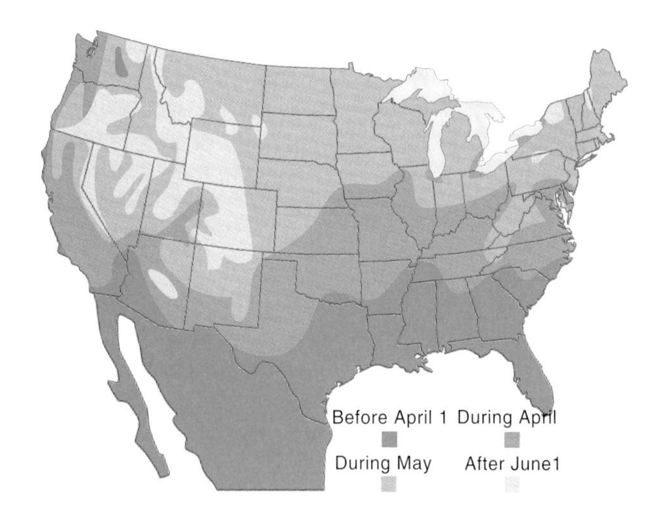


Figure 9.4 Average date of the last occurrence of preezing temperatures in spring. Source: Data from *The Weather Almanac*, 1994, p. 364, Vintage Press.

## *Springtime Conditions*

Early spring finds the high polar regions still locked in midwinter cold and darkness. However, with the vernal equinox comes the return of sunlight to these locales. Once the sun rises there, it doesn't set again until September. Although the polar sun is always low in the sky, the presence of continuous sunlight produces significant warming in the Arctic; the polar spring warmup is much greater than any seasonal changes in the equatorial region.

Although the intensity of the cold in cP air masses wanes with the march of spring, a late-spring intrusion of fresh cP air still is capable of spelling disaster to agricultural interests. Peach and apple blossoms, as well as other vulnerable plants and flowers, can be killed by a late-season frost. Figure 9.4 gives the average date of the last occurrence of freezing temperatures in each region. However, these can be deceiving if the user forgets that "average date" is close to the midpoint of the range of actual last frost dates. Many gardeners wait until their odds are better.

### *Summertime Conditions*

During the Northern Hemisphere summer, only the partially thawed tundra zone of the far north serves as a true cP air mass source region. The presence of prevailing westerly or southwesterly winds at mid-latitudes in summer, plus the absence of strong frontal storms, means that summertime incursions of cP air into middle latitudes are much less common than in winter. But when they do arrive, they bring welcome relief from the stifling, spirit-sapping, will-withering heat and humidity of midsummer. In fact, the drop in dew-point temperature accompanying arrival of a summertime cP air mass is often far greater than the temperature change. Relief is usually fleeting, however, for once the cooler, drier air has left its source region, it is quickly moderated by the high sun and warm underlying surface of mid-latitudes, and eventually, it blends in with the overall lazy, hazy warmth and humidity of a temperate region summer.

### *Autumn Conditions*

During the autumn, winter's foot soldiers become ever bolder as the polar regions chill. Continental polar air masses bring decisive changes to cooler weather. The most intense cooling occurs in the northernmost Arctic, where the sun is now absent. As temperature gradients increase between the frigid polar regions and the still mild mid-latitudes, more intense frontal storms develop; these storms hasten the change toward winter by pulling cold air progressively farther south in their wake.

Fresh cP air masses of autumn are usually associated with clear and dry weather on first arrival. The longer nights promote pronounced radiational cooling near the ground. However, as you read in Chapter 3, such cooling progresses more slowly over water, with its high specific heat capacity and tendency to mix, than on surrounding land. Thus, rivers and lakes remain warm despite the cooler air, and evaporation proceeds vigorously from the warm water surface to the dry, overlying air. Soon, fingers of fog spread through the valleys and expand their reach toward sunrise. The fog may persist well into the morning, because it blocks out the very sunlight that would enhance evaporation through heating.

Air travelers with tight schedules might choose to avoid early morning flights in the fall; the chances for early morning delays caused by fog reach their peak in midautumn. On the other hand, afternoon flights are usually smoother in fall than in summer. Why? Because surface heating by the sun is much reduced compared to summer; therefore, surface-based convection is much feebler.

In the western parts of each time zone, where sunrise is almost an hour later than in eastern parts of the same zone, the combination of early morning fog and darkness encroaching on school bus runs often leads to the same kind of school delays caused by morning snowfall. Fog is a major contributor to accidents, perhaps receiving less than due respect because it is less dramatic than tornadoes, lightning, or floods.

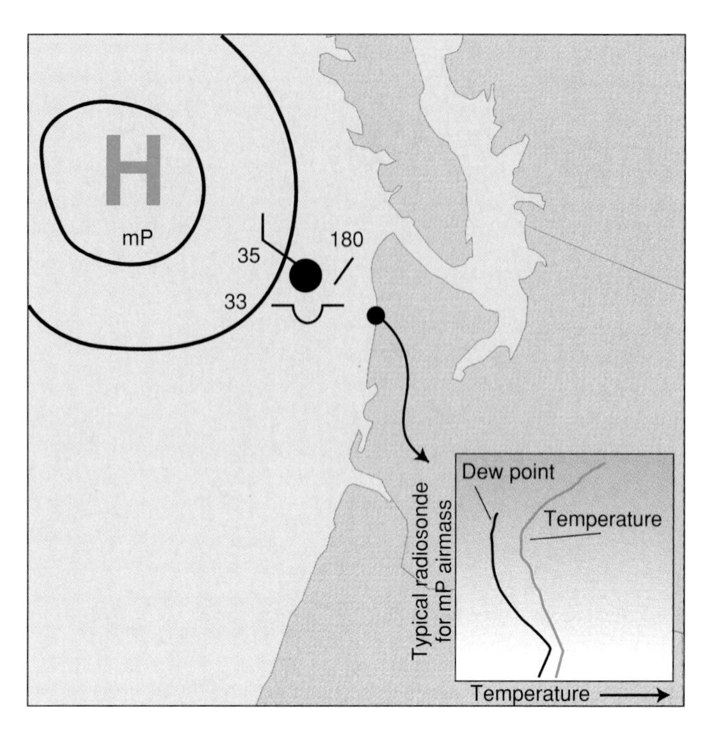


**Figure 9.5** Maritime polar air mass conditions. Note the greater moisture values (both dew points and relative humidity) than in cP air.

## Maritime Polar Air Masses

Maritime polar (mP) air originates over cold ocean waters of high latitudes. From passage over the water surface, it acquires the temperature characteristics of that water. Figure 9.5 depicts a maritime polar air mass. Notice the low-level humidity: dew-point temperatures are close to air temperature in the lowest kilometer, a situation markedly different from that in continental polar air masses. The low-level air is much less stable than in cP air, as well.

Since the oceans change temperature much more slowly than land and never reach the extremes in temperature found on land, temperatures in mP air tend to be more moderate than in cP air masses. With high values of relative humidity and relatively unstable low-level air, low clouds (particularly stratus and stratocumulus) commonly accompany mP air. The combination of cold temperature, high humidity, and heavy cloudiness earns mP air masses a reputation for bringing raw, disagreeable weather.

## Continental Tropical Air Masses

Look at the surface temperature and dew point in Figure 9.6. The combination of high temperatures and large temperature-dew point spreads indicates that we are dealing with a continental tropical (cT) air mass. Continental tropical air masses form when a large air mass resides over dry tropical lands for an extended period. The air becomes hot and dry like the underlying surface.

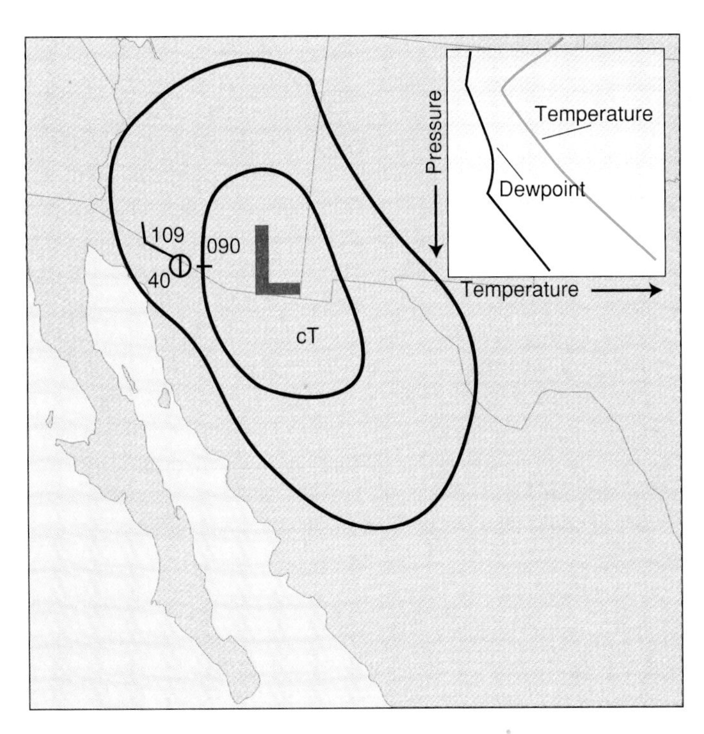


Figure 9.6 Continental tropical air mass conditions. Note the extreme heat and low relative humidity (large temperature-dew point spread).

Notice that the surface pressure system in Figure 9.6 is not a high, as in the case of cP air. Intense heating of the ground and the lower atmosphere in cT air masses causes rising air and low-level convergence, often creating "thermal lows." Despite the convergent, upward air flow in the lower atmosphere, the air generally is too dry for significant cloudiness or precipitation to form. Furthermore, high pressure dominates at upper levels, which usually squelches deep convection.

Where are the source regions for cT air masses? A look back at Figure 9.1 shows there are no truly large tropical land areas in North America. The interior of Mexico and the desert Southwest (in summer), although sufficiently dry and subtropical to serve as cT air mass sources, are somewhat small in area to spawn synoptic scale air masses. Questions arise: How large does an area have to be to qualify as an air mass source region? If an air mass possesses characteristics of a given air mass type but did not originate in a recognized source region, what do we call it? How do we distinguish it from other air masses? In fact, the answer to all of these questions is, "It depends." The air mass classifications we are examining represent scientists' attempts to bring neat order to a range of objects (air mass types) that are not always as distinct as our labels for them suggest.

Perhaps the most extensive source region for cT air masses is the desert lands of northern Africa and the Middle East. Over this large, arid region, surface air temperatures greater than 40°C (104°F) and relative humidities less than 10% are common. The world's highest recorded air temperature, 58°C (136°F), occurred in this region, in Libya.

When hot dry air sweeps from its source region into surrounding areas, the results can be devastating. For example, an

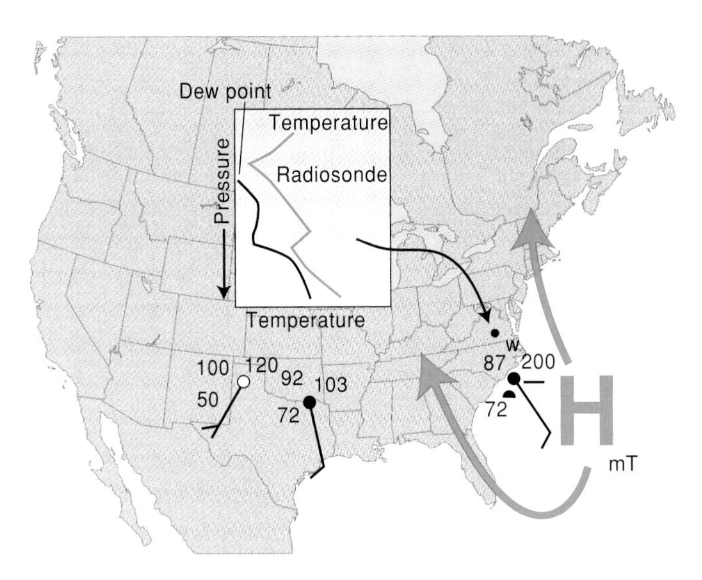


Figure 9.7 Maritime tropical air mass conditions. Notice the slightly lower temperatures but much higher dew points than in cT air.

easterly wind called the Santa Ana sometimes carries cT air from its source region in the desert Southwest westward to the California coast. Coastal California's most devastating fires have occurred when the normally dry undergrowth is ignited and the hot dry winds fan the flames and make them spread like . . . wildfire. You will study Santa Ana winds further in Chapter 11.

## Maritime Tropical Air Masses

Examine the weather conditions in Figure 9.7. Notice the very high dew-point temperatures, a signature of maritime tropical (mT) air masses. Maritime tropical air originates over large, warm bodies of water such as the Gulf of Mexico, the Caribbean Sea, and portions of the Atlantic Ocean. Surface water temperatures in these regions, sometimes as high as 28°C (83°F), lead to high evaporation rates, causing the high dew points. Many people suffer more discomfort in the high-humidity mT air masses than in cT air, which typically has higher temperatures but far lower dew points.

### *Summertime Conditions*

During the warmer part of the year, mT air flows northward through the eastern states with regularity along the path shown in Figure 9.7. Moving over land, an mT air mass may be heated further by contact with the solar-heated ground, producing the withering heat waves that are common in summer.

In the East, mT air is the most common resident among summer air masses. The picture of a hazy blue sky dotted with puffy cumulus clouds adorns advertisements for everything from sunscreen to iced tea. Since surface-based convection is the primary cause of daytime summer cloudiness, the cumulus cloud is the predominant cloud form.

In the nation's midsection, the summertime balance between mT air and cT air from the desert Southwest plays a crucial role in the success of crops. Note in Figure 9.7 that dew points vary greatly with wind direction and thus with air mass type. When mT air covers the region, summer rain, in the form of showers and thundershowers, can be plentiful. In fact, sometimes an easterly flow draws mT air west into Arizona, causing dramatic thunderstorms that make the desert bloom and produce sudden flash floods through the washes.

### *Wintertime Conditions*

During winter, the impact of mT air on the continental United States is limited. While the warmth and moisture of mT air are vital fuel for the development of the winter storms common in the middle latitudes, it often comes and goes quickly. The warm air, with its low density, is easily displaced by denser, polar air masses. When mT air does spread northward through the United States in winter, it is often a contributor to the development of fog—if the wind flow is sufficiently weak. Why? The mT air begins its drift northward as a warm and humid air mass. Moving northward away from its source region, it reaches regions where surface temperatures are progressively colder. Conductive and convective heat losses from the air to the underlying surface cause the air temperature to decrease. Radiational cooling overnight also contributes to cooling the surface air. During the long nights, the air near the ground can easily cool to the saturation point, with the result that fog forms. The fog then reflects the next morning's sunshine, thus shielding the surface air from an energy source that might have warmed it. Much of the mild and muddy, moist and murky weather in winter can be traced in some way to the intrusion of mT air masses.

## Checkpoint

***Review***

This section explained that air masses acquire the moisture and temperature properties of the underlying surface of the area over which they form. There are four major types of air masses whose names are derived from where each originates: continental polar, continental tropical, maritime polar, and maritime tropical.

As an air mass moves, its original properties are modified as it passes over land and water surfaces that have different properties from the region in which the air mass was formed. Knowing the original properties of an air mass and its movement can be an important aid to the weather forecaster. Often, however, the same type of air mass can result in different weather conditions, depending on the season of the year. Thus, the forecaster must have a detailed knowledge of air mass variations by season and region.

*Questions*

1. On the North American map on page 244, indicate the locations of where each type of air mass (continental polar, continental tropical, maritime polar, maritime tropical) is likely to develop.

2. Air masses are distinguished by their moisture and temperature properties. Rank order the four air masses studied in this section from least humid to most humid (in terms of surface dew-point temperatures) and from coldest to warmest. Explain your rankings.

| | Moisture | | Temperature |
|---|---|---|---|
| Least humid | 1. | Coldest | 1. |
| | 2. | | 2. |
| | 3. | | 3. |
| Most humid | 4. | Warmest | 4. |

3. Using the properties of continental polar air masses, explain why residents in the interior of Alaska sometimes have to cope with episodes of smoke pollution.

# Fronts

In Chapter 1 and again in Chapter 8, you studied fronts. Now, it is time to look at these important features in greater detail. We'll see their relationship to and role in synoptic scale cyclones, the "lows" that frequent the middle latitudes.

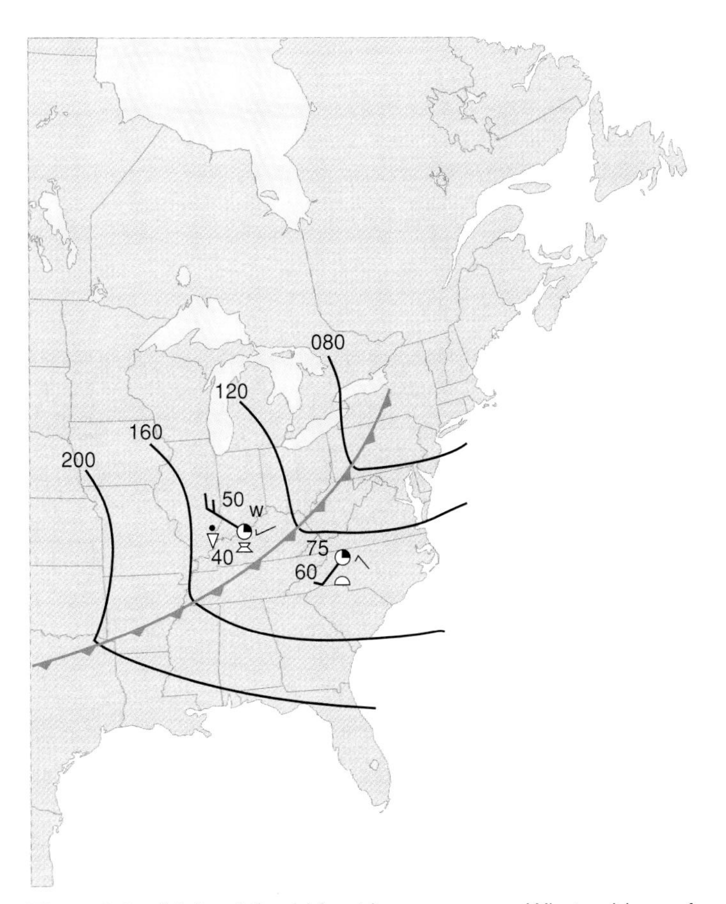


Figure 9.8 A "classic" cold front in many ways. What evidence for the front can you point to?

## What Is a Front?

To review quickly, a front is a boundary between two air masses. The term may have had its origins as an analogy to military operations: the armies in battle encountered each other at the front. A front's movement, if any, is governed by the movement of the air on the colder side of the front. If cold air advances, the front is termed a cold front; if cold air is retreating, it is called a warm front; and if the cold air is neither advancing or retreating, it is a stationary front. A fourth type of front is called an occlusion. An occlusion occurs when one of the other types of front is overtaken by a cold or warm front.

Some fronts, like the one in Figure 9.8, are easy to find on the weather map. They are marked by spectacular changes in temperature and weather. At other times, finding frontal locations is an exercise in frustration. Even a group of trained meteorologists will argue about where a front is located and how it is oriented. In this book, we tend to deal only with textbook cases because this is, of course, a textbook. You will learn to identify these exemplary fronts by referring to a standard frontal model. After you have had some experience finding fronts, you might want to try locating them in more difficult situations. It is important to realize that we study fronts because they are tools that help us explain how the atmosphere is behaving in a certain region. In the real atmosphere, air masses do not always remain distinct, so the concept of a front is an idealization, a model, like many others in atmospheric science, that is not literally true but is nonetheless useful.

Properties of moisture, pressure, and wind all help in the identification and location of fronts. It may seem odd that we use properties other than temperature to locate fronts, since fronts essentially delineate locales of sharp temperature contrast. In fact, fronts are much more than just temperature change zones. Considering all the variables mentioned above, one can construct a more complete and accurate picture of a front's location and structure than would be possible from the temperature pattern alone.

## Cold Fronts

Think of a cold air mass building in the polar night. If this intensely cold air begins moving south, it will encounter air that is less cold. The cold front's position is defined as the leading edge of the change to colder weather. Until the front reaches a given location, it will have little or no effect on the temperature at that place. However, once the front's leading edge passes, it will begin to turn colder. Depending on the contrast between the two air masses and the speed of the cold air's advance, the temperature fall might be barely perceptible or as much as 10°C (18°F) or more in one hour.

Examine the atmospheric pressure pattern surrounding the front in Figure 9.8. The fact that a cold front is found in a low pressure trough accounts for the change in wind direction at the front. This difference in wind directions has great significance: it means that the differing air trajectories draw in air from different air mass source regions, depending on which side of the front you are on. For example, if air on one side of a front came from the Gulf of Mexico and the air on the other side originated in Arctic Canada, it is easy to imagine how different they would be in terms of wind direction, moisture, stability, and so forth. Thus, the front actually is a zone of contrast between air masses whose properties differ in many ways besides temperature. Since these properties often are related, we are given an important tool for analyzing and identifying of fronts. Sometimes one variable or another has relatively weak contrast across a front. One of the other related parameters may then be used to determine the front's location. We can use variables other than temperature to identify and locate fronts because of the strong interrelations among temperature, wind direction, humidity, and so on within a given air mass.

The advancing, subsiding cold air generally makes quickest progress at ground level, where it spreads out beneath the retreating warmer and therefore less dense air. Thus, we would expect the cold frontal surface to slope back toward the core of the cold air mass with increasing elevation, a feature you can observe in Figure 9.9. Note also in the figure that the frontal surface is steepest near the ground. This curvature is due to friction, which exerts a drag on the advancing cold air and causes the cold air mass to bulge forward just above ground level.

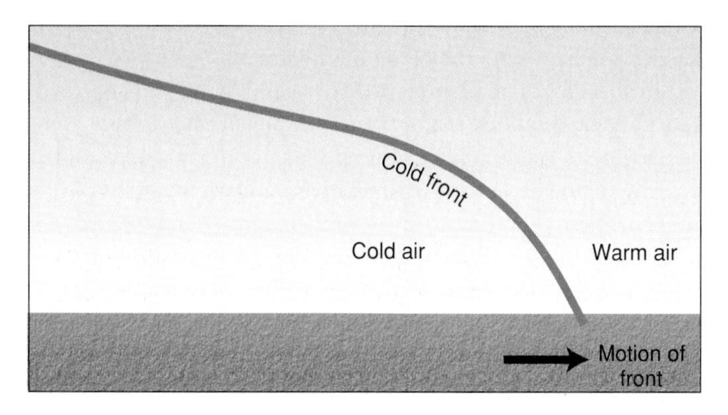


Figure 9.9 A cold frontal surface is steepest at the ground, shallower aloft.

If you carefully observe your own weather during the hours preceding the passage of a cold front, you might notice that some changes occur not at the time of frontal passage but in advance. What is the explanation for this surprising situation? An analogy may be useful: if you push a snow shovel along a driveway after a light snowfall, you can observe that the snow pushed by the shovel pushes snow in front of it, which can result in an area of moving snow several feet in advance of the shovel itself. Similarly, as a cold front advances, the warm air it encounters pushes on other warm air lying farther downstream from the front. The warm air may begin rising and moving out of the way 60–120 kilometers in advance of the front. If there is sufficient moisture in the warm air, only a limited amount of lift will be required to initiate the growth of cumulus clouds. If this development continues, a line of thunderstorms can develop well in advance of the cold front. Such thunderstorm lines are often called squall lines, and they are frequent visitors to the central and eastern United States during the warmer months of the year. You will read more about squall lines in Chapter 12.

An interesting problem for meteorologists arises when a prefrontal squall line produces rain on a summer day. Evaporational cooling causes the temperature to fall with the thunderstorm's passage. Now, you may wonder if that advance drop in temperature should be called the cold front. After all, a front is the leading edge of a change to colder weather. In fact, meteorologists studying thunderstorms and squall lines often observe these small-scale fronts and identify them as gust fronts or mesoscale cold fronts. Figure 9.10 shows such a feature on the Doppler radar. The line of thunderstorms extends in a north-northeast, south-southwest line. There is a brief decrease in surface temperatures in the rain area but a gradual recovery after the squall line goes by. Although on a surface weather chart the temperature contrasts associated with the feature might suggest

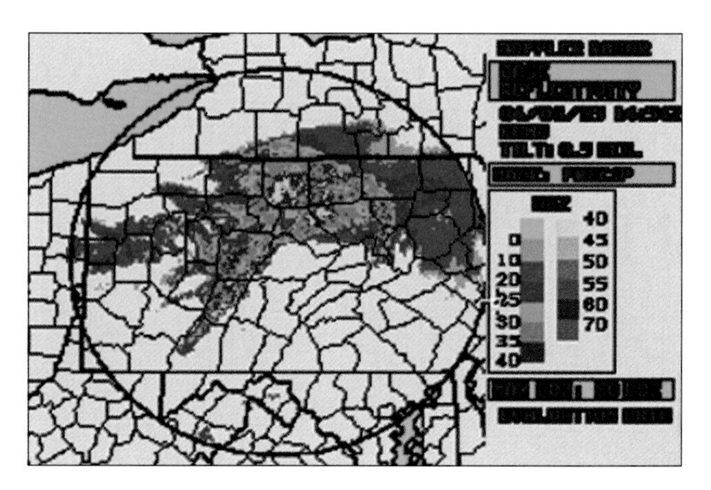


Figure 9.10 Doppler radar image of showers and thunderstorms developing just south of a front lying across Pennsylvania. These shower areas complicate temperature patterns, making it difficult to locate frontal positions.

the feature is a cold front, it is not. It does not represent a contrast between air masses but rather a local modification; further, it does not possess the vertical structure typical of a cold front, as shown in Figure 9.9, and it lacks connection to other fronts and low pressure systems.

### *Backdoor Cold Fronts*

Normally in the middle latitudes, weather systems move from west to east. Cold fronts are no exception: most cold fronts will arrive at a given place from some westerly direction (usually west or northwest). Occasionally, however, a cold front arrives from the opposite direction—the east or northeast—in which case it is called a **backdoor cold front**.

What causes a backdoor cold front? Study the example in Figure 9.11. Note that east of the Appalachian Mountains lies the piedmont, the coastal plain, and then the Atlantic Ocean. From March through late May, the ocean off the northeast coast is much slower to warm than the adjacent land. A shallow maritime polar air mass moving southward from Canada may make limited progress west of the Appalachians. However, east of the mountains, it is sustained by the influence of the chilly Atlantic. The leading edge of the change to colder weather proceeds south and southwest from New England, eventually backing up against the mountains. Since the cold air is shallow, the mountains act as a dam, restricting its further advance; we call this situation **cold air damming.**

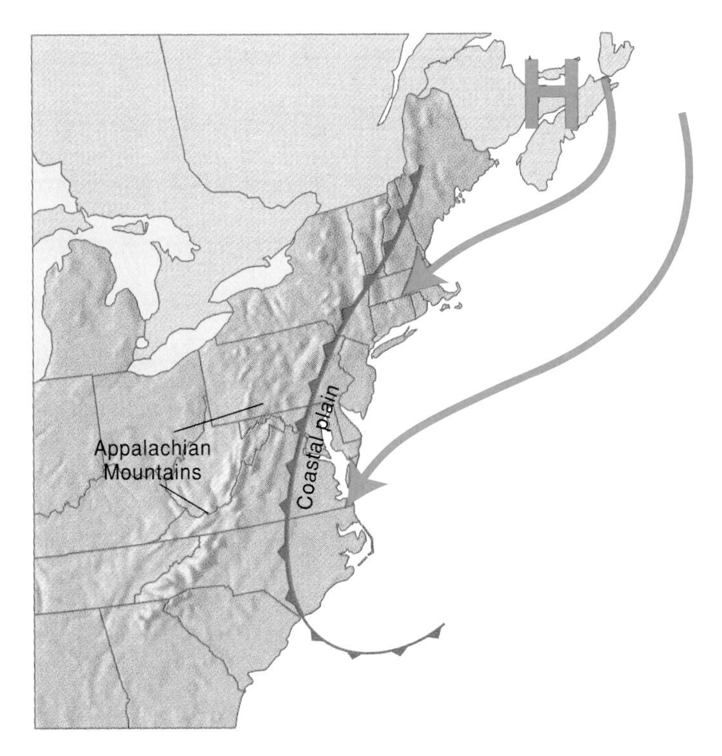


Figure 9.11 Backdoor cold fronts come from the east. What air mass type do they usher in?

Once the chilly air is in place over the costal plain, it is difficult to dislodge. Warm air approaching from west of the Appalachians is less dense than the shallow chilly air mass to the east and may have little success in pushing the mP air offshore. With continued east-to-northeast winds from the Atlantic at ground level, the chilly air mass may grow increasingly moist due to ongoing evaporation from the Atlantic. Clouds and drizzle often result, leading to days of dreary, damp, and dim weather in cities from Boston to Raleigh. Such episodes are a chief source of embarrassment to East Coast meteorologists, because backdoor fronts can take them by surprise.

Often, the days before a backdoor front makes its move are exceptionally warm. The weather feels like early summer, even though it is spring. But heating over the land makes the air less and less dense. Meanwhile, over the ocean, it is still cold. If the winds shift to northeast, the cold mP air begins to move inland, and in a matter of hours, the weather over thousands of square miles can turn from sunny and mild to drizzly and damp. Residents of the East Coast are all too familiar with these fronts as would-be sunny warm days suddenly turn intractably sullen. Computer models are becoming increasingly skillful in handling these fronts but still are far from perfect.

### *Cold Front or Dry Line?*

Examine the contrasts in temperature and in dew points across the dashed line in Figure 9.12. As you can see, temperatures vary little from one side to the other, whereas dew points are sharply lower to the west of the line. Such features are called **dry lines**. The dry line in Figure 9.12 marks the boundary between cT air from the desert Southwest and an mT air mass from the Gulf of Mexico. A low pressure center traveling eastward from Colorado may draw the dry line eastward on its southern flank, triggering some of the most spectacular thunderstorms and tornado situations in the United States.

Is the dry line some special feature, or is it a front? If you compare temperatures on both sides of the front, it is hard to justify calling it a cold front. In fact, some locations west of the dry line are actually warmer than others to the east.

Recall, however, that atmospheric water vapor represents a potential heat source, through release of its latent heat. If this energy were to be released, the moist air mass would be considerably warmer than the dry air mass. Thus, the moist air mass "potentially" is the warmer. But should we call the dry line a cold front on the basis of "potential" warming? This question is debated by meteorologists, and you will see the dry line portrayed in various ways (as a cold front or not) on weather maps.

Dry lines are not the only situations in which temperature contrasts do not follow the normal frontal model. Clouds and showers in the warm sector ahead of a summer cold front may prevent sunlight from reaching the earth's surface, with the result that daytime surface temperatures warm up little from their overnight values. Once the cold front passes and a drier, cloud-free air mass settles in, sunlight may reach the ground,

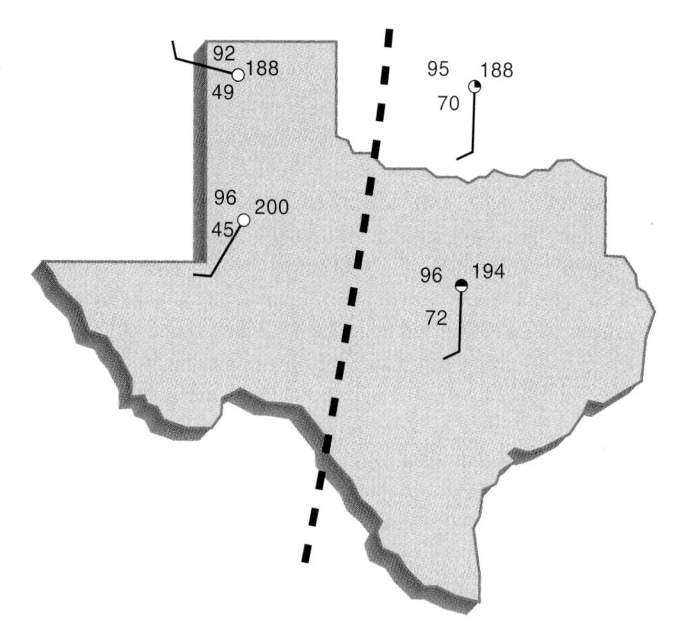


**Figure 9.12** The dry line. In what ways do the data here suggest a cold front? In what ways do they not?

sending surface temperatures to higher values than before the cold front arrived. Thus, the cold frontal passage apparently has ushered in warmer, not colder, temperatures. However, if we consider the potential temperature rise due latent heat release in each air mass, we find the dry air behind the cold front is indeed potentially cooler. Meteorologists use this as justification for drawing cold fronts on their maps even when the temperature pattern alone would not support such a choice. In support of their decision, they might point out that such fronts behave in other respects like cold fronts. It turns out that most of these "cold fronts" do, in fact, exhibit characteristic changes in pressure tendency, wind shifts from some southerly component to more westerly or northerly directions, decreases in dew points after the front passes, the presence of thunderstorms along and/or ahead of the front, and so forth, that are typical of cold fronts. Furthermore, temperature contrasts across the front at upper levels of the atmosphere typically conform to the cold front model in Figure 9.9; only near the surface is the pattern anomalous.

Backdoor cold fronts, dry lines, and low-level temperature variations are just three of many phenomena that raise questions about the standard frontal model. In the end, our concept of frontal classification is seen to be not a rock-solid, explain-everything, clear-cut system. Instead, it is one of many analysis tools, each of which is used thoughtfully and with respect for its limitations.

The idea that scientifically developed models may lack rigid classification criteria and may be imperfect can be discomforting. However, we need to remember that fronts represent a concept humans have developed to understand our environment. In nature, there is no long blue line with triangular teeth in it that sweeps across the land; we have created the cold front model to represent a cluster of atmospheric properties tending to occur together and act in a particular way. As we pointed out at the end of Chapter 1, our models are only models, and not nature itself, and therefore, they inevitably will be imperfect. While this circumstance may make it more difficult for us to understand our world, it doesn't make our study any less scientific.

In one sense, these imperfections are the most interesting aspects of a scientific model. They focus researchers' attention on new questions and problems to solve; thus, they point the way for new research and for improvements in the models themselves.

### *Cold Fronts: A Case Study*

The illustrations in Figure 9.13 show a cold front moving through the southeastern United States in September 1997. First (Figure 9.13A) is the highly simplified, made-for-TV graphic showing the front with various low pressure centers along it.

Figure 9.13B gives the cloud patterns associated with the weather system, as observed by satellite. Note that the greatest concentration of clouds lies not at the front but somewhat in advance of it, as a squall line. Other images reveal features in finer detail. Consider the radar image in Figure 9.13C, for example. The reddest sections show locations of thunderstorms. Evidently the precipitation associated with this front is not widespread, but is confined to just a few regions.

How do we know the objects in Figure 9.13C are thunderstorms? One way to confirm them is to examine the plots of lightning strikes. Figure 9.13D shows such a plot for the same time as the radar image.

Study the surface map with plotted data in Figure 9.13E. Note the wind shifts across the front: in advance (to the east), winds are mainly from the southeast; west of the front, they are west, northwest or north. Temperatures are only slightly cooler behind the front; dewpoint temperatures, however, are ten to 15 degrees Fahrenheit lower in the cooler air.

You have read that fronts lie in troughs of low pressure. A distinct low pressure trough is not evident in Figure 9.13E; observe that pressure values east of the front do not differ much from those along it. Behind the front, however, there is a slight rise, with several stations reporting sea level pressures over 150 (1015.0 millibars). This cool high is of small size, however; notice that pressures on the western border of Figure 9.13E are lower, winds are southerly, and temperatures are much warmer than elsewhere on the map.

Recall from Chapter 8 that the upper air cold front and pressure trough generally are located some distance behind the surface cold front and low pressure center. The jet stream map in Figure 9.13F, depicting strongest windflow high in the troposhere, clearly shows this westward displacement. The trough of lowest pressure and coldest air lie several hundred miles to the west of the surface cold front. Note the distinct counterclock-

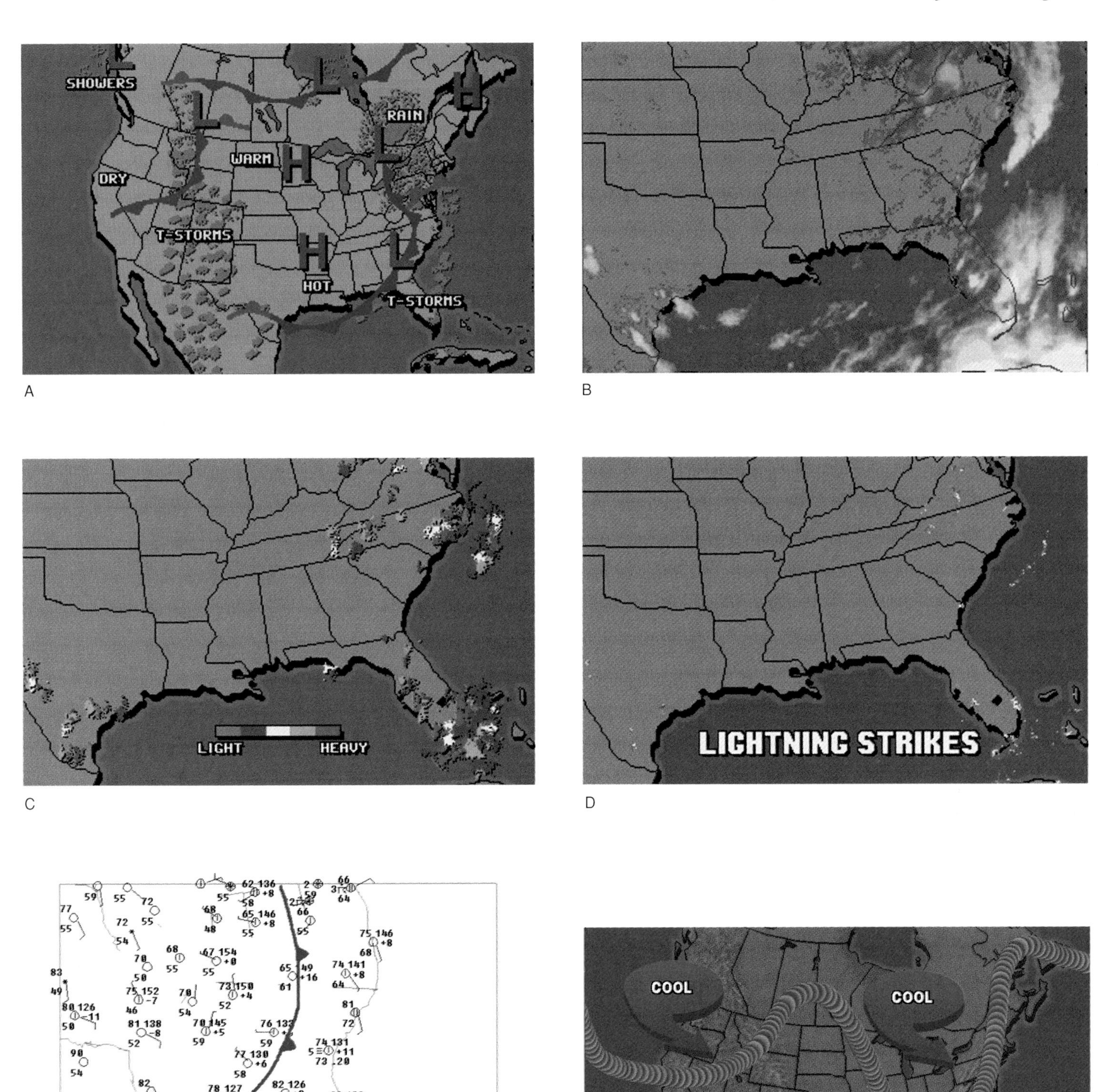


Figure 9.13 Cold front case study.

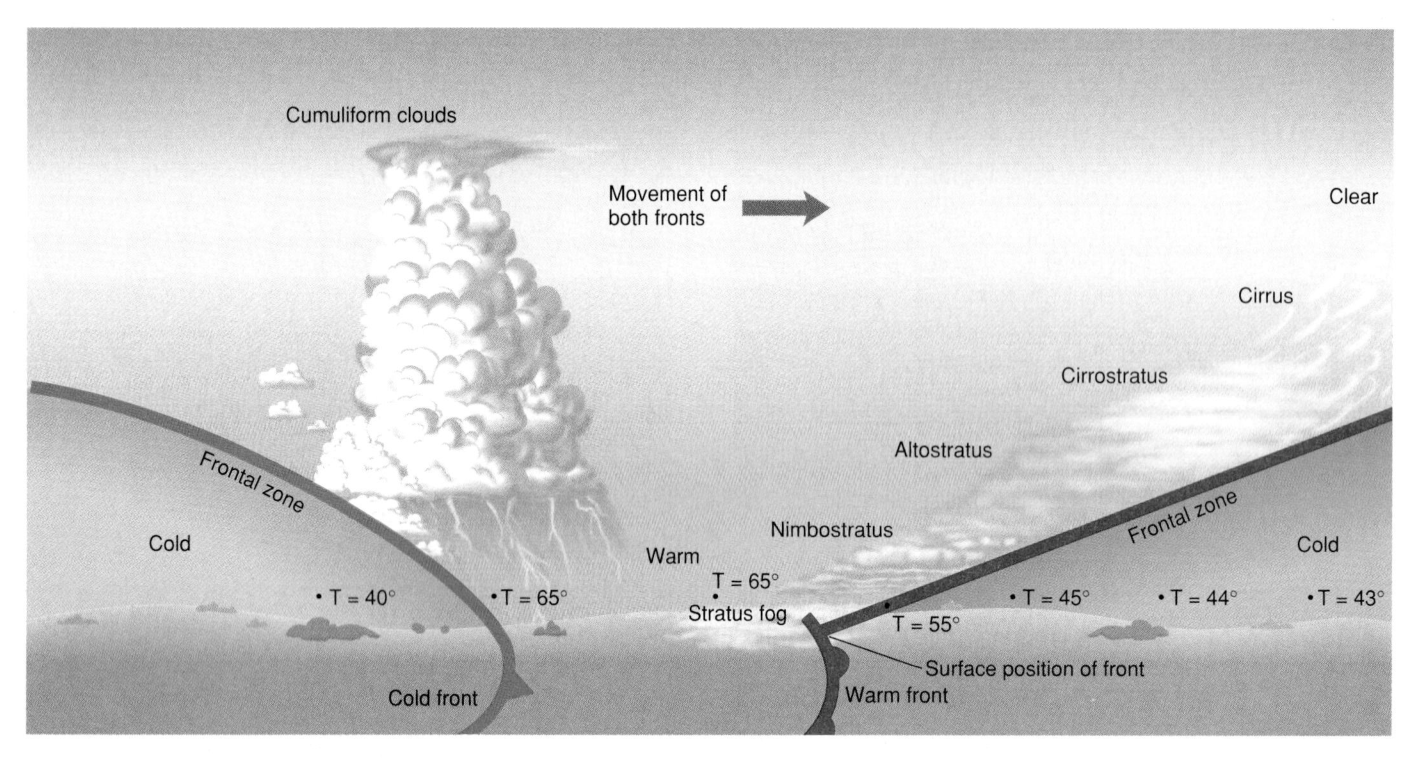


Figure 9.14 Cold and warm fronts in profile, with typical surface air temperatures.

wise turning as the jet stream air moves southeastward, then northeastward, from Tennessee to Alabama and Georgia, then over the Carolinas.

## Warm Fronts

A warm front is one whose movement results in warmer air replacing colder air locally. Study Figure 9.14, which shows an atmospheric cross section through a cold front and warm front, each traveling from left to right. To visualize the effect of each front on your local weather, imagine traveling along the ground from the *right-hand* side of the figure to the *left*.

Notice the placement of the warm front in Figure 9.14. Perhaps the hardest thing to remember about warm fronts is that their position on a weather chart does not represent the leading edge of a change to warmer weather. Instead, as you can see in the figure, the front doesn't come through at a given locale until the temperature change is virtually completed. Thus, in both cases, the front's position is defined as the warm edge of the frontal mixing zone. This arrangement implies that a cold frontal passage occurs at the beginning of the transition from one air mass to another, whereas a warm frontal passage occurs at the end of the transition.

Figure 9.14 illustrates important information about cloudiness accompanying fronts. Note that dense cold air bulges eastward behind the cold front at the left. This cold air blasts into warmer air, pushing it up and out of the way. The lift cools the air to saturation, whereupon violent thunderstorms may develop. Within the warm front, however, the slope is more gentle and the lift more benign. Thick clouds and steady rain just ahead of the warm front's surface position are preceded by thin high clouds as much as 24–48 hours before the front arrives. In the figure, you can see that the temperature transition zone is on the cold side of each front. In both cases, the movement of the front is determined by the movement of the cold air because the cold air is more dense than the warm air. Cold air rules.

The motion of ocean waves at the beach offers an interesting analogy to frontal movements. As each breaking wave charges up the beach (like a cold front), the surging bulge of water pushes into a space formerly occupied by a less dense fluid—the air. As the water retreats down the beach and air reoccupies the space, the motion is smoother, sheetlike, like that in a warm front. Note that the movement of the water, not the less dense air, determines whether the water front is advancing or retreating.

The weather associated with a warm front depends on several factors. If the cold air ahead of it is sluggish, the front will make slow progress. The warm air will glide upward over the stubborn colder air, a process called **overrunning**. If the warm air is moist, it won't take much lifting to cause extensive clouds and precipitation. However, if the cold air ahead of the warm front retreats quickly and/or the warm air mass is dry, there may be very little in the way of clouds or precipitation. As with other weather features, a wide spectrum of possibilities can exist for features that look similar when drawn on the weather map. Still, the model of frontal clouds and weather shown in Figure 9.14 provides a useful frame of reference. Notice that according to the model, when the front is far away (perhaps 1000 miles distant), the sky may be clear. The first sign of the approaching front may be high, thin cirrus clouds. Then, as the front approaches, the cold air in the lower levels of the atmosphere becomes more shallow, while clouds overhead thicken to cause precipitation. In winter, snow may fall ahead of a warm front, though it often changes to rain as the air warms with the front's close approach. From Figure 9.14, you can see that once the warm front passes a locale, clouds and precipitation dissipate as the warm air mass settles in.

The approach of a warm front in winter often causes havoc with transportation. Snow and ice, usually well ahead of the warm front, can be major impediments to travel, while low clouds and fog can shut down flight operations at dozens of airports at once. Also, when the warm air behind the front passes over a snow cover, the air is cooled from below, perhaps to its dew point. If this happens, persistent fog results, and the clear conditions that might otherwise be present in the warm sector do not develop.

## Formation and Dissipation of Fronts

Like many other atmospheric structures, fronts are not permanent features. Fronts form and intensify, a process known as **frontogenesis** (front-o-GEN-esis), when and where air masses of different properties collide. Thus, convergent air flow in the lower and middle troposphere is an important prerequisite for frontogenesis. Frontogenesis commonly happens when a trough or center of low pressure strengthens, thereby intensifying the convergent airflow in the region. Frontogenesis also occurs when an air mass forms and then begins moving, eventually colliding with another air mass of different properties. Along the line of collision, a front forms.

Under different circumstances, fronts undergo **frontolysis** (fron-TOL-ysis), or weakening and dissipation. As you might expect, frontolysis occurs in regions of divergent low-level wind flow, which spreads out and weakens the contrasts between conditions on opposite sides of the front. Another way in which frontolysis occurs is through modification of an air mass by the underlying surface. Consider, for example, a cold front that moves southward to the Gulf of Mexico and out to sea. Over the warm Gulf waters, cP air on the cold side of the front gradually becomes warm and moist like the mT air south of the front. After a while (typically a few days), there is no apparent distinction between the air on opposite sides of the old frontal boundary. At this point, we say the front has ceased to exist, and it is no longer shown on weather charts.

The dissipation of a cold front is often aided by the "bridging" of high pressure across the front. Responding to changes aloft (such as an increase of upper-level convergence), the atmospheric pressure may rise behind, over, and ahead of the front. Soon, what had been a pressure trough with a cold front is just part of a large high pressure area. It is said then that the high pressure area bridged the front.

Bridging typically occurs when the jet stream shifts away from its former location and the flow aloft weakens over the front. The weakening jet indicates that the temperature contrast below is also weakening. An interesting question arises as to whether the temperature contrast or the jet stream initiated weakening in the other. A weaker temperature gradient creates a weaker jet stream, which can cause an increase in upper-level convergence, which further weakens the temperature gradient, in a feedback loop.

## Stationary Fronts

You learned earlier that the motion of a front is determined by air movement on the cold side of that front. Suppose the cold air is neither advancing nor retreating. In such a case, the front doesn't move and is called a stationary front. Figure 9.15 shows a typical stationary front. Notice the cold high pressure area situated well to the north and the warm high pressure area to the south. Since the cold air is not advancing, warm air south of the front is not moving. This means there is little or no mechanism for thunderstorms and showers to form in the warm air south of the front. This situation contrasts with that on the warm side of a cold front, in which squall lines can form, as you read earlier.

What about the weather on the cold side of the stationary front? If you inspect the winds near the front in Figure 9.15, you can see that warm air is approaching the front from the southwest. If warm air is blowing against the front, then why isn't the front moving northward as a warm front? To answer this, recall that warm air is less dense than cold air. The cold air,

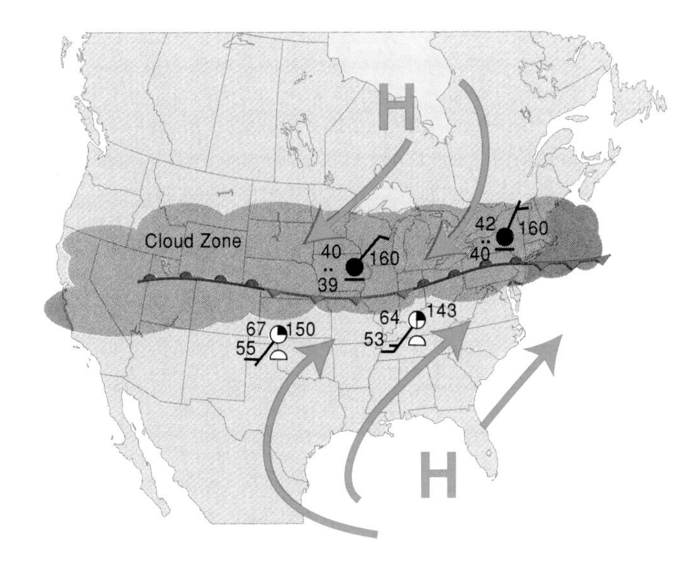


Figure 9.15 Weather conditions accompanying a typical stationary front. Clouds and steady precipitation lie to the north, fair skies and mild temperatures to the south.

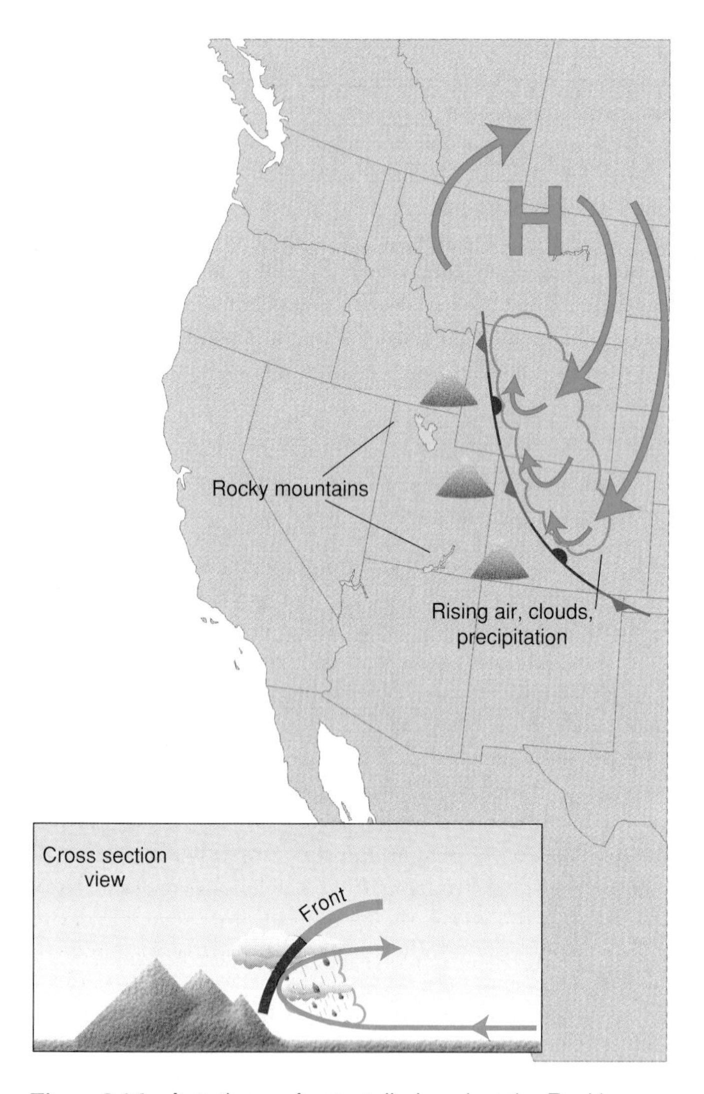


Figure 9.16 A stationary front, stalled against the Rockies.

because it is denser, controls the front's motion. As long as the cold air doesn't give ground, warm air can blow toward the front all day and still not be effective at moving the front. Instead, the warm air overruns the chilly air, just as in the case of a warm front. Knowing this, you can predict likely weather conditions on the cold side of a typical stationary front: layers of stratiform clouds and periods of steady precipitation. Because the front is stationary, these conditions may persist, giving a region several consecutive days of gray, dull weather. In summary, then, stationary front weather resembles warm frontal more than cold frontal weather.

## Special Cases of Stationary Fronts

If a front stops moving, it becomes a stationary front, by definition. If its movement resumes, it is named by whether colder air is advancing (cold front) or retreating (warm front). Usually, the movement of fronts is determined by the behavior of passing high and low pressure systems. You'll see this more clearly when we talk about frontal cyclones later in this chapter. But sometimes, topographic barriers force fronts to stall. For example, when cold air blasts southward through Great Plains in winter, the Rocky Mountains act as a barrier to the westward expansion of the cold air. We already discussed this in a different context when we were looking at air masses. Here, we observe that many cold fronts lose their forward motion and become stationary fronts when they reach a mountain range. This situation is shown in Figure 9.16. Notice that cold air may still be blowing toward the front at ground level, but the mountains deflect the air upward, whereupon clouds and precipitation develop because of the lift-induced cooling. When viewed in cross section, the circulation may look like a slowly swirling horizontal tube. Air moves toward the mountains at ground level, moves up along the mountain slope, then turns back away from the mountains when it reaches the top. When this setup continues for several days along the eastern Rockies, the region can have an extended bout of precipitation.

Another preferred location for stationary fronts is just offshore from the northeastern states in winter. Warm Gulf Stream waters maintain moist, mild mT or mP air offshore, creating sharp contrasts with the cold, dry cP air that prevails over land. When the cP air moves offshore, it is rapidly warmed and moistened. The coastline, then, often is the locale of a front separating the mainland's cP air masses from mP or mT air masses offshore.

## Occluded Fronts

You have seen that warm fronts and cold fronts tend to circulate cyclonically (i.e., counterclockwise in the Northern Hemisphere) around the low pressure center with which they are associated. You may also recall reading (for example, in Chapter 1) that cold fronts commonly move faster than warm fronts. When a cold front overtakes a warm front, an **occluded front,** or **occlusion,** is formed. To picture this event, refer back to Figure 9.14. Imagine the cold front moving from left to right, closing in on the warm front. The warm-sector air between fronts becomes narrower, and the warm air is pushed aloft. Occlusion begins when the cold front reaches the warm front at the earth's surface. At that point, only the two cold air masses are in contact with the earth's surface; the warm air has been displaced upward.

An occluded front has a structure and associated weather that depends on the nature of the two original fronts. Figure 9.17 shows the possibilities. Let's examine them in detail, keeping this basic concept in mind: cold air rules.

### *Cold Occlusions*

In Figure 9.17A, notice that the air behind the cold front is colder than both the air in the warm sector and that east of the warm front (labeled "cool air"). This coldest air mass, being densest, pushes under both the warm air mass and the receding cool air. This type of occluded front is called a **cold occlusion,** because at the surface its structure most resembles that of a cold front. Notice that the warm front has not disappeared; however,

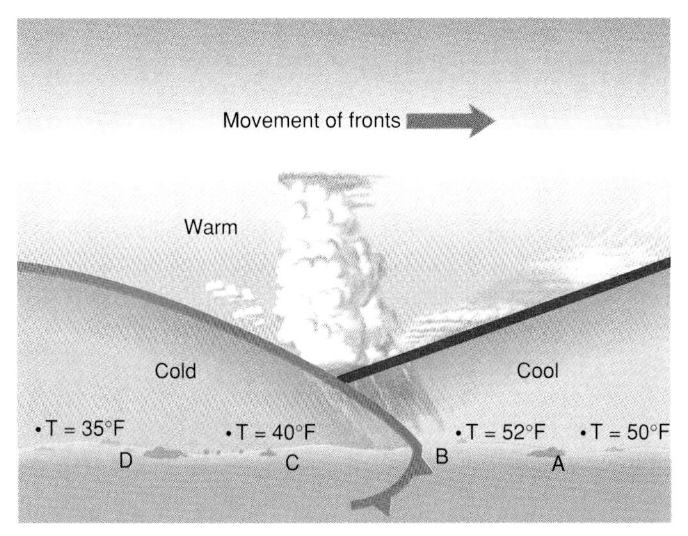


A Cold occlusion

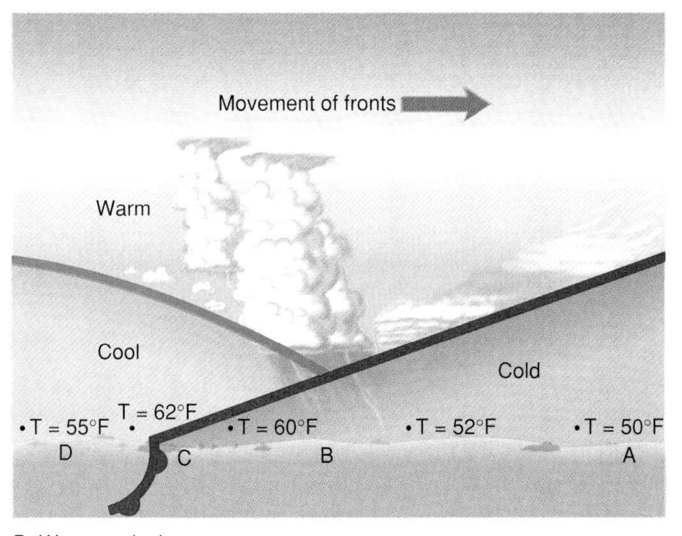


B Warm occlusion

Figure 9.17 Cross-sectional views of cold and warm occluded fronts with typical surface air temperatures. Note the vertical stacking of the three air masses in each case: cold on bottom, cool in middle, and warm on top.

it now lies some distance above the ground, pushed aloft by the cold front.

What sort of weather would you experience with the approach and passage of a cold occlusion? Imagine the weather you would experience if you moved from point A to D in Figure 9.17A. At the start, you are at location A and the fronts are approaching from the west (left). High above your location warm air is gliding up and over a retreating cool air mass. Thus, your weather will resemble that associated with the approach of a warm front. You can expect to observe thickening and lowering stratiform clouds, perhaps followed by periods of steady precipitation. However, as the fronts move past, you will not experience a warm frontal passage. Why not? Look at the figure again; notice that the warm front does not reach the ground, having been pushed aloft by the cold occlusion. Thus, the prewarm frontal weather is terminated by the passage of a cold occlusion, at B, accompanied by a wind shift (from southeasterly to southwesterly or westerly, perhaps) and a change in pressure tendency from falling to rising.

In theory, observers in the zone from B to C experience the effects of both warm frontal uplift (aloft) and cold frontal lifting (from the surface occlusion). Thus, the model suggests that the weather in this zone should be a mixture of clouds of all types, plus steady precipitation punctuated with showery periods.

Moving from C toward D, the warm front aloft no longer is a factor. Overhead, the dome of cold air continues to build, and the observer experiences typical postcold-frontal weather: falling temperatures and dew points, west or northwest winds, and dissipating cumuliform clouds.

While the above sequence actually does occur in some cases, the variety of conditions in and surrounding individual fronts causes considerable departures from the model. Thus, accompanying a given cold occlusion may be considerable precipitation or none at all, a sharp temperature fall following frontal passage or a barely discernible drop; and so on.

### *Warm Occlusions*

Figure 9.17B illustrates a **warm occlusion,** an occlusion in which the cold air receding in advance of the warm front is colder than the "new" cool air approaching from the rear (i.e., the left in the figure). Notice that the new cool air mass is dense enough to push the warm air up and out of its way. However, being less cold—and thus less dense—than the cold air it encounters, it cannot push the cold air away. Instead, the cool air climbs over the cold air, from C to B, much as it would if we were dealing with a warm front.

To understand the sequence of weather observed with a passing warm occlusion, imagine moving from points A to D in Figure 9.17B. At A, you are deep in the cold air in advance of the warm front. High above you, however, warm air glides upward along the frontal surface, causing cirrus and cirrostratus clouds. With time (moving toward B), the clouds thicken and lower in typical warm front fashion; eventually, steady precipitation develops.

When the cold front passes overhead (at point B), precipitation may become heavier, as it would with cold frontal showers and thunderstorms. Often the shower clouds can not be seen from the ground because they are hidden by lower stratiform clouds caused by the overrunning of warmer air over cold.

The passage of the cold front aloft at B provides little if any clearing. This is because the new cold air is forced to rise over the retreating colder air; this lifting leads, as usual, to clouds and precipitation. Finally (at C), the warm occlusion passes, and we find ourselves in milder air than before the front's passage. However, as the new air becomes further entrenched (moving toward D), temperatures fall somewhat. Thus, temperatures typically rise until the surface occlusion passes, then fall.

Because the occluded front model is so complex, individual occluded fronts rarely, if ever, match the model exactly. This has led some meteorologists to abandon use of occluded fronts entirely. They simply consider whichever front can be discerned at the ground. If you look at the diagrams, the observer at the ground sees only a cold front (in the case of a cold occlusion) or a warm front (in a warm occlusion). The other fronts are no longer at the earth's surface, and sometimes their effects are be apparent to only the most astute weather watcher. On the other hand, because the weather surrounding an occluded front will be different, and generally more complex, than in either a cold front or a warm front, many meteorologists believe it is important to indicate occlusions as occlusions, and not as cold or warm fronts, on their maps. Occlusions are an important component of the Norwegian cyclone model, which you will study later in this chapter.

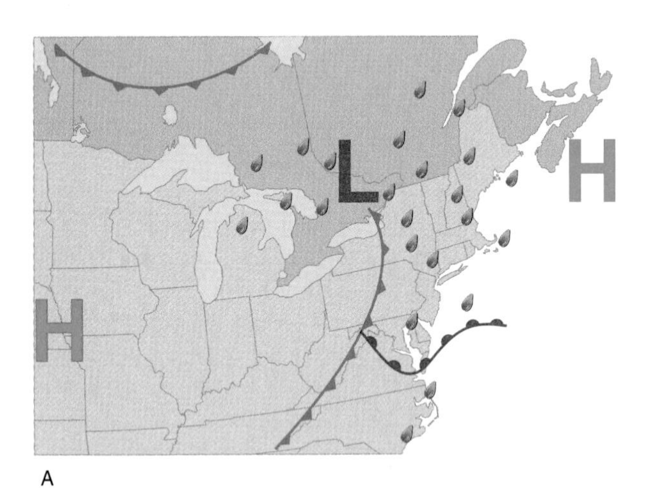


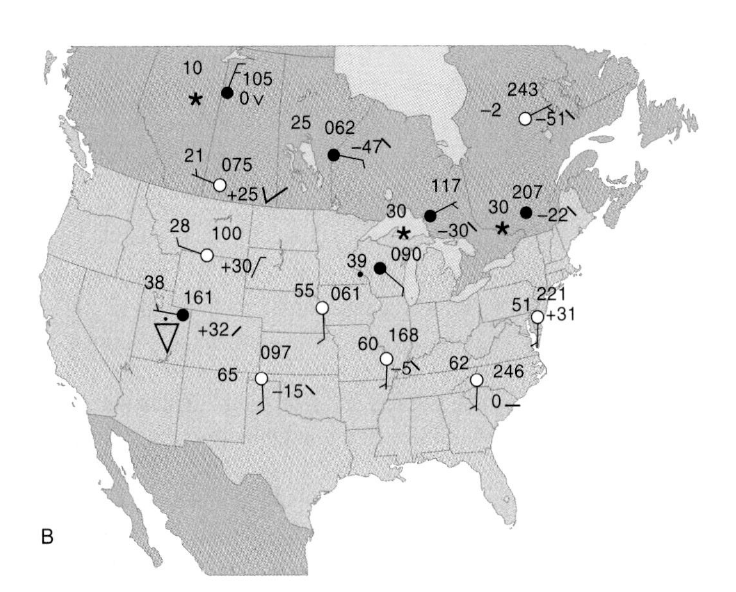


Figure 9.18 (*A*) Examine the plotted data to understand why the fronts were placed as shown. (*B*) Where would you locate fronts in this map?

**Table 9.2 Guidelines for Locating Fronts on Weather Charts**

(1) Fronts separate different air masses.
(2) Fronts lie in troughs of low pressure and/or traverse low pressure centers.
(3) Fronts often mark a shift in wind direction.
(4) Fronts often mark a change in pressure tendencies.
(5) Fronts often mark a change in dew-point temperatures.
(6) Fronts are drawn at the warm air boundary of the frontal mixing zone.
(7) Temperatures show little variation from place to place in the air mass on the warm side of a warm or cold front.
(8) Temperatures decrease steadily, sometimes sharply, on the cold side of a warm or cold front.
(9) Temperatures rise with the approach of an occluded front, then fall with its passage.
(10) Most cloudiness and precipitation occur on the cold side of warm and cold fronts.
(11) Warm frontal clouds tend to be stratiform, with steady precipitation; cold frontal clouds are more cumuliform, with more showery precipitation.

## Fronts on Weather Charts

We conclude this section on fronts with a study of the weather maps in Figure 9.18. In Figure 9.18A, note that the frontal system associated with the low pressure center over Ontario contains cold, warm, and occluded fronts. South of the warm front is a uniform warm air mass. In New England, which is north of the approaching warm front, temperatures are rising. The warming will end, however, with the arrival of the occluded front. This front marks the leading edge of a change to colder weather (since there are no temperatures on the map, you have to trust the analyst). The prewarm-frontal weather occurs north of the warm front and east of the cold occlusion.

The point at which the warm front cold front, and occluded front meet is called the **triple point**. As you can see, the triple point is located at the junction of three air masses: warm, cool, and cold. Often the weather is particularly active in this area because of the close proximity of and interaction between cold and warm frontal effects.

Now consider Figure 9.18B, which contains only the plotted raw data, with no frontal positions shown. Are fronts present in this weather system? If so, where are they located? Table 9.2 gives some guidelines, collected from earlier discussions, to direct your search.

You probably required considerable time to find and refine your frontal positions. Locating the positions of fronts is often as difficult as it is important. Is it any easier for professional meteorologists? Consider the rather disconcerting situation illus-

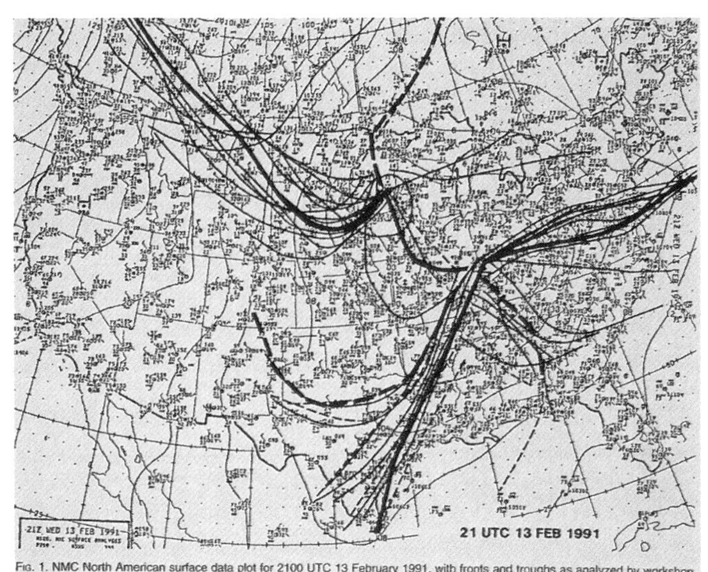


Figure 9.19 Surface weather conditions at 21Z, February 13, 1991, with frontal positions according to meteorologists participating in a professional workshop. Note the wide variety of opinions as to the fronts' positions!

trated in Figure 9.19, which shows the frontal positions drawn by meteorologists attending a workshop on fronts and storms. As you can see, these professionals, all of whom were studying the same weather system, had many different opinions about where the fronts were. Each individual could probably support her or his interpretation with a set of seemingly cogent arguments. The implications of the different interpretations are far-reaching: a forecaster who thinks a front is in a certain location will make a forecast based on this position. Another meteorologist, starting with a different frontal position, may produce a very different forecast.

Figure 9.19 is a timely reminder not to feel frustrated or discouraged if you have a difficult time locating fronts. Clearly, professionals have the same difficulties at times as well! The important thing at this point is that your decisions are consistent as far as possible with the data and with frontal models presented in this section.

As we have seen in this section, fronts often represent concentrations of weather conditions such as cloudiness, precipitation, wind shifts, and temperature changes. Therefore, fronts, like air masses, are of basic importance in weather forecasting in middle and high latitudes.

## Checkpoint

### *Review*

This section reminded you that cold fronts and warm fronts are models that have been developed to explain a cluster of atmospheric phenomena that tend to occur together and act in fairly predictable ways. A front marks the boundary between two air masses. The movement of cold air determines the type of front. A cold front is determined by advancing cold air and a warm front by retreating cold air.

Cold fronts are not only marked by temperature differences across the frontal boundary but also by differences in dew point, pressure, and wind direction. Backdoor cold fronts, dry lines, and low-level temperature variations are phenomena that raise questions about the classical textbook model of a cold front.

Because of the density differences between warm air and cold air, warm and cold frontal surfaces are quite different. The properties of the warm and cold air masses and the season of the year determine the weather associated with an advancing warm front. Several factors that cause frontal formation and strengthening (frontogenesis) and frontal weakening or dissipation (frontolysis) were discussed.

Geographical features that cause fronts to stall, such as mountains or land-ocean boundaries, can contribute to a front's becoming stationary. Occluded fronts occur when a cold front overtakes a warm front. Depending on the relative temperatures of the retreating and advancing cold air, the occlusion may be of the cold or warm type.

### *Questions*

1. Typically, cold fronts are defined and explained by the temperature differences that exist across the frontal boundary. However, dew-point, pressure, and wind direction changes are also often observed. Explain how these other contrasts develop.

2. Briefly describe the three cold front-related weather phenomena listed below and explain how each differs from the textbook cold front model.

| Weather Phenomenon | Description |
|---|---|
| Backdoor cold front | |
| Dry line | |
| Surface temperature variations | |

3. Draw a diagram of the frontal surfaces of a cold front, a warm front, a warm occlusion, and a cold occlusion and explain the differences.

# Frontal Cyclones

Fronts are so useful in organizing and explaining various weather phenomena and figure so prominently in TV and radio weathercasts that it might seem as if they have always been part of meteorology. After all, other physical laws you have learned about were developed hundreds of years ago. However, even as recently as World War I, the concept of fronts was largely unknown to the world's meteorologists. The discovery of fronts was intimately linked to development of a model describing cyclones that form on fronts.

## Historical Background

A number of circumstances converged in the early years of the 20th century that changed the theory and practice of meteorology forever. The explosive growth in powered flight greatly increased the need for accurate weather information. At the same time, the development of rapid wireless telecommunication methods allowed meteorologists to gain a grasp of how the weather was behaving in time to help make a forecast. The development of this modern communications capability cannot be overestimated. Before the wireless, years would pass between the occurrence of a weather event and the preparation of weather charts depicting the event. Once radio communication provided a means of keeping up with the weather, considerable interest developed in improving weather forecast accuracy.

Vilhelm Bjerknes (1861–1952) is widely regarded as the founder of modern meteorology. He and his colleagues in Bergen, Norway, pioneered the concept of storms that develop, grow, mature, and decay in relation to fronts. In the period from 1918 to 1920, Bjerknes's group postulated the existence of the **polar front** as a hemisphere-encircling boundary between warm and cold air masses. Storms came to be seen as waves that developed and grew along these boundaries.

Bjerknes's new theory, like any other scientific proposal, required extensive testing before its acceptance by the scientific community. Unfortunately, times were hard in Norway, and little money was available for continued research. Then, on the night of January 14, 1920, the weather itself intervened. An unpredicted storm sank dozens of fishing boats along the Norwegian coast, resulting in the loss of 21 lives. The storm was a grim reminder of the importance of the weather, and of weather forecasting, in Norwegian life. Funding was maintained, and in 1922, Bjerknes published his landmark paper with H. Solberg of the Bergen School, "Life Cycle of Cyclones and Polar Front Theory of Atmospheric Circulation." To this day, the **Norwegian cyclone model** remains the foundation of our understanding about synoptic scale cyclones and fronts.

## The Norwegian Cyclone Model

The stages of the Norwegian cyclone model, as described by Bjerknes and Solberg in 1922, are shown in Figure 9.20. The drawings are based on those in their work. Note that the basic structure of the model is familiar, being much the same as the model you studied in Chapter 1. We recommend that you spend a few moments at this time reviewing the relevant section of Chapter 1.

The following discussion of the Norwegian model is based on typical Northern Hemisphere conditions. Thus, we assume that winds circulate cyclonically (counterclockwise, in the Northern Hemisphere) around low pressure centers and that colder air is found to the north (top of the charts), warmer air south.

### *Growth to Maturity*

Many frontal cyclones form on a more or less straight stretch of stationary front, such as that shown in Figure 9.20A. Although low-level winds converge toward the front, initially no center of

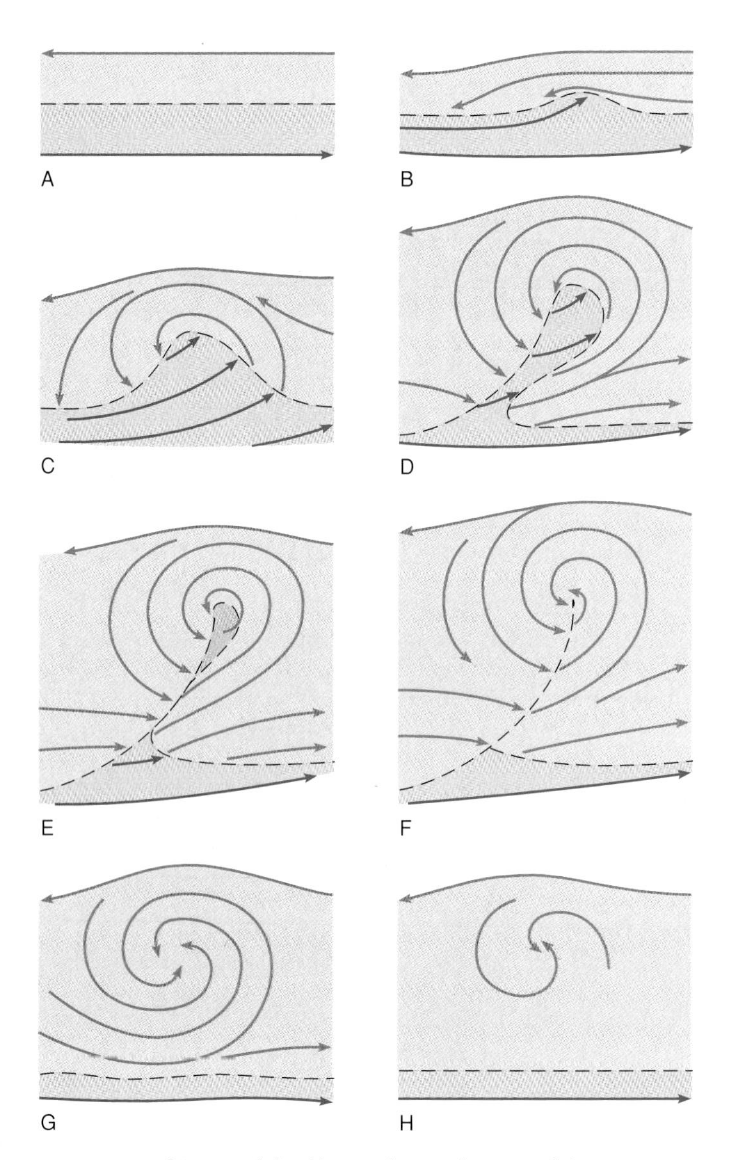


**Figure 9.20** Stages of the Norwegian cyclone model.

rotation may be identifiable. Notice, however, that since the front, like all fronts, lies in a trough of low pressure, cyclonic rotation of the air is present.

What are the possible futures for the front in Figure 9.20A? It may persist in its present form indefinitely; it may weaken and dissipate; or it may intensify. In the latter case, the front may spawn a frontal cyclone. What determines the front's fate? Recall from Chapter 8 that conditions aloft greatly influence surface weather features. Specifically, for cyclonic development at low levels, a divergent wind flow pattern aloft is required. Divergence removes air that converges on the front at lower levels and then rises. (Refer to Figure 8.17 to remind yourself of this process.) If these conditions are met, then low-level convergence is enhanced. As the air spins in toward the front, it acquires additional rotation, which deforms the stationary front, creating a slight wave in it, as illustrated in Figure 9.20B. The wave now deflects wind flow along the front, creating a more obvious cyclonic rotation. Therefore, more focused convergence of low-level air occurs around the wave, accentuated as the surface winds are deflected toward the low pressure center by friction.

If divergence aloft continues to be greater than low-level inflow, cyclonic rotation and distortion of the wave may grow to stage C. Notice that the wave now has a distinct crest; it is at this crest that the center of lowest air pressure is located.

What happens to the fronts and opposing air masses at this point? East of the low pressure area in stage C, cold air retreats and warm air from the south advances. The front in this region is now a warm front. Behind the low pressure area, the circulation has caused the cold air to start advancing south and east. Thus, the polar front acts as a cold front west and southwest of the low pressure area.

The temperature contrasts between air masses on opposite sides of the polar front represent an energy source for the developing circulation. Colder, denser air on the polar side of the front sinks and slides beneath the warmer, lighter air. Thus, thermal energy contrasts are converted to kinetic energy and wind speeds increase. Another source of energy is the heat released when moisture condenses. This heat adds further impetus for the air to rise, which in turn produces more cooling and more condensation. Latent heat energy can be a powerful fuel for storms.

At (D), the storm is nearing its maximum intensity. The cold front sweeps around the south side of the storm, propelled by gusty blusters of cold, dense air.

### *Occlusion and Weakening*

As the storm at the surface reaches its maximum strength, the cold front quickly closes in on the warm front. When the cold air finally overtakes the warm air, the fronts are occluded, as shown in E and F. The onset of occlusion (Figure 9.20E) means that the low pressure center no longer has access to the warm air that helped fuel it. Cold air wraps around the entire storm system. The storm doesn't dissipate immediately; in fact, it may remain quite strong for several days after the onset of occlusion. After all, there is a warm sector aloft for some time after it disappears at the ground. The interaction between warm and cold continues until this situation no longer exists.

Eventually, however, the spinning storm is removed from the main contrast zone between warm and cold (in Figures 9.20E and F). Soon we have a much weakened low pressure area and a stationary front some distance away (F), on which a new cyclone may form. Along many fronts, we see a series of storms, each in different stages of growth and decay.

## What Causes Cyclone Development?

Why does cyclone development occur at one time and place, and not in others? And why do some cyclones develop into only minimal circulations, while others intensify explosively into giant, destructive storms? These are crucial questions for the weather forecaster; we will tackle them next.

### *Short Waves and Long Waves*

Figure 9.21 shows a typical upper-tropospheric circulation pattern. Notice that the flow consists of a relatively smooth, mainly westerly flow with waves embedded, in contrast to the closed circulations typical of surface weather like those in the previous figure. The waves in upper-level flow patterns are of two main types. **Short waves** are associated with individual synoptic scale cyclones and have wavelengths on that scale: typically 1000-2000 kilometers. They travel along through the larger-scale flow pattern, in the direction of that flow. Short wave troughs are of particular interest to meteorologists because they are often associated with individual frontal cyclones. Because temperature variations are responsible for most of the variation in upper-level pressure patterns like that in Figure 9.21, short wave troughs of lower pressure are also generally associated with pools of cold air.

**Long waves,** or **Rossby waves,** have considerably longer wavelengths; four or five long waves commonly span the entire globe at middle latitudes. While long waves move slowly or not at all, short waves travel with the upper-level flow, thus passing through the longwaves. As you will see, such interactions can be important events for cyclone development.

### *The Role of Divergence*

The presence of upper-level divergence is crucial to the health of a cyclone. If upper-level divergence ceases, the converging low-level flow quickly "fills" the low pressure depression, vertical motion ceases, and the cyclone weakens. On the other hand, vigorous upper-level divergence may remove air from the cyclone faster than the converging low-level winds can replace it, causing the air pressure to decrease and thus the surface storm to intensify. Therefore, if we can find regions indicating divergent upper-level wind flow, we will in so doing identify those regions most favorable for cyclone development.

Divergence can take either of two forms, as illustrated in Figure 9.22. Nonparallel wind-flow causes the air to converge or

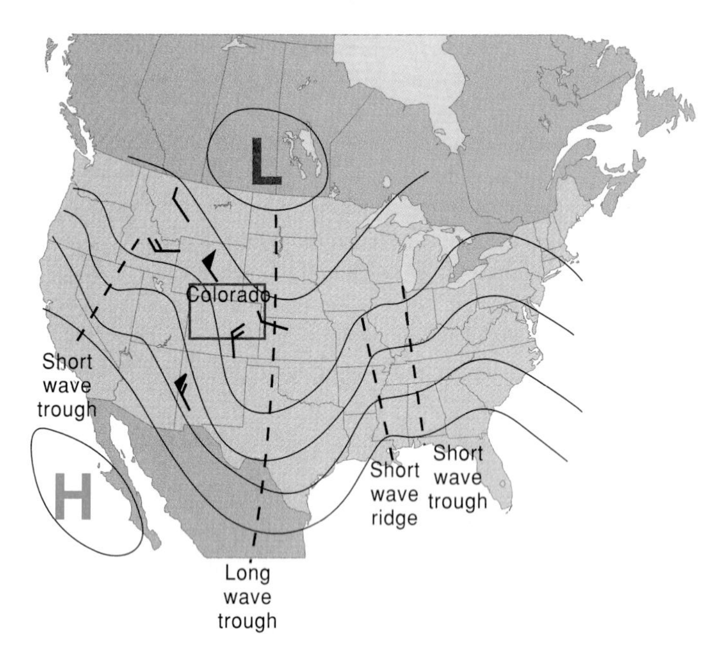


Figure 9.21 Three-hundred-millibar flow pattern showing typical long wave and short wave troughs.

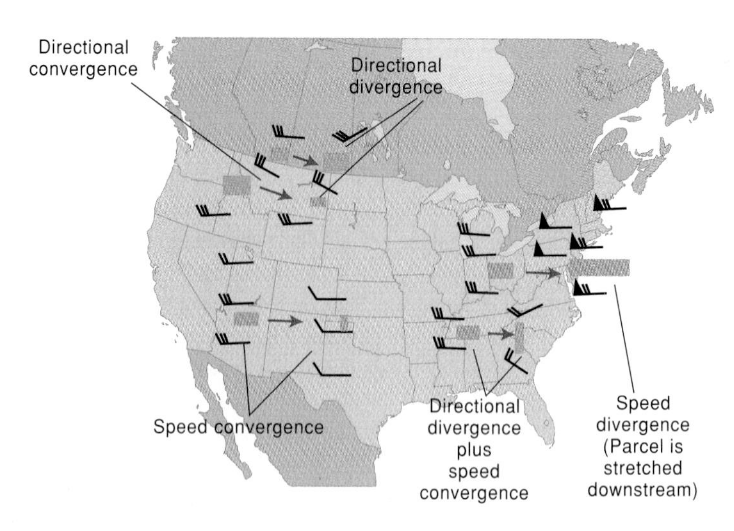


Figure 9.22 Types of horizontal divergent and convergent flow. The amount of divergence of convergence is shown by the change in each air parcel's area.

diverge ("directional convergence or divergence"), and changes in wind speed can also result in convergence or divergence ("speed convergence or divergence").

From Figure 9.22, you can see that divergence is a measure of the change in horizontal area occupied by an air parcel. If the area increases, divergence is considered positive. If the area decreases, the divergence is said to be negative. When used in this way, the term *divergence* can stand for either divergence or convergence; convergence is simply a negative amount of divergence. (Strictly speaking, we should always refer to this type of divergence as horizontal divergence, since air may converge or diverge vertically as well. Since here we are considering only horizontal divergence, however, for brevity we will omit the word *horizontal*.)

Let's look for regions of divergence in the 300-millibar chart shown back in Figure 9.21. Note that over Colorado, the wind directions are clearly divergent. However, in the same region, speed convergence is occurring. The net amount of convergence or divergence, then, equals the difference between the directional divergence, by which air is removed from the region, and speed convergence, by which air is added. Thus, the net divergence is the small difference between two inexactly known quantities. Thus, it is unclear whether the air over Colorado is converging or diverging. This circumstance is typical of fluid flow: divergence of one type is nearly always accompanied by convergence of the other. As another example, consider how water from a garden hose slows down (speed convergence) as it spreads out over a driveway (directional divergence).

If the net divergence is so difficult to evaluate on weather charts, can it be of any value to us in analyzing weather systems? The answer is yes, thanks to its relation to another property of fluid flow, which we consider next.

## *Vorticity*

**Vorticity** is a measure of the amount of rotation present in a fluid. It should come as no surprise, then, that cyclones, being rotating storms, possess considerable vorticity. An air parcel may rotate around its vertical axis (see Figure 9.23) like a top, around a horizontal axis like a car wheel on its axle, or around an axis oriented in any other direction. But because our present topic is frontal cyclones whose rotation is mainly around a vertical axis, we are concerned here with only the component of air's vorticity about the vertical axis.

In Figure 9.24, you can see that the total amount of vorticity a parcel of air possesses, known as **absolute vorticity,** is the sum

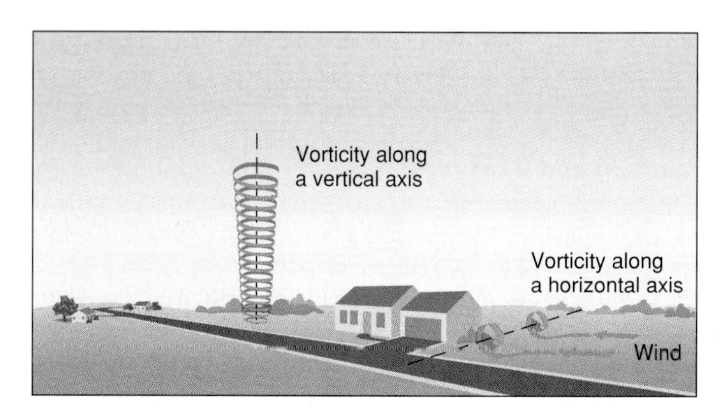


Figure 9.23 Air may possess vorticity about a vertical or a horizontal axis.

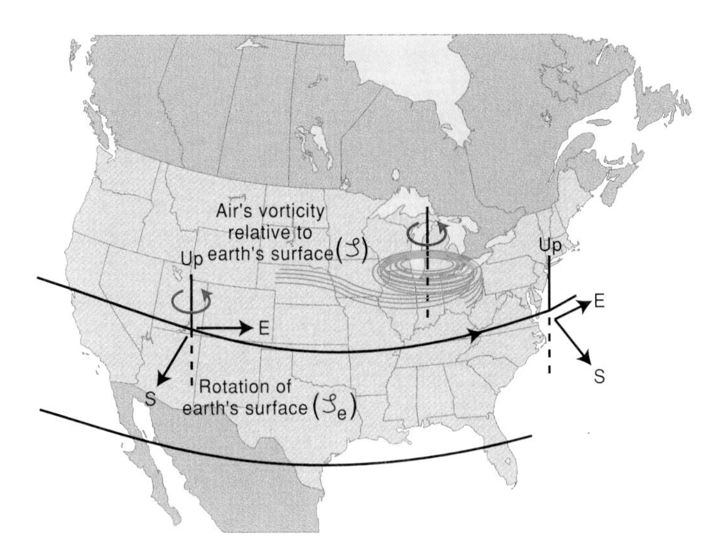


Figure 9.24 Air possesses vorticity of two sources: its rotation relative to the earth's surface, plus the rotation of the earth itself.

of two terms. These are the **relative vorticity,** a measure of the air's rotation with respect to the underlying surface, and the earth's rotation—the rotation of that surface itself (about the vertical axis). In formula form,

absolute vorticity = relative vorticity + earth's rotation

or,

$$\zeta_a = \zeta + \zeta_e \tag{1}$$

(The symbol $\zeta$ is the Greek letter *zeta*.)

Notice in the figure that positive vorticity is taken to be counterclockwise. Thus, in the Northern Hemisphere, the earth's surface possesses positive vorticity, and the relative vorticity of Northern Hemisphere cyclones, which rotate counterclockwise, is also positive. The relative vorticity in anticyclones (high pressure systems) is negative.

As in the case of divergence, vorticity may manifest itself in either of two ways. Figure 9.25 illustrates these two ways. Flow patterns in which air parcels change direction with time possess a form of vorticity known as **curvature**. On the other hand, when wind speeds vary perpendicular to the line of flow, a type of vorticity known as **shear** is present. Unlike the two forms of divergence, which largely cancel each other out, the curvature and shear forms of vorticity are not linked in such a way. At any given locale, curvature and shear terms may be both positive, both negative, or mixed.

With this background, let's see how we can use vorticity to determine patterns of divergence, as promised.

## *Divergence and Vorticity*

Consider the ice skater going into a spin in Figure 9.26. She begins slowly, arms spread apart. But as she draws her arms in close to her body, her spin (vorticity) increases to the point that she may become an indistinguishable blur. Spreading her arms out again, she slows down. (You can demonstrate the same effect on a spinning lab stool. Begin with arms in and have someone give you a gentle spin. The effect will be greater if you hold weights in your hands.)

The skater's spin reveals a key relation between vorticity and divergence. As the horizontal area of her spin *decreases,* her rate of spin *increases*. Now, a "decrease in the horizontal area" is what we have been calling horizontal convergence, and a faster rate of spin is just an increase of absolute vorticity. Thus, we see that convergence is accompanied by an increase in absolute vorticity and divergence by a decrease. This relation can be expressed in a formula, as follows:

$$\frac{\text{rate of change of } \zeta_a}{\text{given period of time t}} = -\text{ divergence}$$

Recall that rates of change are represented by $\Delta$ (Greek *delta*) in many mathematical relations. Using the delta notation, we can rewrite the above formula more succinctly as

$$\Delta\zeta_a/\Delta t = -\text{ divergence} \tag{2}$$

Thus, as an example, in situations where the absolute vorticity increases rapidly (large $\Delta \zeta_a / \Delta t$), the divergence must also be large and negative (and therefore converging, because convergence is negative divergence). Where the absolute vorticity is not changing with time ($\Delta\zeta_a / \Delta t = 0$), the divergence must also be zero.

The significance of this relation can hardly be overstated. It says that we can identify regions of diverging air aloft, which are locales of likely storm intensification, by locating regions of decreasing vorticity aloft. Thus, the vorticity-divergence relation (expression 2) is a basic weather prediction tool.

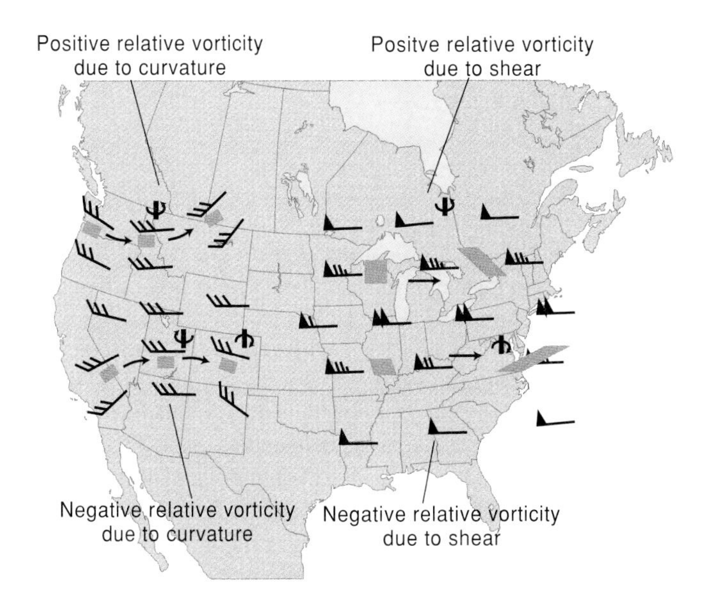


Figure 9.25 Examples of vorticity patterns. Positive vorticity is represented by the ⟲ symbol (counterclockwise). (Negative vorticity is represented by the ⟳ symbol (clockwise).

Figure 9.26 A spinning skater illustrates the relation between divergence and vorticity change.

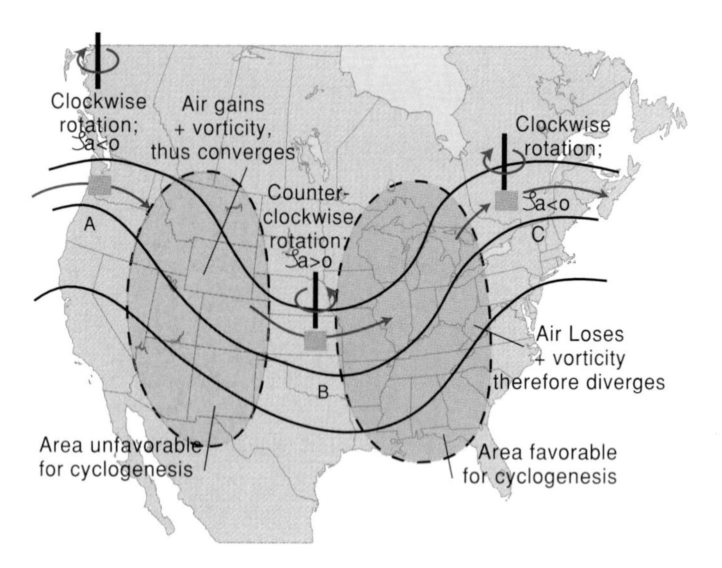


Figure 9.27 Divergence-vorticity relations on an upper air chart.

## *Examples of Vorticity Change*

Let's examine some flow patterns that illustrate the vorticity-divergence relation in action.

### *In Long Waves*

In Figure 9.27, you can see how vorticity change and divergence are linked on a typical upper air chart. Consider the air moving southeastward from the long wave ridge at A to the the trough of the wave at B. How is its vorticity changing? At A, the air is turning clockwise, with respect to the earth's surface, and therefore its relative vorticity ($\zeta$) is negative. Moving into the straight flow region between A and B, $\zeta$ becomes zero and then increases to a positive value as it reaches the trough (B), where it is rotating counterclockwise (again, with respect to the earth's surface). Thus, its relative vorticity increases from a negative to a positive value. Meanwhile, the vorticity of the earth's surface is slightly less at B than at A (the earth rotates more slowly around a vertical axis in lower latitudes) but by an amount generally less than the increase in relative vorticity. The net result is that the parcel's absolute vorticity increases from A to B. Applying expression (2), we see that this region of increasing absolute vorticity must be one of negative divergence (i.e., convergence) aloft, sinking air, and thus divergence at the ground—conditions favorable for anticyclone formation.

Now, consider the vorticity changes to the east of the trough. Through reasoning like that in the previous paragraph, you should be able to determine that the air's absolute vorticity is decreasing from B to C. Therefore, the air aloft is diverging, which promotes upward vertical motion below 300 millibars, which are conditions favorable for cyclogenesis.

### *In Jet Streaks*

Another source of vorticity change is found in the wind shear associated with a jet stream feature called a **jet streak**. A jet streak is a zone of extra-strong winds located within the jet stream. Jet streaks' locations generally change more slowly than the winds themselves; thus, as shown in Figure 9.28, individual particles of jet stream air accelerate into the jet streak region and then decelerate beyond it.

Jet streaks often are caused by short waves aloft. Recall that a short wave itself is associated with a pool of cold air. As the short wave moves through the mean flow, its cold air acts to intensify local temperature gradients, thus causing a local increase in wind speed: a jet streak.

Consider the changes in vorticity an air parcel experiences as it passes through the jet streak in Figure 9.28. Notice that in the northwest quadrant, winds entering the jet streak accelerate more than those to the north, creating an increase in cyclonic (counterclockwise) shear. (To picture this, imagine an inner tube floating in the air in the middle of this wind field. The

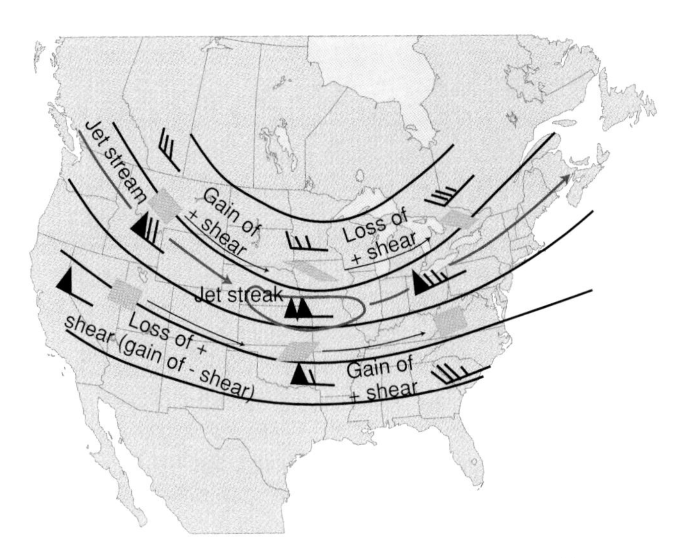


Figure 9.28 Air moving through a jet streak gains and loses vorticity due to changes in shear.

stronger winds at the bottom of the tire would soon start it spinning counterclockwise.) Thus, the air's relative vorticity increases. As this air exits the jet streak in the northeast quadrant, it slows down, resulting in a reduction of shear and therefore a decrease in relative vorticity. South of the jet streak, the situation is reversed. You should be able to verify that air passing through these quadrants first gains negative (clockwise) shear, then loses it.

### *In Cross-Mountain Flow*

In Figure 9.29A, note that the airflow on the windward side of the mountains is deflected upward considerably, while the flow above the ridge level is deflected only slightly. As a result, the parcel of air shown changes shape, becoming shorter vertically but diverging horizontally. The divergence requires a decrease of vorticity at any given level; hence, the initially straight flow acquires a negative (clockwise) rotation, as shown. Once past the ridge line, the opposite conditions occur: the parcel stretches vertically, which is compensated by horizontal convergence, which in turn causes an increase in vorticity. The result is a **lee trough,** a trough of low pressure caused by cross-mountain wind flow. The lee trough is sometimes called a dynamic trough, because it is formed by the dynamic forces associated with the flow over and past the mountains.

Because the trough was set up by a stationary feature—the mountains—it too may remain stationary downwind from the mountains for some time. It often requires an additional upper disturbance to kick it out into the downstream flow. Predicting when such a trough will be ejected is a key challenge to weather forecasters concerned with the central and eastern parts of the United States. Similar challenges exist in other parts of the world where dynamically induced low pressure areas form.

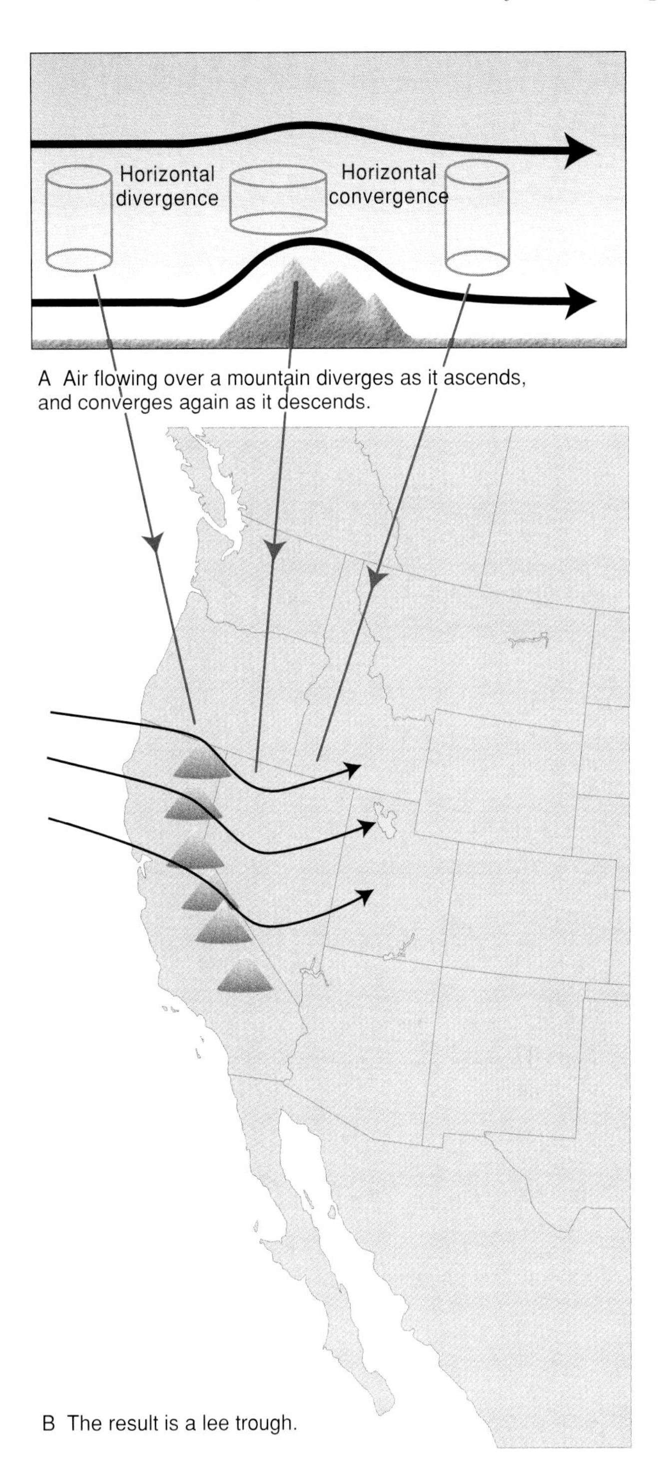


Figure 9.29 Convergence and divergence patterns inducing a lee trough.

### *Baroclinic Flow*

One important additional factor often promotes cyclogenesis. Examine the top two charts in Figure 9.30. Notice in (A) that the patterns of winds (and contour lines) differ from that of the isotherms. As a result, the winds blow across isotherms, not parallel to them. This circumstance causes cold air advection west of the trough and warm air advection to the east. In regions

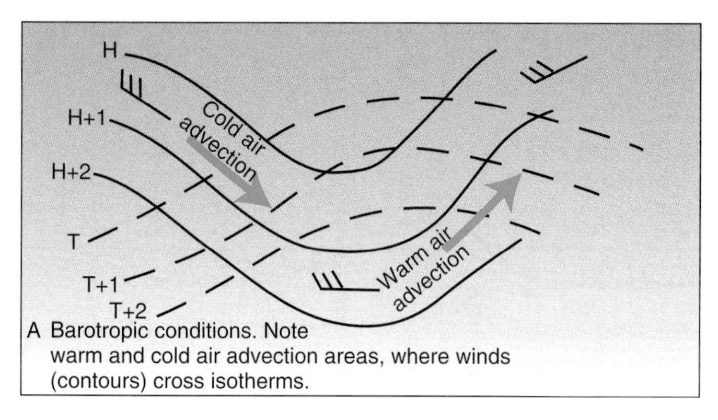

A Barotropic conditions. Note warm and cold air advection areas, where winds (contours) cross isotherms.

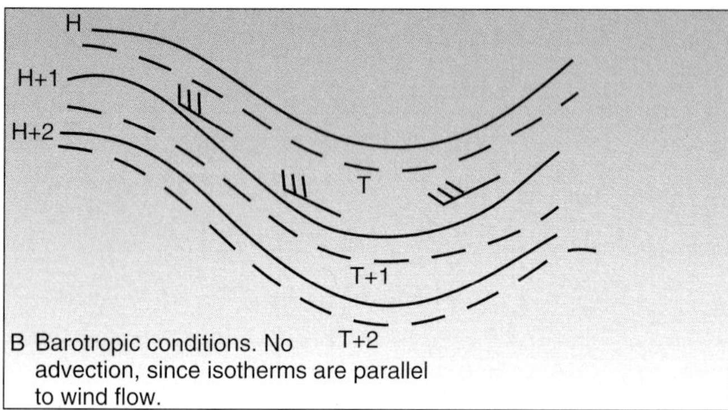

B Barotropic conditions. No advection, since isotherms are parallel to wind flow.

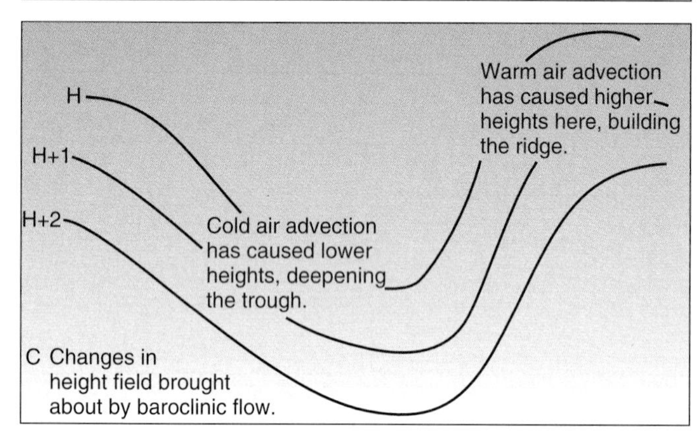

C Changes in height field brought about by baroclinic flow.

**Figure 9.30** Baroclinic (*A*) and barotropic (*B*) flow patterns in the upper atmosphere. The advection of warm and cold air in (*A*) can alter that flow pattern to (*C*).

where temperature advection is occurring, the atmosphere is said to be **baroclinic**. In contrast, Figure 9.30B illustrates a pattern in which isotherms are parallel to the wind flow. In such a case, no temperature advection occurs, a situation known as **barotropic** flow.

In Figure 9.30C, you can see how baroclinic flow has the potential to influence cyclone development. Note that in this baroclinic case, advection of warm and cold air has caused changes in the upper level height field, which alters the wind flow at that level, since winds aloft tend to blow parallel to contours. This example demonstrates how baroclinic flow can change the upper-level wind flow patterns and hence the patterns of upper-level divergence and convergence that are crucial factors in low-level cyclogenesis.

## A Three-Dimensional Cyclone Model

Now we have assembled the key tools needed to probe cyclone development. Let's see how divergence, vorticity, jet streaks, and baroclinicity contribute to cyclone development and relate to the stages in the Norwegian cyclone model. Rarely, if ever, do all these factors come into play together, so for the sake of illustrating their effects, we will consider a hypothetical case of storm development.

The stationary front in Figure 9.31A lies below an upper-level region of tight pressure and temperature gradients, the typical configuration you learned about in Chapter 8. This particular pattern is equivalent to stage A in the Norwegian cyclone model (Figure 9.20A). The lower values of pressure and temperatures aloft to the northwest of the surface front are just what one would expect from considerations of frontal properties. Note also the pool of cold air aloft, in the vicinity of the trough.

The presence of the trough means that upper-level flow is divergent in the region to the east of the trough line, thus promoting upward vertical motions and low-level convergence. Surface air pressures in this region fall, due to the divergence aloft. Since the air in the vicinity of the front already possesses cyclonic shear, enhanced convergence of this air accentuates the cyclonic motion as well. Wherever the upper-level divergence happens to be greatest, surface pressure falls are also greatest, creating a center of cyclonic motion (the "low") on the front.

West of the trough line, the opposite conditions apply. There, the upper winds gain vorticity as they flow southeastward, implying upper-level convergence, subsidence, and low-level divergence.

Suppose now that a short wave moves into the long wave trough, "digging" it deeper and intensifying the gradients of pressure and temperature, as in Figure 9.31B. Note the jet streak associated with the short wave, located at the "bottom" of the trough. The added upper-level divergence due to the jet streak helps to accentuate surface pressure falls within a limited region and hence the formation of a discrete cyclone center.

Conditions in Figure 9.31B correspond approximately to stages B and C in the Norwegian cyclone model. Notice that this cyclonic motion around the low center creates baroclinic conditions: east of the storm center, warm air advection occurs as the front moves northward as a warm front. Meanwhile, west of the low center, cold air advection occurs as the front in that region pushes southward as a cold front. This advection has important consequences of the storm's development.

Warm air advection in advance of the low causes an increase in upper-level height contours, while cold air advection behind the storm has the opposite effect. As a consequence, the upper-

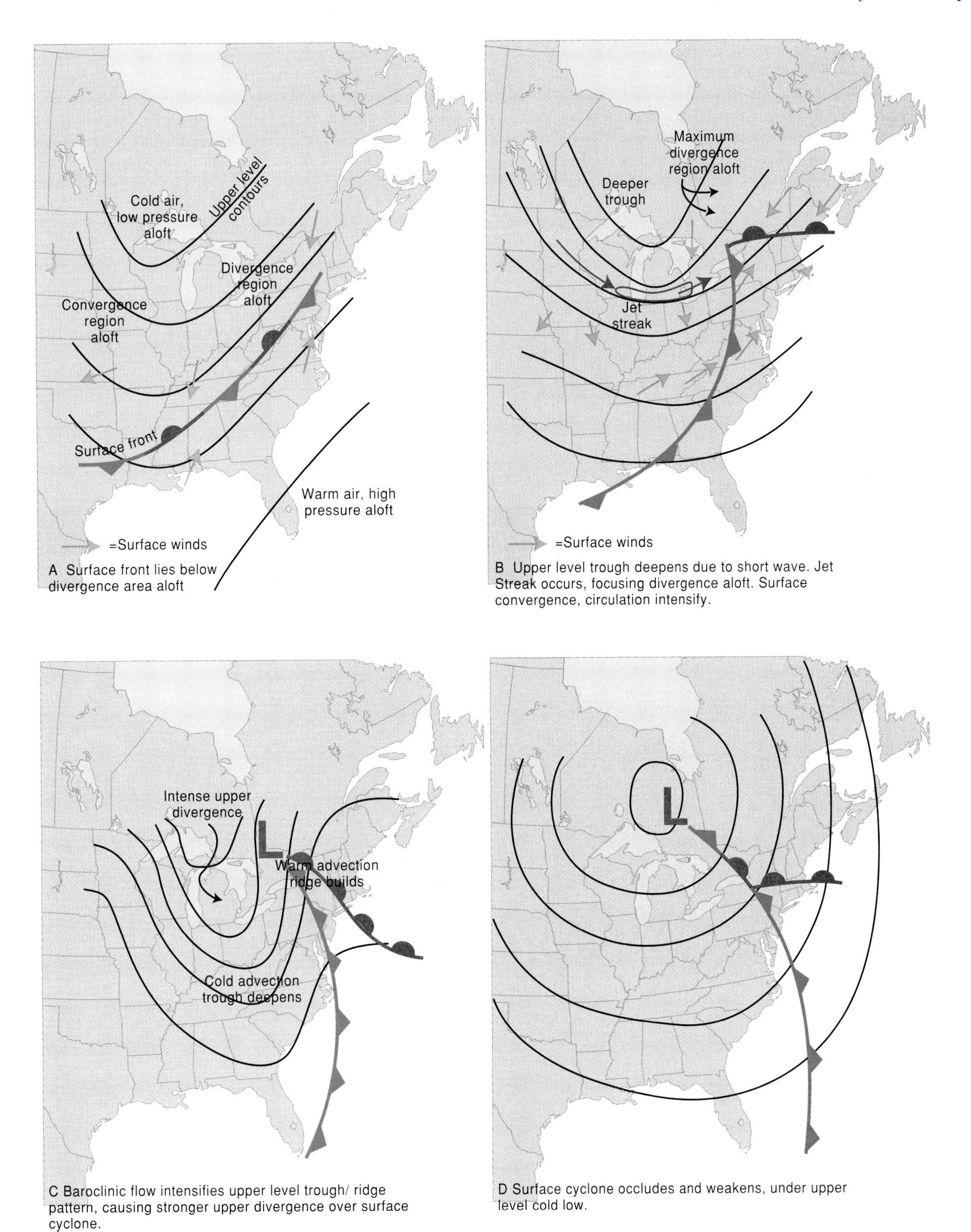


Figure 9.31 A three-dimensional frontal cyclone development model.

level trough becomes deeper. This, in turn, causes even greater changes in vorticity as the air moves into and out of the upper-level trough, which accentuates the divergence, which in turn increases the baroclinic circulation around the low center, which further accentuates the upper-level trough, and so on. This cycle, known as **baroclinic instability,** causes the cyclone to intensify to the stage shown in Figure 9.31C, which is equivalent to stage D of the Norwegian cyclone model.

As the storm develops through stages A to C, cloudiness and precipitation patterns change as well. In stage A clouds and precipitation lie in a rather narrow band north of the stationary front. As the circulation develops, these patterns spread around the low center, particularly to the north of the warm front. Warm air advection to the east of the low center causes pressure falls in that region to be greatest, and the low center moves in that direction, as explained in Chapter 8.

The feedback loop described above cannot last indefinitely, of course. Cold air continues to sweep cyclonically around the low center, eventually surrounding it, which is shown by the occluded front in Figure 9.31D. And cold and warm advection alters the upper-level flow pattern so that the greatest vorticity changes (and divergence) are no longer positioned over the surface low but along the front south of the occluded front. Deprived of the frontal temperature contrasts and upper-level divergence, the surface low ceases to intensify and begins to weaken. However, a new cyclone may develop subsequently at the triple point, where thermal and divergence support are strong.

## Retrospective: The Norwegian Cyclone Model

The Norwegian cyclone model contains some oversimplifications and suffers from its sparse treatment of what is happening aloft. Remarkably, however, the theory has retained its basic form despite advances in meteorological knowledge over the past 70 years. For a great number of the changes we see on daily weather maps, and the types of weather we associate with fronts and storms, the model works. We shall use the model often in the next chapter as we venture into the unknown—the realm of forecasting.

***Review***

The Norwegian cyclone model was used as background for discussion of the development and weakening of cyclones along a frontal boundary. Understanding cyclonic development requires familiarity with two fluid flow properties: divergence and vorticity.

Divergence is a measure of the change in horizontal area occupied by a parcel of air. If the area increases, there is positive divergence; if it decreases, there is convergence or negative divergence. Divergence is a product of nonparallel wind flow (directional convergence and divergence) and/or changes in wind speed (speed convergence or divergence).

Vorticity is a measure of the amount of rotation in a column of air. Like divergence, vorticity can be manifested in two ways. A change in an air parcel's direction of motion exemplifies the form of vorticity known as curvature. Shear is vorticity that occurs when wind speeds vary perpendicular to the line of flow.

Using an ice skater as an example, it was demonstrated that an increase in absolute vorticity is associated with convergence and a decrease in absolute vorticity is associated with divergence. Thus, by locating areas of decreasing vorticity in the upper atmosphere, you can identify areas of divergence, which are needed to maintain and strengthen surface low pressure areas.

Using vorticity changes that occur in long waves, jet streaks, and cross-mountain flow and divergence changes in baroclinic flow, the potential for low-level cyclogenesis was analyzed. Finally, a three-dimensional cyclone model was developed and compared to the classical Norwegian cyclone model.

### *Questions*

1. Explain why an increase in vorticity is associated with convergence and a decrease in vorticity is associated with divergence. (Hint: The example of the ice skater may be helpful.)

2. If you want to identify potential areas where cyclones are likely to develop, why is it important to be able to identify regions where upper-level divergence is occurring?

3. Given the upper air flow diagram below, describe the changes in vorticity and divergence at points A, B, and C and identify the area where low-level cyclonic development is likely.

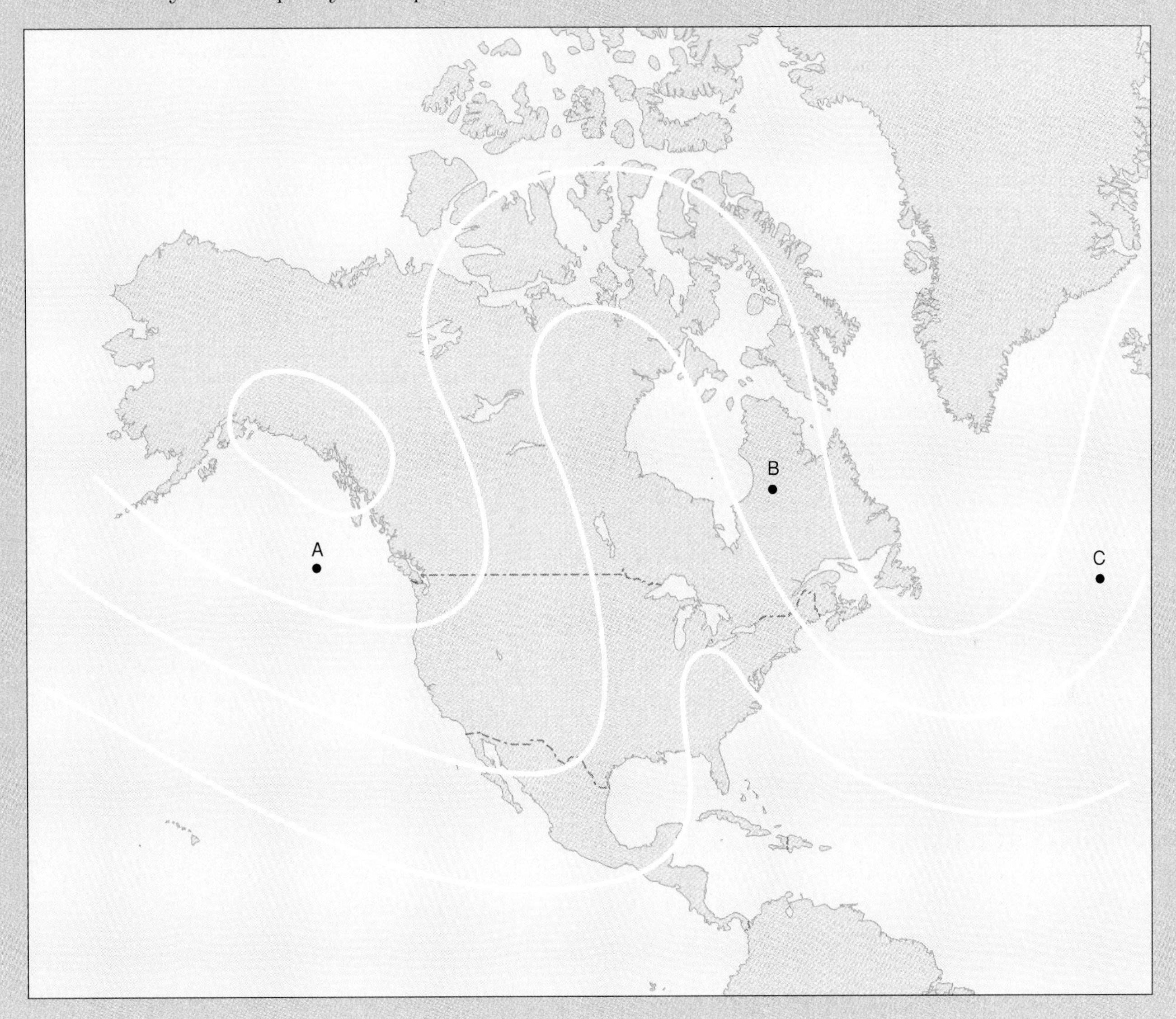

Study Point

## Chapter Summary

Fronts are boundaries between air masses possessing different temperature characteristics. Other contrasts exist in the vicinity of fronts, including wind shifts, pressure troughs, moisture (dew point) discontinuities, cloudiness, and areas of precipitation with varying characteristics. On weather charts, cold fronts and warm fronts are placed at the warm edge of the temperature transition zone. Thus, the change in temperature occurs on the colder side of the front.

The movement of the air on the cold side of any front determines the front's movement. If cold air advances, the front is called a cold front. If cold air retreats and allows warm air to advance, it is a warm front. If the cold air is neither advancing nor retreating, the front is stationary. When a cold front overtakes a warm front, an occluded front results. Occlusions are the most complex fronts to study because the weather at the ground may be caused by hidden frontal locations aloft.

Cold fronts have relatively steep slopes and commonly cause convective precipitation as cold air pushes retreating warm air aloft. In contrast, warm fronts typically have gentler slopes, which fosters development of layered clouds that often produce steady rain or snow. Stationary fronts are often associated with clouds and precipitation due to overrunning, even though there is little or no frontal motion. While fronts may remain stationary for an extended period, they eventually move or dissipate.

Frontal cyclones are large low pressure systems that form on the polar front and are major producers of clouds and precipitation in middle latitudes. The Norwegian cyclone model describes the life cycle of these storms from formation and development through occlusion and dissipation. The development stage is one of special interest because of its relevance to weather forecasting. Divergence and vorticity changes related to upper-level long wave and short wave flow patterns are important indicators of storm development, intensification, and dissipation.

## Key Words

synoptic meteorology
maritime polar (mP) air
maritime tropical (mT) air
backdoor cold front
dry line
frontogenesis
cold occlusion
continental polar (cP) air
continental tropical (cT) air
Arctic (A) air
cold air damming
overrunning
frontolysis
warm occlusion
triple point
Norwegian cyclone model
long wave
vorticity
relative vorticity
shear
lee trough
barotropic
polar front
short wave
Rossby wave
absolute vorticity
curvature
jet streak
baroclinic
baroclinic instability

## Review Questions

1. Explain why a cold front's slope is steepest (most vertical) near the ground and less steep aloft (see Figure 9.9).
2. Explain the differences in structure and associated weather of cold and warm occluded fronts.
3. What type of weather would you predict if a maritime tropical air mass moves into the eastern United States during the summer? Explain your answer.
4. Express the formula $\Delta\zeta_a/\Delta t = -$ divergence in words.
5. Why is fog a frequent occurrence in autumn?
6. What is the difference between a jet streak and a jet stream?
7. What is the weather map feature known as the "triple point"? What is its significance?
8. At what stage in the Norwegian cyclone model does a frontal cyclone reach its greatest intensity? Why does the storm weaken thereafter?
9. In Figure 9.3B, locate (a) speed divergence; (b) directional convergence; (c) positive relative vorticity due to curvature; and (d) negative relative vorticity due to shear.

## Problems

1. Approaching warm fronts can cause various types of weather that depend on the properties of the retreating cold air and the replacing warm air. Using the different properties of the air masses below, describe the types of weather that might be expected.

| Properties of Air Masses | Expected Weather |
|---|---|
| Sluggish cold air plus moderately moist warm air | |
| Sluggish cold air plus very moist warm air | |
| Quickly retreating cold air plus dry warm air | |

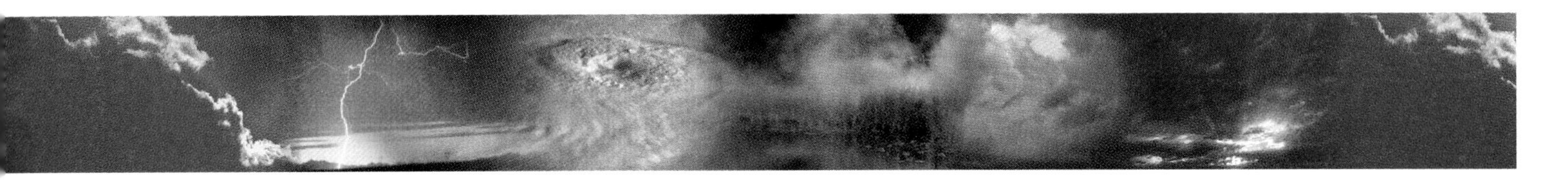

2. What air mass type would follow the passage of a backdoor cold front on the East Coast? A "normal" cold front in Indiana? A warm front in Tennessee? A dry line in New Mexico?
3. Compose a series of weather observations that typify the approach and passage of a cold occlusion. Do the same with a warm occlusion.
4. Broadcast meteorologists often make statements such as, "The jet stream will steer the low to the northeast. . . ." In what way is such a statement valid and useful? How is it incorrect?
5. Upper-level divergence supports cyclone development, so it is important to locate such divergence areas. Why is vorticity involved in the search for divergence areas?
6. Explain how air flowing through a jet streak can establish regions of divergence and convergence aloft.
7. What is meant by baroclinic instability? What is the connection between baroclinic instability and cyclogenesis?

## Explorations

1. Correlate the movement of highs, lows, and fronts through your locale as shown on weather maps with actual conditions you observe. For example, take hourly observations, photos, or both of the sky on a day when a cold frontal passage is expected. Document the sky's appearance within different air mass types. (In many locales, the difference between cP and mT skies can be dramatic.)
2. Download from online sources (such as those mentioned in Resource Links below) charts showing conditions at the 500-millibar or 300-millibar level. On these charts, locate regions of upper-level divergence using the vorticity-divergence relation. Track the movement of these divergence regions, and see under which circumstances cyclogenesis occurs in their vicinity.
3. There is no end of possible projects if you have access to weather graphics and data on the Internet: track motions of highs, lows, and fronts across the country; compare their motions to steering winds, correlate their positions with weather changes reported at the ground; compare weather map positions of fronts, highs, and lows with satellite images of clouds; and so on.

## Resource Links

The following is a sampler of print and electronic references for some of the topics discussed in this chapter. We have included a few words of description for items whose relevance is not clear from their titles. More challenging readings are marked with a C at the end of the listing. For a more extensive and continually updated list, see our World Wide Web home page at www.mhhe.com/sciencemath/geology/danielson.

Carlson, T. N. 1991. *Mid-Latitude Weather Systems*. London: Harper Collins, 507 pp. C Although this book is well beyond the level of this course, it contains much of interest for any reader.

Saunders, F., and C. A. Doswell III. 1995. A case for detailed surface analysis. *American Meteorological Society Bulletin* 76 (4) : 505–21. C

AccuWeather's home page at http://www.AccuWeather.com is an extensive source of current weather conditions; it also provides visible and infrared satellite images, and Doppler radar images, and current and forecast weather maps and discussions.

For readers with access to AccuData, SURMAP offers over one hundred different current and forecast maps showing highs, lows and fronts. SJET plots the jet stream. SIP allows the user to plot data at any geographical location. SATMAP provides a variety of worldwide satellite images; and UIP enables the user to plot upper air data, including heights of various pressure levels, and vorticity.

The Weather Channel's home page at http://www.weather.com/images/curwx.gif offers a current U.S. surface map with isobars, highs, lows, and fronts.

The weather server at Lyndon State College in Vermont offers a beautiful Northern Hemisphere jet stream map at http://apollo.lsc.vsc.edu/gopherdata/Pictures/wxgifs.hem/nhemjet.gif

Ohio State University's weather page at http://asp1.sbs.ohio-state.edu/ offers current vorticity values of the 500-millibar wind pattern.

The National Aeronautics and Space Administration's GOES satellite home page at http://climate.gsfc.nasa.gov/~chesters/goesproject.html offers past and present satellite images, including movies. This site is convenient for following the development of weather systems.

# Chapter 10

# Weather Forecasting

## Chapter Goals and Overview

At 2 o'clock on a recent June afternoon, a group of golfers could be found focused on the Weather Channel® on a TV on Hilton Head Island, South Carolina. They had little better to do since the Oyster Reef golf course had been closed and cleared when a National Weather Service severe thunderstorm warning had been issued for nearby Savannah, Georgia, a half-hour earlier. They could hear reverberating booms of thunder off in the distance, and the sun was a pale disk through the fringe of the storm's anvil top.

One golfer went to the clubhouse desk and observed to the manager that the golf course was in Beaufort County, South Carolina, and not Savannah. The manager retorted that the warning was for Savannah and nearby coastal waters, and in his opinion, the golf course abutted such coastal waters; the course would stay closed until the warning ended at 2:30 P.M. or was called off.

Back at the TV, a consensus based on Weather Channel radar animations was growing that the feared storm appeared to be going by to the south and missing the golf course. It had blown out windows and flattened trees in downtown Savannah, but the only effects at Oyster Reef so far were that muffled thunder could be heard and no golf could be played. Within a few minutes, the dozen or so golfers watching the TV had informally agreed that the short-term forecast for Oyster Reef was "no thunderstorms." Although a meteorologist (one of your authors) happened to be present, the forecast was made strictly by nonmeteorologists.

Like the golfers, watchers of today's weather reports on TV, with sophisticated satellite pictures and extensive radar coverage, can decide on their own what to expect in the next few hours. All that is needed is to determine where you are on the map, then see where the precipitation is moving.

But surely there is more to forecasting than watching where something is moving now and extrapolating to some time in the future. This chapter's topic is Unit 4's central question—"How are weather forecasts made?"

In this chapter, you will learn how you can participate in the kind of forecasting done at the golf course. You will also discover how a far wider range of forecasts is made, and you will be in a position to judge the likelihood of success of most any forecast you hear or see. In addition, you will learn why some forecasts turn out spectacularly accurate while others seem so far off the mark that they appear to have been made for another planet.

### Chapter Goals

By the end of this chapter, you will be able to:

describe and compare several basic forecasting methods;
use some of these methods to make a defensible forecast of weather map features and local weather;
explain how storm development can add compl the forecasting process;
specify what mix of data and other information orologist considers before issuing a forecast;
correctly interpret a weather forecast, and discuss the differences in approach taken by meteorologists in the public and private sectors; and
explain the roles of the public and private sectors in forecasting.

## Forecasting Techniques

As you will see presently, there is no one method of forecasting the weather. Meteorologists employ a variety of methods; in fact, a given forecast is generally the product of several methods. In this section, we will survey the most basic methods.

### Persistence Forecasting

Suppose it is raining at the moment. What is the best forecast for five minutes from now? Almost certainly, it is "rain." For an hour from now? Still, in many cases, "rain." **Persistence forecasting** is based on the idea that the weather will stay the same. Persistence forecasts are highly accurate over short time periods.

In some locales and weather situations, persistence forecasting can be successful for periods of as much as several days. Eventually, however, the weather will change, and by its very nature, a persistence forecast will not predict that change. Persistence forecasts over a period of more than a few hours, however, generally contain some recognition of diurnal changes. After all, local temperatures respond to daily cycles of solar heating and darkness, and these changes set in motion other changes. Thus, if today's weather consists of early morning fog that burns off by 9 A.M., followed by clear skies except for afternoon cumulus, a persistence forecast for tomorrow would incorporate those same changes. In this sense, a persistence forecast means that conditions at a given time tomorrow will parallel those today at the same time.

### Trend Forecasting

Consider the situation in Figure 10.1: a cold front 310 miles to your west is causing a 20-mile-wide band of thunderstorms as it moves eastward at 30 miles per hour. The center of the band of thunderstorms coincides with the actual front. This means the leading edge of the thunderstorms is 300 miles to your west.

When will it start to rain at your location? If you said 10 hours, you would be correct (10 hours × 30 miles per hour = 300 miles).

This example is a simple application of **trend forecasting**. Trend forecasting is based on the assumption that weather features such as fronts, pressure centers, and their associated weather will persist and will move in the future as they have in the past. **Extrapolation** is a more general term for projecting past trends into the future.

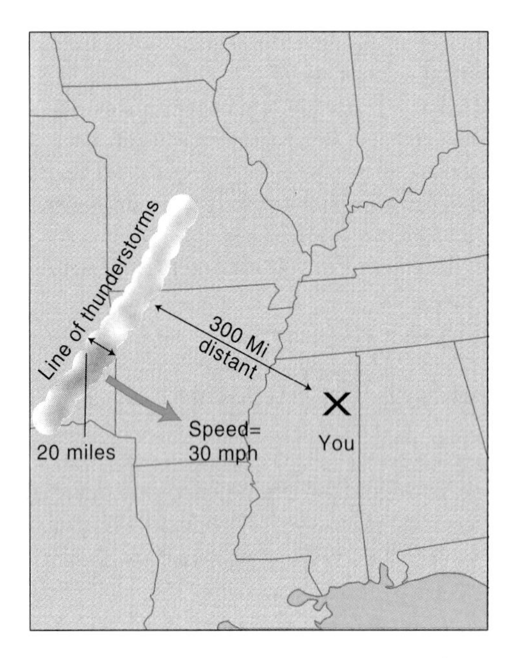


Figure 10.1 Knowing the distance to a thunderstorm line and its forward speed, you can predict its time of arrival.

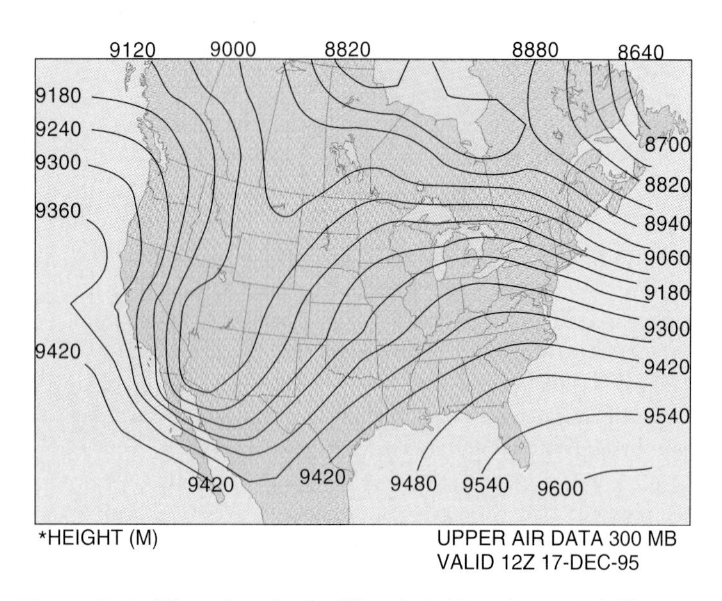


Figure 10.2 Three-hundred-millibar height contours at 12Z, December 17, 1995. Note the long wave trough over the Rocky Mountain states and the ridge over the Midwest/Ohio Valley.

Trend forecasting need not be confined to steady-state weather systems, as long as whatever is changing does so at a constant rate. Consider, for example, a storm that is causing a 100-mile-diameter area of rain in the morning, a 150-mile-diameter area of rain in the afternoon, and a 200-mile-diameter rainfall area later that night. In this case, there is a clear trend in the system's change: it is increasing in size by 100 miles per 12-hour period. Thus, a trend forecast would suggest that the rain area would be 300 miles in diameter by the next morning.

Of course, trend forecasts will be accurate for only so long. Eventually, a steadily moving system will change speed or direction, or will stop growing. Nevertheless, trend forecasts often provide useful information for the forecaster. If you know where the features are, along with their size, direction, and speed of movement, you can lay out a schedule of expected weather events for the region of interest. Then comes the hard part: determining when and how weather systems will change and thus throw the trend-based forecast off track. You'll learn some of the ways this is done later in this chapter.

## Cycles and Trend Forecasting

### *Diurnal Cycles*

We briefly mentioned diurnal variations in connection with persistence forecasting. We concluded that even in a persistent weather pattern, it is important to consider diurnal variations. Most such variations can be traced directly to the sun's daily heating cycle. In addition to the sun's direct effects, it also sets other cycles in motion. Relative humidity decreases during the day and increases overnight because of the changes in the difference between the temperature (which changes considerably) and the dew point (which is relatively conservative).

The wind often follows diurnal cycles. Near the seashore, daytime heating over land causes the air to rise. Cool, dense air from over the ocean moves in to replace the air that left. Predicting when and to what degree this diurnal change will occur can help boaters and those heading for the beach. Elsewhere, when solar heating occurs at the ground, it remains cooler aloft. After a while, the air becomes unstable and starts to overturn. If the winds aloft are stronger than they are near the ground (as is usual), some of the higher speeds are transferred to the ground during the overturning process. Thus, wind speeds follow a diurnal cycle.

One semidiurnal (every 12 hours) effect you could notice if you have a home barometer is the slight twice-daily rise and fall of barometric pressure that accompanies atmospheric tides. Typically, the pressure rises during the morning and falls during the afternoon; it rises in the evening and falls later at night. Someone unaware of these changes might mistakenly believe that a high or low pressure area is approaching when what is happening is just the passing of the tides. Forecasters need to be careful not to be duped by this effect.

### *Three-to-Four-Day Cycles*

Notice the long wave trough and ridge in the contour pattern of the 300-millibar chart in Figure 10.2. You might recall from Chapter 9 that long waves (or Rossby waves) like these generally move slowly or may become stationary for up to weeks at a time. For a locale under the influence of a stationary Rossby wave, the weather may seem to be in a rut wherein it rains at some regular interval, such as every weekend. In the case of Figure 10.2, New Mexico and western Texas lie just east of the long

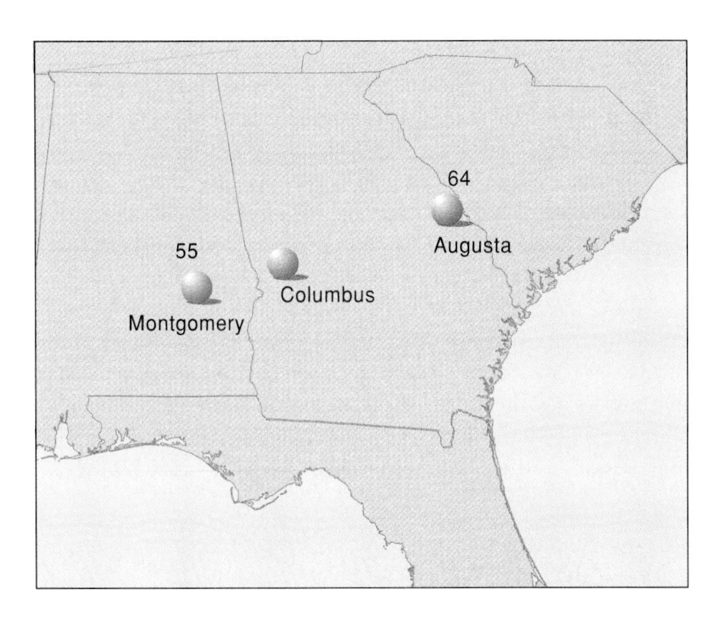


Figure 10.3 Interpolation allows us to estimate values between two (or more) data points. What is a likely temperature in Columbus?

wave trough in a region of upper-level divergence, which you learned in Chapter 9 is a preferred locale for storm development. Thus, short waves, which commonly recur every three or four days, may intensify regularly in this region until the long wave pattern shifts. Double the three-to-four-day time interval and you can see that a storm might visit every six to eight days as well. If this cycle begins on a weekend, a whole series of wet weekends may follow. Forecasters sometimes are blamed for this type of thing, but it is really the fault of those who create our calendars. Why would anyone want to schedule a weekend when it's going to be miserable outside?

## Interpolation

Suppose you know the temperature at Montgomery, Alabama, and Atlanta, Georgia (see Figure 10.3), but you are interested in the temperature in Columbus, Georgia, which lies about one-third of the way from Montgomery to Augusta. Can you predict the temperature in Columbus? You can, if you have reason to believe that temperatures change in some consistent way across the region. Since Columbus lies one-third of the way from Montgomery to Augusta, it would be reasonable to expect its temperature would also be one-third of the way, or, in this case, 58°F. This process of estimating conditions between two locales, or between two times, is known as **interpolation**. Note the difference between interpolation and extrapolation. Interpolation, because it involves predicting conditions between two known data points, is generally more reliable than extrapolation, which requires estimating conditions into space or time for which we have no limiting values on the far end.

The ability to interpolate accurately can be important. On November 15, 1995, a heavy snowstorm dumped 16 inches of snow on State College, Pennsylvania, but only 4 inches at Lewistown, 30 miles away. A major highway runs between the two towns. Accurate interpolation between official observation points was useful in advising motorists where to expect trouble and in more effectively deploying crews to plow the roads.

## Steering

A forecasting technique similar to trend forecasting is based on a convenient fiction: that winds aloft "steer" surface weather patterns. For example, you may recall being told on TV weathercasts that a low pressure area at the surface will be steered in a certain direction by the upper air jet stream. You are left with the impression that the upper winds are going to physically blow the storm one way or another. If you think about that idea, it doesn't really make much sense. How can a wind at an altitude of 10 kilometers (about 30,000 feet) move something that is on the ground? Instead, what happens, as you saw in Chapter 9, is that surface patterns of pressure continually redevelop as dictated by upper-level convergence and divergence patterns. Now, it turns out that this continual redevelopment often occurs in the direction of the upper-level wind flow, in part because of the role played by jet streaks (recall Figure 9.28 in Chapter 9). Thus, although it does not happen literally, it *appears* that the upper-level winds steer surface weather features, as river currents sweep leaves and logs along in their path. So although, meteorologists realize that the physical mechanism is not actually "steering," nonetheless, they use the fiction in preparing trend forecasts.

Of course, as the upper airflow changes, those changes will affect the future movement of the storm. For this reason, **steering** generally is a reliable indicator of storm movement for only relatively short time periods (less than a day).

## Weather Types

**Weather types** are a set of recurrent patterns with recognizable characteristics. For example, most major East Coast snowstorms come from the eastern Gulf of Mexico and strengthen along the North Carolina coast. This could be called an East Coast snowstorm type. Storms that move inland through the Pacific Northwest often cross the Rockies and then reorganize in the Plains. They then track northeastward through the Great Lakes into eastern Canada. Such storms cause wind-whipped snow on their northern and western sides, while spawning showers and thunderstorms along the cold fronts that whip around their south sides. A third weather type involves a storm that moves from the Pacific into southern California. It moves toward the southern Plains and then heads east to northeast.

Forecasters who can recognize that a given weather system is of a certain type are better equipped to make an accurate prediction. At one time, many meteorologists believed that the main use of computers in weather prediction might be to automate and refine the picking of weather types. It has turned out that computer-based approaches that calculate future atmospheric conditions from basic physical and mathematical equations, and not databases of weather types, have become the dominant method of computer-based weather prediction. However, weather types forecasting is still being explored in the development of so-called expert systems, a type of artificial intelligence: a group of experts on a particular topic (weather system types, in this case) input their expertise into a computer database to develop a set of rules the computer uses to predict outcomes.

## Numerical Weather Prediction

Forecasting has changed radically over the past 40 years. Much forecasting effort once involved manual analysis of surface and upper air charts, plus a variety of extrapolation techniques and general rules of thumb to project where weather features would move in the future and what effects they would have. While meteorologists had to learn a great deal about atmospheric physics as part of their training, the actual application of this work was much less formal and took on the characteristics of an art form.

Today's numerical models follow algorithms similar to the one explained in Chapter 2. They begin with a set of initial weather conditions and then use the equations of atmospheric motion and/or an analysis of atmospheric wave forms to project how the situation is expected to change. Once the future atmospheric pressure and moisture patterns are numerically predicted, other programs translate that prediction into actual weather conditions at specific sites. For example, suppose the computer model predicts moderately unstable conditions, with a relative humidity of 65% in the lowest level of the atmosphere and 55% at higher elevations. What are the chances of precipitation there? It is one thing to have a forecast of relative humidity or stability or temperature but quite another to predict how much cloudiness there will be, whether thunderstorms will develop, or what the temperature at the ground will be each hour. The prediction of those actual components of the weather, called the sensible weather, is the primary forecast product we get from "Model Output Statistics," or MOS, prepared every 12 hours by the National Weather Service.

## Climatology

The climatic data for any location should always be consulted before a forecast is made. This is because the data contain the long-term averages and extremes for that site. How should the information be used? Its best use is to establish a frame of reference. If a very cold air mass is coming, the forecaster may wish to find out what the record low temperature is for that time of year. While new records can always be set, it generally pays to be cautious about predicting them until a complete study of the maps, charts, and other weather factors has been completed. If a given locale has 125 years of records and has never been colder than 10 degrees on November 1, you know the odds of it dropping below that level on the next November 1 are quite low.

When using the climate data in a "typical" weather pattern, you can expect temperatures and precipitation to approximate the long-term averages. An additional use for climatological data in forecasting concerns the long range—the period beyond 10–15 days. While some trends or deviations from normal can sometimes be identified even up to a year ahead, daily details cannot be forecast in this manner. Therefore, the best forecast for any given place a year in advance is to "go with climatology."

Once the forecast is disseminated, the job of monitoring and updating it begins. Forecasting is a constant challenge, but it is the variety and ever-changing nature of the weather situation that makes weather forecasting a passion rather than just a job to many of its practitioners. As a further reward, forecasters know that a good forecast can help to save lives and avoid injuries.

***Review***

This section provided an introduction to specific types of forecasting techniques. Persistence forecasting is based on the assumption that the weather will remain the same as it has been. This technique is quite reliable in the very short term, but if projected for more than a few hours, it must take into consideration diurnal changes.

Trend forecasting is based on the idea that weather features such as fronts and pressure centers with their associated weather will persist and move in the future as they have in the past. Since all weather patterns eventually do change, the challenge in trend forecasting is to predict when, how, and where the change will occur.

Interpolation is a technique of estimating conditions between two locations or two times. The underlying assumption of this technique is that the weather varies somewhat consistently between two locations or times.

Steering forecasting techniques rely on the idea that upper air jet stream steers the surface weather as a river steers a floating log. In actuality, the atmospheric physics is not one of steering but instead one of surface pressure patterns continually redeveloping as dictated by upper air convergence and divergence, often in the direction of the upper wind flow.

Weather type forecasting is based on the idea that certain weather systems or patterns with recognizable characteristics tend to recur.

Today's numerical forecasting models use powerful computers to solve equations of atmospheric motion and thus predict atmospheric pressure and moisture patterns. These predictions are then translated by other programs into forecasts of actual weather conditions at specific sites.

Climatological data includes the long-term averages and extremes for given locations.

Short-term forecasts tend to rely on persistence, while very long-term forecasts rely on climatology. Neither of these involves much skill or creates much of a challenge for forecasters. The intermediate times present the greatest challenges and require the greatest skill to analyze the interactions of trends, weather types, and steering influences.

***Questions***

1. List the different weather forecasting techniques mentioned in this section, and give an example of each in practice.
2. Some meteorologists consider trend and persistence forecasting to be one and the same. In what ways are they similar? How do they differ?

## Forecast Case Study: A Summer Night in Pittsburgh

To introduce you to the wide range in techniques used to forecast the weather, we'll use a case study from a summer evening in Pittsburgh in July 1995. After proceeding through the case, we'll back up and look at forecasting in a systematic way. We focus on one meteorologist as she moves through the process of reviewing the relevant data and applying it to produce a forecast.

### First Step: Checking Present Conditions

Before you can predict future weather, it's essential to become familiar with current conditions. Since our meteorologist is making the forecast around 11 P.M. she will have difficulty getting good visual clues about the weather. However, she can look at the hourly weather observations for the last 24 hours or so, which are shown in Table 10.1. In the absence of any other

*Table 10.1* **Observed Weather Conditions at Pittsburgh on July 22, 1995**

| Time | Temperature (°F) | Relative humidity (%) | Wind direction/ speed (mph) | Pressure (inches) | Visibility (miles) | Weather |
|---|---|---|---|---|---|---|
| 12am | 71 | 84 | W/6 | 29.99 | 6 | Haze |
| 1am | 71 | 81 | WSW/3 | 30.00 | 4 | Haze |
| 2am | 71 | 84 | W/5 | 29.98 | 4 | Haze |
| 3am | 71 | 84 | NW/6 | 29.97 | 4 | Haze |
| 4am | 71 | 84 | N/7 | 29.97 | 4 | Haze |
| 5am | 70 | 87 | W/7 | 29.99 | 3 | Light fog |
| 6am | 70 | 87 | NNW/5 | 29.99 | 2 | Light fog |
| 7am | 69 | 93 | S/5 | 29.99 | 2 | Light fog |
| 8am | 70 | 87 | SW/6 | 30.00 | 1 1/2 | Light fog |
| 9am | 72 | 81 | S/6 | 30.00 | 1 1/2 | Light fog |
| 0916 | | | S/3 | 30.00 | 2 1/2 | Light fog |
| 10am | 74 | 75 | WSW/7 | 30.00 | 3 | Haze |
| 11am | 76 | 76 | WSW/3 | 30.00 | 4 | Haze |
| 1127 | | | SSW/6 | 30.00 | 4 | Haze |
| 12pm | 80 | 59 | S/5 | 29.99 | 4 | Haze |
| 1pm | 82 | 58 | SSE/5 | 29.98 | 4 | Haze |
| 2pm | 79 | 64 | SW/8 | 29.97 | 4 | Haze |
| 3pm | 82 | 60 | WSW/6 | 29.96 | 4 | Haze |
| 4pm | 83 | 58 | SW/9 | 29.93 | 4 | Haze |
| 5pm | 85 | 52 | W/8 | 29.92 | 4 | Haze |
| 6pm | 84 | 54 | WSW/13 | 29.91 | 4 | Haze |
| 7pm | 83 | 56 | SW/10 | 29.91 | 5 | Haze |
| 8pm | 80 | 59 | SSW/7 | 29.90 | 5 | Haze |
| 9pm | 78 | 63 | SSE/7 | 29.90 | 5 | Haze |
| 10pm | 75 | 70 | S/7 | 29.89 | 5 | Haze |
| 11pm | 75 | 73 | S/6 | 29.91 | 6 | Partly cloudy |

information, this would allow her to make a forecast based on persistence and local trends.

Actually, one type of observation is easier to make at night than during the day: the observation of lightning. If you are away from the greatest concentration of urban lights, lightning shows up quite well. If you follow the shifting locations of lightning for a period of time, you can usually get a sense of where individual thunderstorm cells are located, which way they are moving, and how quickly they are advancing. Flash frequency can also suggest whether the storms are intense. If the flashes are diffuse, it may be because rain is hiding the individual streaks. Sometimes the flashes illuminate the clouds long enough for you to discern some pattern to the cloudiness.

Notice in Table 10.1 that the most recent barometer reading is lower than it was yesterday at this time. This suggests that a high pressure area is leaving and a low pressure area is drawing closer. The observed south wind is consistent with this idea. Recall Buys Ballot's law, which states that when the wind comes from your back (in this case, the south), lower pressure is to your left (the west, in this case). Since most mid-latitude weather systems move from the west toward the east, the south wind suggests the presence of a low "upstream."

The greatest pressure fall, according to Table 10.1's data, occurred during the afternoon, but since then, the values have leveled off. Does that mean the pace of change is decelerating? Perhaps, but recall that air pressure tends to fall during the afternoon hours due to tidal effects.

## Climatology

Another potentially useful piece of information is knowledge about the area's climate at the time of year for which you are forecasting. Table 10.2 shows the relevant climatological data for Pittsburgh. In this case, long-term records (called normals when they apply to the 30-year base period ending in the year of the last national census) suggest a daily high temperature of 83°F and a low of 62°F. Inspection of the prior day's hourly data shows the high temperature was very close to normal, but the morning low temperature was well above normal. The haze and relatively high humidity help explain the warmer-than-normal low temperature. Recall that water vapor is a greenhouse gas and that high concentrations will tend to prevent long wave radiational heat loss, thus keeping nighttime temperatures from falling.

Based only on persistence and climatology, a reasonable forecast at this point might be: Continued hazy, warm and humid tonight through tomorrow. Low tonight about 70; high tomorrow 83.

## The Radar Image

Moving outward from the local Pittsburgh data, our meteorologist wants next to see the latest Pittsburgh-area Doppler radar scan. If thunderstorms were imminent, it would dictate a different order of work than if storms were absent or a distance away. As you can see from Figure 10.4, there are storms in Ohio, but

Table 10.2 **Observed Weather for July 15–21, 1995, and July Climatological Data for Pittsburgh**

| | TEMPERATURE DATA | | | | | | | | | | | |
|---|---|---|---|---|---|---|---|---|---|---|---|---|
| | ACTUAL | | | NORMAL | | | | | RECORD | | | |
| DAY | HI | LOW | AVE | HI | LOW | AVE | DEPT | HDD | HI | YEAR | LOW | YEAR |
| 15 | 100 | 71 | 86 | 83 | 62 | 72 | +14 | 0 | 100 | 1995 | 49 | 1960 |
| 16 | 91 | 70 | 81 | 83 | 62 | 72 | +9 | 0 | 103 | 1988 | 48 | 1960 |
| 17 | 87 | 72 | 80 | 83 | 62 | 72 | +8 | 0 | 101 | 1887 | 52 | 1970* |
| 18 | 84 | 68 | 76 | 83 | 62 | 73 | +3 | 0 | 101 | 1878 | 49 | 1971 |
| 19 | 82 | 62 | 72 | 83 | 62 | 73 | −1 | 0 | 97 | 1894* | 52 | 1909 |
| 20 | 86 | 62 | 74 | 83 | 62 | 73 | +1 | 0 | 100 | 1878 | 45 | 1965 |
| 21 | 80 | 68 | 74 | 83 | 62 | 73 | +1 | 0 | 99 | 1885 | 46 | 1966 |
| 22 | M | M | M | 83 | 62 | 73 | M | M | 96 | 1933* | 50 | 1965 |
| 23 | M | M | M | 83 | 62 | 73 | M | M | 98 | 1933 | 50 | 1985* |
| 24 | M | M | M | 83 | 62 | 73 | M | M | 98 | 1934 | 52 | 1977 |
| 25 | M | M | M | 83 | 62 | 73 | M | M | 96 | 1934* | 54 | 1911 |
| 26 | M | M | M | 83 | 62 | 72 | M | M | 96 | 1934* | 51 | 1977* |
| 27 | M | M | M | 83 | 62 | 72 | M | M | 96 | 1941* | 44 | 1977 |
| 28 | M | M | M | 83 | 62 | 72 | M | M | 96 | 1993 | 47 | 1962 |
| 29 | M | M | M | 83 | 62 | 72 | M | M | 96 | 1901 | 52 | 1982 |
| 30 | M | M | M | 83 | 62 | 72 | M | M | 94 | 1988* | 50 | 1968* |
| 31 | M | M | M | 83 | 62 | 72 | M | M | 90 | 1975 | 50 | 1965 |

Note: M = Missing Data; Dept = temperature departure from normal for data; HDD = heating degree days; * = and other years.

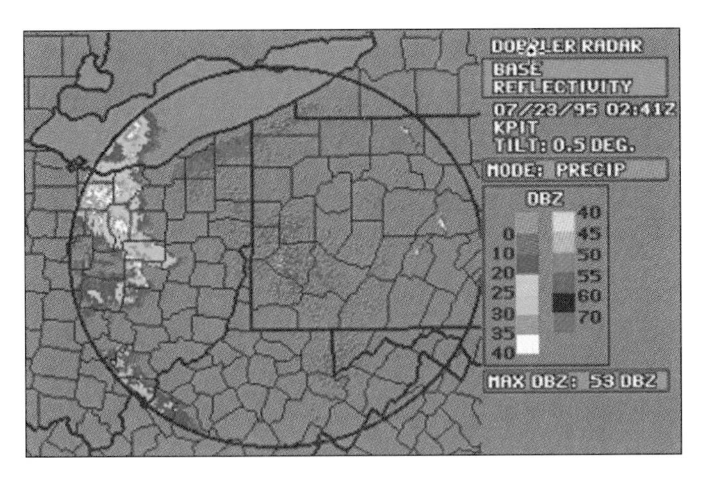


Figure 10.4 Doppler radar map, 02:41Z, July 23, 1995 (10:41 P.M., July 22, 1995), Pittsburgh. Note the strong echoes on the western perimeter of the display.

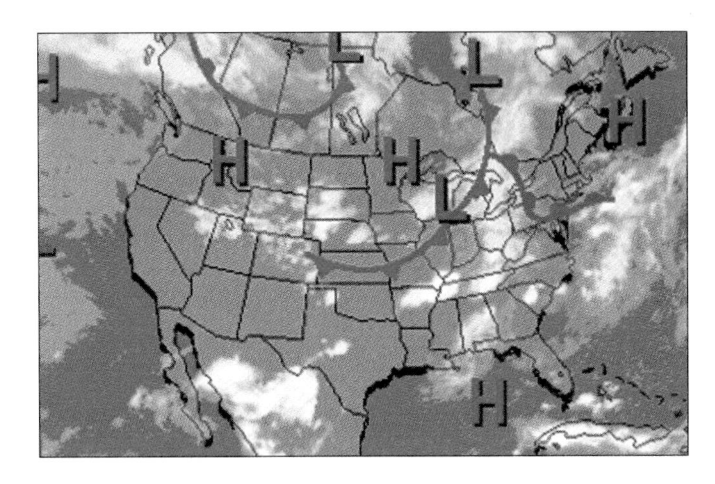

Figure 10.5 Satellite view of clouds, with weather system symbols superimposed.

they are still some distance from Pittsburgh. The precipitation is painted in green and yellow at the left edge of the radar screen. You can also see some precipitation straddling the Ohio River at the southwest edge of the picture. What about the blue pattern scattered near Pittsburgh and just inland from the Lake Erie shoreline in northeast Ohio and northwest Pennsylvania? That kind of return is seen commonly in the summertime and may represent insects or pollen. If you look at the color scale that shows the intensity of radar returns, the blue dots all fall within the range of the first three color blocks, representing 20 dbz or less. Dbz values are decibels and indicate the strength of the reflected radar energy. The higher the value, the stronger the return. In the summertime, 20 dbz is considered to be about the threshold between actual precipitation and other atmospheric reflectors. In winter, signals in the 20 dbz range are sometimes indications of light snow.

You might notice in Figure 10.4 that a thin band of rain seems to be jutting out ahead of the main body of precipitation. If the precipitation is following the customary west-to-east movement we often see at the latitude of Pittsburgh, this will be the first precipitation received by places downwind. The meteorologist will want to track the rain to see if she can time the onset of the rain for various places. At this stage, we cannot say how wide the rainfall area is. We would have to look at radar images made at places farther west to determine this important fact. Why is the rain area's width important to determine? People need to know not only when rain will start but how long it will likely continue. Knowing the rain area's size and its velocity, a forecaster can calculate the expected duration of precipitation.

## Synoptic Scale Maps

After checking the local radar, the meteorologist takes a broader view of the overall pattern across North America. One way of doing this is by analyzing a detailed map showing plots of current weather at the earth's surface at this stage. Another is to examine a chart like that in Figure 10.5, an infrared satellite image upon which fronts and pressure systems have been drawn. Strictly speaking, infrared pictures show areas with different temperature, not clouds. Fortunately for meteorologists, however, the temperature changes as you ascend through the atmosphere. Recall what you learned in Chapter 3 about radiation and how objects radiate in relation to their temperature. If we know what temperature each radiation (brightness) level on the picture corresponds to and the temperature profile of the atmosphere at the time of the picture, then we can determine how high the cloud tops are. As a general rule, the greater the vertical extent of cloudiness, the more potent the weather system and the greater the chance precipitation is being produced.

On the picture this evening (Figure 10.5), there is a band of cloudiness near the cold front extending from Quebec to Wisconsin and another band of cloudiness from Michigan and southwest Ontario south across Ohio and down to Tennessee. Close inspection shows a little finger of thicker (higher) clouds extending from eastern Ohio to western Pennsylvania. This corresponds to the finger of rain we saw on the radar image. However, you cannot easily see where it is raining just from the satellite picture. One very obvious problem with trying to do so is that the satellite is looking at the scene from above, whereas the rain falls out of the hidden bottoms of the clouds.

Meteorologists can infer precipitation locations from the structures they see in satellite images. Radar images, however, give much more accurate indications of precipitation's extent and intensity. For this reason and others, a composite of all radar reports across much of the country is useful. Such a composite, shown in Figure 10.6, makes it easy for the forecaster to relate areas of precipitation to the map and satellite views.

In addition to these images, there are a variety of satellite views, each designed to focus closely on one or more specific atmospheric parameters. Water vapor images look like giant rivers in the sky. Various infrared views allow a detailed look at clouds at different levels in the atmosphere. During the daytime, we

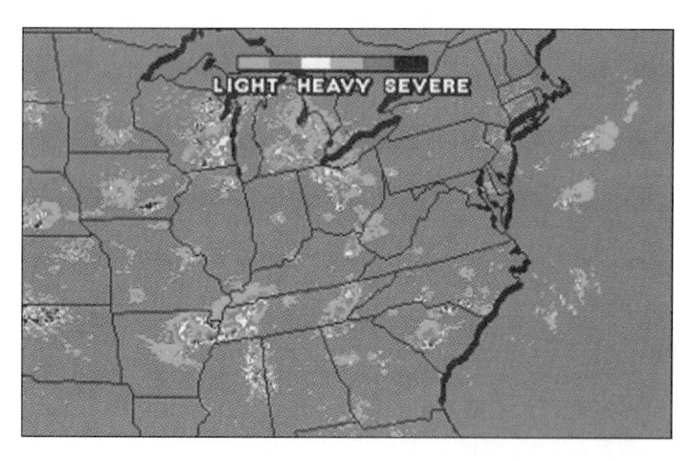


Figure 10.6 Doppler Radar Composite at 02Z, July 23, 1995. Compare this map with the lightning strike map in Figure 10.7 to see which radar echoes represent thunderstorms.

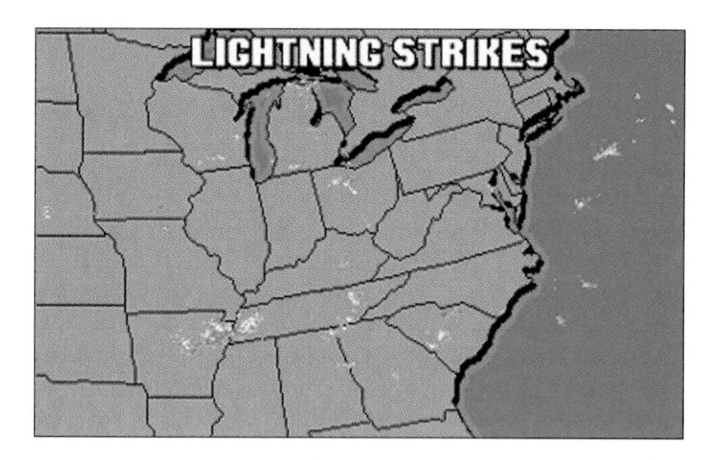


Figure 10.7 Lightning strike map, 01Z-02Z, July 23, 1995.

receive satellite images taken in visible wavelengths as well as the infrared.

The meteorologist is curious about which radar precipitation areas represent showers and which include thunderstorms. The lightning strike map, Figure 10.7, solves this riddle. She notes the storms in Ohio have a concentration of lightning that will have to be tracked during her shift.

Our meteorologist might consult the jet stream map shown in Figure 10.8 or upper air charts (at the 500- and 300-millibar levels, for example) to look for signs of "upper-level support" for shower and thunderstorm development. Recall from Chapter 9 (see Figure 9.27) that regions of decreasing vorticity, found typically to the east of upper-level low pressure troughs, are also regions of upper-level divergence and upward vertical motion, which promote convective development. Remember also (from Figure 9.29, for example) that decreases in shear as air moves through a jet streak are also divergence regions.

From the data considered so far, our meteorologist knows that a cold front is approaching, that it seems to be setting off thunderstorms ahead of it, and that the jet stream map shows

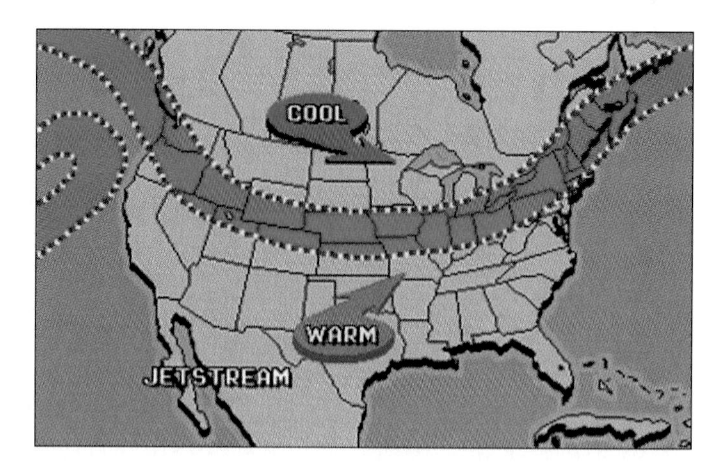


Figure 10.8 Jet stream map at 00Z, July 23, 1995.

that Pittsburgh is east of an upper trough, a likely place for clouds and precipitation. She will also study temperature maps and temperature change maps (such as 24-hour changes) to see how strong the frontal effect is and how much the temperature falls behind it.

Before proceeding, the meteorologist wants to return to the climatology data for this time of year. This establishes a good reference frame in which to consider the current pattern. If the upper airflow deviates in some way from normal, it will affect the temperatures. By assessing the amount the pattern deviates from long-term averages, the meteorologist can make a good estimate about how the temperatures will depart from normals as well. Table 10.2 shows values of normal and record temperatures at this time of year. The meteorologist summons this data to the screen, repeating a step she took earlier. Forecasting often involves mulling over the same information time and again in the hope of ferreting out more information that will help with the prediction.

## Computer Guidance

Now, it is time to look at some of the "numerical guidance," as computer-based analyses and forecasts are called. The meteorologist wants to assess of how long the threat period will be for showers and thunderstorms. Figure 10.9 is a 12-hour computer forecast (valid at 8 A.M. tomorrow, Sunday, Pittsburgh time) of relative humidity at the 850-millibar level, which is about 5000 feet overhead. Note that a concentration of high moisture values is centered in eastern Ohio and western Pennsylvania. It is reasonable to conclude that this moisture is contributing to the thunderstorms that are showing up on radar. At this point, the forecaster may want to get a map showing the initial moisture distribution (12 hours earlier than this map). This would help her get a sense for how quickly the model was predicting the moist area would move. By matching up the moist values on the maps with the actual arrangement of clouds and precipitation, the forecaster can make an assessment about how well the model is handling the weather system and what problems or deviations to look for. In any case, keep in mind that the 12-hour forecast

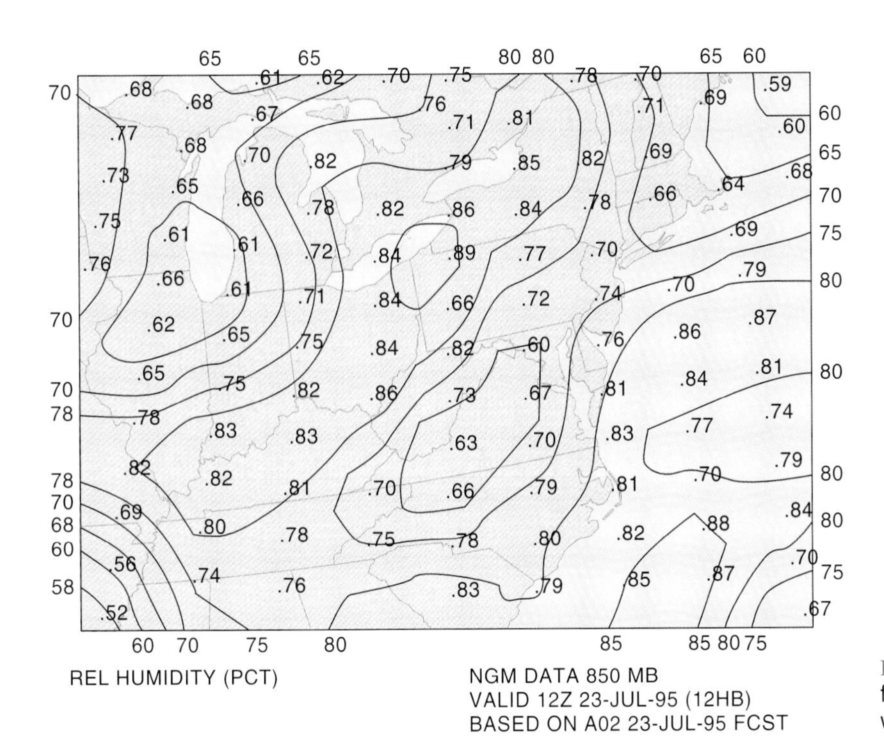


Figure 10.9 Twelve-hour, 850-millibar relative humidity forecast for 12Z, July 23, 1995. What are the values in western Pennsylvania?

applies at 12Z on July 23. This is about nine hours after the time when the forecaster is doing her work.

The computer predictions are based on conditions observed at 00Z on July 23 (8 P.M. Pittsburgh time on July 22). When the computer makes the forecast, it is solving complex, second-order nonlinear differential equations. Since these equations do not have exact solutions, the computer program actually makes a series of approximations to represent the various meteorological parameters. The projections are made in very small increments of time, perhaps in steps of five minutes each. As of the mid-1990s, it still took some of the world's most powerful computers to handle the volume of data and calculations these forecasts require—if there was to be any hope of completing the forecast in a timely fashion.

The output from computer models is in several forms. The maps in Figure 10.9 are one example. Another are the specific data predicted for a given locale, shown in Table 10.3. Notice that the data come from three different computer models: the European model, the ETA model, and the Nested Grid Model (NGM). Note also that the three different models' output do not exactly agree! The human forecasters must choose which model, if any, to believe.

The statistical output in Table 10.4 is based on NGM model output but was designed specifically by Accu-Weather Inc., a private forecasting company, to enhance its forecasts for various clients. The table lists excerpts of the predicted values of various weather parameters for each hour out to 48 hours.

One interesting aspect in Table 10.4 is the rainfall prediction. The model predicts 0.08" of rain by 12Z (8 A.M.) Sunday, July 23. However, the rainfall is not assessed as a certainty. In the row labeled "PRB LIQUID (%)", there is an estimate of the probability of getting 0.01" or more of rain during the period specified. This probability is 53%. The duration (next row) is predicted to be one hour. This type of information can be quite useful to some weather forecast consumers. For example, it can be helpful if a meteorologist can tell a client that yes, it is going to rain, but no, it shouldn't last more than an hour.

## Moisture and Stability

Stability often is a crucial variable in summertime forecasting. Our forecaster spends some time reviewing stability information.

### *Moisture Aloft*

After studying conditions at the 850-millibar level, the meteorologist looks at the moisture situation aloft. She knows if she finds low-level moisture topped by drier air, the atmosphere may become more unstable and chances of severe thunderstorms would increase. Why? Recall from Chapter 6 that the stability of a lifted air layer decreases rapidly when lower parts of the layer cool at the moist adiabatic rate, while upper levels cool at the greater, dry adiabatic rate. The resulting increase in the difference between lower- and upper-level temperatures makes the air even more likely to overturn or become unstable. This is a classic setup for thunderstorm development. Figure 10.10 shows the relative humidity forecast at 12 hours for the 500-millibar level. This 500-millibar forecast shows drier air to the west. However, information about the winds at each level will be needed before it can be determined whether the dry air is likely to intrude over the moist low-level air.

Table 10.5

## Three Different Computer Model Forecasts for Pittsburgh on July 22, 1995

EUROPEAN MODEL FORECAST

| | 850 TMP (DEG C) | SFC PRS (MB) | 500 HGT (DM) | 1000-500 THK (DM) |
|---|---|---|---|---|
| SAT 12Z 22-JUL | 15.4 | 1015 | 584 | 572 |
| SUN 12Z 23-JUL | 17.0 | 1011 | 582 | 573 |
| MON 12Z 24-JUL | 14.8 | 1012 | 581 | 572 |
| TUE 12Z 25-JUL | 15.6 | 1016 | 587 | 574 |
| WED 12Z 26-JUL | 16.6 | 1012 | 584 | 574 |
| THU 12Z 27-JUL | | | | |
| FRI 12Z 28-JUL | 13.6 | 1019 | 586 | 571 |
| SAT 12Z 29-JUL | | | 585 | |
| SUN 12Z 30-JUL | | | | |

ETA FORECAST

| | 850 TMP (DEG C) | 1000 TMP (DEG C) | SFC PRS (MB) | 700 RHU (PCT) | SFC PCP (IN) | 1000-500 THK (DM) |
|---|---|---|---|---|---|---|
| SUN 12Z 23-JUL | 17.2 | 25.7 | 1009 | 81 | 0.10 | 576 |
| SUN 18Z 23-JUL | 17.8 | 27.4 | 1007 | 55 | 0.08 | 575 |
| MON 00Z 24-JUL | 17.9 | 28.3 | 1010 | 55 | 0.08 | 575 |
| MON 06Z 24-JUL | 17.1 | 26.3 | 1011 | 63 | 0.07 | 575 |
| MON 12Z 24-JUL | 16.6 | 25.2 | 1013 | 68 | 0.28 | 574 |
| MON 18Z 24-JUL | 16.7 | 27.0 | 1013 | 68 | 0.17 | 575 |
| TUE 00Z 25-JUL | 18.2 | 27.9 | 1014 | 62 | 0.21 | 576 |
| TUE 06Z 25-JUL | | | | | | |
| TUE 12Z 25-JUL | | | | | | |

NGM FORECAST

| | 850 TMP (DEG C) | 1000 TMP (DEG C) | SFC PRS (MB) | 700 RHU (PCT) | SFC PCP (IN) | 1000-500 THK (DM) |
|---|---|---|---|---|---|---|
| SUN 12Z 23-JUL | 16.4 | 23.5 | 1008 | 70 | 0.28 | 575 |
| SUN 18Z 23-JUL | 16.9 | 28.6 | 1009 | 49 | 576 | |
| MON 00Z 24-JUL | 17.4 | 29.5 | 1008 | 41 | 0.00 | 577 |
| MON 06Z 24-JUL | 16.7 | 28.2 | 1010 | 47 | 576 | |
| MON 12Z 24-JUL | 15.8 | 26.0 | 1011 | 52 | 0.12 | 574 |
| MON 18Z 24-JUL | | | | | | |
| TUE 00Z 25-JUL | 17.8 | 29.9 | 1012 | 43 | 0.04 | 577 |

Note: Most abbreviations in the table are evident. Thus, 850 TMP is the temperature at the 850-millibar level. Less obvious abbreviations are:

| | |
|---|---|
| 500 HGT (DM) | Height of the 500-millibar surface in decameters (tens of meters). Thus, 584 is 584 decameters, or 5840 meters. |
| 1000–500 THK (DM) | Thickness in decameters of the 1000–500 millibar layer. |
| RHU | Relative humidity. |
| SFC PCP | Precipitation predicted for the earth's surface. |

**Table 10.4** Accu-Weather Forecast Data for Pittsburgh on July 22, 1995

| Data | 00Z/23-12Z/23 | 12Z/23-00Z/24 | 00Z/24-12Z/24 | 12Z/24-00Z/25 |
|---|---|---|---|---|
| MX/MN TMP (deg F) | 68 | 84 | 68 | 88 |
| AVG DEW (deg F) | 67 | 71 | 67 | 65 |
| AVG RHU (%) | 84 | 78 | 82 | 58 |
| AVG WND SPD (kts) | 5 | 12 | 7 | 8 |
| AVG WDR (deg) | 150 | 250 | 270 | 280 |
| AVG GUS (kts) | 8 | 18 | 12 | 14 |
| MAX APT (deg F) | 84 | 87 | 81 | 90 |
| AVG CLD (%) | 87 | 84 | 42 | 31 |
| AVG CIG (100 ft) | 129 | 47 | 135 | 158 |
| MAX UVI (0-10) | 0 | 4 | 0 | 7 |
| MIN VIS (mi) | 3 | 2 | 4 | 4 |
| THUNDER (%) | 24 | 37 | 15 | 5 |
| AMT RAIN (in) | 0.08 | 0.00 | 0.00 | 0.00 |
| PRB RAIN (%) | 53 33/53 | 37 22/22 | 15 14/5 | 6 1/4 |
| DUR RAIN (hr) | 1 0/ 1 | 0 0/ 0 | 0 0/ 0 | 0 0/ 0 |
| AMT SNOW (in) | 0.0 | 0.0 | 0.0 | 0.0 |
| PRB SNOW (%) | 0 0/ 0 | 0/ 0 0 | 0/ 0 0 | 0 0/ 0 |
| DUR SNOW (hr) | 0 0/ 0 | 0 0/ 0 | 0 0/ 0 | 0 0/ 0 |
| AMT ICE (in) | 0 0/ 0 | 0 0/ 0 | 0 0/ 0 | 0 0/ 0 |
| PRB ICE (%) | 0 0/ 0 | 0 0/ 0 | 0 0/ 0 | 0 0/ 0 |
| DUR ICE (hr) | 0 0/ 0 | 0 0/ 0 | 0 0/ 0 | 0 0/ 0 |
| AMT LIQUID (in) | 0.08 | 0.00 | 0.00 | 0.00 |
| PRB LIQUID (%) | 53 33/53 | 37 22/22 | 15 14/ 5 | 6 1/ 4 |
| DUR LIQUID (hr) | 1 0/ 1 | 0 0/ 0 | 0 0/ 0 | 0 0/ 0 |

Note: Most abbreviations in the table are evident. Others are:

MAX APT Maximum temperature at the airport for the period.

AVG CIG Average ceiling; the height at which someone rising through the atmosphere would first reach broken or overcast clouds.

MAX UVI Maximum ultraviolet sunlight index.

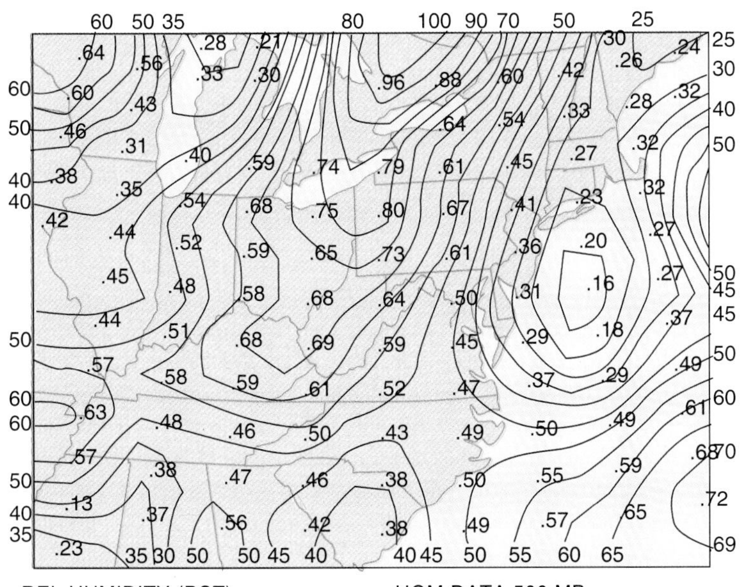


Figure 10.10 Twelve-hour 500-millibar relative humidity forecast for 12Z, July 23, 1995. Notice the region of high relative humidity values running through western Pennsylvania and eastern Ohio.

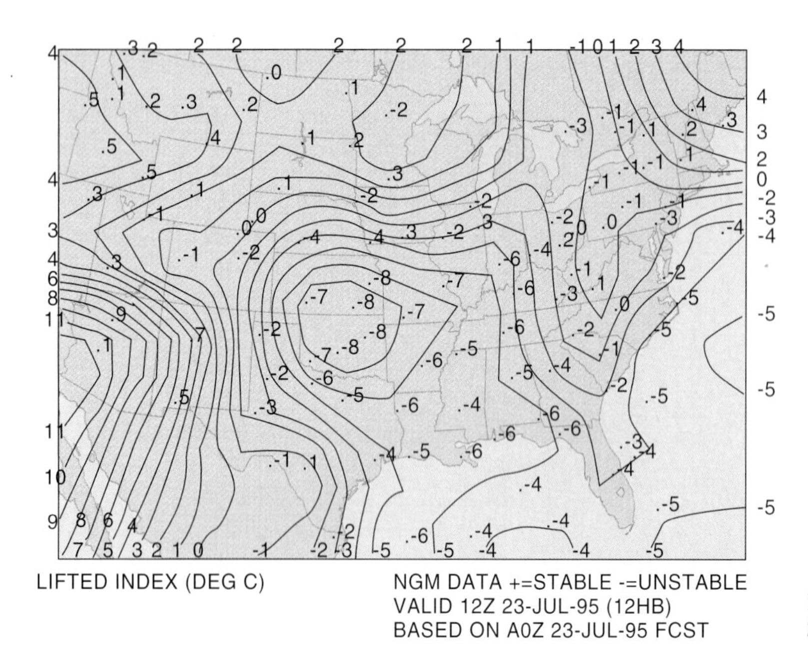


Figure 10.11 Twelve-hour lifted index forecast for 12Z, July 23, 1995. Where are the most unstable values expected?

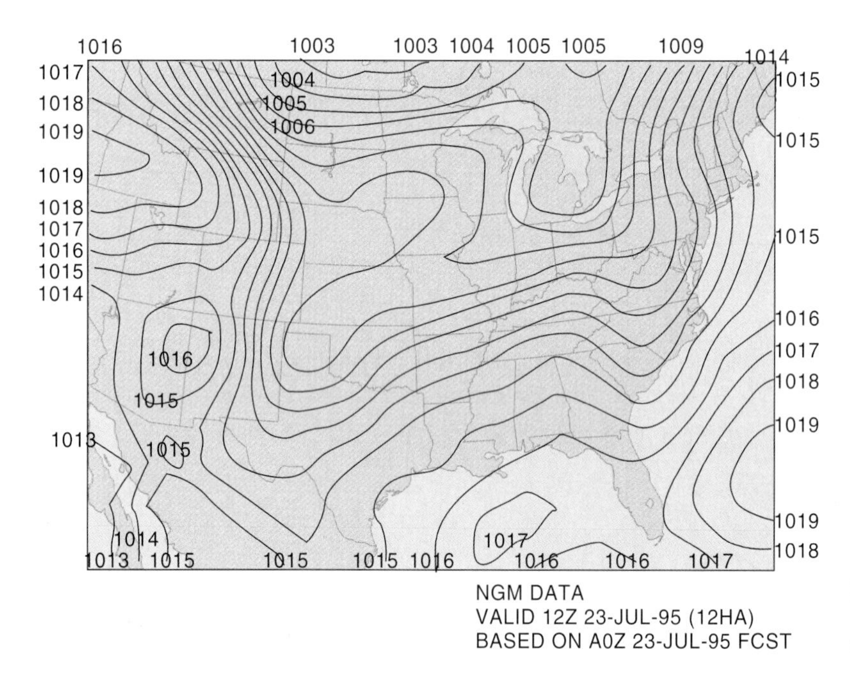


Figure 10.12 Twelve-hour sea-level pressure forecast for 12Z, July 23, 1995. Compare this pattern with the positions of fronts in Figure 10.5.

### *The Lifted Index*

Before going further with this thought, the meteorologist wants to look at the computer forecast of instability as expressed by a calculated value called the **lifted index.** The index is computed by examining the temperature and moisture differences between two atmospheric levels known to be important in determining stability (850 and 500 millibar). The more negative the lifted index values, the greater the probability of thunderstorm formation. Normally, lifted index values below zero reflect sufficient instability for thunderstorm development.

A map of predicted lifted index values (Figure 10.11) shows lowest values (greatest instability) of −8 occurring over Kansas and Oklahoma. That is where the most intense thunderstorms are likely to occur. In fact, on this particular night, there were devastating windstorms in the Oklahoma City area. The lifted index values near Pittsburgh were predicted to be about −1 or −2, a moderately unstable value. There is no set rule that says if the stability index reaches a certain value, there will or will not be a thunderstorm; many other factors are involved. For example, if there are strong thunderstorms about 50–100 miles away from you, the air closer to your location may be subsiding (which would act to suppress thunderstorm formation) in reaction to the thunderstorm updrafts. Furthermore, recall that the lifted index was based on values at just two atmospheric elevations. You

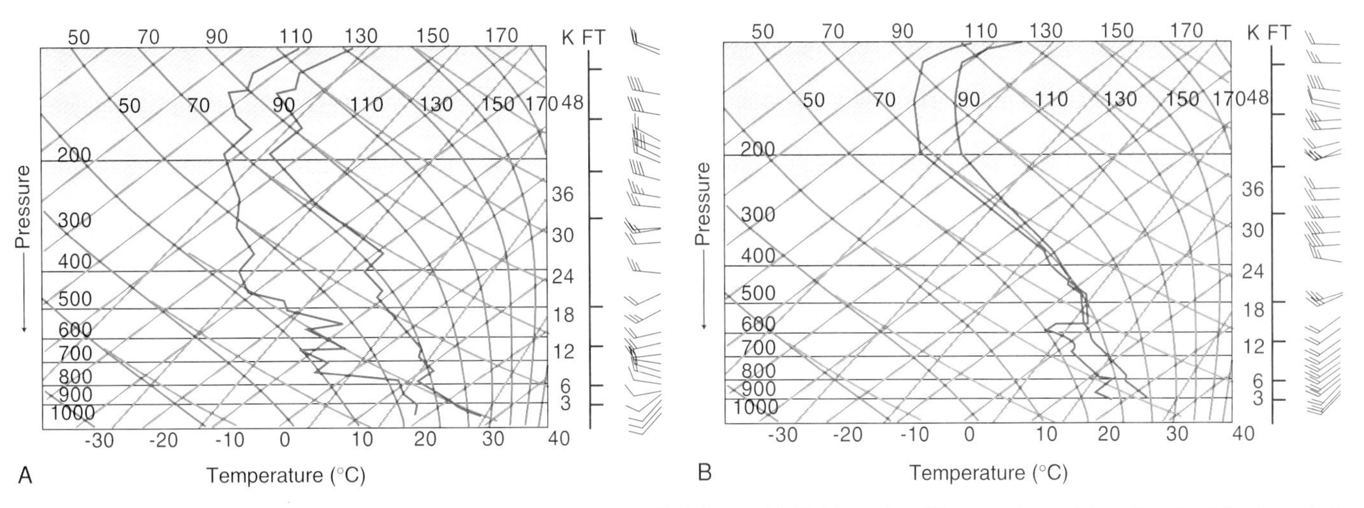


Figure 10.13 Radiosonde plots for Pittsburgh (*A*) and Dayton (*B*) 00Z, July 23, 1995. Note the difference in moisture levels at the two stations. Note: KTF 5 Thousands of feet above sea level. Thus 36 means 36,000 feet.

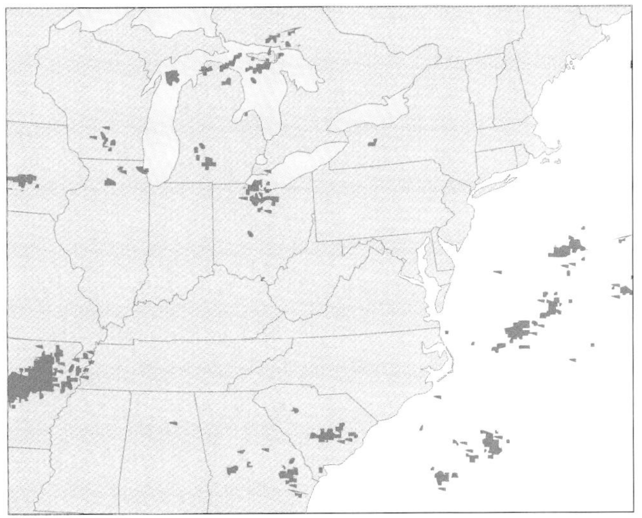


Figure 10.14 Lightning strikes recorded between 02:35Z and 03:357, July 23, 1995. Compare with the earlier lightning map (Figure 10.7). How are the lightning strikes in Ohio moving?

have seen a number of vertical temperature and moisture profiles; clearly, they are much too complicated and intricate to expect one index value to tell you everything about the nature of that profile. Having used caution in interpreting the lifted index forecast, it would be prudent to believe there was a reasonable chance of thunderstorms in the forecast area. The forecaster decides she will have to look at more information before deciding how to phrase the forecast.

The meteorologist turns next to Figure 10.12, the computer projection of the surface pressure pattern at 12Z (8 A.M. tomorrow, Sunday). Note the pressure trough extending from Lake Huron to northern Illinois and out through Kansas. Could this be a cold front?

### *Plotted Soundings*

To assess further the moisture and stability of the regional atmosphere, the forecaster inspects soundings of temperature, moisture, and wind. The Pittsburgh and Dayton, Ohio, soundings are shown in Figure 10.13. Keep in mind that these soundings represent actual conditions at 00Z on July 23. They are not forecasts. Note how much moister the layer from about 10,000 feet to 40,000 feet is at Dayton than at Pittsburgh. This may account for why there were thunderstorms in Ohio at 00Z but not in Pittsburgh. However, a little detective work is needed. Where exactly in Ohio does the moisture transition zone from Dayton to Pittsburgh conditions lie? By looking at satellite and radar pictures, it should be possible to make that determination. Thick clouds, showers, and thunderstorms would be found in the region that most closely matches the Dayton conditions.

The winds aloft show where the showers and thunderstorms should move and even give a clue about how strong the wind gusts might be in thunderstorms. Here again, there are no magic formulas. Gust forecasts were developed empirically. This means many actual cases were studied. Researchers noted what wind gusts resulted with various wind profiles aloft. Tables were then constructed to give a reasonable estimate of the expected maximum wind gust for any given set of circumstances. However (there are a lot of "howevers" in forecasting, aren't there?), remember that the soundings show conditions in only two locations. Big differences may exist between sampling sites. When you consider that Dayton and Pittsburgh are more than 200 miles apart but the area affected by severe thunderstorm winds may be less than one square mile, you can appreciate the danger of putting too much faith in any index value or guidance table.

At this point, an updated and more detailed look at the lightning pattern is warranted. Examine Figure 10.14. Notice the lightning plots in Ohio. There is also considerable lightning near Lake Erie but little in the storms in central Ohio.

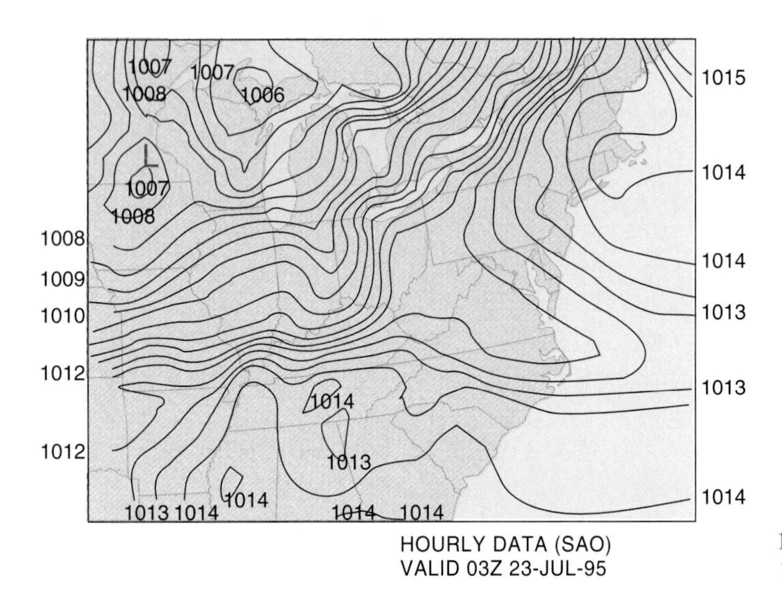


Figure 10.15 Regional sea-level pressure analysis 03Z, July 23, 1995. Note the trough of low pressure over Ohio.

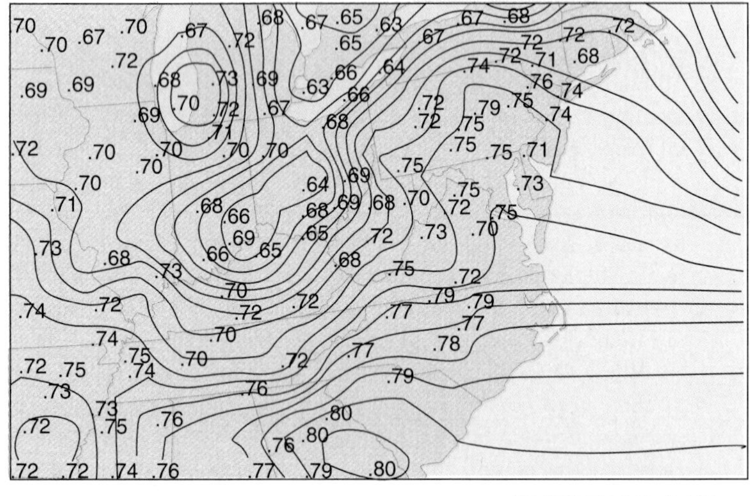


Figure 10.16 Regional dew-point analysis, 03Z, July 23, 1995.

## Regional Weather Maps

Next, the meteorologist examines a map showing the regional pressure pattern in high resolution. This is shown in Figure 10.15. Notice that a low pressure trough in Ohio matches well with the area of showers and thunderstorms back in Figure 10.4. A look at the dew-point map, Figure 10.16, suggests why. The dew points surge above 70 in the same area where the thunderstorms are breaking out. The thunderstorms are probably being forced by the influx of warm, humid air from the southwest cutting under an air mass that is beginning to dry out aloft. Note that for our discussion here, we are skipping a step. The forecaster would need to look at additional charts to come to the conclusion just presented.

The forecaster also notes a dew-point surge east of the Appalachians. She makes a note to check out the Philadelphia radar after she has finished her other work.

Based on what she has seen so far, the forecaster has decided to predict thunderstorms for early tomorrow morning. The air mass will be moderately unstable, and both models indicate at least some rain. The timing is a compromise between the two models, along with an awareness of how fast the real storms have been moving in Ohio. But what about later tomorrow and out through the following week? The forecast must cover all these time periods. To get a grasp on this challenge, the forecaster reads the National Meteorological Center (NMC) discussion for the short to medium range. In this connection, "medium range" extends out about five days. (Note: In the NMC discussion, POP means probability of precipitation; NRN, ERN, etc., mean northern, eastern, etc.; MRF is the medium range forecast computer model.)

*Extended Forecast Discussion for Jul 24 Thru Jul 26 1995 Meteorological Operations Division..NMC..NWS..Washington DC 2:15 pm edt Sat Jul 22 1995*

*Complex and low amplitude flow across SRN Canada and the NRN half of the lower 48 should persist thru the medium range period. A series of short waves will track rapidly ewd over the top of a pesky srn U.S. upper ridge. Some trof energy will penetrate esewd into the ridge creating a weakness e of the ms river. Expect generally above normal pops across the NRN tier states and the ERN half of the nation. Expect hot summertime temps from the plains EWD and seasonal/below normal temps from the Rockies WWD.*

*All models are slowly coming around to the more progressive flow regime depicted by the preferred mrf over the last few runs. Accordingly..MRF guidance was in the most part used as a guide for my manual progs. However..I have low confidence that any of the models have a strong handle on the timing and strength of individual shortwaves and associated surface systems.*

Once she has digested all this material and discussed the weather situation with her colleagues in Pittsburgh and in other nearby weather offices, the meteorologist writes a "regional discussion." A regional discussion identifies the main weather systems affecting the area and describes in general terms the weather these systems are expected to bring to the region.

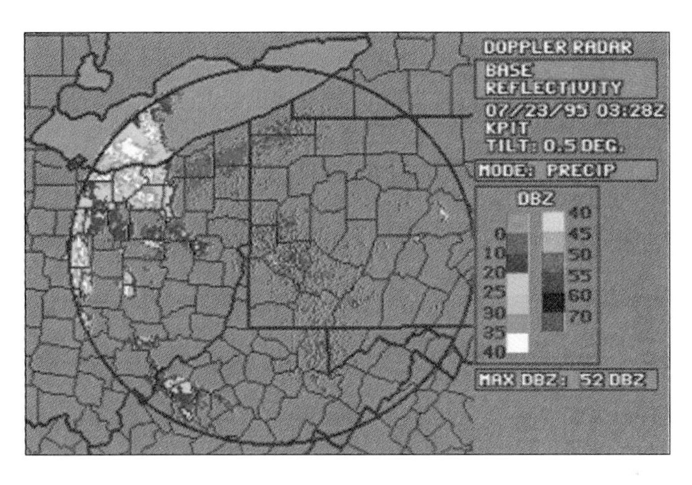


Figure 10.17 Doppler radar image at 03:28Z, July 23, 1995. Compare this image with that an hour earlier (in Figure 10.4).

## The Forecast

Here is what our meteorologist decided to write:

*State Forecast Discussion*
*National Weather Service Pittsburgh PA*
*1017 pm EDT Sat Jul 22 1995*

*Isolated thunderstorms over the central mountains should continue to linger. Showers and thunderstorms entering central Ohio should cross the WPA border at or around midnight then spread eastward after midnight. Will continue the chance of showers and thunderstorms in central zones tonight and after midnight in western zones as new area of precip approaches. Hazy and humid conditions should also continue overnight into Sunday.*

*On Sunday with the approach of a cold front showers and thunderstorms are a good bet across all zones. Thunderstorms across southern zones may be mainly during the afternoon however with the increase in instability as the front approaches will keep mention of them the entire day. Have made some small adjustment to temps and pops as well as wind speeds for overnight and Sunday. No other major changes.*

The weather study and forecast deliberations have gone on for about an hour. Perhaps the radar has changed enough that a pattern can be seen to be emerging. Examine Figure 10.17. Note that the thunderstorms and showers are moving eastward, but the main area of activity is still well west of Pittsburgh.

At last, the meteorologist is ready to compose the forecast. For the general population, there are three components for each forecast zone:

- the **nowcast,** a prediction for the next few hours
- the **short-term forecast,** applying to the next 48–60 hours
- the **extended forecast,** which goes out five days.

*.Now . . .*
*Showers and thunderstorms will be moving into the region anytime after midnight. Temperatures will fall into the mid to upper 60s by 4 am.*
*State forecast for western Pennsylvania and the northern panhandle of West Virginia . . . updated*
*National Weather Service Pittsburgh PA*
*900 pm EDT Sat Jul 22 1995*

*.Tonight . . . hazy warm and humid. Scattered thunderstorms east while more likely west of the mountains toward or after midnight. Low in the 60s. .Sunday . . . warm and humid with showers and thunderstorms likely. High in the upper 70s to mid 80s.*
*.Sunday night . . . partly cloudy. A chance of thunderstorms southeast early. Low in the 60s.*
*.Monday . . . partly sunny. High in the upper 70s to mid 80s.*

*.Extended Forecast . . .*
*.Tuesday . . . a chance of thunderstorms. low 65 to 70. High in the 80s. .Wednesday . . . partly sunny. Low in the 60s. High in the 80s. .Thursday . . . humid with a chance of thunderstorms. Low 65 to 70. High in the 80s.*

## Verifying the Forecast

So, what actually happened?

Here is the sequence of hourly weather observations that followed the issuance of the evening forecast.

The process of comparing forecast conditions with the weather that actually occurred and scoring the forecast is called **verification.** A forecast that comes true is said to verify. As you can see, no rain was reported on any of the hourly observations.

Table 10.5 Observed Pittsburgh Weather, July 23, 1995

| Location * | Time | Temperature °F | Relative humidity % | Wind direction/ speed (mph) | Pressure inches | Visibility miles | Weather |
|---|---|---|---|---|---|---|---|
| Pittsburgh | 3z | 75 | 73 | S/6 | 29.91 | 6 | Partly cloudy |
| | 4z | 73 | 81 | SE/6 | 29.89 | 6 | Partly cloudy |
| | 5z | 73 | 73 | ESE/6 | 29.87 | 6 | Cloudy |
| | 6z | 73 | 75 | SSE/8 | 29.85 | 6 | Partly cloudy |
| | 7z | 73 | 75 | SE/8 | 29.83 | 6 | Cloudy |
| | 8z | 73 | 81 | WSW/7 | 29.83 | 6 | Cloudy |
| | 9z | 72 | 87 | W/7 | 29.84 | 6 | Cloudy |
| | 10z | 71 | 87 | SSW/5 | 29.84 | 4 | Haze |
| | 11z | 71 | 87 | S/7 | 29.82 | 4 | Haze |
| | 12z | 73 | 84 | SW/9 | 29.83 | 5 | Haze |
| | 13z | 77 | 73 | WSW/16 | 29.86 | 6 | Haze |
| | 14z | 77 | 76 | WSW/15 | 29.87 | 7 | Mostly cloudy |
| | 15z | 78 | 66 | WSW/19 | 29.88 | 10 | Cloudy |
| | 16z | 76 | 70 | WSW/16 | 29.90 | 10 | Cloudy |
| | 17z | 78 | 68 | W/16 | 29.91 | 10 | Cloudy |
| | 18z | 81 | 57 | WSW/16 | 29.90 | 10 | Partly sunny |
| | 19z | 83 | 58 | W/14 | 29.90 | 12 | Mostly sunny |
| | 20z | 83 | 54 | W/14 | 29.90 | 12 | Mostly sunny |
| | 21z | 84 | 56 | W/14 | 29.88 | 12 | Mostly sunny |
| | 22z | 85 | 52 | WSW/14 | 29.86 | 12 | Mostly clear |
| | 23z | 83 | 56 | SW/13 | 29.87 | 12 | Mostly clear |
| | 0z | 82 | 58 | W/13 | 29.87 | 12 | Mostly clear |
| | 1z | 79 | 66 | SW/8 | 29.88 | 12 | Partly cloudy |

PIT 23 temps: high = 85 at 22z low = 71 at 10z mean = 77.0 precip = Trace

The ending summary shows a trace of rain (less than 0.01 inches), so there must have been a brief shower in between two of the hourly observations. It was so minor that no special observations were taken to report it. In the end, all the consternation about which model was timing the precipitation correctly came to naught. Neither forecast verified! The ETA model had predicted some precipitation for each six-hour time period. The NGM had predicted more than a quarter inch of rain by morning but then nothing during the day. In terms of helping people plan their outdoor activities on a Sunday, the NGM would have been more useful. Yet, when the forecaster made her prediction, rain was approaching and all the computer guidance predicted at least some rain. In essence, the forecast of showers and thunderstorms for Sunday missed the mark; it turned out to be a good summer day for outdoor activities.

Before simply chalking up this case to error, we should consider one other aspect. How do we know whether it rained heavily a few miles away from the official observation site, and does that matter in verifying the forecast? If a forecaster makes a prediction for an area the size of a city, it is reasonable to expect the weather to vary across the area. The official determination of accurate or inaccurate is often made on the basis of one observation point, but that may not tell the whole story. If rain occurred within the forecast area, many people will believe the forecast was correct. By using Doppler-derived rainfall measurements in concert with the network of rain gauges, it is possible to reconstruct a record of where it did and did not rain. How should this information be integrated into forecast verification systems? That's an issue for the future.

Such questions are important not only to the forecaster but for other purposes as well. For example, private weather service firms do a brisk business in weather reconstructions for the legal profession. Suppose Bill was in an accident on the night of our case study, and he claimed a giant puddle caused his car to careen out of control. If only the data from the official observing point are used, it will suggest that there was no rain and the puddle story might be all wet. However, with data sources such as archives of weather radar images, a meteorologist may be able to corroborate Bill's story.

Perhaps this case study has given you some appreciation for the complexity of weather forecasting, even when meteorologists are surrounded by data and provided with detailed guidance.

Hidden behind the computer models and all the careful detective work is the fact that subtle change can have big effects. Until thunderstorms break out, there is nothing but theory to track. Once storms form in a particular area, they influence the atmosphere around them. Unfortunately for forecasters, often by the time this occurs, most people have made their plans and are executing them.

## Updating the Forecast

For weather-sensitive interests, the key is not always what the original forecast said. Rather, it's the updating that matters. As long as money and resources have not been spent on a particular activity, the latest forecast can have economic value. For example, if an electric company had been planning line work for Sunday to repair damage from previous storms, the forecast in Table 10.5 may have caused it to cancel its plans. However, if the company used the services of a private forecasting company that had been instructed to watch the weather constantly, the utility could have continued working with confidence that it would be alerted as soon as any adverse weather was approaching. In this case, the same private meteorologist whose forecast went wrong for the general public could get credit for issuing useful updates and helping weather-sensitive interests, all with the same weather pattern on the same work shift! This apparent paradox brings up an important point: weather forecasts are most useful when the meteorologist knows the specific use for which the forecast is intended. We'll look at some more examples of this later in the chapter.

### Checkpoint

***Review***

This section introduced you to the complexities of weather forecasting. Even though modern forecasters are surrounded by data and computers, it is still up to the individual forecaster to analyze the data and make the final decision as to the issued forecast. Thus, forecasting is an ever-changing, complex piece of detective work aided by modern technology.

In making a forecast, a meteorologist consults numerous data sources. These include present conditions, climatology, radar images, synoptic scale maps, computer guidance, moisture and stability indices, regional weather maps, and other meteorologists. After the forecast is issued, the process of comparing the forecasted conditions with the actual weather, or verification, is undertaken. Besides the obvious benefits to the forecaster, verification is important to private weather services, which are often asked to reconstruct weather events for legal purposes.

Finally, the importance of updating the forecast is critical, especially for clients whose services are weather-sensitive and who require specific and detailed weather forecasts.

***Excercise***

Below are the data sources commonly used by meteorologists in preparing a forecast. For each, describe the type of information it will provide and how the information can be useful to the forecaster in making an accurate forecast.

| Data Source | Used by Forecaster |
|---|---|
| Present conditions | |
| Climatology | |
| Radar images | |
| Synoptic scale maps | |
| Computer/numerical models | |
| Moisture aloft | |
| Lifted index | |
| Soundings | |
| Regional surface maps | |
| Consultation with other forecasters | |

# National Weather Service Forecasts and Services

The activity you read about in the last section is just one small part of the daily business of the National Weather Service (NWS), an agency of the U.S. government. In this section, you will learn about other types of NWS forecasts and services.

## The National Weather Service

The annual National Weather Service budget in the mid-1990s was approximately $5 billion. In 1995, the service was in the middle of an ambitious transition to a completely modernized operation. The agency received the go-ahead on this in part because of a 1992 study by the National Institute of Standards and Technology that estimated that modernization would result in improved weather information services that would benefit the U.S. economy by approximately $7 billion *per year*. If realized, this will be an excellent return on the public's investment.

The mission of the National Weather Service is:

*To provide weather and flood warnings, public forecasts and advisories for all of the United States, its territories, adjacent*

Table 10.6

## Extended and 6–10-Day Forecasts

**AccuWeather for Windows (Record Start)**
**Saturday April 19 1997, 19:39:41**

STATE FORECAST FOR OREGON
NATIONAL WEATHER SERVICE PORTLAND OR
[ISSUED AT] 300 PM PDT SAT APR 19 1997
ORZ001>009-011-012-201000-
STATE FORECAST FOR WESTERN OREGON
.EXTENDED FORECAST. . .
.TUESDAY. . . RAIN CHANGING TO SCATTERED SHOWERS. BREEZY NEAR THE COAST. LOWS 45 TO 50. HIGHS COAST IN THE UPPER 50S. . . INLAND IN THE LOWER 60S NORTH TO UPPER 60S SOUTH.
.WEDNESDAY. . . SCATTERED SHOWERS. LOWS IN THE UPPER 30S TO MID 40S. HIGHS COAST IN THE UPPER 50S. . . INLAND 60 TO 65.
.THURSDAY. . . PARTLY CLOUDY. CHANCE OF SHOWERS. LOWS IN THE UPPER 30S TO MID 40S. HIGHS COAST NEAR 60. . . INLAND IN THE MID 60S.

tenday
# of Reports [RETURN] for most Recent:

NMCEONUS
FEUS 40 KWBC 181916
6-TO-10-DAY OUTLOOK FOR APRIL 24 - 28 1997
CLIMATE PREDICTION CENTER NCEP NATIONAL WEATHER SERVICE WASHINGTON DC 3 PM EDT FRI APRIL 18 1997

THE NATIONAL WEATHER SERVICE 6 TO 10 DAY OUTLOOK FOR THURSDAY APRIL 24 1997 TO MONDAY APRIL 28 1997 CALLS FOR ABOVE NORMAL TEMPERATURES OVER WESTERN AND CENTRAL WASHINGTON. . . THE NORTHWESTERN HALF OF OREGON. . . AND COASTAL AREAS OF NORTHERN AND CENTRAL CALIFORNIA.

NEAR MEDIAN PRECIPITATION TOTALS ARE INDICATED FOR EXTREME SOUTHWESTERN WASHINGTON. . . WESTERN OREGON AND THE WESTERN HALF OF NORTHERN CALIFORNIA. . .

| STATE | TEMP | PCPN | STATE | TEMP | PCPN | STATE | TEMP | PCPN |
|---|---|---|---|---|---|---|---|---|
| WASHINGTON | A | A | OREGON | A | A | NRN CALIF | N | N |
| SRN CALIF | N | NP | IDAHO | N | A | NEVADA | B | A |

. . . . . . . . . . . . . . .

LEGEND

TEMPS WITH RESPECT TO NORMAL
NA - MUCH ABOVE    A - ABOVE
N - NEAR NORMAL    B - BELOW
MB - MUCH BELOW

PCPN WITH RESPECT TO MEDIAN
A - ABOVE    N - NEAR MEDIAN
B - BELOW    NP - NO PCPN

THE FORECAST CLASSES REPRESENT AVERAGES FOR EACH STATE. NORMAL VALUES—WHICH MAY VARY WIDELY ACROSS SOME STATES—ARE AVAILABLE FROM YOUR LOCAL WEATHER SERVICE FORECAST OFFICE.

*waters and ocean areas, primarily for the protection of life and property. NWS data and products are provided to private meteorologists for the provision of all specialized services.*

To achieve the mission, the NWS will:

coordinate its activities with state, local, and federal agencies;
provide a range of weather services to the private hydrometeorological community;
fulfill international hydrometeorological obligations;
provide data and products to the private sector;
work closely with the mass media because it is the chief means of communicating weather and flood warnings and forecasts to the public;
conduct applied research with other agencies and scientists to improve warnings and forecasts based on the latest scientific and technological advances;
devise better ways to distribute and exchange information; and
help with the emergency management decision process.

## NWS Forecast Products

### *Routine Forecasts*

For very short time periods starting with "now" and extending out several hours, the National Weather Service issues a kind of "nowcast." It is a narrative that expresses what should happen during the next few hours in a defined coverage area.

The NWS provides routine forecasts for areas comprising one to several counties four times per day; these are intended for general use. Typical issuance times are prior to 6 A.M., 11 A.M., 5 P.M., and 11 P.M. These forecasts typically cover the next 24–48 hours and are updated, if needed.

Beyond 48 hours, the National Weather Service provides an extended forecast. This general and often sparsely worded forecast covers the period from three to five days from now. In addition, a 6–10-day outlook is issued three times a week. It predicts departures from normal temperature and precipitation on a state-by-state basis. In Table 10.6, you can see examples of these forecasts.

All of these public forecasts are issued from a network of 100 forecast centers scattered strategically across the country. Viewers of the Weather Channel are quite familiar with these products, for that service is a conduit for getting these products from the NWS to you without alteration.

### *Products Involving Dangerous Weather*

The National Weather Service has as one of its core responsibilities the warning of the public about potentially dangerous weather conditions. These conditions include floods, severe thunderstorms, tornadoes, snow and ice storms, and hurricanes. Special environmental prediction centers within the NWS handle specific categories of hazards. For example, the National Hurricane Center (NHC) coordinates the tracking and prediction of and the issuance of watches and warnings for all tropical-origin storms. The Atlantic and Gulf coasts are served from the NHC headquarters in Miami; storms in the eastern Pacific are handled through an office in San Francisco, and mid-Pacific storms are forecast through the office in Honolulu.

Severe weather advisories are of two basic forms: watches and warnings. A **watch** is an alert: it says conditions are ripe for the possible development or occurrence of the named event (tornadoes, floods, etc.) in a specified area during a defined time period. In the case of hurricanes, a watch may apply if a storm is more than 24 hours from landfall in the watch area. For tornadoes, the danger may exist for the next 4–10 hours in the specified area. This difference in times is caused by the nature of the events themselves. Hurricanes typically are tracked over long distances. Severe thunderstorms and tornadoes tend to develop quickly, sometimes in a matter of minutes.

As you may already have surmised, a **warning** indicates severe weather is imminent. It is a direct call to action: protect life and property because they are in the path of peril. One of the great promises of Doppler radar is its ability to detect potentially severe storms further in advance than is possible using other methods. In some cases, warnings extend as much as 30 or 45 minutes ahead of the arrival of a tornado or severe thunderstorm.

In some cases, a forecast office may choose to issue a **special statement** instead of a watch or warning. Such statements are informational or advisory in nature. They describe an unusual or perhaps significant event but stop short of putting a watch or warning in effect. They may be issued at any time of day and may cover expected or ongoing situations. People involved in weather-sensitive activities should always be aware of whether special statements are in effect and their potential implications.

The ability to predict accurately severe weather raises interesting questions about appropriate responses. What should people do with the increased lead time or advance notice? If storm shelters are available, the answer may be obvious. But how will people react when a warning catches them far from home, family, or safe shelter? Will people panic, running off to pick up children at school or dashing home from the office? Will the warning trigger more problems than the hazard itself? Can these problems be managed? Will our increased warning capabilities lead to improved emergency facilities and plans? Our hope is that it will not require a tragedy for the appropriate steps to be taken.

Flood forecasts are coordinated and issued by a group of River Forecast Centers, each responsible for monitoring and predicting conditions on watersheds in the region. In addition, these offices prepare summaries of temperature and precipitation observations for numerous substations throughout their coverage area. Many of these stations are in the backyards of the homes of volunteer private citizens. Some of these "cooperative observers" have been assisting the NWS for decades.

Figure 10.18 Snow and ice buildup on aircraft is just one of a host of weather-related hazards facing the airline industry. Here, a crew de-ices an airplane prior to takeoff.

Severe thunderstorms and tornadoes have been centrally monitored and predicted by the Storm Prediction Center in Norman, Oklahoma. Plans are unfolding for decentralizing this function and placing more decision making in the hands of the individual regional forecast offices. These offices already are responsible for issuing watches and warnings for winter storms and their hazards.

### *Forecasts for Special Interests*

Aviation forecasts comprise a major portion of the National Weather Service's forecasting effort. Weather conditions affect aircraft, pilots, and ground crews in a multitude of ways (see Figure 10.18). Even with instrument landing systems, aircraft are not permitted to take off or land if the ceiling and visibility decrease below certain thresholds. A common limit is a 400-foot ceiling and one-quarter-mile visibility.

Sophisticated computer models are used to predict winds at a multitude of places and altitudes. Using this information and aircraft performance data, airlines determine the most economical route for each flight. If you have flown, you may remember the pilot announcing the expected flight time to your destination, a value that has been calculated using this method.

If icing is encountered en route, special measures are taken to shed the ice. If the methods fail, the plane's wings may lose the design characteristics that permit flight. Showers and thunderstorms present additional hazards. Turbulence can cause great discomfort for passengers, and planes are put through vigorous wind tunnel tests before they are certified for flight. Extremely heavy precipitation has been known to interfere with engine operation, and hail can cause considerable damage. Winds affect planes while cruising and also during takeoff and landing. Wind shear, the rapid change in wind direction and/or speed over a short distance, is a serious threat. Doppler radars installed at airports promise to reduce the hazard by identifying downbursts and other small-scale wind shear events before pilots encounter them.

The NWS also makes forecasts for various agricultural and forestry interests. These include frost advisories for fruit growers and fire weather forecasts. Provision of these specialized services has been a subject of controversy over the years. On one hand, the public has an interest in the protection of the nation's food supply. On the other hand, spending taxpayer dollars for the specific benefit of one group of businesses raises questions.

## NWS Research Activities

The National Oceanic and Atmospheric Administration (NOAA), the parent agency of the National Weather Service, collaborates with universities and other agencies to develop and exchange scientific and technical knowledge about the atmosphere. These projects range from basic meteorological and hydrological research to developing methods and products to improve forecasts and warnings. In addition, research in computing systems is underway to provide better ways of assimilating, visualizing, and using the vast amount of incoming data and computer model outputs to make better numerical predictions and short-term local and regional forecasts.

Much NOAA research is carried out through the Environmental Research Laboratories (ERL), the National Weather Service, and the National Environmental Satellite, Data and Information Service (NESDIS).

### *Environmental Research Laboratories*

At the ERL Forecast Systems Laboratory in Boulder, Colorado, (Figure 10.19), research is underway to:

find better algorithms for Doppler radar data (for example, refining the relationship between rapid velocity changes over short distances and severe storms and tornadoes);
use sounding data from geostationary satellites and ground-based atmospheric profilers, replacing twice-a-day, balloon-launched radiosondes with continuous, comprehensive data sources;
develop computer workstations that will enable forecasters to assemble more information faster and more clearly to help with decision making; and
develop local data networks, Doppler wind data, and profiler output to provide detailed, three-dimensional hourly analyses of wind, temperature, and moisture. In the fore-

Figure 10.19 NOAA's Environmental Research Laboratories in Boulder, Colorado.

cast example earlier in this chapter, such tools would have allowed the forecast to be updated and corrected earlier than otherwise. It might have been possible to discover that the ingredients for the predicted showers were not coming together as originally thought.

ERL's National Severe Storms Laboratory in Norman, Oklahoma, is a research site for severe weather phenomena and forecasting. In 1994 and 1995, a major effort called VORTEX (Verification of the Origins of Rotation in Tornadoes Experiment) was launched to look at questions related to the origin of tornadoes and the way tornadoes operate. A major aim of the project is to be able to distinguish between those storms that produce tornadoes and those that do not.

The ERL Atlantic Oceanographic and Meteorological Laboratory performs research aimed at improving our understanding and forecasting of hurricanes. In 1995, for example, a new model that replicates the central vortex of a tropical storm was introduced; it was developed in association with the Geophysical Fluid Dynamics Laboratory (GFDL) in Princeton, New Jersey.

GFDL is perhaps best known for its general circulation models employed in climatic change study (see Chapter 4). However, the laboratory also develops, tests, and evaluates numerical mesoscale and synoptic scale models that generate forecasts from a week to a season ahead.

### *National Meteorological Center*

The National Meteorological Center constantly refines and improves the operational forecast models on which all practicing meteorologists rely. Current research is concentrated in three main areas: the development of models and techniques to forecast regional weather events and features, such as thunderstorm outbreaks; global weather and climate modeling, including numerical techniques for modeling the interactions between the oceans, the atmosphere, and land surfaces; and ocean modeling research, particularly computer models of ocean waves and sea ice.

### *NESDIS Research Programs*

Efforts at the National Environmental Satellite, Data and Information Service are aimed at providing data from satellite sensors to improve all aspects of weather analysis and prediction. Work is underway to provide better moisture and atmospheric stability products. One obstacle to the preparation of better satellite-derived temperature and moisture profiles has been the problem of detection within and below clouds. Since this is just the area where the most detailed information is needed, much work has been done to overcome the problems.

## Checkpoint

***Review***

The mission of the National Weather Service is to provide weather and flood warnings, public forecasts, and advisories for all of the United States, its territories, adjacent waters, and ocean areas, primarily for the protection of life and property. To meet this mission, the NWS routinely issues short-term forecasts for the next few hours, forecasts for the 24–48-hour period, extended forecasts for 3–5 days, and a 6–10-day extended forecast.

Specialized forecasts for aviation, forest, and agricultural concerns are also issued by the NWS. Such forecasts have become controversial over the years because of concerns about spending taxpayer dollars for the specific benefit of one or another group of businesses. Many argue that these special interests may be better served by private weather forecasting services.

The NWS also meets its responsibility for warning the public about potentially dangerous weather conditions by operating the National Hurricane Center, River Forecast Centers, and the National Severe Storms Forecast Center. Severe weather information is communicated to the public as either a watch or a warning. A watch is an alert that is issued when conditions are ripe for the possible development of specific severe weather. Warnings are issued when such weather is imminent.

Through its parent agency, the National Oceanic and Atmospheric Administration, the NWS collaborates with universities and other scientific agencies on research and development projects intended to improve our understanding about the atmosphere and provide better forecasting models and communication networks.

***Excercices***

1. State in your own words the mission of the National Weather Service.

2. Name four government-sponsored atmospheric science laboratories and describe briefly the nature of the work done in each.

## Private Weather Forecasting Services

A thriving demand exists for forecasts specifically tailored for special applications. Here are some examples of the services offered by private meteorological firms.

### Forecasts for Utility Companies

Figure 10.20 shows the relation between daily temperature and natural gas demand by customers in Connecticut. If you were in charge of maintaining gas supplies at a utility company, you would need to have accurate estimates of future customer demand so you could match them with your supply. If a gas utility runs short, it must suddenly buy supplies at higher prices on the open market, raising costs. Conversely, any oversupply must be either stored or sold, at lower profit.

This is just one example of electric and gas utilities' needs for customized weather information. Utilities have developed sophisticated statistical models for predicting their customers' energy usage based on temperature, like the relation shown in Figure 10.20. Some of these models are so accurate that a utility company may be able to use the relation in reverse and determine within a degree or two what the temperature is just based on the current demand for gas or electricity.

A private forecast for a utility generally includes a prediction of the temperature to the nearest degree for every hour or two out to 48 hours, with a more general outlook beyond that. The meteorologists on duty must monitor the progress of the forecast on a continuous basis, for many utilities require an immediate update any time it is determined that one of the hourly values will be off by more than a degree or two. Utilities also use forecasts for scheduling installation and repair work.

A relatively new use of forecasts in the electric industry is watching for lightning strikes. Detection systems can tell where lightning strikes occur and transmit that information to forecasters in a matter of seconds. Lightning location and progress maps (which show the pathways of a succession of lightning strokes as a storm moves) are matched with locations of work crews in order to warn them of approaching peril. In addition, these stroke maps allow utilities to judge where most of their service calls are likely to originate.

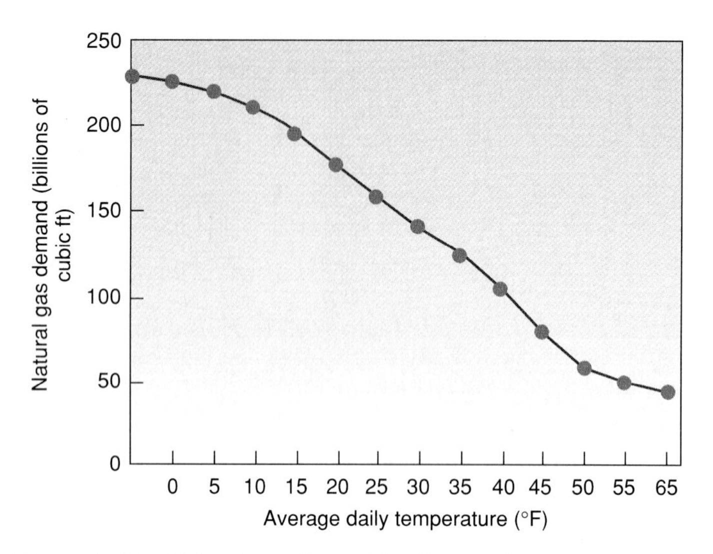


**Figure 10.20** Natural gas demand by Connecticut customers varies with air temperature.

### Forecasts for the Media

With NWS forecasts available at no cost, you may wonder why the broadcast or print media outlet would pay for private forecasts. Again, customized service is the key. Private meteorologists are able to tailor their forecasts specifically to a given audience. Such specialized service is not part of the National Weather Service's mission, nor is it something most people would wish to pay for with their tax dollars.

#### *Newspapers*

Many newspaper weather pages like those in Figure 10.21 are prepared by private meteorological firms. These services collect a variety of climatological information from the client's service area and add forecasts for areas as the paper specifies. They draw national, regional, and local weather maps using design schemes developed in collaboration with the paper's graphics department and management. A separate look can be developed for each newspaper.

#### *Radio*

Private meteorologists also make forecasts for radio and television stations. In each case, the station is looking for products specifically designed to serve its target market. A radio station in farm country may want to offer special agricultural forecasts, while one near ski country will want ski reports. Further, the private service may be able to prepare a forecast that spells out in greater detail the differences expected in different parts of the coverage area.

Radio stations typically do not have the resources to hire their own staff of meteorologists, but they can obtain the services of a private meteorological company. Since each broadcast meteorologist may serve 15 to 25 stations each day, the costs are spread out among the stations. The stations with good advertising sales staffs will turn around and sell sponsorships at a premium, making a solid profit on a service it pays for. This is not a service that the NWS is commissioned to handle. Some of the NWS meteorologists do go on the air, but if one station began to use the NWS forecasters as its own weather team, the service would have to be offered to all other stations. There is no provision for such services, and if the agency tried to provide it with existing staff, its mission might be compromised.

#### *Television*

Private weather firms provide television weather services in many different forms. On cable, the Weather Channel (Figure

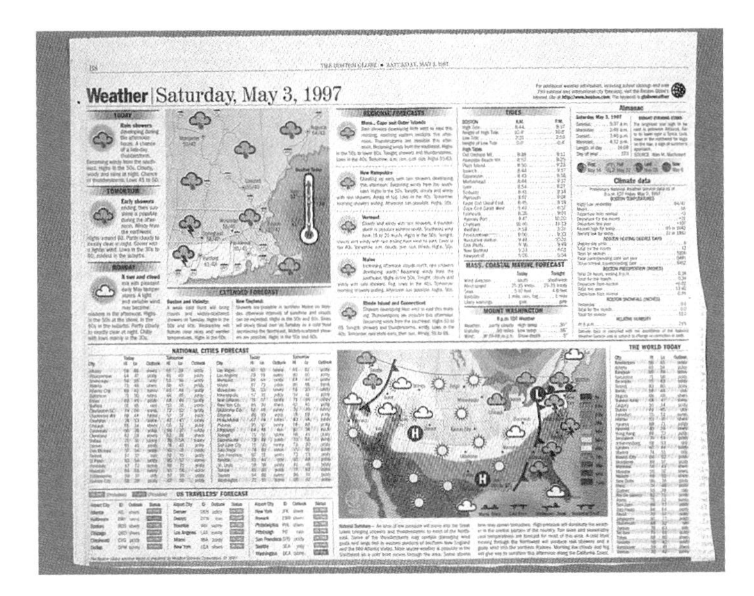

Weather | Saturday, May 3, 1997

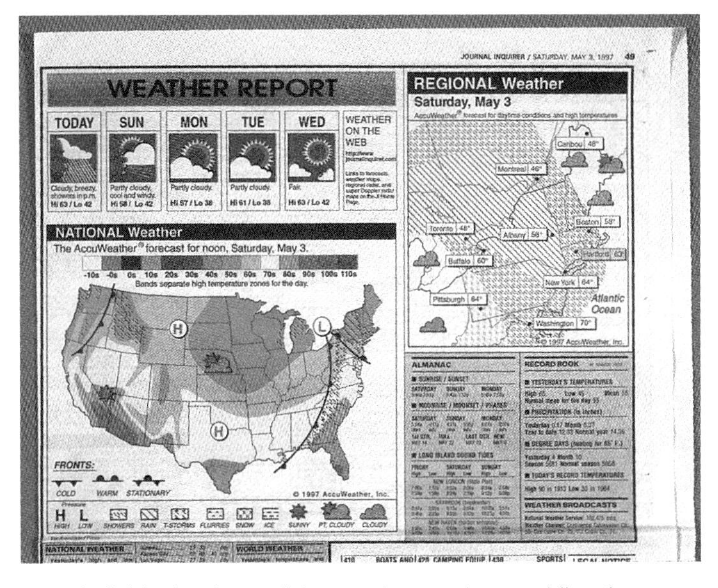

WEATHER REPORT

| TODAY | SUN | MON | TUE | WED |
|---|---|---|---|---|
| Cloudy, breezy, showers in p.m. Hi 63 / Lo 42 | Partly cloudy, cool and windy. Hi 58 / Lo 42 | Partly cloudy. Hi 57 / Lo 38 | Partly cloudy. Hi 61 / Lo 38 | Fair. Hi 63 / Lo 42 |

WEATHER ON THE WEB

REGIONAL Weather

Saturday, May 3

NATIONAL Weather

The AccuWeather® forecast for noon, Saturday, May 3.

FRONTS:

COLD WARM STATIONARY

© 1997 AccuWeather, Inc.

Figure 10.21 Newspaper weather maps vary considerably due to newspaper individual styles and the weather service providing the map.

Figure 10.22 The Weather Channel devotes 100% of its broadcasting to weather information.

10.22) is by far the biggest private supplier of comprehensive weather information. Its on-air meteorologists show an extensive array of graphics and video footage of actual weather events to explain all the weather systems affecting the nation. For local coverage, the Weather Channel arranges with local operators to provide a custom selection of products. These include local and regional observations, forecasts, and Doppler radar displays. Recall this chapter began with a group of golfers standing around watching the Weather Channel. That type of scene is repeated around the country as people often take short-term forecasting into their own hands. Using what you have learned in this course, you are equipped to try it yourself. As use of the Internet expands, it will be interesting to see whether specialized TV channels hold their own or expand, or whether people turn to potentially more flexible, do-it-yourself information-gathering systems online.

On traditional broadcast television stations, there is wide diversity in the weathercasters' backgrounds. Some have bachelor's, master's, or even doctoral degrees in meteorology; others have no formal meteorology training at all. A number of weathercasters with weak meteorology backgrounds use the services of one of the private weather firms to obtain forecasts and weather briefings. A typical briefing might be a personal phone conversation with the meteorologist who specializes in the particular region and is familiar with each client's market and special needs. The briefer walks the presenter through the upcoming weather scenario, suggesting how to interpret and explain the relevant weather factors. A person with good broadcasting skills can take this information and present a weather show that will rival the product of the best professionally educated meteorologists.

Broadcast meteorology can be a lucrative business. The top meteorologists who present weather shows on commercial television are the highest paid members of the meteorological profession. In the top 10 broadcast markets, salaries frequently topped a quarter of a million dollars in 1995, and a few individuals make salaries comparable with all but the best known sports figures.

Many meteorologists who prepare their own shows use a variety of graphics systems to create the stunning effects seen on TV. They are able to incorporate a wide variety of data types and maps to create fluid animations and representations of that day's weather. In other cases, the TV station subscribes to a private meteorological company that provides on-air graphics from a central location. WSI in Massachusetts and AccuWeather in Pennsylvania are two of the largest suppliers. AccuWeather is believed to have more meteorologists on its staff than any other private weather service. These services employ computer graphics specialists or artists to help in the creative process. For art majors, weather graphics design is an employment field that did not exist before the 1990s.

Figure 10.23 These workers will be idled in rainy weather. Accurate forecasts can reduce losses in wages and time.

### *Certification of Broadcast Meteorologists*

To recognize the training and skill that contribute to good weather presentations, meteorological organizations have established certification programs for their members. The American Meteorological Society is the largest and most respected of these organizations, and its Seal of Approval program has existed since 1959. Applicants submit a college transcript and tapes of their weather shows covering a series of consecutive days. This allows the review board, composed of sealholders, to assess the quality of the presentation. Successful applicants are awarded the Seal of Approval of the American Meteorological Society.

## Forecasts for the Construction Industry

Weather impacts construction every day (see Figure 10.23). If it is too cold, concrete does not cure properly, and concrete pours must be postponed. Rainy weather halts most outdoor work and causes problems at the job site. When it is too dry, dusty areas must be treated to prevent dust particles from the construction vehicles filling the air. The possibility of weather delays is built into the contracts for many projects, but a serious delay can mean big penalties for the construction company as well as disruption for those waiting for completion.

Many of the weather problems must simply be dealt with as they occur. Specifically tailored weather forecasts are feasible from an economic sense only if the affected people can take some action that will save money or prevent damage and extra costs. For example, some labor agreements call for the automatic payment for four hours of work if the crew reports for work. If the crew can be contacted in advance, work can be called off and that money is saved. On a large project, just a few such days can add up to big savings.

On paving jobs, concrete supplies are often trucked in from distant sites by convoy on days scheduled for major pours. If it rains, the already mixed concrete may be wasted and need to be disposed of. A good forecast can prevent the loss. A relatively new idea is for the meteorologist to be in constant communication with project managers during the workday. In the past, if a thunderstorm approached, the managers might call off work for the rest of the day. But suppose a meteorologist could tell them the shower would last only 20 minutes and would be the only one that day. If so, work could proceed. Asphalt changes character in very hot weather, and steps must be taken to avoid the bleed-through of oils that would render the surface too slick for safe travel and thus substandard when road inspectors check the work. A good temperature forecast can help management plan the workday around whatever weather is coming.

## Forecasts for Government Agencies

Why would a government unit pay for private forecasts if the federal government provides them for free? There are two main reasons: first, the individual government unit has special needs not addressed by existing forecasts. This means a request would have to be made for specialized services, which in turn would require extra appropriations. And second, there may be a need for redundancy. Agencies like the Nuclear Regulatory Commission have to be able to respond to a problem quickly and effectively. It makes sense for them to have access to multiple information sources and pathways, as well as second opinions.

For states and local governments in colder climates, snow removal can be a big expense. Private meteorologists provide specific forecasts of when snow or ice will begin and how much will accumulate on various surfaces in specified time periods. This helps save money and allows for work to be scheduled effectively.

## Surface Transportation Forecasts

While the NWS devotes considerable resources to making forecasts for aviation, it does not give the same attention to surface transportation interests. Instead, private meteorologists provide maps and forecasts to trucking companies, railroads, and transit facilities to help them deal with the weather. Consider a trucking company on a day when ice will affect part of the roadway between its terminals, which are eight hours apart. Based on the private service's forecasts, the trucking company may select different routes or alter distribution points.

Intriguing possibilities will arise with the increased use of satellite-based navigation systems and communications. These devices can determine a vehicle's location to within a few feet. Imagine a trucker or other motorist pressing a button to indicate icy conditions or any other problem; the condition and the location of the problem could be communicated to a weather service and plotted on a map. Such capabilities will greatly enhance the ability of all forecasters to provide weather alerts for users of the nation's highways with a specificity never before possible.

Railroad operations have their own special weather sensitivities. Ice can freeze switches. Tidal flooding at coastal rail yards can introduce saltwater into sensitive wheel bearings, requiring expensive overhaul. Intense storms can interfere with freight and passenger handling. Strong cross winds can affect high-profile rolling stock (a problem for truckers as well). Cold weather reduces the effectiveness of hydraulic systems in a train's brakes. In some mountainous areas of the West, the length of freight trains is dictated by the lowest expected temperature in the mountains. If the temperature drops too low, braking can be compromised or fail if the load is too great.

## Forecasts for the Skiing Industry

In many parts of the country, especially the East, skiing is a viable business only with snowmaking. Snowmaking equipment (Figure 10.24) uses a variable mix of water and air. Compressed air is forced out of a nozzle into a fine mist of water droplets. The air's rapid expansion as it leaves the nozzle causes its temperature to plunge, and the droplets turn to fine particles of ice. Evaporation helps with the chilling process. This means that the colder and drier the air is, the more water and less air are needed to produce snow. At temperatures in the teens with dry air, a good system can produce huge volumes of snow. Private meteorologists provide forecasts of temperature and relative humidity to ski areas. Based on this information and wind forecasts, snowmaking managers can schedule crews and estimate how much snow can be made and where to make it. They can answer such questions as: "Can we open for the weekend? Should we build up core slopes to prepare for a coming rainstorm, or can we spread out to open more slopes?" The wind forecast enables operators to aim the snow guns where they will do the most good. Joel N. Myers, president of AccuWeather, invented this aspect of the private forecasting business in the 1960s, and his firm has served some ski areas for more than three decades.

Figure 10.24 Successful snowmaking depends in many ways on the right weather conditions.

## International Forecasts

Since national meteorological services in each country naturally focus on forecasting within their nation's boundaries, it has made sense for the private sector to branch into the international field. High-speed computers and international communications networks mean a forecaster anywhere can find out what is happening virtually anywhere else. This data accessibility means that accurate forecasts can be produced anywhere and transmitted instantly.

***Review***
Private-sector meteorologists provide a wide variety of services to business, industry, and some public agencies. While relying on centrally collected weather data and extensive, computer-based weather predictions generated by federal agencies like the National Weather Service, they provide specialized forecasts for which public funds are not available or the expenditure of tax dollars cannot be justified.

***Excercise***
Listed below are some clients who require special weather forecasts and information. For each, list the types of weather information that might be helpful to them and why the special weather information is needed.

Construction companies; public utility companies; insurance companies; professional athletic teams; schools; city governments; television stations; transportation companies; golf courses; hotels and motels.

Study Point

## Chapter Summary

How are weather forecasts made? You have seen in this unit that this question has no single answer. The approach depends on factors such as the time span and region for which the forecast is issued, and the particular weather elements that are predicted. A person can make a reasonable forecast by paying attention only to local weather elements such as air pressure changes, wind direction, and cloud patterns. A professional forecasting team, on the other hand, might base its prediction on persistence, climatology, trend forecasting, steering, and dynamic clues such as divergence, not to mention a good look out the window, all within a single forecast.

Familiarity with the unique properties of atmospheric features such as fronts, highs, and lows guides the forecaster. For example, knowledge of the structure of a typical warm front can be applied to predict the weather changes in store when you know a warm front is on the way.

In a more physics-oriented vein, you saw that the atmosphere's behavior rests on a few basic scientific principles and laws. Thus the geostrophic relation, which can be used to predict wind speeds and directions in the free atmosphere, is an expression of the balance between pressure gradient and Coriolis forces; and the concept of force itself takes us back to Newton's second law of motion. As another example, the intensification of weather systems can be predicted by using the divergence-vorticity relation, a process that generally is carried out by computer. Computer-based forecasting, in fact, rests almost entirely on relations such as these.

You have seen that weather prediction, like the weather itself, is a highly complex subject. This complexity, of course, adds to the weather's interest. It explains in part the fascination many of us feel when we look at the sky or at a satellite or radar map and ask, "I wonder what it will do next?"

## Key Words

persistence forecasting
trend forecasting
extrapolation
interpolation
steering
weather types forecasting
lifted index
nowcast
short-term forecast
extended forecast
forecast verification
watch
warning
special statement

## Review Exercises

1. Below are listed several different forecasting techniques discussed in this chapter. For each, list some of the advantages and limitations of its use.

| Technique | Advantages | Limitations |
|---|---|---|
| Persistence | | |
| Trend | | |
| Cycles and trend | | |
| Interpolation | | |
| Steering | | |
| Weather types | | |
| Numerical | | |

2. For each of the weather situations below, suggest a reliable forecasting technique and explain the reason for your selection.

| Weather Situation | Forecasting Technique |
|---|---|
| Weather in the next hour | |
| Weather next month | |
| Weather 30 miles away | |

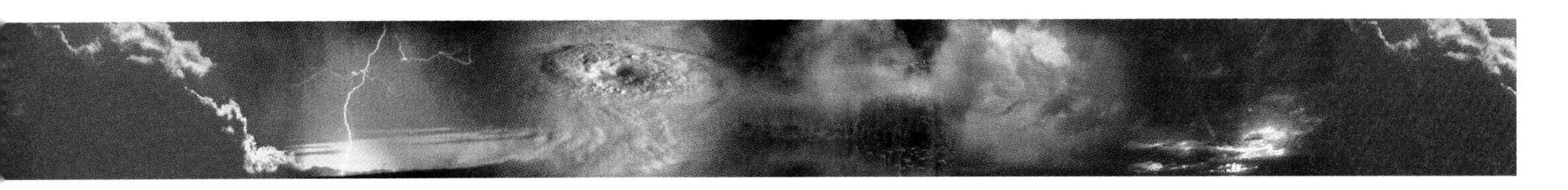

A low pressure area begins
to form as usual off the North
Carolina coast
Weather in two days with a
persistent jet stream pattern
A forecast for record-breaking
cold and snow
A forecast three days in advance
Sunny and warm today, sunny
and warm tomorrow; what will the
temperature be tonight?

## Problems

1. Controversy exists over whether the NWS should provide specialized forecasts for certain businesses such as agriculture. Some argue that taxpayers' money should not be spent to support a particular industry and such forecasting should be left to the private weather services. Others argue that the nation's food supply is of critical public interest and warrants the spending of tax money to protect it. What is your opinion?

## Explorations

1. The most obvious project to pursue after studying this chapter is to prepare your own weather forecasts. Use as many of the sources described in this chapter as you can find (check the Internet). After you have made your prediction, compare it with other local forecasts. Then follow the weather over the forecast period and see if it verifies.
2. Collect forecasts for your local area from as many different sources as possible, such as NOAA weather radio, various radio and TV stations, newspapers, Internet sources (see suggested sources in Resource Links below.) Compare the different forecasts in terms of the type of information provided (high and low temperatures? Tides? UV index?).

## Resource Links

The following is a sampler of print and electronic references for some of the topics discussed in this chapter. We have included a few words of description for items whose relevance is not clear from their titles. More challanging readings are marked with a C at the end of the listing. For a more extensive and continually updated list, see our World Wide Web home page at www.mhhe.com/mathscience/geology/danielson.

Baumann, W. H. III, and S. Businger. 1996. Nowcasting for space shuttle landings at Kennedy Space Center, Florida. *American Meteorological Society Bulletin 77* (10):2295–2305. C A detailed look at "nowcasting" for a specific customer.

AccuWeather's home page at http://www.AccuWeather.com is an extensive source of current weather conditions, and forecasts for US cities. It also provides visible and infrared satellite images, and Doppler radar images, and current and forecast weather maps and discussions.

For readers with access to AccuData, SURMAP offers over one hundred different current and forecast maps. FST provides short range and extended forecasts, including computer model forecasts. AVIPAC, AGPAC and MARPAC provide specialized data for aviation, agricultural and marine interests respectively.

The National Weather Service's home page is at http://www.nws.noaa.gov/ .

For West Coast weather, an excellent source is the San Francisco State University web site at http://tornado.sfsu.edu/geosciences/geosciences.html .

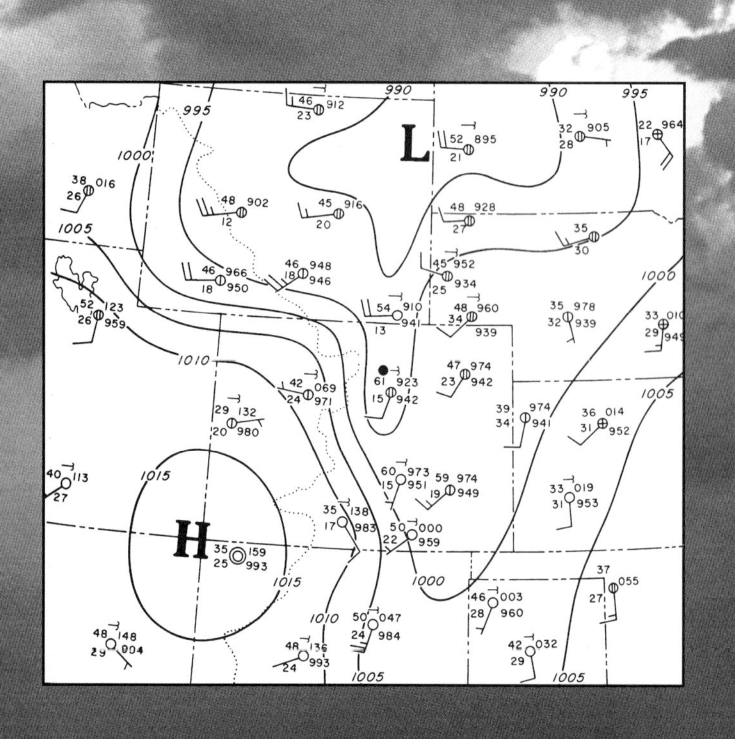
L
H
990
995
1000
1005
1010
1015

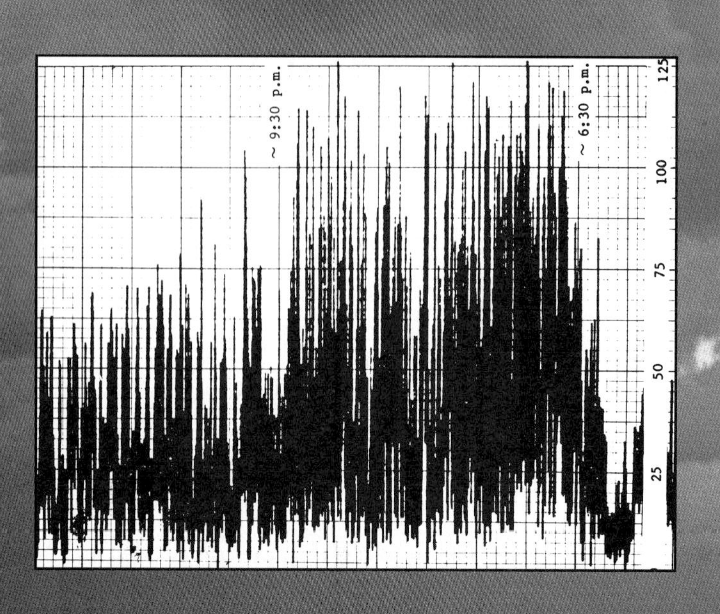
~ 9:30 p.m.
~ 6:30 p.m.
125
100
75
50
25

# Are There Limits to Atmospheric Predictability?

The weather map in the facing figure shows surface weather conditions over Colorado on January 7, 1969. Note the location of the city of Boulder, slightly northwest of Denver. If you had to estimate conditions prevailing at Boulder, what would you conclude? Based on observations at nearby stations, a reasonable estimate might be "partly cloudy skies, temperature around 60, and wind speed around 15 knots."

As you can see from the anemometer trace, the weather actually occurring at Boulder on this day was vastly different. Winds sweeping down from the Rocky Mountains produced spectacular winds in the city, causing considerable damage. One resident reported the winds in the evening were so strong that they broke windows and tore the roof off his house. "We heard the windows crashing, and looking up, we saw the stars in the sky," he said.

How could the weather at Boulder that day bear so little resemblance to that in nearby Denver? Or, stated from another point of view, how could the weather map be so misleading?

This example illustrates well that the earth's weather is composed of events occurring simultaneously on a number of different scales.

In this unit of three chapters, you will explore atmospheric circulations from the very small to the very large. As you will see, the sheer number of small-scale events and their potential to interact in complex ways can make precise weather forecasting seem unattainable. Thus, the unit's central question is, "How does the vast variety of interacting atmospheric events affect the atmosphere's predictability?" This topic is the focus of considerable research at present, leading to fascinating new areas of science such as chaos.

# Chapter 11

# Circulations Large and Small

## Chapter Goals and Overview

In this chapter, you will survey atmospheric patterns ranging from molecular to worldwide. Many of these phenomena, such as wind gusts, sea breezes, and monsoons, are basic components of the world's weather. Thus, to understand the atmosphere and its weather, you need to understand these features.

An important theme linking the topics in this chapter is the fact that atmospheric circulations are linked to each other through chains of cause and effect. None of them exists in isolation; rather, each shapes and is shaped by features on many different scales of action. This linkage leads to some surprising circumstances, as you will see in the chapter's final section.

### Chapter Goals

By the end of this chapter, you will be able to:

compare and contrast laminar (smooth) and turbulent flow, and explain how fluid motion may change from one to the other;
give examples of atmospheric circulations on five different scales of size and time;
make a sketch showing typical flow patterns for each of these: dust devils, clear air turbulence, land and sea breezes, mountain and valley winds, lee waves, Hadley cells, monsoon circulations, and the general circulation;
describe the factors and forces that are important in the formation of each of the circulations listed above;
give examples of how circulations of different scales interact; and
define *chaos* and describe the "butterfly effect."

# Forms and Scales of Atmospheric Motion

This section is a brief survey of the most common atmospheric flow patterns, their basic shapes, and their dimensions in space and time.

### Forms of Fluid Motion

Examine for a moment the fluid motions in Figure 11.1. If you made an inventory of flow forms in these and other images, you might note the following features:

smooth flow in parallel lines or sheets; such flow is called **laminar flow;**
wave motions; and
vortices, or eddies.

You also might list various combinations of these features, such as

straight flow made up of smaller-scale features;
waves made up of smaller-scale features;
straight flow that is part of larger-scale waves; and
complex combinations of many forms. Complex flow forms that tend to change rapidly with time are known as **turbulent flow,** or **turbulence.**

Notice that the list above progresses from the simple to the complex. Laminar flow, the least complex, is smooth and predictable: knowing the speed of the flow, you can calculate where a given fluid parcel will be at a later time. Moving down the list to wave and eddy motions, the flow becomes more complex and prediction becomes more difficult. In these more complex patterns, two parcels at only slightly different starting points may end up in very different locations later on. Fully turbulent flow, consisting of rapidly changing waves and eddies of many different sizes, is the most complex. Generally, it is not possible to predict exactly the future location of a particle, or conditions at some future moment at a given location, in turbulent flow, because the flow pattern itself is constantly changing. The problem is a bit like trying to predict the position of a moving automobile some time in the future when you don't know where the road goes.

These flow patterns are not restricted to the earth's atmosphere, as the images of clouds on Jupiter indicate. Notice that you can also find them in other fluids, such as water.

In summary, atmospheric flow may be laminar, fully turbulent, or somewhere in between. Its appearance ranges from straight or gently curving smooth flow to waves, eddies, and finally to turbulent combinations of these features. Waves and eddies may be oriented vertically or horizontally and may look very different from different perspectives, as you can see in Figure 11.1.

### Scales of Atmospheric Circulations

The images in Figure 11.1 illustrate striking likenesses between some flow patterns of very different sizes. It's hard not to notice the structural similarities between a hurricane and a frontal cyclone, for example. Although these similarities can be instructive, crucial differences remain. A hurricane is not just a small-diameter version of a frontal cyclone. Hurricanes form under different circumstances and mixes of pressure gradient, Coriolis, and frictional forces than frontal cyclones. Thus, scale is an important property of atmospheric flow features.

Another observation evident from the illustrations is that the form of flow pattern you recognize depends to a considerable extent on the scale on which you observe. Thus, viewed from space, a frontal cyclone might appear as a wave; however, a meteorologist analyzing one region of the same cyclone might consider the flow straight and laminar, while an observer experiencing the wind's moment-to-moment gustiness would probably describe it as turbulent. Therefore, defining the scale on which you are working can be an important step toward understanding.

Considerations such as those above have led atmospheric scientists to develop a simple classification system for circulations of

A
B
C
D
E

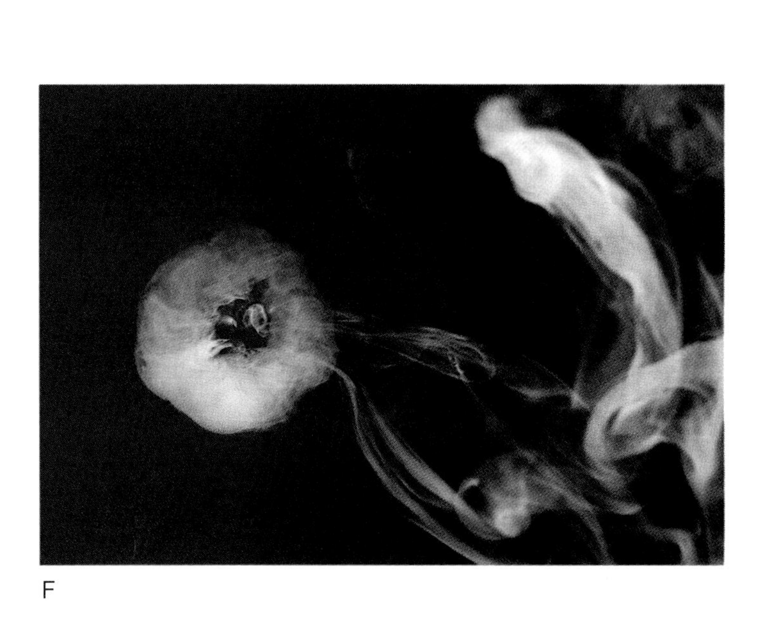

F

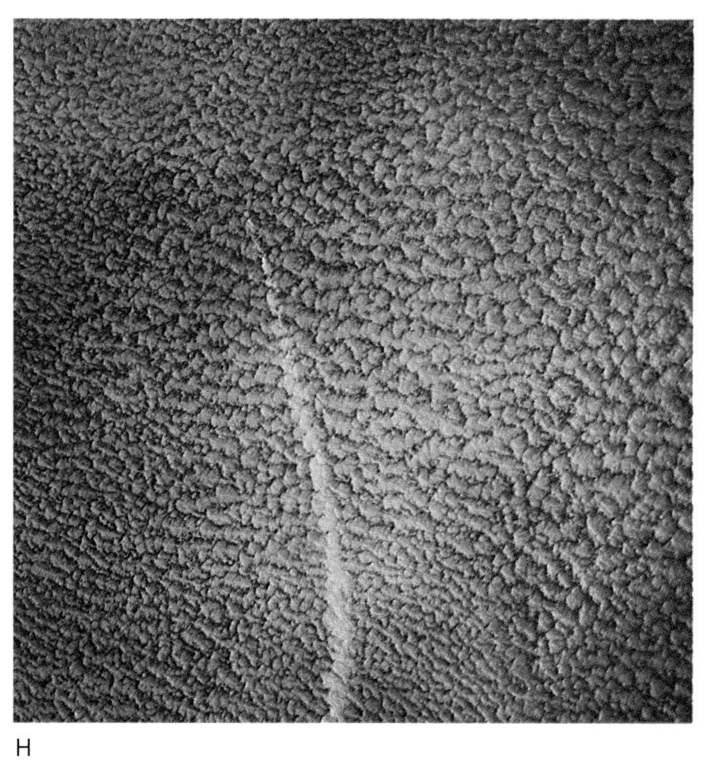

H

G

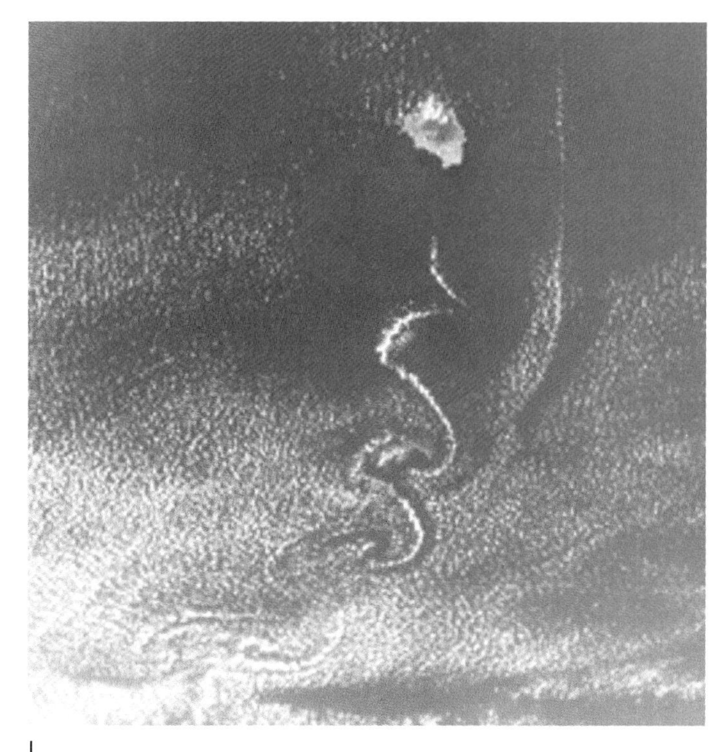

I

Figure 11.1 Atmospheric flow patterns exhibit great variety in scale but some similarities in form.

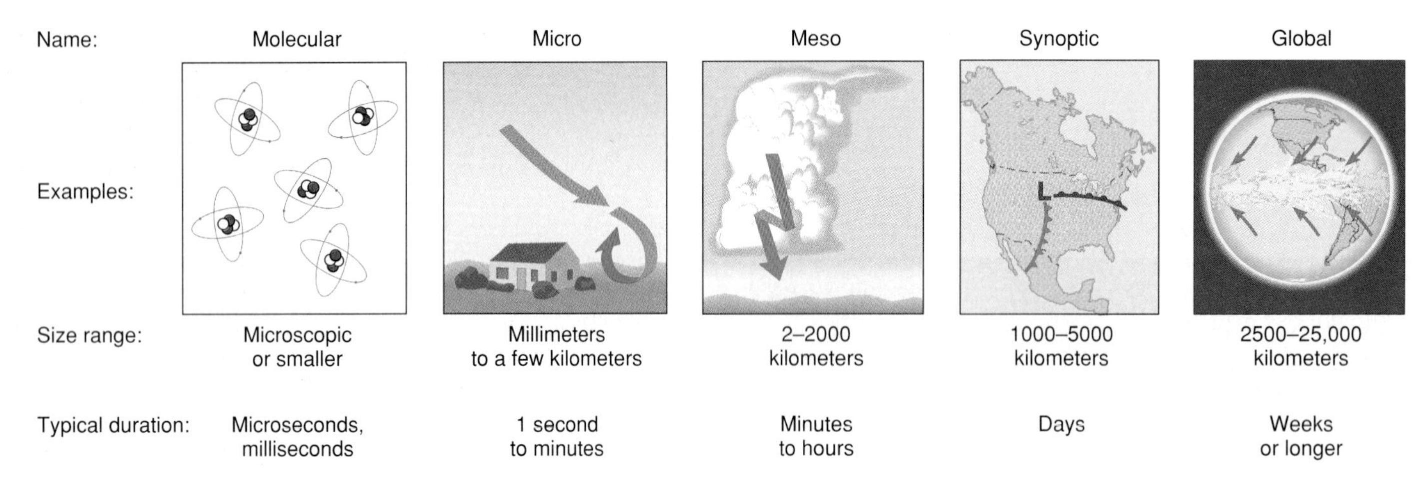


Figure 11.2 Scales of atmospheric flow, from molecular to global.

different dimensions. The system, shown in Figure 11.2, consists of five categories: **molecular scale** (involving individual molecules, in processes such as in heat conduction); **microscale** (such as wind gusts); **mesoscale** (thunderstorms); synoptic scale (frontal cyclones); and **global scale** (features affecting large portions of the earth's atmosphere). In the following sections, you will study important features on each of these scales, proceeding from smallest to largest. Note that neighboring categories in Figure 11.2 do not have precise boundaries; instead, some overlap exists between, for example, microscale and mesoscale dimensions.

# The Molecular Scale

Molecular scale events are too small to observe directly. However, we can employ the atomic and kinetic models developed in earlier chapters to help understand events at this scale. We begin by examining molecular behavior in laminar flow.

## Laminar Flow

Although atmospheric motions are never perfectly laminar, sometimes they come close enough so that they may be assumed to be so. For example, geostrophic flow, which frequently is a good approximation of large-scale motions, is laminar.

Figure 11.3 shows a cross-sectional view of the atmosphere exhibiting laminar flow near the earth's surface. Notice the wind speed profile in this layer. Speeds are constant at upper levels but decrease as you approach the surface. Surface frictional forces would explain the decrease at the very interface; but why is the speed reduced at intermediate levels as well?

To answer this question, let's shift scales downward for a moment. If you could observe Figure 11.3's airflow at the molecular level, you would see that the air molecules at any one level do not always remain strictly at that level, as the arrows in the figure suggest. Instead, the molecules' random thermal motions carry them into neighboring layers as they move to the right with the flow. Near the surface, downward migrating molecules may strike the ground, losing speed and therefore transferring kinetic energy to the earth's surface. Similarly, molecules in the next higher layer lose energy by colliding with the slower-moving layer below, and so on. Thus, random motions generate molecular collisions between laminar air layers, exerting a type of mutual friction. As a result, molecules in the faster-moving layers lose speed and thus kinetic energy, conveying the energy to lower, slower-moving levels.

Such frictional forces within a fluid are called **viscous forces.** In the example we have considered, the viscous forces are due to the air's **molecular viscosity,** which is a measure of

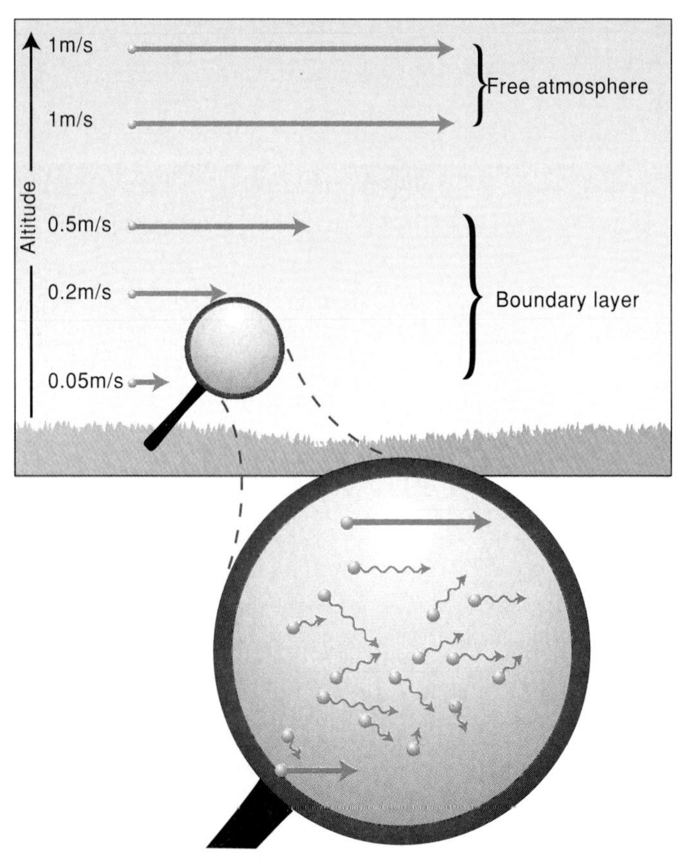


Figure 11.3 Laminar wind flow near the earth's surface. Notice that the air in different layers does interact through random molecular migration.

how effectively the air's random molecular motions convey viscous forces. The greater a fluid's viscosity, the greater the frictional resistance it offers. Thus, thick, gluey fluids that resist changes in motion are more viscous than thin ones. You may be familiar with the term **viscosity** from its use in connection with motor oils. Although molecular viscosity in liquids is due more to intermolecular attractions than in gases, its effect—resistance to changes in motion—is the same in both.

Notice that the net effect of viscous and surface frictional forces in Figure 11.3 is to remove kinetic energy from the atmosphere by passing it to the earth's surface as heat energy, a process is known as "frictional dissipation." Without a supply of new energy, frictional dissipation at the earth's surface eventually would bring all atmospheric motion to a halt. This surface layer in which the atmosphere's kinetic energy is moved downward and out to the earth by frictional dissipation is known as the **boundary layer.** The atmosphere above the boundary layer, which is free of surface frictional effects, is referred to as the **free atmosphere.**

## The Transition to Turbulence

What happens to the flow if the wind shear across the boundary layer in Figure 11.3 increases? Greater shear means stronger viscous forces between layers, as molecules of radically different speeds collide. Eventually, these interactions alter the laminar nature of the flow, much as a smooth flow of highway traffic is disrupted when slower vehicles try to merge with the flow. In the atmosphere, eddies and waves begin to develop in the previously smooth flow. You can see this transition in several of the images in this chapter's opening figure.

Depending on factors such as the degree of wind shear and the atmosphere's stability, one of three things will occur: (1) the eddies and waves will dissipate as the flow becomes laminar once more; (2) eddies and waves of all sizes will proliferate, as the flow becomes turbulent; or (3) the flow will remain in a transitional state, as a combination of smooth flow and scattered eddies and waves. Several examples of transitional regimes between laminar and turbulent flow are pictured in the collage of photographs opening this chapter.

The breakdown of laminar flow in the above example is due to what meteorologists call "mechanical "means, as air molecules of different speeds collide with each other or the earth's surface. Turbulence triggered in such a way is therefore called **mechanical turbulence.** Common examples of mechanical turbulence (see Figure 11.4A) are the gustiness of the wind due to collisions with trees, buildings, and so on, and turbulence produced by wind flow over mountain ranges. Mechanical turbulence can also occur in the free atmosphere, as you will see presently.

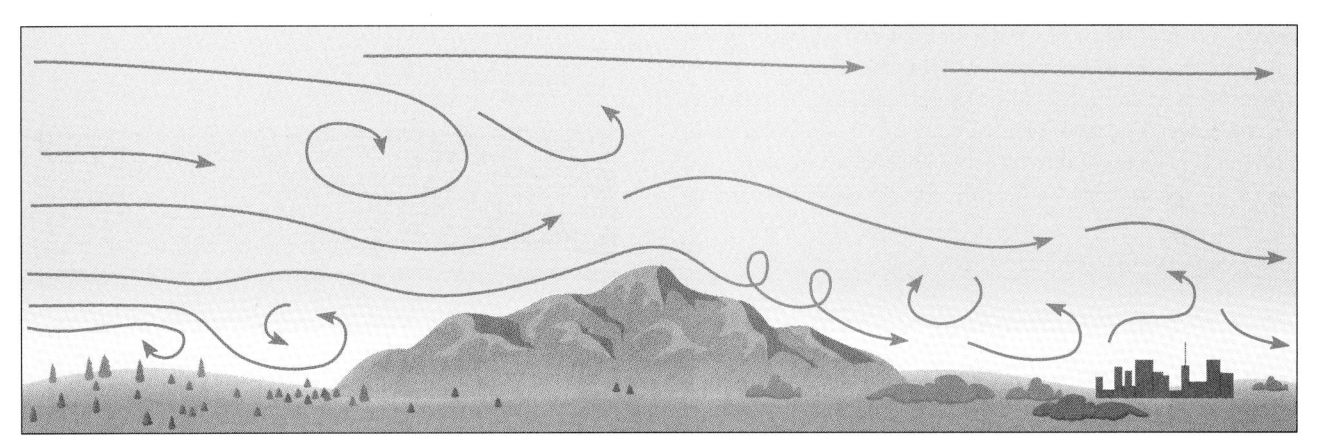

A Mechanical turbulence

B Thermal turbulence

Figure 11.4 Turbulence may develop by mechanical or thermal means—or a combination of both.

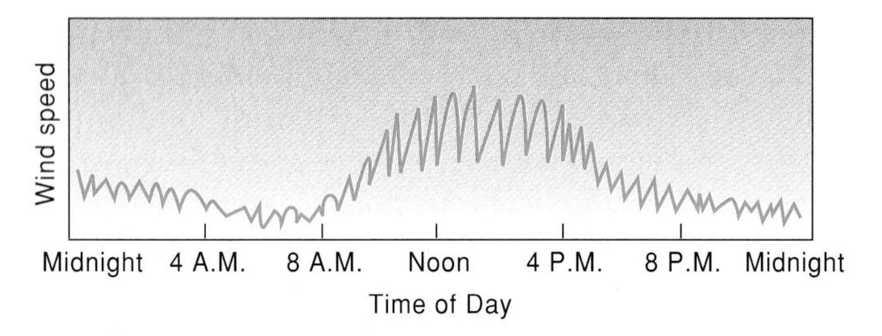


**Figure 11.5** The wind speed record for a spring day at Caribous Maine. Why is the wind so much more variable during the day?

Not all turbulence is mechanical in origin, however. Buoyant forces can also induce a transition from laminar to turbulent flow. Thus, **thermal turbulence** is produced by convective air currents (Figure 11.4B) and is exemplified in cumulus cells, as well as in a number of other circulations you will study later in this unit.

Generally, atmospheric turbulence is neither purely mechanical nor purely thermal but a combination of both factors. The wind speed trace in Figure 11.5 shows how changes in static stability, due to surface heating and cooling, typically amplify mechanical turbulence by day and suppress it by night.

## Turbulent Flow

Suppose that the breakdown of laminar flow in Figure 11.3 proceeds all the way to turbulence, which is the normal form of boundary layer flow. How does the breakdown alter the wind profile and energy transfer, compared to the case of laminar flow?

As you can see in Figure 11.6, energy differences between colliding molecules and frictional dissipation are far greater in turbulent than in laminar flow. Turbulent eddies mix air from different layers much more efficiently than molecular viscous forces do. The situation is closely analogous to heat transfer by conduction versus convection; recall (from Chapter 2) that conduction occurs on a molecule-to-molecule basis, while convection involves eddy transport of whole parcels of air at once.

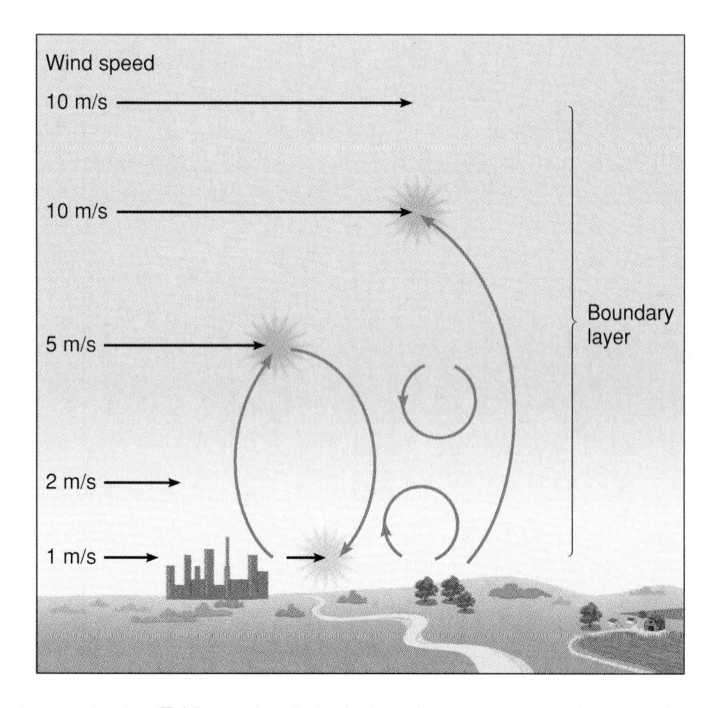


**Figure 11.6** Eddy motion in turbulent flow causes surface winds to mix directly with much stronger winds high above the ground.

Since energy is dissipated rapidly in turbulent flow, it takes a greater supply of new energy to maintain the flow than when the flow is laminar. In a sense, the air is more viscous when turbulent. This greater viscosity due to eddy motion in turbulent flow is known as eddy viscosity. The air's eddy viscosity may be as much as 1 million times greater than its molecular viscosity.

Note also in Figure 11.6 that more efficient mixing due to eddy motion causes the boundary layer to be far deeper in turbulent than in laminar flow. In turbulent low-level flow typical of the earth's boundary layer, it is on the order of a kilometer in depth; thus, winds up to a kilometer above the ground show some effect of surface drag. In pure laminar flow, the boundary layer would extend no more than a few meters above the surface.

Thus, eddy viscosity is a far more effective agent than molecular viscosity in removing energy from faster-moving, larger circulations and dissipating that energy downward to the earth through the planetary boundary layer. The English meteorologist L. F. Richardson aptly summarized this process of turbulent energy exchange as follows:

*Big whirls have little whirls*
*that feed on their velocity*
*and little whirls have lesser whirls*
*and so on to viscosity.*

### Checkpoint

***Review***
Atmospheric motions may be laminar, turbulent, or transitional between these two forms. The table below compares the extremes of pure laminar and pure turbulent flow.

| | Laminar Flow | Turbulent Flow |
|---|---|---|
| TYPICAL FEATURES | Parallel flow; straight or gently curving | Eddies and waves of many sizes |
| PERSISTENCE OF PATTERN | Highly persistent | Not persistent; continually changing |
| MOTIONS OF FLUID PARTICLES | Highly predictable, along streamlines | Unpredictable due to changes in flow pattern |
| INTERNAL FRICTION DUE MAINLY TO | Molecular viscosity | Eddy viscosity |

*Questions*

1. Locate examples of laminar flow, waves, eddies, and turbulent flow in Figure 11.1.
2. Of which scale of atmospheric circulations is a frontal cyclone an example? A swirl of wind gusting down a city street?
3. How does molecular viscosity differ from eddy viscosity?

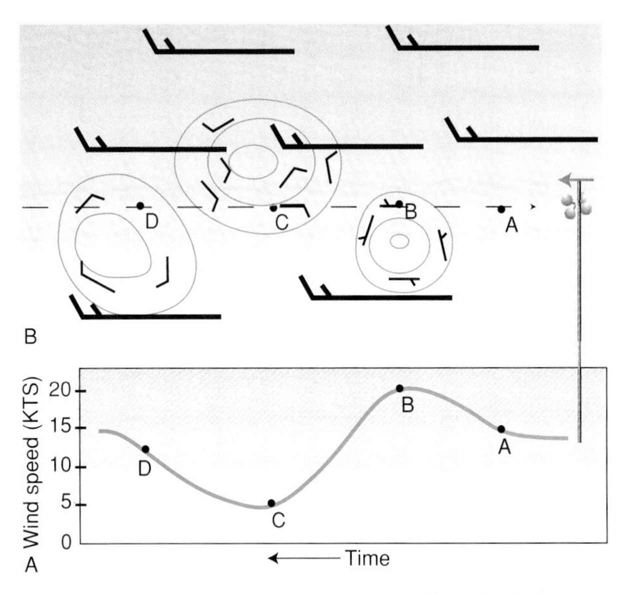


Figure 11.7 The passage of a series of eddies within the mean flow causes surface winds to vary.

# Microscale Circulations

Step outside and feel a gust of wind brush across your face: you have just had an encounter with a microscale circulation. Wind gusts are examples of microscale eddies, typically spun off from larger-scale circulations interacting with surface objects.

## Forces on the Microscale

What forces shape microscale circulations such as wind gusts? Pressure gradient force is the major motive force, as it is for the synoptic scale features you studied in the last unit. You learned there that in large-scale circulations, pressure often is balanced by Coriolis force, resulting in geostrophic flow. On the microscale, however, geostrophic balance does not occur. What would account for the difference?

The answer relates to the small time scale of microscale circulations: during their brief lifetimes (a few seconds or minutes), the earth does not rotate enough for Coriolis force to have a noticeable effect. As a result, geostrophic balance does not occur on the microscale.

Frictional forces often are particularly important on the microscale. Microscale circulations generally consist of light winds, which may be greatly modified or even extinguished entirely by surface friction or viscous forces.

## Surface Turbulence

Observe the record of wind speed and direction in Figure 11.7A, a typical signature of surface turbulence. We have stated that such turbulence results from eddy motions. But note the presence of waves, not eddies, in the wind pattern. Where do the waves come from?

Suppose that the eddies are embedded in a prevailing wind that is part of some much larger circulation, a normal circumstance. Figure 11.7B shows the situation, with some idealized, circular eddies shown in blue in a constant flow from the left (shown by the black wind arrows) and an anemometer and wind vane to measure the airflow. Consider what happens as these eddies move past the anemometer. The air at A is not affected by an eddy; therefore the anemometer records its passage with a reading of 15 knots, the speed of the larger-scale wind. However, the eddy at B adds 5 knots to the mean flow; in response, the anemometer records a gust to 20 (15 + 5) knots. The eddy at C moves against the mean flow at 10 knots, causing a decrease in wind speed to 5 (15−10) knots. At D, eddy motion both changes the speed and deflects the direction of the flow. In this way, wind gusts and lulls can be interpreted as the passage of a succession of eddies embedded in the main flow.

Although microscale eddy motions often are positioned at random in a larger flow, as in the example shown in Figure 11.7, surface obstructions may induce more predictable microscale patterns. For example, the deflecting effect of a building may cause a persistent eddy in its lee, which in turn causes surface winds to flow in a direction opposite to that of the main flow. Snowdrifts around buildings, trees, and fences (Figure 11.8) often reveal the presence of eddy flow during a storm.

Figure 11.8 Snowdrifts are records of eddy motion that persisted during the storm.

**Figure 11.9** A dust devil interrupts a Baltimore Orioles, Pittsburgh Pirates exhibition baseball game in Miami on March 13, 1960. How might the geometry of the stadium have influenced the dust devil's formation?

## The Dust Devil: A Microscale Circulation

Not all microscale structures are lumped together under general headings like turbulence, eddies, and so forth. Some have distinctive characteristics and names of their own. The dust devil (Figure 11.9) is a well-known example. Dust devils are centers of warm or hot rising air into which surrounding air spirals to replace the upward-moving flow. They resemble small-scale tornadoes but are not associated with thunderstorms. Although most dust devils are not dangerous, the largest and most powerful ones have been known to cause structural damage.

At the core of a dust devil is a region of warm, ascending air. As it rises, surface air spirals inward to replace it. This should seem reasonable and familiar to you: by now, you have seen many examples of convergent, rotating motion on larger scales. However, in the case of the dust devil, questions arise: why does the air spiral inward, rather than flowing straight to the center of circulation? If Coriolis force is insignificant on the microscale, what causes the dust devil to rotate?

The answer lies in two laws. The first is the law of probability. If you flip a coin 1000 times, chances are you will get a slight preponderance of heads or tails and not exactly 500 or each. Similarly, the air molecules surrounding the dust devil, like those in any air parcel you choose, are likely to have some tendency, however slight, to rotate one way or the other and not be free of all spin. Sometimes the air need not depend on the law of probability at all: local obstructions (such as the stadium in Figure 11.9) may systematically deflect the airflow, imposing a rotation on it.

The second law is best illustrated by asking our spinning ice skater of Chapter 9 (Figure 9.26) for a repeat performance, as illustrated in Figure 11.10A. Recall that her "spin up"as she pulls in her arms illustrated an increase of vorticity related to a decrease in the area through which she was spinning, a process called convergence. A concept related to the vorticity-divergence relation is a law known as the **conservation of angular momentum.** This law states that as an object rotates, the product of its mass (m), its rotational speed (v) around the center of rotation, and its distance from the center of rotation (r) tends to remain constant: thus,

$$\text{angular momentum} = mvr = \text{constant}$$

Note in Figure 11.10B that the rotational speed (v) is just part of the parcel's motion: it is the circular part, exclusive of any inward or outward speed. The law applies only to this circular, rotational speed.

Applying the law of conservation of angular momentum to the inward spiraling air in the dust devil in Figure 11.10C, you see that the air's rotational speed increases in proportion to its decreasing distance from the center of the vortex. Just how much rotational speed does the air acquire? Suppose a 1-kilogram air parcel 100 meters from the dust devil's center has a rotational speed of 0.5 meters per second ($ms^{-1}$). Then, the parcel's angular momentum is

$$\begin{aligned} mvr &= (1\ kg) \times (0.5\ ms^{-1}) \times (100\ m) \\ &= 50\ kg\ m^2s^{-1} \end{aligned}$$

Now, as the 1-kilogram parcel spirals toward the center of the dust devil, its mass remains unchanged. However, the product of (v) × (r) must remain equal to 50, according to the law. You can use the expression (v) × (r) = 50, or v = 50/r, to calculate the value of the wind's speed at various distances from the center of rotation. When the parcel is 3 meters from the center, you should find it is rotating at a speed of about 16.7 meters per second (roughly 30 mph), a respectable rate for a dust devil.

## Turbulence Aloft

Microscale turbulence is not restricted to near-surface regions; it may occur high above the ground as well. Thermal updrafts in cumulus clouds are obvious visible examples. Another, invisible form of microscale turbulence is known as **clear air turbulence,** or **CAT.** On an airplane in clear skies, miles above the ground, you may have experienced CAT as light turbulence, dubbed "choppy air," or in a more severe form known colloquially as an "air pocket."

CAT typically forms in the zone of strong shear between a pronounced jet stream and lighter neighboring winds, as you can see from Figure 11.11. In both the jet and the lighter wind zone, speeds vary little and the flow may be laminar or nearly so. In the transition zone, however, the shear is too great for laminar flow to persist, and it is here that turbulent eddies form. Since the difference in wind speed across the jet may be 50 meters per second (roughly 100 knots) or more, the resultant eddies may be very pronounced. An airplane suddenly encountering a 100-knot downdraft will experience a sudden drop in altitude, which can cause unsecured objects to be thrown about violently as well as structural damage to the aircraft itself.

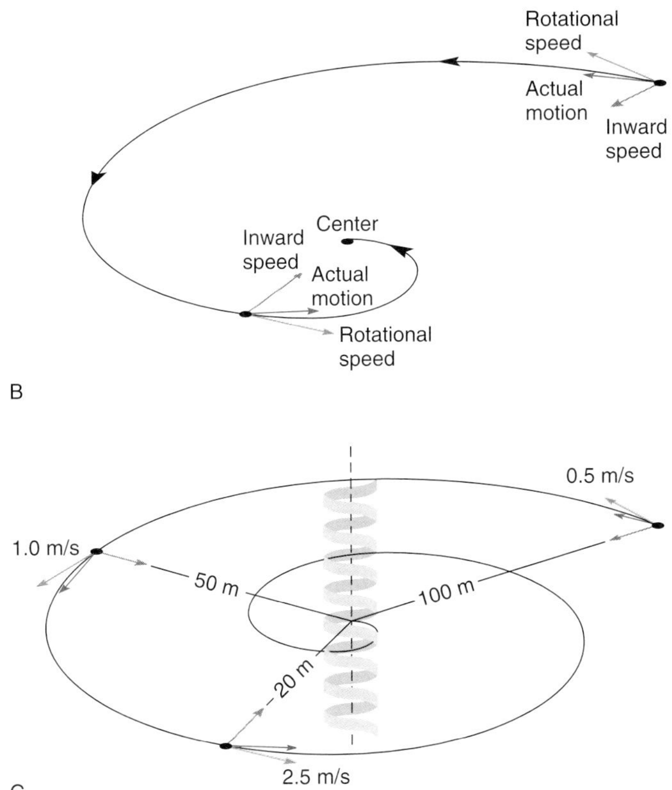


Figure 11.10 (*A*) Ice skater demonstrating conservation of angular momentum. (*B*) Rotational speed is just that component around the center, not toward or away from it. (*C*) As the parcel approaches the center, its rotational speed increases.

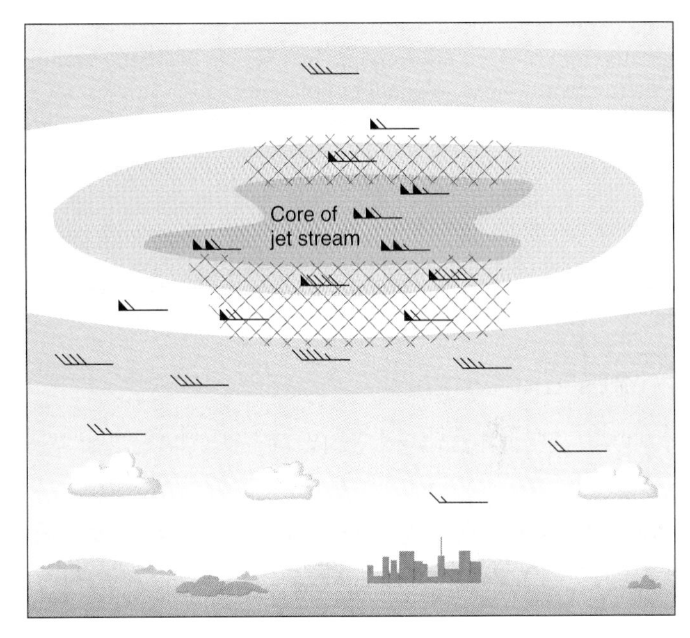


Figure 11.11 Regions most vulnerable to CAT are cross-hatched. Why is CAT less likely in the jet stream core, where winds are strongest?

## Checkpoint

***Review***

You have seen in this section that microscale circulations often are constituent eddies of pure mechanical or thermal turbulence. Microscale circulations are more vigorous near the ground, because of the role of the surface in breaking down laminar flow. However, turbulent microscale circulations do occur above the planetary boundary layer, in convective cells and as clear air turbulence.

***Questions***

1. Why is Coriolis force insignificant on the microscale?
2. Suppose the initial wind speed in the dust devil example was 2 meters per second, while other values remained the same. What would the rotational wind speed be at 5 meters from the center of the vortex?
3. Would you expect low-level air turbulence for aircraft to be greater early in the morning or late in the afternoon? Why?

# Mesoscale Circulations

As you saw back in Figure 11.2, mesoscale circulations are those that persist for minutes to hours and have dimensions from 2 to 1000 kilometers. Unlike smaller-scale features, mesoscale circulations are large enough and persist long enough that many characteristic types can be identified and studied. This contrast with microscale events is largely a matter of perspective; if we were microscopic in size, we probably would be able to identify many distinct circulations within microscale turbulence and perhaps recognize their counterparts on larger scales as well.

In any case, the mesoscale boasts a rich variety of circulation types. Three particularly violent forms—the thunderstorm, tornado, and hurricane—are so important that two subsequent chapters are devoted entirely to them. In this section, you will survey several other common mesoscale examples.

## Forces and Origins

In smaller mesoscale circulations such as tornadoes and some isolated thunderstorms, Coriolis force is relatively unimportant and the flow is nongeostrophic. (Recall from Chapter 8 that Coriolis force is also not a factor at or near the equator.) In longer-lasting mesoscale circulations, Coriolis assumes a more important role. Friction often has a significant influence in mesoscale systems; but the pressure gradient force remains the fundamental driving force, as it is on all scales of atmospheric motion.

Many mesoscale features also contribute to larger-scale wind flow, by passing energy "upward" to large circulations, rather than only drawing kinetic energy downward and out of the atmosphere, as in the case of microscale turbulence.

Often variations in the earth's surface properties (land to water, mountain to valley, etc.) are influential in causing and shaping mesoscale circulations. Notice the important role these features play in the examples below.

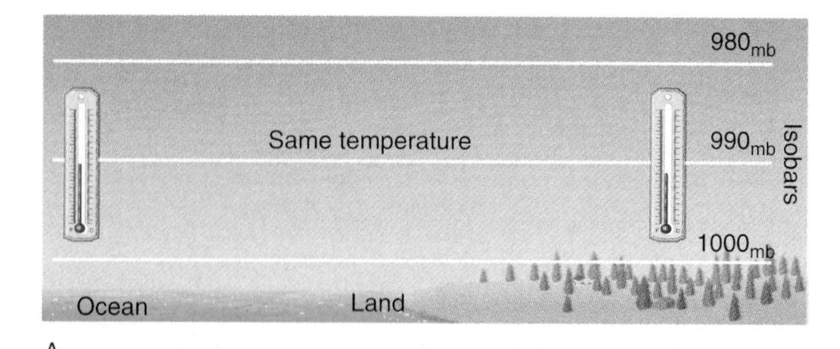


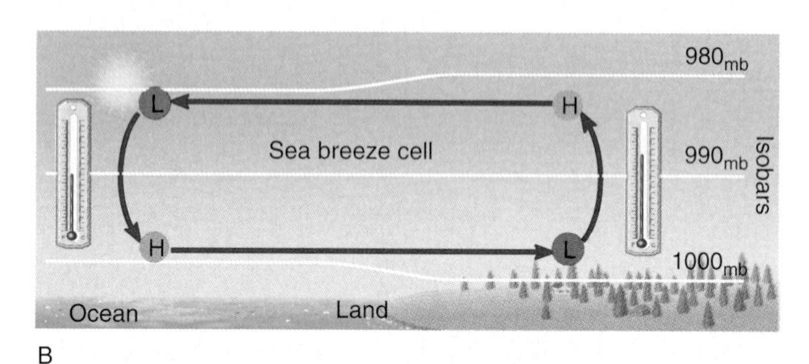


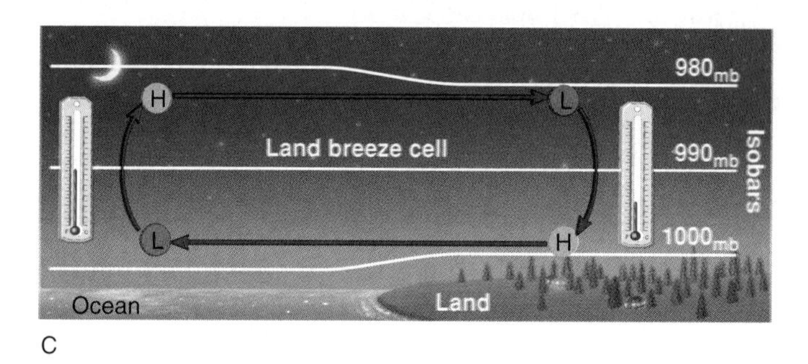


**Figure 11.12** The land and sea breeze circulation. (*A*) In mid-morning or early evening. (*B*) Afternoon. Note the surface pressure is lower over land, higher over water. Aloft, the pressure pattern and wind flow are reversed. (*C*) Late evening. How has the pressure pattern changed from daytime?

## Land and Sea Breeze Circulations

Look at the cross-sectional views of the lower atmosphere at the seacoast in Figure 11.12. Notice that in (A), which represents conditions in mid-morning, temperatures and air pressures are more or less uniform horizontally. What are the wind speed and direction in Figure 11.12A? Since the isobars are horizontal, there is no horizontal pressure gradient force, so there is no horizontal wind.

In (B), the sun has warmed the earth's surface for several hours. As you can see, this heating causes a greater temperature rise over the land than over the water (recall discussion of this topic in Chapter 2), leading to important changes in the near-surface pressure and wind regimes. The flow of surface air from the water onshore to the thermal low pressure center is known as a **sea breeze.** Notice that the sea breeze is just part of a mesoscale convection cell that includes warm air rising over land, returning to the ocean aloft, then sinking. A typical sea breeze cell is 30 to 100 kilometers in diameter. The sea breeze cell strengthens during a sunny day, typically reaching its peak strength and size in mid-afternoon.

Through the evening and night, the earth's surface cools. Again, the land changes temperature more rapidly than the ocean surface, so that some time later (early evening), the ocean and land surface temperatures are equal, and the sea breeze circulation ceases. Once again, conditions resemble those in (A). However, by late evening (C), the water surface is relatively warmer than the land. The circulation redevelops, but in the opposite sense, with a **land breeze** carrying cool air offshore to a weak thermal surface low. The nighttime temperature contrasts between land and water are not so extreme as during the day; as a result, the land breeze circulation generally is weaker at night than the daytime cycle.

The daily land and sea breeze cycle is a regular feature of many coastal regions. People living within 40 kilometers or so of the coast depend on the arrival of the cooling sea breezes to cap

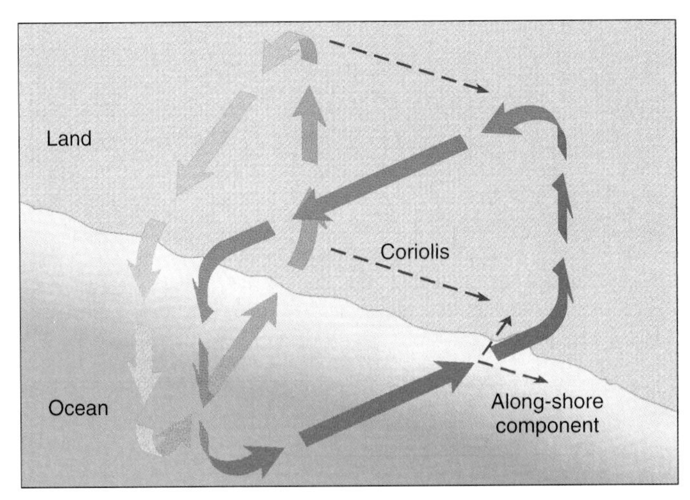


A

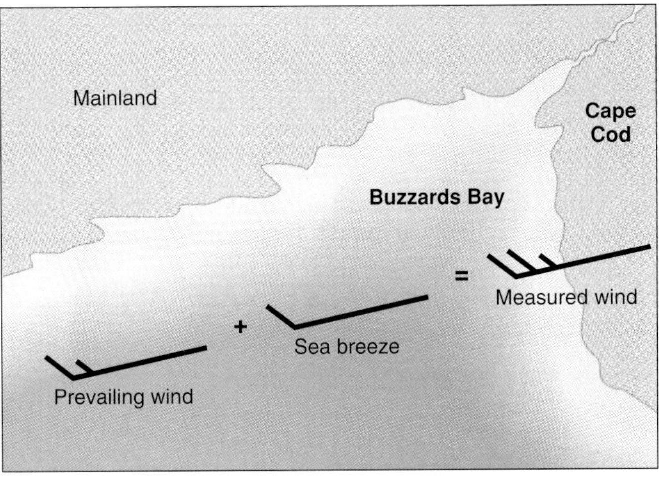


B

**Figure 11.13** (*A*) Coriolis force may deflect the sea breeze circulation, giving it an along-shore component. (*B*) Afternoon winds on the east shore of Buzzards Bay, Massachusetts, often are strong due to the combination of sea breezes and a prevailing westerly flow.

the day's temperature rise, a form of air conditioning people living further inland do not experience.

Like other atmospheric circulations, land and sea breezes rarely exist in isolation; generally, they show the effects of other influences. For example, because land and sea breezes persist for hours at a time, they often show the effects of Coriolis force. Figure 11.13A shows how Coriolis can cause an along-shore component to the sea breeze.

Often winds of larger-scale circulations also modify the land and sea breeze circulation. The resultant wind is a combination of the large-scale prevailing wind and the mesoscale land and sea breeze circulation (plus microscale turbulence, perhaps). When these circulations flow in the same direction, the resulting wind is particularly strong. Figure 11.13B gives an example.

To recapitulate, land and sea breeze circulations are driven by differential heating from land to water over the period of one day-night cycle of solar heating. The land and sea breeze is an example of a convective, or thermal, circulation; it is driven solely and directly by pressure changes due to surface heating and cooling. You will see several other important examples of thermal circulations later in this chapter.

## Mountain and Valley Winds

On a clear, calm September night in 1984, scientists in Colorado injected smoke into the surface air in a high mountain valley and then photographed by moonlight the subsequent movement of this smoky air. Some of their photos are reproduced in Figure 11.14. From the sequence of photos, what can you say about the air's movement?

These photos illustrate one portion of a circulation known as **mountain and valley winds.** Mountain and valley winds develop when diurnal heating and cooling combine with topography to create a circulation system that, like the land and sea breeze, reverses from day to night.

Figure 11.15A shows the night phase of the mountain and valley circulation. Under clear skies, the surface cools radiatively, chilling the air in contact. This cooler air, being denser, flows downhill and is called a mountain breeze. It is this branch of the circulation that the photos in Figure 11.14 illustrate so dramatically. Unlike the land and sea breeze circulation, mountain and valley winds do not comprise a closed cell. Instead, the downward flowing air is replaced by inflow from the surrounding regions. During the day, the circulation flows in the opposite direction. As you can see in Figure 11.15B, surface heating causes the surface air to move upslope (a valley breeze).

Remember that the flow patterns illustrated in Figure 11.15, like the land and sea breeze model, are idealizations; many factors, including the orientation and steepness of the valley, the presence of a prevailing wind, and cloud cover, can modify the mountain and valley circulation substantially. To cite just one example, the west side of a valley that runs north to south receives morning sun earlier than the east side; therefore, uphill valley winds may commence on the west slope, while mountain breezes continue to drain cold air downhill on the east.

## The Effects of Mountain Ranges

You have read already about ways in which surface obstacles may generate or modify microscale circulations. Large mountain ranges, particularly those aligned perpendicular to the prevailing winds, such as the Rocky Mountains, can cause circulation features on scales from the mesoscale all the way to global.

Examine the diagram of wind flow across a range of mountains such as the Rockies in Figure 11.16. Note that the surface air acquires an upward component as it ascends the windward side. What happens when this air crosses the ridge?

The answer depends largely on the air's static stability. If the ascending air is unstable, you know from Chapter 6 that it will continue rising; depending on the degree of instability, convective clouds, showers, or thunderstorms may result.

Figure 11.14 Note the progress of the cold air (marked by smoke) as it drains downhill at night.

A Mountain winds

B Valley winds

Figure 11.15 Low-level winds on mountain slopes reverse direction from day to night. What causes this change in flow?

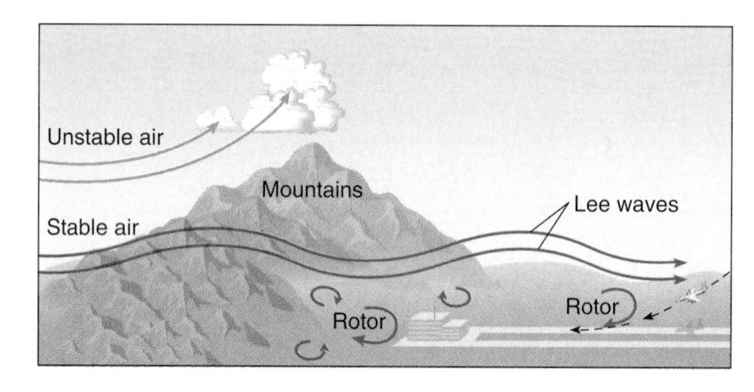


Figure 11.16 Waves and eddies induced by flow over a mountain ridge.

If the air is stable, however, then it will sink on the lee side. This rising and sinking may generate a series of waves downstream, known as **lee waves,** as the air gradually settles at its eventual equilibrium level. As you saw in Chapter 6, lenticular clouds may form on the crests of the lee waves, making the wave crests visible.

Closer to the ground, other events may occur. The waves may "scrape bottom," encountering the greater friction at the earth's surface, and degenerate into turbulent eddies that rotate on horizontal axes, called **rolls,** or **rotors,** plus smaller-scale turbulence. Low-altitude rotors can pose severe hazards to aircraft on takeoff and landing. A number of accidents have been attributed to this form of turbulence, and now at some airports, low-level wind shear is routinely monitored.

Mountain topography may funnel the airflow in unexpected ways, causing much stronger winds in some locations than in others. Boulder, Colorado, which lies just to the lee of the continental divide and happens to be the site of the National Center for Atmospheric Research, falls victim to such winds every decade or so. (Recall this unit's opening discussion on p. 297.)

Many other mesoscale wind systems develop as a result of synoptic scale circulations interacting with surface topography. You will study several important examples in the next section.

## Checkpoint

***Review***

Land and sea breezes and mountain and valley winds are sister circulations in several ways. Both are reversing, diurnal, thermal circulations formed and shaped by topography: reversing because the circulations move in opposite directions at different times and diurnal because the cycle of reversal is a daily one. (Why "thermal" and why "shaped by topography"?)

Mountain ranges may cause other mesoscale circulations, such as lee waves, large turbulent eddies, and dramatic local variations in wind flow.

***Questions***

1. What are the dimensions of mesoscale circulations? How long do they persist?
2. Where would you expect clouds or showers to form during daytime in a land and sea breeze circulation? Where would they form at night?
3. A large-scale weather system is causing a southerly 10-knot wind flow over the Florida panhandle. Forecasters expect a southerly, 5-knot sea breeze circulation to develop over the region during the afternoon. What should they predict for afternoon coastal wind speeds?

# Synoptic Scale Circulations and Local Winds

Synoptic scale circulations are best exemplified by the frontal cyclones and anticyclones of middle and high latitudes you studied in Chapters 8 and 9.

Unlike microscale and many mesoscale features, which often are at least as tall as they are wide, synoptic scale features are distinctly flat. Because of their great horizontal extent and their prevalence (outside of the tropics), synoptic scale circulations tend to dominate wind flow patterns in the lower atmosphere.

Like circulations on other scales, synoptic scale circulations constantly interact with features on larger and smaller scales. For example, recall from the last unit that frontal cyclones form in, and draw their energy from, zones of strong thermal contrast, that is, along the polar front. They "feed on" this thermal potential energy of the larger flow, like the whirls in Richardson's poem, converting this energy to the kinetic energy of the cyclone's circulation.

In this section you will study a number of examples of synoptic scale circulations' interaction with the surface environment. These interactions produce a variety of smaller-scale flow patterns and notable local winds. You will see that many of these local winds are not complete circulations like the land and sea system but are simply one portion of the larger system, within which distinctive weather conditions prevail.

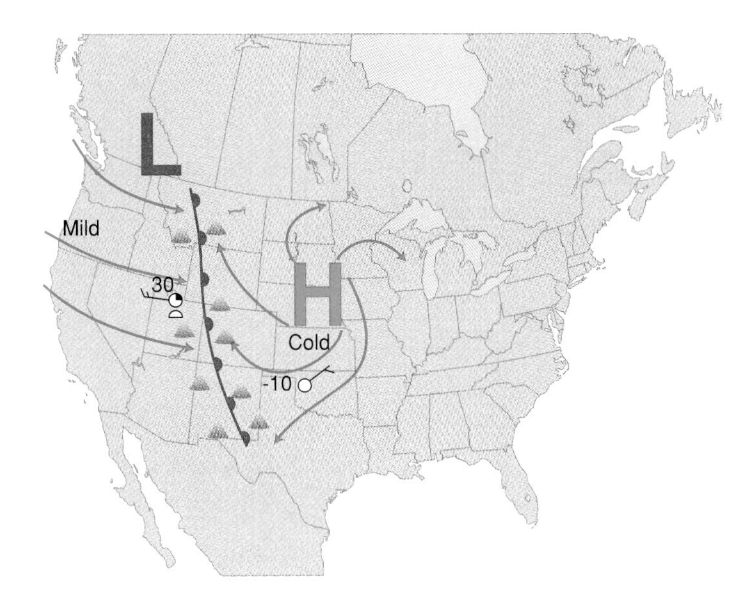


Figure 11.17 Synoptic scale situation that generates Chinook winds. What locations will experience the Chinook? Why?

## Chinook Winds

Note the synoptic situation in Figure 11.17. A cell of Arctic high pressure lies just east of the Rockies; meanwhile, warmer air of Pacific origin is moving eastward behind a warm front. How will the weather change in the lee of the mountains as the Pacific air advances?

Any location that experiences a change from an Arctic air mass to a modified maritime one is in for a substantial warming. However, in the western Plains states, the warming may be intensified for two reasons. First, exceptionally cold air often pools in the lee of the Rockies, so that the return to "normal," milder air is a more dramatic change than in many other places. Second, the mild air from the west often is warmed by ascending moist adiabatically to the continental divide and then descending dry adiabatically to the plains. As you read in Chapter 6, the difference results in a net warming due to latent heat release.

The resulting warm, dry downslope wind is known in the United States as a **Chinook wind,** named after an intense version that originates in the former Chinook Indian Territory in Oregon. Chinook winds are also known as "snow eaters"; it is said that a strong Chinook can melt more than a foot of snow overnight. In parts of Europe, the Chinook wind is called a foehn (pronounced like fern without the "r") wind.

One of the most impressive aspects of the Chinook wind is the abruptness of the warming it brings. The residents of Spearfish, South Dakota, were certainly surprised on January 22, 1943, when a Chinook lifted the temperature from −20°C to +7°C (−4°F to +45°F) in a period of only two minutes. No wonder that to many Westerners waiting out a cold snap, the

Figure 11.18 Waiting for a Chinook, by Charles M. Russell.

Chinook is the most welcome of winds. It is immortalized in the title of a painting by Charles Russell (Figure 11.18).

## Santa Ana Winds

Try to imagine 100 mph winds bringing 100-degree temperatures and humidities almost too low to measure. This blast furnace is known as the **Santa Ana,** a searing wind that carries desert air southwestward to portions of the southern California coast in the vicinity of the city of Santa Ana. A severe Santa Ana wind can uproot trees, blow vehicles off roads, and, worst of all, fan fires that spring up in the tinder-dry canyons, reducing vegetation and residences to ashes.

What is the synoptic scale setting in which Santa Ana winds occur? For winds to persist from the northeast, strong high pressure must lie to the north, with lower pressure to the south. As this desert air makes its way to the coast, it experiences a Chinook-like warming in the process of descending to sea level. Topography of the local canyons acts to funnel the winds to great speeds in places. The Santa Ana winds illustrate well how a synoptic scale circulation can interact with local and regional topography to create spectacular microscale and mesoscale flow patterns.

## Nor'easters

The most intense storms to affect the eastern seaboard of the United States during the cold months are frontal cyclones that move northeastward alongshore or just offshore. Although the storms themselves move from the southwest, they are known to coastal residents as "**nor'easters**" (northeasters) because they produce northeasterly surface winds along the coast (see Figure 11.19). The northeast inflow feeds the storm with cold, moist air; heavy snowfalls often result. The blizzard of March 1993, discussed in Chapter 7, almost qualified as a nor'easter; however, its track was slightly too far west (over land) to be considered a model example.

Early in the chapter, we commented that predicting future conditions in turbulent flow is more difficult than in laminar flow. Nor'easters illustrate the point well. Not only is it difficult to predict where and when a potential nor'easter will form and where it will move, but the prediction must be highly accurate if forecasts of local weather conditions are to be accurate at all. For example, suppose you predict that a nor'easter will form and move northward just offshore, like the example in Figure 11.19. Consistent with this prediction, you forecast below-freezing temperatures and traffic-paralyzing snow along the coast. If, instead, the storm tracks just inland, coastal locations may experience rain, a warm frontal passage, and a period of south winds with temperatures of +10°C (50°F)! Thus, a small error in predicting the storm's path has led to a giant error in the local forecast.

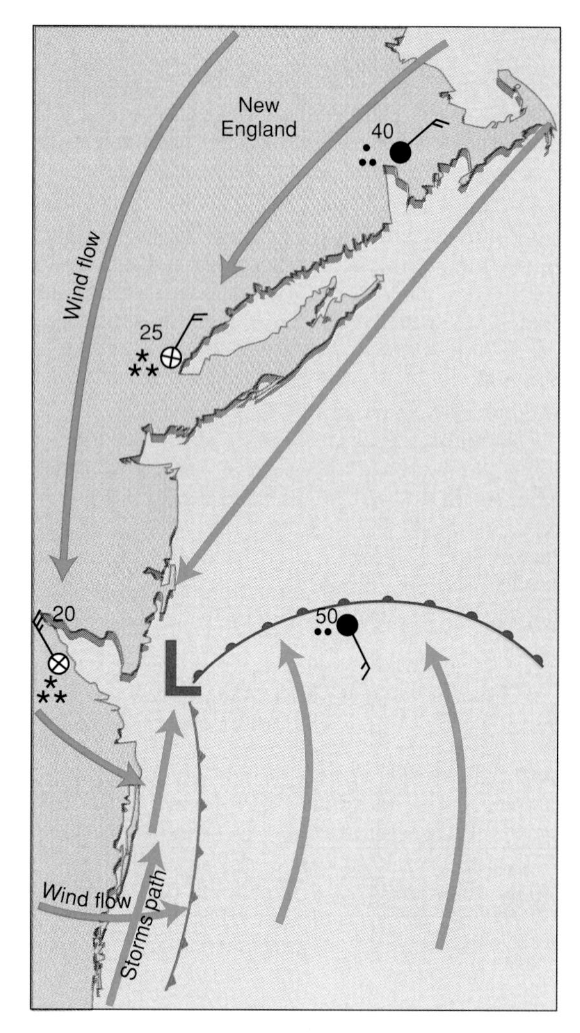


Figure 11.19 A typical "nor'easter." How would the coastal weather differ if the storm's track were 100 miles to the west?

## Lake Effect Snowstorms

One of the snowiest regions in the United States (outside of mountain regions) includes parts of New York, Pennsylvania, and Michigan. Figure 11.20 should satisfy you that snowfall in these regions can be substantial, even overwhelming. Perhaps just as remarkable as the quantity of snow is the fact that these deluges of white come not with the approach of a frontal cyclone but with the northwest flow of cold air in its wake.

Figure 11.21 shows synoptic scale conditions favorable for such **lake effect snowstorms**: a brisk cyclonic flow of cold, neutrally stable or unstable air across the relatively warm open water of the Great Lakes. The surface air gains heat and moisture over the water, leading to convection, heavy cumulus development, and then to snow flurries and squalls. Orographic lifting as the air reaches land intensifies the development and subsequent precipitation.

Lake effect snows occur in the lee of many large water bodies including Great Salt Lake and, until they freeze over, Lake Winnipeg and Great Bear Lake in Canada. Hudson Bay, although not a lake, also causes lake effect snows until it freezes and thus ceases to act as a source of heat and moisture to the low-level air. Look at the Hudson Bay snowfall records in Figure 11.22. Can you determine the approximate date of freeze-up from the snowfall data?

Figure 11.20 Clearing a roof of lake effect snow

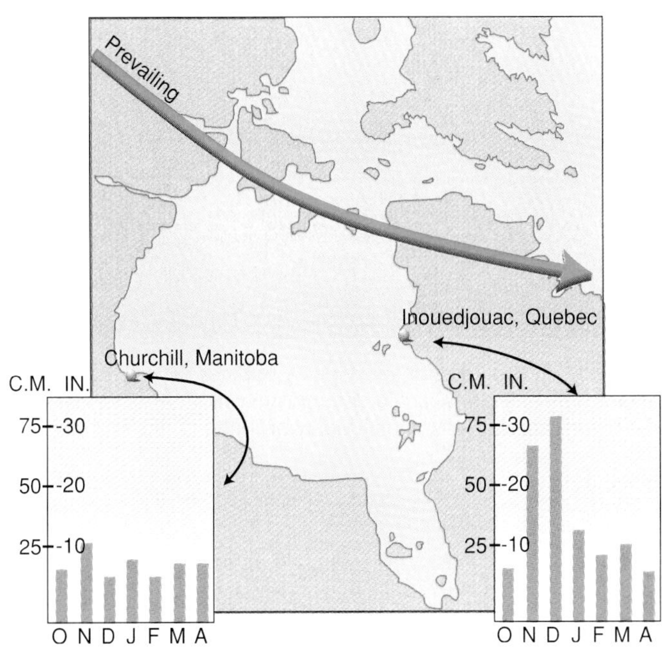


Figure 11.22 Inouedjouac is subject to lake effect snow until Hudson Bay freezes. In what month is the freeze-up?

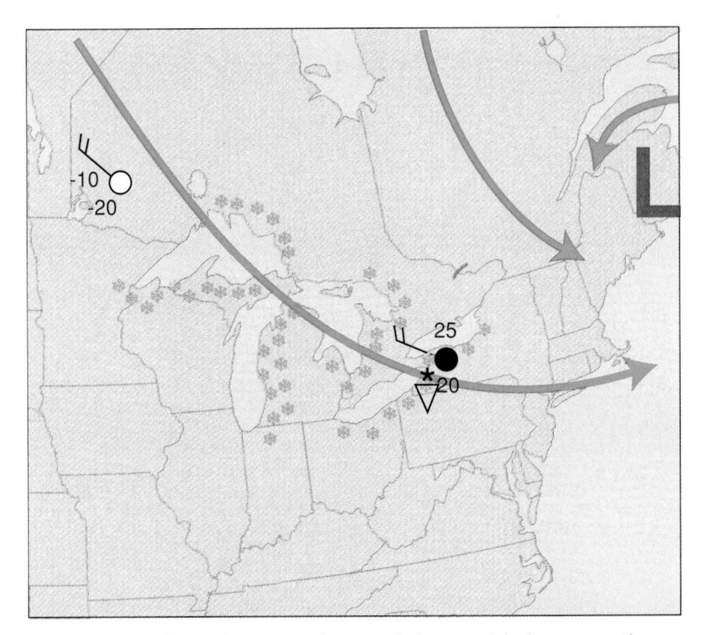


Figure 11.21 A northwest trajectory brings cold air across the waters of the Great Lakes and produces lake effect snows. Note the changes in air temperature and dew point.

## Checkpoint

***Review***

The main point of this section is that synoptic scale circulations, like those of other scales, are linked to larger and smaller features. Rather ordinary synoptic scale circulations may be influenced by local topography to produce extraordinary weather conditions on a small scale. The Chinook and Santa Ana Winds, nor'easters, and lake effect snowstorms are just four of many possible examples of these notable local systems.

***Questions***

1. In what ways is the Santa Ana similar to the Chinook wind?

2. Only 300 miles separate Buffalo, New York, and New York City, yet their heaviest snowfalls occur under different synoptic scale weather conditions. Why?

Figure 11.23 The equatorial band of cloudiness in this satellite image is part of a global scale circulation system. At times it can be followed more or less continuously around the entire earth.

# Global Scale Structures and the General Circulation

Look at the large-scale patterns of clear skies and cloudiness (such as the band of clouds along the equator) in Figure 11.23. These features are global scale features; they comprise the atmosphere's largest and longest-lived flow structures. Collectively, they form a planetary flow pattern known as the **general circulation.** A typical global-scale feature has a lifetime of weeks or more (some are essentially permanent) and may cover up to a tenth or more of the planet. Let's see how they can be understood in relation to the discussion of fluid flow in earlier sections. We begin with tropical circulation patterns.

## The Tropics

Perhaps the world's most persistent winds are the **trade winds,** which flow toward the equator from the northeast (Northern Hemisphere) or southeast (Southern Hemisphere). The trades' very name is a monument to their reliability: early mariners sailing in this zone knew they could depend on the trade winds to power their ships from port to port.

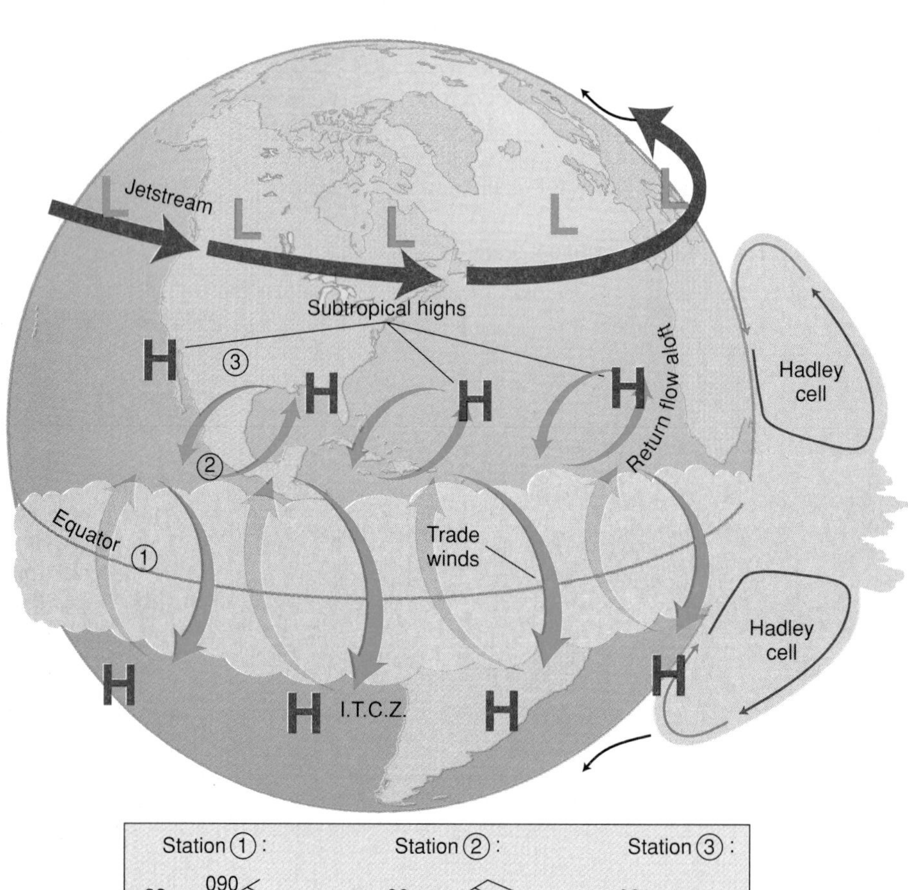


Figure 11.24 Global scale equatorial weather systems. Notice the thermal cells are tilted (differently in each hemisphere) due to Coriolis force. Weather stations illustrate typical daytime conditions.

What is the explanation for such a consistent wind pattern over a large part of the globe? Since the trades converge onto the equator, where incoming solar radiation is greatest, you might hypothesize that the trades are part of a thermal circulation centered on or near the equator.

How might you test this hypothesis? If the trades are part of a thermal circulation, you would expect to find other thermal circulation components as well. For example, the zone of rising air should be one of low surface pressure but relatively high pressure aloft. You would look for a return flow aloft and sinking air—with high pressure and generally clear skies—on the poleward sides of the trades. Further, you might expect to find that being "solar powered," the entire circulation pattern migrates seasonally, shifting northward with the sun during the Northern Hemisphere's summer and southward during the Southern Hemisphere's summer. Finally, your hypothesis would have to accommodate the fact that the trades do not flow directly toward the equator but instead have an east-to-west component.

To test the thermal circulation hypothesis, examine the observed large-scale equatorial circulation patterns, shown in schematic form in Figure 11.24. Compare these features and the satellite image in Figure 11.23 with the list of expectations in the last paragraph. Are you inclined to accept the thermal hypothesis or reject it? What might be an explanation for the westward component of the trade winds, and the eastward component of the return flow aloft? (Hint: Refer back to Figure 11.13A, which shows a similar effect operating on a circulation of smaller scale.)

In the 18th century, the English meteorologist George Hadley proposed a thermal explanation for the trade wind circulation. In his honor, the circulation is called the **Hadley cell.** The Northern and Southern Hemispheres' Hadley cells dominate global weather patterns from nearly 25 degrees north latitude to 25 degrees south latitude, almost half of the surface area of the earth.

The region of low surface air pressure where the trades converge is known as the **intertropical convergence zone,** or simply the **ITCZ.** The ITCZ is the equatorial zone of cloudiness visible in Figure 11.23. Air in the ITCZ derives its warmth and buoyancy in part from sensible heat flux due to solar heating of the earth's surface but also from latent heat fluxes. Vast amounts of energy are released as water vapor condenses in the rising air.

Colloquially, the ITCZ is known as the **doldrums.** A region of variable (sometimes calm) winds and frequent showers, the doldrums posed difficult conditions for early mariners. Trade wind sailors who ventured too close to the ITCZ found themselves "down in the doldrums," which is now an expression for being in a melancholy mood.

What is weather like in the Hadley cell? It depends on your location within the cell and the season. Figure 11.24 shows representative daytime weather conditions at several sites.

The return flow in the Hadley cell is from southwest to northeast (Northern Hemisphere) and reaches great speeds at the northern edge of the cell. This zone of strong winds aloft is known as the **subtropical jet stream.** What causes this wind speed maximum?

The answer lies in the conservation of angular momentum. Figure 11.25 is a reminder that objects on the earth's surface, including the atmosphere, are rotating around the earth's axis once per day at hundreds of kilometers per hour. If their distance from the axis changes, then their speed must change as well so that their angular momentum is conserved.

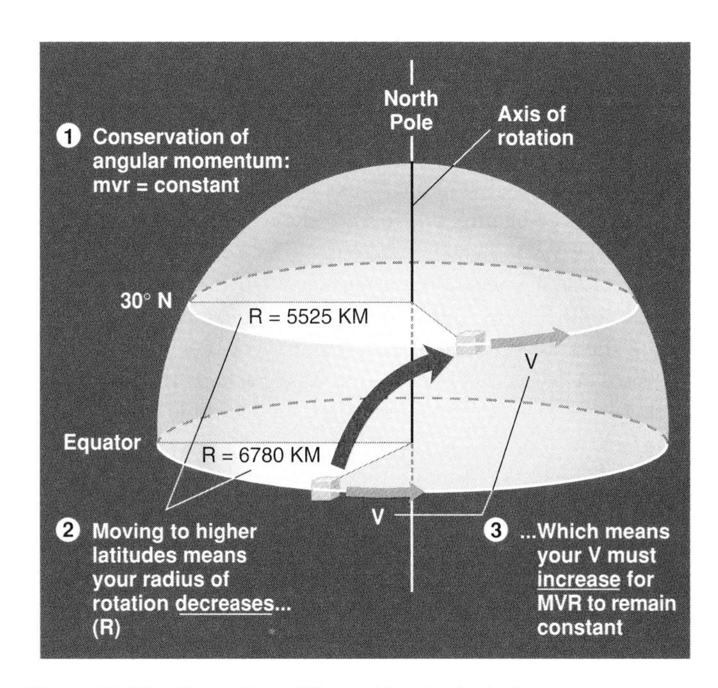


Figure 11.25 Formation of the subtropical jet stream.

How does this rotational motion explain the subtropical jet stream? Figure 11.25 shows that as tropical air moves poleward from the ITCZ, it moves closer to the earth's axis. Its radius decreases, so its speed of rotation must increase correspondingly. As a result, the poleward-moving air stream acquires a strong eastward component as it moves to higher latitudes. In this way, the Hadley cells feed the subtropical jet stream with fast-moving equatorial air.

## The Subtropics

The subtropics are those regions between about 20 and 35 degrees of latitude, depending on season; meteorologically, they lie just poleward of the trade winds. Look back at the satellite photo of earth in Figure 11.23 for a moment. What are the prevailing weather conditions in the subtropics?

The answer should be clear: the subtropics enjoy some of the most cloud-free weather on earth. Many of the world's largest deserts, including the Sahara, Arabian, Mojave, and Australian, are found in this zone. Knowing that clear skies occur in regions of subsiding air and high pressure, you should expect this region to be dominated by high pressure. The cross-sectional view of the Hadley cell in Figure 11.24 confirms this expectation; this zone is known as the **subtropical high pressure belt.**

What causes the poleward-moving flow in the Hadley cell to subside in the subtropics? Why doesn't it continue aloft all the way to the pole? As it moves away from the equator, the upper-level air cools by radiating long wave energy to space. This cooling makes the air denser and initiates its descent in the subtropics.

Like all atmospheric circulations, the subtropical high pressure belt does not exist in isolation but is shaped by other factors, particularly geographical ones. In summer, subtropical land surfaces become warmer than ocean surfaces at the same latitudes. (Recall the discussion of heat capacity and mixing in Chapter 3.) Strong surface heating encourages the formation of thermal low pressure currents, which weaken or disrupt the flow of subsiding air. Over the ocean, however, where surface heating is much less pronounced, the subsiding air meets no such obstacle. As a result, the subtropical belt in summer resembles a series of high pressure cells located over the oceans. Perhaps the best known of these cells is the "Bermuda High," which you met in earlier chapters.

In winter, the temperature contrast between subtropical land and water surfaces is much less than in summer. Consequently, individual cells of high pressure within the subtropical high pressure zone are not so strongly developed then.

What are winds like near a subtropical high pressure center? Recall from Chapter 8 that wind speeds are proportional to the horizontal pressure gradient and that at the centers of any pressure system, high or low, the gradient is zero. Thus, winds near the center of the subtropical high pressure cells are light—often too light for sailing. Sailors traveling between Europe and America who ventured into the Bermuda High were becalmed and spent weeks adrift, sometimes resulting in the death of their cargoes of livestock by starvation or dehydration. These regions became known as the "**horse latitudes**" for the dead horses thrown overboard from these drifting merchant ships.

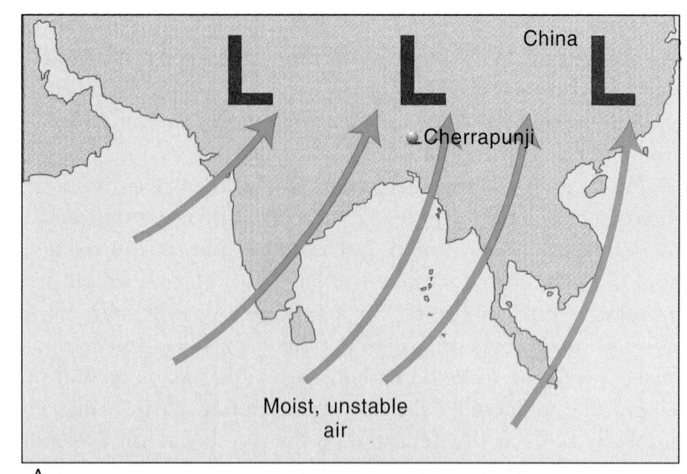


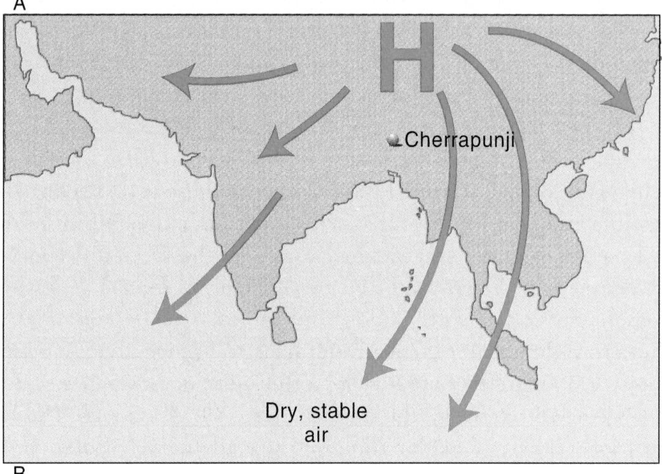


**Figure 11.26** The Asian monsoon. (*A*) The southwest, or wet, monsoon (May–September). (*B*) The northeast, or dry, monsoon (November–March).

## Monsoon Circulations

In southern Asia and a few other locations, the configuration of land and oceans causes a global-scale, seasonal thermal circulation known as a **monsoon**. Think of a monsoon system as a land and sea breeze circulation of enormous proportions that alternates from summer to winter instead of from day to night.

Figure 11.26 shows the flow patterns associated with the Asian monsoon. Summertime heating of the continent generates onshore winds, laden with moisture from the tropical oceans, which falls as rain as the air moves onshore and upslope toward the Himalayan mountains. The volume of water carried inland is difficult to imagine: in Cherrapunji, India, where the most extreme monsoonal precipitation occurs, the average rainfall for the month of July alone is nearly 250 centimeters (100 inches).

During winter, the monsoon circulation reverses direction (Figure 11.26B). Then, the land is cooler than the ocean to the south, and the flow is downslope and offshore. Under clear skies and high pressure of the dry monsoon, Cherrapunji's January precipitation is just 0.2 centimeters (0.1 inches). In spring, the wet monsoon returns on southwest winds. Its onset has great religious and economic significance in many south Asian countries.

The Asian monsoon is by far the largest monsoon circulation. A smaller, less persistent version exists along the gulf coast of the United States, where winds tend to be moist and onshore during summer, dry and offshore during winter.

## Middle and High Latitudes

Many features of the general circulation of middle and high latitudes are familiar to you already. For example, in Figure 11.27, you can follow some of the low-level outflow from the subtropical high pressure belt as it moves into middle latitudes as the **prevailing westerlies** (marked by "A" in the figure).

Notice that this poleward-moving, tropical air collides with air masses of polar origin moving equatorward (at B) at the polar front. Therefore, you might expect the general circulation to exhibit higher surface pressure in high latitudes, and lower pressure at the polar frontal convergence zone. Is this what is observed?

The answer is a qualified yes. The polar front is indeed observed to be a region of low pressure and rising air that extends

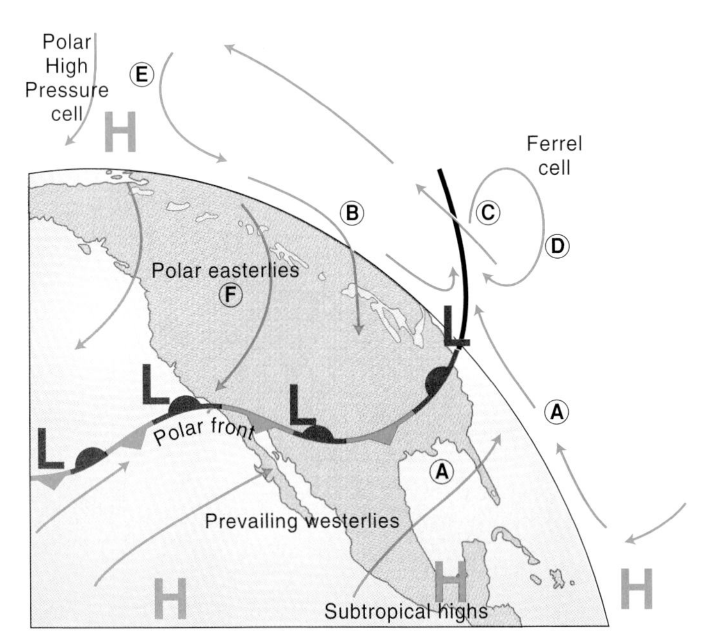


**Figure 11.27** Model of general circulation features of middle and high latitudes.

around large portions of the globe, somewhere between 35 and 60 degrees of latitude. Some of the rising air at the front (at C) is observed to turn equatorward but soon sinks (at D), completing a circulation known as the Ferrel cell. The Ferrel cell, named for American meteorologist William Ferrel, is much less extensive in area and less persistent than the Hadley cell. Persistent **polar high pressure cells** (at E), strongest in winter, are observed to dominate the general circulation in the coldest locations. Surface outflow from these cells is deflected by Coriolis force to form the **polar easterlies** (F). The polar easterlies meet the **prevailing westerlies** along the polar front. In all these respects, then, the general circulation scheme of Figure 11.27 is validated.

However, many discrepancies exist between the model sketched above and observed weather conditions for two reasons, at least. First, the daily circulation patterns at middle and high latitudes are so strongly dominated by migrating frontal cyclones and anticyclones that larger-scale features often are not apparent except as long-term averages. This situation differs greatly from that at low latitudes, where the general circulation features are so strong and persistent that the flow on any given day generally resembles that of the general circulation. Second, land and sea temperature contrasts impose enormous variability on the circulation at middle and high latitudes, considerably altering the model of Figure 11.27. The maps in the next section illustrate these deviations.

## Observed General Circulation at the Earth's Surface

Figure 11.28 shows the observed global sea-level pressure and wind patterns for January and July. You can use these charts to compare circulation features actually observed with those suggested by the model of the last section.

Consider January conditions first, beginning with polar conditions. You can see several departures from the model. For example, the polar high pressure cell is located not near the North Pole but appears as two distinct cells, much farther south, over the colder landmasses of North America and eastern Asia. On the other hand, the relative warmth of the open oceans in high latitudes causes the polar frontal zone to be far more intense in these regions than over the cold continents. As a result, two intense low pressure cells exist: the Aleutian Low and the Icelandic Low.

Note the subtropical high pressure belt, at around 30 degrees north latitude, and the ITCZ, which has followed the sun south of the equator in January. Also note the strength of the subtropical high pressure belt in the Southern Hemisphere. (Why is it so much stronger over the oceans?)

In July, Northern Hemisphere temperatures are colder over the icy Arctic Ocean than over the continents, which lose their snow cover; therefore, the polar high retreats toward the pole. The polar frontal zone appears, weakly, along the northern coasts of North America and Eurasia. Clearly, the Northern Hemisphere's circulation is dominated at this time by the Bermuda and Pacific subtropical highs and the Asian monsoon.

Thus, the observed general circulation conforms to the model in general, if we allow for variations due to topographic effects. It is interesting to notice that the correspondence is better in the Southern Hemisphere, where the surface is more uniformly oceanic and thus land and ocean contrasts are less significant influences.

## Observed 500-Millibar General Circulation

The general circulation at the 500-millibar level is depicted in Figure 11.29. Study these flow patterns for a moment. How do they differ from the surface circulation of Figure 11.28?

After examining the charts, you might feel that a better question would be, "Can you find anything in common?" The obvious differences seem to far outweigh any similarities. Except for the wintertime Aleutian and Icelandic Lows and the summertime subtropical highs, no features match from one level to the other. In fact, the 500-millibar general circulation appears to be a one-cell system in each hemisphere, with high pressure at the equator, low pressure at the pole, and a westerly flow in between. This is quite a contrast to the surface circulation, which consists of three zones in each hemisphere, each of which is broken further into cells.

Several factors contribute to these differences. First is the obvious fact that the earth's surface has less influence on the flow aloft than at the surface. Therefore, the 500-millibar flow does not reflect the complex distribution of oceans, continents, and mountain ranges so obviously as the surface circulation.

Another factor is that wind flow patterns high in the troposphere are determined almost entirely by the temperature of

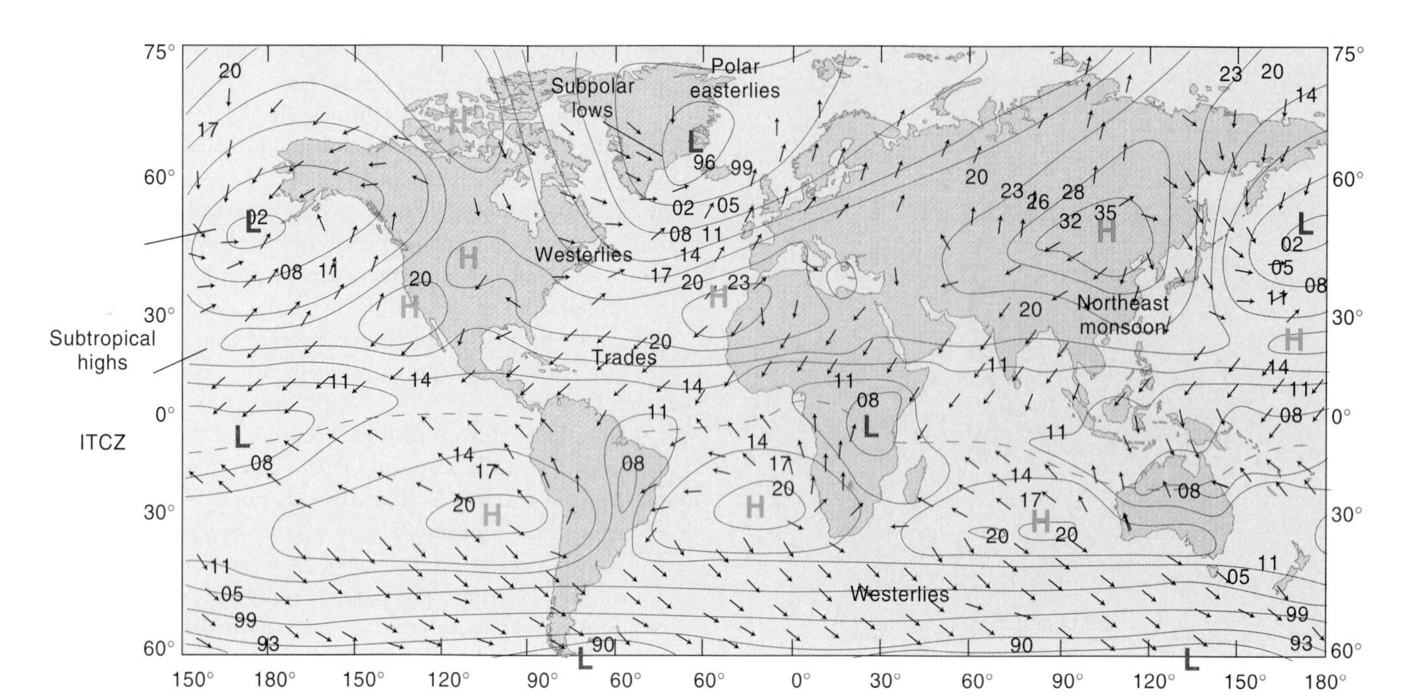


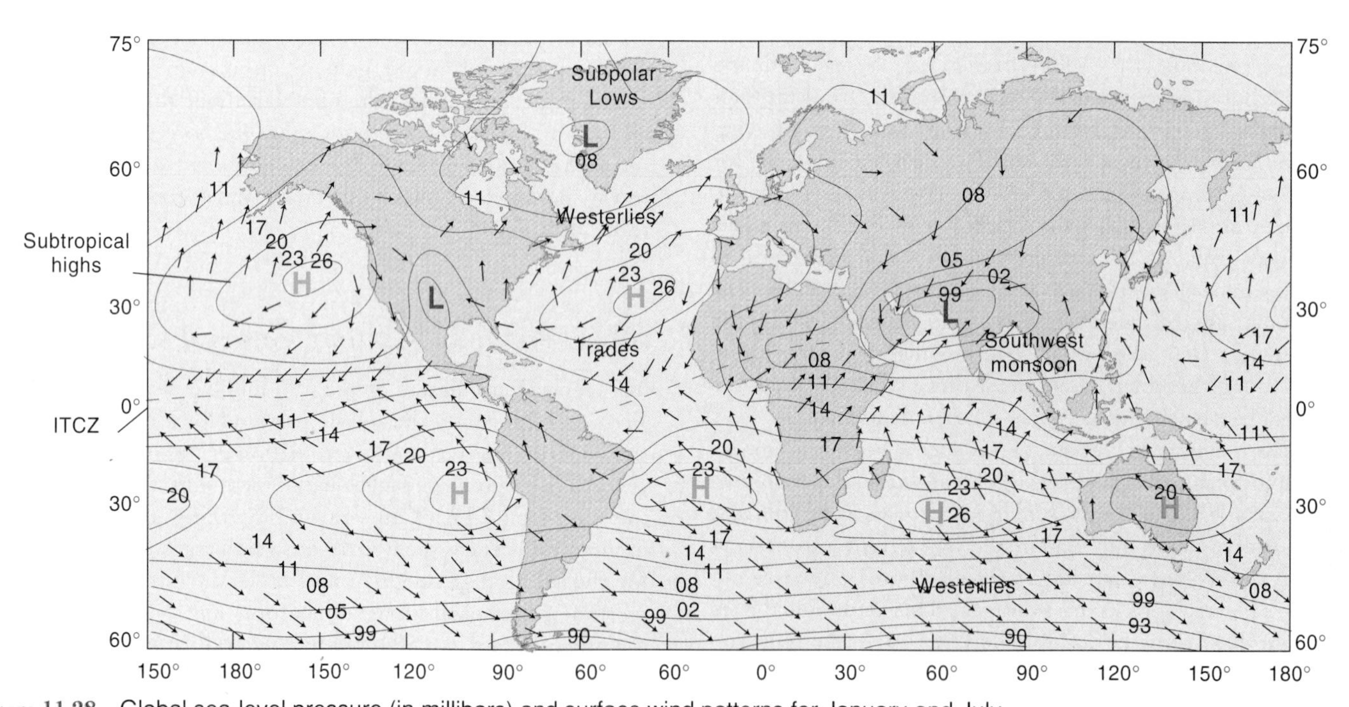


**Figure 11.28** Global sea-level pressure (in millibars) and surface wind patterns for January and July.

the underlying atmosphere. (Recall the discussion of the thermal wind in Chapter 8.) For this reason, low pressure is found aloft over the cold, polar regions, while high pressure prevails over the warm, equatorial lower atmosphere. And since surface differences are less important at high altitudes, the temperature, and therefore wind flow pattern, is determined mainly by the pole-to-equator temperature pattern, a truly global scale pattern. Therefore, wind flow features at this level are primarily global as well.

A final reason for the smoother pattern aloft is a more mundane one: the network of observations at the earth's surface is much finer than at 500 millibars. There well may be upper-level features, especially on the mesoscale, of which we have little idea at present.

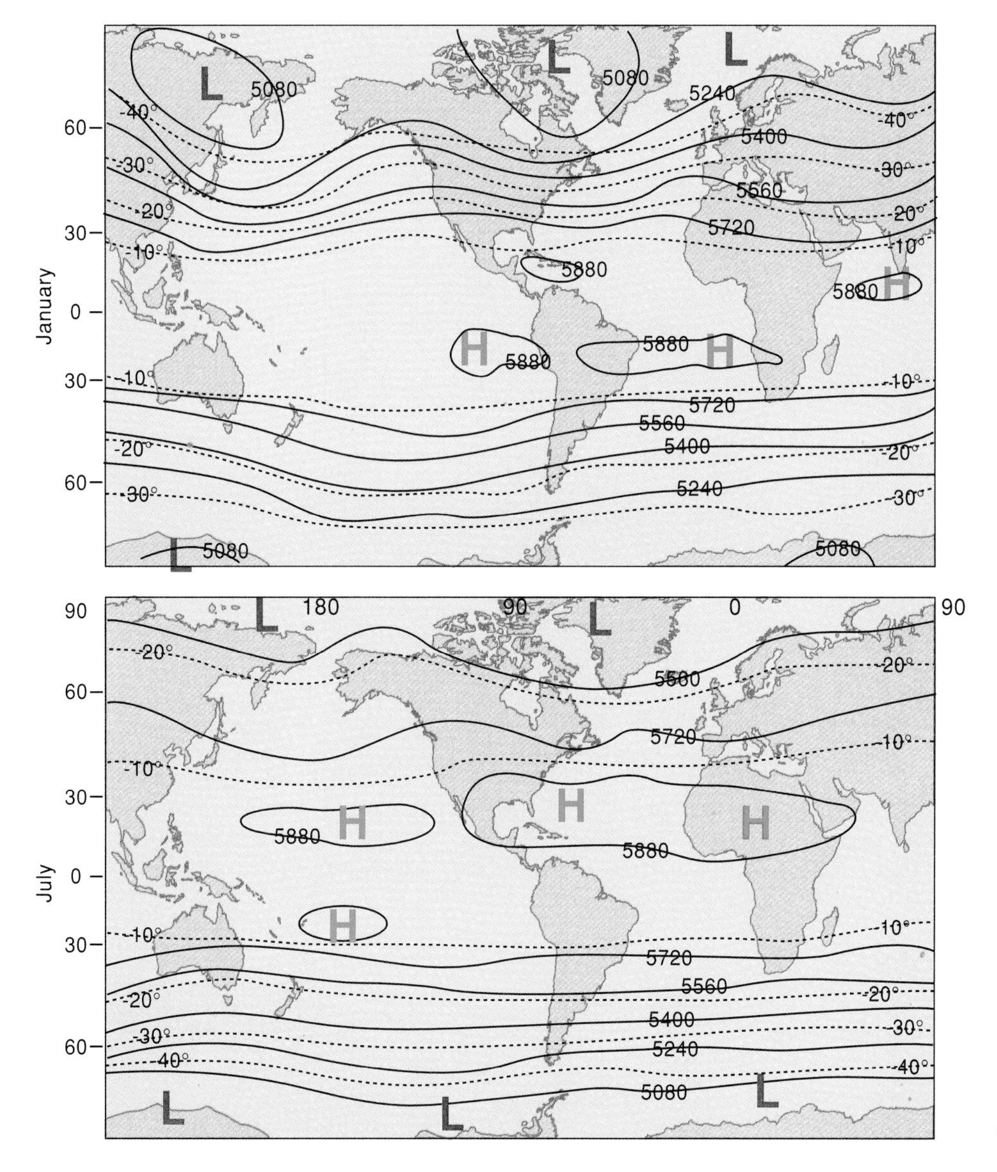


Figure 11.29 Global 500-millibar, contours (in meters), temperatures (°C), and wind flow for January and July.

## Checkpoint

***Review***

Remember that the general circulation is a response to an excess of solar energy in low latitudes and a corresponding energy deficit in polar regions. The visual depictions in Figures 11.28 and 11.29 summarize the general circulation's key features.

***Questions***

1. What is the main attribute shared by Hadley cells, monsoon winds, and land and sea breezes?

2. Use the general circulation model to estimate the location of each of these weather stations:

92 227
73
A.

8 073
**
6
B.

110 122
45
C.

3. Why does the general circulation appear more laminar at 300 millibars than at the surface?

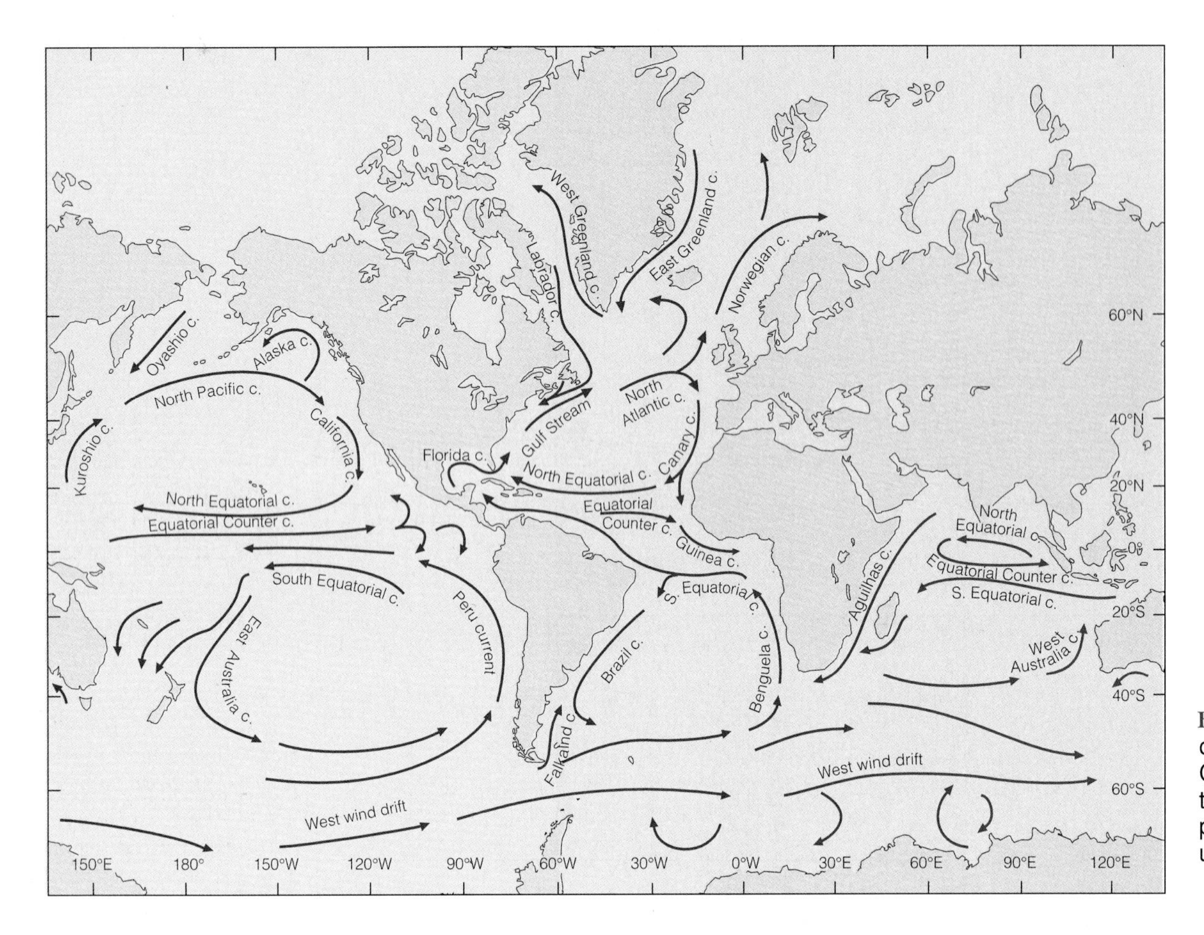


Figure 11.30 Major ocean surface currents. Compare flow patterns in this figure with global air pressure patterns in Figure 11.28.

# Ocean-Atmosphere Interactions

Roughly two-thirds of the earth's surface is covered by water. Since water, like air, is a fluid, ocean waters exhibit many of the same fluid flow features you have read about in earlier sections. In this section, you will examine three important ways in which ocean and atmospheric circulations interact to produce distinctive weather conditions.

## Global Scale Ocean Currents

Examine the chart of major ocean currents in Figure 11.30. Compare it to the atmospheric flow patterns in Figure 11.28. What similarities do you notice?

Perhaps the most striking correspondence occurs over the subtropical oceans; beneath each subtropical high pressure cell is a similar **gyre,** or oceanic circulation, of surface ocean waters. This correspondence is no coincidence. Winds of the general circulation exert frictional forces on the surface waters, causing ocean currents to flow in much the same patterns as global scale winds.

Do the ocean currents thus formed have any reciprocal influence on the atmosphere? Indeed, they do; the effects are large and varied. For example, currents that flow from low to high lat-

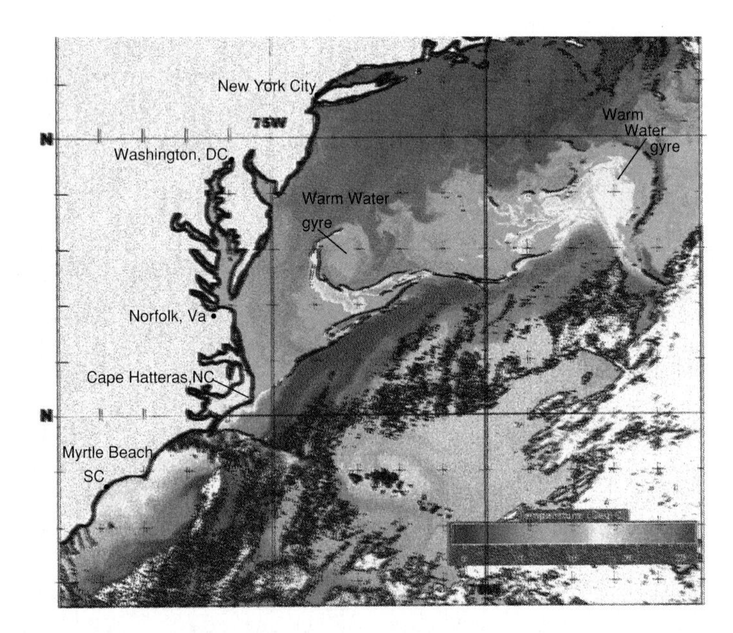


Figure 11.31 This chart of ocean surface temperatures was produced from satellite measurements. Warmest waters are shown in red, coldest in blue. Notice the waves along the northern border of the Gulf Stream and gyres of warm water that have spun off from the Gulf Stream.

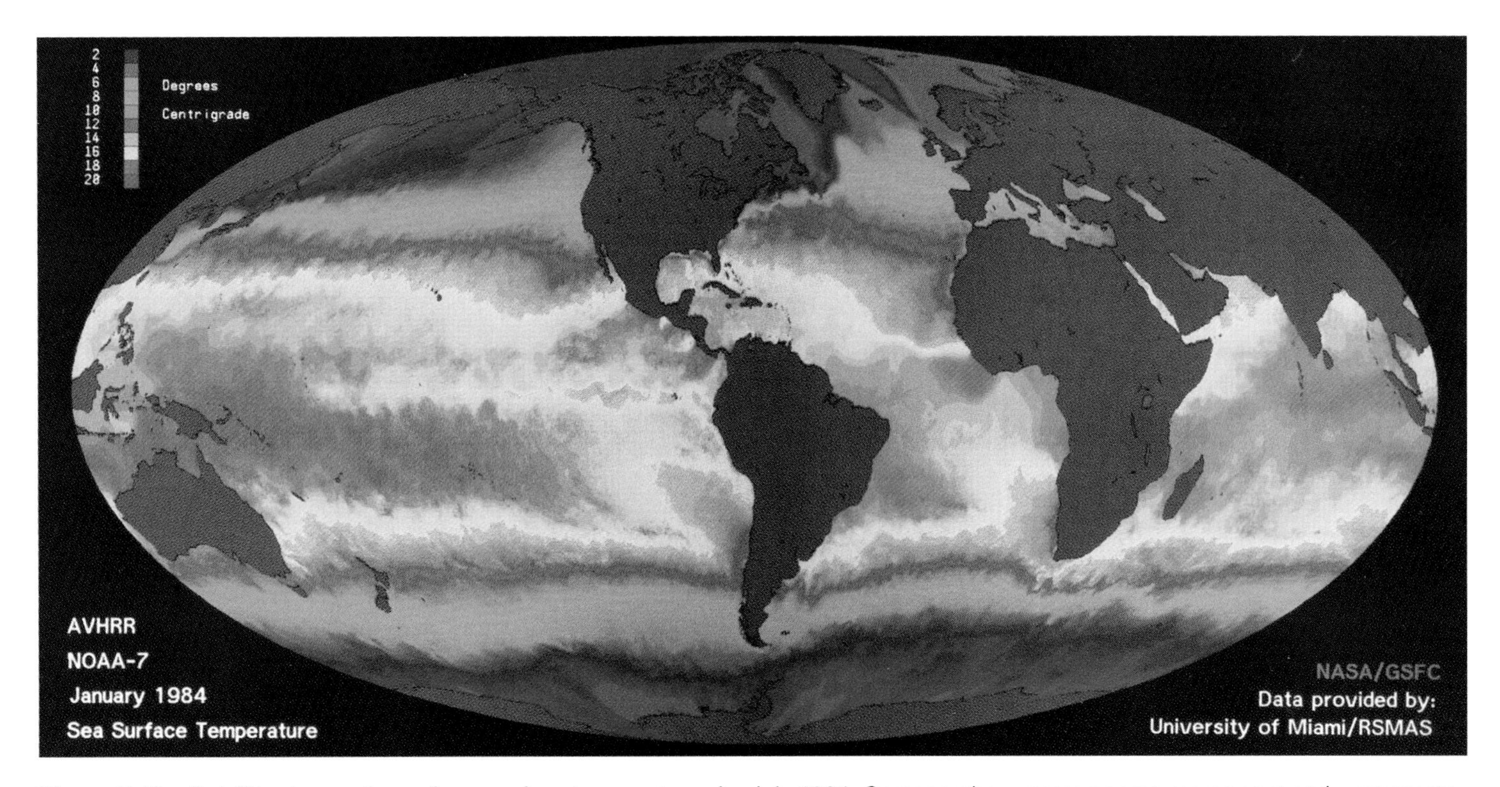


Figure 11.32 Satellite observations of sea surface temperatures for July 1984. Compare the water temperatures on east and west coasts at similar latitudes.

itudes, like the Gulf Stream along the east coast of North America, transport vast amounts of heat from the tropics to cooler regions. In Figure 11.31, you can see gyres of warm water transported by the Gulf Stream. Without heat transport by oceans currents, temperature contrasts between equatorial and polar regions would be more extreme, which could lead to more intense atmospheric circulations such as frontal cyclones and hurricanes. As it is, the Gulf Stream and North Atlantic Current and the Japan and North Pacific Currents keep the western coasts of Europe and North America considerably warmer than they would be otherwise.

## Upwelling

Study the sea surface temperature charts in Figure 11.32. In particular, note the variations in temperature along lines of latitude. The chart indicates that waters along western coasts (such as California) are colder than that on eastern seaboards (Virginia, for example) at the same latitude.

This contrast is due to a phenomenon called **upwelling.** Upwelling is the rising of cold, deep water to the surface. Upwelling causes more than cold water for bathers and surfers. It also cools the summertime climate of coastal locales and can cause

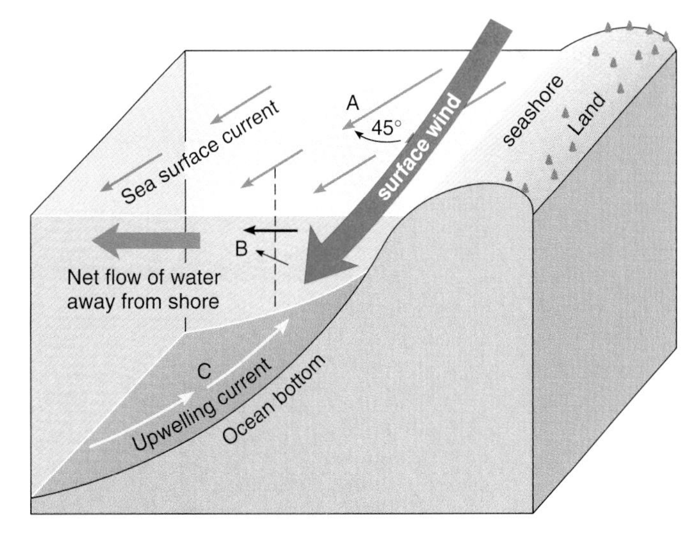


Figure 11.33 Upwelling. Because of Coriolis and frictional effects, surface ocean currents (*A*) move 45 degrees to the right of the surface wind flow. The same two forces cause subsurface water to be deflected further to the right (*B*), resulting in a net flow of water 90 degrees to the right of the wind flow. Deep water upwells (*C*) to replace the water that moves away at upper levels.

low-level temperature inversions, which in turn may cause fog or air pollution episodes.

Upwelling occurs when surface and near-surface water moves away from the coastline, as illustrated in Figure 11.33. In response, deep water rises to "fill the gap."

Notice that contrary to what you might expect, upwelling occurs in regions where low-level winds tend to blow parallel to the coastline, rather than across it. This is due to the fact that Coriolis and frictional forces deflect the water's motion, causing each layer of water to move slightly to the right (Northern Hemisphere) of the layer above. The sum effect is that the upper-level water moves 90 degrees to the right of the surface wind. Therefore, you should expect upwelling to be most significant where prevailing winds blow parallel to the coast with land to the left and ocean to the right of motion. In the Southern Hemisphere, where the Coriolis effect works in the opposite direction, land should lie to the right and ocean to the left of the wind. You can check this conjecture by comparing sea surface temperatures (Figure 11.32) with sea-level wind flow (Figure 11.28).

## ENSO

Perhaps the most complex and fascinating example of the coupling between atmospheric and oceanic circulations is found in the tropical southern Pacific Ocean and atmosphere, and is known as **El Niño Southern Oscillation,** or **ENSO.** The main circumstances leading to and including ENSO are summarized in Figure 11.34.

Notice that under normal conditions in this region (Figure 11.34A), the trade winds cause westward-moving surface currents on each side of the equator. Continued action of the trades and currents causes this warm surface water to "pile up" in the western Pacific. This warm water helps to create anomalously warm, moist, and unstable air masses over the western Pacific, which leads to convective cloud formation and to rainfall totals greater than anywhere else in the world for an area of similar size.

On the circulation's eastern edge, however, conditions are very different. As you can see from the figure, winds curling around the subtropical high pressure cell lead to upwelling of cold water to the surface along the west coast of Peru. As a result, conditions there are cool and dry. A further consequence of the upwelling is that great quantities of organic nutrients, the remains of plant and animal life that have settled to the ocean's depths, are carried upward, where they nourish surface marine life. In this way, the upwelling water supports a fishing industry in Peru that is among the most productive in the world.

Notice also in Figure 11.34A that surface air pressure is higher in the eastern Pacific than in the west. This pressure

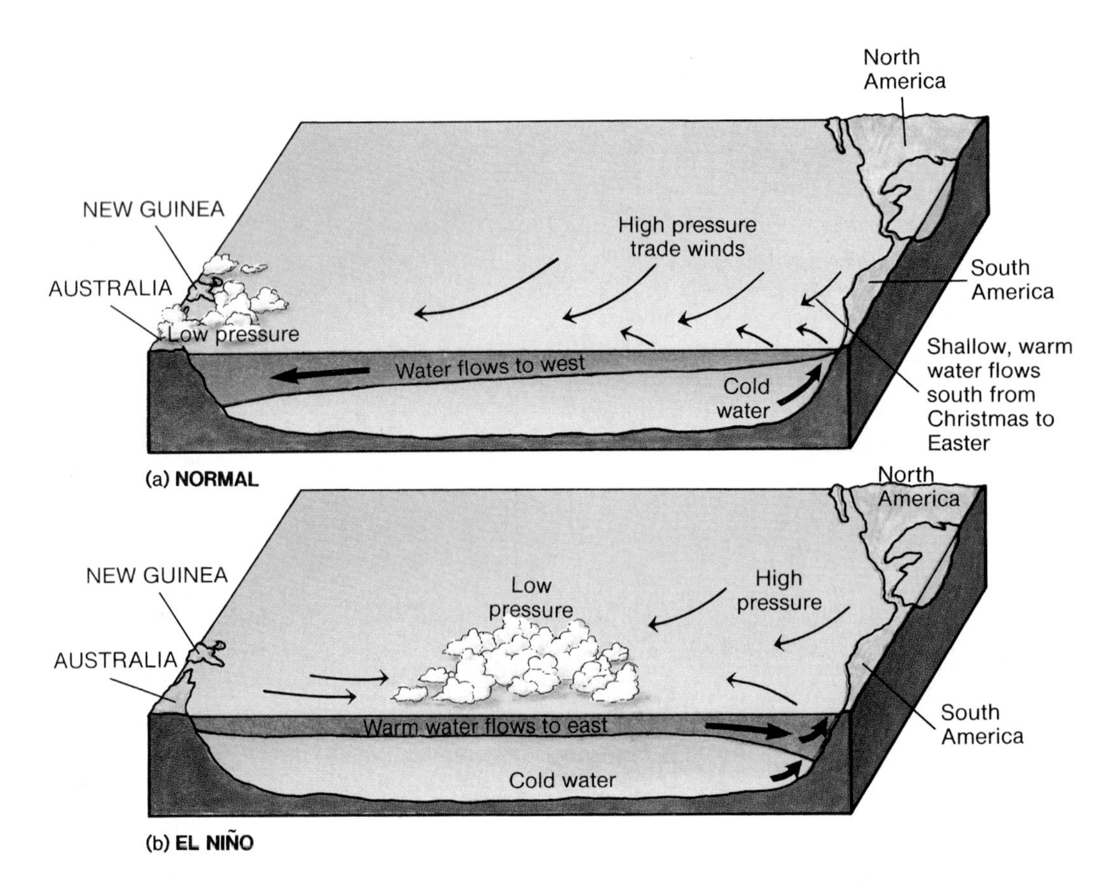


Figure 11.34 ENSO circulations. (*A*) Under normal circumstances, high sea-level pressure over the eastern Pacific causes surface winds and water to move westward. Note upwelling along eastern ocean boundary. (*B*) Periodically, the pressure difference weakens or reverses, resulting in major shifts in atmospheric and oceanic patterns.

difference is the driving force behind the east-to-west wind flow.

The situation described in the previous paragraphs is not a permanent one, however, as you can see from Figure 11.34B. For reasons not yet fully understood, the surface air pressure difference periodically declines, then reverses: higher pressure forms in the western Pacific, while the eastern high weakens. This alternation in surface air pressure regimes is known as the **southern oscillation.** In Figure 11.35, you can trace the history of the southern oscillation over a number of years.

Notice in Figure 11.34B that in response to the reversal in the air pressure pattern, the trade winds weaken; in some cases, they disappear altogether and are replaced by a westerly wind flow. The warm surface waters of the western Pacific travel slowly eastward, carrying with them a region of cloudiness and precipitation. Off the coast of Peru, the lack of strong trade winds and the arrival of the warm surface water cause a temporary end to coastal upwelling—and drastic declines in the fishing industry.

The intrusion of warm surface water as shown in Figure 11.34B recurs over periods of three to seven years. Because it tends to arrive during or near the Christmas season, the warming is known as "El Niño," or "the (Christ) child." A typical El Niño episode lasts for several months (although some recent ones have persisted for much longer). Eventually, surface air pressures rise once more over the eastern Pacific, and conditions return to the more normal ones depicted in Figure 11.34A.

ENSO is particularly interesting to researchers because its influence apparently extends far beyond its own region. For example, during El Niño episodes, droughts occur in a region from Australia to southern Africa, and severe winters are the rule for eastern North America.

***Review***

Surface wind and air pressure patterns appear to be responsible for a number of oceanographic circulation features, such as the major ocean currents, upwelling, and El Niño episodes. Closer inspection, however, reveals that each of these marine phenomena in turn affect the atmosphere in a variety of ways. Thus, oceanic and atmospheric circulations are intimately connected. ENSO is the most complex, varied, and far reaching of these connections.

***Questions***

1. Compare Figures 11.28 and 11.30. What atmospheric circulation feature appears to be responsible for the Alaska Current? The West Wind Drift? The Brazil Current?

2. The coastline of the Florida panhandle runs approximately from east to west. From what direction must the wind blow to cause upwelling?

3. How are the southern oscillation and El Niño related?

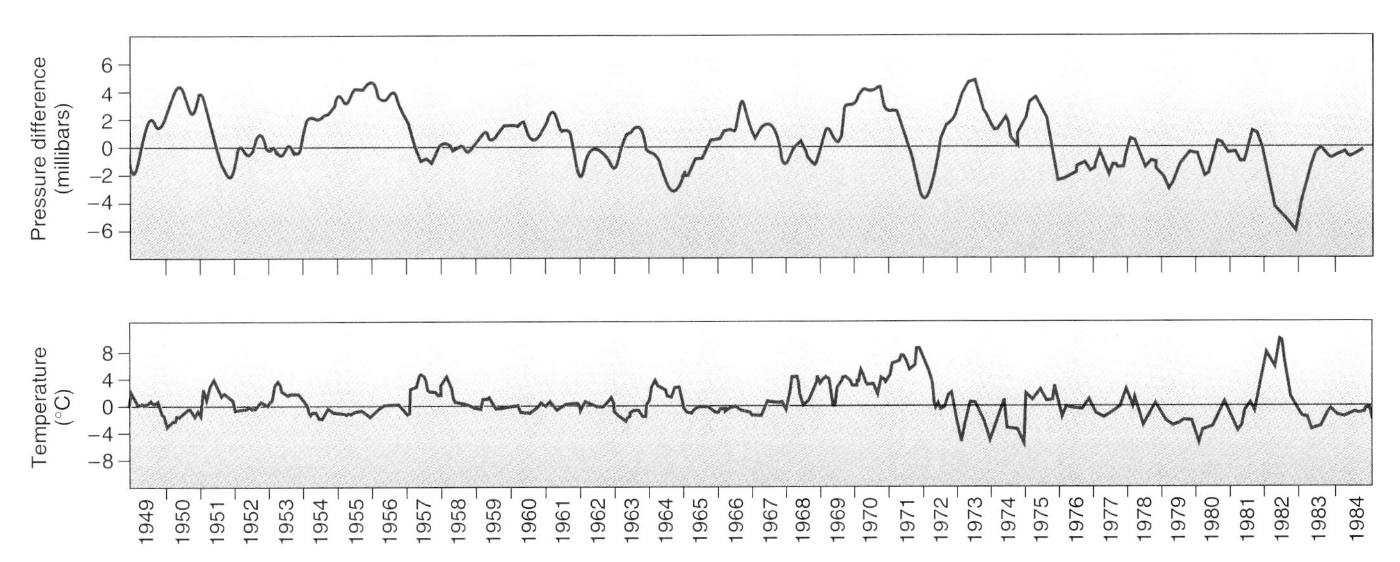


**Figure 11.35** The southern oscillation index and sea surface temperatures off the coast of Peru. The index is calculated by subtracting sea-level air pressure at a western locale (Darwin, Australia) from one in the east (Easter Island). Thus, positive index values mean higher pressure in the east, the situation you saw in Figure 11.34A. Negative values mean lower pressures to the west, as in Figure 11.34B. Can you see a relation between the index and the temperature graphs? Source: C.S. Ramage, in *Scientific American*. Copyright © 1986 by Scientific American, Inc. All rights reserved.

## Chaos and the Butterfly Effect

Suppose that tomorrow morning, your alarm clock goes off one second late. Will that tiny perturbation on your daily schedule change the entire course of your life? You never know for sure, since you can't live your life over again, as a control case, to see the effect of that one extra sleepy second. But most people would dismiss the proposition as far-fetched: we save and waste far more time every day in a hundred different ways. Besides, events like the beginning of a class or a TV program act to resynchronize us periodically with the rest of the world. Surely, the one-second perturbation on your life would soon disappear in the "noise"of everyday living.

Until recently, many meteorologists believed the same of small perturbations in the atmosphere; surely, microscale disturbances like wind gusts must dissipate quickly and leave no lasting mark on larger-scale circulations.

In the early 1960s, meteorologist Edward Lorenz inadvertently performed an experiment that suggested the atmosphere might behave otherwise. Lorenz had written a computer model that simulated basic atmospheric flow patterns, like those graphed in Figure 11.36. One day, intending to duplicate a run he had just made, he restarted the program with initial values he believed were the same as in the previous run.

Lorenz was surprised to see that although at first the two runs gave nearly identical output, after a while they began to diverge, and eventually they bore no resemblance to each other whatsoever. Investigating, Lorenz found the initial conditions he had entered for the second run differed by a tiny amount from those of the first. Rather than disappearing with time, Lorenz's accidental perturbations propagated, eventually taking his model atmosphere into a future that was totally different from that of the first run.

Lorenz's program outputs exhibited what has become known as chaotic behavior, or, simply, **chaos**. Chaotic behavior has several distinguishing properties. It is very sensitive to initial conditions: as you saw above, tiny differences in starting values may lead to large changes later on. Chaotic behavior is nonperiodic: that is, it never repeats itself exactly. And chaotic behavior is deterministic: it is not random or capricious but follows physical laws like other phenomena.

Figure 11.37 shows a beautiful example of chaotic fluid motion in a convection cell. The white lines show information about the fluid's speed of circulation (horizontal axis) and vertical temperature conditions (vertical axis) for two different experiments with nearly identical initial conditions. Notice that the two paths are nearly identical for some time, then they abruptly diverge. In one experiment, the circulation switches to a completely different range of speeds and vertical temperatures. Running this simulation for a longer time reveals a number of such switches from one mode to another. It also reveals that the trajectories never exactly repeat each other; they are nonperiodic.

Are the fluids' trajectories in Figure 11.37 deterministic? Certainly they are; after all, they were calculated, in the simulation, from the laws of physics. Then why can't we just use these laws, as used in the simulation, to determine actual atmospheric convection, and to see which future will occur? For brief periods into the future, this approach is possible, if the convection in the actual atmosphere is as simple as that in the simulation. For longer periods of time, however, the calculation inevitably will be inaccurate, due to its sensitivity to initial conditions.

To illustrate the problem, suppose you measure wind speed in a convection cell at 10 meters per second. You enter this value into the computer simulation, begin the program, and receive a prediction for the cell's future like one of the trajectories in Figure 11.37.

Now, no measurement of any natural phenomenon is perfectly accurate, and your 10 meters per second value is no exception. Suppose it is off by just 0.01 meters per second a very

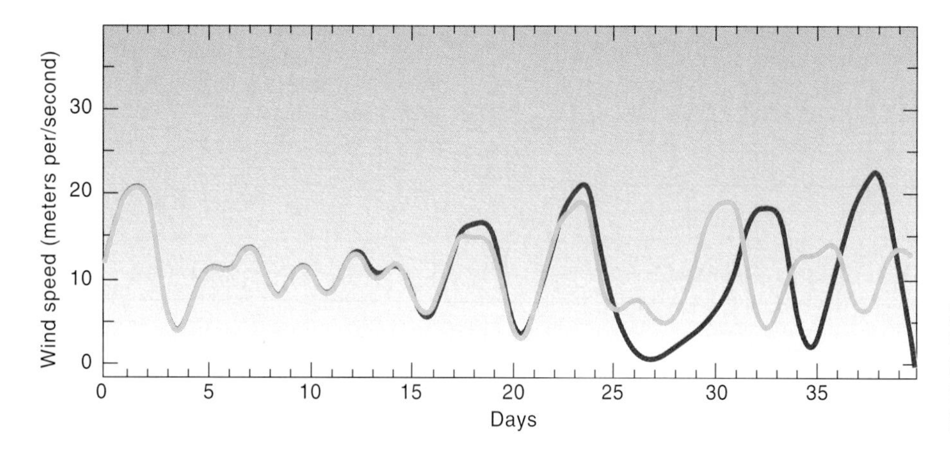


**Figure 11.36** Two of Lorenz's simulations of wind speeds for a 40-day period. Initial conditions for the two runs were nearly identical; note the divergence in predicted speeds by the end of the period.

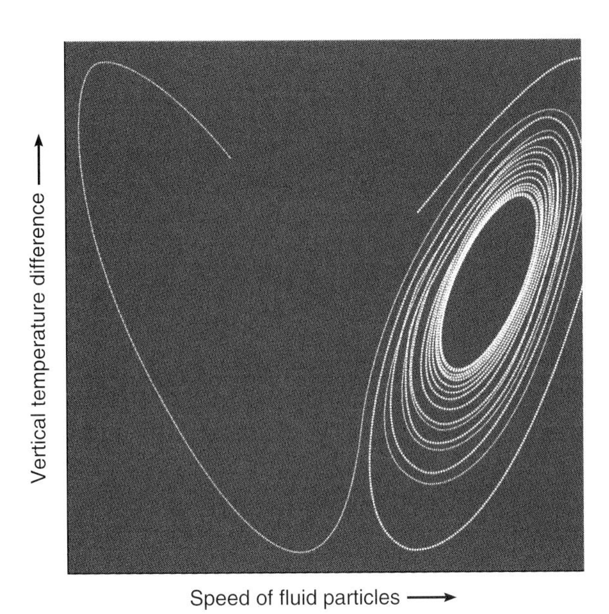


**Figure 11.37** Two neighboring particles with nearly identical initial conditions eventually have very different futures.

small error by meteorological standards. This error—any error, in fact—is enough to ensure that your calculation will diverge at some future time from the trajectory based on the correct value and that eventually the two sets of trajectories will diverge like those in Figure 11.37. Specific conditions at a given moment in the predicted future, based on an initial value of 10.00 meters per second, will be completely different from those that actually occur based on the correct starting value of 10.01 meters per second. And since it is impossible to know if the initial wind speed was 10 or 10.01 meters per second, it is impossible to predict which trajectory the cell will follow.

Thus, chaos theory predicts that chaotic systems like the one you have been studying, and the actual atmosphere as well, are unpredictable in detail due to their sensitive dependence on initial conditions. Speaking more practically, in a chaotic atmosphere, we can never expect to have exact forecasts of weather conditions a month or a year ahead, for example.

Lorenz has suggested that chaos may generate a more active form of unpredictability through what is known as the **butterfly effect.** This term comes from a question Lorenz posed in a lecture: Could the flap of a butterfly's wings over Brazil cause a tornado over Texas?

The argument runs like this: Suppose a butterfly flaps its wing. The air moved by its wing causes a very small microscale eddy. Eventually, this perturbation interacts with another, causing a larger wave to form, which may be just the boost needed to cause some other, still larger-scale change, and so on. In an atmosphere sensitive to initial conditions, a butterfly may be all that's needed to change the future of atmospheric events; and two weeks after the original perturbation, a thunderstorm occurs in Texas that wouldn't have happened without that Brazilian butterfly flapping its wings.

The extent of the butterfly effect is still a matter of conjecture. The more significant it turns out to be, the less predictable the atmosphere will be in the long term. (Imagine having to monitor butterflies, birds, door closings and openings, campfires, and all the other things that perturb the atmosphere before being able to make a forecast.)

Chaos and the butterfly effect are not about to end progress in understanding and predicting the atmosphere's behavior, however. On the contrary, viewing the atmosphere as a chaotic system provides an entirely new perspective on atmospheric events, and chaos theory offers researchers new analytical tools with which to approach old problems.

Thus, at this point, we have a partial answer to the main question for this unit: the tendency of the atmosphere's component parts and circulations to interact does indeed lead to problems of unpredictability. In the next two chapters, you will explore the problem further, concentrating on three specific circulations: thunderstorms, tornadoes, and hurricanes.

# Study Point

## Chapter Summary

The circus of atmospheric events we call weather results from the interplay of flow phenomena on all scales, from molecular viscosity to the turbulence of microscale circulations, to the thermally driven mesoscale land and sea breezes and mountain and valley winds, to synoptic scale cyclones and anticyclones, to global scale features of the general circulation. The flow at each scale may be laminar (smooth), turbulent (characterized by rapidly changing eddies and waves of all sizes), or transitional (a blend of laminar and turbulent) properties. The nature of the flow often varies with location, being generally more turbulent close to the earth's surface, for example. The character also varies with scale: thus, sometimes small-scale regions of laminar flow are present within large-scale turbulence patterns; other times, turbulence may be found embedded within laminar flow.

Some forces, particularly Coriolis, are more influential at some scales than at others. At every scale, thermally driven circulations are evident. These include the land and sea breeze, mountain and valley winds, the monsoon, and Hadley cell circulations. Other notable winds are the result of large-scale systems interacting with surface topographical features. In this category are lee waves, Chinook and Santa Ana winds, nor'easters, and lake effect snowstorms.

Interactions between atmosphere and oceans affect the general circulation of the atmosphere in a variety of ways. Large-scale ocean currents, upwelling circulations, and ENSO are just three examples of interactions that have major impacts on airflow patterns.

The atmosphere demonstrates elements of chaotic behavior, which means that future conditions are sensitively dependent on earlier ones. A related idea is that the smallest atmospheric disturbances may have major influences on much larger systems, a phenomenon known as the butterfly effect. These considerations suggest there may be limits to the atmosphere's long-term predictability.

## Key Words

laminar flow
molecular scale
mesoscale
viscous forces
boundary layer
mechanical turbulence
eddy viscosity
law of conservation of angular momentum
turbulent flow or turbulence
microscale
global scale
molecular viscosity
free atmosphere
thermal turbulence
clear air turbulence (CAT)
land and sea breeze circulation
mountain and valley winds
roll or rotor
Santa Ana wind
lake effect snowstorms
trade winds
intertropical convergence zone (ITCZ)
subtropical jet stream
horse latitudes
prevailing westerlies
polar easterlies
upwelling
southern oscillation
chaos
lee waves
Chinook wind
nor'easter
general circulation
Hadley Cell
doldrums
subtropical high pressure belt
monsoon
polar high pressure cell
gyre
El Niño
ENSO
butterfly effect

## Review Exercises

1. Compare the path of a balloon released into laminar flow with one released into turbulent flow.
2. List the different scales of atmospheric motion, with ranges of size and time for each.
3. Give an example of an eddy on each scale from global to microscale.
4. At what time of day is the surface wind flow most likely to be laminar? Why?
5. Compare molecular and eddy viscosity. Why is transfer via eddy viscosity so much more efficient?
6. Give an example of pure mechanical turbulence; of pure thermal turbulence.
7. What conditions can lead to clear air turbulence aloft?
8. Using Figure 11.13, explain the daily cycle of the land and sea breeze circulation.
9. Why is it inevitable that synoptic and global scale circulations are much broader horizontally than they are tall?
10. Fabricate representative weather observations for each of these locations: in the ITCZ, in the mid-trades, in the center of the subtropical high pressure belt, just north of the polar front, and in a polar high pressure cell. Make sure wind directions, air pressures, temperatures, and so on are consistent.
11. What is the butterfly effect? How is it related to chaos?

## Problems

1. How do different layers of air interact with each other in laminar flow? What evidence exists that they do interact, in fact?
2. A 1-kilogram air parcel rotates at 3 meters per second around the center of an eddy 10 meters away. The air spi-

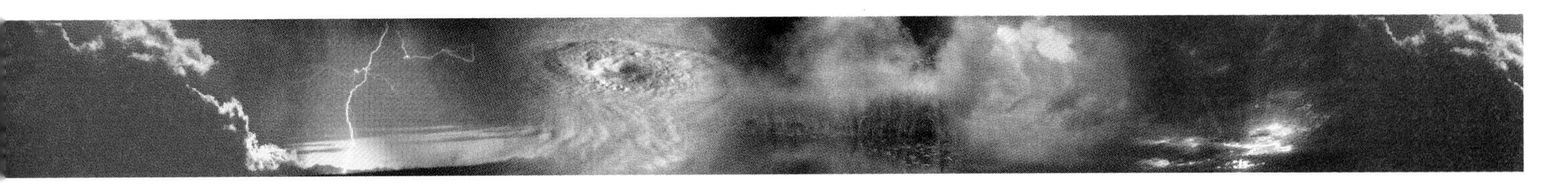

rals inward, gaining speed due to conservation of angular momentum. What is its speed when it is 2 meters from the center of the eddy?
3. You are at the beach for an all-night party. On which side of the campfire should you sit to avoid the smoke? Why?
4. Suppose the wind flow at the seashore is the sum of a land and sea breeze that produces a 7-meters per second onshore wind by day and a 3-meters per second offshore flow at night, plus a larger-scale (i.e., synoptic) wind flow that contributes a steady 2-meters per second onshore flow throughout the period. What is the net wind during the day? During the night?
5. How might the winds in a valley that slopes downward to the east differ from those in one that slopes downward to the south?
6. Great Lakes cities that receive excess winter snowfalls do not observe enhanced precipitation in summer. Why?
7. On a blank map of the United States, position synoptic scale highs and lows that would simultaneously cause a Chinook in the Rockies, a Santa Ana in southern California, and lake effect snow in Salt Lake City.
8. Someone commented once that "the ITCZ is the atmosphere's version of the equator." What does the comment mean?
9. You saw (in Figure 11.24, for example) that clouds form in the warm, rising air at the ITCZ. These clouds reflect sunlight, reducing the solar income in the region. What additional energy source helps to maintain the low pressure and rising air currents at the ITCZ?
10. The doldrums and the horse latitudes are both regions of light or calm winds; in nearly every other way, however, weather conditions in the two regions bear no resemblance to each other. Why is their weather alike in this way but not in others?
11. At the surface, the general circulation is dominated by low pressure at the equator and high pressure at the poles. Aloft (500 millibars), this arrangement is reversed. Why?

# Explorations

1. Pour a dollop of milk into a mug of coffee or tea and observe the circulation that develops. Repeat the experiment a number of times to see the effect of imparting some rotation on the coffee or milk, changing the temperature contrast between coffee and milk, heating the mug from below, adding a mountain range (rock), and so forth. Record your observations in writing, or use a video camera.
2. Many local winds, such as the Chinook, are vividly described in historical writings. Make a study of such eyewitness accounts; look for good observations of the phenomenon, as well as its impact on the region and its people.
3. If you have access to a meteorological database like AccuData or to Internet sites (see Resource Links below), collect data showing land and sea breezes, monsoon winds, mountain and valley winds, general circulation features such as the Hadley cell or subtropical highs, or other circulations discussed in this chapter. Innumerable such studies can be made. For example, compare the onset of the sea breeze at various coastal locations from day to day, depending on weather conditions; or look for evidence of land and sea breezes on hour-to-hour satellite images.

# Resource Links

The following is a sampler of print and electronic references for some of the topics discussed in this chapter. We have included a few words of description for items whose relevance is not clear from their titles. More challenging readings are marked with a C at the end of the listing. For a more extensive and continually updated list, see our World Wide Web home page at www.mhhe.com/mathscience/geology/danielson.

Gleick, J. 1987. *Chaos*. New York: Viking Press. 352 pp. A popular exposition of chaos; Chapter 1 deals with E. Lorenz and the butterfly effect.

Kocin, P. J., and L. W. Uccellini. 1990. *Snowstorms Along the Northeastern Coast of the United States: 1955 to 1985. Meteorological Monograph Number 44.* Boston: American Meteorological Society. 280 pp. C (but worth the effort).

Lorenz, E. N. 1993. *The Essence of Chaos*. Seattle: University of Washington Press. 227 pp.

Simpson, J. E. 1994. *Sea Breeze and Local Wind.* Port Chester, NY: Cambridge University Press. 220 pp.

Watson, B. 1993. New respect for nor'easters. *Weatherwise* 46 (6): 18–23. Easy-to-read review of eastern coastal storms.

For readers with access to AccuData, SJET gives maps of the North American jet stream, and UIP provides the user with maps of worldwide upper air windflow.

NOAA's Geophysical Fluid Dynamics Laboratory offers "Mesoscale studies at GFDL," with movies, lectures, graphics, and more, at http://www.gfdl.gov/ . A report on El Niño is also available at GFDL, at http://www.gfdl.gov/brochure/2El_Nino.doc.html .

An instructional unit on El Niño is available through the University of Illinois Daily Planet web site at http://covis.atmos.uiuc.edu/guide/El-Nino/ . Another unit, at the same web site, is "Coastal Weather" at http://covis.atmos.uiuc.edu/guide/rsteve/html/lakewx.html .

# *Thunderstorms and Tornadoes*

## Chapter Goals and Overview

On May 11, 1992, meteorologists monitoring the NEXRAD radar screen in Norman, Oklahoma, observed echoes they believed indicated a developing tornado. They promptly issued a tornado warning for the town of Kingston, Oklahoma. Alarms were sounded, and people took cover. Twenty minutes later, a tornado scored a direct hit on the town, causing considerable damage but little human injury—thanks to the warning.

What grade would you give the forecasters in Norman for their forecast? On the time span of 20 minutes, you would probably agree they deserve an A+ for a perfect prediction based on the latest theory and technology. But a resident of Kingston who drove the 100 miles to Oklahoma City that day might complain that a 20-minute warning was of little use; there was insufficient time to return home and make sure that the children were safe. The forecasters might point out in response that they had issued a "severe weather watch" several hours before the tornado struck, alerting people that tornadoes were possible that day within a large area of southern Oklahoma that included Kingston.

This chapter is devoted to study of the thunderstorm and its violent offspring, the tornado. You will learn about the formation, structure, and distribution of thunderstorms and tornadoes. You will consider some of the problems involved in observing their behavior. A number of times throughout the chapter, you will apply this information to the unit question, "How predictable are small-scale circulations?"

### Chapter Goals

By the end of the chapter, you should be able to:

- sketch the stages in development of a typical thunderstorm, showing patterns of wind flow, temperatures, and precipitation;
- list the characteristics that distinguish thunderstorms of various intensities and aggregations;
- describe two hypotheses that address the formation of electric charge in thunderstorms;
- describe the conditions under which tornadoes form;
- list safety precautions appropriate for lightning and tornadoes; and
- discuss problems and successes in forecasting thunderstorms and tornadoes.

# Thunderstorm Structure and Life Cycle

What is a thunderstorm? From its name, it is evidently a storm producing thunder and thus its companion, lightning. In addition to thunder and lightning, a typical thunderstorm produces showers, sometimes of torrential intensity, and occasionally hail or, more rarely, snow, along with gusty winds, sometimes in the form of dust storms or tornadoes. Ounce for ounce, a vigorous thunderstorm generates a greater variety of violent weather than any other atmospheric circulation. A fully developed thunderstorm is also a beautiful and majestic sight, as the photo opening this chapter illustrates. Perhaps the photograph by itself is the best definition of "thunderstorm."

## Life Cycle of a Thunderstorm

It may seem incredible that an object as large as the storm in the opening photo, containing several cubic kilometers of cloud, millions of gallons of precipitation, plus thunder, lightning, and violent winds, can develop "out of the blue" in a period of just an hour or so. How does it happen?

Research conducted in Ohio and Florida in the late 1940s led to a three-stage model of thunderstorm development. The model applies to **ordinary** or **air mass thunderstorms,** which form individually, due to local heating, and not due to large-scale lifting mechanisms such as fronts or orography. The three stages are diagrammed in Figure 12.1. Although more recent models describe certain types of thunderstorms in greater detail (as you will see later in this chapter), the three-stage model continues to serve as a useful general description.

### *The Cumulus Stage*

The first stage in thunderstorm development is known as the **cumulus stage** and is illustrated in Figure 12.1A. Examine the airflow pattern in this stage. Notice the converging and rising air flow, typical of any growing cumulus cloud. Why is this air rising? Follow the isotherms (dashed lines in the figure); if you compare the temperature inside the cloud with that outside at the same altitude, you can see that the air in the updraft is warmer than in the surrounding air; hence, it is buoyant.

Recall from Chapter 6 that buoyant, rising air cools adiabatically; if it rises high enough, condensation occurs. Condensation releases latent heat, which warms the air within the cloud, adding to its buoyancy. Thus, the thunderstorm begins as a typical cumulus cloud: it becomes larger and more buoyant than most cumuli, but its basic properties are the same.

And where does the growing cumulus come from? Perhaps you feel the three-stage model should include an earlier, "zero stage," to span the period from clear skies to the cumulus stage. Mechanisms leading to the cumulus stage, like that stage itself, in part are just typical cumulus-forming processes you read about in Chapter 6: warm, moist, conditionally unstable low-level air is heated to buoyancy. However, clouds large enough to become thunderstorms often grow due to other causes as well. For example, under the right conditions, neighboring cumuli sometimes merge, creating a single, much larger and more potent cell. We will return to this subject later, when we consider specific types of thunderstorms.

### *The Mature Stage*

If conditions for growth remain favorable, the updraft in the cumulus cloud strengthens, drawing in more moist air at low levels, which in turn releases more latent heat, strengthening the

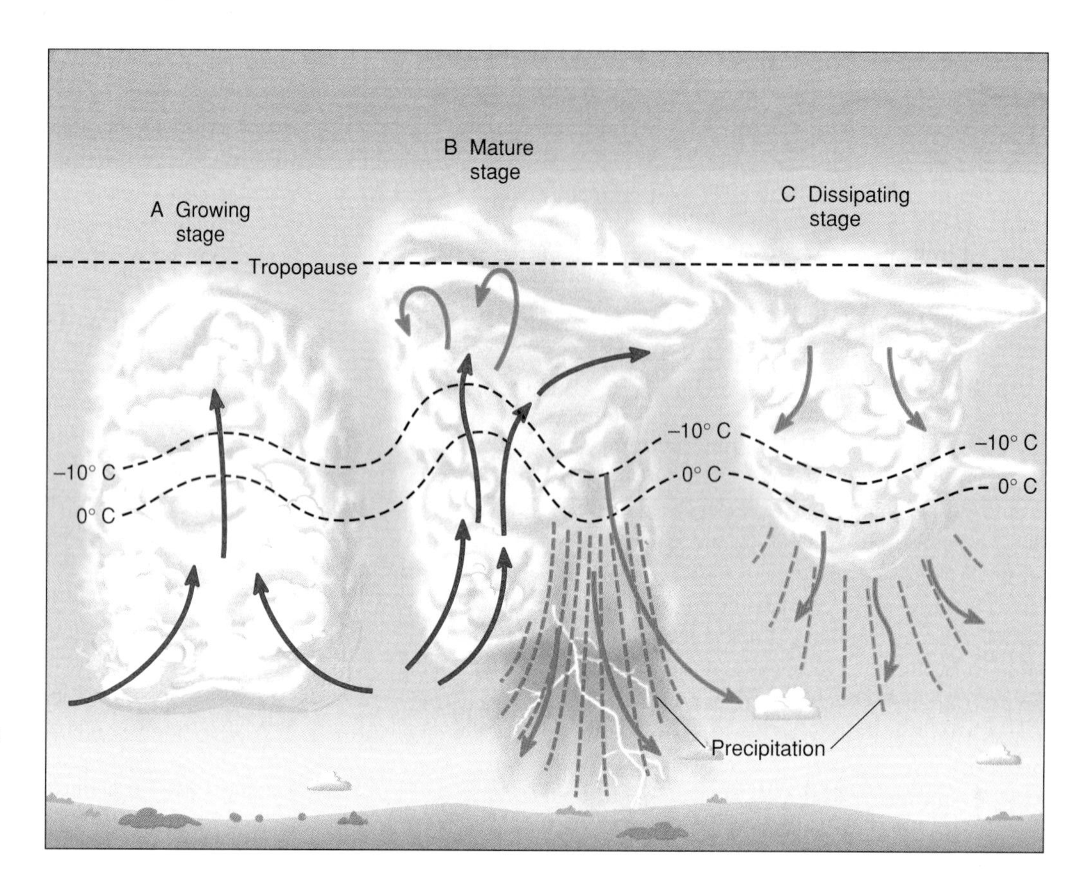


Figure 12.1 Stages in thunderstorm growth. Follow the evolution of updrafts, downdrafts, and temperature patterns through the three-stage sequence.

updraft. Fifteen to 30 minutes after having reached the cumulus stage, precipitation begins falling from the storm's base, which marks its transition to the **mature stage**. Compare this stage, in Figure 12.1B, with the cumulus stage. What changes do you see?

Most immediately evident is that the storm is now clothed in full thunderstorm regalia, with thunder, lightning, heavy precipitation of various kinds, and turbulent winds. A strong, deep updraft may lift air as high as the tropopause, whose stability caps the upward motion and causes an anvil to begin forming. Notice in the figure, however, that the air in a vigorous updraft may be rising so fast it "overshoots" the tropopause, creating a cloud dome that bulges upward from the anvil into the lower stratosphere.

Typical updraft speeds in the storm's mature stage are on the order of 10 to 20 meters per second (20 to 40 mph). Thus, in one minute, the ascending air may rise as much as a kilometer or more. (Traveling for 60 seconds at 20 meters per second, the distance covered is $60 \text{ s} \times 20 \text{ m/s} = 1200 \text{ m} = 1.2 \text{ km}$). At these speeds, it takes only 10 minutes or so for updraft air to ascend from ground level to the tropopause.

This rapid lifting has important implications for precipitation formation. Recall from Chapter 6 that rising air cools adiabatically at 10°C per kilometer if unsaturated and somewhat less rapidly (6°C per kilometer, as a typical value) if saturated. Thus, the air in a strong updraft may cool by 10°C or more per minute for several minutes of ascent. This sudden cooling causes most of the air's water vapor to change phase to liquid water or ice, which has two important consequences: first, vast amounts of latent heat are released as the water vapor changes phase; this heating enhances the buoyancy of the rising air. Second, the sudden copious supply of water in liquid and frozen form leads to the rapid growth of precipitation-sized particles.

Strangely, the powerful updraft that generates vast numbers of precipitation particles also tends to prevent them from actually falling. Recall the discussion of terminal velocities in Chapter 5. Table 5.2 indicated that terminal velocity for a large raindrop is approximately 10 meters per second. If the updraft is stronger than the drop's terminal velocity, the drop will continue to ascend. How, then, does the precipitation make its way out of the cloud?

The answer is, "In downdrafts, which sweep the precipitation downward and out through the cloud base." In Figure 12.2A, you can see that if the updraft is not exactly vertical but is tilted with height, its precipitation may fall out of the updraft and into regions of more slowly ascending air. (Updraft tilting is a more important feature of some other thunderstorm types, as you will read later.) Other precipitation particles rise in the updraft and then spread into parts of the cloud lacking strong updrafts. As they fall, they begin to drag the surrounding air downward with them, initiating a downdraft. The descending air may be joined by downdrafts of drier, environmental air that was entrained into the cloud top or sides.

Falling farther into lower, warmer air, the precipitation cools this lower air in two ways. First, since this cold rain (and melting snow and ice) is colder than the low-level air, it cools the air by contact. Second, both within the cloud and especially below the cloud base, a substantial amount of the precipitation evaporates into the surrounding air. The latent heat required for these phase changes comes from the air itself, cooling it further.

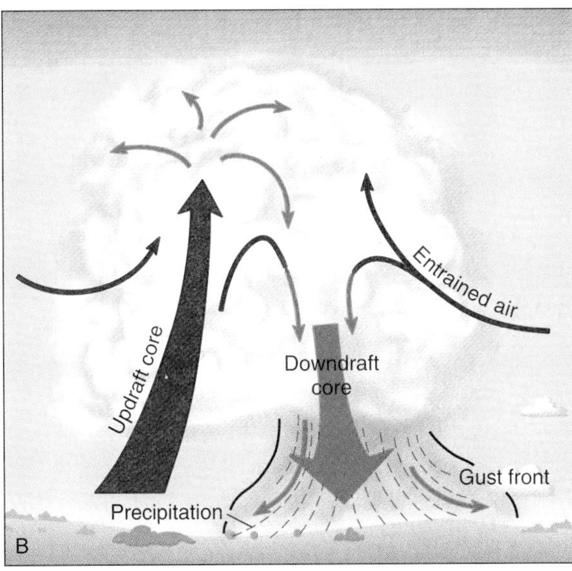


**Figure 12.2** Development of a downdraft and gust front. (*A*) Precipitation particles falling out of or rising through the updraft initiate downdraft current. (*B*) The mature downdraft, cooled by its precipitation, spreads out at ground level behind the gust front.

The cooling increases the air's density in the downdraft, causing it to sink faster. Through these processes, downdrafts form and intensify. The presence of vigorous updrafts and downdrafts is characteristic of mature stage thunderstorms.

Upon reaching the ground, the dense downdraft air spreads outward as a mesoscale cold high pressure cell. The boundary between this cold air and the warm ascending air, called a **gust front** (see Figure 12.2B), is often very distinct, like a sharp cold front. Sometimes the gust front is so pronounced, it appears on weather radar images, as in Figure 12.3. As the gust front sweeps along the ground, it lifts the warmer, inflowing air, thus providing an extra lift to the air headed for the updraft.

You can probably recall experiencing the passage of a gust front. Just before the rain arrives, the wind strengthens and switches direction, coming straight from the darkest part of the sky; also, the temperature may fall sharply. The downdraft air has reached you; the heaviest rain and lightning are close at hand. Time to take cover, fast.

### *The Dissipating Stage*

Fifteen to 30 minutes after a typical thunderstorm has entered its mature stage, the cold downdraft air has spread out beneath the cloud base and has cut off the storm's supply of warm, humid air (see Figure 12.1C). Deprived of its source of heat energy and moisture, the storm soon weakens. Now it is in its final, or **dissipating, stage.** Weak downdrafts, light precipitation, and temperatures colder than those at the same level outside the cloud characterize the lower regions of the cloud in this stage. Thunder and lightning diminish, then cease altogether.

Although a single thunderstorm typically moves through all three stages in an hour or so, some effects are more long-lasting. For example, in the upper regions of the cloud, weak updrafts may persist for some time. The anvil becomes an extended deck of cirrostratus and altostratus clouds that may stretch over hundreds of square miles by the time the thunderstorm dissipates. Drifting through the atmosphere, these layer clouds may cause light precipitation several hours after the disappearance of the low-level updraft that gave them birth. Much of the middle- and high-level cloudiness observed in warm climates consists of the remnants of dissipating thunderstorms.

## Thunderstorm Frequencies

Where and when are thunderstorms likely to form? Since they begin as growing cumulus clouds, the conditions under which thunderstorms form are similar to those for cumulus cloud formation you read about in Chapter 6. These conditions include the presence of warm, moist, unstable air, particularly at low levels, and some lifting mechanism. Thunderstorms grow to greater heights than other cumulus clouds; therefore, the atmosphere must be free of major inversions that might act as a cap to vertical growth.

Figure 12.4 shows a typical sounding on a day air mass thunderstorm development might be expected. Note the presence of conditionally unstable air at lower levels and the absence of any extensive inversion in the troposphere. Strong solar heating will cause the surface air to become buoyant, providing the lift to initiate deep convection and leading to thunderstorm formation.

The requirement for warm air suggests that thunderstorms, like cumulus clouds, will be more common in the warm months of the year, in the warmest time of day, and more in tropical than high latitude locations. The humidity requirement suggests locales that experience maritime air masses. Finally, the need for a lifting mechanism suggests that locales experiencing local winds (mountain and valley winds, sea and land breezes, etc.) or frequent frontal passages will see more thunderstorm development.

In Figure 12.5, you can check these suppositions. Note that in the United States, the maximum occurs in the southeastern states, a region dominated by maritime tropical air masses and abundant sunlight to heat the lower atmosphere. Cold frontal passages in the Midwest and Plains states make these regions

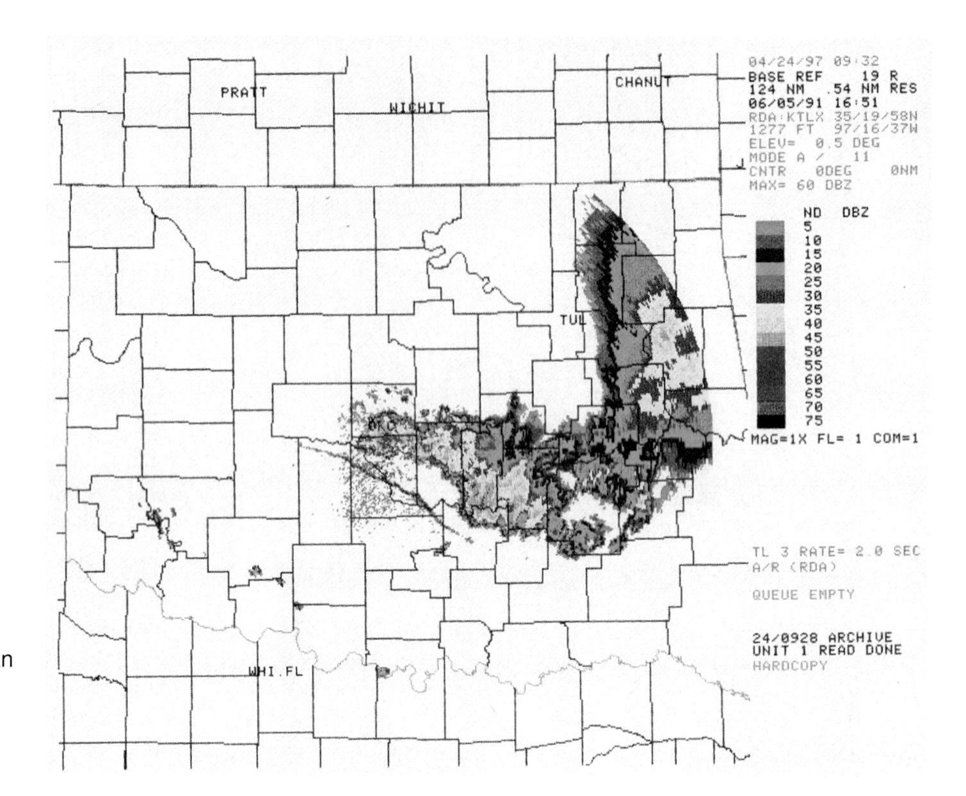


**Figure 12.3** NEXRAD Doppler radar image of an Oklahoma thunderstorm (shown in red, yellow, and green) and its gust front (the thin blue line). Successive radar images show the gust front spreading outward from the thunderstorm cell.

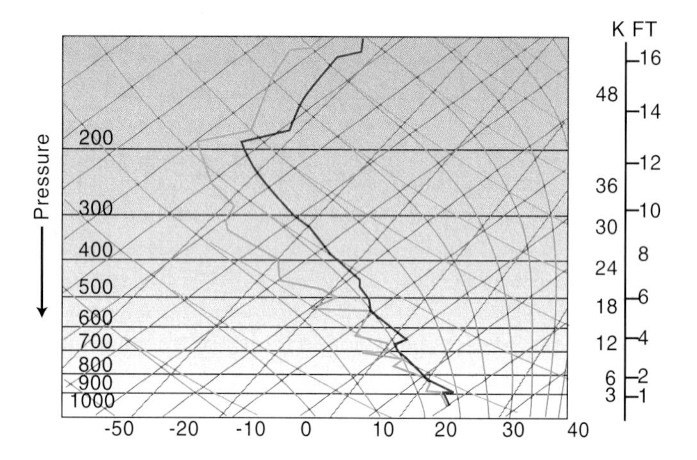


**Figure 12.4** Atmospheric sounding for Tallahassee, Florida, on a day when air mass thunderstorms were expected.

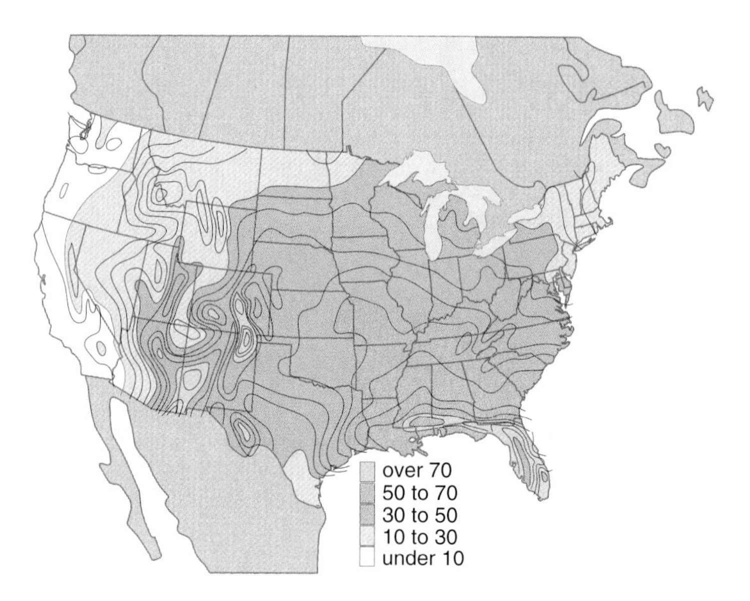


**Figure 12.5** Average number of days per year with thunderstorms. What conditions account for the maximum in Florida?
Source: Data from NOAA *Thunderstorms and Lightning Pamphlet*, NOAA/PA 92053, ARC 5001.

subject to large numbers of thunderstorms as well. Other maxima from Colorado southward to New Mexico and Arizona are the result of orographic lifting. Note the relation between thunderstorm frequency and latitude: in general, thunderstorms are more common in lower latitudes. Note also the minimum along the Pacific Coast states, where cool ocean water and a persistent inversion tend to create stable low-level conditions, thus suppressing thunderstorm development.

In most states, thunderstorms occur most frequently in the warmest months of the year and during the warmest hours of the day. The major exception is in some Midwestern and Plains states, where thunderstorm activity is greatest at night. It is thought that convergent low-level wind flow, which often forms at night in these regions, is responsible for the nighttime thunderstorm maximum.

On a global scale, thunderstorms are most common over the tropical oceans, particularly in the vicinity of the intertropical convergence zone. Convergence along the ITCZ provides a lifting mechanism for the very warm, moist air. It is estimated that at any given moment, several hundred thunderstorms are in progress at various places along the ITCZ.

# Checkpoint

### *Review*

The concept map in Figure 12.6 summarizes the main concepts in this section. It is by no means a complete map for thunderstorms. In later sections, we will be adding considerably to the "other thunderstorms" and "thunder and lightning" branches.

### *Questions*

1. Describe the changes in updrafts and downdrafts as a thunderstorm cell evolves from growing to dissipating stage.

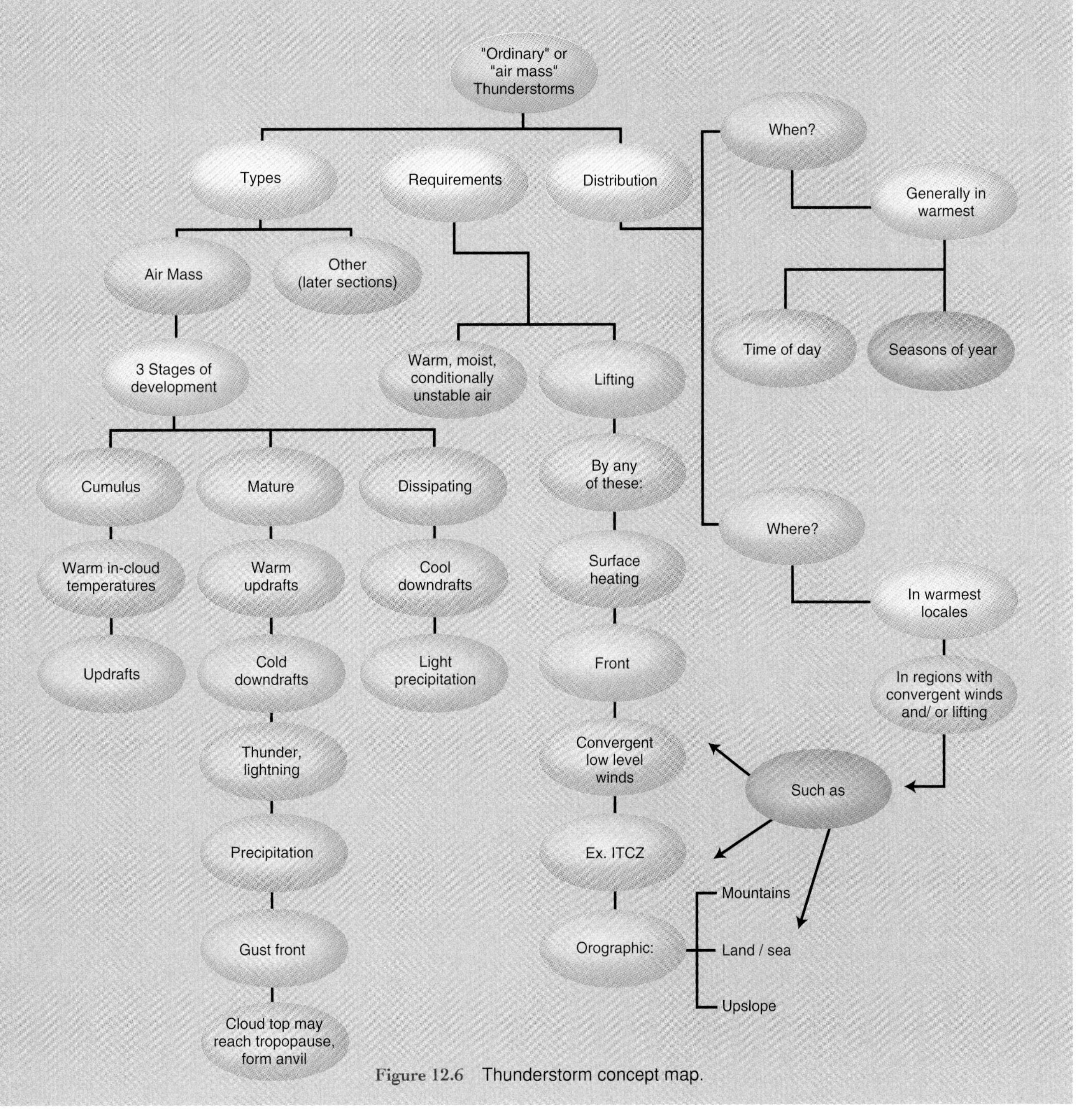


**Figure 12.6** Thunderstorm concept map.

2. What is the fall rate of a drop with a 9-meters per second terminal velocity, caught in a 12-meters per second updraft? In a 5-meters per second updraft?

3. Los Angeles and Charleston, South Carolina, lie at nearly the same latitude. In which city are thunderstorms more common (see Figure 12.5)? Why?

## Lightning and Thunder

Certainly the most distinguishing feature of thunderstorms is their electrical activity. Lightning is a gigantic spark discharge: a sudden, intense flow of electricity along a narrow atmospheric path. And thunder? Explosive heating of the air along the lightning bolt's path produces the rolling rumble (if distant) or ear-splitting crash (if nearby) called thunder.

### Electrical Properties of Matter

How does lightning form in a thundercloud? Surprising as it might seem, more than 200 years after Benjamin Franklin's supremely dangerous experiments that demonstrated lightning was a form of electricity, we still don't know for certain. In the next section, you will consider some important facts known about lightning and then examine several hypotheses that attempt to explain its occurrence. To do justice to this fascinating subject, however, you must be familiar with a few basic electricity concepts. This section is devoted to that task.

#### *Electric Charges*

In Chapter 2, you read that electric charge is a basic property of matter. Recall that every electron has one unit of negative electric charge, and every proton has one unit of positive charge. The magnitudes of positive and negative charges are equal but opposite in sign. Therefore, a molecule with equal numbers of protons and electrons has a net charge of zero and is electrically neutral. On the other hand, if you ionize an atom by removing one of its electrons, you have created two tiny charge centers: a negative one centered on the electron and a positive one surrounding the ionized molecule. Ionization of molecules is the primary way in which electric charge centers form.

#### *Charge Separation and Electric Forces*

Let's consider an example of charge separation, or ionization, from everyday life, which is illustrated in Figure 12.7. When you run a comb through clean, dry hair in dry weather, you experience electric charge separation and electric forces firsthand. The combing action ionizes some of the molecules in the comb, depositing electrons in your hair. As a result, the comb, now having lost some electrons, has a net positive charge, and the hair, having gained electrons, has acquired a negative charge of the same magnitude.

You probably have noticed that when you charge your hair in this way, it tends to stick out in all directions. In Figure 12.7C, you can see that each strand of negatively charged hair experiences a repelling force between it and its neighbors; by sticking out, it forces itself as far as possible from the other hairs. On the other hand, if you move the comb back and forth a small distance above your hair, you may notice the hair rise toward the comb as it passes. The negatively charged hair experiences an attractive force toward the positively charged comb.

Experiments such as these illustrate that oppositely charged objects (one positive, the other negative) experience an attractive force; similarly charged objects (both positive or both negative) experience a repulsive force. In short: opposites attract,

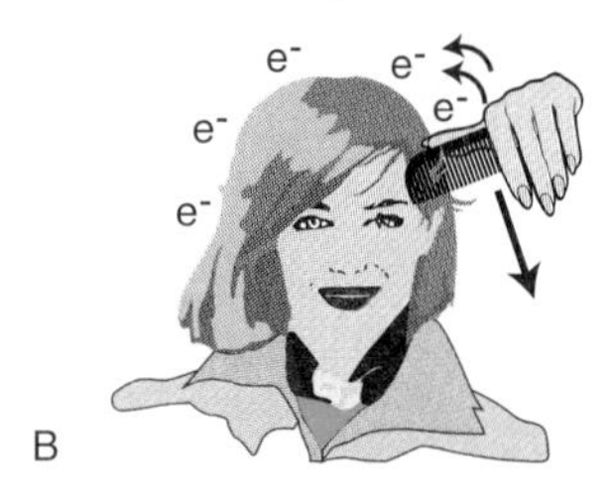

**Figure 12.7** A common example of ionization.

likes repel. The forces' strength varies with the amount of charge (i.e., the number of ionized particles) and with the distance between charges: the farther the charge centers are separated from each other, the less the force.

### *Induction*

If you charge a comb as in the previous example, you can use it to attract small bits of paper. This situation differs from the last in one important way: the paper bits have no net charge. Why, then, are they attracted to the comb?

The explanation goes as follows: As you move the comb close to the paper, the comb's positive charge draws negative charge centers (electrons) to the paper's surface nearest the comb, while repelling positive charges. As a result, the part of the paper closest to the comb becomes negatively charged. Notice that as a whole, the paper remains electrically neutral: no charges have been added to it or removed. However, its charges are rearranged in response to the approaching positively charged comb. Charging caused in this way, by the approach of a charged object, is known as **induction.** Charging by induction plays an important role in cloud-to-ground lightning, as you will see presently.

### *Electric Discharge Currents*

We offer one last, small-scale example before we turn to thunderstorm lightning. Picture yourself walking across a deep pile rug on a winter day. As you go, you build a charge (typically negative) on your body, as shown in Figure 12.8A. Reaching for the door handle, your charged hand induces a positive charge on the handle. The closer your hand comes to the handle, the stronger the electric attraction. At some small distance, you experience an unpleasant "zap" as a spark flies between your hand and the handle. The spark is a momentary flow of charged particles between oppositely charged objects and is called an **electric discharge current.** The current brings opposite charges together, where they recombine (de-ionize), thus eliminating or greatly reducing the charge differences between your body and the surface of the door handle.

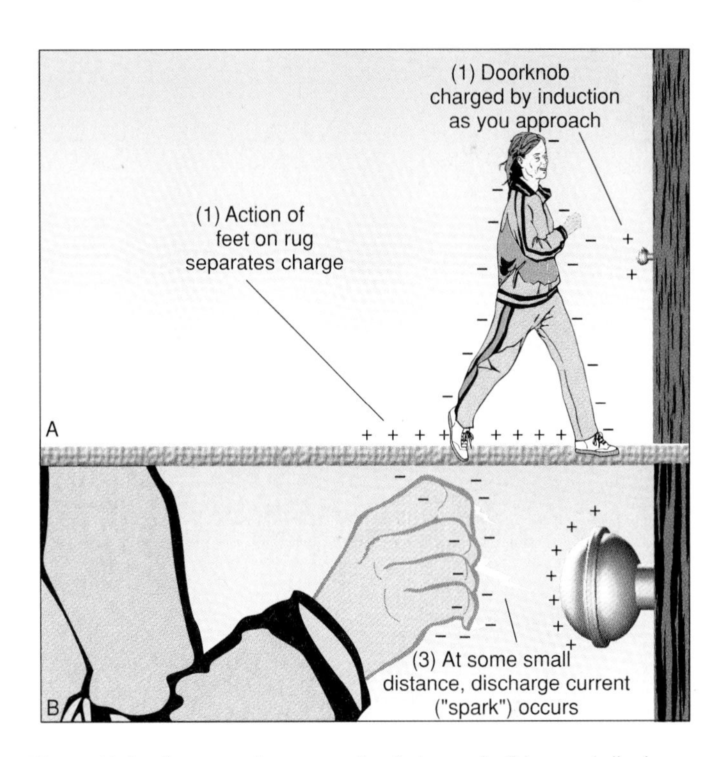


Figure 12.8 An everyday example of charge buildup and discharge.

## Properties of Thunderstorm Electricity

With this background, we are now prepared to apply these electric charge concepts to lightning. As in the human-scale examples above, formation of regions of opposite electric charge is the first step in creating lightning. Initially, the air, which is an excellent electric insulator, prevents electric currents from flowing between the charge centers. However, if the charges increase beyond certain values, or if the charge centers move close to one another (as they did in Figure 12.8), or if the atmosphere's conductivity increases (precipitation substantially improves conductivity), a brief electric discharge current flows through the air. Over short distances, such as from your finger to the doorknob or across the gap of a spark plug, the discharge is called a spark. From a cloud base to the ground, the discharge is called lightning.

There are, of course, vast differences of scale between a spark discharge from hand to doorknob and a lightning bolt. Lightning is an immensely energetic electric current driven by voltage differences of as much as 100 million volts, flowing through the atmosphere along a narrow path typically a few centimeters wide, over distances as great as 30 kilometers.

Atmospheric scientists have measured electric charge concentrations in and around a number of thunderstorms. Figure 12.9 shows a typical pattern of electric charge distribution. Notice that negative charge typically accumulates in the central and lower cloud levels, while upper regions are positive. The most intense negative charge is located near the −15°C level, which happens to be a region of active precipitation formation in many thunderclouds. Near the cloud base, the charge generally is negative; however, as you can see from the figure, positive charge often occurs in the vicinity of precipitation. Note also that the very lowest levels of the atmosphere and the ground itself generally are positively charged.

To understand how lightning forms, then, it is necessary to understand how electric charges develop in the patterns shown in Figure 12.9. We turn to this topic next.

## How Does Charge Separation Occur in a Thunderstorm?

Most atmospheric scientists believe the distribution of thunderstorm charge regions as depicted in Figure 12.9 to be basically correct. There is much less agreement, however, on the important question of how these charge centers form. Current hypotheses fall in two main categories.

### *Convective Hypotheses*

Figure 12.10 illustrates the main features of a typical **convection hypothesis.** Note the sequence of steps: (1) In fair weather, the

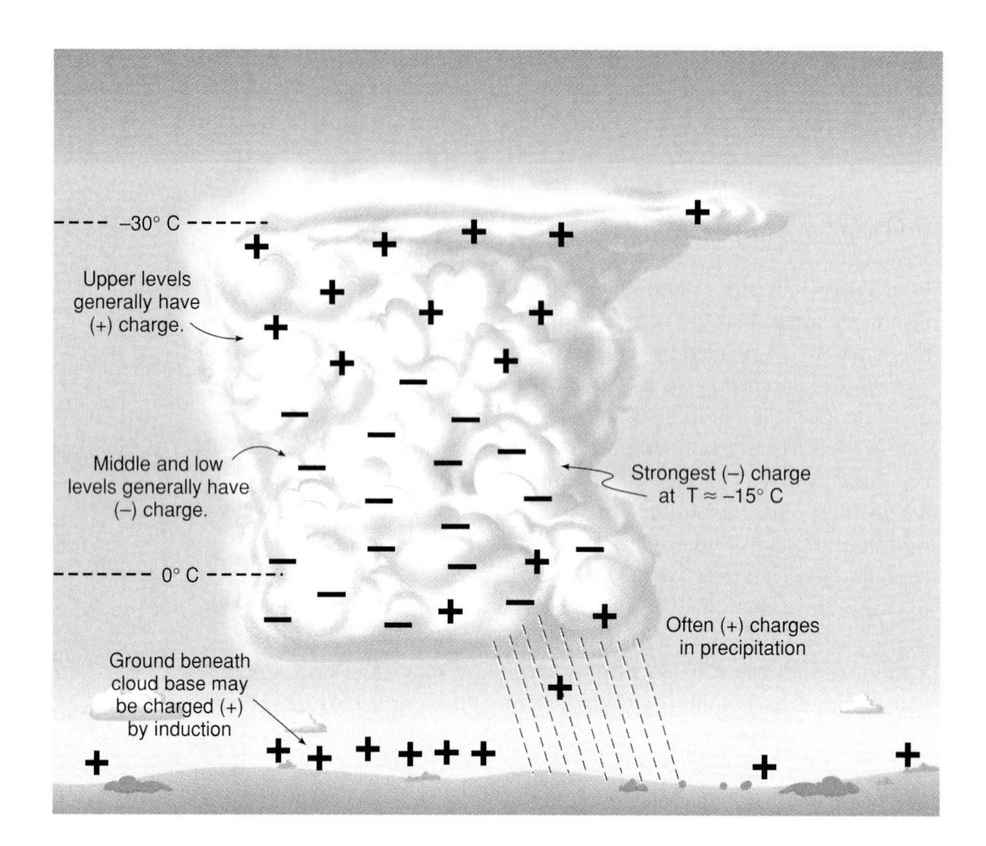


**Figure 12.9** Charge separation regions in and around thunderstorms.

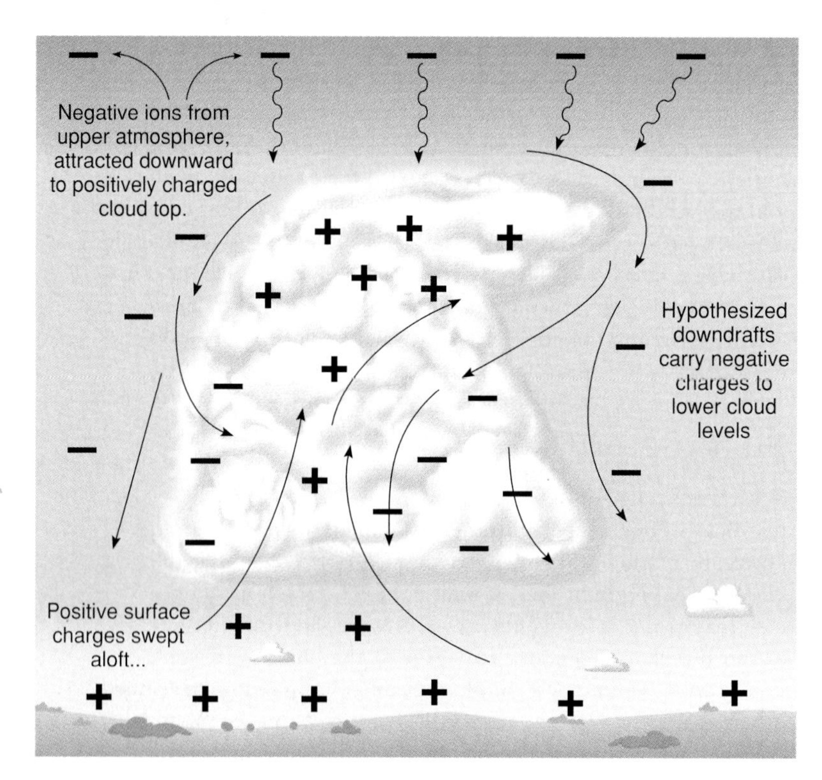


**Figure 12.10** The convection hypothesis for charge separation in thunderclouds. Convection currents move positive ions upward, negative ions downward.

lower atmosphere and the ground possess a slight surplus of positive charge. (This surplus is a well-established fact.) (2) Convective currents within a vigorous thundercloud sweep some of this low-level positive charge to upper cloud regions. (3) This positive charge center aloft attracts negative ions from the upper atmosphere to the cloud boundaries. (4) The downward moving negative charges encounter downdrafts on the periphery of the cloud, which carry them to middle and lower levels of the cloud.

How does the convection hypothesis fare in the face of observational evidence? The hypothesis explains the general configuration of positive charge aloft and negative charge at lower levels in thunderclouds. However, the convection hypothesis has

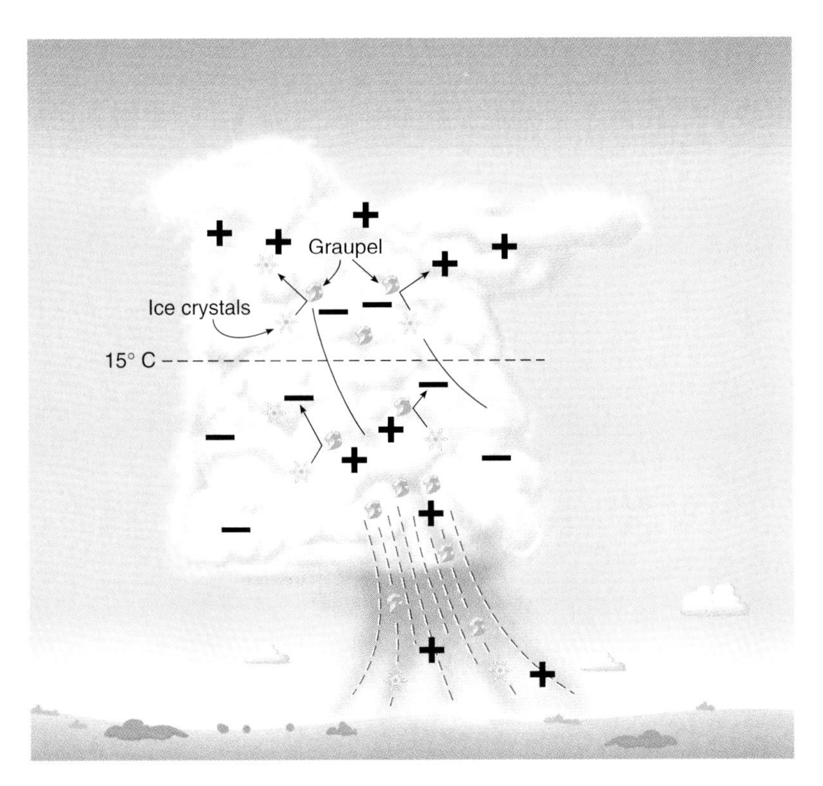


**Figure 12.11** A precipitation hypothesis for thunderstorm charge separation. When ice crystals and graupel collide at temperatures colder than about $-15°C$, the ice crystals acquire positive charge. Being lighter, they tend to remain at higher levels than the heavier, negatively charged graupel particles. Thus, the cloud acquires positive charge at upper levels and a negative charge below.

no explanation for the fact that negative ions are observed to accumulate at the same time and location that precipitation begins to form in the cloud.

### *Precipitation Hypotheses*

A typical **precipitation hypothesis** is outlined in Figure 12.11. In this example, the process is the following: (1) Graupel particles and ice crystals collide within the cloud. The collision causes electric charges to become separated. Typically, the ice crystals acquire a positive charge, while the graupel becomes equally but negatively charged. (2) Updrafts carry the lighter, positively charged particles higher in the cloud than the negatively charged graupel.

What is the evidence for and against the precipitation hypotheses? Many researchers have verified in laboratory experiments that charge separation like that drawn in Figure 12.11 does occur when precipitation particles collide. Generally, the smaller particles acquire positive charges at cold temperatures. Also, the fact that lightning often occurs at the onset of precipitation is evidence in support of a hypothesis that links precipitation and lightning.

On the other hand, the charge separation process remains mysterious at the present. Why does the graupel acquire a negative charge at low temperatures but a positive charge at temperatures closer to freezing? And why does the charge distribution seem to depend also on the amount of moisture in the air? Until these perplexing questions are answered, precipitation hypotheses will remain in doubt.

Currently, most scientists consider the precipitation hypotheses to be the more convincing, although it is possible that both mechanisms play some role, depending on conditions prevailing in a given storm. The uncertainties surrounding both precipitational and convection hypotheses highlight the difficulties involved in observing charge separation in an actual thunderstorm. This complex and important topic of research may occupy atmospheric scientists for years to come.

## The Lightning Stroke

Lightning is possible between any two charge regions of opposite sign in and around a thunderstorm. Figure 12.12 summarizes some of the possibilities. Most common are cloud-to-ground and in-cloud lightning. The "vertical upward" stroke has been seen only rarely, by aircraft-based observers.

Let's consider a typical cloud-to-ground lightning stroke in some detail. A buildup of negative charge near the cloud base induces a strong positive charge on the ground, particularly on tall objects such as trees and buildings. When the charge difference becomes sufficiently great, the attractive electrical force between the charge regions overcomes the air's electrical resistance. A flood of electrons rushes earthward in a series of segments known as a **stepped leader.** When the stepped leader approaches the ground, it is met by the **return stroke,** which carries the main flow of electricity, a positive current from the ground to the cloud, along the channel that was prepared by the leader. The return stroke typically consists of several surges of electricity, each just a few thousandths of a second in duration but altogether carrying a flood of electrical energy sufficient, if harnessed, to light a city of 100,000 people for several minutes. Figure 12.13 illustrates a stepped leader and its return stroke, based on high-speed photography.

### *Thunder*

Thunder is the sound made by a lightning bolt. Just how does lightning generate sound? Picture a lightning bolt ripping

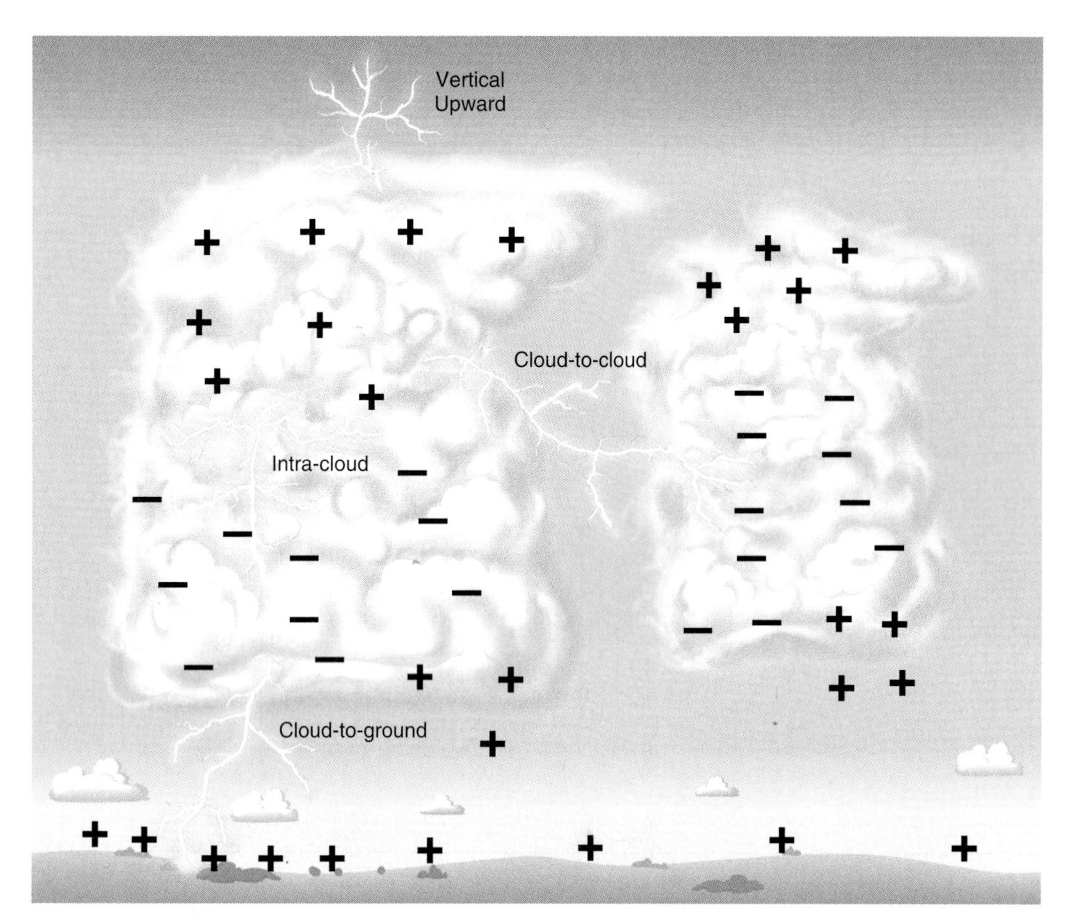


Figure 12.12 Lightning can join any two centers of opposite charge within the vicinity of the thunderstorm.

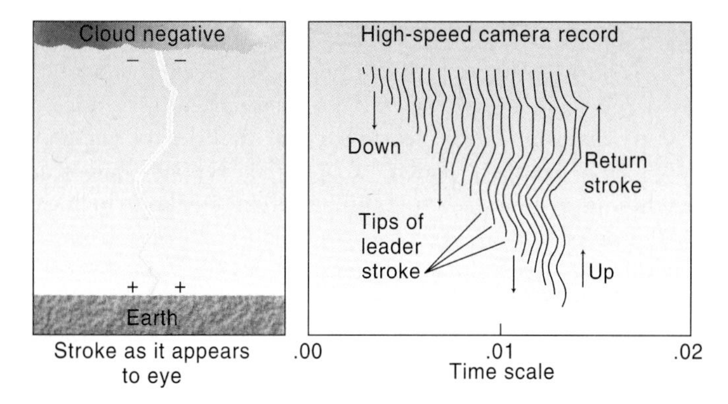


Figure 12.13 These drawings are traces of an actual lightning bolt, based on photographs made with a high-speed camera. Notice the stepped leader proceeding downward, followed by the return stroke (the darker line on the extreme right). Time scale in seconds.

through the air, heating it to as much as 30,000°C (50,000°F) in just a fraction of a second. This extraordinary heating causes a corresponding explosive increase in the air pressure within the channel. This high pressure air expands violently, "slamming" against unheated air just beyond the channel and causing higher pressure at the point of collision, lower pressure in its wake. Thus, thunder, like other sounds, is a "compression wave," a sequence of microscale shells of high and low pressure that propagate outward through the air at approximately 350 meters per second ("the speed of sound").

Observers close to lightning strikes have noticed that different parts of cloud-to-ground strokes generate different sounds. They report hearing a tearing sound as the stepped leader begins its descent, followed by a sharp smacking sound when the leader and the ground stroke make contact. The return stroke provides the full-fledged crash.

### *How Far Away Was That Lightning?*

The difference in travel time between the flash of lightning and the sound of the associated thunder clap provides a simple method of determining the distance to the lightning. Light from the lightning bolt travels outward at over 100,000 miles per second; thus, the flash from any lightning strike within 10 miles will reach you in less than one-ten thousand of a second—instantaneously, for all practical purposes. Thunder, on the other hand, travels at the speed of sound, which is approximately 350 meters per second or, very roughly, 1000 feet per second. Therefore, every second that elapses between when you see the lightning and when you hear the thunder means the lightning is another 1,000 feet distant. So, multiply the time in seconds by 1000, and you have a rough estimate of the distance in feet to the lightning bolt. If you don't have a watch, begin counting when you see the lightning: "A-thousand-one, a-thousand-two, a-thousand-three," and so on. The number you reach when you hear the thunder will be the approximate distance (in thousands of feet) to the bolt. As an example, if you begin counting the instant you see a lightning bolt and reach "a-thousand-six" before hearing the thunder, the thunder must have traveled for six seconds before reaching you. Therefore, the bolt must have struck approximately $6 \times 1000 = 6000$ feet away.

## Lightning Safety

At least 100,000 thunderstorms occur in the United States each year. Lightning from these storms kills approximately 100 people annually. If you are caught outdoors in a thunderstorm, find a low spot away from trees, electric wires, and fences. If you feel your skin tingle or your hair stands on end (see Figure 12.14), squat low to the ground but do not lie down flat; make yourself as small a target as possible. Cars (not convertibles!) generally afford good protection from lightning because they conduct most or all of the bolt harmlessly to the ground. Indoors, stay away from windows during thunderstorms. Do not use plumbing or telephones, as pipes and wires are good electrical conductors. Unplug TV sets and computers to protect them from electrical surges.

Lightning rods protect buildings and their occupants from lightning. A lightning rod is a short metal probe installed at the peak of a building and connected to a cable that is sunk deeply into the ground. In the event of a lightning strike, the electrical energy is carried harmlessly to the ground, sparing the building. Tall buildings such as the Empire State Building have been struck by lightning thousands of times and have suffered little or no adverse effect.

Figure 12.14 No laughing matter. Five minutes after this picture was taken and the young woman had left, the spot where she had been standing was struck by lightning. One person was killed, one permanently disabled, and six others injured.

## *Checkpoint*

***Review***
The concept map in Figure 12.15 summarizes the concepts from this section on lightning.

***Questions***

1. Before a cloud forms, the atmosphere has little or no net electric charge. Yet minutes later, large centers of charge are present. Where were the electric charges before the cloud forms?
2. Define charging by induction, then give an example.
3. Describe two hypotheses advanced to explain how electric charge centers are created in thunderstorms. What are their strong and weak points?
4. You observe a flash of lightning and measure the time until you hear the thunder to be 12 seconds. Approximately how far away was the lightning?

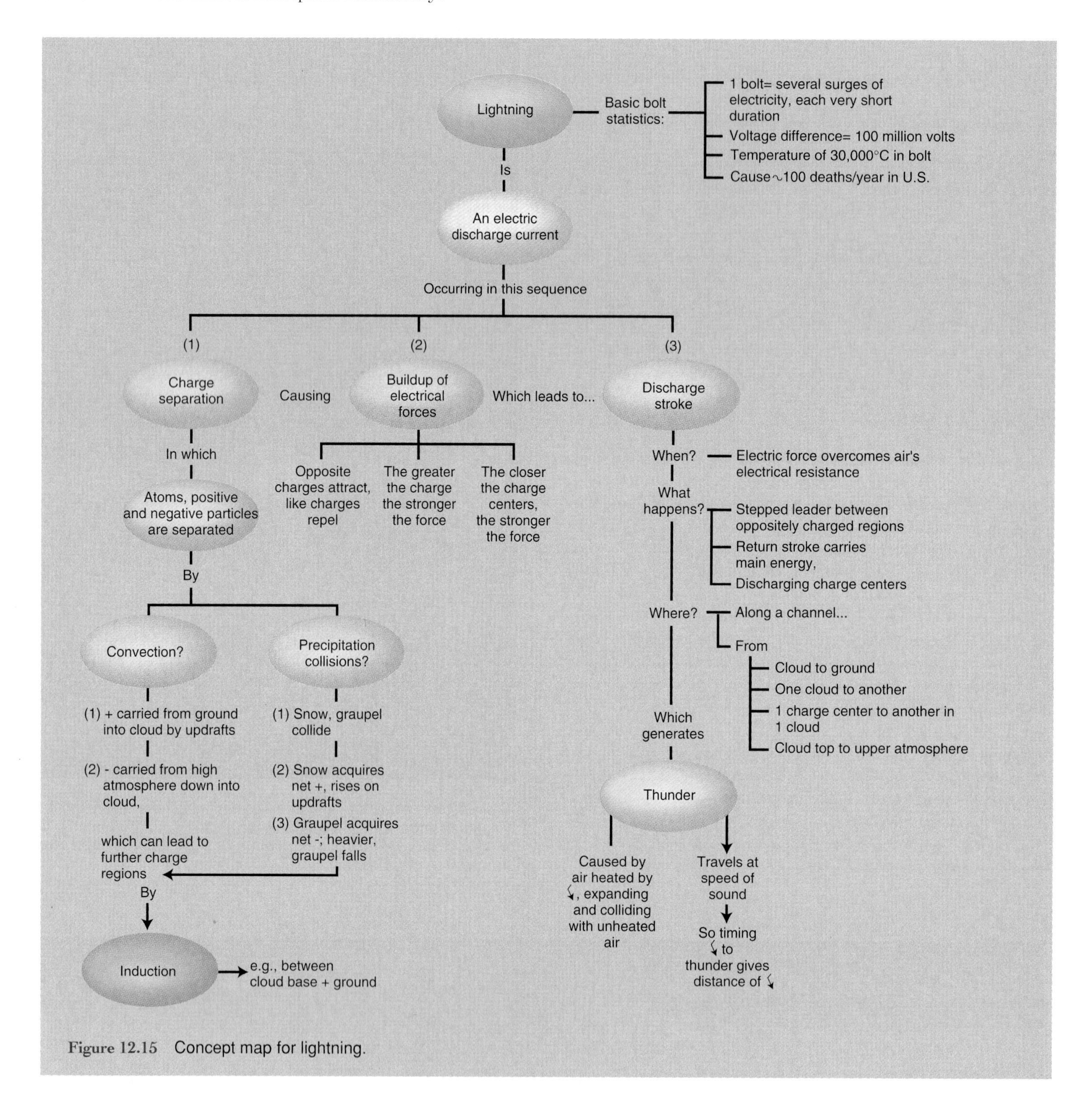


**Figure 12.15** Concept map for lightning.

## Multicell and Severe Thunderstorms

We've described something of a fiction in the first section of this chapter: a model, single cell thunderstorm that grows, flourishes and fades in isolation from its meteorological and geographical context. However, in the last chapter, you saw that flow patterns on all scales interact with each other in diverse and complex ways. Thunderstorms are no exception: a given thunderstorm may differ considerably from the model of Figure 12.1, depending on the interactions it experiences with features like fronts, wind shear, and even other thunderstorms. In the following two sections, you will see the importance of these influences and the effect they can have on thunderstorm size and intensity.

## Multicell Thunderstorms

Examine the thunderstorm pictured in Figure 12.16. Notice that there seems to be not one cell but several in various stages of development in this **multicell thunderstorm.** Many thunderstorms exhibit multicell structure. What causes this clustering?

Part of the explanation is a general one: if atmospheric conditions are suitable for one cell to form, they are likely to be right for other cells to form in the same region. Other factors include the influence the thunderstorm itself exerts on the surrounding air and effects of topography. Notice in Figure 12.16B that the gust front emanating from the downdraft in a mature cell can lift warm, moist air drawn toward the cell's updraft. This is just one way in which a mature cell may stimulate the development of

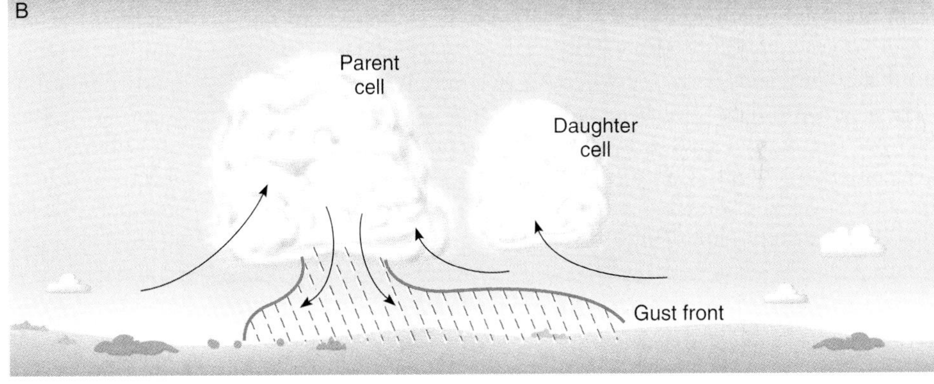


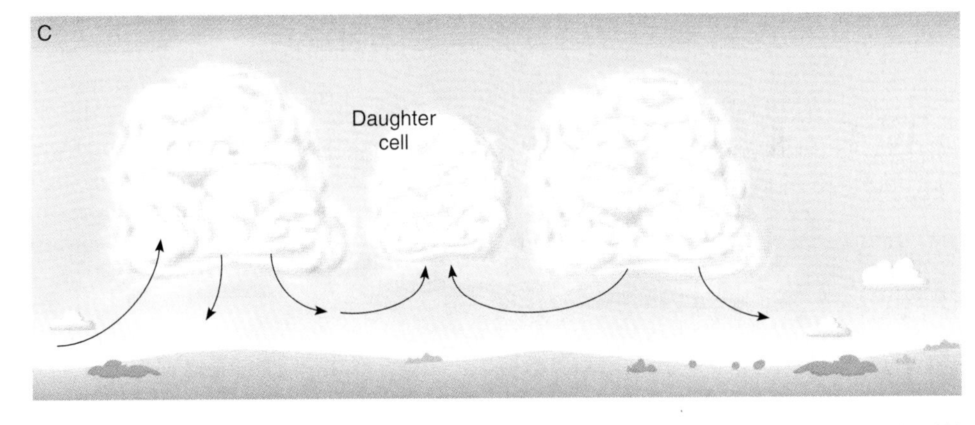

**Figure 12.16** Multicell thunderstorms. Try to identify the parent and daughter cells in (*A*).

"daughter cells." As the mature cell dissipates, a daughter cell may reach the mature stage, at which point it may generate offspring of its own. Repeated generation of daughter cells can extend the life span of the multicell thunderstorm far beyond that of a single cell, similar to successive generations of humans extending the existence of a family for centuries.

It is interesting to consider how thunderstorms' tendency to cluster affects their predictability. On the one hand, the large size and longevity of multicell storms gives a multicell group more continuity than an individual cell that grows from clear skies or a field of cumulus clouds. As an example, in the spring and summer of 1993, multicell thunderstorms formed regularly in the upper Mississippi Valley. Their persistence and slow rate of movement meant that a general forecast of "frequent showers and thundershowers throughout the day" often turned out to be accurate in this region. On the other hand, the complex interactions among different cells made specific predications of shower onset or end, wind speed and direction, and cloudiness much more difficult than on a day when an isolated afternoon thunderstorm is predicted. Because one cell exerts an immediate effect on its daughter cell(s), multicell storms present us with an accelerated version of Lorenz's butterfly effect. An error in predicting the onset of the first cell likely will generate errors in predictions of the daughter cells also.

## Larger Associations of Thunderstorms

At times, thunderstorms cluster in groups even larger than multicell storms. Look at the thunderstorm anvil clouds (the white oval masses) in the satellite image in Figure 12.17. These cells stretch across hundreds of kilometers. Notice that they are not randomly scattered but tend to lie in lines running from northeast to southwest. What would cause the cells to become aligned in this way? One possible mechanism is a cold front, which provides the requisite lifting. Thunderstorms that form along a front are known, logically enough, as **frontal thunderstorms.**

**Figure 12.17** A prefrontal squall line may lie hundreds of kilometers in advance of a cold front.

### *Frontal and Squall Line Thunderstorms*

Is frontal lifting responsible for the line of thunderstorms in Figure 12.17? Compare the positions of the front and the thunderstorm line in the figure. You can see the thunderstorms lie hundreds of kilometers downwind from the front! How could the front possibly cause these storms located so far in advance of its position?

Figure 12.18 shows two important ways in which this situation can occur: (1) thunderstorms may form at the front due to frontal lifting and then move forward faster than the front, generating daughter cells as they go; and (2) wave motions caused by air flowing over the front may induce a region of rising air downstream; if the air is unstable, this lift may be sufficient to initiate thunderstorm formation. In either case, the result is a **prefrontal squall line,** a line of thunderstorms generally caused directly or indirectly by the front and lying 100–500 kilometers ahead of it.

Other factors besides fronts may generate squall lines. They may form wherever lifting occurs along a line, such as along mountain ridges, seacoasts, and lines of wind shear.

In an environment of contrasting air masses, fronts, and substantial vertical shear of the horizontal wind, thunderstorm structure often differs significantly from the three-stage model outlined in the first section of this chapter. A typical frontal thunderstorm in its mature stage is shown in Figure 12.19. Compare this with the mature stage of the air mass thunderstorm in Figure 12.1B. Notice that large quantities of warm, dry environmental air are entrained into the frontal storm at low and middle levels. As it penetrates the cloud, this warm air is cooled and moistened, and plunges downward, strengthening the downdraft. The down rush can cause particularly strong winds at the gust front, as it brings mid-level environmental winds to the surface. The powerful downdraft also enhances the updraft, thus intensifying the storm.

Vertical shear of the horizontal wind is often more pronounced in frontal thunderstorms than in air mass storms. Wind shear can affect thunderstorm development in several ways. For example, an increase in wind speed with altitude can cause greater tilting of the main thunderstorm updraft. The greater the tilting, the more quickly precipitation falls from the updraft, thus sparing the updraft from the precipitation's retarding frictional effects. This leads to stronger, more persistent updrafts in storms with some wind shear. (However, very great wind shear can disrupt the updraft altogether, weakening the storm.) Wind shear is also responsible for the asymmetry in the thunderstorm's anvil top, elongating it in the downwind direction.

Wind shear in the form of changes in wind direction affects thunderstorm development as well. Shifts in wind direction with height can cause the circulation in and around the thunderstorm to be varied and complex. Further, winds aloft may transport air of different properties into the thunderstorm's environment. Advection of cold air aloft lowers the air's stability and promotes further storm growth. (Recall the discussion on this subject in Chapter 6.)

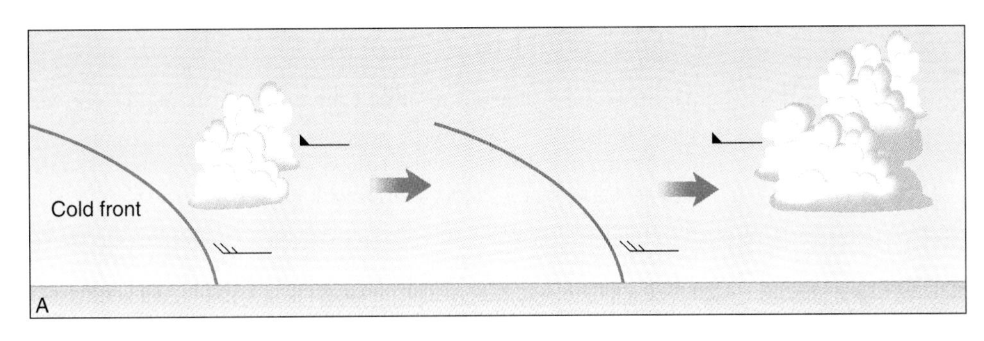


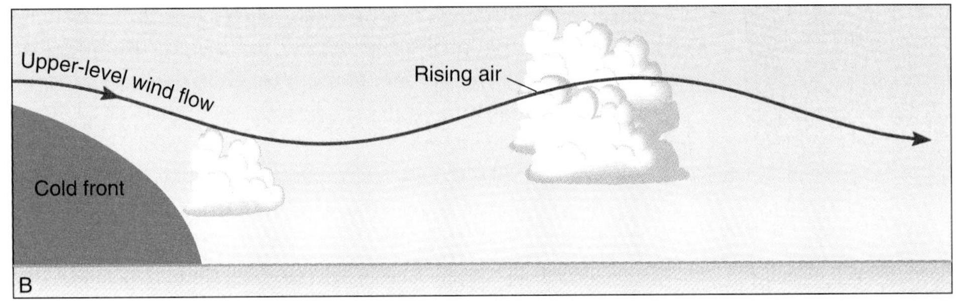


Figure 12.18 Two mechanisms of squall line formation. (*A*) Strong winds aloft move frontal thunderstorms in advance of cold front. (*B*) A wave, induced by upper level wind flow over the cold front, may stimulate thunderstorm growth downstream due to lifting.

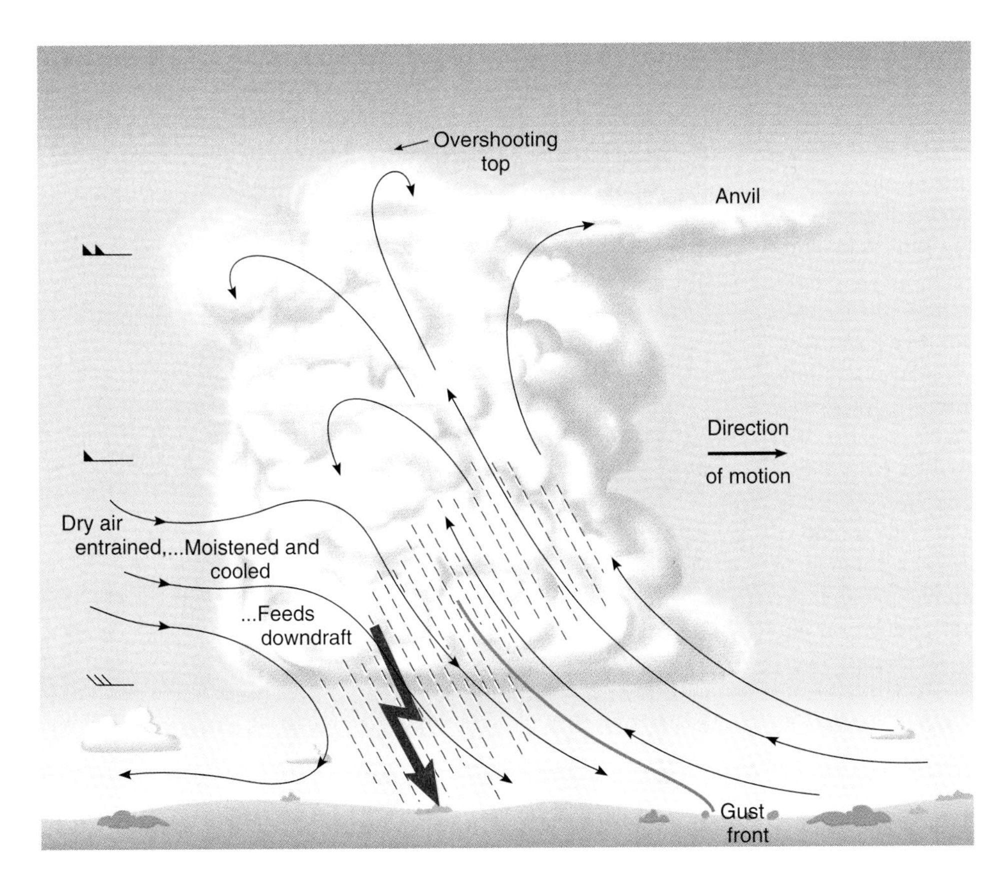


Figure 12.19 Structure of a mature frontal thunderstorm. How does this differ from the structure of an air mass thunderstorm? (See Figure 12.1*B*).

Rotation has additional effects on thunderstorm development. It is an important ingredient of the most severe thunderstorms, as you will see in a later section.

## *Mesoscale Convective Complexes*

The nearly circular cloud mass covering western Pennsylvania in Figure 12.20 is a region of concentrated thunderstorm activity known as a **mesoscale convective complex,** or **MCC.** Dozens of cells in this complex produced heavy rainfall for a period of several hours, leading to devastating, fatal floods in Johnstown.

Unlike other weather systems (such as synoptic scale cyclones, hurricanes, and individual thunderstorm cells) that have been observed for centuries, MCCs are relatively "new" objects: they were discovered only with the advent of satellite images like that in Figure 12.20. That their discovery is so recent is somewhat surprising, considering their importance: MCCs generate

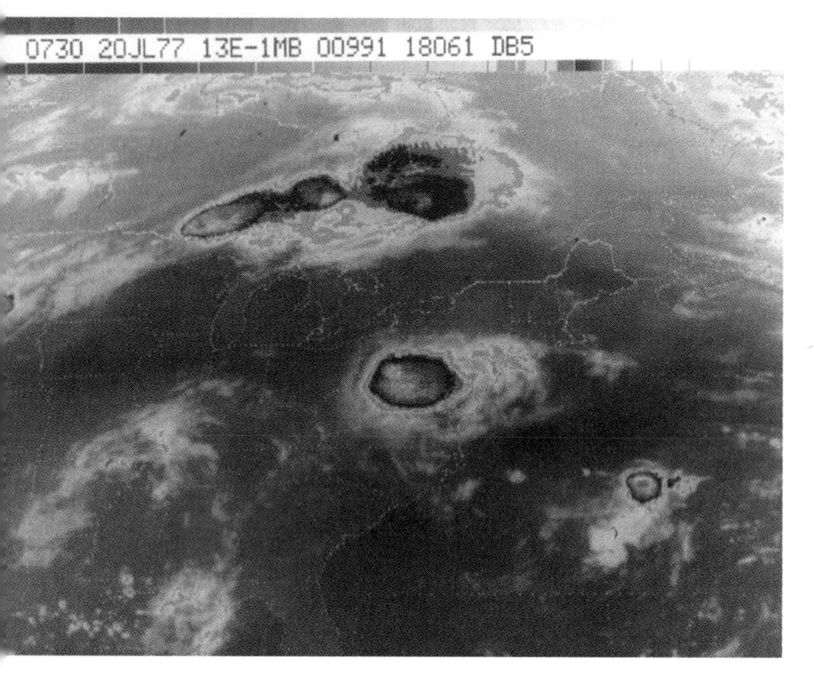


**Figure 12.20** Mesoscale convective complex over Pennsylvania. This storm brought devastating floods to Johnstown, causing $100 million in damages and taking at least 76 lives.

most of the precipitation during the growing season in many agricultural regions of the Midwest. Unfortunately, they also can produce a variety of severe weather, including hail, damaging winds, torrential rain, and tornadoes.

MCCs are even larger associations of thunderstorms than squall lines. In fact, many MCCs contain multicell thunderstorms and squall lines, perhaps even forming on or around them.

How does a MCC become organized, combining to create a structure on a larger scale? This is a topic of active research at present. (In other words, we don't know the answer yet!) However, researchers have collected considerable data about conditions surrounding MCCs. They have found that MCCs tend to form within regions of deep and extensive warm, humid air. Fronts and air mass contrasts, although sometimes present, are not necessary components of MCCs. The main circulation features are a slightly cyclonic convergence at low levels and divergence aloft. Heaviest precipitation falls from powerful convective cells, but light rain falls from stratiform clouds that provide a more or less continuous cover throughout the MCC. A typical MCC is on the order of 400 kilometers (250 miles) in diameter. They tend to be slow moving, to be most active at night, and to reform the following night.

The MCC is a prime example of how circulations at different scales of motion can interact. Typically, synoptic scale flow patterns establish a convergent, low-level flow of warm, moist air and divergence aloft, conditions conducive for MCC formation. The most active elements in the MCC are individual thunderstorms, which are small-scale mesoscale events. These individual thunderstorm cells combine to produce a much larger circulation feature. The result is the MCC, a circulation smaller than synoptic scale but considerably larger than thunderstorm scale.

## Supercell Thunderstorms

Thunderstorms vary not only in the number of constituent cells; they also vary considerably in intensity. The most intense thunderstorms are called **severe thunderstorms.** A severe thunderstorm is one that is capable of producing heavy downpours, flash flooding, strong and gusty winds, frequent lightning, hail, and/or tornadoes. What kinds of thunderstorms generate severe weather? Actually, any thunderstorm is capable of reaching the severe level. However, multicellular storms produce severe weather more often than individual, single cell thunderstorms. By far the most prolific producer of severe thunderstorm weather, however, is the monster storm known as the supercell.

Examine the thunderstorm pictured in Figure 12.21. If you compare it with images of other thunderstorms, like those sketched in Figure 12.1 for example, you can see that this is one extraordinarily large thunderstorm. The life span of this particular storm matched its prodigious size, as the cell persisted for several hours. Such blockbuster storms are known as **supercells.** A supercell thunderstorm is one of exceptional size and intensity,

**Figure 12.21** This supercell thunderstorm raged over Cheyenne, Wyoming, for 3 hours in 1985. Flood waters piled up hail to a depth of 8 feet.

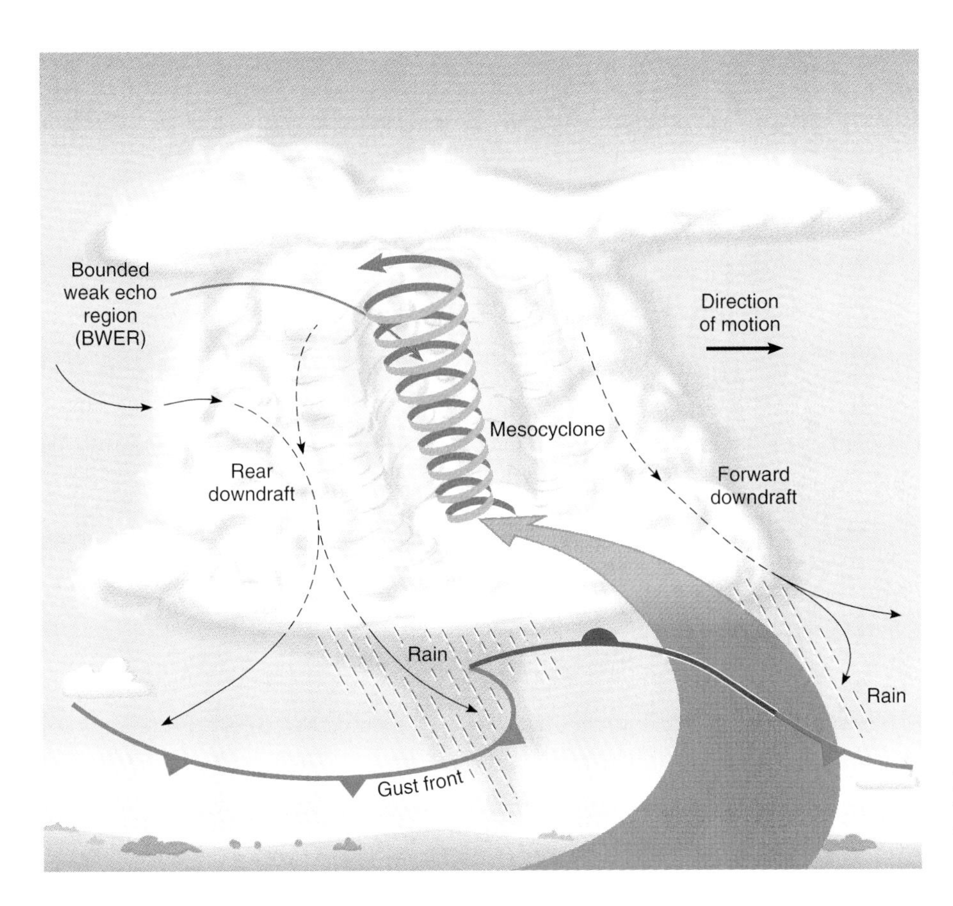


**Figure 12.22** Wind flow in a typical supercell thunderstorm. How is the circulation similar to that in a normal size, frontal thunderstorm? (Compare with Figure 12.19.) How does it differ?

that persists in the mature stage for considerably longer than most thunderstorms.

The structure of a typical supercell thunderstorm is shown in Figure 12.22. Notice its resemblance to a synoptic scale low pressure center, but on a much smaller scale. The supercell's gust front, coming from the downdraft in the rear of the storm, resembles the cold front in a frontal cyclone, and the inrushing warm, moist air is like the circulation's warm sector. Many supercell storms contain a second downdraft, in the forward-most part of the storm, which carries precipitation out of the cloud.

## *Bounded Weak Echo Regions*

Updraft and downdraft velocities in a mature supercell are as awesome as other aspects of the storm. In the most intense supercells, updraft speeds can reach 50 meters per second (over 100 mph). It would seem that precipitation particles would grow to enormous size in these updrafts, as water vapor rapidly cools and condenses in its upward rush. Surprisingly, radar images often record weaker precipitation echoes in the cores of these strongest updrafts than in surrounding regions within the cloud. These weak echo zones are so distinctive they have become known as **bounded weak echo regions (BWER)**. Figure 12.23 shows an example.

What is the explanation for a BWER? Atmospheric scientists believe that moisture particles pass through the strongest supercell updrafts so rapidly there is not time for them to grow to be strong radar targets. Thus, the presence of a BWER on a radar screen is an important signal to the meteorologist, an ominous indication of severe updrafts and therefore a powerful storm.

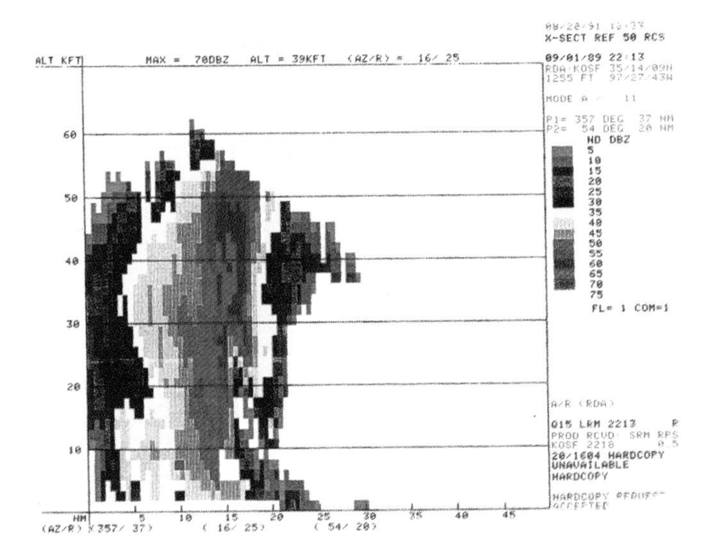


**Figure 12.23** A bounded weak echo region. This NEXRAD Doppler radar image represents a vertical slice through a thunderstorm. The BWER is the region in green and yellow, surrounded by stronger (orange and red) echoes, up to approximately 28,000 feet at a distance of 13–14 nautical rules.

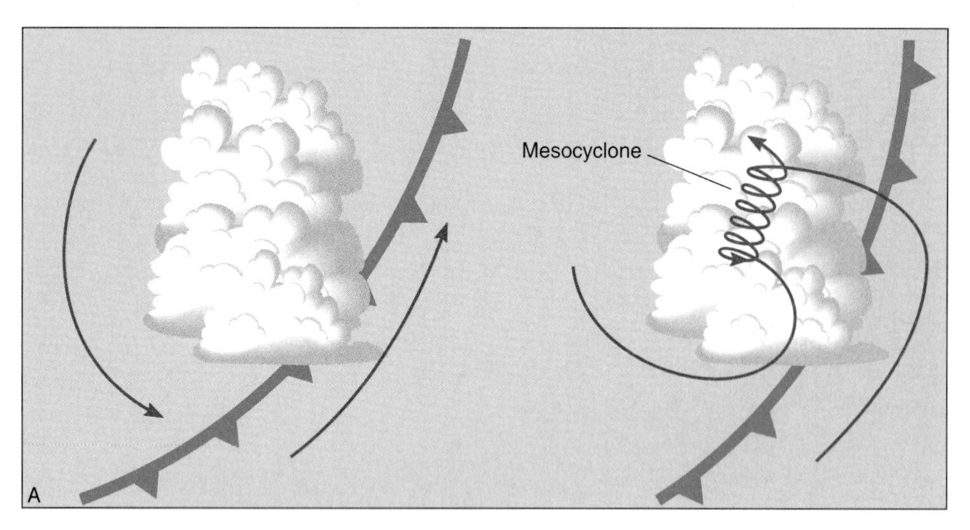


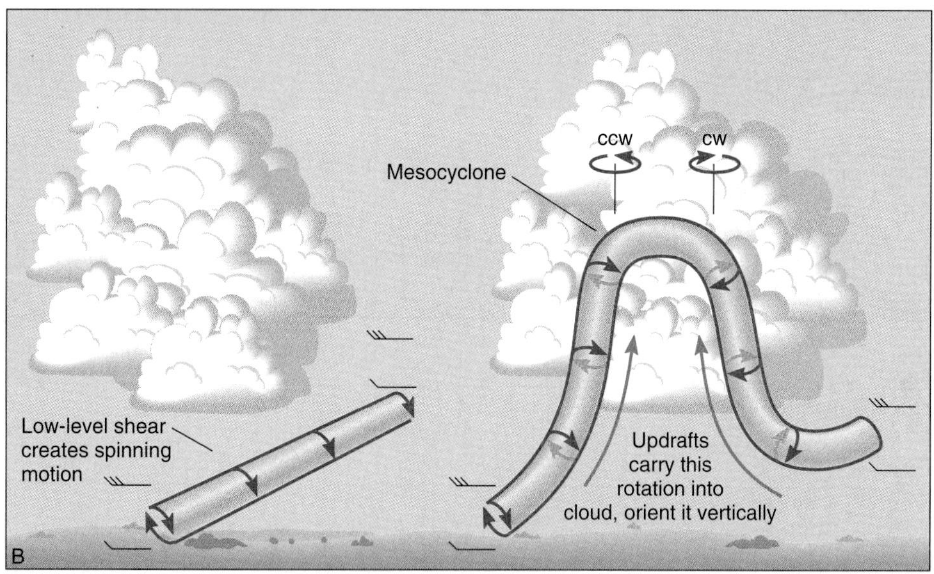


**Figure 12.24** Sources of rotation in thunderstorms. Notice in (*B*) how low-level wind shear can be swept up into the thundercloud, contributing vortex motion to the mesocyclone.

### *Mesocyclones*

Although rotation certainly is present in other thunderstorms, particularly frontal thunderstorms, it appears to play a special role in supercells. Inflowing warm, moist surface air spirals cyclonically as it is drawn toward the updraft, where a small-scale, counterclockwise turning, low pressure center called a **mesocyclone** forms.

The mesocyclone's rotation helps to maintain the supercell's great intensity. Recall from Chapter 9 that vortex motion tends to be conserved; vortexes persist for long periods of time. (If you induce a rotation in a hot drink by stirring it in a circle, the motion will persist much longer than if you just slosh the spoon back and forth.) The persistent, organized vortex motion centered on the mesocyclone helps to maintain a flow of energy (warm, moist air) into the updraft for a longer period of time than in a cell in which rotation is less prominent.

Where does the supercell acquire its rotation? Figure 12.24 suggests two possibilities. In case (A), the thunderstorm has formed in a region of preexisting rotation, such as near a front or the center of a frontal cyclone. In the second case (B), wind shear provides the vortex motion. Notice that the wind in the lowest kilometer changes significantly with height. (Fronts often are a source of such shear.) With help from the arrows in the diagram, you can visualize this shear as a rotation. Thus, the near-surface air is rotating—but along a horizontal axis, not a vertical one. However, as this air is drawn into the updraft, the axis of rotation is swept upward as well, thus changing its orientation from horizontal to vertical.

You can see in Figure 12.24 that this process generates regions of both clockwise and counterclockwise rotation in the supercell thunderstorm. Generally (but not always), the main updraft cell grows in the region of counterclockwise rotation, that is, in the mesocyclone. With rotation acting to stabilize the circulation and to feed the updraft with a source of warm, humid air, the supercell can grow to extraordinary size and intensity. Strong surface winds, frequent lightning, softball-sized hail, torrential rain, and flash flooding (see Figure 12.25) make an encounter with supercell weather an unforgettable experience. Supercells are also a major source of tornadoes, which are the subject of the next section.

Figure 12.25 Ten seconds after the car (top) crossed the culvert, a flash flood turned the road into a potential deathtrap.

## Checkpoint

### *Review*

The concept map on the next page (Figure 12.26) outlines the main differences among nonair mass thunderstorms. A word of caution, however: the different thunderstorm categories are not so distinct as the concept map may suggest. Various combinations and mixing of properties occur. For example, some supercell thunderstorms are multicellular, and some thunderstorms in all categories exhibit rotation, though generally it is not as intense as in supercells.

### *Questions*

1. Distinguish between air mass, multicell, frontal, and supercell thunderstorms.
2. Give two examples of thunderstorm cells combining to create larger structures.
3. What is a BWER? What does it indicate?
4. Explain how vertical shear of the low-level horizontal wind can contribute to rotation in a thunderstorm.

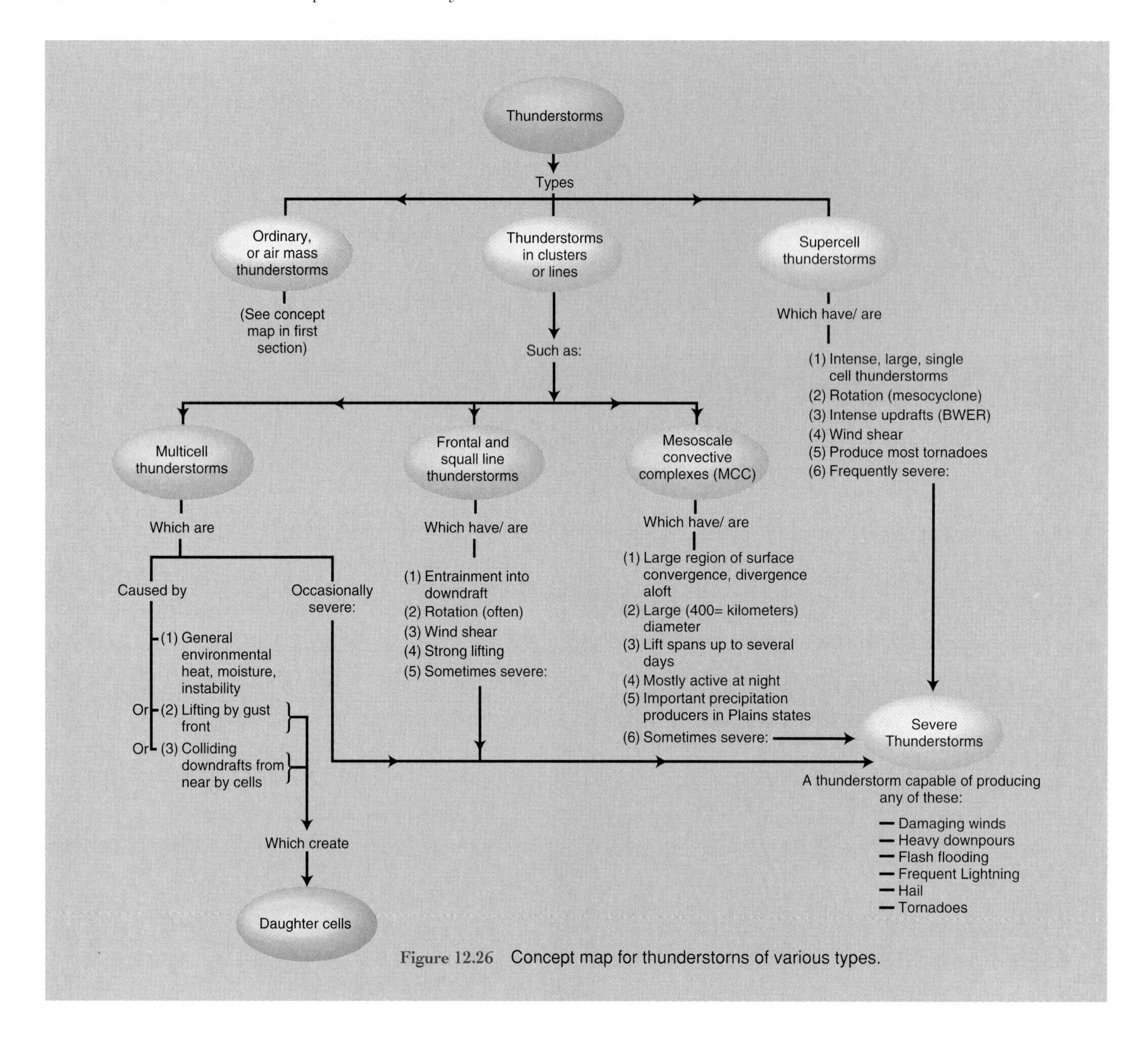


**Figure 12.26** Concept map for thunderstorms of various types.

# Tornadoes

Anyone who thinks scientific research is a dull affair should consider tornado research. Few spectacles in nature can equal the elemental fury of a powerful tornado like the one shown in Figure 12.27. Try to picture conditions within that small but incredibly intense circulation: winds strong enough to hurl trailer trucks hundreds of meters through the air; the air so thick with dust and larger pieces of airborne debris that visibility is nil, except within the "eye" at the circulation's center; surface air pressure changes so great that buildings are blown apart as the storm passes overhead; and a sound that has been likened to "the roar of a thousand freight trains."

## Observing Tornadoes

Suppose you were an atmospheric scientist interested in studying tornadoes. How would you proceed? The first step is the hardest: "Get one tornado. . . ."

This gag vividly brings to mind various problems in tornado research: tornadoes are small, short-lived phenomena; they are relatively rare; and they are incredibly dangerous to confront. Only rarely does a tornado pass over or close to a weather-observing station, and when it does, it generally destroys the observing instruments. So, how do you study tornadoes?

**Figure 12.27** A violent tornado roars through Hesston, Kansas, in March 1990.

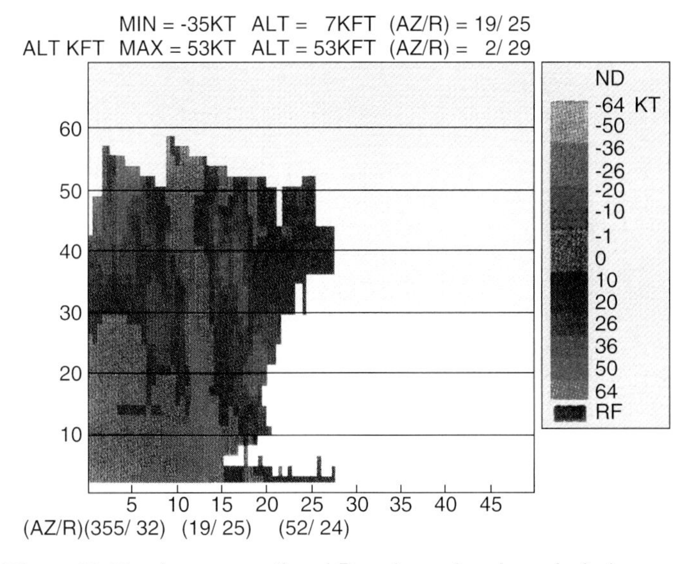


**Figure 12.28** A cross-sectional Doppler radar view of winds within a severe thunderstorm. Green indicates winds directed out of the page, toward the reader. Red depicts winds directed into the page, away from the reader. Note the close proximity of the green and red areas, indicating sharp contrasts in wind flow.

## *Field Measurements*

Despite the risks, a number of atmospheric research teams do take to the field with the intent of measuring tornado circulations directly. Closely monitoring the weather, especially radar images, on a day of expected severe weather, the team locates itself where it believes tornado development is most likely. When tornado formation occurs, they position themselves as close to the storm as is feasible and proceed to observe, photograph, and measure conditions in the storm's vicinity. The researchers may place specially designed instrument packages in the tornado's expected path in hopes of obtaining a continuous record of conditions as the tornado passes. They also may operate a portable

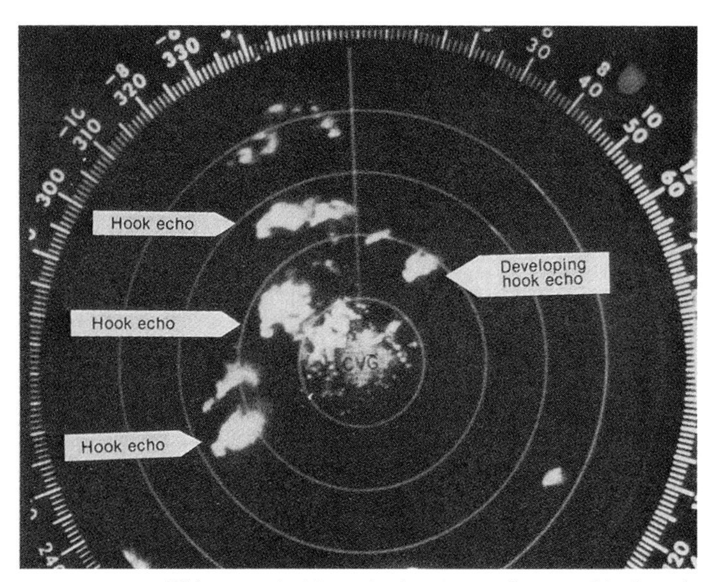


**Figure 12.29** This remarkable radar image made near Cincinnati, Ohio, on April 3, 1974, shows four hook-shaped echoes associated with severe thunderstorms in various stages of development. All four of these thunderstorms produced tornadoes on Ohio's worst tornado day ever.

Doppler radar system. Most tornado research, however, is conducted at safer distances.

## *Weather Radar Observations*

The recently installed NEXRAD Doppler radar network is one of the most important components of tornado research at present. Doppler radar images like that in Figure 12.28 present the patterns of wind flow within thunderstorms, including tornado-bearing ones.

The presence of a BWER is evidence of powerful updrafts, as you read in the last section, and strong updrafts are thought to be necessary for tornado formation. Even more diagnostic is the "**hook echo,**" shown in Figure 12.29, a small, highly reflecting area shaped like the numeral 6. Meteorologists have long recognized that hook echoes often are associated with tornado formation.

## *Poststorm Damage Assessments*

Another important source of data about tornadoes comes from damage analyses. Look at Figure 12.30. What can you tell about the wind flow from the figure? Theodore Fujita of the University of Chicago has specialized in studying tornadoes through records of storm damage. His research and that of others have led to a better understanding of tornado wind strength, which is summarized in the **Fujita scale** of tornado intensity, in Table 12.1.

As you can see from the table, weak tornadoes are far more common than more severe types. However, because of their severity and often greater size, the rarer F4 and F5 tornadoes cause by far the most damage and personal injury.

Fujita also recognized from storm damage patterns that many tornadoes are not single vortexes but have an internal structure as shown in Figure 12.31. Notice that within the tornado are smaller-scale circulations called **suction vortexes,**

A F1

B F2

C F3

D F4

which rotate within the main vortex. These suction vortexes contain the strongest winds and most severe pressure changes, and therefore cause the greatest damage. An early eyewitness account of possible suction vortexes comes from a man who looked directly up into a tornado as it passed overhead:

"Everything was as still as death. There was a strong gassy odor, and I could hardly breathe. There was a screaming, hissing sound coming directly from the end of the funnel. I looked up and to my astonishment, I saw right into the heart of the tornado. There was a circular opening in the center of the funnel, about fifty or a hundred feet in diameter and extending straight upwards for at least half a mile, as best I could judge. The walls of this opening were rotating clouds and the hole was brilliantly lighted with the constant flashes of lightning which flashed from side to side. . . .

Around the rim of the great vortex, small tornadoes were constantly forming and breaking away. These looked like tails as they writhed their way around the funnel. It was these that made the hissing sound. I noticed the rotation of the great whirl was anticlockwise, but some of the small twisters rotated clockwise. . . ."[1]

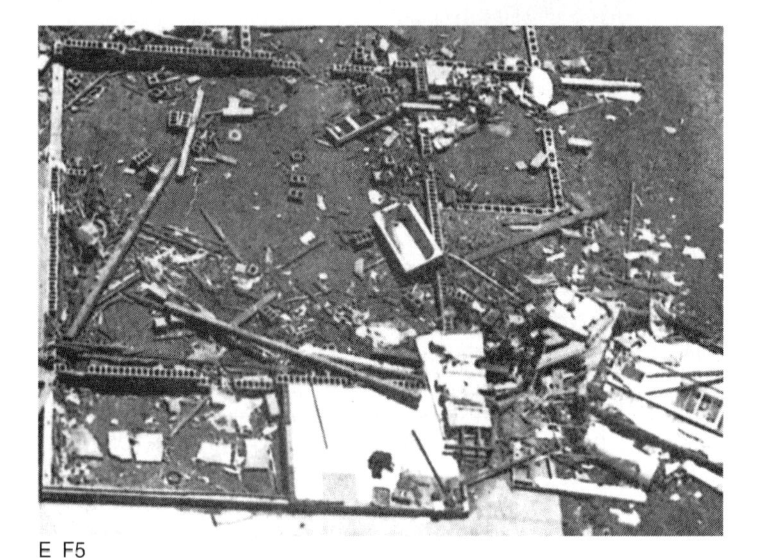
E F5

**Figure 12.30** Damage wrought by different classes of tornadoes on the Fujita scale.

[1]Monthly Weather Review, May 1930.

Table 12.1 Fujita Scale for Damaging Wind

| Class | % of All Tornadoes | Fujita Category | Wind Speed | Damage |
|---|---|---|---|---|
| Weak | 69% | F0 | 35–2 knots<br>40–72 mph | Light; small trees uprooted |
| | | F1 | 63–97 knots<br>73–112 mph | Moderate; trailer homes damaged |
| Strong | 29% | F2 | 98–136 knots<br>113–157 mph | Considerable; roofs torn off houses |
| | | F3 | 137–179 knots<br>158–206 mph | Severe; cars lifted off the ground |
| Violent | 2% | F4 | 180–226 knots<br>207–260 mph | Devastating; houses destroyed |
| | | F5 | 227–276 knots<br>261–318 mph | Incredible; steel structures destroyed |

Figure 12.31 Suction vortexes within a tornado cloud. These smaller-scale swirls are thought to cause much of the damage in severe tornadoes.

## How Do Tornadoes Form?

Tornadoes are offsprings of thunderstorms. The majority of tornadoes come from intense thunderstorms, such as supercells, but they are produced in other storms as well, especially in squall line and cold frontal thunderstorms. Even air mass thunderstorms occasionally generate tornadoes, although such events are relatively uncommon.

Given the variety of parents a tornado might have, it should come as no surprise that no one model of tornado development and structure fits all cases. Here, we will concentrate on the source of the most numerous and powerful tornadoes, the supercell. Figure 12.32 presents one possible sequence of events.

Notice that the overall circulation in all three stages is similar to that in the supercell model shown back in Figure 12.22. In stage (A), the mesocyclone is wider, shorter, and rotating more slowly than in later stages. In the transition to stage (B), the mesocyclone is stretched vertically, and narrowed horizontally, with the result that rotation around its core becomes much faster as it conserves its angular momentum. The cause of this stretching and accelerating is not known; perhaps it results from the advection of vorticity into the storm by one of the processes outlined in Figure 12.24. In any case, a very dark, slowly rotating cloud called a **wall cloud** develops and extends below the cloud base below the mesocyclone (see Figure 12.33). The wall cloud normally forms toward the rear of a typical, northeastward- moving thunderstorm, in the southwest quadrant of the storm.

The rotating mesocyclone and wall cloud are regions of intense low pressure. Air below the cloud base rushes into this cyclone, and, conserving angular momentum, it accelerates to great speeds. As its pressure decreases to match that of its surroundings, the air expands and cools. The cooling may lead to condensation, making the air visible as a **funnel cloud.** A funnel cloud is a vortex of spinning air extending downward from a thunderstorm (typically from a wall cloud) but not reaching the

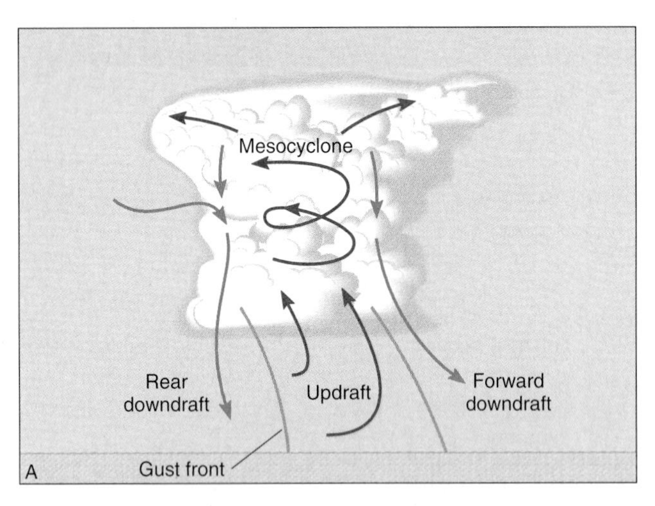


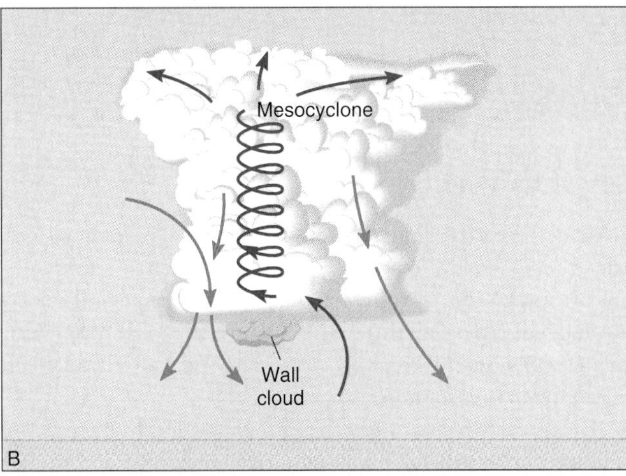


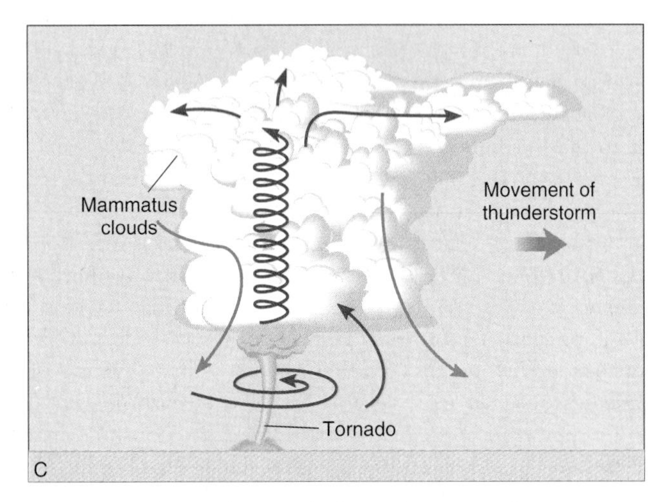


**Figure 12.32** Stages in tornado formation, according to one model. Note that the tornado builds downward, towards the surface, from the mesocyclone in the thunderstorm.

ground. If the funnel cloud makes contact with the earth's surface, it is classified as a **tornado.** The great majority of tornadoes in the Northern Hemisphere rotate counterclockwise, although clockwise-rotating examples have been confirmed.

**Figure 12.33** (*A*) A wall cloud protrudes from the base of a supercell thunderstorm. (*B*) A funnel cloud descends from a wall cloud over Eldorado, Kansas.

What large-scale conditions would favor tornado formation? The synoptic situation shown in Figure 12.34 is a particularly favorable one as well as a rather common one. As you see, the frontal cyclone system provides a general environment of convergent, cyclonic flow. It also stimulates the flow of warm, moist air from the Gulf of Mexico, providing the low-level heat and humidity necessary for thunderstorm development. At higher levels, however, the wind is from the west or northwest and has properties of a drier and generally cooler air mass. Thus, we have warm, moisture-laden, low-level air lying beneath cold, dry air aloft, a highly unstable configuration. A temperature profile typical of this situation is shown in the insert in Figure 12.34. Notice that if the surface air were to be lifted, it would quickly reach condensation and then would cool at the relatively slow moist adiabatic rate as it continued upward through the cold, dry air aloft. This air, then, has the potential to be highly unstable, given surface heating or some other lifting mechanism. Lifted indices in the range of −6 or

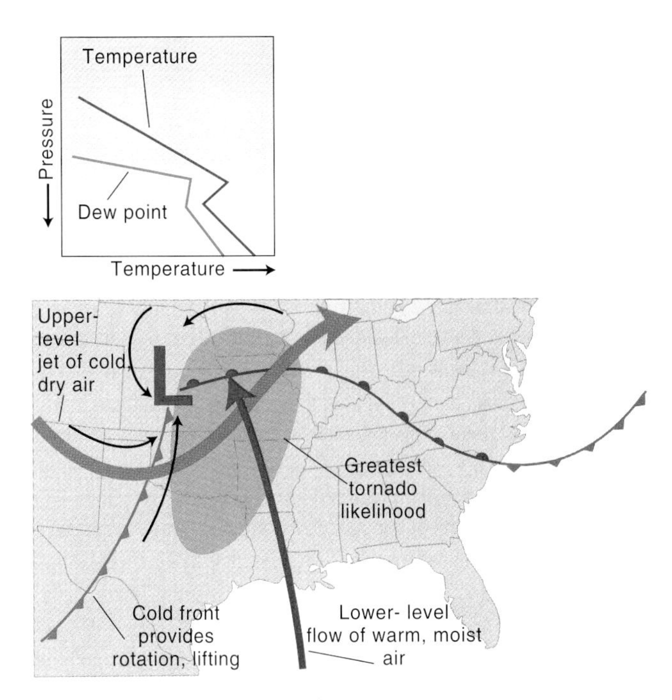


Figure 12.34 Ideal conditions for tornado formation. The green area marks the most likely region. Why does cold, dry air aloft promote tornado formation?

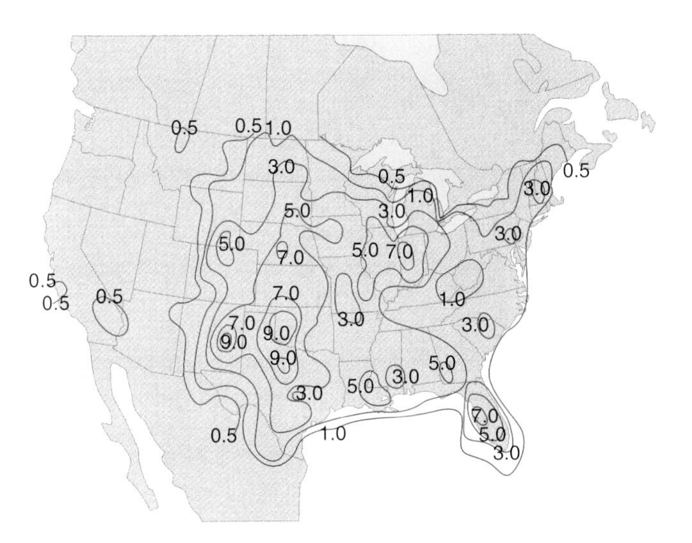


Figure 12.35 Average annual tornado incidence per 10,000 square miles, 1953–1980. Ten thousand square miles is equal to a square 100 miles on a side. Thus, the figures on the chart give the average number of tornadoes per year in a 100 × 100 mile area. Source : Data from Anonymous, *Tornado Safety*, NOAA Pamphlet, NOAA/PA 82001. January 1982, U.S. Government Printing office, 1989-244-143, page 8. U.S. Department of Commerce, National Weather Service.

greater occur in situations, like this one, that are most likely to spawn tornadoes.

The inversion in the sounding, marking the transition from warm, moist surface air to drier air aloft, plays an important role. The inversion limits the growth of smaller convective cells; as a result, low-level heat energy is vented upward only through the strongest cells that are able to penetrate the capping inversion. Once past the inversion, however, these strong cells blossom great cauliflower heads in the colder and drier air aloft.

The final player in Figure 12.34's drama is the cold front. As it sweeps eastward, the front provides an extra lift to the warm, unstable surface air; furthermore, squall lines may develop in the warm sector ahead of the front, as discussed earlier.

Figure 12.35 depicts tornado frequencies throughout the United States. Note the consistency of this chart with the discussion in the previous paragraphs, with the greatest frequency of tornadoes occurring in the south-central and southeastern states. In fact, the region's unique geography makes it the most tornado-prone territory in the world.

The model described in this section describes reasonably well the development of many tornadoes, particularly the most intense ones. However, many questions remain. For example, why don't all supercell thunderstorms generate tornadoes? And why do tornadoes develop from nonsupercell thunderstorms, even (occasionally) from growing cumulus clouds with no apparent downdraft structure? Clearly, there is much yet to learn about tornado formation.

## Tornado Forecasting and Safety

There's an unfortunate irony in the fact that tornadoes, which are nature's most destructive storms, are also in many ways the most difficult to predict. Forming on short notice, lasting typically for just minutes, and growing to diameters of only 100 meters or so on the average, individual tornadoes exist on a scale that is simply too small to be observed or predicted using the methods employed on larger, less violent circulations. Given these difficulties, it is remarkable that tornado forecasting is as successful as it is. Furthermore, the prospects for improvement are good, as you will see in this section.

### *Tornado Forecasting*

How do you predict where and when a tornado will appear? Think of tornado prediction as a two-step process. The first step is to identify large-scale (synoptic scale) conditions favorable for tornado development, such as those you considered in Figure 12.34. When conditions seem favorable, forecasters at the National Severe Storms Forecast Center in Kansas City, Missouri, issue a **severe storm watch** or **tornado watch** for the region threatened. A watch means that severe weather has not been reported, but conditions are such that there is a significant chance of its occurrence. Typically, severe weather watches are issued several hours before severe weather is likely to occur.

Once the severe weather watch has been issued, meteorologists pay special attention to the watch area. If visual or radar data indicate that severe weather is occurring or is imminent, meteorologists at the local National Weather Service office take the second step and issue a **severe weather warning** or **tornado warning.** A warning refers to much more specific

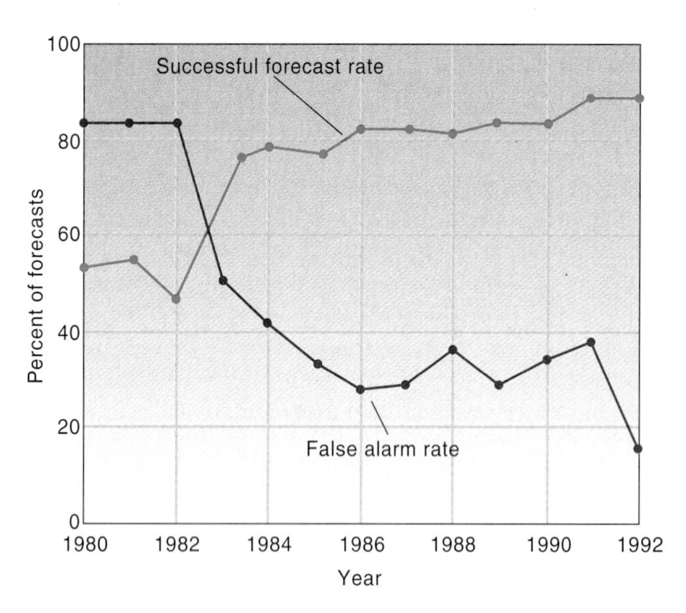


**Figure 12.36** The successful forecast rate (green) and false alarm rate (red) for tornado and severe thunderstorm warnings issued by the National Weather Service office in Norman, Oklahoma. Source: Data from Polger, et al., *Bulletin of the American Meteorological Society.* February 1994

events, generally within narrower limits of time and space, than a watch.

Why this two-step process? Why not skip the watch and issue the warning, say, six hours ahead of time? The answer is that no one can tell hours ahead exactly where severe weather or a tornado will develop. Six hours is many times the life span of the typical tornado. Over that period, far too many atmospheric changes and interactions occur on the scale of a tornado for such a forecast to be accurate. It would be a bit like trying to forecast the positions and intensities of synoptic scale lows and highs six weeks in advance. Therefore, severe weather and tornado forecasting proceeds by successive approximation, from the watch category to the warning.

We conclude this discussion by returning to the Kingston, Oklahoma, tornado cited at the beginning of this chapter. The forecast was an extraordinarily accurate one. Can we claim it as an example of improved forecasting technique, or was it just a lucky guess? How often do the forecasters fail to predict a tornado occurrence?

Figure 12.36 shows the "successful forecast rate" at Norman, Oklahoma, where the Kingston forecast was issued. Note the steady improvement over the years, to the point where at present the success rate is near 90%.

You might wonder how often these forecasters issue false alarms, warning people of severe weather that never materializes. After all, someone who is constantly predicting that severe weather is coming won't miss any severe weather events—although, like the boy who cried wolf, his false alarm rate will be near 100%, and no one will take his forecasts seriously. In Norman, where many meteorologists are highly experienced forecasters and experts with the NEXRAD Doppler radar system, the false alarm rate for severe weather warnings is only 20% (refer again to Figure 12.36). Thus, only one time in five is a tornado warning issued when no tornado actually occurs.

In summary, the Kingston tornado forecast seems to have been no lucky guess. Thanks to many years of studying conditions attending tornado formation, capped by the recent installation of the NEXRAD network, tornado forecasting is showing solid improvement.

The Kingston forecast illustrates another important aspect of tornado forecasting: although the 20-minute lead time between the warning and the tornado's actual occurrence may sound uncomfortably short, for many practical purposes it is quite enough. Twenty minutes is long enough for most people to take basic, potentially life-saving precautions, as long as they hear the warning promptly. The perfect tornado forecast of the future may be a 20-minute warning, coupled with a communication system that instantly informs everyone in the warning area.

### *Tornado Safety*

Just what precautions do you take if you see a tornado or hear a tornado warning? The National Weather Service offers the advice presented in Table 12.2.

Above all, take a tornado threat very, very seriously. Tornadoes kill hundreds of people every year. Respect the tornado's power; take cover without delay.

*Table 12.2* **Tornado Safety**

If a warning is issued or threatening weather approaches:
In a home or building, move to a predesignated shelter, such as a basement.
If an underground shelter is not available, move to an interior room or hallway on the lowest floor and get under a sturdy piece of furniture.
Stay away from windows.
Get out of automobiles.
Do not try to outrun a tornado in your car; instead, leave your car immediately.
If caught outside or in a vehicle, lie flat in a nearby ditch or depression.
Mobile homes, even if tied down, offer little protection and should be abandoned.

# Checkpoint

***Review***

The concept map in Figure 12.37 summarizes much of the material presented in this section.

***Questions***

1. What is a "hook echo"? Why is its shape so significant?
2. Investigators inspecting a tornado's damage observe mobile homes destroyed and roofs removed from houses (but the houses are otherwise intact) but no evident movement of cars or damage to steel-reinforced buildings. What category on the Fujita scale was the tornado? What were its likely maximum winds?
3. Explain the role of the mesocyclone in tornado formation.
4. You are driving a car when you sight a tornado. What safety precautions should you take?

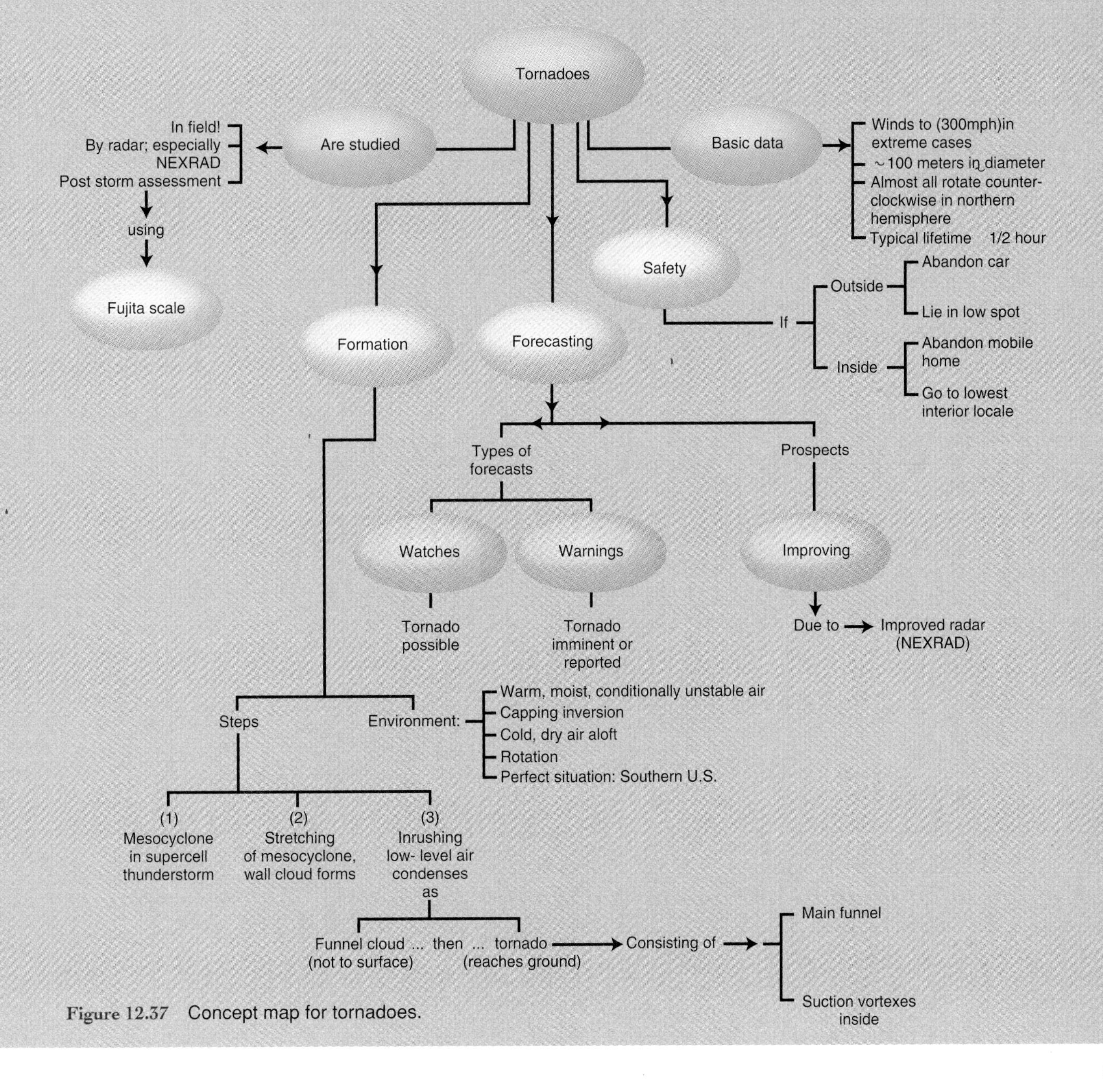


**Figure 12.37** Concept map for tornadoes.

Study Point

## Chapter Summary

An air mass, or ordinary, thunderstorm evolves through three stages over a period of about an hour. Updrafts and warm in-cloud temperatures of the cumulus stage give way to up- and downdrafts, precipitation, thunder, and lightning in the mature stage. The dissipating stage is characterized by downdrafts and light precipitation.

Regions of intense electric charge develop in a mature thunderstorm through processes not fully understood. Hypotheses include convection currents that carry charge from the ground and the upper atmosphere into different regions of the cloud, and collisions of precipitation particles within the cloud. Lightning is a sudden flow of electricity between two charge centers, discharging them. Thunder is the sound made by the violently heated, expanding air along the lightning's channel.

Multicelled, frontal or squall line, and supercell thunderstorms all show significant differences from the basic, air mass thunderstorm model. Interactions among neighboring cells, coupled with the influence of large-scale convergence and divergence patterns, may stimulate dense regions of thunderstorm development called mesoscale convective complexes. Severe thunderstorms tend to form where shear and/or rotation, dry air aloft, and frontal or orographic lifting supplement the basic requirements of all thunderstorms—warm, moist, and unstable air.

Tornadoes form most frequently from supercell storms, in which a rotating air column called a mesocyclone exists. Under some circumstances, the mesocyclone is stretched vertically upward and downward, intensifying as it does so, and extending its influence below the cloud base, where the tornado forms. Tornado winds may reach 300 mph in the most violent cases, causing total destruction of objects in their path. The southern United States is the most tornado-prone region in the world.

Thunderstorm and tornado forecasting is showing substantial improvement, thanks in large measure to the newly installed NEXRAD Doppler radar network. However, lead times for specific thunderstorm and tornado warnings are likely to remain on the order of minutes to an hour or two, like the life spans of these storms.

## Key Words

ordinary, or air mass, thunderstorm
gust front
thunderstorm stages: cumulus, mature, dissipating
electric discharge current
precipitation hypothesis
return stroke
frontal thunderstorm
mesoscale convective complex (MCC)
supercell thunderstorm
mesocyclone
Fujita scale
wall cloud
tornado
severe weather warning
convection hypothesis
stepped leader
multicell thunderstorm
prefrontal squall line
severe thounderstorm
bounded weak echo region (BWER)
hook echo
suction vortex
funnel cloud
severe storm or tornado watch

## Review Exercises

1. How does the mature stage of an air mass thunderstorm differ from the cumulus stage? From the dissipating stage?
2. Why are thunderstorms more frequent in southern states than in northern? Why are they more common along the eastern seaboard than in the western coastal states?
3. Name several agents that can provide the lifting that enhances thunderstorm formation.
4. Explain how an atom that has no net charge can be altered to create centers of both positive and negative charge.
5. Give an example of charging by induction in a thunderstorm.
6. During a thunderstorm, you see lightning strike a hilltop you know to be 2 miles away. How long will it be before you hear the thunder from that bolt?
7. What is a daughter cell? How is it formed? Give two examples.
8. How does the wind flow in a mature frontal thunderstorm differ from that in an air mass thunderstorm?
9. Describe two ways in which a prefrontal squall line may form in advance of an advancing cold front.
10. Where and when do mesoscale convective complexes form most commonly? What benefits do they bring? What problems can they cause?
11. Draw the numeral 6 on a piece of paper; as you do so, consider the trajectory of your pencil point. Now explain why a weather radar echo in the shape of a 6 is considered a tornado echo.
12. What are the wind speeds in an F2 tornado? What sorts of damage occur?
13. What is the difference between a funnel cloud and a tornado?
14. Distinguish between a tornado watch and a tornado warning.

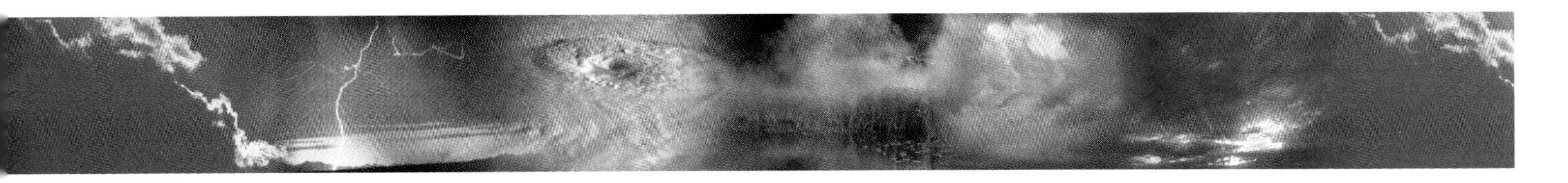

## Problems

1. Suppose a hailstone with a terminal velocity of 15 meters per second is embedded in an updraft of 15 meters per second. What will be its rate of fall with respect to the ground?
2. Suppose eddies within the thunderstorm cell carry the hailstone in the previous example into a 20- meters per second downdraft. What will be its rate of fall with respect to the ground? How long would it take the hailstone to reach the ground from an altitude of 3 kilometers (roughly 10,000 feet)?
3. Describe the weather conditions (wind direction and speed, cloud cover, precipitation, lightning, thunder) you might observe as the thunderstorm in Figure 12.2B passes over you from left to right.
4. Two opposite charge centers in a thunderstorm are separated by a distance of 1 kilometer. (a) Is the electric force between the centers an attractive or a repelling one? (b) Which of the following would increase the electric force: (1) the charge centers move closer; (2) they move farther apart; (3) the charge on one center increases; (4) the charge on one center decreases.
5. Explain how negative charge accumulates in the middle and lower regions of a thundercloud, according to (a) the convection hypothesis and (b) the precipitation hypothesis.
6. You are out on a golf course, far from the clubhouse, when a thunderstorm with frequent lightning overtakes you. What measures should you take to avoid becoming a lightning statistic?
7. How are multicell thunderstorms similar to MCCs? How are they different?
8. A supercell thunderstorm interacts with its surroundings in ways an air mass thunderstorm does not. Describe some of these ways.
9. Why is rotation an important component of severe thunderstorms, particularly supercells?
10. Precipitation particles form and grow in thunderstorm updrafts. Why, then, do radar images show little or no precipitation particles in the strongest thunderstorm updrafts?
11. Compare the maps of thunderstorm frequency (Figure 12.5) and tornado frequency (Figure 12.35). In what ways are they similar? How do they differ, and why?
12. Explain how the geography of the southern United States and the structural elements of a typical frontal cyclone can combine to create ideal conditions for severe thunderstorm and tornado formation.

## Explorations

1. Construct a "tornado in a box," following Figure 1 (below). Observe the structure and behavior of the vortex as the water warms. In what ways is this a valid model of an actual tornado? In what ways is it not?
2. Bernard Vonnegut suggested that atmospheric physicists studying charge separation in thunderstorms are paying too much attention to precipitation hypotheses and too little to alternative explanations. Read his argument in the January 1994 issue of the *Bulletin of the American Meteorological Society* (see Resource Links below) and write a paper on your findings.
3. If you have access to real-time weather data, make a "multimedia" study of a group of thunderstorms (see Resource Links below for data sources). Collect satellite and Doppler radar images; radiosonde soundings; surface, winds aloft, and lightning stroke charts; and surface observations. If the thunderstorms are in your vicinity, include photographs and observations of your own. How complete a picture of the thunderstorms and their environment are you able to assemble?

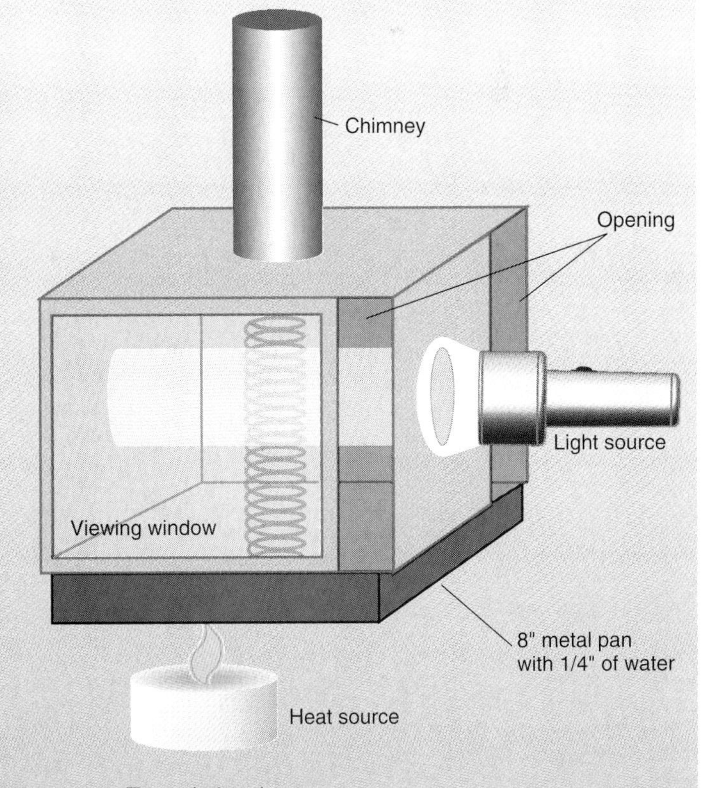


**Figure 1** Tornado in a box.

## Resource Links

The following is a sampler of print and electronic references for some of the topics discussed in this chapter. We have included a few words of description for items whose relevance is not clear from their titles. More challenging readings are marked with a C at the end of the listing. For a more extensive and continually updated list, see our World Wide Web home page at www.mhhe.com/sciencemath/geology/danielson.

Bentley, Mace. 1996. A midsummer's nightmare. *Weatherwise* 49 (4):13–19. A discussion of severe nocturnal thunderstorms.

Cotton, W. R. 1990. *Storms*. Fort Collins, CO: Aster Press, 158 pp. Excellent presentations of thunderstorm and tornado development.

Davies-Jones, R. 1995. Tornadoes. *Scientific American* 273 (August): 48–57. A review of tornado anatomy and formation.

Livingston, E. S., J. W. Nielson-Gammon, and R. E. Orville. 1996. A climatology, synoptic assessment, and thermodynamic evaluation for cloud-to-ground lightning in Georgia: A study for the 1996 Summer Olympics. *American Meteorological Society Bulletin* 77 (7): 1483–95. C

Vonnegut, B. 1994. The atmospheric electricity paradigm. *American Meteorological Society Bulletin* 75 (1): 53–61. C

AccuWeather's home page at http://www.AccuWeather.com is an extensive source of current weather conditions, satellite images, and NEXRAD Doppler radar images.

For readers with access to AccuData, NEXRAD provides an extensive array of Doppler radar images and data including reflectivity, rainfall accumulations, cloud tops, and radial wind velocity. LIGHT and LGTA plot the locations of lightning strokes seconds after they occur. DAMG and DAMSUM report damage as a result of severe weather. BULSUM and CANADV summarize all current severe weather watches and warnings in the U.S. and Canada respectively. CVC1A shows forecast thunderstorms, plotted by risk and severity. CVCV is a detailed discussion of the potential for severe thunderstorms.

The University of Oklahoma's CAPS home page at http://www.uoknor.edu/tornado/CAPS.WWW/tornado.html offers a wide variety of data and images on thunderstorms and tornadoes for people of all ages.

The NOAA National Severe Storms Laboratory in Norman, Oklahoma, is at http://www.nssl.uoknor.edu/ .

The National Severe Storms Forecast Center in Kansas City, Missouri, at http://www.awc-kc.noaa.gov .

# Chapter 13

## Hurricanes

## Chapter Goals and Overview

Hurricane! To those who have experienced its fury, the mere mention of the word arouses strong emotions and vivid, sometimes painful memories. Hurricanes are among the most destructive storms on earth. Hurricanes' violent weather, their long life spans and trajectories, and their propensity for sudden changes in intensity and movement combine to make hurricane forecasting a difficult and important problem.

In this chapter, you will study one particular Atlantic hurricane from its genesis on an August day to its dissipation two weeks later. You will observe the storm's formation, development, and eventual decay. You will probe its structure and witness its destructive might. And every day, for as long as the storm exists, you will grapple with crucial forecast questions: Where will it go next, and how fast? How severe will weather conditions be within the storm?

### Chapter Goals

By the end of this chapter, you will be able to:

- name the conditions required for a hurricane to form;
- locate on a map the regions of most frequent hurricane formation and indicate typical hurricane paths;
- sketch a cross section of a hurricane, indicating structural components (eye, eyewall, etc.) and general patterns of wind flow, air pressure, temperature, clouds, and precipitation;
- identify energy sources that power a hurricane and explain how the flow of energy influences its growth and decay;
- describe weather and sea conditions attending a hurricane's landfall and the kinds of resulting damage; and
- identify the various data and techniques used in hurricane forecasting, how accurate the forecasts are, and the prospects for improving their accuracy.

## Hurricanes and Tropical Weather

To study the entire life of a hurricane, you need to be there at the beginning. This section provides data on where, when, and under what conditions hurricanes form.

### What Is a Hurricane?

We begin by defining a few important terms. A **hurricane** is a severe tropical cyclone on the order of 500 kilometers (300 miles) in diameter, with winds of 65 knots (74 mph) or higher, which forms over warm ocean waters in maritime tropical air. The name is derived from a Caribbean Indian word meaning "evil wind."

As you can see from Figure 13.1, hurricanes are by no means unique to the Caribbean. Every tropical ocean except the South Atlantic experiences them, although in different locations they are called by different names, such as **typhoons** (in the western Pacific) and cyclones (in the Indian Ocean). Nonetheless, the term *hurricane* is used as a general one; when we speak of hurricane structure or hurricane development, we are referring to hurricane-like storms anywhere—not just those of the Atlantic or Eastern Pacific.

Another useful term is **tropical cyclone.** Tropical cyclone is the most inclusive of hurricane terms; it refers to any member of the hurricane family, whatever its location or strength. Thus, hurricanes, typhoons, and hurricane-like storms with winds less than 64 knots are all examples of tropical cyclones.

### When and Where Do Hurricanes Form?

Figure 13.1 shows the major regions of tropical cyclone formation. Notice that they develop only over the oceans and only between the "Tropics," that is, between the Tropic of Cancer and the Tropic of Capricorn. Note also, however, that the South Atlantic and eastern South Pacific oceans are relatively immune to tropical storm formation, as is a narrow band centered on the equator. Because we are interested in studying hurricanes that have the greatest potential to affect the United States, we will focus on the North Atlantic formation region, between the northeast coast of South America and west Africa. Storms in this region tend to travel westward or northwestward, as indicated in Figure 13.1.

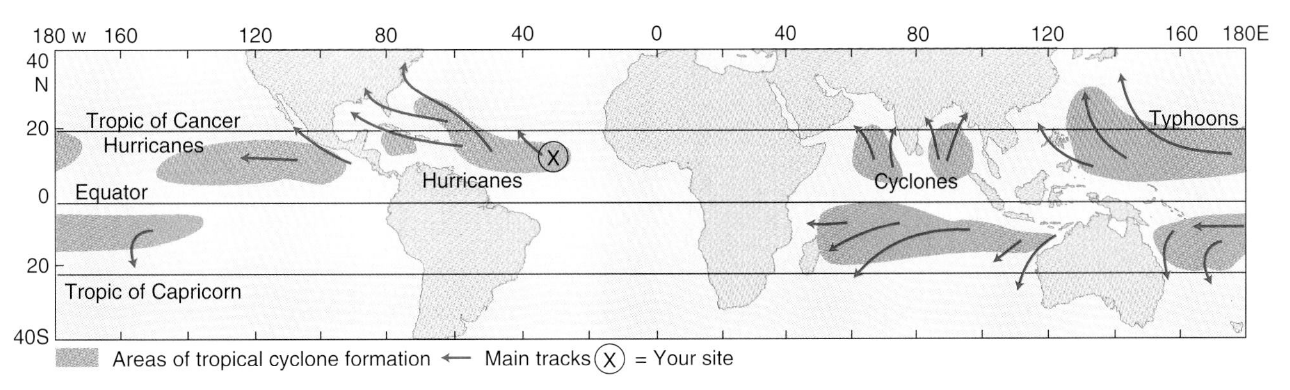


**Figure 13.1** Tropical cyclone formation regions, typical paths, and names. What continents are almost free of the threat of tropical cyclones?

Atlantic hurricanes have a distinct season, which is evident in Figure 13.2. Notice the pronounced tropical cyclone maximum in August and September, and the nearly total absence of these storms from December through April.

Based on the data of Figure 13.2, beginning in August, a hunter of Atlantic hurricanes should focus attention on the hatched regions of Figure 13.1. Imagine, then, that you have joined a research team that hopes to witness the formation of a hurricane and follow its development and movement throughout its lifetime. Your team stations itself in the eastern North Atlantic Ocean in early August and hopes for the worst possible weather.

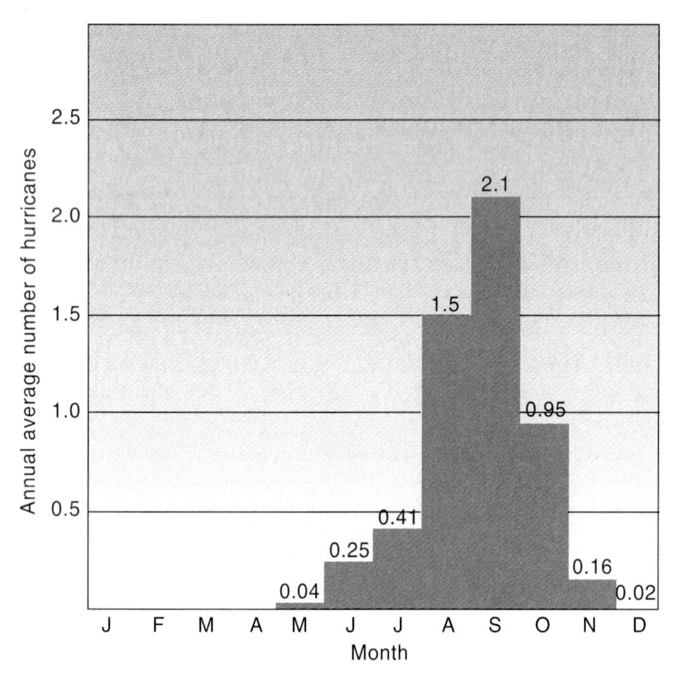


Figure 13.2 The frequency distribution of Atlantic tropical cyclones by month of the year.

## Tropical Weather

What are normal weather conditions at your site? A nontechnical description would be "delightful." Day after day, the sun shines, interrupted only by occasional cumulus buildups and passing showers. The showers bring a slight temporary drop in temperature but no change of air mass; you are deep within a dome of maritime tropical air. The air and sea surface temperatures hover around 27°C (81°F). The high relative humidity (75%) makes the air a bit sticky, but a steady trade wind breeze from the east keeps you comfortable. Each day is much like the last; absent are the passing synoptic scale events that bring major day-to-day changes to mid-latitude weather.

Figure 13.3 shows the relation of your local weather conditions to large-scale features of the tropical atmosphere. Note that the amount of cloudiness and shower activity depends to a large extent on your proximity to the region of warmest sea surface temperatures, which generally lie near or slightly poleward of the intertropical convergence zone (ITCZ). Far from this region (i.e., on the left side of Figure 13.3), weather is dominated by a strong trade wind inversion and subsiding air associated with the subtropical high pressure belt. Moving over progressively warmer waters, the surface air becomes warmer, moister, and less stable. The trade wind inversion is less significant, and deep convective currents are more common.

Note also the scales of the weather features in Figure 13.3. The intertropical convergence zone, the trade winds, and the

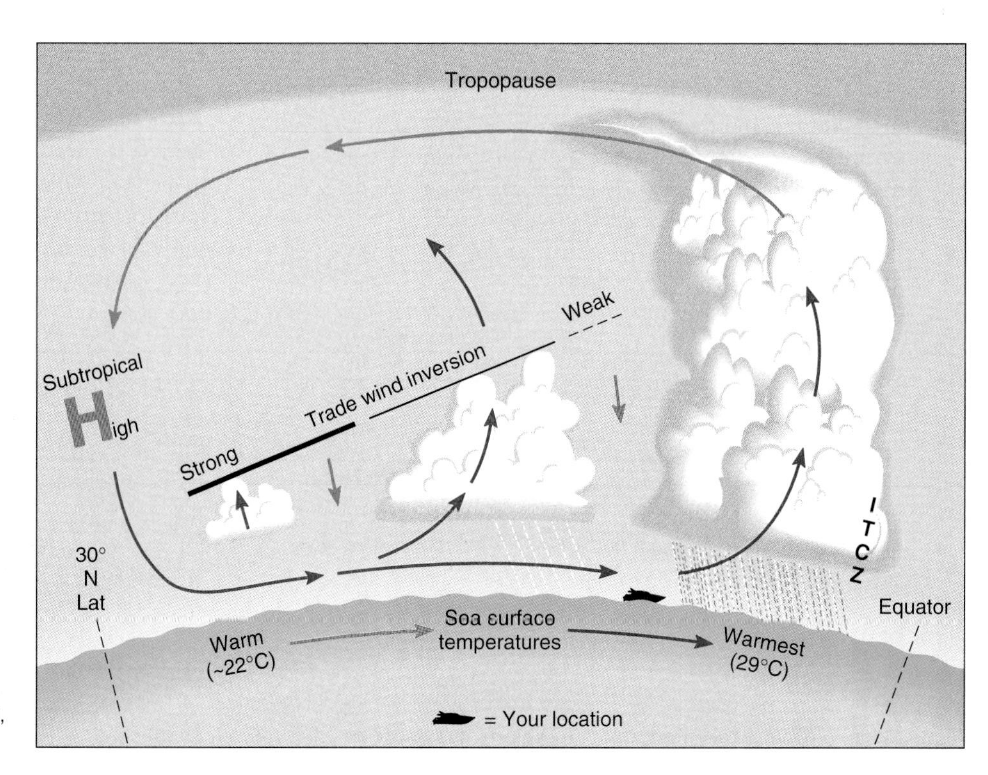


Figure 13.3 Large-scale circulation features in the tropics. Source: Data based on K. Emanuel, *American Scientist*, Volume 76, July/August 1988.

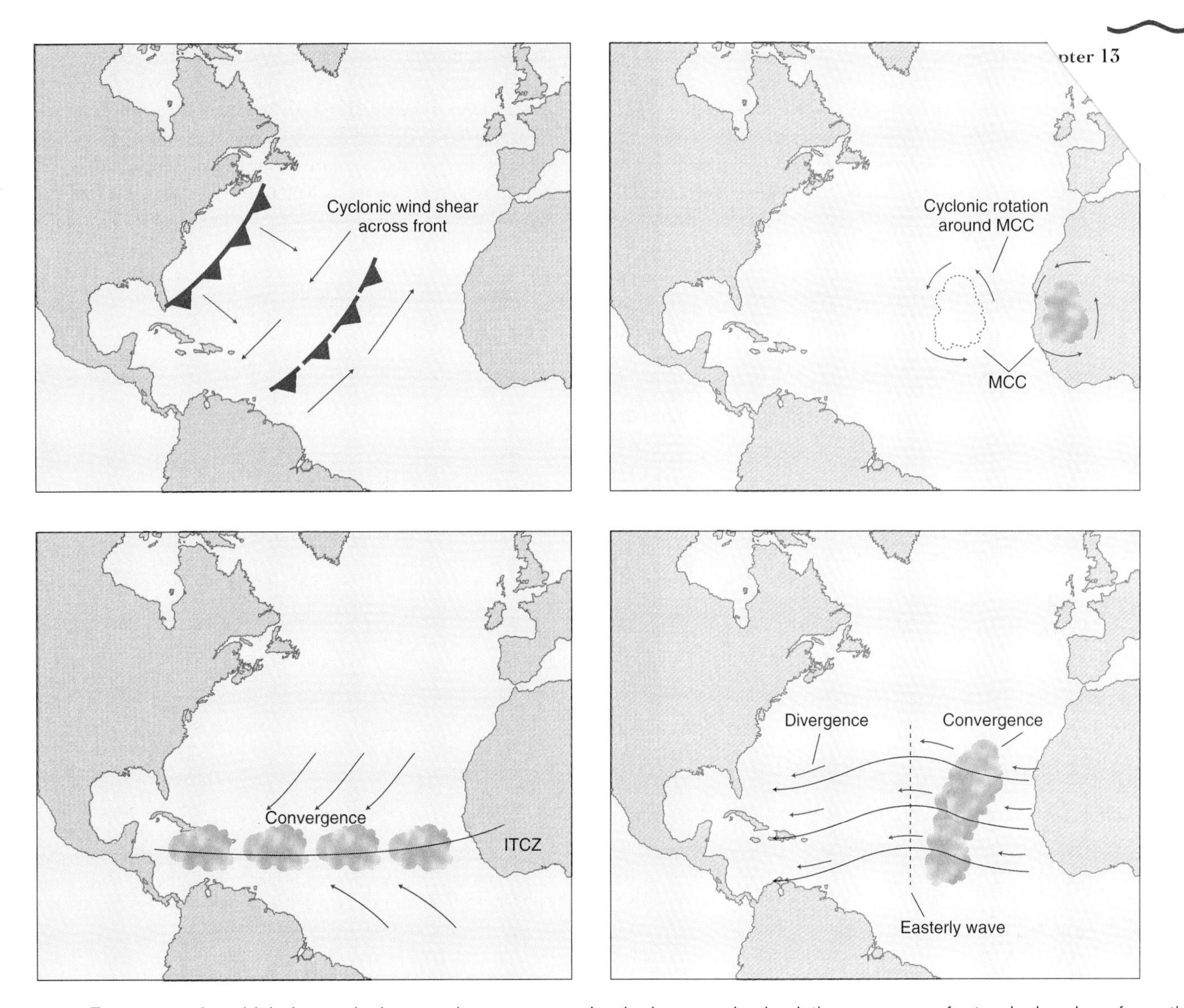


**Figure 13.4** Four means by which the tropical atmosphere may acquire the large-scale circulation necessary for tropical cyclone formation.

subtropical highs pressure belt are global in scale. However, the clouds (and therefore tropical precipitation) generally show little organizational structure larger than that of individual clouds. In particular, large mesoscale or synoptic scale features, which are so characteristic of the weather in mid-latitudes, are much less common in the tropics. This absence of synoptic scale circulation accounts for the lack of day-to-day variability in tropical weather—and thus in the weather you observe at your site.

## Requirements for Tropical Cyclone Formation

It may seem remarkable that large and violent hurricanes can develop within the tranquil tropical weather regime described above. In fact, hurricanes do not just spring from these quiet conditions; a number of factors must be present for them to develop. Although at present they are unable to predict in detail the when and where of tropical cyclone development, atmospheric scientists have identified several special conditions that seem to be required before such development will occur. Let's examine those conditions.

### *A Preexisting, Large-Scale Flow Pattern*

Unlike frontal cyclones, which can develop from small-scale instabilities on the polar front, tropical cyclones almost always form within preexisting, large-scale flow patterns. These patterns must contain low-level convergence and upper-level divergence. Most such structures also possess some cyclonic rotation or shear, another important property. You read in the last section that large-scale structures are not typical features of the tropical environment. Where, then, do they come from?

Any one of several agents may impose the necessary large-scale flow pattern on the tropical atmosphere, as shown in Figure 13.4. Remnants of fronts or mesoscale convective complexes may drift into the region, carrying the requisite motions. The ITCZ can provide the low-level convergence and upper-level divergence. Another important source is the feature known as the **easterly wave.**

As its name suggests, an easterly wave travels from east to west, embedded in the trade winds, and is a wave disturbance—not a cyclone with closed isobars. Easterly waves are on the order of 2000 kilometers (1300 miles) in length, which puts them ynoptic scale category. As you can see from Figure 13.4, rly waves differ from mid-latitude low pressure systems in ier ways besides their direction of movement and open strucure. Notice that with lower pressure toward the equator, the wave looks "inverted," compared to waves embedded in the westerlies, which commonly have lower pressure in a poleward direction. Also note that the region of greatest cloudiness and precipitation lies on the trailing side of the wave (its east side), rather than in advance of it. The clouds and precipitation are caused by a region of low-level convergence, which you can see in the figure. It is in this convergent wind pattern that tropical cyclones are sometimes born. In part because easterly waves are common features of the tropical atmosphere, they play a major role in hurricane formation. In fact, most Atlantic hurricanes are spawned in easterly waves.

### *Moist, Unstable Air*

Thunderstorms are the pistons that power the hurricane's engine. Therefore, conditions that favor thunderstorm formation favor hurricane development as well. From Chapter 12, you know that favorable conditions include warm surface temperatures, a deep layer of moist air, and the absence of any strong inversions that would limit cloud growth.

### *Absence of Strong Wind Shear*

Compare the two situations shown in Figure 13.5. Notice that when strong shear is present in the horizontal wind from lower to upper levels, as in (A), upper regions of convective cells are torn from lower parts. Through this action, strong wind shear inhibits deep convection. If the shear is minimal, however, as in (B), then the rising currents of warm, moist air can grow to great heights.

Wind shear also inhibits convection by causing air of different origins to be advected into the region at different levels. This action removes the local very warm, humid air and replaces it with air that is likely to be cooler and drier.

### *Warm Ocean Surface Water*

If thunderstorms are the hurricane's pistons, then the warm ocean surface provides the fuel. Hurricanes typically form over water whose surface and near-surface temperature exceeds 26°C (79°F) to a depth of at least 60 meters (200 feet). Why must the warm water extend to such depths? As a tropical cyclone forms, it generates huge ocean waves that mix the water to great depths. If the warm water layer were shallower, mixing would bring colder water to the surface, dissipating the storm's warm-water energy supply.

### *Nonequatorial Location*

You may have noticed in Figure 13.1 that tropical cyclones do not seem to form along the equator. These regions certainly have ample supplies of warm water and warm, humid air. Thus, the absence of tropical cyclones there seems to require an explanation.

Recall from earlier chapters that Coriolis force, derived from the earth's rotation, is responsible for most of the rotation of mesoscale and larger circulations. On the equator, however, there is no rotation about the local vertical. Thus, Coriolis force is absent on and near the equator, and rotating storms such as hurricanes cannot form there.

The absence of Coriolis force near the equator also helps to explain why the ITCZ is not more often a factor in tropical cyclone development. Although large-scale convergence is a permanent feature of the ITCZ, whenever the zone lies near the equator, it lacks the requisite rotational motion. The ITCZ is an effective stimulus only when it lies well to the north or south of the equator. In the Atlantic, only one hurricane in six arises in the ITCZ.

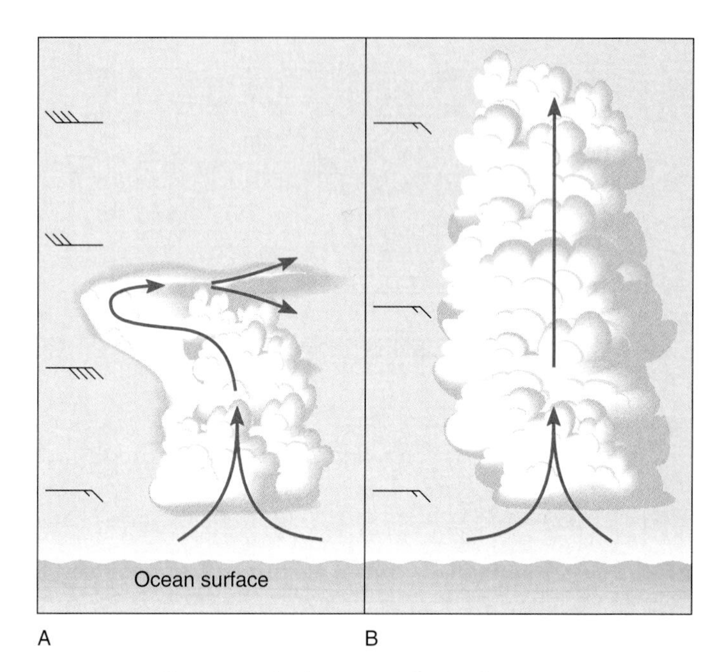


Figure 13.5 Strong vertical shear of the horizontal wind can inhibit convection. (*A*) Strong vertical shear, as horizontal wind changes radically with altitude. Result: Convection is inhibited. (*B*) No vertical shear, as the horizontal wind is the same at all altitudes. Result: Deep convection develops.

***Review***

The concept map in Figure 13.6 summarizes hurricane terminology and the specifics of hurricane formation. Notice the numerous references to "tropical" and "maritime" in the concept map, reminders of the importance of heat and humidity in hurricane formation.

***Questions***

1. Why are Atlantic hurricanes more common in August and September than in other months?
2. Imagine you are located on a ship in the tropics during the passage of an easterly wave. Describe the changes in weather that might occur. (Refer to Figure 13.4.)
3. List four conditions necessary for a tropical cyclone to develop.

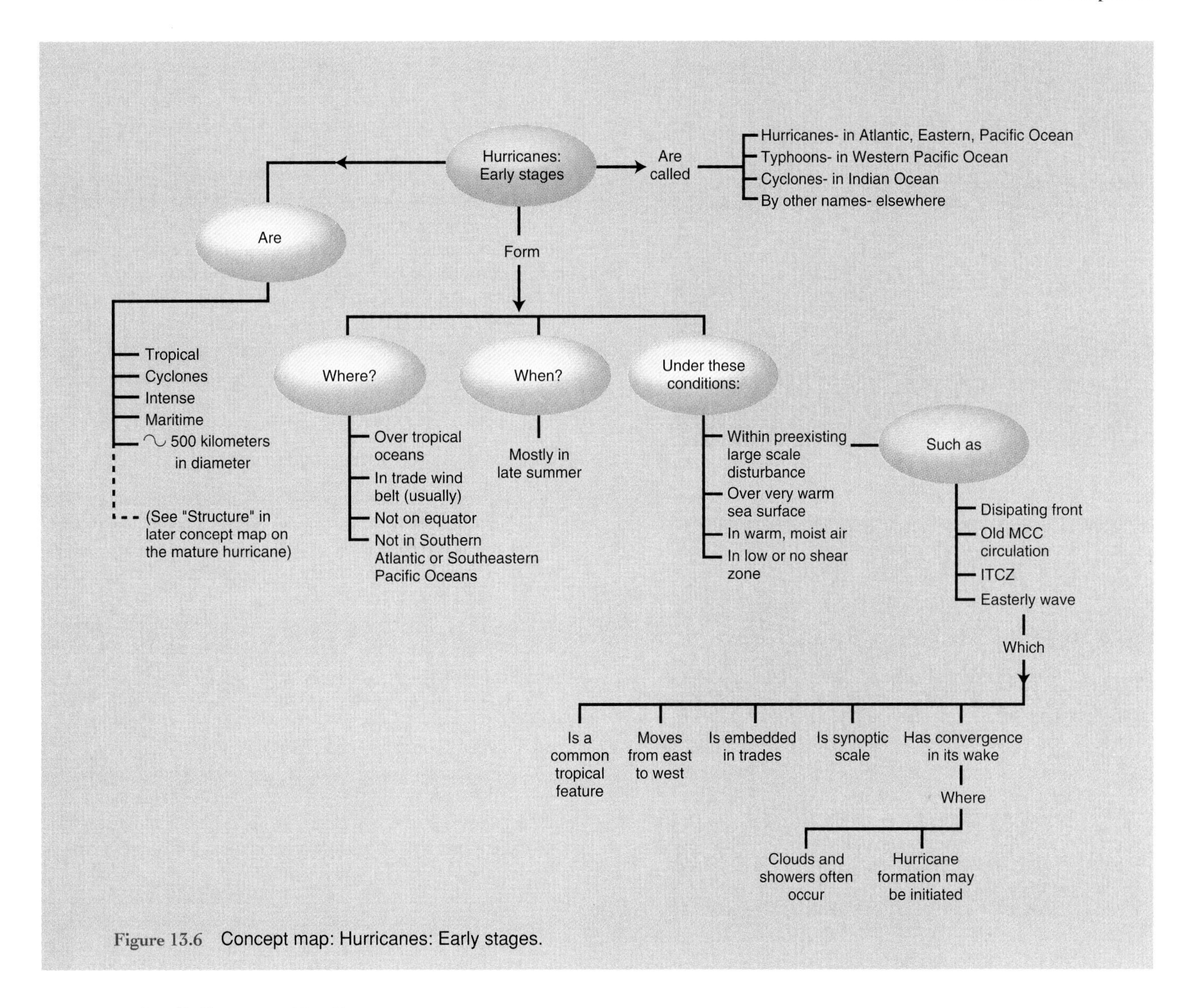


**Figure 13.6** Concept map: Hurricanes: Early stages.

# A Tropical Storm Is Born

For a number of days, your observing team sits in position, taking measurements of atmosphere and ocean and searching for signs of tropical cyclone formation. The ocean temperature is warm, the atmosphere is warm and humid, and the trade wind inversion is weak. But the weather remains exasperatingly pleasant, typically tropical, day after day. Then, near the end of your second week on station, things begin to happen.

## Tropical Storm Genesis (Days 1–3)

On August 14, an easterly wave sweeps off the African continent and through your region. The wave is embedded in a deep trade wind flow and moves steadily westward at 20 knots. Cloudiness begins to concentrate in the zone of convergence around the east flank of the wave (see Figure 13.7A).

The wave continues westward. On the next day, August 15, you observe it passing south of the Cape Verde Islands. You study satellite images of the storm (Figure 13.7B) and note that considerable cloudiness persists in the vicinity of the wave. Weather reconnaissance aircraft begin a schedule of regular observation flights into the young storm.

The satellite images for day 3, in Figure 13.7C, reveal that the cloud mass has acquired a distinct cyclonic structure, with spiral bands. Weather reconnaissance aircraft crews report the system's central pressure to be 1010 millibars—a remarkably high value for a tropical cyclone. However, sustained wind speeds within the disturbance have increased to 20 knots. Therefore, meteorologists at the National Hurricane Center in Coral Gables, Florida, announce the storm has reached the category of "tropical depression." You record the latitude and longitude of the storm's center and plot it on your hurricane tracking chart (Figure 13.8).

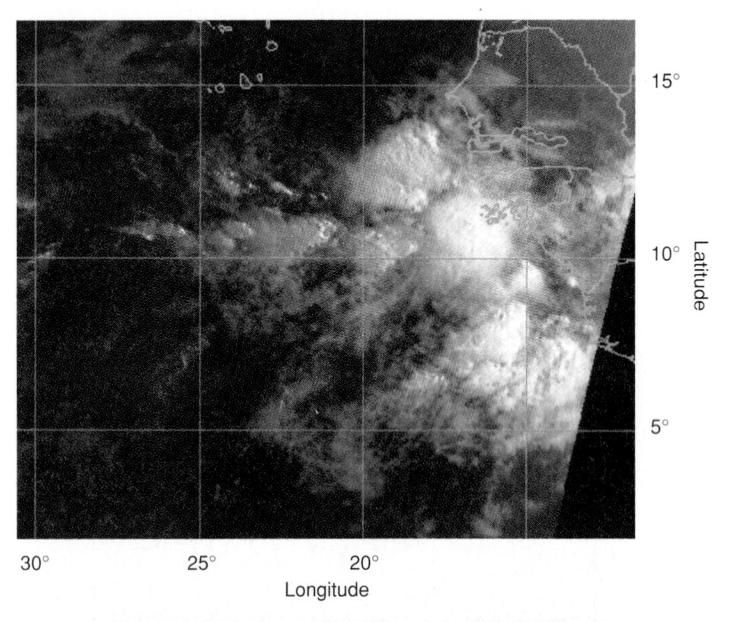

**Figure 13.7** Satellite images of an incipient tropical cyclone on days 1–3. The storm moves westward along the 10° latitude line. The west African coastline is marked in green on the right edge of the first two images.

## Tropical Cyclone Categories

Meteorologists group tropical cyclones into three main categories according to the strongest sustained surface winds in the storm. (A sustained wind speed, as opposed to a momentary gust, is the average speed over a period of one minute, measured 10 meters above the earth's [or ocean's] surface.) In Table 13.1, you can see that a tropical disturbance must contain sustained winds of at least 20 knots to be classified as a **tropical depression.** It must also exhibit some cyclonic structure.

Note that this classification system applies only to tropical cyclones. Other circulations such as frontal cyclones or tornadoes may contain winds in excess of 64 knots, but then are not referred to as hurricanes because they do not possess tropical cyclone characteristics.

## Slight Strengthening (Days 4–5)

Will the storm grow or dissipate? The odds are against growth: in the Atlantic, about one tropical depression in five reaches tropical storm status, and only half of those become hurricanes. One of the remarkable things about tropical cyclones, given that they occur at all, is their relative rarity.

What is required for a tropical depression to mature into a full-blown hurricane? The same conditions needed to create the storm (warm surface water temperature, absence of strong shear, etc.) must persist long enough for it to reach maturity—typically, for several days at least. The loss of any one of these factors can be enough to stem further growth.

At the moment, wind shear variations turn out to be a major player in the evolution of this particular tropical disturbance. For its first three days, vertical shear within it is substantial, and the disturbance's growth, though steady, has been slow. Late on day 3, however, the depression moves into a region of reduced wind shear. Convection builds to greater heights, the inflowing wind accelerates, and on day 4, maximum wind speeds almost double from the previous day, surpassing 35 knots. The disturbance now has reached **tropical storm** intensity and receives a name: Andrew.

## Tropical Cyclone Names

Since the early 1950s, each tropical cyclone that reaches tropical storm intensity has been given a name. Lists of names are drawn up in advance by the World Meteorological Organization, with input from countries affected by tropical cyclones. Then names are assigned alphabetically from the list as cyclones reach tropical storm status. Different lists of names are used for Atlantic, Eastern Pacific, and Western Pacific storms. Table 13.2 gives the names of Atlantic and Eastern Pacific hurricanes through 1999.

The lists of names are recycled over six-year periods, with minor changes. However, the names of particularly devastating

**Figure 13.8** Standard tracking chart for tropical cyclones of the Atlantic Ocean. Source: Data from NOAA, January 1982, U.S. Government Printing Office, 0–364–750.

*Table 13.1*

## Tropical Cyclone Categories

| Name | Maximum Sustained Winds (Knots) | Maximum Sustained Winds (Miles per hour) |
|---|---|---|
| Tropical depression | 20–34 | 23–39 |
| Tropical storm | 35–64 | 40–74 |
| Hurricane (Atlantic, Eastern Pacific) } Typhoon (Western Pacific) } Cyclone (Australia, Indian Ocean) } | 65 + | 74 + |

Table 13.2 Six-Year List of Names for Atlantic Storms

| 1994 | 1995 | 1996 | 1997 | 1998 | 1999 |
|---|---|---|---|---|---|
| Alberto | Allison | Arthur | Ana | Alex | Arlene |
| Beryl | Barry | Bertha | Bill | Bonnie | Bret |
| Chris | Chantal | Cesar | Claudette | Charley | Cindy |
| Debby | Dean | Dolly | Danny | Danielle | Dennis |
| Ernesto | Erin | Edouard | Erika | Earl | Emily |
| Florence | Felix | Fran | Fabian | Frances | Floyd |
| Gordon | Gabrielle | Gustav | Grace | Georges | Gert |
| Helene | Humberto | Hortense | Henri | Hermine | Harvey |
| Isaac | Iris | Isodore | Isabel | Ivan | Irene |
| Joyce | Jerry | Josephine | Juan | Jeanne | Jose |
| Keith | Karen | Kyle | Kate | Karl | Katrina |
| Leslie | Luis | Lili | Larry | Lisa | Lenny |
| Michael | Marilyn | Marco | Mindy | Mitch | Maria |
| Nadine | Noel | Nana | Nicholas | Nocole | Nate |
| Oscar | Opal | Omar | Odette | Otto | Ophelia |
| Patty | Pablo | Paloma | Peter | Paula | Philippe |
| Rafael | Roxanne | Rene | Rose | Richard | Rita |
| Sandy | Sebastien | Sally | Sam | Shary | Stan |
| Tony | Tanya | Teddy | Teresa | Tomas | Tammy |
| Valerie | Van | Vicky | Victor | Virginie | Vince |
| William | Wendy | Wilfred | Wanda | Walter | Wilma |

Note: Names of particular individuals have not been chosen for inclusion in the list of hurricane names.

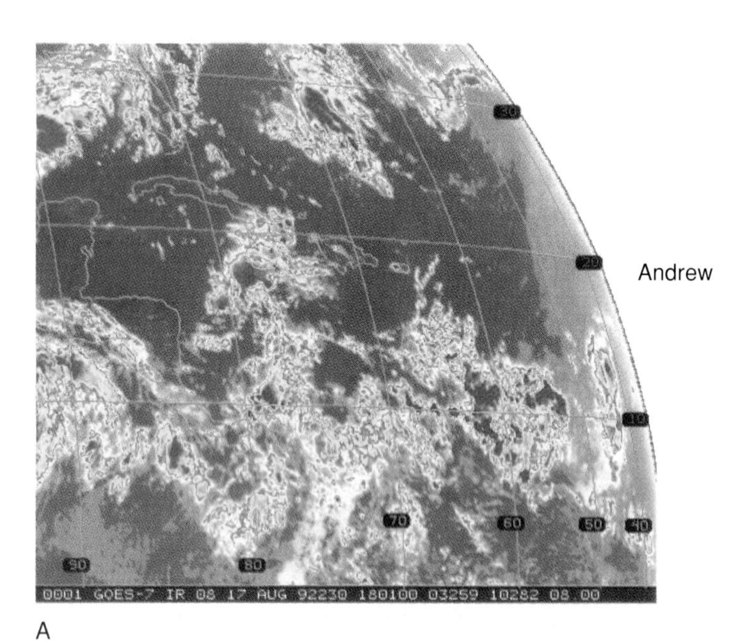


A

B

Figure 13.9 Satellite images of tropical storm Andrew on days 4 and 6.

or otherwise noteworthy hurricanes are retired from the list. Carol, Edna, Hugo, Camille, and Bob are names that will not be used again for Atlantic hurricanes.

Since the storm you are following is the first of the year to reach tropical storm intensity, it receives the first name on this year's list of Atlantic storms: Andrew. In Figure 13.9A, you can see Andrew's portrait (satellite image) on its first day as a named storm.

## Andrew Tracks Northwestward (Days 6–8)

Study the satellite images of tropical storm Andrew taken on days 4 and 6 (Figures 13.9A and B). Note that the circulation appears to be becoming more organized. However aircraft reconnaissance late on day 6 reports that "only a diffuse low-level circulation center remains," with a surprisingly high central pressure of 1015 millibars.

Is Andrew in the process of dissipating? Will it weaken and evaporate like most tropical storms, or will it reintensify, perhaps to become a hurricane? The answers to these questions depend to a great extent on the environment through which Andrew moves in the next days. If the storm tracks into regions of warm surface waters, divergence aloft, and minimal wind shear, its chances of intensification are improved. Thus, the storm's track will affect its future intensity.

## Paths of Tropical Cyclones

For the next two days, Tropical Storm Andrew progresses steadily northwestward. Figure 13.10 presents the situation through day 7, along with a record of the storm's maximum wind, central pressure, and track to date.

First, examine the data on Andrew's central sea-level pressure and maximum winds in Figure 13.10B. Notice the inverse relation between these two curves. Increases in wind speed occur simultaneously with decreases in air pressure. As in all atmospheric circulations, a deepening of central pressure increases the horizontal pressure gradient, which causes stronger winds.

Note in particular the trend of pressure and wind speed for days 6 and 7. As you see, Andrew is weakening. Its central pressure rises to 1015 millibars, while its strongest sustained winds decrease to around 40 knots. However, a storm with sustained winds of 40 knots is still a powerful cyclone and requires close watching.

At this stage, meteorologists are particularly concerned about the potential impact of larger-scale flow features surrounding Andrew. Consider Figure 13.10C. Notice that Andrew is embedded in the easterly trade wind flow around the subtropical high pressure cell located near Bermuda (the Bermuda High). The trades, dominant through the lower troposphere, have been sweeping Andrew westward at a steady pace. This path, if maintained, should keep the storm over very warm water for the next several days. Observe, however, that the upper-level winds are blowing from the southwest. They have the potential to affect Andrew in two ways. First, they are imposing considerable wind shear over the storm. Forecasters suspect this wind shear is inhibiting Andrew's growth. And second, the winds may steer Andrew onto a more northward track.

## Predicting Andrew's Course

Despite Andrew's relative weakness at present, forecasters at the National Hurricane Center in Coral Gables are intensely interested in the storm. A glance back at Figure 13.10C shows you why: Andrew is traveling uncomfortably close to Puerto Rico and a number of other Caribbean islands, and is headed in the general direction of the southeastern United States. Will the storm affect these islands? Will it strike the U.S. mainland, and if so, where?

Hurricane forecasting, like most meteorological forecasting, begins properly with careful observations of present conditions. Probably the most important forecasting tool is an accurate fix

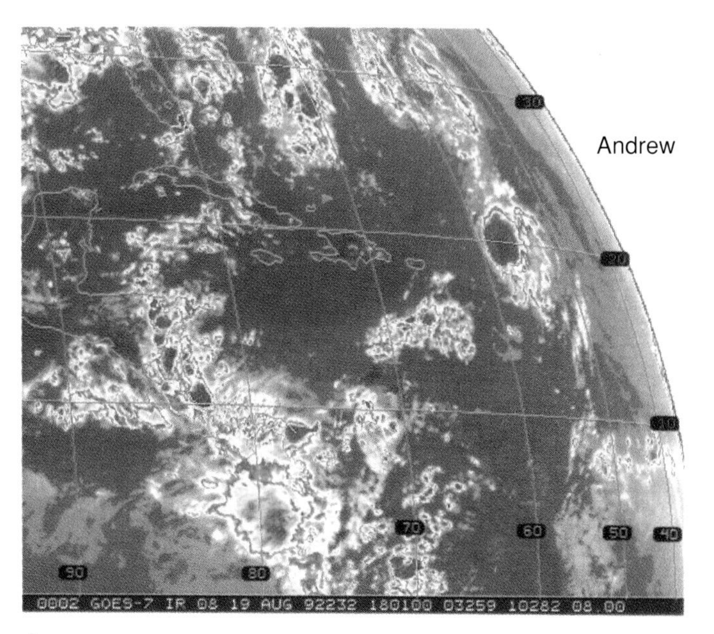


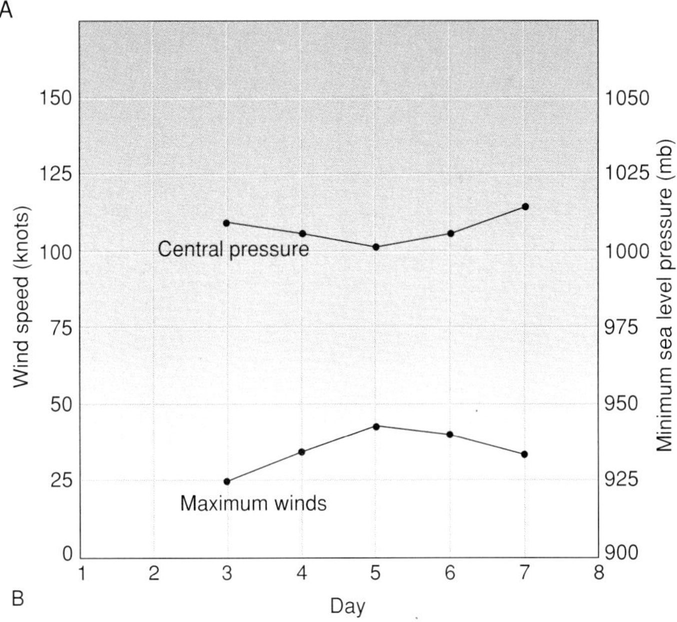


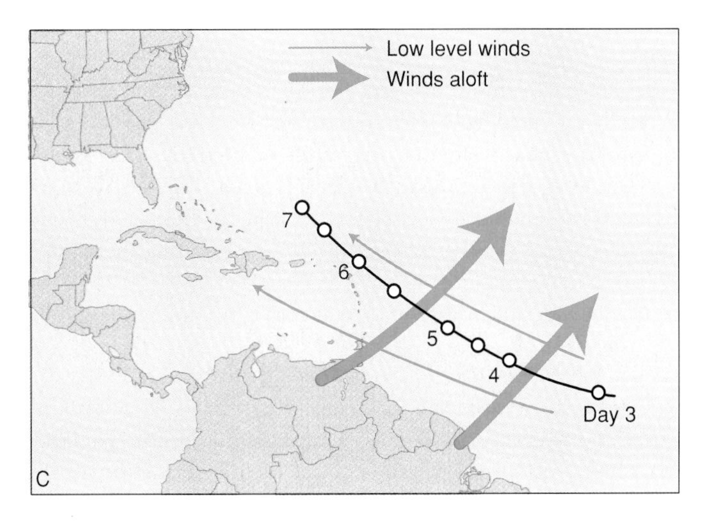


**Figure 13.10** Date for Andrew through day 7.

A

B

**Figure 13.11** Hurricane aircraft reconnaissance. (*A*) Manned "weather recon" aircraft penetrate hurricanes, often to their centers, directly measuring weather conditions and pinpointing the storm's locations. (*B*) *Perseus* is an unmanned experimental reconnaissance aircraft that can be directed into hurricanes.

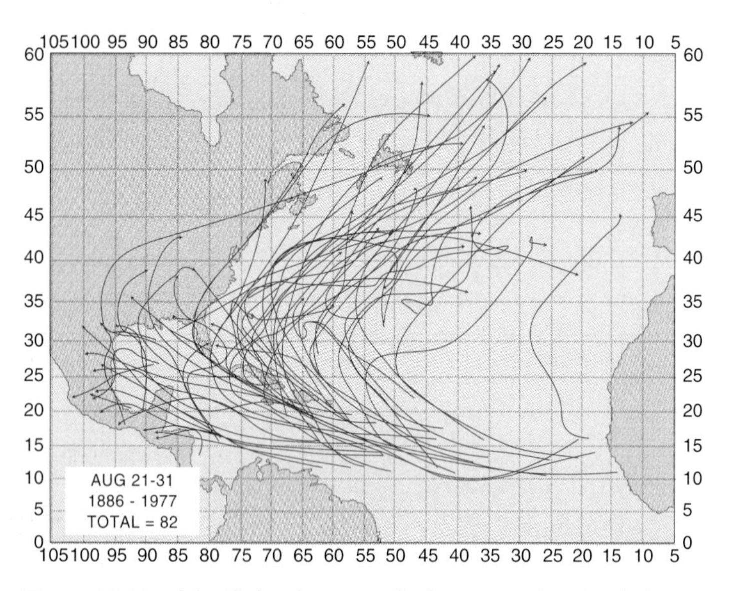


**Figure 13.12** Atlantic hurricane tracks for storms forming in late August. Based on data from 1900–1978. Source: Data from C. J. Neumann, et al., *Tropical Cyclones of the North Atlantic Oceans, 1971–1977*, 1978 page 161. Government Printing Office, NOAA, Washington, D.C.

on the storm's location, present motion, and attendant weather. For hurricane meteorology, standard surface, radiosonde, radar, and satellite measurements are supplemented by those made from weather reconnaissance aircraft (see Figure 13.11A). These aircraft penetrate hurricanes, often to their centers, directly measuring weather conditions and pinpointing the storm's location. Unmanned aircraft such as *Perseus* (Figure 13.11B) are recent experimental additions to weather reconnaissance.

## *Guidance from Tracks of Past Hurricanes*

Once they have accurate data on the storm's location and weather, forecasters can turn to several techniques for guidance in predicting its future. The historical record of past hurricane tracks is one such technique, a form of the "weather types" forecasting you read about in Chapter 10. Compare the tracks shown in Figure 13.12 with Andrew's path to date in Figure 13.10C. Notice that Andrew is moving along a well-traveled route.

What guidance about future motion does Figure 13.12 offer about tropical storm Andrew? The tracks seem to suggest two main possibilities: one, that Andrew will continue on a mainly westward course, striking the Bahama Islands, Cuba, or southern Florida, then possibly moving into the Gulf of Mexico. The second is that Andrew might **recurve,** turning northwest, north, and eventually northeastward, either striking or remaining offshore from the U.S. East Coast. Based on the tracks shown, the chances of Andrew moving southward into the Caribbean Islands are low. A caution, however: careful examination of Figure 13.12 shows hurricanes can take erratic paths. Clearly there are exceptions to the overall patterns shown in the figure.

## *Guidance from Extrapolation*

Suppose that you use Andrew's past path as a guide to predict its future track—an example of trend forecasting by extrapolation. Figure 13.13 shows the projected tracks resulting from three of many possible extrapolation methods. Extrapolation works best when the storm's past track is **linear,** that is, it lies on a straight line. The three most recent fixes on Andrew's position are nearly linear, and the first track in Figure 13.13A shows linear extrapolation based on these three positions.

Take a ruler and try to position it so every point on Andrew's track in Figure 13.13B lies on the ruler's edge. What do you find? Clearly, Andrew's past track is nonlinear. How do you extrapolate a nonlinear path? The general idea is to project forward the curved shape of the storm's past path, much as you projected the straight line forward in case (A). Difficulties arise, however, in deciding just which curve to project. Notice in (B) that Andrew's entire track from day 1 to the present can be represented as a single, gently upward-curving arc. All points of the storm's path lie on or close to this curve, although some recent points lie some distance off. Extrapolating Andrew's position along this curve suggests the storm will recurve, missing Florida but possibly posing a threat to points farther north.

If you consider only the most recent four points (in Figure 13.13C), the storm seems to be following a trajectory that turns

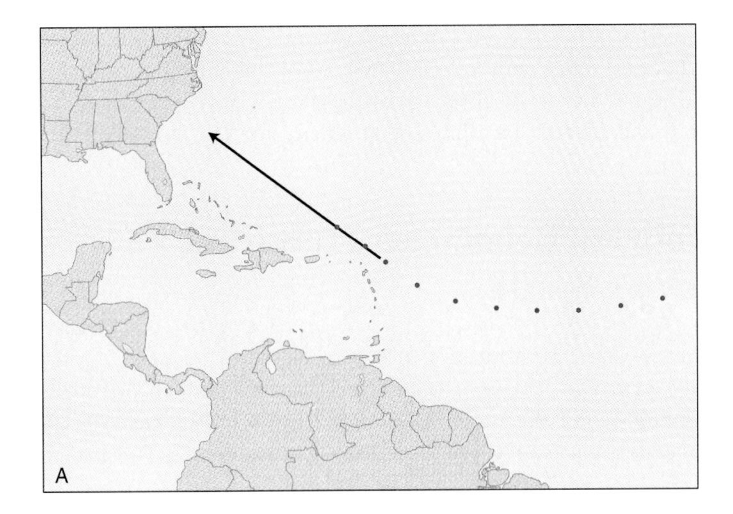


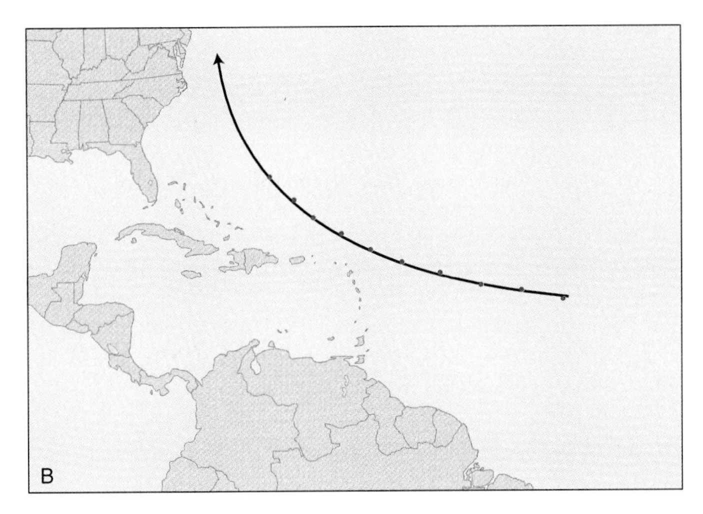


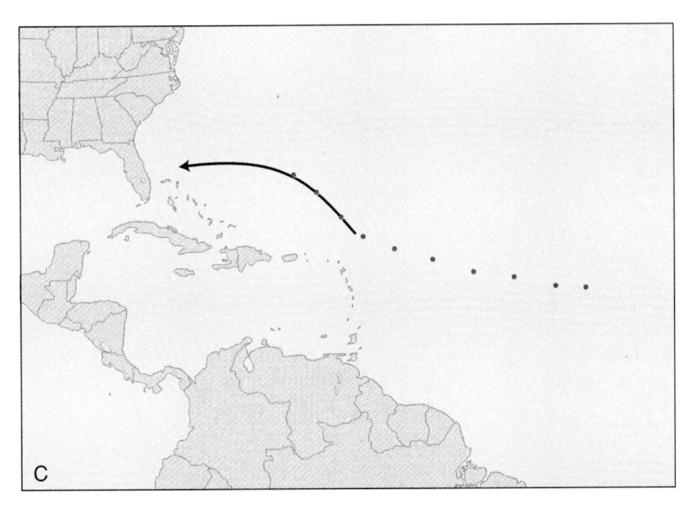


Figure 13.13 Extrapolating Andrew's future track from positions up to day 8. (*A*) Linear extrapolation from three most recent positions. (*B*) Curved extrapolation based on all positions. (*C*) Curved extrapolation based on four most recent positions. Source: U.S. Government Printing Office, NOAA.

west rather than north. Extrapolating along this curve moves Andrew toward central Florida.

Thus, the three methods illustrated give very different results, projecting landfall anywhere from Florida to New England. Which, then, is the correct method to use?

Unfortunately, no one method is invariably best or worst. Linear extrapolation often is safest for short periods of time (24 hours or less); but you saw in Figure 13.12 that few hurricanes travel in straight lines their entire lives. Using most or all of the storm's positions to derive a curve (Figure 13.13B) helps to reveal the large-scale features of its track and small, random point-to-point variations. However, sometimes those small variations are important indicators of a real change in the storm's direction. Thus, extrapolation, like the historical record of past storm tracks, is useful in showing several possibilities but generally does not yield a single, unequivocal projection for a given storm.

### *Guidance from the Storm's Interactions with Its Environment*

Like mid-latitude frontal cyclones, tropical cyclones interact in complex ways with larger-scale atmospheric systems in which they are embedded. These larger systems, such as the trade winds, steer the hurricane's motion and affect its development. As the hurricane moves and develops, it in turn affects the surrounding large-scale circulations.

In recent years, computer-based forecast models have become increasingly skillful in forecasting movement and intensity changes in tropical cyclones. The National Hurricane Center uses several different computer models to predict hurricane development and motion.

Based on data collected through the evening of day 7, three different computer models make the predictions shown in Figure 13.14. Compare these forecasts with the extrapolations of the previous figure. Notice that all three models are in fairly close

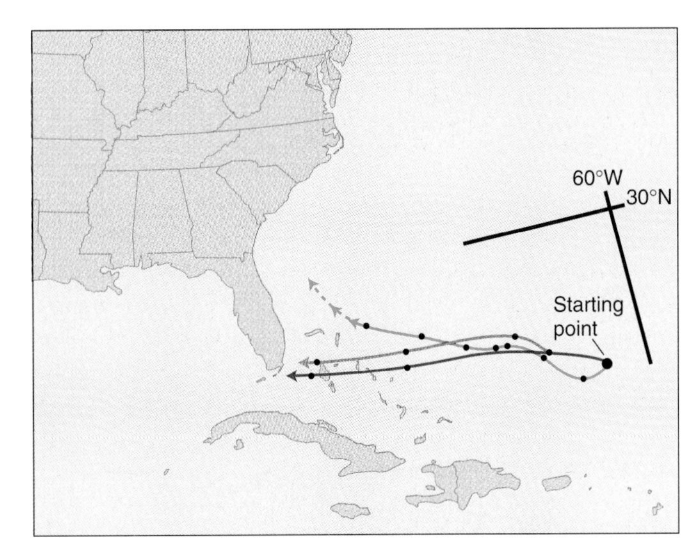


Figure 13.14 Seventy-two-hour forecast trajectories (96 hours for the track marked in blue) for Andrew according to three computer models. How do these models compare to the extrapolated positions in Figure 13.13?

agreement for the first 48 hours, indicating a basically westward motion. However, by the end of their 72-hour forecast period, they begin to diverge, leaving forecasters with a familiar dilemma: Will Andrew recurve or not?

## The Forecast

Choosing the correct track and getting the timing right becomes increasingly important with time: as Andrew approaches land, people must be warned and precautions taken. Thus, forecasters at the National Hurricane Center (NHC) critically evaluate each numerical prediction for reasonableness and consistency with other forecasts, observed data, the historical record, and extrapolated positions. They also monitor closely the latest observations that pour in from satellites, reconnaissance aircraft, ships, radar, and other sources. Based on all of this information, every six hours the NHC forecasters produce a new forecast of the storm's projected location and intensity for the next 48 hours.

# Andrew Becomes a Hurricane (Days 8–10)

The question of Andrew's future track takes on new urgency on days 8 and 9. Upper-level wind patterns shift dramatically. Moving out of the region dominated by low-level easterly and upper-level westerly winds, Andrew now enters a region of

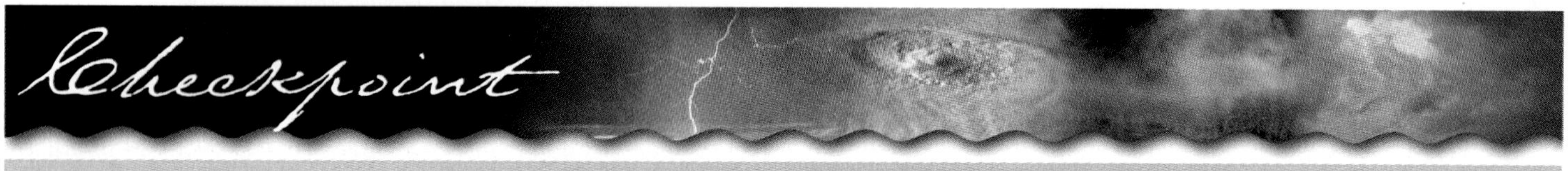

## Checkpoint

***Review***

The concept map in Figure 13.15 reflects the content of the past two sections. To review this information, move through the concept map considering how each item relates or might relate to Tropical Storm Andrew.

***Questions***

1. What is the designation (tropical depression, storm, etc.) for a tropical cyclone with maximum sustained winds of 50 knots? 30 knots? 70 knots?
2. Why do tropical cyclones generally travel from east to west?
3. Explain why extrapolation (Figure 3.13) can yield a variety of different forecast paths.

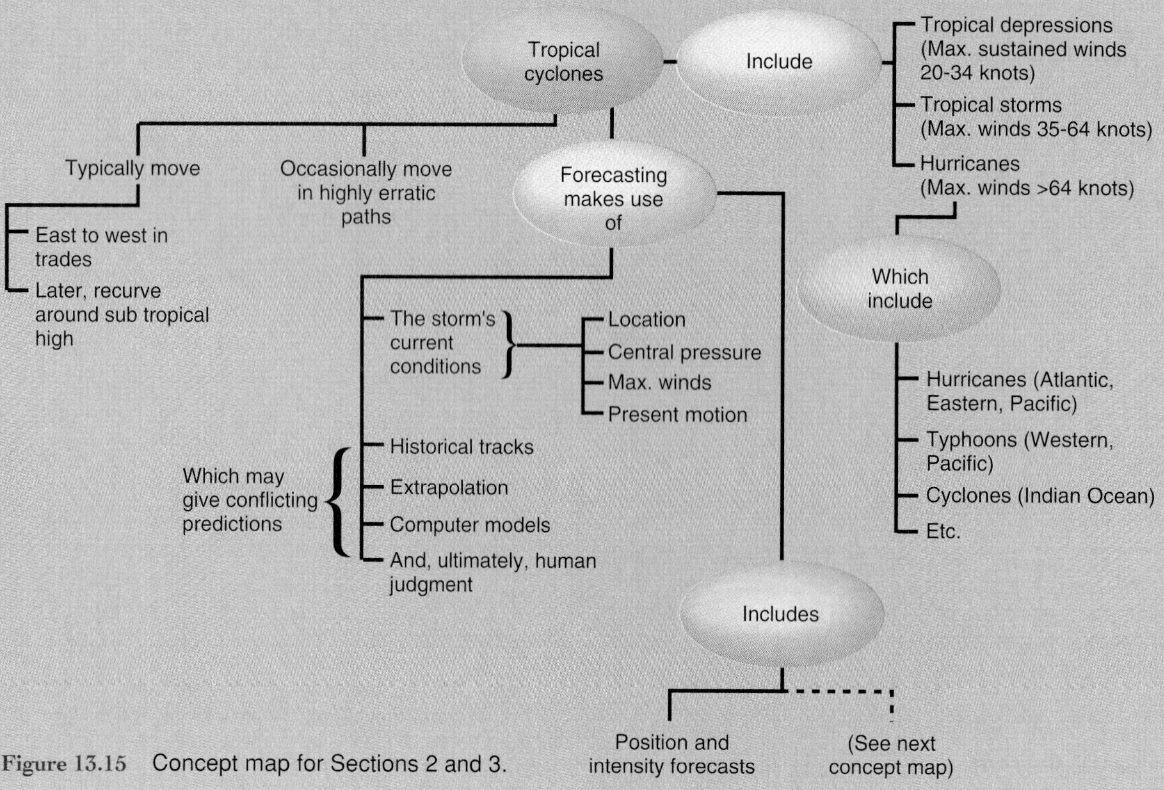

**Figure 13.15** Concept map for Sections 2 and 3.

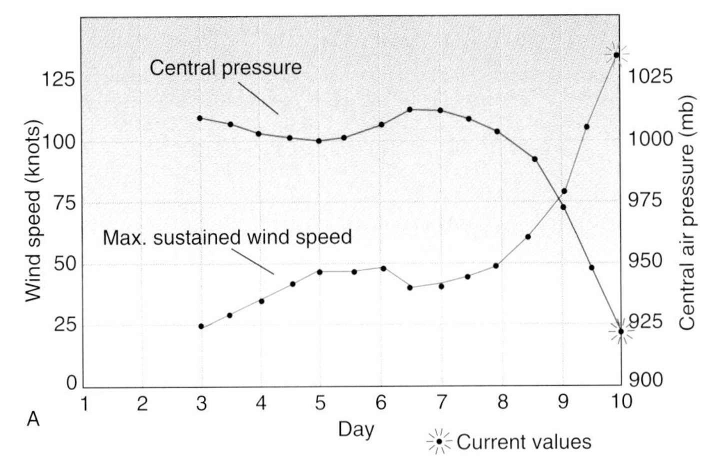


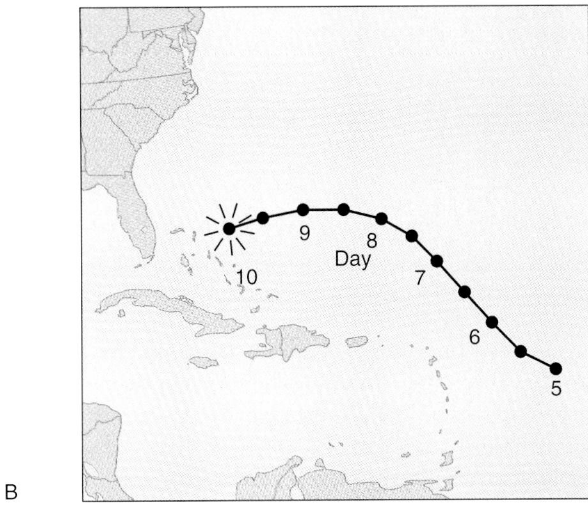


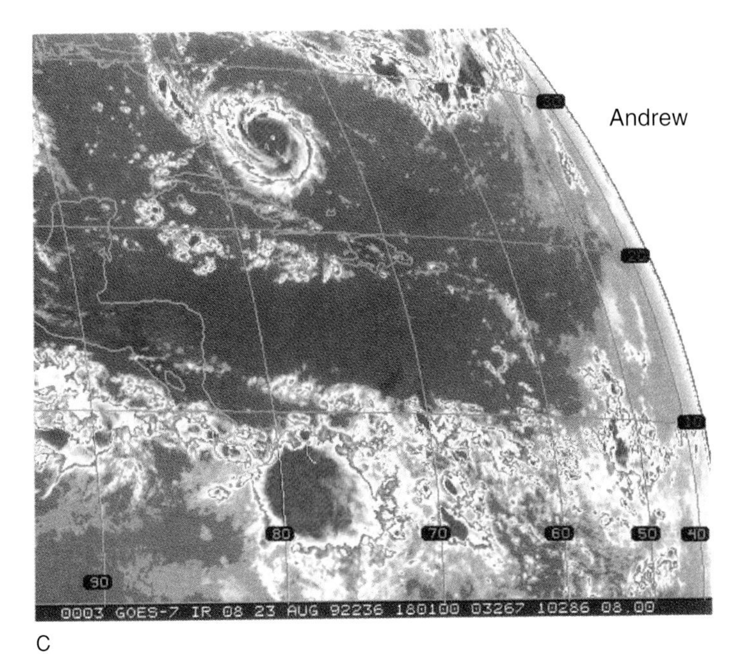


**Figure 13.16** Conditions through day 10. How great was the increase in wind speed from day 9 to day 10?

easterly flow at nearly all levels. In the absence of wind shear, deep convective currents begin to develop. The new upper-level wind pattern is also a divergent one, which further encourages vertical motions. Suddenly, all conditions are suitable for intensification.

Figure 13.16 shows Andrew's extraordinary response to its new environment. From day 7 to day 10, central sea-level pressure decreases from 1015 to 922 millibars—a change of 93 millibars! This precipitous decline has a correspondingly dramatic effect on Andrew's winds. From the figure, you can see that maximum sustained wind speeds increase to over 130 knots on day 10.

## Structure of Mature Hurricanes

Andrew's structure now resembles that of a typical intense hurricane whose important characteristics are summarized in Figure 13.17. Notice the general wind flow pattern: convergent at low levels, rising in convective clouds, and divergent at upper levels. The **eye**—a distinctive region of clear or nearly clear skies, subsiding air, and light winds—is the center of the circulation. Surrounding the eye is the **eyewall,** a ring of towering thunderstorms that typically contain the storm's heaviest rain and strongest winds. Note also that the clouds and rain are not distributed evenly throughout the cloud mass. For reasons not understood at present, the most vigorous thunderstorms and the heaviest rain are concentrated in narrow bands. Between these bands, precipitation is less intense.

Beyond the hurricane's cloud mass is a region of subsiding air. As the air descends, it warms adiabatically, and clouds evaporate. Hurricanes tend to stand out prominently in satellite images (look at Figure 13.18, for example), surrounded as they are by this ring of subsiding clear air.

Figure 13.17 reveals additional important structural details. For example, trace the isotherms (dashed lines) on the chart. Notice that they bulge upward at the storm's center; hurricanes have warm cores. By comparing the two isotherms, you can also see that the warmth is more pronounced at lower levels. The hurricane's warm core is due to the release of vast amounts of latent heat in showers and thunderstorms and to adiabatic heating through subsidence in the eye.

Now examine the isobars in Figure 13.17. Notice the striking "V" shape of the 920-millibar isobar, indicating the dramatic drop of air pressure as you approach the storm's center. This pronounced pattern does not persist aloft, however. Recall from Chapter 8 that low pressure centers centered in warm air weaken with height. Thus, horizontal pressure gradients in the hurricane become progressively flatter at higher altitudes. To verify this, trace the 300-millibar isobar. You can see that near the tropopause, almost all signs of the low have disappeared. Except for a small region over the storm's very center, pressure aloft tends to be higher within the hurricane than in the surrounding atmosphere.

Because pressure gradients in the hurricane are greatest at low altitudes, wind speeds are strongest there as well. (Surface friction reduces speeds somewhat in the very lowest levels, of course.)

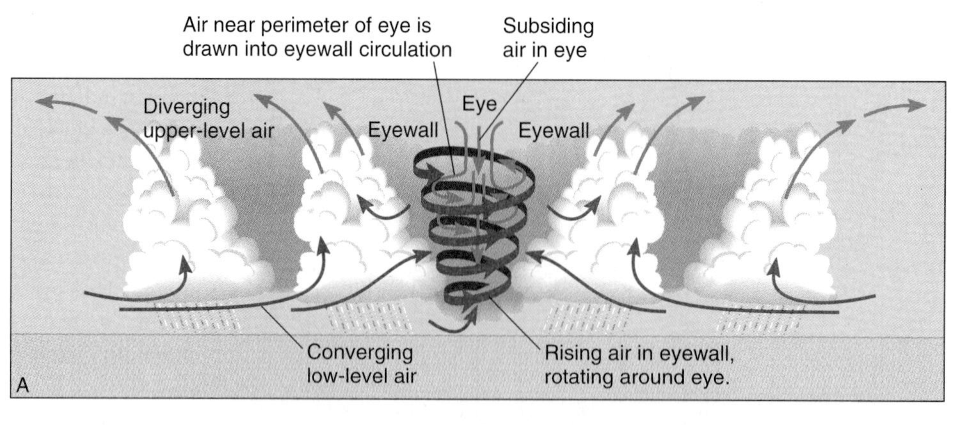


**Figure 13.17** Structure of a mature hurricane. (*A*) Basic wind flow, cloud, and precipitation patterns. (*B*) Profiles of temperature (dashed), pressure (black), and wind (blue). Note the warmth at the storm's center. Also note that the gradients of temperature and pressure, and the speeds of the winds, decrease upward.

**Figure 13.18** From space, hurricanes are prominently visible because they tend to be surrounded by clear air. Note the spiral cloud bands and eye in this image.

## The Hurricane's Eye

How remarkable it is that at the very center of the hurricane's fury is an oasis of quiet weather! To anyone directly in the hurricane's path, the eye is a momentary, if ominous, respite from the storm's fury; it is the center of a towering coliseum of eyewall clouds (see Figure 13.18); it is the half-time break in the action, before the tempest continues. The eye is also refuge for sea birds swept into the hurricane's tumult. They fly imprisoned within the eye, traveling wherever the storm moves. As a result, tropical birds are sometimes found as far north as New England and eastern Canada in the aftermath of an Atlantic hurricane.

Why does a hurricane have an eye? Why doesn't the inflowing low-level air circulate all the way to the hurricane's very center? Like many other aspects of the hurricane, there is no completely satisfactory explanation for the eye's formation and persistence. A partial explanation can be found by comparing the hurricane's eye to the small "eye" at the center of a tornado vortex. Recall from Chapter 11 that the angular momentum of the inward spiraling air tends to be conserved. Thus, as an air parcel's distance from the center decreases, its rotational speed increases proportionally. Remember also that there are limits to this process. If the air were to spiral all the way to the vortex center, its speed would increase beyond limit. Like the racing driver who turns too sharply at great speed and drifts outward, at some radius the inward spiraling air reaches a speed beyond which the pressure gradient force is insufficiently strong to maintain such a sharply curved trajectory. Thus, at some distance from the center of the vortex, typically 5 to 15 kilometers, the low-level air's inward motion ceases. The air rises in the hurricane's eyewall.

Follow the motion of the air in the eye itself, in Figure 13.17. Observe how air near the perimeter of the eye is drawn outward and spirals upward into the ascending eyewall. To replace this outward and upward flow, air descends throughout the central region of the eye. This subsidence causes the air to become warmer and clouds to evaporate, hence, the region of reduced cloudiness in the eye.

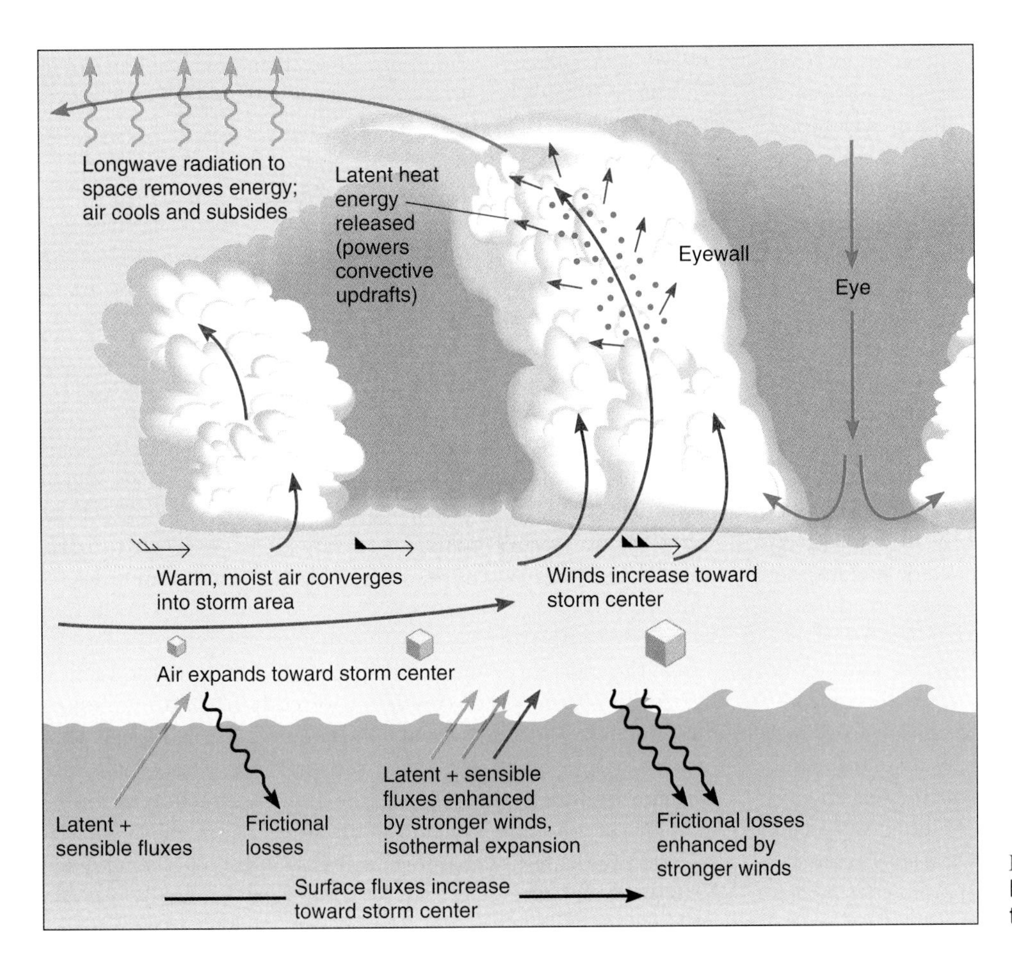


**Figure 13.19** Energy flow within a mature hurricane. Through what mechanisms does the storm lose energy?

## What Sustains the Hurricane?

How does the hurricane maintain its towering structure and awesome power? What are its energy sources, how does the storm's energy relate to its structure, and how is the storm's energy consumed? We will tackle these questions with the help of Figure 13.19, a vertical cross section through the hurricane center, which illustrates the primary energy fluxes within the hurricane.

Consider first the large-scale surface wind flow, shown by the long, red horizontal arrow. This surface air, already warm and moist due to prolonged contact with the ocean surface, gains even more heat from the ocean (represented by the red arrows pointing upward from the ocean) as it is swept toward the storm's center. Note that the surface air also loses energy to the sea surface through friction.

Reaching the convective cloud bands that comprise the eyewall, the air rises and cools moist adiabatically. Latent heat release from the condensing water vapor heats the air, adding to its buoyancy and therefore its upward velocity.

At the top of the hurricane, the rising air diverges from the storm center. Note that in this region, it loses energy through longwave radiation loss upward. Eventually, at some distance from the center, the air subsides, perhaps all the way to the sea surface, where it may repeat the loop, gaining heat and moisture from the sea surface, converging toward the storm center, and so on.

Let's take a closer look at processes at work at the sea surface in Figure 13.19. You can see the low-level wind accelerating, due to angular momentum conservation and the increasing pressure gradient as it converges on the storm center. This acceleration affects the rate of energy transfer between sea surface and air, in a manner you know about if you have ever stood outside in a strong wind after a swim: namely, the stronger the wind, the greater the fluxes of sensible and latent heat from a wet surface—the ocean or your body. Furthermore, over the open ocean, stronger wind speeds also mean a rougher ocean surface. Heat and water vapor are transferred readily into the air from the towering waves' ragged, wind-blown crests.

Stronger wind speeds and greater surface roughness mean that frictional dissipative fluxes increase as well. Thus, notice in Figure 13.19 that all these sea-to-atmosphere fluxes increase in magnitude as the air converges toward the hurricane's center.

You might be skeptical about the increase in sensible and latent heat fluxes in Figure 13.19. After all, you learned in Chapter 5 that a (net) flux of water vapor occurs only when the air is unsaturated. Surely, in the raging tempest of a hurricane, the near-surface air must be as close to saturated as air can become. How, then, could there be a significant net flux of sensible or latent heat?

Part of the explanation lies in turbulent mixing. Eddies of all sizes carry surface air aloft and replace it with slightly less-saturated air. Another factor is illustrated by the expanding boxes in Figure 13.19 and operates as follows.

As surface air moves from the perimeter of a mature hurricane toward its center, it experiences a substantial pressure decrease, typically on the order of 50 millibars. As you know from Chapter 6, air expands as its pressure decreases. In the absence of outside energy sources, the air's internal (thermal) energy fuels this expansion. Hence, the air temperature decreases, at the adiabatic rate, as the air moves toward the storm center. For the air parcel in the figure, which undergoes a 50-millibar decrease in air pressure, adiabatic cooling would be approximately 10°C.

In fact, the in-rushing surface air doesn't experience this degree of cooling, at least not all at once. As soon as the air pressure drops slightly, the resultant cooling lowers the air temperature slightly below sea surface temperature. This temperature difference generates a flux of sensible and latent heat from sea to air, to bring the air back up to equilibrium with the sea surface. This expansion, followed by cooling, followed by flux of heat from the sea surface, occurs continually as the parcel is swept inward toward the hurricane's center. The result is that the air undergoes what is known as **isothermal expansion**—expansion at constant temperature—in which any temperature decrease is quickly compensated by a flow of energy into the air from the sea surface. In summary, as the air moves toward lower pressure, some of its heat energy is used to drive the air's expansion. This energy consumption is quickly made up for by a flux of energy from the ocean surface. In a sense, the expansion provides an extra niche in which the air can remove and store energy from the sea, energy it releases as it rises in the eyewall, powering the hurricane. The effect is significant only because a hurricane's sea-level pressure is substantially lower at the storm's center than on the perimeter.

Finally, note that the lower the hurricane's central pressure, the greater the effect. On day 10, Andrew, an intense hurricane with very low central pressure, is receiving a considerable energy boost from isothermal expansion.

## Andrew Approaches Land: Warnings Are Issued

Turn back for a moment to Figures 13.14 and 13.16. Comparing Andrew's path to the forecasts made on day 8, you can see that the predictions of due westward movement turned out to be the most accurate. By day 9, computer models and human forecasters all predict that Andrew will continue moving westward toward the Bahama Islands and southern Florida.

Ever since Andrew's birth, forecasters have kept residents of the Bahamas and the southeastern states informed of its progress. By day 9, however, more specific forecasts are in order. The National Hurricane Center issues a **hurricane watch** for the northwest Bahamas and the southeast coast of Florida. A hurricane watch means that a particular region faces the threat of hurricane conditions within 24 to 36 hours. It is a call to residents of the region to watch and listen for further bulletins and to consider what steps to take should a hurricane warning be issued; but a hurricane watch is not a call to evacuate.

The hurricane watches for Andrew are followed a few hours later by **hurricane warnings** for the Bahamas and most of the southern Florida coast. A hurricane warning means that hurricane conditions are expected and that residents should begin immediately to take precautions, including evacuation from low-lying areas subject to flooding. Over 1 million residents of the southeast Florida coast are told to evacuate and seek shelter out of the hurricane's path.

## Hurricane Conditions: A Triple Threat

What can residents within the hurricane warning area expect from the storm? Most hurricane damage is caused by one or more of three elements: strong winds, surging ocean waters, and heavy rain.

### *Hurricane Winds*

Despite a hurricane's rather symmetrical, often circular shape, winds are not equally strong in all directions from the center. Two factors that cause hurricane winds to be stronger to the right of the storm's motion than elsewhere are shown in Figure 13.20. The first results from adding the circulation around the storm's center to the forward motion of the storm itself. In regions where the storm's circulating winds and its forward motion are in the same direction, wind speeds will be strongest. (This effect is similar to the combination of eddy motion and smooth flow, which you read about in Chapter 11; see Figure 11.7.) From Figure 13.20, you can see that this effect will cause strongest winds to the right of the storm's direction of motion.

The second effect occurs as the storm approaches land. Notice that to the left of the storm center, winds blow from the land into the storm. Greater surface friction retards these winds compared to those to the right of the storm, which have remained over open water.

Now apply these ideas to Hurricane Andrew. You can see that the strongest winds should be expected slightly to the north of landfall. If the forecasts are correct, the Homestead-to-Fort Lauderdale region of Florida is at greatest risk. South of the expected landfall, winds should be less.

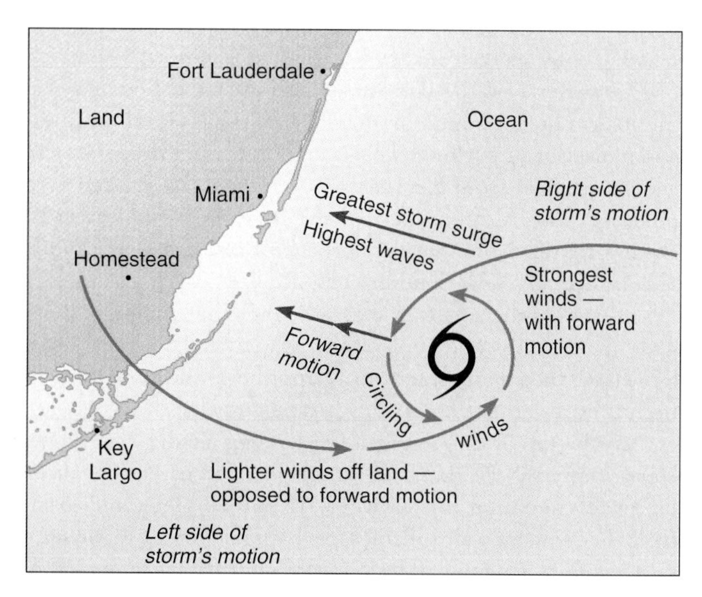


**Figure 13.20** Relationships between wind circulation, storm movement, and coastline make the right side of a hurricane (following its motion) the most destructive one.

Tornadoes and other small-scale circulations, which occasionally are spawned in the eyewall when a hurricane moves overland, cause some of the most extreme wind damage. These small-scale vortexes add their rotational speed to or subtract it from the hurricane's rotation and forward speed. The result is the possibility of brief, local winds well in excess of the already powerful hurricane winds.

### *Storm Surges*

As a hurricane churns across the ocean, it shapes the ocean surface in distinctive ways. Notice in Figure 13.20 that winds on the right side of the storm's track push surface water forward, along with the storm. This action causes a **storm surge,** a rise of the sea surface along the hurricane's path (see Figure 13.21). Generally, the storm surge is magnified in shallow coastal waters as the hurricane approaches land.

Low air pressure at the center of a hurricane also contributes to the storm surge. The sea surface acts something like a barometer, rising 1 centimeter for every millibar of surface air pressure difference between the storm's center and its periphery. This process is similar to what happens if you press a balloon against the floor with your foot: the part of the balloon not under your foot, subject to less downward pressure, rises higher. Hurricane Andrew's extraordinarily low central pressure causes a rise of nearly a meter in the sea surface near its center.

The combined effects of winds and low pressure can generate coastal storm surges that are as great as 8 meters (25 feet) high and are the cause of the greatest number of hurricane casualties. In 1900, a hurricane-related storm surge claimed over 6,000 lives in Galveston, Texas. A cyclone (hurricane) in the Indian Ocean caused over 200,000 fatalities in November 1970, as the storm surge rolled across highly populated, low-lying coastal plains.

**Figure 13.21** The storm surge in the great 1938 hurricane flooded downtown Providence, Rhode Island, to a depth of 10 feet.

**Table 13.3 Saffir-Simpson Scale of Hurricane Intensity Maximum**

| Category | Central Pressure | Sustained Winds | Storm Surge | Damage |
|---|---|---|---|---|
| 1 Weak | > = 980 mb<br>> = 28.94 in | 74–95 mph<br>65–82 kts<br>20–42 m/s | 4–5 ft<br>1.2–1.6 m | Mobile homes, trees damaged; minor flooding |
| 2 Moderate | 965–979 mb<br>28.50–28.91 in | 96–110 mph<br>83–95 kts<br>43–49 m/s | 6–8 ft<br>1.7–2.6 m | Some damage to houses; small boats break moorings |
| 3 Strong | 945–964 mb<br>27.91–28.47 in | 111–130 mph<br>96–113 kts<br>50–68 m/s | 9–12 ft<br>2.7–3.7 m | Mobile homes destroyed; flooding destroyed coastal residences |
| 4 Very strong | 920–944 mb<br>27.17–27.88 in | 131–155 mph<br>114–135 kts<br>59–69 m/s | 13–18 ft<br>3.8–5.5 m | Extensive damage to houses; major beach erosion |
| 5 Devastating | < 920 mb<br>< 27.17 in | > 155 mph<br>> 135 kts<br>> 69 m/s | >18 ft<br>>5.5 m | Many buildings destroyed; massive evacuation |

*Heavy Rainfall*

Although normally less of a threat than strong winds and storm surges, hurricane rainfall can produce considerable damage under the right conditions. Typical rainfall totals during a hurricane's passage are 10–15 centimeters (4–6 inches), but they can be far greater in a slow-moving storm. The remains of Tropical Storm Alberto dumped 2 feet of rain on parts of southern Georgia in a three-day period in 1994. And in July 1911, a typhoon moving at snail's pace through the Philippines produced 224 centimeters (88 inches) of rainfall in the town of Baguio over a four-day period. In deluges of such magnitude, destructive flooding is almost a certainty.

### The Saffir-Simpson Scale

The **Saffir-Simpson scale** of hurricane intensity, shown in Table 13.3, classifies hurricanes according to central pressure, maximum sustained winds, storm surge, and attendant damage. The scale is useful for gauging a storm's overall intensity. Remember, however, that even a Category 1 hurricane is a dangerous storm whose destructive potential must be respected.

Use the latest wind and pressure data for Andrew (in Figure 13.16) to determine Andrew's intensity on the Saffir-Simpson scale. You should find that on day 10, Andrew is a Category 4 hurricane.

## Landfall and Dissipation (Days 10–14)

Here is the current situation in brief: on day 10, powerful Hurricane Andrew, a Category 4 storm, is moving steadily on a track that would take it across the Bahamas and into southern Florida. What will conditions be like at landfall? Imagine that your observation team is located in Homestead, Florida, on high ground in a building reinforced to withstand hurricane conditions. You are ready to observe Hurricane Andrew firsthand.

### Observing the Hurricane Come Ashore

Day 10 dawns bright and clear in Homestead. Temperature, dew point, air pressure, and wind are all in normal ranges. As the morning progresses, you notice wisps of cirrus advancing from the east.

In the afternoon, you observe cumulus clouds forming, a common occurrence at Homestead. Twelve hours before the hurricane's expected landfall, the sky holds few clues of what is to come. (Recall from Figure 13.17 that a ring of subsiding air and clear skies surrounds a typical hurricane.) Around sunset, however, heavier cumulus clouds and thunderstorms move in from the northeast. Wind speed increases sharply, and the air pressure begins to nose-dive. Rather abruptly, Andrew has arrived.

By mid-evening, conditions have deteriorated considerably. The first of Andrew's cloud bands sweep over your observing site, raking it with scattered showers, some of moderate to heavy intensity, on blustery winds. You observe a steady increase in wind speed during those evening hours, coupled with a steady decrease in air pressure.

Conditions worsen throughout the night. For several hours, you observe a truly precipitous decline in air pressure, accompanied by winds that reach average speeds of 140 mph, with gusts perhaps 30 mph stronger. The rain is torrential. Even more frightening than the meteorological data, however, are the sounds that reach you. The wind shrieks and wails as it rips around buildings and trees. Many of these objects are incapable of withstanding such battering and collapse or blow away, punctuating the wind's howls with an assortment of crashes, thuds, and booms. Your building shudders and shakes as if a continuous

### Checkpoint

***Review***

This section has added a great deal of information to the hurricane concept map (Figure 13.22). Included are major new branches on hurricane structure, energy, and weather conditions, as well as additions to the branches on hurricane forecasts and classifications.

One effective way of reviewing hurricane structure is to compare their structure to that of frontal cyclones. You will find many interesting and instructive contrasts.

***Questions***

1. In a frontal cyclone, wind speeds generally increase with altitude. In a hurricane, however, winds decrease upwards (above the friction layer). Why?
2. Make a cross-sectional sketch of the wind flow in a hurricane, showing typical motions in the outer reaches of the storm, the spiral cloud bands, the eyewall, and the eye.
3. When a northward-moving hurricane makes landfall on a coast that runs from east to west, the greatest damage generally occurs on the east side of the storm's path. Why?
4. Suppose winds circulate at 110 knots around a hurricane that is moving westward at 25 knots. What will the maximum winds in the storm be? If a tornado, with eddy winds of 100 knots, forms in the region of strongest hurricane winds, what is the maximum speed the wind might reach? Draw a sketch to illustrate this situation.

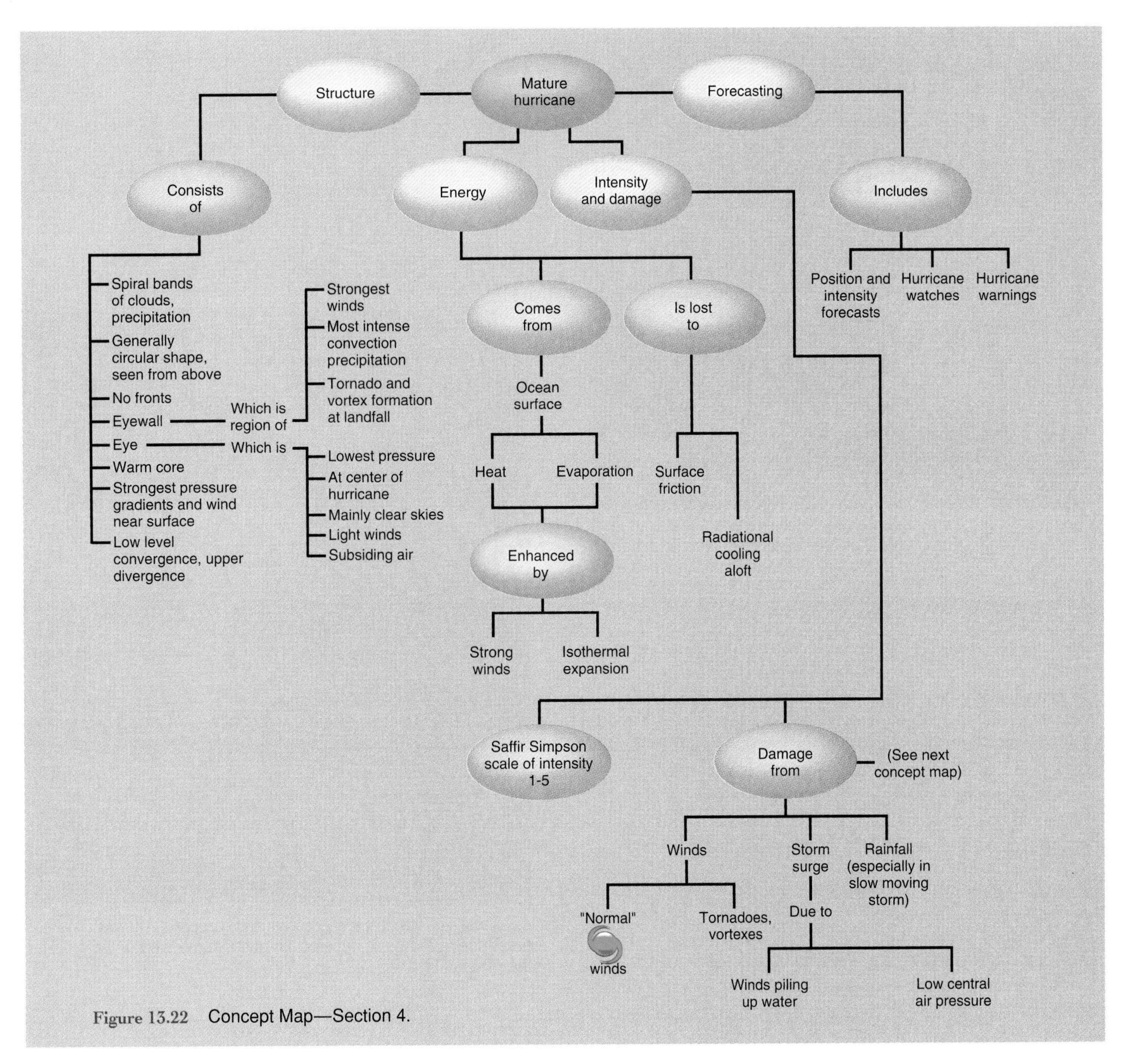


**Figure 13.22** Concept Map—Section 4.

earthquake were in progress. Glasses and books jiggle, dance, and tumble off tabletops.

Observers in other locations have their own stories to tell. For example, at Burger King International Headquarters on the shore of Biscayne Bay, the storm surge reaches 5.15 meters (16.9 feet). Huge ocean waves whipped up by the winds carry destruction to much greater heights along the immediate coast. At 4:35 A.M., wind gusts over 100 knots topple the weather radar dome at the National Hurricane Center, just seconds after it records its final image of Andrew (see Figure 13.23).

Suddenly, around 7 A.M., you notice that all is quiet outside. The rain has ceased, the wind is nearly calm, and the air pressure has ceased falling, although it sits at the incredibly low value of 922 millibars (27.23 inches of mercury). You are in the eye of the storm. You venture outside and see almost clear skies overhead. In the distance in all directions, however, you see walls of cumulonimbus clouds—the eyewall clouds. You spend a few minutes photographing the scene of destruction (see Figure 13.24). In every direction, trees, utility poles and wires, cars, boats, and other objects lie in tangled heaps. Of the 1,176 mobile homes in Homestead, all but 9 are destroyed. But you cut your damage survey short and dash for cover because the wind is freshening, this time from the southwest; the second half of the storm is about to begin.

In a few minutes, the hurricane's fury returns. Again, wind speeds exceed 100 knots, and the rain flies past in horizontal

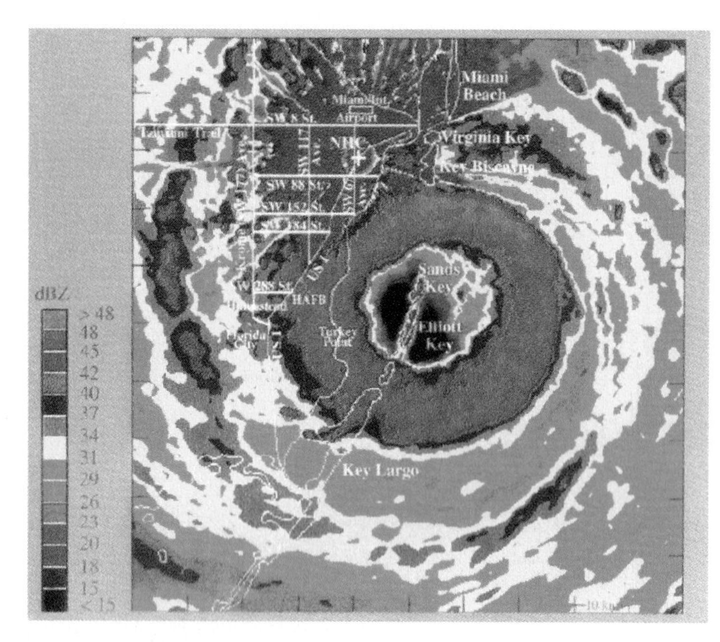


**Figure 13.23** The last full sweep of the weather radar at the National Hurricane Center before the radar dome toppled in Andrew's winds. Locate NHC on the map. What was its distance from the eye at the time this image was made? (Distance scale is in lower right corner.) Where are the most intense echoes? Where are they weakest, and why?

torrents. However, you observe important differences in wind direction and in pressure change compared to those in the first half of the storm. From these changes, you realize (Figure 13.25) that the hurricane is moving away toward the west. Indeed, by mid-morning, the rain ends, the winds lighten, and the skies begin to clear a final time. The hurricane is gone, but the damage it caused will take many months to repair.

## Andrew Weakens, Then Regenerates

Just as landfall generally marks the beginning of a hurricane's most destructive phase, it also marks the onset of the storm's decline. Andrew is no exception to this rule; as you can see in Figure 13.26, the storm's maximum winds decrease significantly, and its central pressure rises, as it crosses the Florida peninsula. In its four hours over Florida, Andrew weakens one category on the Saffir-Simpson scale, to a Category 3 storm.

### *Causes of Hurricane Weakening*

Why do hurricanes begin to dissipate as soon as they move over land? Recall from an earlier section that for a hurricane to sustain itself, there must be a deep layer of warm water, deep enough so that evaporation does not cool the water significantly. Also required are strong winds to maintain a continuous, massive transfer of heat and water vapor from the water surface to the atmosphere.

As the hurricane moves inland, both of these sustaining factors are diminished. Objects such as buildings, trees, and hills are obstacles to the wind flow and reduce wind speeds and therefore turbulent energy fluxes from the surface. And while heavy rains certainly have made the land's surface wet, evaporative energy loss from the earth's surface is not replaced, as it is over the

**Figure 13.24** Destruction caused by Hurricane Andrew. The presence of the boat in the photo gives some idea of the extent of the storm surge.

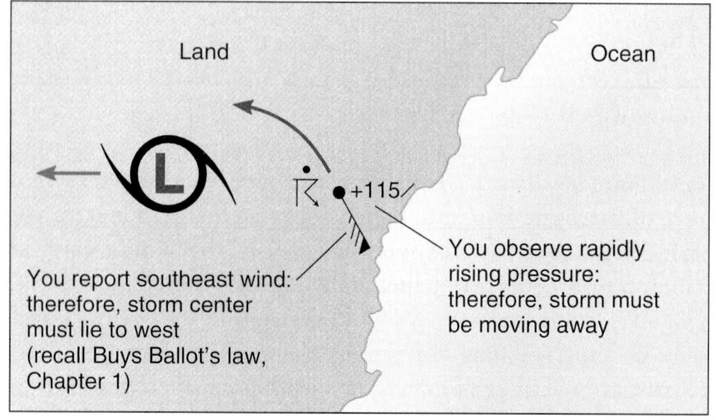


**Figure 13.25** South winds at your location imply that the hurricane's center must lie to your west. Rapidly rising air pressure indicates the storm is moving away.

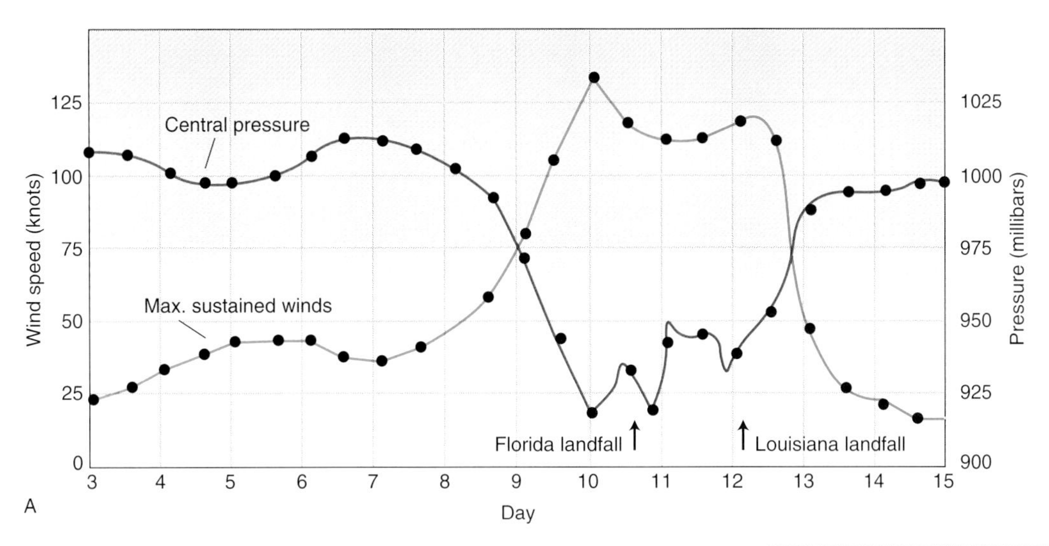


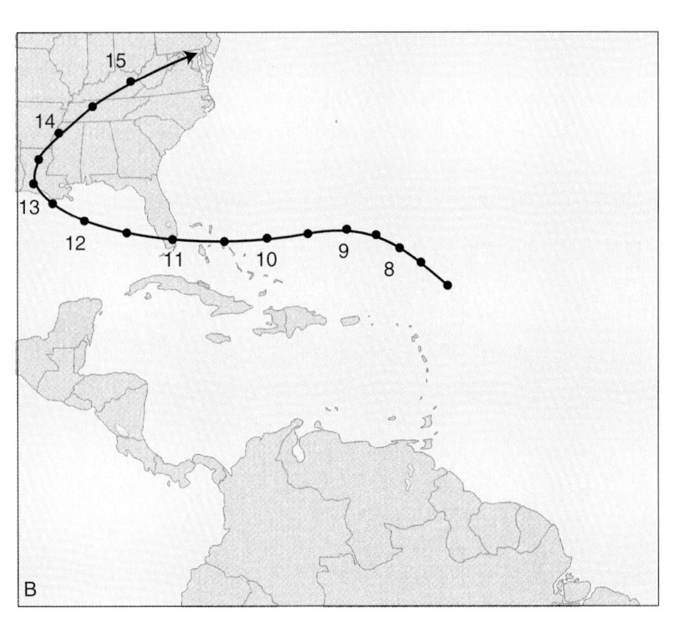

**Figure 13.26** Hurricane Andrew data through day 14 and satellite image near landfall on day 12. Note the rapid rise in central pressure and decrease in wind speed at landfall, particularly in Louisiana.

ocean, by energy from a fresh supply of warm water from below. Thus, fluxes of heat and water vapor are less over land than over the ocean. With reduced inputs of latent and sensible heat energy, the storm's warm core is cooled by as much as 10°C in its upper levels in the first 24 hours following landfall. The presence of cooler, drier (and therefore denser) air in the storm's center causes a rise in its central pressure, resulting in weaker horizontal pressure gradients and diminishing winds.

### *Reintensification over the Gulf of Mexico*

Although weakened during its traversal of Florida, Andrew is far from finished. Around 8 A.M. on day 11, the hurricane reaches the warm waters of the Gulf of Mexico. Examine Figure 13.26 to see the effect this event has on the storm's intensity. Observe that Andrew has survived its encounter with Florida, has reintensified, and now possesses the potential to strike the U.S. coast a second time.

A hurricane in the Gulf of Mexico is a particularly dangerous circumstance. Like the proverbial bull in a china shop, a Gulf hurricane cannot move without crashing into something. And Andrew does continue to move, west-northwestward. On the afternoon of day 11, the National Hurricane Center issues a new hurricane warning, this time for portions of the Louisiana and Texas coasts. Approximately 1.5 million people in these regions evacuate their homes for safer locations inland.

## A Second Landfall, Then Dissipation

Inspecting Andrew's trajectory for day 12 in Figure 13.26, you can see the first signs of an important change. Influenced by a change in direction of the steering winds aloft, Andrew turns more northward, apparently initiating its recurvature. Recurvature is good news to residents of coastal Texas, but not to Louisianians, who now lie directly in the storm's path.

### *Louisiana Landfall*

Shortly after midnight on day 13, Andrew makes landfall for the last time, in a sparsely populated section of the Louisiana coast. Its forward movement slows to 8 knots as steering winds from the southwest stall the storm and force it to recurve. Moving at this slower speed, Andrew ravages each place along its path for a longer time. One consequence is that rainfall amounts in Louisiana exceed those in Florida: Hammond, Louisiana, records nearly a foot of rain (11.92 inches). Although the storm's intensity is not quite so severe as it was in Florida, the combination of storm surge, rainfall, and wind—this time including 48 tornadoes from Louisiana to Tennessee—delivers a second round of destruction.

### *Final Decay*

Most hurricanes decay to tropical depression intensity or less as they recurve or move inland. Moving inland increases surface frictional dissipative effects, as discussed earlier. Recurving also means movement toward mid-latitudes, where the storm is liable to encounter air masses that are not tropical, along with more pronounced wind shear than in the tropics. Under these conditions, the storm loses its warm-water energy source, its vertical continuity, and its tropical cyclone properties—its warm core and its symmetry.

Over the next several days, you observe (Figure 13.26) that Andrew follows the above script rather closely. Moving inland over Louisiana and recurving, Andrew weakens dramatically. Notice the rapid increase in the storm's minimum air pressure and the decrease in maximum wind speed during days 13 and 14. On day 15, with sustained winds of just 20 knots and central pressure of 1000 millibars, the remnants of once mighty Hurricane Andrew merge with a frontal system over the mid-Atlantic states. As this weather system moves through Pennsylvania, Maryland, and Delaware, heading once again toward the open water, it generates 13 more tornadoes—a final destructive spasm.

## Andrew's Legacy

Yes, the Hurricane Andrew of this chapter was *the* Hurricane Andrew of August 23–28, 1992. Although now the storm is an event in history, its legacy will be with us for years. The families of the 26 people—3 in the Bahamas, 15 in Florida, and 8 in Louisiana—whose deaths were directly caused by the storm will never forget Hurricane Andrew. Nor will those who lost a home or other property in Andrew's $30 billion rampage—the costliest natural disaster in American history. Because of the ferocity and damage record of the storm that bore it, Andrew has been retired from the list of names for future Atlantic hurricanes.

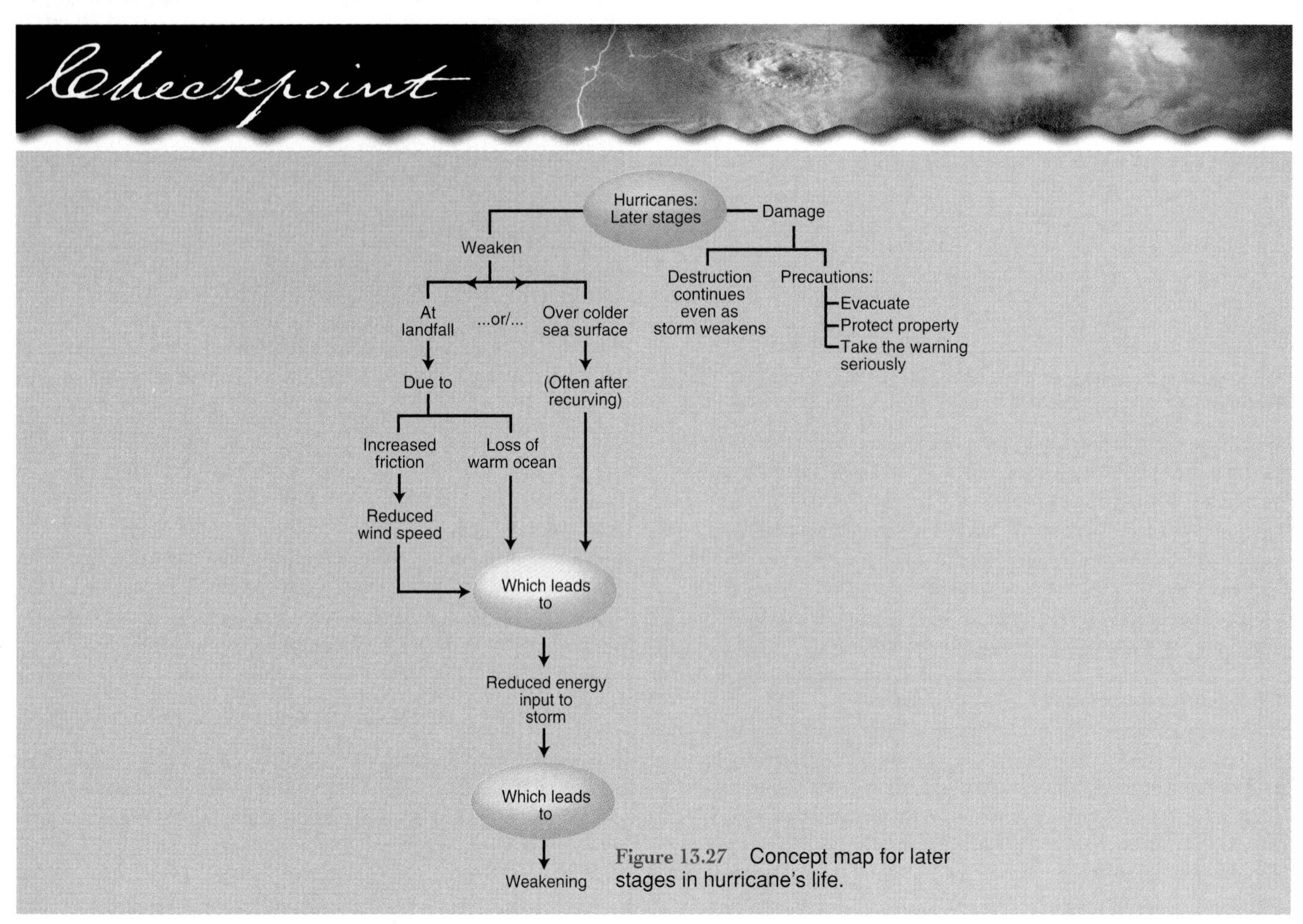


**Figure 13.27** Concept map for later stages in hurricane's life.

***Review***

Weather conditions at hurricane landfall are impossible to describe adequately in words; even the most graphic photographs cannot capture the fury and devastation of these storms. An important objective of this section is that you acquire a healthy, perhaps life-saving, respect for hurricane conditions.

This section's concept map (Figure 13.27) deals with later stages in a hurricane's life span. Note that although the branch shows that weakening can occur at landfall or after recurving, the weakening's ultimate cause is the same in each case: a reduction in the fluxes of sensible and latent heat energy from the earth's surface.

***Questions***

1. From Figure 13.26, determine the change in Andrew's central sea-level pressure (a) while crossing Florida, (b) upon reaching the Gulf of Mexico, and (c) while moving inland over Louisiana. Explain why the pressure changed as it did at each of these times.

2. Andrew was a slightly more powerful storm when it struck Florida than at landfall in Louisiana. Why, then, were rainfall totals greater in Louisiana?

## Retrospective: Are Hurricanes and Tornadoes Predictable?

Now you have studied two of nature's most destructive weather systems, tornadoes and hurricanes. Here, we return briefly to the central question of this unit: Are these storms predictable?

Certainly, many forecasts continue to fall short of perfection, as you have seen in this unit. Critics point out that lead times for tornadoes are often given in terms of minutes and that hurricane forecasts often mislocate storms by dozens of miles. They might add that hurricane warning zones typically are three times the size of the areas that actually experience hurricane conditions; thus, two-thirds of the region is "overwarned." Clearly, if *predictable* means pinpoint accuracy, hours or days ahead of time, then these circulations currently are unpredictable.

A more optimistic person, on the other hand, might cite data like those in Figure 13.28, which show improvement in forecast skill and dramatic reductions in the number of deaths due to these storms. Obviously, forecasts are better now than ever, both in mathematical terms such as mean errors, and in practical terms such as fatalities. Are the forecasts good enough to allow us to say we can predict small-scale circulations like tornadoes and hurricanes? If the answer is no, then what level of accuracy *is* required for the answer to be yes?

The issue of predictability offers an important perspective on our relation to our natural world. Thunderstorms, tornadoes, and hurricanes are just three of many complex natural processes that are predictable only within some significant range of uncertainty. Furthermore, according to chaos concepts such as the butterfly effect, we will never be able to predict such phenomena exactly. In addition to the limits imposed by chaos and complexity, there are limits in terms of the resources we as a society can afford to devote to making tornado or hurricane forecasts arbitrarily accurate.

Thus, we will always live with some level of uncertainty about our future. We will have to tolerate some degree of ignorance about what the natural system of which we are a part has in store for us. A pessimist might receive such news with a despairing "I *knew* it." To the optimist, uncertainty can be a frontier to explore, a challenge to meet, a breath of fresh air in the drama of life on earth.

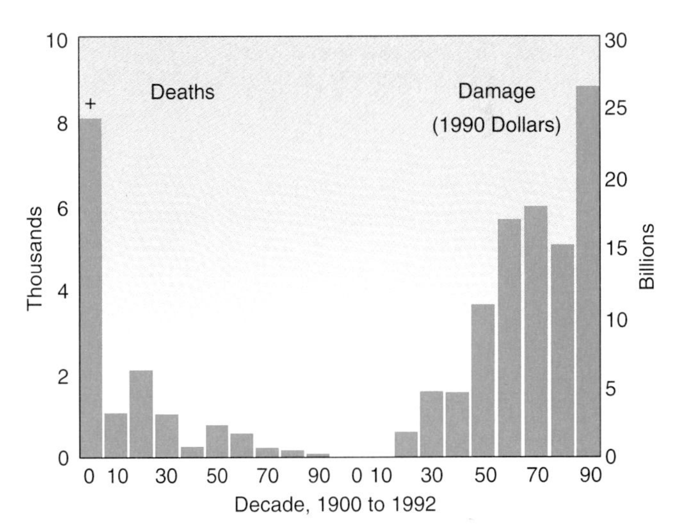


**Figure 13.28** Deaths and damage from hurricanes. Source: Data from *Hurricane!* NOAA Pamphlet 91001, p. 15. Revised April, 1993. NOAA

# Study Point

## Chapter Summary

Hurricanes are intense tropical cyclones of large mesoscale proportions that develop over warm ocean water in many parts of the world. A deep layer of warm ocean water provides the storm with its energy source in the form of large fluxes of sensible and (especially) latent heat. Also required are a preexisting, large-scale wind flow system that imposes a low-level convergent and/or upper-level divergent flow on the region and an air mass that is warm, humid, and free of strong inversions or wind shear. Due to an absence of Coriolis force on and near the equator, hurricanes do not form in this region.

To qualify as a hurricane, a tropical cyclone must have sustained winds in excess of 64 knots. Hurricanelike storms with lesser winds are designated tropical depressions (maximum winds between 20 and 34 knots) or tropical storms (maximum winds between 35 and 64 knots). Hurricanes are known by different names in different parts of the world: *typhoons* (in the Western Pacific) and *cyclones* (in the Indian Ocean) are two examples.

Hurricanes differ from frontal cyclones in many ways. They are smaller than frontal storms, are roughly circular, consist of a single (mT) air mass, and have no fronts. Being warm-cored, the hurricane's circulation is strongest near sea level and weakens with height. As air spirals toward the storm's center, it gains speed, sometimes reaching well over 100 knots. Tornadoes and smaller vortexes spawned in the hurricane's eyewall create even stronger, short-duration winds and gusts. One of the hurricane's most distinctive features is its eye, a region of relative calm at the very center of the storm.

Hurricanes generally move westward, steered by the trade winds on the equator side of the subtropical high pressure belt. Reaching the west side of the subtropical high cell, many storms recurve northward. If a hurricane makes landfall, it can cause widespread devastation due to a combination of strong winds, storm surges of ocean water, and heavy rainfall. Moving over land or colder ocean waters, hurricanes lose energy and dissipate, often merging with mid-latitude weather systems.

As with other circulations from tornadoes to frontal cyclones, we have much to learn before we can predict accurately just when and where hurricane formation will occur. Once a hurricane has formed, however, forecasts generally are much more accurate. With guidance from satellite imagery and aircraft reconnaissance to provide an accurate record of the storm's past, charts of past hurricane tracks, and computer modeling, forecasts have shown considerable improvement in skill in recent years. The dramatic reduction in hurricane fatalities in recent years is important evidence of this improvement.

## Key Words

hurricane
tropical cyclone
tropical depression
recurvature
eye (of hurricane)
isothermal expansion
hurricane warning
Saffir-Simpson scale
typhoon
easterly wave
tropical storm
linear
eyewall
hurricane watch
storm surge

## Review Exercises

1. A tropical cyclone has maximum sustained winds of 40 knots. What is its designation (tropical depression, etc.)? Suppose it weakens, its peak winds dropping to 25 knots. Now what is its designation?
2. Suppose you were planning a two-month-long, transatlantic voyage on a small boat. When would be the best time of year to make the voyage?
3. Some hurricanes, spawned in the Eastern Pacific, move into the Western Pacific. What effect would such movement have on their names?
4. What percentage of Atlantic tropical depressions become tropical storms? Hurricanes?
5. Sketch and describe the wind flow in the hurricane's eyewall and eye.
6. Why is the weather generally more severe in the forward right quadrant of a hurricane?
7. Describe two factors that contribute to a storm surge.
8. Hurricane Bonnie (1992) reached a minimum central pressure of 965 millibars and maximum sustained winds of 95 knots. What category of hurricane was Bonnie at this time? If Bonnie made landfall at this time, how great a storm surge might you expect?
9. Despite breaking all records for costliness, Hurricane Andrew caused relatively few fatalities. Why?

## Problems

1. Compare the processes of isothermal expansion as surface air spirals toward a hurricane center with adiabatic expansion as it rises. Imagine in each case that the air begins at a temperature of 30°C and is at its saturation mixing ratio. Suppose the air pressure decreases by 100 millibars (an extreme but not impossible value). What will be the final temperature and dew point in each case?

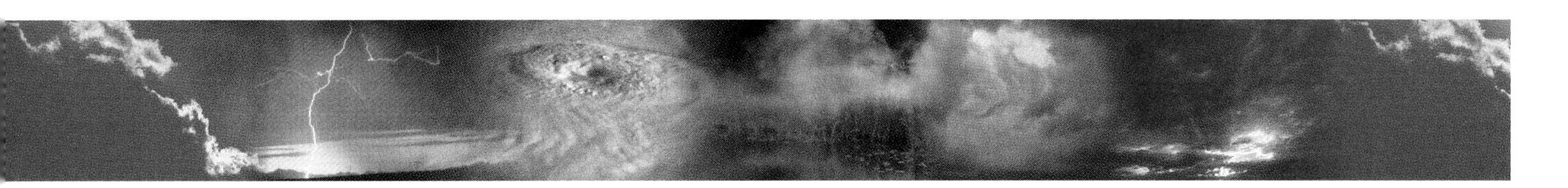

2. Why do different computer forecast models, based on the same observational data, give different forecasts?
3. A tropical cyclone near Japan has maximum sustained winds of 110 knots and a central sea-level pressure of 953 millibars. What name and classification category apply to this storm?
4. Why is the East Coast of the United States far more susceptible to hurricanes than the West Coast?
5. Suppose you live in Corpus Christi, Texas. One day, you hear there are two hurricanes on the map: one centered at latitude 30° north latitude and 120° west longitude. The second is located at 23° north and 50° west. Which storm is closer to you at present? Which is more likely to affect your weather? Why?
6. Compare the conditions described in the Saffir-Simpson scale with those in the Fujita scale of tornado damage and those of the Beaufort wind scale. At equal wind speeds, do the descriptions of damage and so on seem to match?
7. What role do you think hurricanes play in the global energy balance? (Hint: Imagine the oceans were warmer than at present. What effect might this have on hurricane frequency or intensity?)
8. Generally, hurricane warnings are issued about 24 hours before expected landfall. The amount of lead time poses interesting questions. Longer lead times (say, 36 or 48 hours) would give residents more time to prepare, to change plans, perhaps to evacuate. On the other hand, a 48-hour warning inevitably will be less accurate. Discuss the pros and cons of longer lead times.

## Explorations

1. If you have access to real-time weather data, establish a tropical cyclone tracking program. Several good Internet sites are listed in Resource Links. Although the best times for Atlantic and Eastern Pacific hurricanes are late summer and early fall, tropical cyclones occur in every month of the year. Follow day-to-day changes in storm position, central sea-level pressure, and maximum winds. Make predictions based on historical data and extrapolation. Verify your predictions with future observed positions.
2. Hurricanes have made intriguing history, from their impact on Columbus's voyages in 1492 to the impact of Hurricane Andrew exactly 500 years later. Other storms of special interest are the Galveston Flood of 1900, the New England Hurricane of 1938 (particularly significant because it took everyone by surprise), and many others: Camille, Hugo, Diane, Carol, and Esther, for example. Each of these has fascinating stories behind it. Write a short research paper or prepare a presentation to your class on one of these storms.

## Resource Links

The following is a sampler of print and electronic references for some of the topics discussed in this chapter. We have included a few words of description for items whose relevance is not clear from their titles. More challenging readings are marked with a C at the end of the listing. For a much more extensive and continually updated list, see our World Wide Web home page at www.mhhe.com/sciencemath/geology/danielson.

Cotton, W. R. 1990. *Storms*. Fort Collins, CO: Aster Press. 158 pp. Good discussion of hurricane formation.

Emanuel, K. A. 1988. Toward a general theory of hurricanes. *American Scientist* 76 (4): 371–79.

Minsinger, W. E., and C. T. Orloff, eds. 1994. *The North Atlantic Hurricane*. Milton, MA: Blue Hill Meteorological Observatory. 64 pp. Photos and vivid descriptions of the September 1944 hurricane.

Wakimoto, R. M., and P. G. Black. 1994. Damage survey of Hurricane Andrew and its relationship to the eyewall. *American Meteorological Society Bulletin* 75 (2): 189–200.

AccuWeather's home page at http://www.AccuWeather.com is an extensive source of current weather and forecast conditions, including hurricane data.

For readers with access to AccuData, HURRG gives a wide variety of hurricane graphics including hurricane position, movement, strength, etc. TRPA, TRPP and TRPI provide tropical advisories, reconnaissance and discussions.

NOAA Geophysical Fluid Dynamics Laboratory has an interesting graphic of circulation within Hurricane Andrew at http://www.gfdl.gov/brochure/Cover.doc.-html and a spectacular image of Hurricane Emily at http://www.gfdl.gov/hurricane.html .

The National Hurricane Center home page at http://www.nhc.noaa.gov/index.html contains a wealth of information about tropical storms worldwide.

The University of Wisconsin (Madison) Tropical Weather Page at http://cimss.ssec.wisc.edu/tropic/tropic.html gives current conditions for hurricanes and other storms from late spring until November 30; it also includes archives and movies.

The University of Illinois at http://www.atmos.uiuc.edu/ is a good source of information about hurricanes and tropical meteorology.

Chapter 14

# How Stable Is Earth's Climate?

## Chapter Goals and Overview

The Montana wheat farmers in the opening figure are a study in people's faith in the stability of their region's climate. Year after year, they bet their livelihoods and their financial resources that rainfall during the growing season will be plentiful (but not too plentiful); that temperatures will fall within a range suitable for wheat to germinate, grow, and mature; and that hail and wind will not level their crops. Most years, their bet is a safe one. Summer weather in Montana tends to fall within rather clearly defined ranges. Almost as though it had memory, the temperature for a given year follows traditions of past years, climbing into a predetermined range by a given date each spring.

But some years, the wheat farmers lose their bet with the climate. A spring too wet or too dry or a half-hour of summer hail can wipe out a year's investment of effort and money. Such anomalous weather conditions may simply be random events lasting only one or a few years before conditions return to normal, or they may herald some long-term change in the region's climate. As you can imagine, the latter possibility has serious implications for both the farmers and the appetites of the world's population.

In this chapter, you will study many issues related to climate. First, you will become acquainted with a few basic statistical tools used to describe a region's climate. After reviewing the factors that determine climate, you will survey the different climatic regions of our planet. Finally, you will consider issues of climate change.

Climatology is a particularly appropriate topic for the final chapter of a meteorology textbook, because climate can be viewed as a synthesis of all weather events. Stated another way, in the previous 13 chapters, we have analyzed specific atmospheric features in some detail. In this chapter, we take a more holistic approach as we consider how these features combine to produce a larger entity known as climate. Because this chapter is based heavily on concepts developed earlier, you will see frequent references to previous sections. We recommend you turn back to these sections often, to refresh your memory.

Finally, we should say that this chapter is just a bare introduction to a fascinating, extensive, and increasingly important subject that is the topic of entire college courses. We hope this taste of climatology whets your appetite and moves you to study the subject in greater depth.

### Chapter Goals

By the end of this chapter, you will be able to:

- define climate and identify types of data commonly used to specify a region's climate;
- explain the concept of climatic stability;
- Identify factors that control a region's climate;
- classify a locale's climate, based on long-term temperature and precipitation data; and
- discuss the roles of astronomical, geophysical, and anthropomorphic factors in climate change.

## Defining and Measuring a Region's Climate

Up to now, we have used the term *climate,* somewhat imprecisely, to refer to long-term weather. Now, it is time to define the term more rigorously. **Climate** is the sum of all weather events affecting a region over a long period of time. Often, a 30-year period is considered standard for defining a region's climate.

### Standard Measures of Climate

From the all-inclusive nature of the definition given above, it may seem that only a moment-by-moment listing of a locale's weather for 30 years, like the sample shown in Table 14.1, would capture the essence of its climate. Such a compilation, although complete in one sense, would be uninformative because of its very completeness. We need to process this volume of data to reduce it to comprehensible proportions and to find patterns and trends. The following statistical measures are routinely used for such reduction and generalization.

#### *Means and Normals*

The mean (or average), the most widely used statistic in climatology, needs little introduction. Climatologists use data such as the mean annual temperature in San Antonio, Texas, (20.4°C, or 68.7°F) or the mean July rainfall in Orlando, Florida, (203.2 millimeters, or 8 inches) to reduce thousands of individual readings to a more meaningful form.

**Normal** values, on the other hand, demand a bit more explanation. You may have heard statements such as "The normal high temperature for July in Boston is 82°F, and the normal low is 65°F." Normals are climatological means, measured and calculated over a standard (usually 30-year) time span. The term *normal* can be misleading because it sounds as though it is the *usual* value. In fact, Boston's daily July maximum temperature only rarely is 82°F exactly; almost every day, it is either higher or lower. Thus, there's nothing particularly normal about a "normal." When you hear talk of "normal" temperatures, think "mean."

#### *Ranges and Extremes*

Examine the two graphs in Figure 14.1. They are examples of the **climograph,** a basic and widely used tool in climatology. As you can see, the climograph presents a considerable amount of information about monthly temperatures and precipitation for a given locale.

Look at the January temperature data (in red) for Eureka, California, in Figure 14.1. The bars represent the mean daily range of temperatures for each month. The **range,** as its name suggests, is simply the span of individual values from highest to lowest: in this case, from the mean daily high to the mean daily low temperature. What is the mean daily temperature range in January for Eureka? From the figure, you can see the mean high is 54°F (12°C), the mean low is 41°F (6°C), and thus the range = 54 − 41, or 13°F (6°C). On an "average January day" in Eureka, then, the temperature varies by 13°F, from a low of 41°F to a high of 54°F. We hasten to remind you, however, that such a

Table 14.1 Decoded Extract from Surface Weather Observations at Windsor Locks, Connecticut, on January 22, 1961

| Time (local std.) | Sky and Ceiling | Visibility (miles.) | Weather and Obstructions to Vision | Sea-Level Pressure (millibars) | Temperature (°F) | Dew Point (°F) | Wind Speed (knots) |
|---|---|---|---|---|---|---|---|
| 0356 | Clear | 15+ | | 1017.2 | −13 | −20 | Calm |
| 0430 | Clear | 15+ | | | | | Calm |
| 0452 | Clear | 15+ | | 1016.9 | −14 | −19 | Calm |
| 0530 | Clear | 15+ | | | | | Calm |
| 0557 | Clear | 15+ | | 1017.2 | −20 | −25 | Calm |
| 0630 | Clear | 15+ | | | | | Calm |
| 0655 | Sctd cirrus | 15+ | | 1017.6 | −21 | −21 | Calm |
| 20000 63020 17679 00901 79303 90423 41774 LOXAT | | | | | | | |
| 0755 | Sctd cirrus | 4 | Ice fog | 1018.2 | −19 | −24 | Calm |
| 0830 | Sctd cirrus | 5 | Ice fog | | | | Calm |

Note: The five-digit numbers in the line following the 6.55 A.M. data contain special data that are reported four times per day. For example, 90423 means that the snow depth was 23 inches. The 41774 group gives daily temperature ranges; the 17 indicates the high was 17°F (−8°C); 74 means (in coded form) that the morning's low was −26°F (−32°C). The letters LOXAT indicate that −26°F was the all-time record low temperature observed at the station. (It occurred between the 6 and 7 A.M. reports and therefore is not shown in the temperature column.)

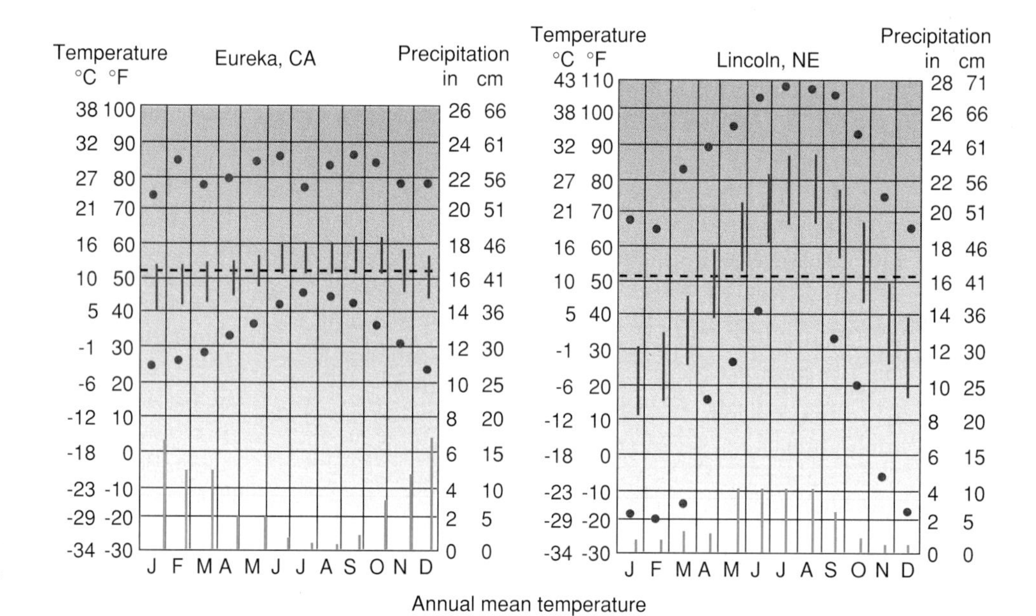


Figure 14.1 Climographs for Eureka, California, and Lincoln, Nebraska. Notice the greater range of temperatures in Lincoln. In what season does each city receive the most precipitation? In what months are the mean temperatures at the two stations roughly equal?

day is a fictional "average day" and that almost all individual days will see temperatures that differ from the mean in one way or another.

Notice from Figure 14.1 that Eureka and Lincoln, Nebraska, have nearly the same annual mean temperature (11.4°C, or 52.5°F). You can see from the monthly temperature bars, however, that month-to-month variations in the temperature record at the two stations differ markedly. A given month's mean temperature in Eureka is likely to be closer to the annual mean temperature than in Lincoln, where large seasonal deviations from the annual mean occur. The amount by which individual data points vary from their mean, sometimes called **scatter,** is another important climatological measure. In this case, we would say that Lincoln's monthly mean data exhibit greater scatter about the annual mean than do Eureka's.

**Extreme values,** as the name implies, are "highest ever," "lowest ever," "fastest ever," "most ever," and so forth. Extreme temperature values are shown in Figure 14.1 as red dots above and below the mean temperature bar for each month. What is the highest June temperature ever recorded in Eureka? In Lincoln? From the climograph you can verify these values are 85°F (29°C) and 103°F (39°C), respectively.

Extreme data are particularly useful to people engaged in planning building construction, highway drainage, agricultural projects, and so on. What is the greatest 24-hour rainfall ever measured in Los Angeles? (It was 155.1 millimeters, or 6.11

inches, in January 1956.) What is the coldest temperature ever recorded in New Haven, Connecticut? (It was −26°C, or −15°F, in February 1934.) Like the examples above, extreme values generally are all-time records, not just the highest or lowest recorded during the past 30-year standard data period.

## Proxy Data

Study the two sites illustrated in Figure 14.2. Why is the vegetation so different in the two scenes? A region's flora is controlled to a large extent by the local climate. In the early years of the 20th century, when long-period weather records were more sparse than they are now, scientists used existing maps of vegetation as indicators of regional climates. Thus, plant distributions have a long history of service as proxy data in place of standard climatological measurements.

Researchers investigating climate change often require data extending much farther back in time than the span of available meteorological data—a period of only 100 to 200 years. For longer periods, they must rely on proxy data of the sort you read about in Chapter 2. Tree rings, lake sediments, and glacial ice layers all contain information about past climates. Two recently completed Greenland ice coring projects have yielded particularly valuable data on climate conditions as much as 120,000 years ago.

## What Is a "Stable Climate"?

Suppose that some natural event (a meteor shower) or some human enterprise (nuclear explosions) caused the Arctic Ocean's ice cap to melt. Replacing millions of square miles of ice with open water would cause a drastic perturbation in the climate of the Arctic—and perhaps of the entire earth.

And then, what would happen? At least four possibilities come to mind: (1) The new climate might persist, with a permanently ice-free Arctic Ocean. (2) The ice might form again, with the climate returning to earlier conditions. (3) The climate might fluctuate, either regularly or erratically. (4) The climate might settle into a pattern different from both preepisode and immediate postepisode conditions.

These scenarios are the responses of unstable (case 1), stable (case 2), and neutrally stable systems (cases 3 and 4) to a perturbation inflicted on the system. You have met these stability concepts in earlier chapters. To cite two examples, recall that in Chapter 6 you saw static stability as an indicator of a perturbed air parcel's tendency to return to its original state. And in Chapter 9, you read that waves may amplify unstably on the polar front, developing into frontal cyclones. In general, stability is a measure of a system's tendency, when perturbed, to restore itself to earlier conditions.

Notice that a stable system is not necessarily an unchanging one. On the contrary, stability is based on the assumption that perturbations do occur but that feedback mechanisms within the system act to return the perturbed system to its earlier state. Thus, in the discussion that follows, we are seeking to learn primarily not whether the earth's climate has changed but whether the present climate, if perturbed, is likely to return to its present condition.

**Figure 14.2** The type of vegetation a locale supports depends on its climate. It should be obvious which of these two places receives more precipitation.

Checkpoint

***Review***

Remember that climate is not merely "average weather." Climate is the sum of all weather events affecting a locale. Thus, while typical values (such as means) are important descriptors of a region's climate, so also are measures of scatter. Proxy data are important supplements to the record of standard weather observations.

***Questions***

1. What is the difference between weather and climate?

2. Arrange these values from largest to smallest: April mean temperature, April mean daily maximum temperature, April all-time low temperature, April mean daily minimum temperature, April all-time high temperature.

3. Which of the following are considered primarily weather data and which are climatological data? Normal April minimum temperature in San Diego; wind speed at 4:00 P.M. on January 6, 1996, in Fairbanks; the all-time low temperature at Duluth, Minnesota; the average number of tropical cyclones to strike the Texas coast each year.

## What Factors Control a Region's Climate?

Like the weather itself whose conditions over the years define a given climate, a region's climate is shaped by influences operating on all scales of space and time from microscale to global. In this section, we divide these influences into two broad categories. First are the more or less permanent features, such as a region's latitude or its location with respect to the ocean. Second are factors that are subject to some variation, such as prevailing wind systems and ocean currents. Some of these factors will be familiar to you from earlier chapters.

### Constant Factors

Perhaps we should qualify the word *permanent* in the preceding paragraph; after all, very little of the physical earth is truly permanent. Over long enough time spans, mountains form and erode, oceans expand and contract, and continents move, collide, and rebound. However, for time periods of tens or hundreds of years, such factors as a location's latitude or altitude generally can be considered permanent. It is in this context that we use the word here.

#### *Latitude*

Every autumn, many millions of butterflies, birds, and humans migrate from higher to lower latitudes, seeking a warmer climate until the end of winter. This massive, once-a-year commute underscores the dominant impact of latitude on climate. In fact, a region's latitude is arguably its most important climatic attribute, due to the effect it has on the amount of solar radiation reaching the earth's surface. As you saw in Chapter 3, low-latitude regions receive more solar energy and therefore are warmer than those nearer the poles. Thus, you might expect climate to change more rapidly from north to south than from east to west. This supposition is correct for changes in temperature, as you can see from the maps of global temperatures in Figure 14.3. Notice that the warmest temperatures lie on or near the equator and the coldest are near the poles, and that isotherms tend to run along parallels of latitude, an indication of latitude's controlling influence.

#### *Altitude*

Based on its latitude (within 3 degrees of the equator) Mount Kilimanjaro, pictured in Figure 14.4, should have a tropical climate. However, its snow-capped summit bespeaks a climate far from tropical.

The peak's great elevation (5895 meters, or 19,335 feet, above sea level) accounts for this patch of Arctic climate on the equator. The greater a locale's altitude, the more its climate is like those at higher latitudes. As a rough rule of thumb, you can figure that in middle latitudes, each 100-meter (330-foot) rise in elevation causes a temperature decrease equivalent to moving poleward one degree of latitude (approximately 100 kilometers, or 70 miles).

As another example, consider the summit of Mount Washington, New Hampshire, at 1917 meters (6288 feet) above sea level. To what sea-level latitude do its summit temperatures correspond?

To find the answer, divide the altitude (1917 meters) by 100 meters per degree of latitude, which gives 19.17 degrees of latitude. Thus, the summit temperatures are typical of those at sea level 19 degrees of latitude farther north. Since Mount Washington lies at 45° north latitude, the rule indicates its summit temperatures are roughly like those at a latitude of $45° + 19° = 64°$, nearly as far north as the Arctic Circle.

#### *Location with Respect to Mountain Ranges*

You do not have to be actually located on a mountain to experience its impact on climate. As you read in earlier chapters, mountain ranges can profoundly affect the climate for considerable distances by stimulating synoptic scale cyclone formation and inducing orographic clouds and precipitation. The data for Seattle and Spokane, Washington, shown in Figure 14.5, offer an interesting example. Seattle, lying between the Pacific Ocean

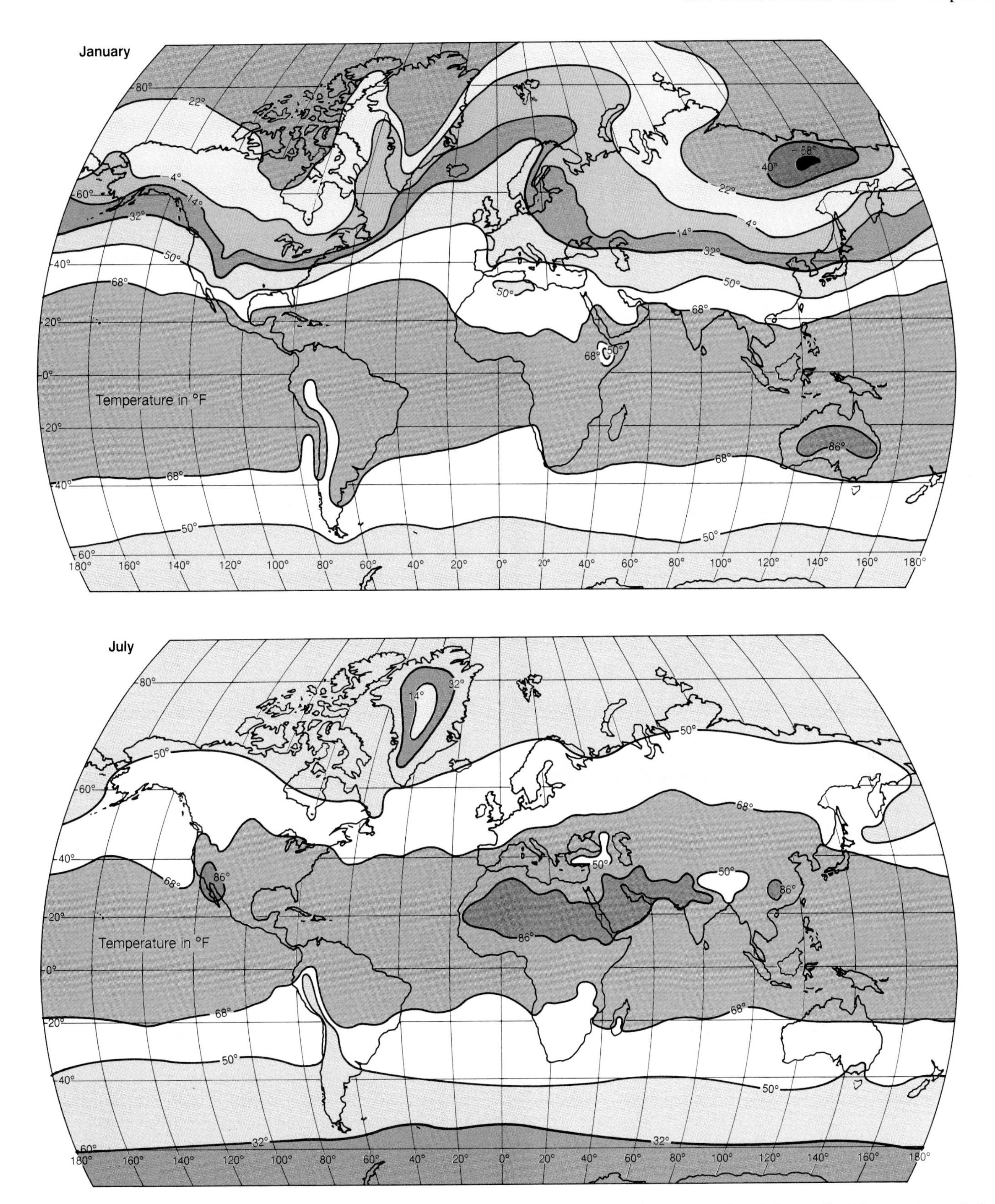


**Figure 14.3** Global mean temperatures for January and July. In which hemisphere (northern or southern) do you find the greater variation in temperatures? Why? (Look for the answer and an explanation in the following pages.)

Figure 14.4 Mount Kilimanjaro, Kenya. Notice changes in the nature of the surface as you move to higher elevations.

and the Cascade Mountains, has a climate that reflects a stronger maritime influence (greater precipitation, less temperature variation) than Spokane's.

### *Location with Respect to Oceans*

Eureka, California, and Lincoln, Nebraska, whose climographs you saw back in Figure 14.1, lie at nearly the same latitude (40.8° north), and neither is influenced by the vagaries of mountain ranges. Notice that despite these similarities, their weather differs dramatically. Why the great difference?

The main difference lies in the degree of maritime influence they experience. Recall the discussion in Chapter 3 that explained why daily and annual temperature variations are far less over the oceans than over land. The ocean's "moderating" influence is an important determinant of a region's climate. You can see this moderating effect in the map of global temperatures in Figure 14.3. Observe that both the hottest summer temperatures and the coldest winter temperatures occur over land, not over oceans.

Earlier in this text, you have read about other ways in which land-ocean contrasts affect weather—and therefore climate. For example, recall that the temperature contrast across the land-sea boundary encourages differential heating, which can lead to increased storminess, from thunderstorms to frontal cyclones, along coastal regions. Further, you read in Chapter 11 that upwelling of deep ocean water, which can strongly influence regional climate, occurs along coastlines under proper wind flow conditions.

### *Local Topography*

Notice the vegetation patterns in Figure 14.6. These patterns are mainly the result of differences in the **microclimate,** the climate over microscale regions. Variations in the slope of the ground can have profound effects on absorbed sunlight, exposure to winds, and runoff.

## Variable Factors

In contrast to the constant factors discussed in the previous section, the factors we consider next affect climate in ways that can vary considerably in time.

### *Global Wind and Pressure Regimes and Ocean Currents*

You studied the general circulation of the atmosphere and global ocean currents in Chapter 11. We encourage you to review this information now (in Figures 11.30 and 11.32, for example), because it plays an important role in determining regional climate.

The atmospheric general circulation features are of special importance; hence, we will consider them in more detail in Figure 14.7. Notice in the first cross section that each of these major features (such as the ITCZ) defines a characteristic climate (e.g., warm and rainy). These circulation features tend to migrate with the sun as it moves north and south of the equator during the course of the year. Thus, as you can verify in Figures 14.7B and C, the climates associated with these features also migrate somewhat with the seasons. The annual cycles of temperature and precipitation at individual stations reflect these seasonal variations. Thus, for example, a locale at 10° north latitude is likely to experience a rainy May through July due to the northward migration of the ITCZ but a dry November through January as the subtropical high moves southward and dominates the region's weather. As you might suspect, these effects are greatest in places where influences of other climate controls such as mountain ranges or ocean-land boundaries are minimal.

### *Nature of the Surface*

Is the ground soggy, moist, or dry? Is it unfrozen, frozen but bare, or snow-covered? Is it covered by grass, leaves, trees, crops, pavement, or buildings? Local surface conditions can shape climate to a considerable extent. For example, as you saw in Chapter 3, conditions at the earth's surface determine its albedo and therefore the proportion of incoming solar energy

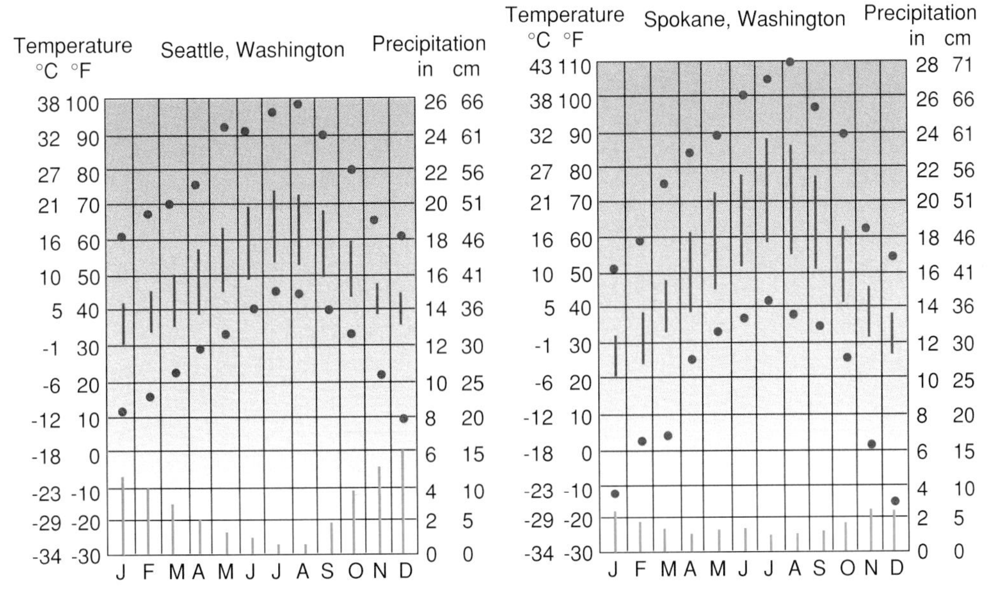


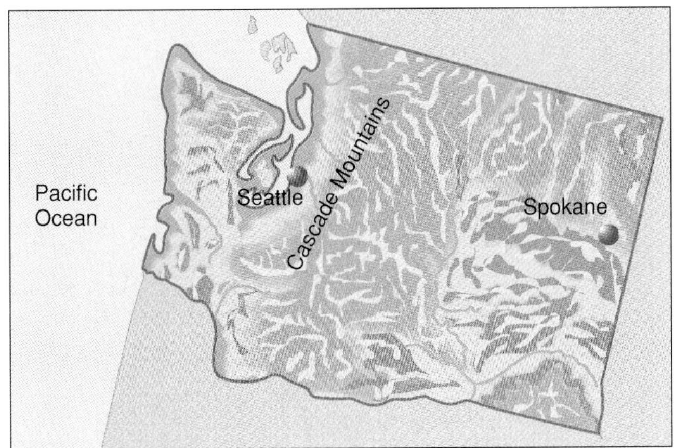


Figure 14.5 Climograph for Seattle and Spokane, Washington. In what ways might the Cascade Mountains be responsible for differences in precipitation between the two stations? For differences in temperature?

that is captured and converted to heat. Thus, temperature, a key element of a region's climate, is determined in part by the local surface albedo. But recall also that albedo may be subject to powerful feedback mechanisms: temperature can mightily influence albedo, as in the case of snow melting and exposing a darker, soil-covered surface.

As another example, compare energy fluxes over a moist surface, such as wet tidal flats at the seashore, with those over dry sand at slightly higher elevations. Over the moist surface, both latent and sensible heat fluxes operate, whereas over the dry surface, where no water is available to evaporate, only sensible fluxes develop. Thus, when exposed to a large energy input such as morning sunlight, the dry surface and the atmosphere in contact with it warm rapidly compared to the moist surface, much of whose energy is removed by latent heat flux. As a result, the tidal flats' climate may differ radically from that of the sand dunes only a short distance away.

A final example of the role the surface plays in climate concerns the effect of surface roughness on convective transfer. Mechanical turbulence over terrain made rough by features such as hills, tall buildings or trees promotes the development of eddy circulations in the lowest atmospheric layers, thus enhancing the

Figure 14.6 Variations in microclimate can lead to great local variations in vegetation.

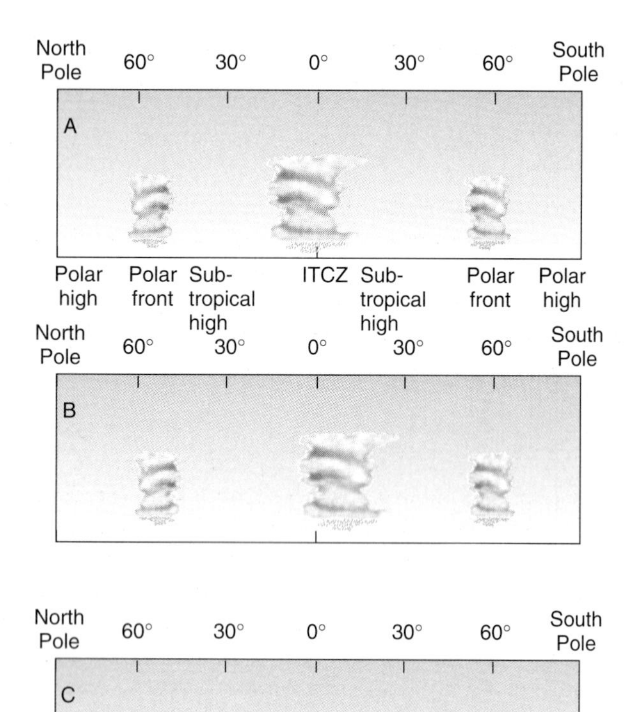


**Figure 14.7** As general circulation features migrate north and south with the sun, they bring seasonal climate changes to many regions.

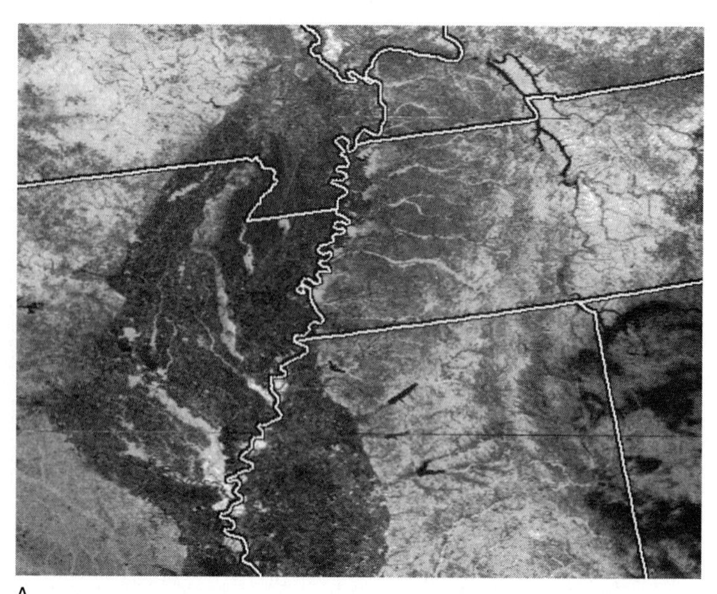


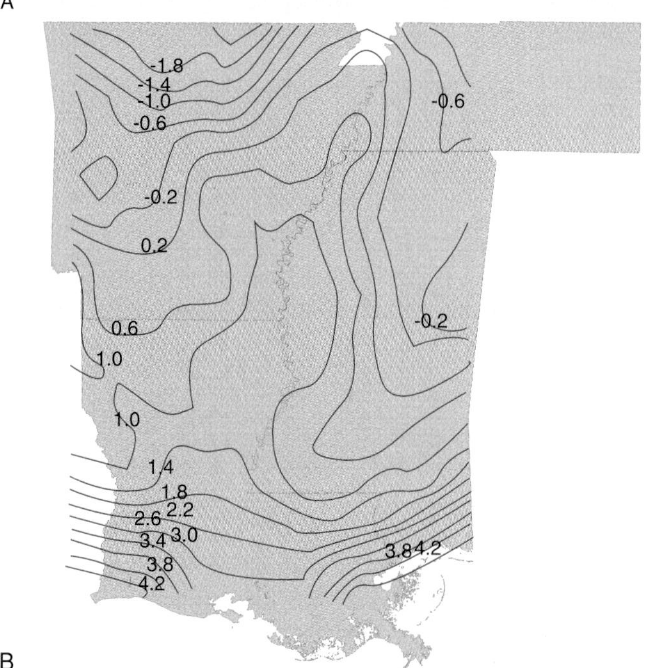


**Figure 14.8** (*A*) A satellite image of vegetation in the lower Mississippi Valley in late spring. Lighter areas have greater amounts of vegetation than darker areas, which are under cultivation. (*B*) Variations in daily minimum surface temperature over the same region at the same time of year.

transfer of energy between the earth's surface and the atmosphere. Such action can have many local effects: it brings stronger winds aloft down to the surface; it mixes heat and moisture within the turbulent layer, thus reducing the occurrence of extreme conditions at the surface; and it increases the formation of stratocumulus cloud layers due to mixing (see Chapter 6).

The next section presents an interesting illustration of the role of surface moisture on local climate.

## *Anthropogenic Effects*

Examine the satellite image of the lower Mississippi River flood plain in Figure 14.8A. Notice the contrast between lands along the river and those farther distant. Agriculture along the flood plain has reduced the amount of vegetation, particularly in spring and early summer when crops are young, causing a darker image than in the forested regions on either side. With less vegetation, the surface has less moisture available for evaporation. As a result, it experiences higher temperatures than the forested region, as you can verify in Figure 14.8B.

Figure 14.8 illustrates just one example of anthropogenic change (change caused by humans) to our climate. Recall that you considered other examples in Chapters 3 and 4; you will study still others in the final section of this chapter.

Variable factors such as anthropogenic ones are particularly significant to the study of climatic stability precisely because they vary. The continual perturbations provide scientists with an ongoing series of experiments on climate stability. We will consider stability issues further, after first becoming acquainted with the various climate regions on our planet.

***Review***

A region's climate is the product of a number of contributing factors. Some, such as the region's geographical setting (latitude, proximity to oceans, etc.), are constants. Others, such as the effects of large-scale atmospheric and ocean circulations, vary over time. These varying factors continually perturb the earth's climate.

***Questions***

1. Why do the isotherms in Figure 14.3 run generally in an east-west direction and not, for example, from north to south?
2. Mount Mitchell, North Carolina, at 36° north latitude, has an altitude of approximately 2 kilometers (6684 feet). Roughly to what sea-level latitude do temperature conditions at its summit correspond?

## Climate Classification

Now we have assembled the basic tools of climatology. In this section, you will examine a wealth of climatological data and study one system of classifying the variety of the earth's climates into a few, more or less distinct, types. The intention here is to survey the earth's climates somewhat briefly in preparation for tackling the chapter's central question, "How stable is our climate?" We hope that you will choose to study climatology and climatological classification systems in more detail in some future course.

### The Köppen System

The most widely employed climate classification system was first developed in 1918 by the German scientist Wladimer Köppen. (To pronounce *Köppen,* think "KER-pen," and then don't quite say the "r".) Köppen's system is still widely used, with only minor modifications, nearly a century after its creation. This durability is remarkable, considering the paucity of data available to Köppen in the early 1900s compared to the wealth of current climatological data that must fit into his system!

Köppen solved the problem of sparse meteorological data by (1) assuming that a given plant species (white spruce, for example) grows in similar climates wherever it is found and then (2) defining climate zones according to distributions of those species. Thus, plant distributions, which were better known than long-term climatological data, served Köppen as proxy data for climatological zones. It is not surprising, then, that Köppen's categories bear names related to plant communities, such as like "tropical rain forest climate," "snow forest climate," and "tundra climate."

The **Köppen system,** slightly modified over the years, is outlined in Figure 14.9 and mapped in Figure 14.10. Notice that it consists of six broad zones, labeled A, B, C, D, E, and H. For reasons you just read about, these zones generally conform to distinct ecological regions, such as tropical rain forest, desert, or ice cap. Note in the figure the rather strong relation of climatic zone to latitude. The exception is the highland (type H) climate type, for which altitude, not latitude, is the dominant factor.

Köppen and others have further divided the main zones according to prevailing moisture and temperature conditions. For example, Figure 14.9 indicates that the letter f (from the German *feucht,* meaning "moist") designates locales with no dry seasons; the w and s designators are applied to places with dry winters and dry summers, respectively. The third character indicates the character of the temperature regime, with the letters a through d representing successively colder conditions. The letters k and h also indicate temperature variations, in this case for desert climates. You can find details of these and other subdivisions in Figure 14.9.

### Large-Scale Features

How do you make sense of a dense, complex pattern like that illustrated in Figure 14.10? Concentrate first on large-scale structure, ignoring details if possible. You might note the tendency in the figure for most climate zones to appear as horizontal bands, stretching farther in an east-to-west direction than from north to south. This is most evident for the northernmost climates in the Northern Hemisphere, but once you are aware of the phenomenon, you can recognize it in many of the other climate types as well. This trend, which is even more pronounced in the global temperature chart you studied back in Figure 14.3, reflects the influence of temperature, through the effects of latitude, on a region's climate according to Köppen's system.

Where do you find exceptions to this pattern? Notice the north-south orientation of climate zones along the western regions of North and South America, largely a response to the mountains that run nearly the length of the Western Hemisphere. Other departures from the east-west alignment are evident in eastern Australia, southern Africa, and southern Asia. (Which climate controls might be responsible for these departures?)

Starting at the equator in Figure 14.10, you can trace the sequence of climates as you move northward: A, B, C, D, E. Moving southward from the equator, you see the same sequence begins: A, B, C. However, D and E climates are nearly absent from the Southern Hemisphere. These cold winter climates occur only over continents at high latitudes. The lack of continents at high southern latitudes (with the exception of Antarctica) is responsible for the absence of D and E climates in the Southern Hemisphere.

| Climate | | Description | Conditions |
|---|---|---|---|
| A | | Tropical Rainy | T for every month >=18°C (64°F) or higher, and climate is not dry (see B criteria below). |
| | | Precipitation-based subdivisions: | |
| | f | wet | No dry season: P for each mont >=6cm |
| | w | wet/dry | Wet summer, dry "winter": P for driest month <6cm and <(10-p/25) |
| | m | monsoon | Short dry season: P for driest month <6cm and >(10-p/25) |
| B | | Dry | Precipitation is less than potential evaporation and transpiration, determined as follows: Drier in summer: If <30% of precip falls in warmest 6 months, then climate type is B if p=<2t; Drier in winter: If <30% of precip falls in coolest 6 months, then climate type is B if p<2t+28; No drier season: If neither half of the year receives <30% of the mean annual precip, then climate type is B if p<2t+14 |
| | | Precipitation-based subdivisions: | |
| | S | Steppe | p=at least 1/2 of values in p, 2t formulas above. |
| | W | Desert | p<1/2 of values in p, 2t formulas above. |
| | | Temperature-based subdivisions: | |
| | h | hot | Mean annual temperature (t) >=16C |
| | k | cool | Mean annual temperature (t) <16C |
| C | | Moist; mild winter | T of coldest month between -3°C (27°F) and +18°C (64°F); climate not dry (see B criteria above). |
| | | Precipitation-based subdivisions: | |
| | w | dry winter | |
| | s | dry summer | |
| | f | moist all year | |
| | | Temperature-based subdivisions: | |
| | a | long, hot summers | T of warmest month >=22°C (72°F); at least 4 months with T >10°C (50°F). |
| | b | long, warm summers | P of wettest winter month <22°C (72°F); at least 4 months with T >10°C (50°F). |
| | c | short, cool summers | T of all months <22°C (50°F); 1-3 months with T> 10°C (50°F). |
| D | | Moist; cold winter | T of coldest month between <-3°C (27°F); T of warmest month > 10°C (50°F); climate not dry (see B criteria). |
| | | Precipitation-based subdivisions: | Precipitation-based subdivisions: |
| | w, s, f | same as in C climates | a, b, c same as in C climates; d severe winters: T of coldest month <=-38°C (-36°F). |
| E | | Polar | T of warmest month < 10°C (50°F). |
| | | Temperature-based subdivisions: | |
| | T | tundra | T of warmest month is between 0° and 10°C (32° and 50°F). |
| | F | ice cap | T of warmest month <0°C (32°F). |

T= mean temperature (°C) for one month
t= mean annual temperature (°C)
P= mean precipitation for one month (cm)
p= mean annual precipitation (cm)

**Figure 14.9** A Köppen climate classification system.

Köppen's climate zones show the effects of still other climate controls. For example, the contrast in climate from coastal southern California to coastal South Carolina is a result largely of differences in global air pressure and wind patterns and ocean currents, a topic we will consider in more detail presently.

Thus, the global climate patterns of Figure 14.10 reflect the influences of the large-scale controls listed in the previous sections, particularly the influence of latitude. Now let's take a closer look at the properties of individual climate regions.

## Moist Tropical (A) Climates

Climates of Köppen type A are known as moist tropical climates. The mean temperature for every month of the year is 18°C (64°F) or more. Further, significant precipitation occurs in at least some months in A climates. Notice in Figure 14.11 where such climates are found: along and near the equator, in places such as the Amazon River basin in South America, central Africa, and Southeast Asia. Further, you can see that the group is divided into three subcategories.

### *Tropical Wet (Af) Climates*

Picture living under the constant threat of showers or thundershowers every day of the year in a place where the temperature never goes below the 60s or above the lower 90s; a place that is humid and sticky, all day, all year; a place where June weather is essentially the same as December weather.

Tropical wet (Af) climates lie throughout the year within range of the ITCZ. As this zone of convergence meanders through the region, it brings spells of showery weather. Af climates are characterized, therefore, by warm temperatures and abundant rainfall every month of the year. In everyday language, these are the steamy tropics, where high heat and humidity are always present. Tumaco, Colombia, whose climograph you can examine in Figure 14.11, is a near-perfect example of a station with an Af climate. Notice that temperatures in Tumaco lie within a few degrees of 25°C (77°F), re-

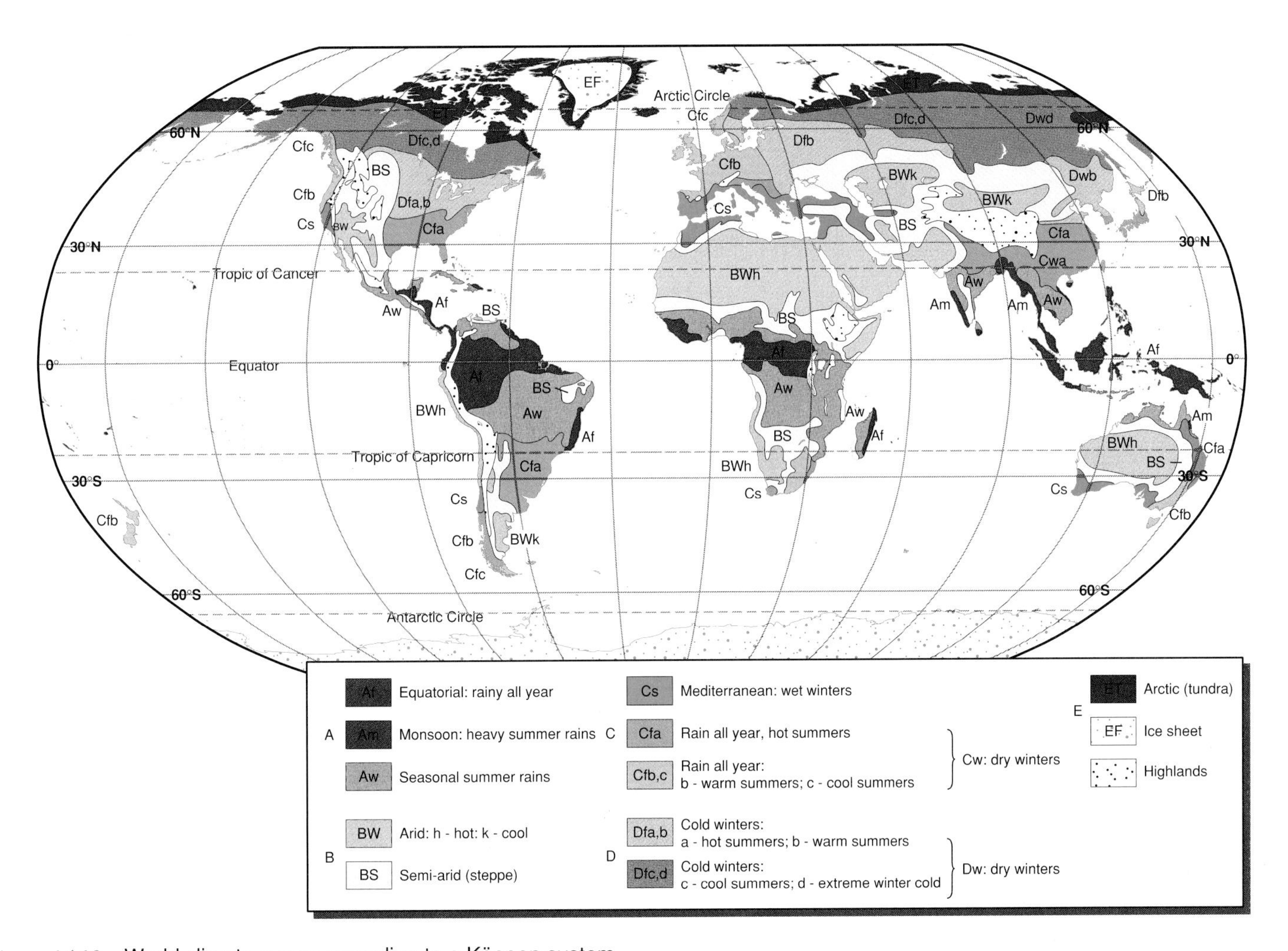


**Figure 14.10** World climate zones according to a Köppen system.

gardless of time of day or season of the year. Rainfall is abundant all year long, although amounts are considerably greater in the first half of the year. Dramatic seasonal changes, to which residents of middle and high latitudes are accustomed, are absent in Tumaco.

Tropical wet climates support rain forests. Figure 14.12A illustrates a typical Af rain forest environment.

## *Tropical Wet-and-Dry (Aw) Climates*

Traveling poleward from the tropical rain forest, you would notice distinct changes in vegetation. The forest begins to thin out, acquiring a parklike appearance, as in Figure 14.12B. Such ecosystems are called **savannas**. Savannas experience unstable, showery weather associated with the ITCZ for the high sun ("summer") seasons. During low sun seasons, however, dry weather is the rule under the dominance of the subtropical high pressure zone. Thus, savanna climates are also known as tropical wet-and-dry (Aw) climates. Notice the alternation between wet and dry regimes in the climograph of Ho Chi Minh City, Vietnam, in Figure 14.11. Also note that the temperatures vary little throughout the year; the seasonal temperature rhythm is nearly absent (except for the trends of extreme monthly values), as it was in the Af climate zone.

## *Tropical Monsoon (Am) Climates*

In some tropical regions, the contrast between land and oceans imposes a monsoon circulation on places that otherwise would experience Af climates. (Recall the discussion of monsoon circulations in Chapter 11.) The monsoon circulation increases precipitation during high sun months but suppresses it during low sun months. The net effect is a long, very rainy season followed by a relatively short dry season. You can see the extent of such monsoon (Am) climates in the climograph for Tavoy, Burma, Figure 14.11. Figure 14.12C illustrates a typical monsoon ecosystem. Such regions experience radical changes in weather from season to season.

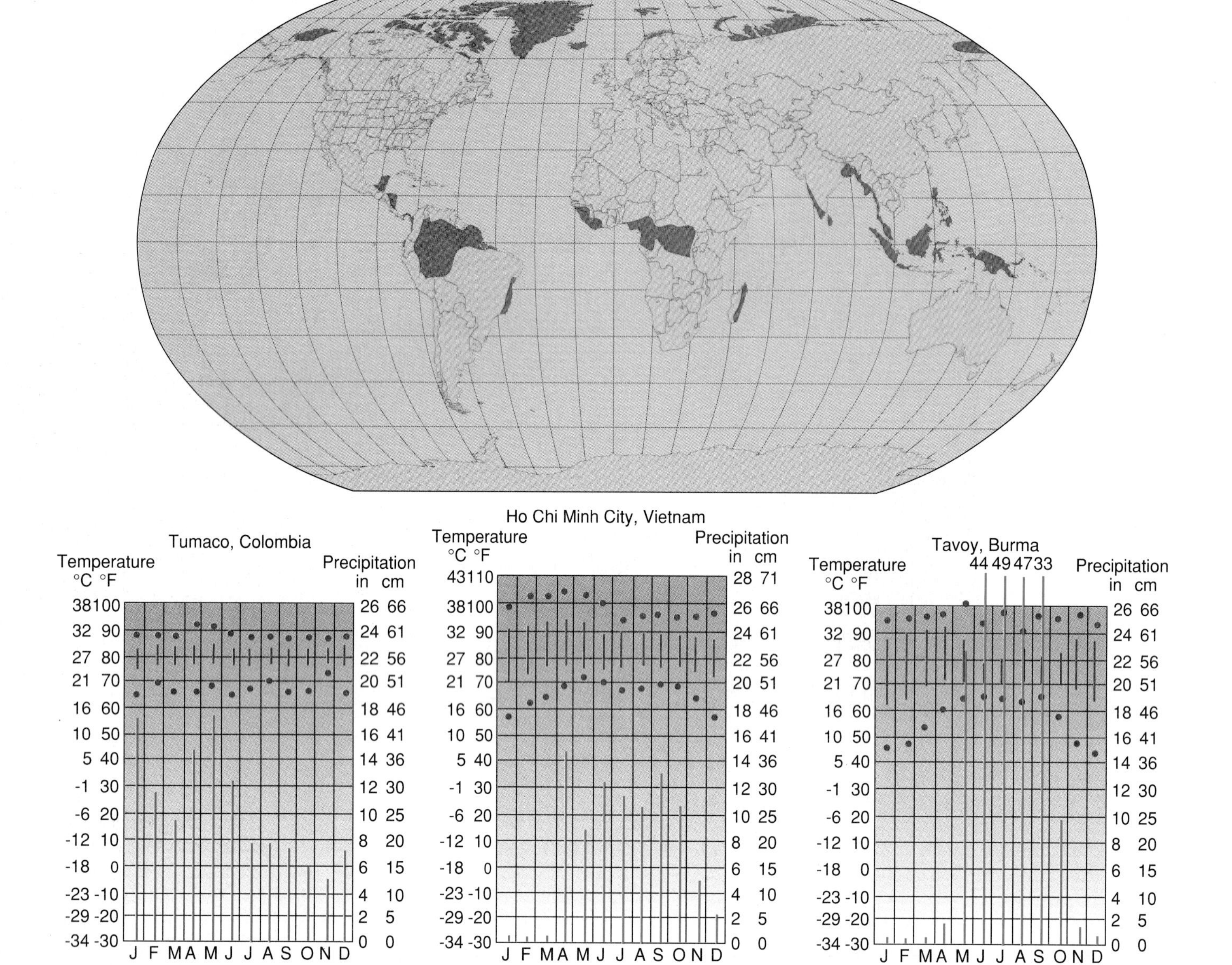


**Figure 14.11** Moist tropical (A) climate zones and representative climographs. Why is there so little range in temperatures from month to month?

## Dry (B) Climates

Moving poleward from the A climates, we enter the territory of the subtropical high pressure belts. In these regions of subsiding air and clear skies, we find dry climates labeled by Köppen as B climates. Figure 14.13 shows the distribution of B climates, which are divided into two main subcategories.

### *Desert (BW) Climates*

Climates of type Bw are true deserts. (The w comes from the German *wüste*, meaning "desert.") Notice the strong tendency in Figure 14.13 for B climates to lie in the 20° to 30° latitude zone, along the subtropical high pressure belt. The exception to this rule, you can see, is in eastern Asia. There, the Himalaya Mountains and the extensive east-west coastline from India to Vietnam strongly affect the climate, displacing Bw climates considerably poleward.

Precipitation is so meager in BW regions that mean monthly data can be misleading. Consider a location that receives no June precipitation whatsoever for 29 years and then experiences a single 60-millimeter (2.5 inch) downpour on one June afternoon in year 30. Under such circumstances, the mean June precipitation is calculated to be 60 millimeters/30 June months = 2 millimeters/month of June, a figure that certainly does not adequately describe the precipitation history at the site!

**Figure 14.12** Vegetation in moist tropical climates, in a (*A*) wet (Af) locale, (*B*) wet-and-dry (Aw) region, and (*C*) tropical monsoon (Am) terrain. Notice the change from the thick, lush growth of the Aw region to the parklike, seasonally dry monsoon regime.

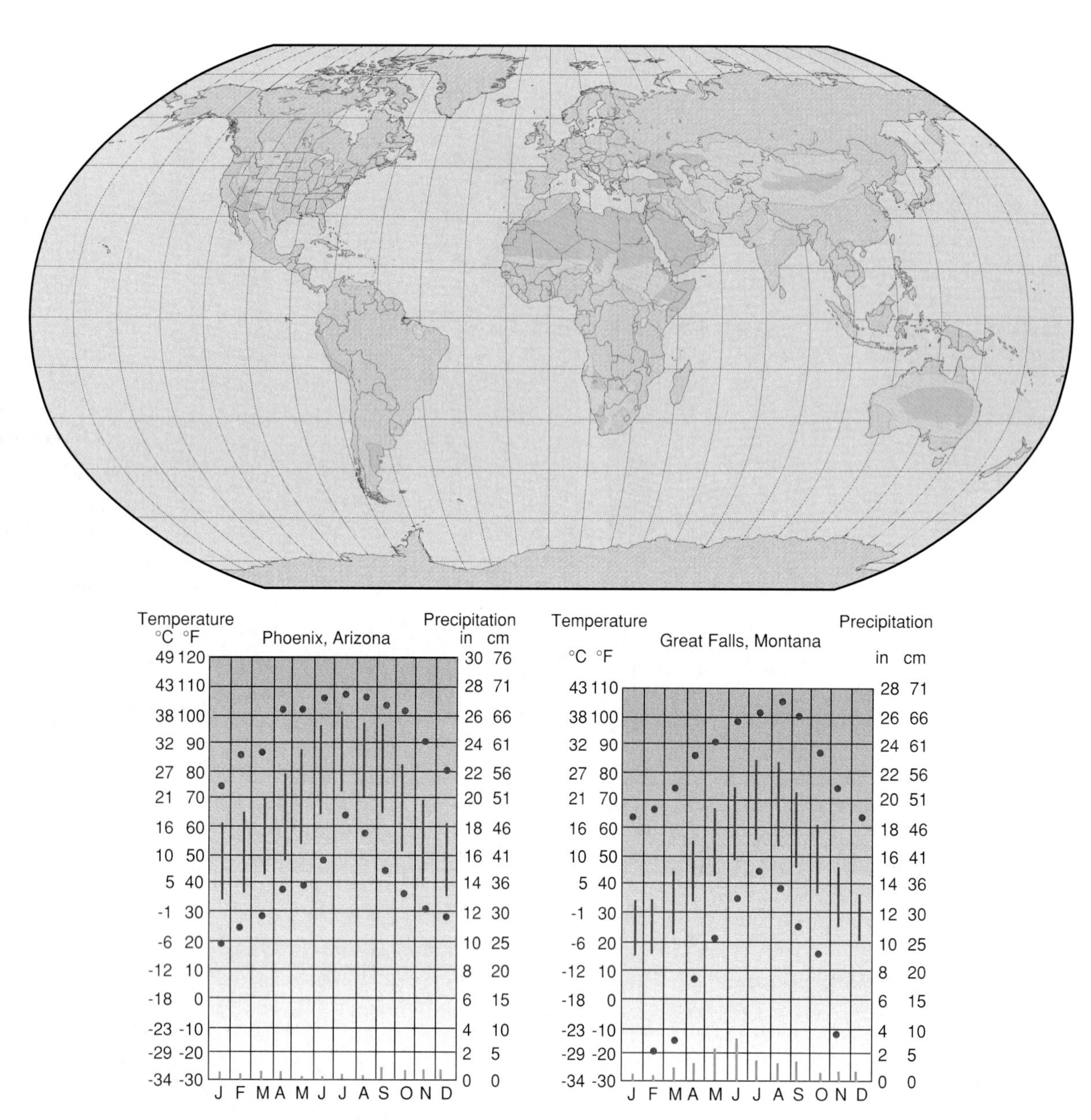


Figure 14.13 Dry climate: (B) zones, and climographs for a typical desert locale (Phoenix, Arizona) and steppe locale (Great Falls, Montana). Compare the annual temperature range and amount of precipitation shown in these climographs to those with those in A climates (Figure 14.11)

## *Steppe (BS) Climates*

Examine the landscape in Figure 14.14B. You can see that, although trees and shrubs are nearly absent, a cover of grass and other small plants thrives. This is typical semiarid, or **steppe,** climate. Steppe climates tend to surround Bw climate zones, as you can see by inspecting Figure 14.13. Regions slightly equatorward of BW climates experience occasional incursions of tropical moistness during the high sun season. On the other hand, locales poleward of the high pressure belt receive occasional precipitation from fronts and frontal cyclones during winter months or from summer showers and thundershowers, as in Great Falls, Montana, whose data are given in Figure 14.13.

A

B

**Figure 14.14** Desert (*A*) and steppe (*B*) landscapes. Limited water makes it impossible for many forms of vegetation to thrive, resulting in plant communities unique to B climates.

## Moist Mild-Winter (C) Climates

Climates of type C comprise a diverse group of several subtypes highlighted in Figure 14.15. Notice that they occur mainly in two locales: on east sides of continents, around 30° latitude (including locales such as the southeastern United States and China), and on west sides of continents at slightly higher latitudes (coastlines of the western states and western Europe). C climates lie near or slightly poleward of the subtropical high pressure belt. As a result, they often bear the imprint of interactions between subtropical highs and frontal systems of the prevailing westerlies.

How do you distinguish C climates from the A and B types? Unlike tropical A climates, locales with C climates experience a distinct cool season. Although winters are mild, at least one month averages below 18°C (64°F), according to Köppen's system. To determine whether a given locale's climate belongs to type B or C, examine its precipitation data. C climates are moist for part or all of the year (refer to Figure 14.9).

### *Humid Subtropical (Cfa) Climates*

The Cfa designation represents mid-latitude (C) climates like that of Charleston, South Carolina; such climates are moist (f) with long, hot summers (a). Note their locations in Figure 14.15. They occur in the subtropics, as their name suggests, on the east sides of land masses in North and South America, Asia, and Australia. Most of the southeastern United States lies in the Cfa region. As you can see from the climographs, temperatures are hot in summer and cool but not cold in winter, and precipitation is abundant all year. In summer, flow of mT air around the west side of subtropical high pressure cells (see Figure 14.16) carries very warm, moisture-laden air into the region, resulting in frequent showers and thundershowers. During winter months, frontal cyclones bring precipitation and periodic incursions of cold cP air.

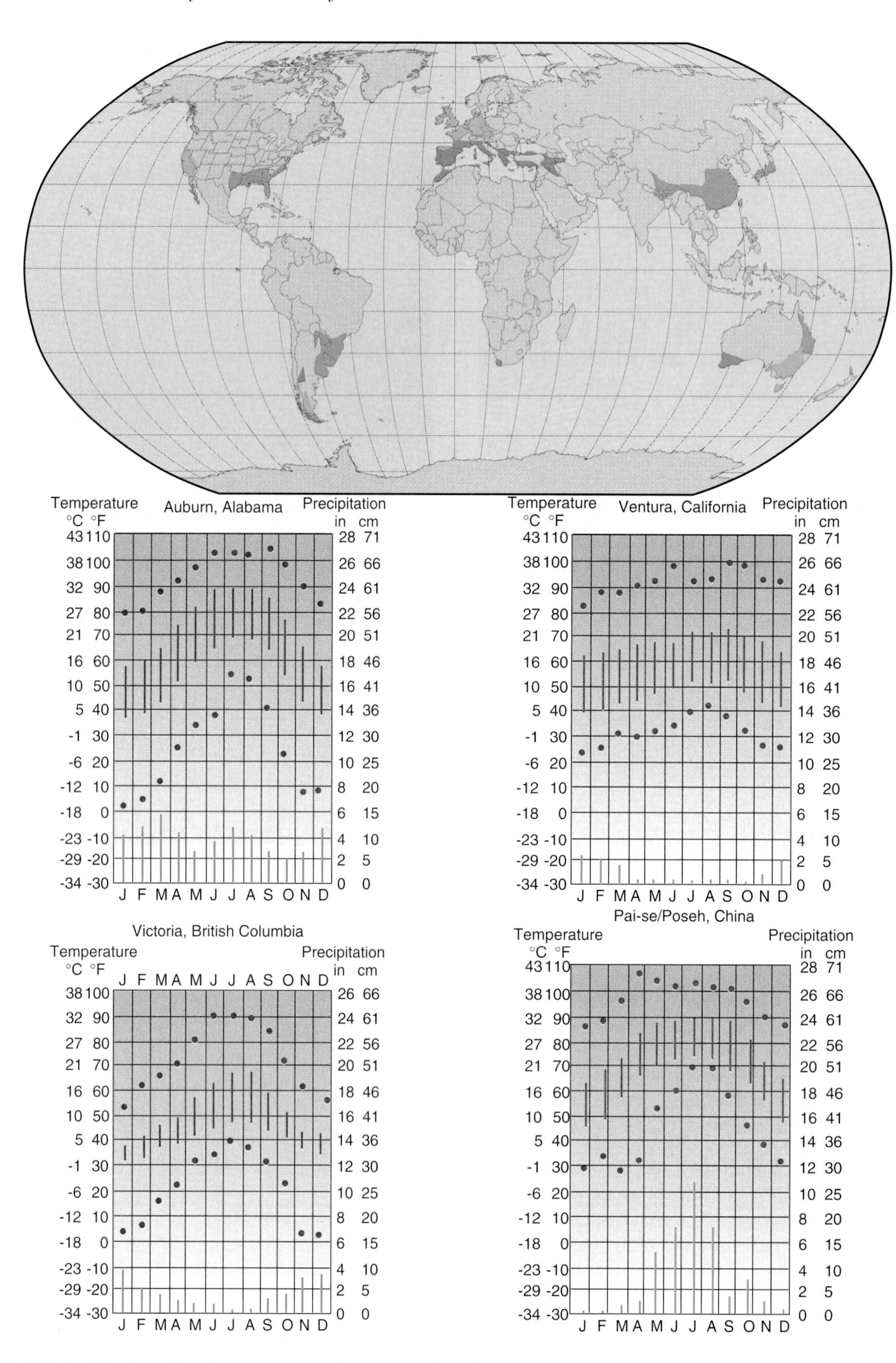


**Figure 14.15** Moist, mild-winter (C) climate zones and representative climographs. Notice the wide variety of temperature and precipitation regimes on these climographs. C climates show more diversity than any other type.

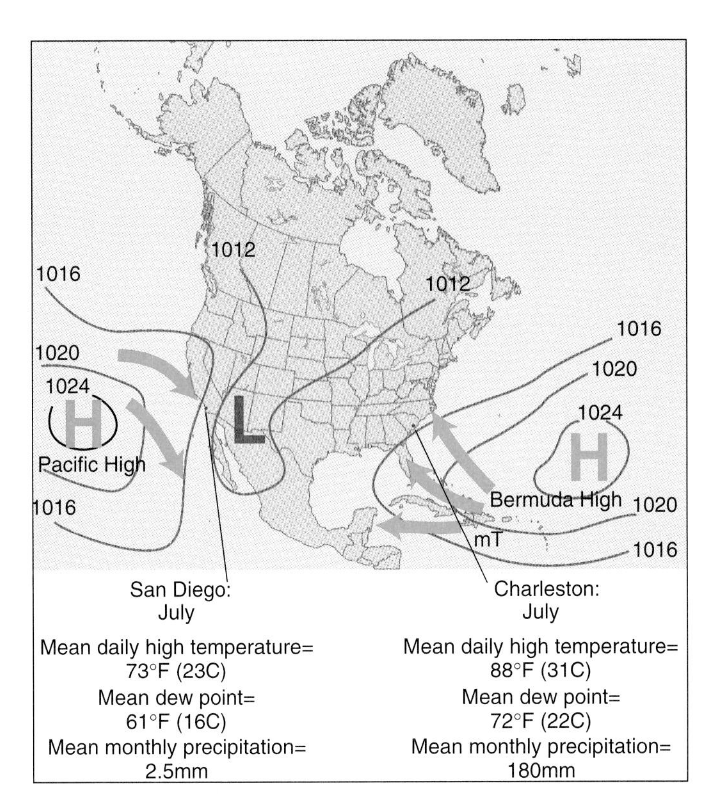


Figure 14.16 Conditions differ sharply from east side to west side of a sub-tropical high pressure cell. Compare trajectories of the wind flow in San Diego, California, and Charleston, South Carolina.

The warm temperatures and abundant precipitation in Cf climate regions support vigorous plant growth. Note the dense, mixed evergreen and hardwood forest in Figure 14.17A.

### *Mediterranean (Cs) Climates*

While Cfa climates are found along the eastern continental regions, the Cs, or Mediterranean, climates occur along continental west shores. Observe in Figure 14.15 that the greatest extent of Cs climates is around the Mediterranean Sea but that smaller regions occur in southern California, southern Australia, and the southern tip of Africa.

The s designator means that summers are dry. To understand the summer dryness in these regions, refer back to Figure 14.16. Note that Mediterranean climates, lying east of subtropical high pressure centers, do not experience the warm, humid flow of tropical air typical of Cfa climates. Instead, summertime wind flow originates at higher, cooler latitudes. Furthermore, these winds typically lead to upwelling of cold, deep water along the coast, causing low-level air to be stable and summer precipitation to be minimal, although fog is common in coastal regions.

In winter, Mediterranean climates lie under a different set of general circulation features. The subtropical highs slide away equatorward, and frontal cyclones move into the region, accounting for wintertime precipitation.

The dry, sunny Mediterranean climates are favorites of many people. These climates also support distinctive plant communities, such as that shown in Figure 14.17B.

### *Coastal Mid-Latitude (Cfb, Cfc) Climates*

Maritime air masses dominate Cfb and Cfc climates, such as that of Seattle. As you can see from Figure 14.15, much of Europe also lies in the Cfb zone. The absence of coastal mountains in western Europe allows maritime air to penetrate throughout much of that continent. This situation contrasts sharply with that in North and South America, where coastal mountains confine Cfb and Cfc climates to a narrow coastal zone. The ocean's moderating effect causes Cfb and Cfc winters to be mild (no monthly mean temperature colder than −3°C, or 27°F), and summer heat to be rare (no monthly mean temperature warmer than 22°C, or 72°F).

What distinguishes a Cfb climate from a Cfc one? By referring back to Figure 14.9, you can see that the distinction is based on summer "warmth." If fewer than four months have mean temperatures greater than 10°C (50°F), the climate is designated Cfc, according to the Köppen system.

The moist, temperate Cfb and Cfc climates support a variety of vegetation, including the dense forest growth illustrated in Figure 14.17C.

### *Subtropical Monsoon (Cwa) Climates*

Earlier, you read that tropical monsoon climates are designated Am climates. There are other locales, represented by the Cwa regions in Figure 14.15, that experience monsoon climates but have temperatures too cool to fall within the Am category. These regions are said to experience subtropical monsoon (Cwa) climates.

Climates of the Cwa type are particularly significant from the point of view of agriculture. The long months of warm, rainy weather provide ideal conditions for growing rice and other staple crops (see Figure 14.17D).

## Moist Cold-Winter (D) Climates

Poleward of B and C climates lie the cool, humid, continental cold-winter regimes. Note in Figure 14.18 the broad extent of D climates; they cover much of the northeastern United States and nearly all of the great continental expanses of Canada and Russia.

Köppen called the D regime "snow forest climates." The name is apt: D climate locales generally are forested regions that experience extended periods of winter snow cover.

Most locales with D climates receive abundant precipitation throughout the year and are therefore labeled Df climates. However, inspection of Figure 14.18 shows a Dw climate region in eastern Asia, which means it experiences dry winters. The dry winters are a result of the influence of the intensely cold, Siberian high pressure center.

**Figure 14.17** Landscapes in various C climate zones. (Note the wide variety of conditions. (*A*) Humid subtropical (Cfa). (*B*) Mediterranean (Cs). (*C*) Coastal mid latitude (Cfb, Cfc). (*D*) Sub-tropical monsoon (Cwa).

**Figure 14.18** (right) Moist, cold weather (D) and polar (E) climates. These regions experience snow that remains on the ground for extended periods. E climate locales experience no months with mean temperatures above 10°C (50°F). Note how closely this line corresponds to the border between northern forest and tundra vegetation.

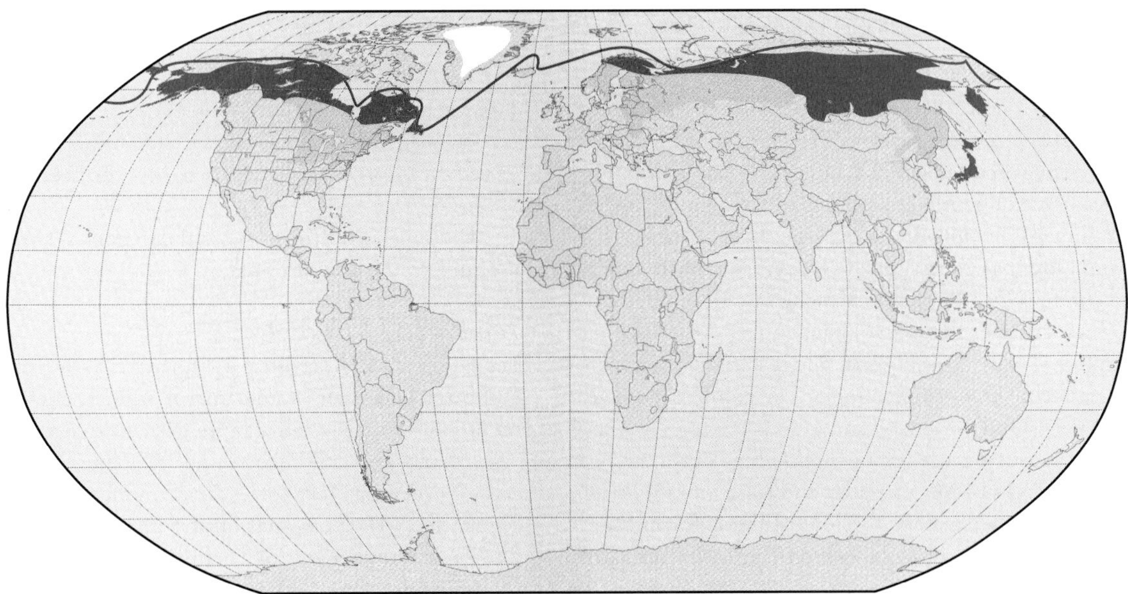

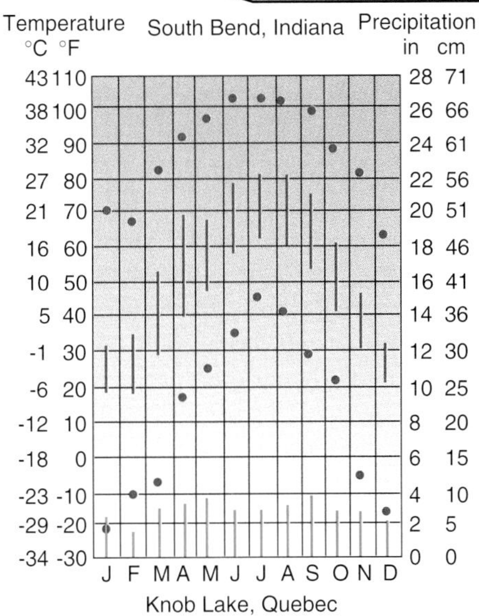
Temperature
°C °F
South Bend, Indiana
Precipitation
in cm
J F M A M J J A S O N D

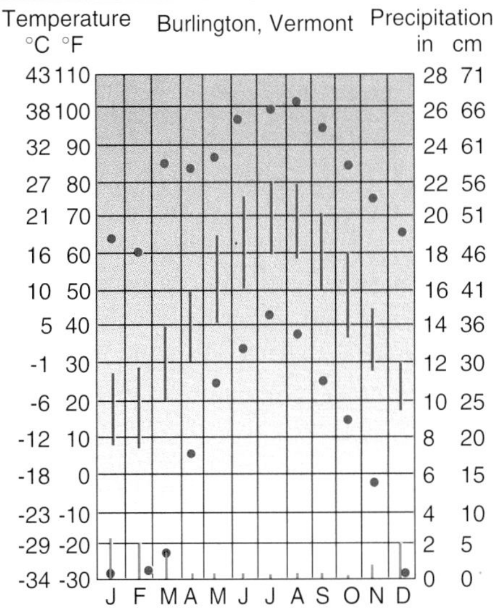
Temperature
°C °F
Burlington, Vermont
Precipitation
in cm
J F M A M J J A S O N D

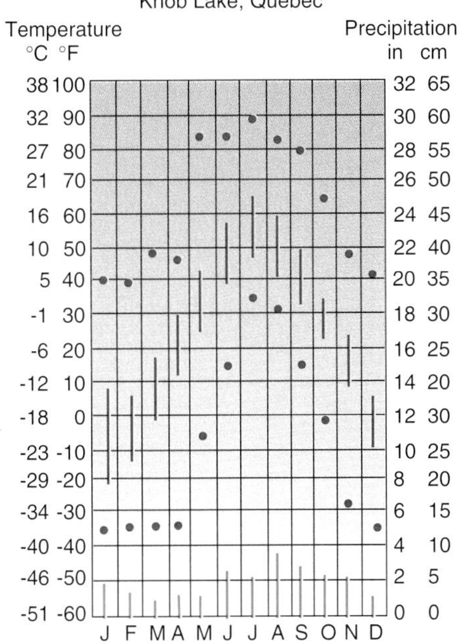
Knob Lake, Quebec
Temperature
°C °F
Precipitation
in cm
J F M A M J J A S O N D

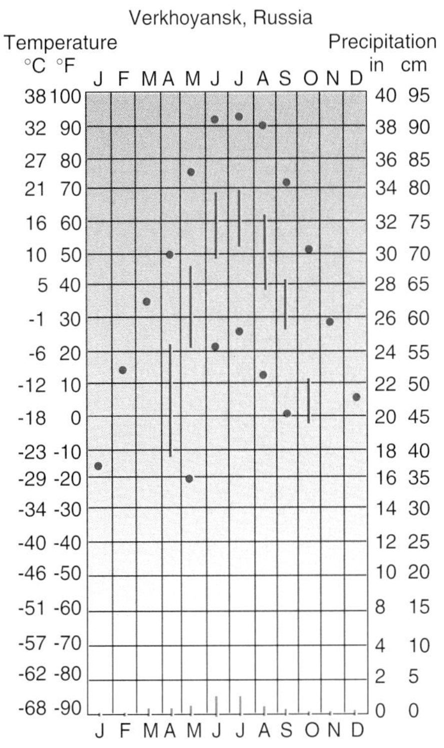
Verkhoyansk, Russia
Temperature
°C °F
Precipitation
in cm
J F M A M J J A S O N D

As you can see in Figure 14.18, D climates are divided into subcategories according to temperature. The coldest of these, labeled Dwd, experience an intensity of winter cold that is difficult to imagine: monthly mean winter temperatures are below −38°C (−36°F). Dwd climates occur only in Siberia.

In the Da and Db zones, forest cover is a mixture of hardwoods (leaf trees) and conifers (needle trees). Moving into the colder Dc climate region, the nature of the forest changes, as you can see in Figure 14.19. In this "boreal forest" zone, where snow persists for many months, coniferous trees predominate. The conifers' conical shape helps them withstand prolonged snow loads.

Approaching the northern edge of the D climate zone, you find the forest thins. Trees become smaller and less common, and finally disappear altogether except for low, stunted forms that hug the ground like a mat. You have crossed from snow forest to polar climate.

## Polar (E) Climates

Examine the position of the D-E climate boundary in Figure 14.18. The boundary is defined by the position of the 10°C isotherm for the warmest month. Notice that it corresponds rather closely to the line that marks the northern limit of trees (the tree line). Thus, trees do not grow in locales where monthly mean temperatures are colder than 10°C throughout the entire year. (Note that this does not mean temperatures never rise above 10°C in these regions but only that the 30-year monthly mean temperature is less than 10°C for every month of the year.)

The close match between the tree line and 10°C isotherm for the warmest month means that the tree line can serve as proxy data for the 10°C isotherm and thus for the D-E climate boundary. As early as 1877, a German climatologist named H. Dörr observed this correspondence between the tree line and the 10°C isotherm.

E climates fall into two subcategories: tundra and ice cap. We turn to these types next.

### *Tundra (ET) Climates*

Figure 14.20A illustrates a typical ET, or **tundra,** landscape. Tundra consists of a mixture of mosses, lichens, shrubs, and small matlike trees, all adapted to survive many months of severe, prolonged cold. Examine the climograph for Cambridge Bay, Northwest Territories, in Canada, a typical ET locale, in Figure 14.20B. Note that all months average colder than 10°C (50°F), a requirement of E climates. But summer temperatures at Cambridge Bay do rise above freezing (0°C), permitting the snow to melt and tundra plants to rouse from dormancy and rush through their brief growing season.

### *Polar Ice Cap (EF) Climates*

The name says it all: polar ice cap (EF) climates are found at high latitudes and are characterized by ice and snow cover throughout the year. Mean monthly temperatures are colder than 0°C for every month. Plant life does not exist. In the most remote of these frozen wastes, only one form of life is sometimes found, a particularly curious species known as the research scientist (see Figure 14.21A).

Despite their remote and forbidding nature, ice cap regions are important and perhaps fragile components of the global climate. We expect you understand by now that large-

A

B

Figure 14.19 Typical D climate landscapes. In milder (Dfa and Dfb) D zones, a wide variety of hardwood and coniferous forests thrives (*A*). In the more harsh Dfc climates (*B*), the forests are almost entirely coniferous with some birch and aspen.

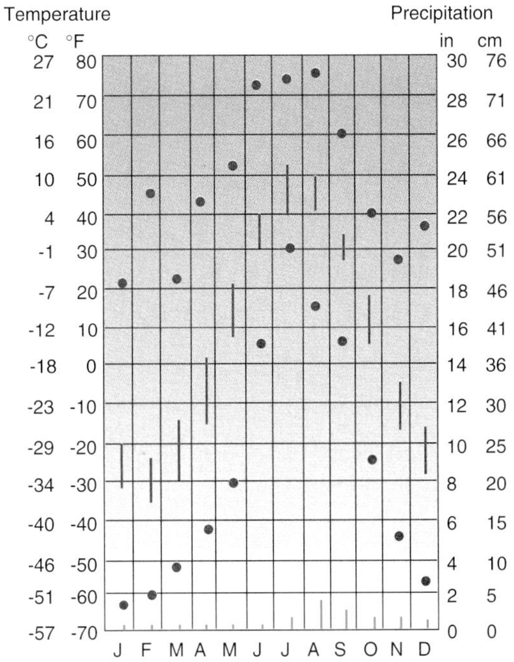


**Figure 14.20** A typical tundra landscape (photo on left). As in desert climates, harsh conditions limit vegetation to certain, distinctive forms. How can you tell from its climograph data that Cambridge Bay, Northwest Territories, lies in the ET zone?

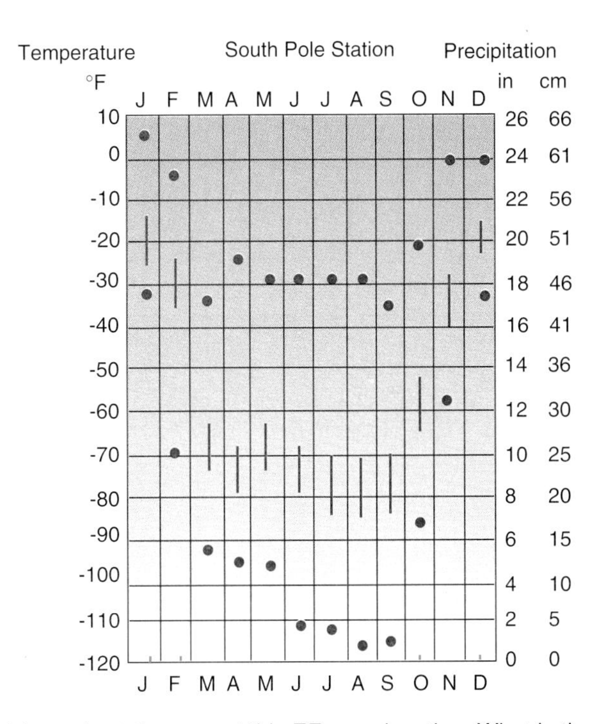


**Figure 14.21** Scott-Amundsen South Pole station. Notice the incredible cold throughout the year at this EF zone location. What is the extreme minimum temperature recorded? The extreme maximum?

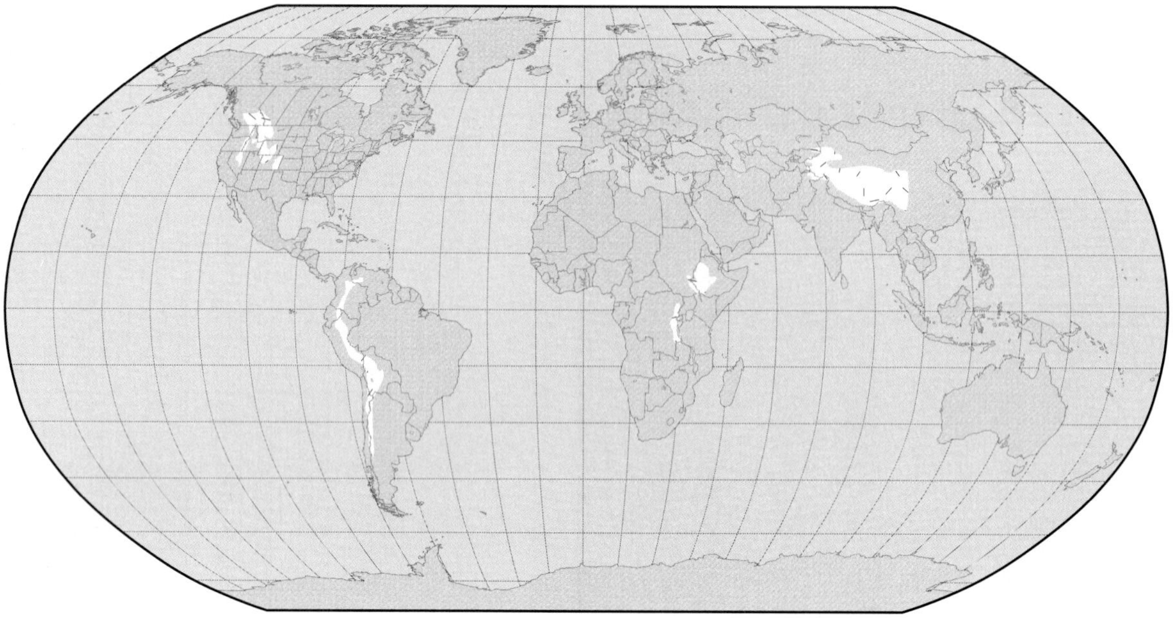

Figure 14.22 Locations of highland (H) climates.

scale alteration of any region's climate—including the polar ice cap climate—could have unexpected and possibly disastrous effects elsewhere.

### Highland (H) Climates

You read earlier in this chapter of the effects of altitude and slope on a locale's climate. In Köppen's system, all mountain climates are grouped together under the rubric "highland (H) climates." Conditions within this category are highly variable, since abrupt local changes in altitude and exposure in mountainous terrain cause correspondingly abrupt changes the climate. Figure 14.22 outlines the regions in which H climates prevail.

## How Stable Is Earth's Climate?

Now that you have surveyed the range of climates on earth and the factors that produce these climates, let us consider the central question of this chapter: How stable is our planet's climate?

### Perturbations and Responses

We are particularly interested in knowing how the climate might respond to influences exerted on it—human influences (air pollution, for example) as well as natural ones (a small change in the sun's energy output, say). Therefore, we restate the main question slightly more specifically: "How does our climate react to perturbations?"

## Checkpoint

***Review***

Use Figure 14.9 as a one-page summary of the large volume of information presented in this section. Try constructing your own version of this table, perhaps as a concept map.

***Questions***

1. Why do the main Köppen climate zones (A, B, etc.) tend to extend farther in east-west directions than from north to south?
2. Locale P has a Cfa climate; locale Q has a Cfc climate. In what ways are their climates alike? How do they differ?
3. The city of Meteoris is located at 50° north latitude, in the interior of a large continent. In which Köppen climate category is Meteoris likely to fall?

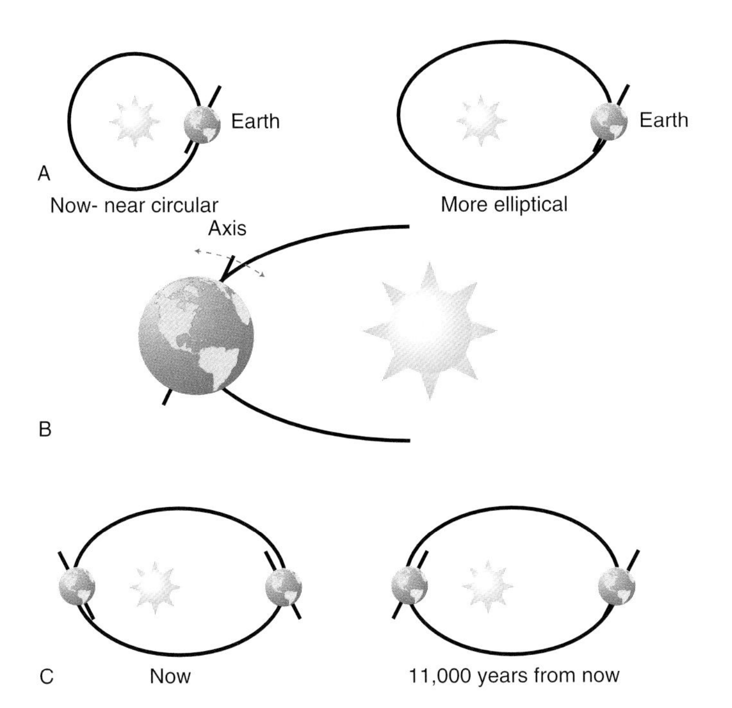


Figure 14.23 Milankovitch cycles. (*A*) The ellipticity of the earth's orbit varies over a period of 100,000 years. (*B*) The tilt of the earth's axis varies from approximately 21 to 24 degrees over a period of 41,000 years. (*C*) The earth's axis wobbles (precesses) over a period of 22,000 years.

We can tell you the answer to that question right now, without killing any of the suspense. The answer is "We don't know." We *do* know that throughout earth's history, a number of different forces have perturbed the climate, and we know that many of these forces are still at work. We also know from proxy data that earth's climate has undergone enormous variations in the past. But we do not know exactly how the atmosphere responds to a specific perturbation—much less to a number of perturbations acting simultaneously.

However, work is in progress on this highly important topic. In this final section, you will briefly consider a number of mechanisms that are known or suspected to perturb earth's climate and review some of the research recently completed or currently underway.

## Astronomical Perturbations

You read in Chapter 3 that the ultimate energy source for all weather (and therefore climate) is the sun and in Chapter 4 that measurements indicate the sun's energy output is quite steady. Unfortunately, accurate measurements span a period of only a century or so, so we don't know what changes the sun may undergo over periods of hundreds or thousands of years. Over much longer time periods, such as billions of years, we are certain that the sun's output will vary substantially, and with it, earth's climate. The sun, like all stars, has a finite life span and eventually will undergo extraordinary changes, but these are not expected for billions of years.

The sun itself need not change for earth to experience variations in insolation, however. In the 1930s, a Serbian astronomer named Milutin Milankovitch pointed out that variations in earth's orbit, over periods of thousands of years, cause significant variations in the sunlight received. Figure 14.23 illustrates the genesis of these variations, which are now known as **Milankovitch cycles.** The most significant cycle is due to changes in the shape of earth's orbit, as shown in Figure 14.23A. Recall from Chapter 3 (and from Figure 3.5) that the earth's orbit is not a perfect circle but is slightly elliptical, or oval-shaped. Over a period of roughly 100,000 years, the shape changes from slightly less elliptical than it is at present to significantly more elliptical.

How would this change of shape in earth's orbit affect incoming sunlight? A more elliptical orbit means that earth's distance from the sun varies more during the course of a year (see Figure 14.23A). As our distance from the sun varies, the amount of solar energy intercepted by earth varies as well.

The other two effects deal with earth's axis of rotation. In Figure 14.23B, you can see that the tilt of the axis varies slightly over a period of about 41,000 years. When the axis is more vertical compared to the plane of earth's orbit, seasonal variations in incoming sunlight are less.

The third Milankovitch cycle results from a slow turning, or wobbling, of earth's axis, known as **precession,** shown in Figure 14.23C. Earth's axis completes one precessional cycle in 22,000 years. Precession, combined with the eccentricity of earth's orbit, causes first the Northern Hemisphere and then the Southern Hemisphere to experience greater seasonal changes. Note from the figure that at present, earth is closest to the sun during the Southern Hemisphere's summer and farthest during the Southern Hemisphere's winter, a combination that accentuates winter cold and summer warmth in that hemisphere. In the Northern Hemisphere, on the other hand, the summertime tilt of earth's axis toward the sun is partly compensated by the greater earth-sun distance. In 11,000 years, the pattern will be reversed (see the figure), with the Northern Hemisphere experiencing more extreme and the Southern Hemisphere less extreme seasonal contrasts.

Numerous studies have shown that earth's climate has indeed fluctuated more or less in phase with Milankovitch's cycles. The explanation seems to be as follows: At times when seasonal changes are small, polar regions tend to experience milder winters. In the less frigid air, more snow falls than under conditions of bitter cold. On the other hand, summers during periods of small seasonal change are cooler, with the result that the snow melts more slowly. The result is that glaciers tend to advance during these periods and retreat when seasonal variations are greatest. At the height of the last glacial epoch, which ended about 10,000 years ago, all of Canada and one-third of the United States experienced ice cap climates (EF climates in Köppen's system). Tundra (ET) climates prevailed over much of the central United States at that time.

The sun is not the only celestial object capable of changing earth's climate. In the early 1970s, a group of scientists led by Louis Alvarez of the University of California advanced the theory that 60 million years ago, a giant meteor struck earth. Alvarez's group theorizes that the impact triggered catastrophic changes in climate that led to the extinction of many forms of life, including most species of dinosaurs.

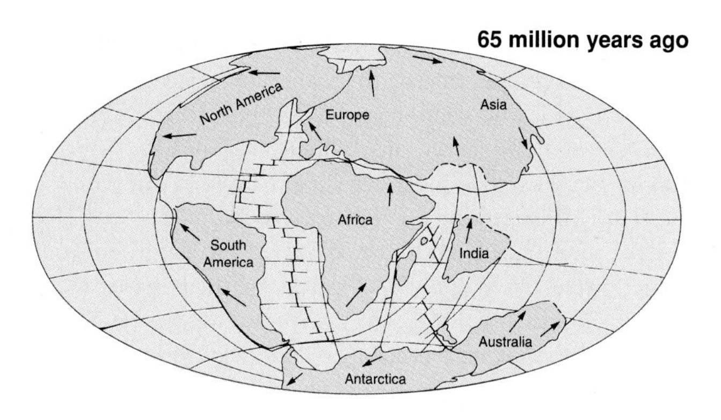


Figure 14.24 Map of the continents' positions 65 million years ago. Notice the position of India, separate from Asia, and the land bridge from Europe to North America. Source: American Petroleum Institute

Climate change through meteor impact may be a much more common event than many scientists once thought. Sediments from British Columbia, Canada, indicate that some sudden climatic change around 250 million years ago caused 96% of oceanic life forms to become extinct in an instant (geologically speaking), according to Kun Wang of the University of Ottawa. Wang speculates that some catastrophic event, such as an asteroid impact, caused a sudden change in global climate, which led to the extinctions. Furthermore, the July 1994 collision of Comet Schumaker-Levy 9 with the planet Jupiter reminded us all that meteor impacts are not just things of the distant past but even now remain within the realm of possibility for earth.

## Geological Perturbations

Examine the world map in Figure 14.24. The distortions you see are not the result of careless cartography but are the actual locations and shapes of continents over 65 million years ago. Confirmation of continental drift and of the more general theory of plate tectonics of which it is a part is one of the most remarkable scientific developments of the past century.

Plate tectonics teaches us that oceans open and close, continents change shape and slide about and sometimes crunch into one another, as India now is plowing into Asia and creating the Himalaya Mountains. It should be obvious that as continents change shape, size, and location the climates on those continents also will change. To cite an example, 200 million years ago, New England is thought to have been situated on the equator. Think of the changes its climate must have undergone since then.

Mountain building in the form of volcanic eruptions can also have great impact on climate. Over earth's history, volcanoes have been important sources of water vapor, carbon dioxide, sulfur, and particulates, all of which affect earth's energy budget and hence its weather and climate. And as volcanoes grow, they can become significant obstacles to airflow, thus affecting atmospheric circulation systems.

## Ocean-Atmosphere Interactions

Ocean currents influence the atmosphere's energy budget in important ways. Recall from Chapters 3 and 11 that a significant

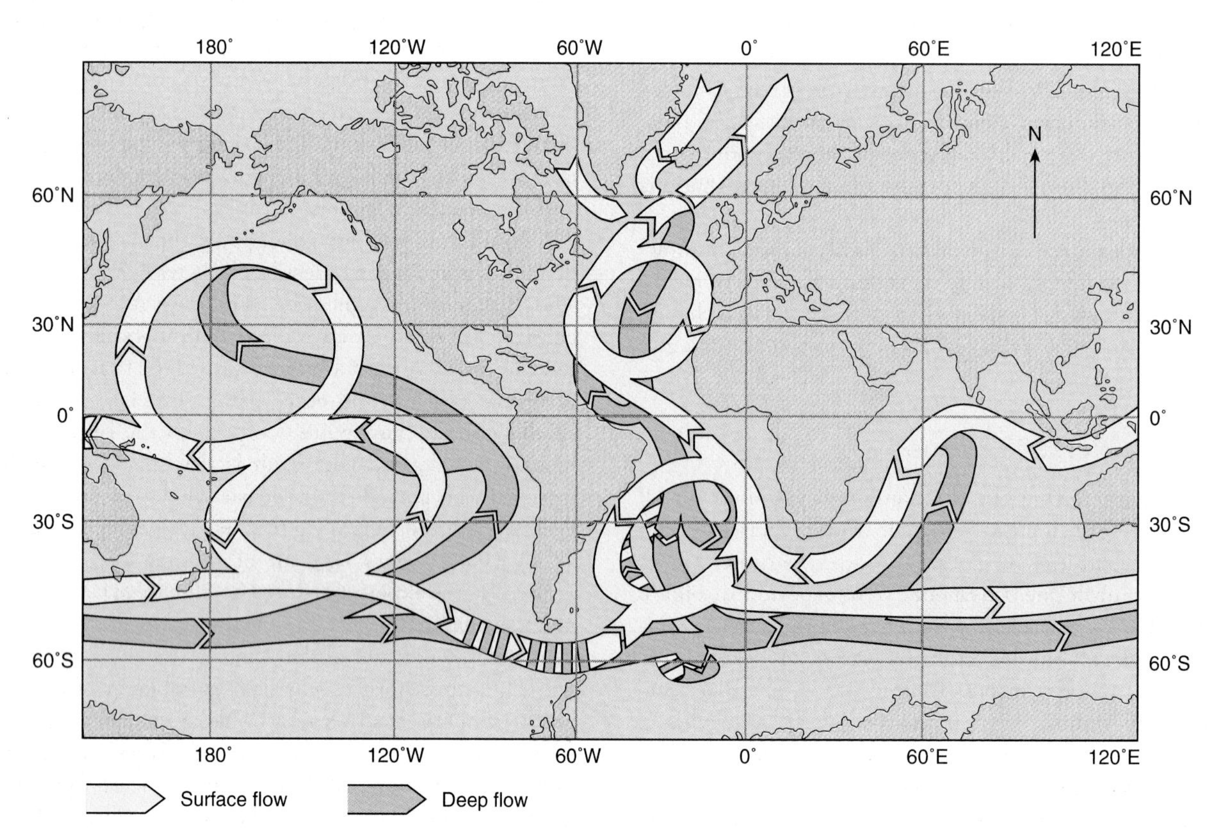


Figure 14.25 The oceanic conveyor belt model. Notice the surface (yellow) circulations in the Atlantic and Pacific oceans. To what atmospheric circulations do these currents correspond?

fraction of the excess solar energy absorbed in low latitudes is transported poleward via ocean currents. Changes in ocean current locations or volumes of flow will lead to changes in this energy transport and thus to changes in atmospheric circulations and earth's climate.

In recent years, oceanographers have developed a new model that describes the global network of ocean currents. The new description is known as the **ocean conveyor belt** model and is illustrated in Figure 14.25. Ocean-atmosphere interactions in the North Atlantic Ocean play a crucial role in this model. South of Iceland, warm surface water carried northeastward in the North Atlantic Drift loses vast amounts of energy through evaporation and contact with cold air masses. Evaporation removes water and heat from the ocean surface but leaves the salt behind, causing the surface water to become both saltier and colder. The decrease in temperature and increase in salinity act to increase the surface water's density; eventually, according to the model, it becomes so dense it sinks to the ocean bottom. The sinking is represented in Figure 14.25 by the change from yellow to green in the North Atlantic Ocean currents. It is primarily this sinking cold water in the North Atlantic Ocean that drives the ocean conveyor.

Now, if some event were to disrupt the conveyor belt circulation and therefore the poleward transport of heat, worldwide repercussions would ensue. What kinds of climatic change might occur?

Canadian scientists Andrew Weaver and Tertia Hughes are using a sophisticated computer model to examine the North Atlantic conveyor system. Their simulations have suggested that if the climate were to warm, the conveyor would become unstable, changing speeds erratically. The result would be sudden swings of climate, such as those that occurred before the last ice age, roughly 100,000 years ago. At that time, scientists at work on polar ice coring operations have discovered, annual temperatures varied by as much as 14°C (25°F) over just a few decades. (Imagine mean winter temperatures in your locale suddenly warming or cooling by 25°F!)

## Anthropogenic Changes

In Chapters 2 and 4, we considered the impact of anthropogenic influences such as increases in airborne particulates and greenhouse gases, and changes to the earth's surface such as cities. These changes have the potential for significant climate change on scales ranging from the mesoscale to the global.

We will not repeat those discussions here, but we recommend you review the relevant sections in Chapters 2 and 4. One might argue that *humans* are the greatest unknown in questions of climate stability: of all possible perturbations, anthropogenic ones are by far the most recent and therefore poorly understood and perhaps the least predictable.

Anthropogenic changes are often assumed to be deleterious. However, such is not always the case. For example, in China, 300 million trees were planted along the China-Mongolia border in the 1950s to reduce airborne dust from the Gobi Desert. Trees, it was theorized, would reduce surface winds and help to retain soil moisture, both of which would tend to reduce dust

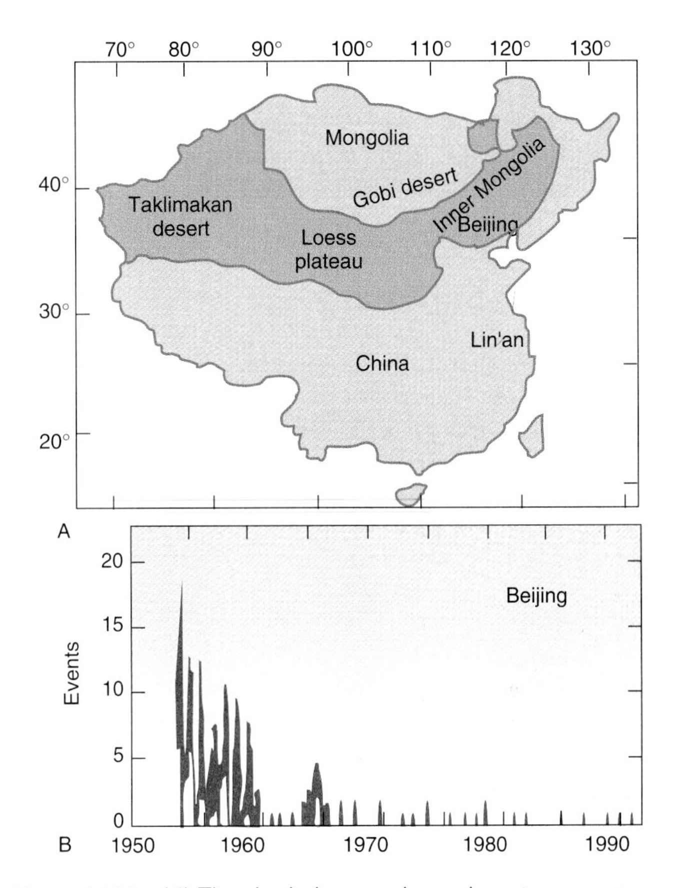


Figure 14.26 (*A*) The shaded areas show where trees were planted in China to reduce the incidence of dust storms. (*B*) Graph showing the number of observed dust storms per month in Beijing. Source: Parungo, *Geophysical Research Letter*, June 1 1994. American Geophysical Union.

levels. Note in Figure 14.26 the dramatic changes in air quality in Beijing since the trees were planted.

## Effects of Chaos

It may turn out that earth's climate doesn't require global scale perturbations like those we have just discussed in order to undergo major changes. It is possible that the climate possesses its own inherent sources of instability and that, like Lorenz's butterfly effect, large climate changes might arise from minor internal fluctuations. The relative stability of earth's climate over the past several thousand years argues against this suggestion. Still, because the atmosphere is a chaotic system, the sobering possibility exists.

# Recapitulation

You have seen that research from all ends of the globe documents the variability of climate over long periods of time. You have also seen that natural and human agents are perturbing the climate in a number of different ways. Thus, our climate is not static, it never has been, and perhaps it never will be. But with deeper understanding, we can plan steps to protect our atmosphere—humanity's most precious resource.

Study Point

## Chapter Summary

Climate is the sum of all weather events affecting a locale. Climatologists use statistical measures such as means (or "normals") and extremes to describe a location's climate and proxy data to extend the record into times and places where data are not available.

Many factors influence a locale's climate. Chief among them are latitude, altitude, location with respect to oceans or mountains, local topographic and surface conditions, global winds and ocean currents, and anthropogenic effects.

According to a climate classification based on the system of Wladimer Köppen, (Figures 14.9 and 14.10), the world's climates fall into six broad categories. The divisions are based on differences in monthly and annual means of temperature and precipitation. Climates of type A are found at and near the equator; climates B, C, D, and E typically occur at successively higher latitudes. The sixth type, labeled H (for highlands), comprises high altitude climates. The six basic types are subdivided according to specifics of monthly mean moisture and temperature data.

It is known that astronomical, geological, oceanic, and anthropogenic factors are likely to have caused or to be causing perturbations in earth's climate. It is also known that in the past, the climate has exhibited instability, changing abruptly from one regime to another. However, in general, we do not know whether a given perturbation will cause the atmosphere to respond in a stable, unstable, or neutral fashion. Many researchers are at work on various aspects of this important problem.

## Review Exercises

1. Use the graph in Figure 14.1 to determine the normal April daily maximum temperature, minimum temperature, and range in Lincoln, Nebraska.
2. Distinguish between climate and weather.
3. What is meant by the term "stable climate"?
4. Why is a locale's latitude an important determinant of its climate?
5. Suppose you climb Black Mountain in Kentucky, gaining 1000 meters (3300 feet) in altitude from the base of the mountain to its summit. To what change in latitude is this altitude change roughly equivalent?
6. Imagine you have before you the climatological data for some mid-latitude, west coast locale. How would you determine whether its climate is of type Cfb or Cfc? (See Figure 14.9.)
7. D climates, which cover vast areas of the Northern Hemisphere, are absent from the Southern Hemisphere. Why?
8. Suppose you were interested in mapping the boundary between D and E climates in a remote, uninhabited northern location. What data would you use to determine the boundary?
9. Identify the climate types associated with the following: steppe, tundra, Mediterranean, ice cap, desert, snow forest.
10. Which of the following are thought to be agents of climate change? Changes in the sun's energy output, changes in earth's orbit, meteor impacts, volcanic eruptions, continental drift.

## Key Words

| | |
|---|---|
| climate | normal |
| climograph | range |
| scatter | extreme value |
| microclimate | Köppen classification |
| savanna | steppe |
| tundra | Milankovitch cycles |
| precession | ocean conveyor belt |

## Problems

1. "Climate is average weather." In what sense is this statement true? In what sense is it false?
2. What advantages do proxy data offer the climatologist over conventional measurements of temperature and precipitation? What are the disadvantages?
3. Of all the factors affecting a region's climate, anthropogenic ones might be the least predictable. Why?

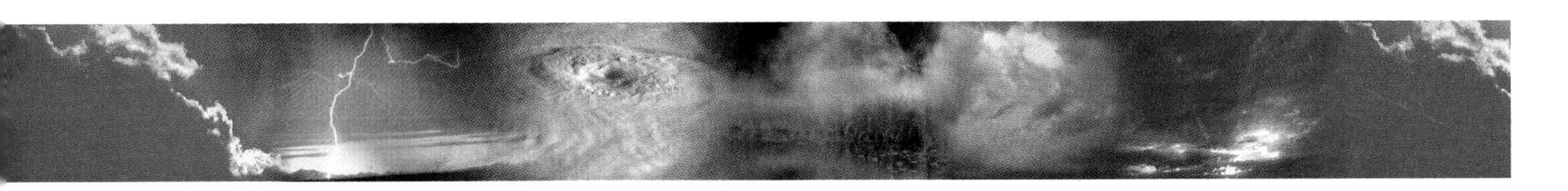

4. When Köppen developed his climate classification some 80 years ago, climate data were sparse or nonexistent over much of the globe. How did Köppen solve this problem? How is his solution reflected in his classification system?
5. "Regions bathed in copious quantities of solar energy have warmer climates than sunshine-deficient locales." How is this statement supported or refuted in the pattern of climate types according to Köppen (see Figure 14.10)?
6. Between roughly 40° and 60° north latitude, the west coasts of North America and Europe experience C climates, whereas the east coasts have D climates (see Figure 14.10). What is the explanation for the difference?
7. Climatological data for Birmingham, Alabama, are listed in Appendix M. Use the data and Figure 14.9 to explain why Macon's climate is classified as Cfa.
8. What causes the dry period in tropical wet-and-dry (Aw) climates? In tropical monsoon (Am) climates?
9. Explain how precession of earth's axis of rotation can lead to long-term changes in climate.

# Explorations

1. If you write computer programs in a language such as Pascal, BASIC, FORTRAN, and so on, or with software such as HyperCard, construct a climate identification key based on the Köppen system. The program should ask the person using it various questions about mean temperature and precipitation data at some locale; then it should determine and display the locale's climate type.
2. Climate change is a topic of very active research; every month, new discoveries are made. Use your college or university library or the Internet to find and review the most recent research on the stability of earth's climate. Then present your findings in an oral presentation or a paper.
3. Using a data source such as AccuWeather or one of the Internet sites listed in Resource Links, find data and construct climographs for locales along the West Coast of the United States. Then do the same for locales along the East Coast. Are your climographs consistent with the Köppen classification system?

# Resource Links

The following is a sampler of print and electronic references for some of the topics discussed in this chapter. We have included a few words of description for items whose relevance is not clear from their titles. For a more extensive and continually updated list, see our World Wide Web home page at www.mhhe.com/sciencemath/geology/danielson.

Graedel, T. E., and P. J. Crutzen. 1995. *Atmosphere, Climate and Change*. New York: W. H. Freeman and Company. 206 pp.

The following four articles by Richard Monastersky give a sample of the current research in climate and climate change:

Ten thousand cloud makers: Is airplane exhaust altering earth's climate? *Science News* 150 (1): 12–13.

The case of the global jitters: Even in seemingly stable times, climate can take an abrupt turn, *Science News* 149 (9): 140–41.

Iron versus the greenhouse: Oceanographers cautiously explore a global warming therapy, *Science News* 148 (14): 220–22.

Staggering through the ice ages: What made the planet careen between climate extremes? *Science News* 146 (5): 74–76.

For readers with access to AccuData, PLAY plots average and record temperatures for North American locations. DAY/R/X gives daily record temperature data. DAY provides daily climatological data for over a thousand worldwide locations. RECD is a listing of record setting weather throughout the U.S.

The Geophysical Fluid Dynamics Laboratory at http://www.gfdl.gov/ offers an interview on global warming, plus information on climate modeling.

NOAA's National Climate Data Center is at http://www.ncdc.noaa.gov/ .

The U.S. Government Climate Page at http://www.cdc.noaa.gov/USclimate/ is an interactive page at which you can request specific climate information for various U.S. locations.

Interactive, worldwide climate information is available at http://grads.iges.org/pix/clim.html .

# Appendices

# Appendix A

# Some Notable Dates in the History of Meteorology

| Dates in History | |
|---|---|
| c 1060 B.C. | Official regular weather records kept, in China. |
| c 400 B.C. | First known use of rain gauges, in India. |
| c 400 B.C. | Hippocrates writes of air pollution in Greek cities. |
| c 340 B.C. | Aristotle writes *Meteorologica,* the first book to deal systematically with atmospheric phenomena. Despite many erroneous explanations, it remains the standard reference for nearly 2000 years. |
| c 50 A.D. | Seneca and Pliny write treatises on the causes of weather; Seneca decries air pollution in Rome. |
| c 140 | Greek scientist Ptolemy develops an astrology-based system of weather prediction. |
| c 703 | Bede the Venerable, an English scholar, writes the meteorology treatise *De Natura Rerum.* |
| c 1000 | Arab scientist Al Hazen explains twilight and other optical phenomena. |
| 1273 | Use of coal prohibited in London due to air pollution. |
| c 1450 | L. Alberti and N. Cryfts invent hygroscopes for observing humidity. |
| c 1593 | Galileo invents the thermometer. |
| 1635 | René Descartes writes *Les Meteores* and proposes that air is made of innumerable tiny particles. |
| 1643 | First barometer is constructed by Evangelista Torrecelli. |
| 1644 | First American weather records are maintained by the Rev. John Campanius, near Wilmington, Delaware. |
| 1648 | Blaise Pascal and Florin Perier measure the decrease of air pressure with altitude. |
| 1661 | Robert Boyle discovers that at constant temperature, an air parcel's pressure is proportional to its density. |
| 1686 | Edmund Halley publishes first weather map of global winds. |
| 1687 | Isaac Newton publishes *Principia,* the foundation of modern mathematical theory for all physical science, including meteorology. |
| 1714 | First mercury thermometers with reliable scales are made by Gabriel Fahrenheit. |
| 1735 | George Hadley publishes the theory of atmospheric circulation, "Hadley cells." |
| 1738 | Daniel Bernouilli develops the kinetic theory of gases. |
| 1752 | Joseph Black isolates and studies $CO_2$, the first atmospheric constituent to be so studied. |
| 1752 | Benjamin Franklin conducts kite experiments, demonstrating "the sameness of electrical matter with that of lightning". |
| 1755 | Leonhard Euler derives equations of fluid flow. |
| 1783 | First balloons are used for weather observing. |
| 1804 | J. L. Gay-Lussac makes balloon ascents to measure upper atmosphere; reaches 7 kilometers (23,000 feet). |
| c 1820 | First daily weather maps are made. |
| 1844 | G. Coriolis gives mathematical explanation of Coriolis force. |
| 1894 | Argon is identified as a constituent of air. |
| 1898 | Inversion at tropopause is discovered. |

**Dates in History** ***continued***

| | |
|---|---|
| c 1900 | Max Planck, Wilhelm Wien, and Ludwig Boltzmann develop radiation laws. |
| 1917 | Polar front theory is formulated by V. Bjerknes. |
| 1922 | L. F. Richardson publishes results of first attempt at numerical weather prediction attempt. |
| c 1927 | First radiosondes, produced by French and Russian scientists. |
| c 1942 | Weather radar is developed. |
| 1945 | Windchill factor concept is introduced. |
| 1946 | Vincent Schaefer and Irving Langmuir conduct first dry ice cloud-seeding experiments. |
| 1947 | First computer-generated weather predictions. |
| 1953 | Computer predictions of storm development are demonstrated. |
| 1959 | Van Allen radiation belts are discovered. |
| 1960 | First weather satellite, TIROS I, is launched. |
| 1964 | Nimbus-1 satellite is launched; always points at earth. |
| 1972 | First computer-based climate models of the atmosphere are developed. |
| 1984 | Launch of the Earth Radiation Budget Satellite, the first satellite devoted to climate research. |
| 1985 | Discovery of the seasonal thinning of the ozone layer. |
| 1993–95 | WSR-88 (NEXRAD) Doppler radar network is installed. |

## *Sources*

Middleton, W. E. K. 1969. *Invention of the Meteorological Instruments.* Baltimore: Johns Hopkins Press.
Frisinger, Howard. 1983. *The History of Meteorology to 1800.* Boston: American Meteorological Society.
Griffiths, J.F. 1977. A chronology of items of meteorological interest. *American Meteorological Society Bulletin.* 58(10): 1058.
Heidorn, K.C. 1978. A chronology of important events in the history of air pollution meteorology to 1970. *American Meteorological Society Bulletin.* 59(12): 1589.

# Appendix B

# Scientific Notation

"A molecule of diatomic nitrogen or oxygen is approximately 0.0000000002 meters in diameter; in a single deep breath, you inhale roughly 50,000,000,000,000,000,000,000 of these molecules."

Numbers with many digits, like those above, are difficult to read and remember, and awkward to use in calculations. Expressing them in a format known as scientific notation reduces these difficulties. Here are the same numbers, and others, expressed in scientific notation:

| Scientific Notation | |
|---|---|
| $2 \times 10^{-10}$ | the diameter of a diatomic oxygen molecule in meters |
| $5 \times 10^{22}$ | the number of molecules in one deep breath |
| $3.65 \times 10^{2}$ | the number of days in a year (365) |
| $3.6 \times 10^{3}$ | the number of seconds in an hour (3600) |
| $-4.682 \times 10^{4}$ | the number −46,820 |

As these examples illustrate, every number has the same format when expressed in scientific notation, specifically:

- a leading plus sign (optional) or minus sign;
- a single digit (in the "ones" place);
- often, a decimal point and one or more digits to its right;
- the symbols "$\times$ 10"; and finally,
- an exponent that tells by how many powers of 10 the n.nnn value must be multiplied (if a positive exponent) or divided (if a negative exponent) to represent the original number.

Recall that multiplying or dividing a number by some power of 10 is equivalent to moving its decimal point to the right or left of its starting position. Thus, expressing the number $3.65 \times 10^2$ means moving the decimal point 2 places to the right:

$$3.65 \times 10^{2} \quad = 365$$

Similarly, the number $6.9 \times 10^{-5}$ can be written as

$$6.9 \times 10^{-5} \quad = 0.000069$$

Notice in the latter example that it was necessary to add zeros as the decimal point moved leftward, in order to represent correctly the number of places it moved.

For practice, work through the following examples. The answers are at the end of this appendix.

1. Express the following numbers in scientific notation:
   149,600,000 (mean earth-sun distance in kilometers)
   12,756 (equatorial diameter of the earth in kilometers)
   0.01 (diameter of typical cloud droplet, in millimeters)
   46.1 (all-time high temperature [°C] in Santa Barbara, California)

2. Express these numbers in standard notation:
   $2.9998 \times 10^{8}$ (speed of light in meters per second)
   $4.2 \times 10^{-3}$ (mass of a typical raindrop, in grams)
   $3.1536 \times 10^{7}$ (number of seconds in one year)
   $9.75 \times 10^{2}$ (mean annual rainfall in Seattle, in millimeters)

*Answers*

1. $1.496 \times 10^{8}$; $1.2756 \times 10^{4}$; $1.0 \times 10^{-2}$; $4.61 \times 10^{1}$
2. 299,800,000; 0.0042; 31,536,000; 975

# Appendix C

# SI Units of Measurement

Most scientific data are expressed in a system of units known as SI (for *Systeme International*). In the SI system, the basic units are:

length—the meter (m)

mass—the kilogram (kg)

time—the second (s)

temperature—the Kelvin (K)

Other units are expressed in terms of these four. Thus, in SI units, speed is expressed in meters per second (m/s, or m $s^{-1}$), density is expressed in kg/$m^3$ (kg $m^{-3}$), and so on.

For many scientific measurements and descriptions, the meter, kilogram, and so on are of inconvenient size. For example, a raindrop's mass is a tiny part of a kilogram, while a hurricane may be hundreds of thousands of meters in diameter. To work with units that are more compatible with the objects under study, scientists often scale the standard units up or down by an appropriate number of powers of 10 and indicate this scaling by adding one of the prefixes listed below to the standard unit.

*Example:* An air pollution particle is reported to be 3.24 micrometers in diameter. Express this length in meters.

**Units of Measurement**

| Prefix | Name | Power of 10 | Example |
|---|---|---|---|
| nano- | 1 billionth | $10^{-9}$ | nanosecond |
| micro- | 1 millionth | $10^{-6}$ | micrometer |
| milli- | 1 thousandth | $10^{-3}$ | millibar |
| centi- | 1 hundredth | $10^{-2}$ | centimeter |
| deci- | 1 tenth | $10^{-1}$ | decibel (sound) |
| deka- | ten | $10^{1}$ | dekameter |
| hecto- | 1 hundred | $10^{2}$ | hectogram |
| kilo- | 1 thousand | $10^{3}$ | kilogram, kilopascal |
| mega- | 1 million | $10^{6}$ | megawatt (electric power) |
| giga- | 1 billion | $10^{9}$ | gigabyte (computer memory) |
| tera- | 1 trillion | $10^{12}$ | terabyte (computer memory) |

*Solution:* Referring to the table above, you can convert micrometers into meters:

$$\begin{aligned} 3.24 \text{ micrometers} &= 3.24 \text{ millionths of a meter} \\ &= 3.24 \times 10^{-6} \text{ m} \\ &= 0.00000324 \text{ m} \end{aligned}$$

# Appendix D

# Other Units, Conversion Factors, and Constants

Atmospheric data are not always presented in SI units and not always in scientific notation—even in this textbook. Terms such as knots, millibars, and degrees Fahrenheit have a long meteorological tradition and remain in use today, particularly by weather observing and forecasting services and the media. With the following tables, you can convert values from one unit system to another.

### Length Conversion

**Length** TO CONVERT FROM

statute miles (mi) to kilometers (km), multiply mi by 1.609
kilometers to statute miles, multiply km by 0.6214

nautical miles (n mi) to kilometers (km), multiply n mi by 1.853
kilometers to nautical miles, multiply km by 0.5396

statute miles to nautical miles, multiply mi by 0.8684
nautical miles to statute miles, multiply n mi by 1.152

feet (ft) to meters (m), multiply ft by 0.3048
meters to feet, multiply m by 3.281

inches (in) to centimeters (cm), multiply in by 2.54
centimeters to inches, multiply cm by 0.3937

### Speed Conversion

**Speed** TO CONVERT FROM

miles per hour (mph) to kilometers (km) per hour, multiply mph by 1.609
kilometers per hour to mph, multiply km per hour by 0.6214

nautical mi per hour (knots) to kilometers (km) per hour, multiply knots by 1.853
kilometers per hour to knots, multiply km per hour by 0.5396

miles per hour to knots, multiply mph by 0.8684
knots to miles per hour, multiply knots by 1.152

knots to meters per second, multiply knots by 0.5148
meters per second to knots, multiply meters per second by 1.943

### Air Pressure Conversion

**Air Pressure** TO CONVERT FROM

inches (in) of mercury to millibars (mb), multiply in by 33.8639
millibars to inches of mercury, multiply mb by 0.02953

inches (in) of mercury to millimeters (mm) of mercury, multiply in by 25.4
millimeters of mercury to inches of mercury, multiply mm by 0.03937

millibars (mb) to Pascals, multiply mb by 100
Pascals to millibars, multiply Pascals by 0.01

millibars (mb) to kilopascals (kPa), multiply mb by 0.1
kilo-pascals to millibars, multiply kPa by 10

### Temperature Conversion

| Temperature | To convert between Fahrenheit (°F), Celsius (°C) and Kelvin (K) scales |
|---|---|
| °F to °C: | °C = (°F − 32) × 5/9 |
| °C to °F: | °F = °C × 9/5 + 32 |
| K to °C: | °C = K − 273.2 |
| °C to K: | K = °C + 273.2 |

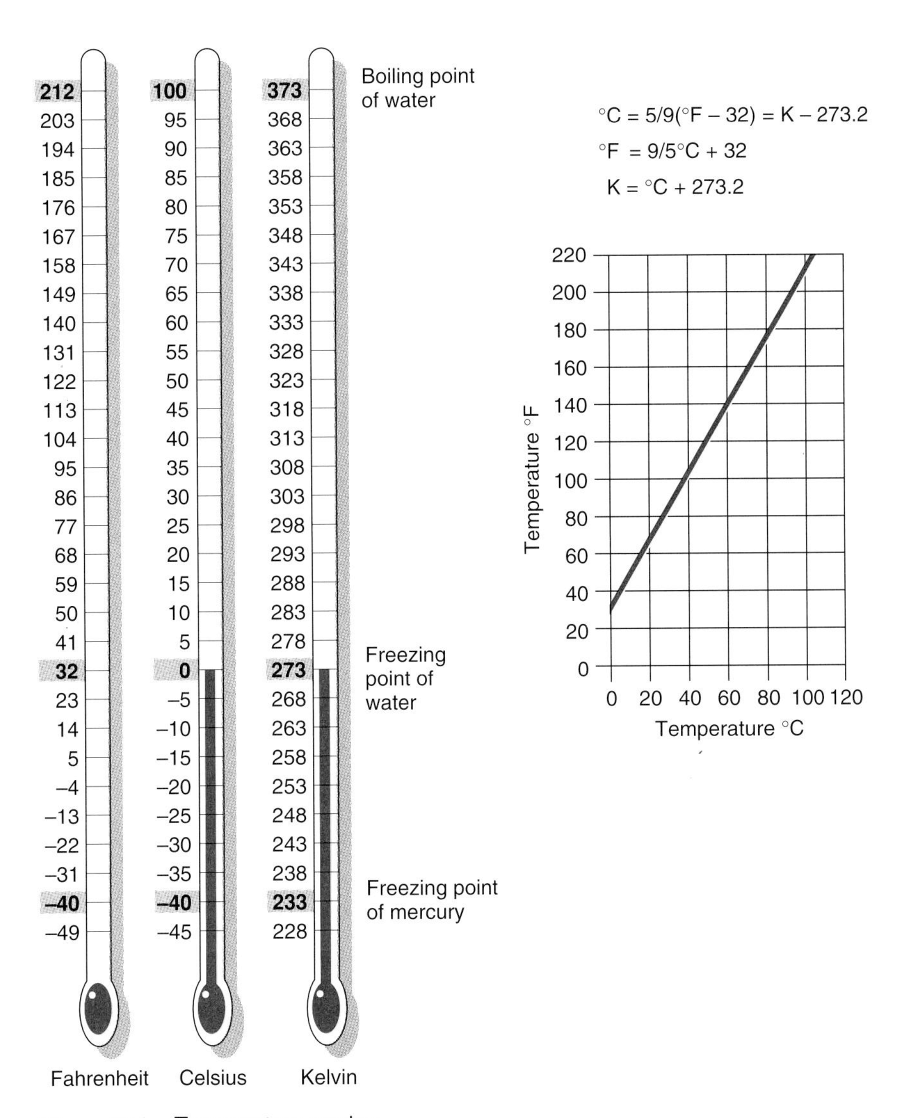


Figure D.1 Temperature scales.

## Constants

### Some Important Constants

Diameter of earth (at equator) = 12,756 km
Standard sea level pressure = 1013.25 mb
Speed of light = $2.9998 \times 10^{8}$ m $s^{-1}$ (meters per second)
Boltzmann constant = $8.132 \times 10^{-11}$ cal $cm^{-2}$ $K^{-4}$ $min^{-1}$
Wien's constant = 0.2897 $\mu$m K
Dry adiabatic lapse rate = 0.98°/100 m
One standard atmosphere = 1013.25 mb
= 101.325 kPa
= 760 mm Hg
= 29.92 in Hg
= 14.7 pounds per square inch

# Appendix E

## List of Elements

**Elements**

| Atomic Number | Name | Symbol | Atomic Number | Name | Symbol |
|---|---|---|---|---|---|
| 1 | Hydrogen | H | 47 | Silver | Ag |
| 2 | Helium | He | 48 | Cadmium | Cd |
| 3 | Lithium | Li | 49 | Indium | In |
| 4 | Berillium | Be | 50 | Tin | Sn |
| 5 | Boron | B | 51 | Antimony | Sb |
| 6 | Carbon | C | 52 | Tellurium | Te |
| 7 | Nitrogen | N | 53 | Iodine | I |
| 8 | Oxygen | O | 54 | Xenon | Xe |
| 9 | Fluorine | F | 55 | Cesium | Cs |
| 10 | Neon | Ne | 56 | Barium | Ba |
| 11 | Sodium | Na | 57 | Lanthanum | La |
| 12 | Magnesium | Mg | 58 | Cerium | Ce |
| 13 | Aluminum | Al | 59 | Praesodymium | Pr |
| 14 | Silicon | Si | 60 | Nyodemium | Nd |
| 15 | Phosphorus | P | 61 | Promethium | Pm |
| 16 | Sulfur | S | 62 | Samarium | Sm |
| 17 | Chlorine | Cl | 63 | Europium | Eu |
| 18 | Argon | Ar | 64 | Gadolinium | Gd |
| 19 | Potassium | K | 65 | Terbium | Tb |
| 20 | Calcium | Ca | 66 | Dysprosium | Dy |
| 21 | Scandium | Sc | 67 | Holmium | Ho |
| 22 | Titanium | Ti | 68 | Erbium | Er |
| 23 | Vanadium | V | 69 | Thulium | Tm |
| 24 | Chromium | Cr | 70 | Ytterbium | Yb |
| 25 | Manganese | Mn | 71 | Lutetium | Lu |
| 26 | Iron | Fe | 72 | Hafnium | Hf |
| 27 | Cobalt | Co | 73 | Tantalum | Ta |
| 28 | Nickel | Ni | 74 | Tungsten | W |
| 29 | Copper | Cu | 75 | Rhenium | Re |
| 30 | Zinc | Zn | 76 | Osmium | Os |
| 31 | Gallium | Ga | 77 | Iridium | Ir |
| 32 | Germanium | Ge | 78 | Platinum | Pt |
| 33 | Arsenic | As | 79 | Gold | Au |
| 34 | Selenium | Se | 80 | Mercury | Hg |
| 35 | Bromine | Br | 81 | Thallium | Ti |
| 36 | Krypton | Kr | 82 | Lead | Pb |
| 37 | Rubidium | Rb | 83 | Bismuth | Bi |
| 38 | Strontium | Sr | 84 | Polonium | Po |
| 39 | Yttrium | Y | 85 | Astitine | At |
| 40 | Zirconium | Zr | 86 | Radon | Rn |
| 41 | Niobium | Nb | 87 | Francium | Fr |
| 42 | Molybdenum | Mb | 88 | Radium | Ra |
| 43 | Technetium | Tc | 89 | Actinium | Ac |
| 44 | Ruthenium | Ru | 90 | Thorium | Th |
| 45 | Rhodium | Rh | 91 | Protactinium | Pa |
| 46 | Palladium | Pd | 92 | Uranium | U |

# Appendix F

# Latitude and Longitude

Locations on earth are specified by the values of a pair of coordinates called latitude and longitude.

Latitude is a measure of a point's distance north (+) or south (−) of a reference latitude line, the equator. The equator itself is formed (as shown in Figure F.1A) by the intersection of the earth's surface and an imaginary plane passing through earth's center, perpendicular to earth's axis of rotation. Values of latitude (see Figure F.1B) range from −90° at the South Pole through 0° at the Equator to +90° at the North Pole. Notice that latitude lines lie parallel to each other.

Longitude (Figure F.1C) is a measure of a location's distance east (+) or west (−) of a reference longitude known as the prime meridian. The prime meridian, whose longitude is defined as 0°, passes through Greenwich, England. Values of longitude range from −180° (or 180°W) through 0° at the prime meridian to +180° (or 180°E). Notice from the figure that lines of longitude are not parallel, as in the case of latitude lines, but radiate from North and South Poles, like sections of an orange.

Figure F.1D shows the earth with both latitude and longitude coordinates superimposed.

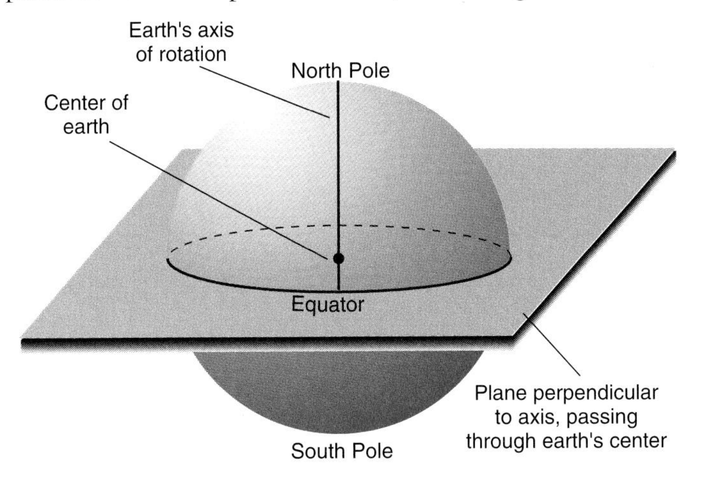


Figure F.1A Earth's axis of rotation, North and South Poles, and Equator.

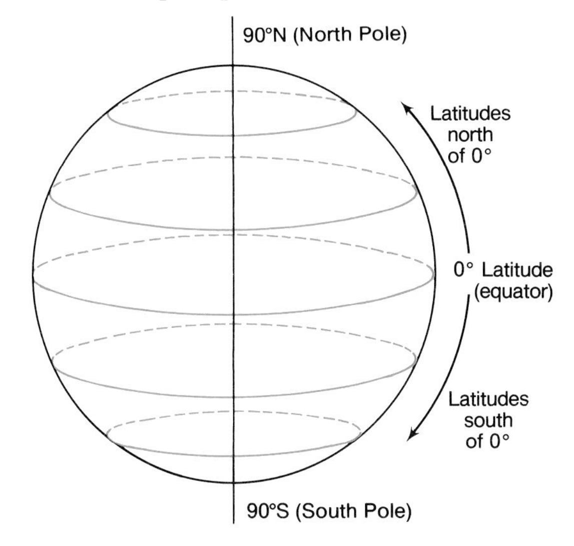


Figure F.1B Latitude ranges from −90° to +90°.

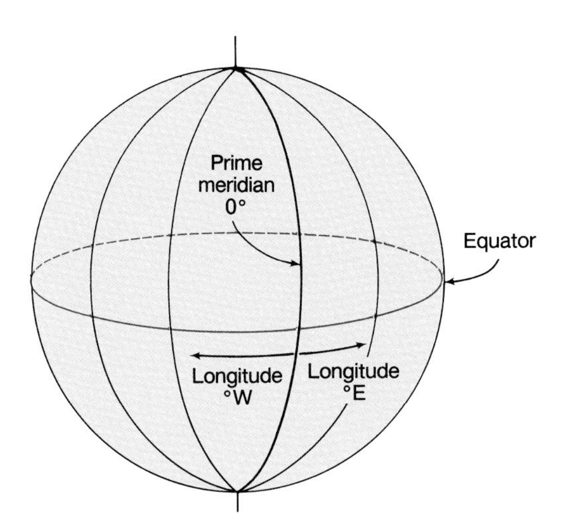


Figure F.1C Longitude is measured from the prime meridian and varies from −180° to +180°.

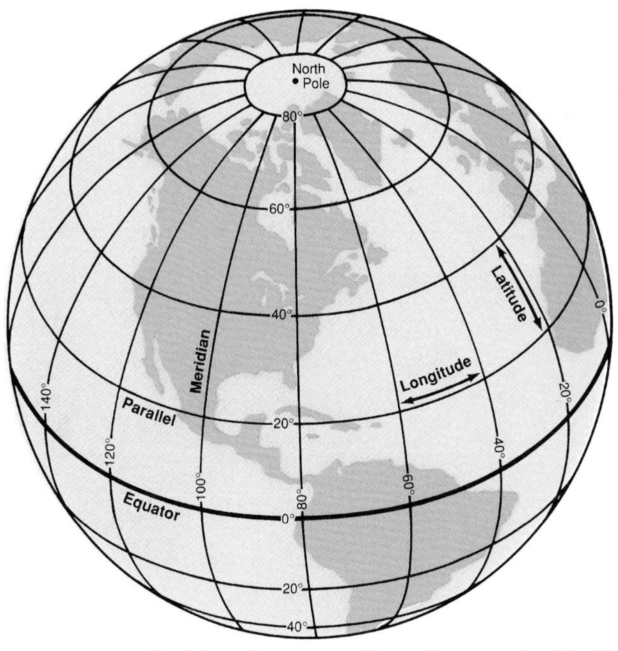


Figure F.1D Each location on earth's surface can be identified by its unique latitude and longitude.

# Appendix G

# Time Zones

As earth rotates eastward on its axis, each location is carried into the sunlight (daytime) and then into the dark (night), as shown in Figure G.1. Local solar time is based on the sun's position at a given locale; thus, local solar noon occurs when the sun is highest in the sky at a given location.

From Figure G.1, you can see that local solar time varies with a location's longitude. Thus, local solar time varies from one town to another—even from one house to its neighbor to the west or east, technically. Consequently, local solar time is impractical for modern timekeeping. Governments have agreed to standardize time within certain regions known as time zones, shown in Figure G.2. Within a given time zone, all clocks are kept on the same standard time. Changing time zones means resetting your clock, generally by an hour for each time zone boundary crossed.

If you traveled constantly in the same direction (westward, say), you would always be resetting your watch back (to earlier times); traveling around the world, you would arrive back home having set your watch earlier 24 times: if a calendar watch, it would indicate the correct time but be off by a day! The International Date Line (see Figure G.2) resolves such problems. Crossing the date line westward, you add a day (24 hours); crossing eastward, you subtract a day (24 hours) from your time.

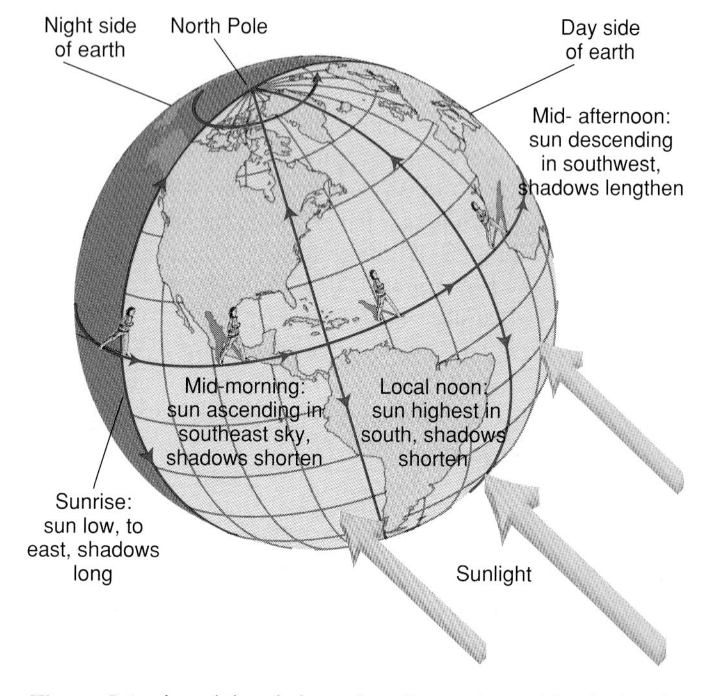


**Figure G.1** Local time is based on the sun's position in the sky.

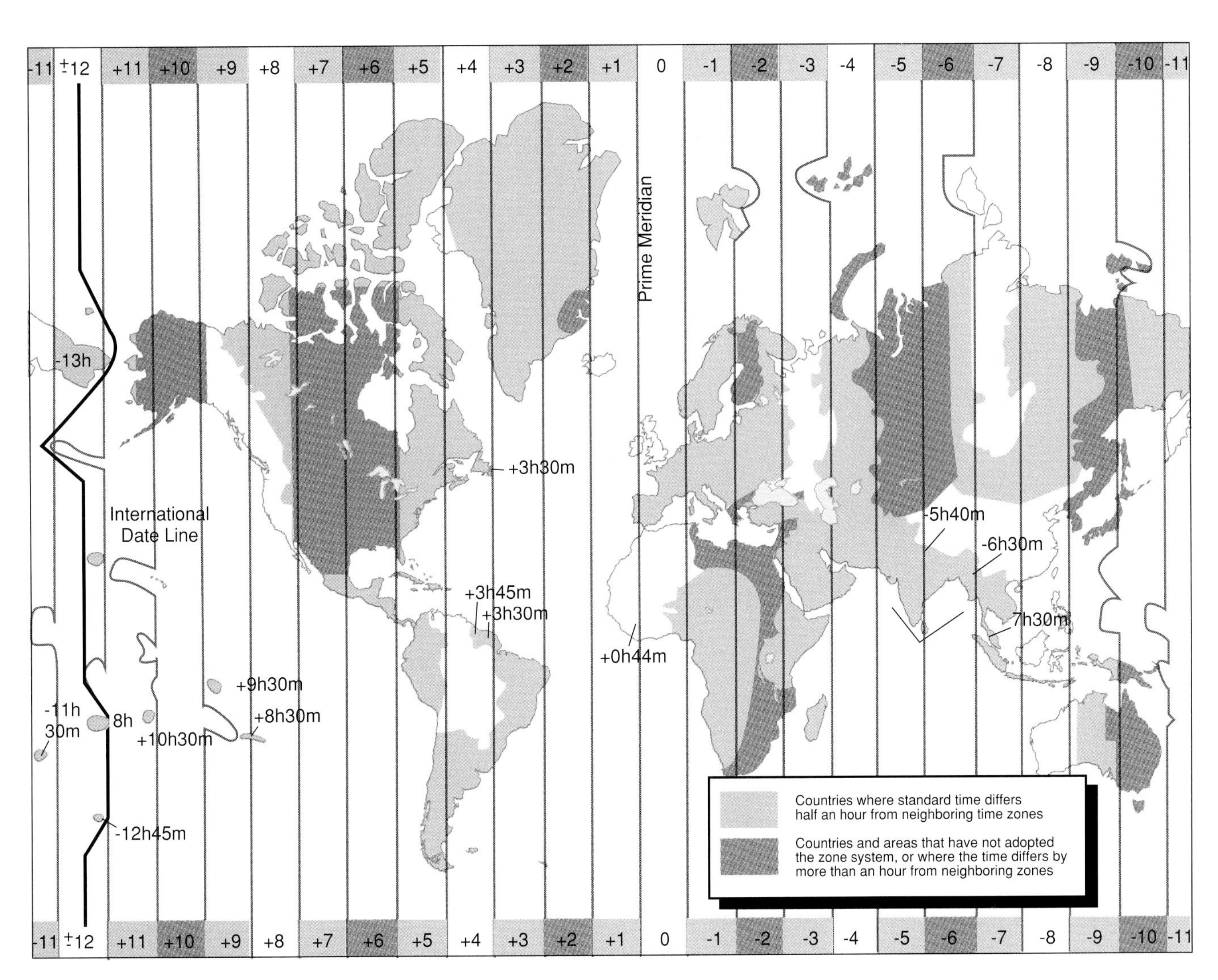


Figure G.2 Time zones and the International Date Line.

# Appendix H

# Weather Plotting Symbols

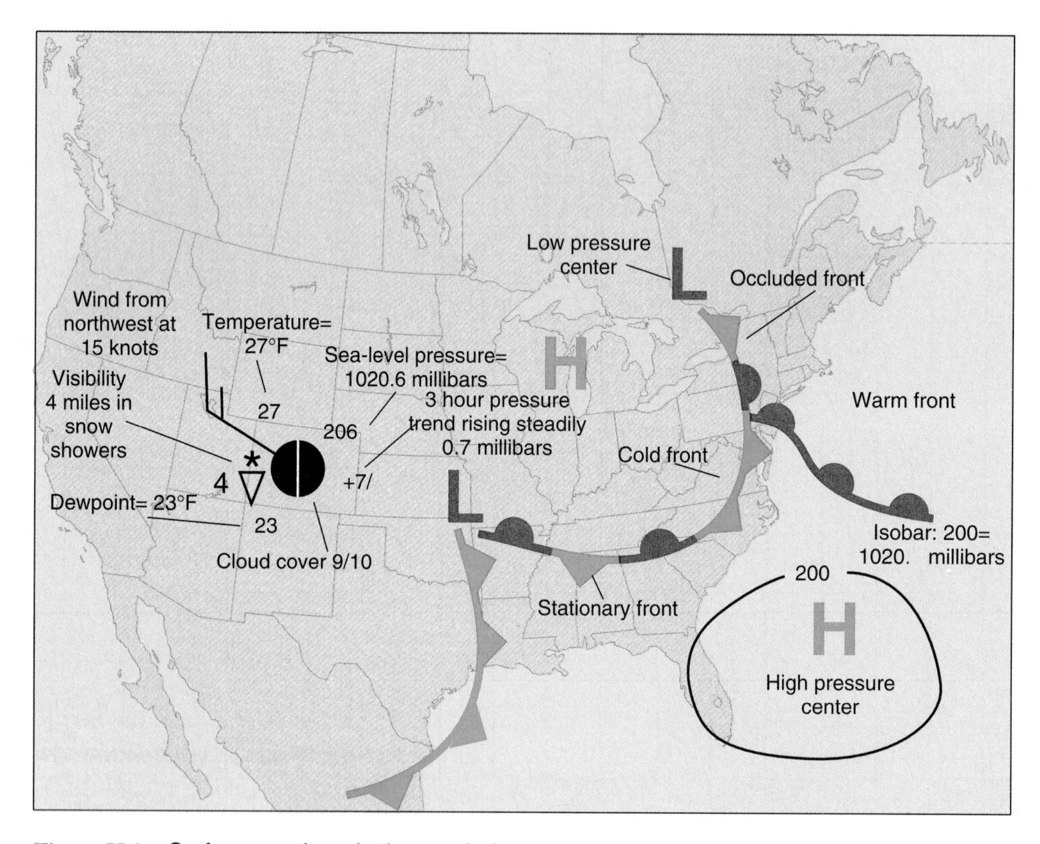


Figure H.1 Surface weather plotting symbols.

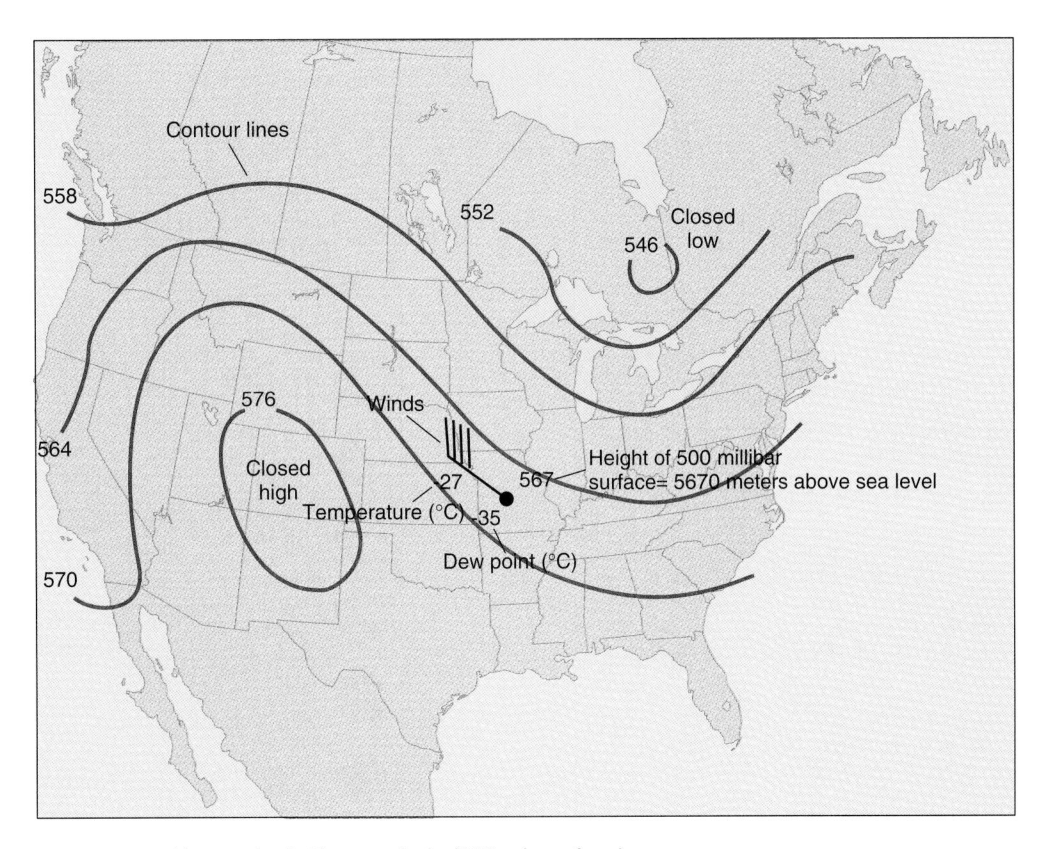


Figure H.2 Upper air plotting symbols (500 mb surface).

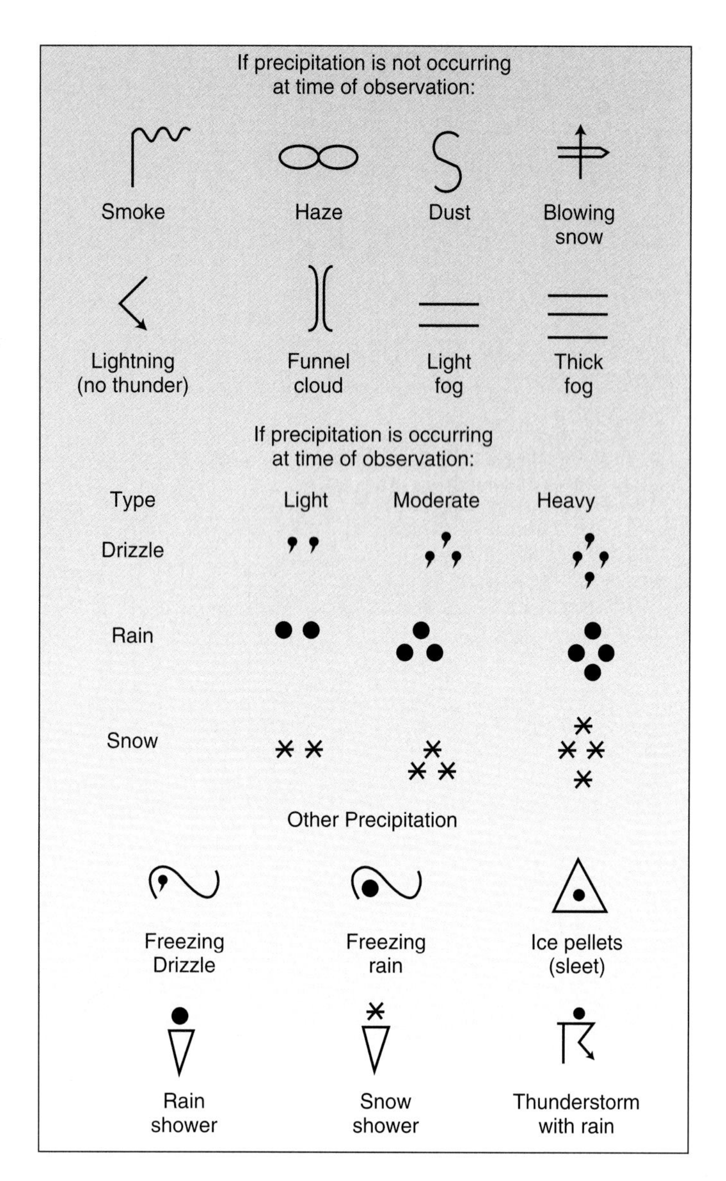


Figure H.3 Some present weather symbols

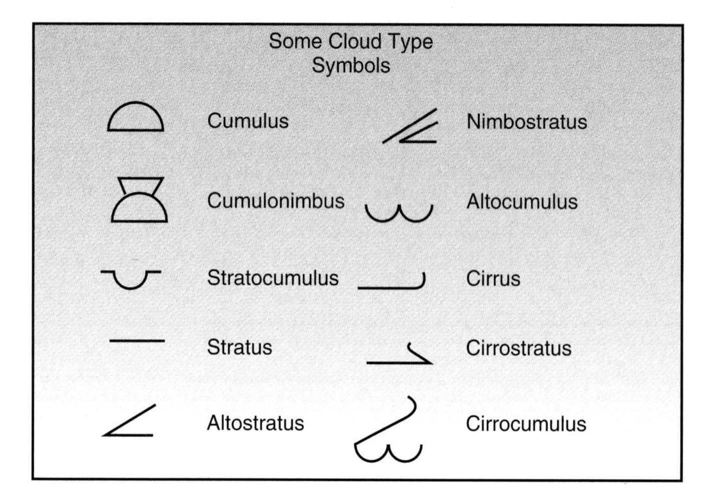


Figure H.4 Some cloud type symbols

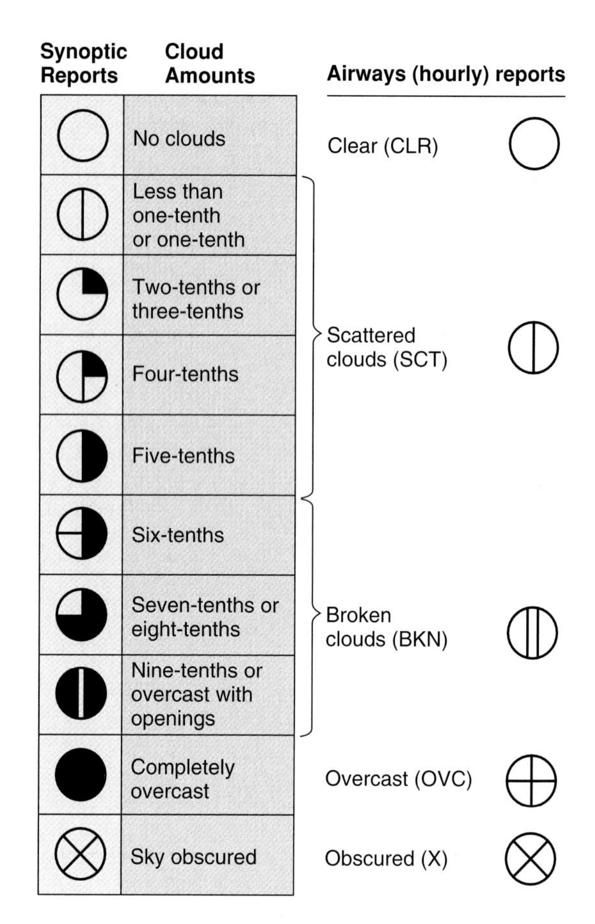


Figure H.5

**Winds**

| Symbol | Knots | Miles (statute) per hour | Meters per second |
|---|---|---|---|
| | Calm | Calm | ≤0.2 |
| | 1 - 2 | 1 - 2 | 0.3 - 1.2 |
| | 3 - 7 | 3 - 8 | 1.3 - 3.8 |
| | 8 - 12 | 9 - 14 | 3.9 - 6.4 |
| | 13 - 17 | 15 - 20 | 6.5 - 9.0 |
| | 18 - 22 | 21 - 25 | 9.1 - 11.5 |
| | 23 - 27 | 26 - 31 | 11.6 - 14.1 |
| | 28 - 32 | 32 - 37 | 14.2 - 16.7 |
| | 33 - 37 | 38 - 43 | 16.8 - 19.3 |
| | 38 - 42 | 44 - 49 | 19.4 - 21.6 |
| | 43 - 47 | 50 - 54 | 21.7 - 24.4 |
| | 48 - 52 | 55 - 60 | 24.5 - 26.7 |
| | 53 - 57 | 61 - 66 | 26.8 - 29.3 |
| | 58 - 62 | 67 - 71 | 29.4 - 31.9 |
| | 63 - 67 | 72 - 77 | 32.0 - 34.7 |
| | 68 - 72 | 78 - 83 | 34.8 - 37.3 |
| | 73 - 77 | 84 - 89 | 37.4 - 39.6 |

Figure H.6

| Code number | a | Barometric tendency | |
|---|---|---|---|
| 0 | | Rising, then falling | |
| 1 | | Rising, then steady; or rising, then rising more slowly | Barometer now higher than three hours ago |
| 2 | | Rising steadily, or unsteadily | |
| 3 | | Falling or steady, then rising; or rising, then rising more quickly | |
| 4 | | Steady, same as three hours ago | |
| 5 | | Falling, then rising, same or lower than three hours ago | |
| 6 | | Falling, then steady; or falling, then falling more slowly | Barometer now lower than three hours ago |
| 7 | | Falling steadily, or unsteadily | |
| 8 | | Steady or rising, then falling; or falling, then falling more quickly | |

Figure H.7

*Appendix I*

# Windchill and Heat Index Tables

## Windchill Table

| Temperature (°F) | Wind Speed (mph) 5 | 10 | 15 | 20 | 25 | 30 | 35 | 40 | 45 | 50 | 55 | 60 |
|---|---|---|---|---|---|---|---|---|---|---|---|---|
| 34 | 31 | 21 | 14 | 10 | 6 | 4 | 2 | 1 | −0 | −1 | −1 | −1 |
| 32 | 29 | 18 | 12 | 7 | 3 | 1 | −1 | −3 | −3 | −4 | −4 | −4 |
| 30 | 27 | 16 | 9 | 4 | 0 | −2 | −4 | −6 | −7 | −7 | −7 | −7 |
| 28 | 25 | 13 | 6 | 1 | −3 | −5 | −7 | −9 | −10 | −10 | −11 | −11 |
| 26 | 23 | 11 | 4 | −2 | −6 | −8 | −11 | −12 | −13 | −14 | −14 | −14 |
| 24 | 21 | 9 | 1 | −5 | −9 | −11 | −14 | −15 | −16 | −17 | −17 | −17 |
| 22 | 18 | 6 | −2 | −7 | −12 | −15 | −17 | −18 | −19 | −20 | −20 | −20 |
| 20 | 16 | 4 | −5 | −10 | −14 | −18 | −20 | −22 | −23 | −23 | −24 | −24 |
| 18 | 14 | 1 | −7 | −13 | −17 | −21 | −23 | −25 | −26 | −27 | −27 | −27 |
| 16 | 12 | −1 | −10 | −16 | −20 | −24 | −26 | −28 | −29 | −30 | −30 | −30 |
| 14 | 10 | −4 | −13 | −19 | −23 | −27 | −29 | −31 | −32 | −33 | −33 | −33 |
| 12 | 8 | −6 | −15 | −22 | −26 | −30 | −32 | −34 | −35 | −36 | −36 | −36 |
| 10 | 6 | −9 | −18 | −25 | −29 | −33 | −35 | −37 | −39 | −39 | −40 | −40 |
| 8 | 4 | −11 | −21 | −27 | −32 | −36 | −39 | −41 | −42 | −43 | −43 | −43 |
| 6 | 2 | −14 | −23 | −30 | −35 | −39 | −42 | −44 | −45 | −46 | −46 | −46 |
| 4 | −0 | −16 | −26 | −33 | −38 | −42 | −45 | −47 | −48 | −49 | −49 | −49 |
| 2 | −3 | −19 | −29 | −36 | −41 | −45 | −48 | −50 | −51 | −52 | −53 | −52 |
| 0 | −5 | −21 | −31 | −39 | −44 | −48 | −51 | −53 | −55 | −55 | −56 | −56 |
| −2 | −7 | −23 | −34 | −42 | −47 | −51 | −54 | −56 | −58 | −59 | −59 | −59 |
| −4 | −9 | −26 | −37 | −44 | −50 | −54 | −57 | −59 | −61 | −62 | −62 | −62 |
| −6 | −11 | −28 | −39 | −47 | −53 | −57 | −60 | −63 | −64 | −65 | −65 | −65 |
| −8 | −13 | −31 | −42 | −50 | −56 | −60 | −64 | −66 | −67 | −68 | −69 | −69 |
| −10 | −15 | −33 | −45 | −53 | −59 | −63 | −67 | −69 | −71 | −71 | −72 | −72 |
| −12 | −17 | −36 | −48 | −56 | −62 | −66 | −70 | −72 | −74 | −75 | −75 | −75 |
| −14 | −19 | −38 | −50 | −59 | −65 | −70 | −73 | −75 | −77 | −78 | −78 | −78 |
| −16 | −21 | −41 | −53 | −62 | −68 | −73 | −76 | −78 | −80 | −81 | −82 | −81 |
| −18 | −24 | −43 | −56 | −64 | −71 | −76 | −79 | −82 | −83 | −84 | −85 | −85 |
| −20 | −26 | −46 | −58 | −67 | −74 | −79 | −82 | −85 | −87 | −88 | −88 | −88 |
| −22 | −28 | −48 | −61 | −70 | −77 | −82 | −85 | −88 | −90 | −91 | −91 | −91 |
| −24 | −30 | −51 | −64 | −73 | −80 | −85 | −88 | −91 | −93 | −94 | −94 | −94 |
| −26 | −32 | −53 | −66 | −76 | −83 | −88 | −92 | −94 | −96 | −97 | −98 | −98 |
| −28 | −34 | −55 | −69 | −79 | −86 | −91 | −95 | −97 | −99 | −100 | −101 | −101 |
| −30 | −36 | −58 | −72 | −81 | −89 | −94 | −98 | −101 | −102 | −104 | −104 | −104 |
| −32 | −38 | −60 | −74 | −84 | −92 | −97 | −101 | −104 | −106 | −107 | −107 | −107 |
| −34 | −40 | −63 | −77 | −87 | −95 | −100 | −104 | −107 | −109 | −110 | −111 | −110 |
| −36 | −43 | −65 | −80 | −90 | −98 | −103 | −107 | −110 | −112 | −113 | −114 | −114 |
| −38 | −45 | −68 | −82 | −93 | −101 | −106 | −110 | −113 | −115 | −116 | −117 | −117 |
| −40 | −47 | −70 | −85 | −96 | −103 | −109 | −113 | −116 | −118 | −120 | −120 | −120 |
| −42 | −49 | −73 | −88 | −99 | −106 | −112 | −117 | −120 | −122 | −123 | −123 | −123 |
| −44 | −51 | −75 | −91 | −101 | −109 | −115 | −120 | −123 | −125 | −126 | −127 | −127 |
| −46 | −53 | −78 | −93 | −104 | −112 | −118 | −123 | −126 | −128 | −129 | −130 | −130 |
| −48 | −55 | −80 | −96 | −107 | −115 | −121 | −126 | −129 | −131 | −133 | −133 | −133 |
| −50 | −57 | −83 | −99 | −110 | −118 | −124 | −129 | −132 | −134 | −136 | −136 | −136 |
| −52 | −59 | −85 | −101 | −113 | −121 | −128 | −132 | −135 | −138 | −139 | −140 | −140 |
| −54 | −61 | −87 | −104 | −116 | −124 | −131 | −135 | −139 | −141 | −142 | −143 | −143 |

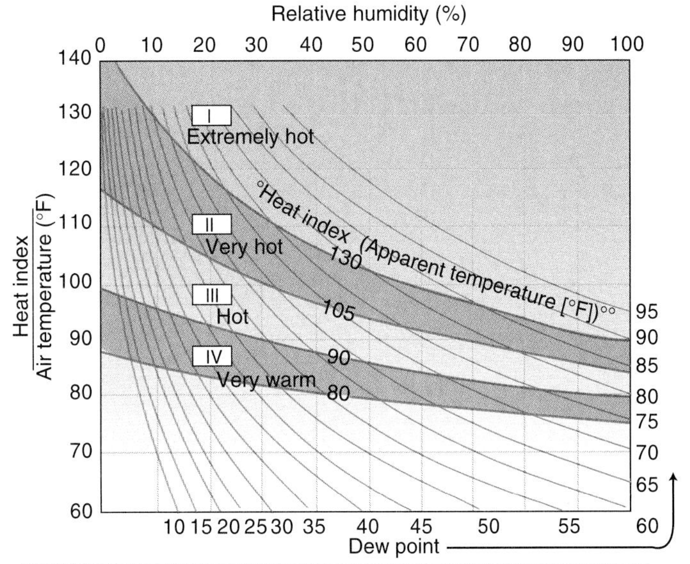


| Category | Apparent temperature (°F) | Heat symptoms |
|---|---|---|
| I | 130° or higher | Heatstroke/ Sunstroke highly likely with continued exposure |
| II | 105° - 130° | Sunstroke, heat cramps, or heat exhaustion likely. Heatstroke possible with prolonged exposure and physical activity. |
| III | 90° - 105° | Sunstroke, heat cramps, and heat exhaustion possible with prolonged exposure and physical activity. |
| IV | 80° - 90° | Fatigue possible with prolonged exposure and physical activity. |

Figure I.1 Source: Data from Lans P. Rothfusz, "The Heat Index Equation" in *Western Region Technical Attachment No. 90-24*, July 10, 1990. National Weather Service, Salt Lake City, UT.

# Appendix J

# Humidity Tables

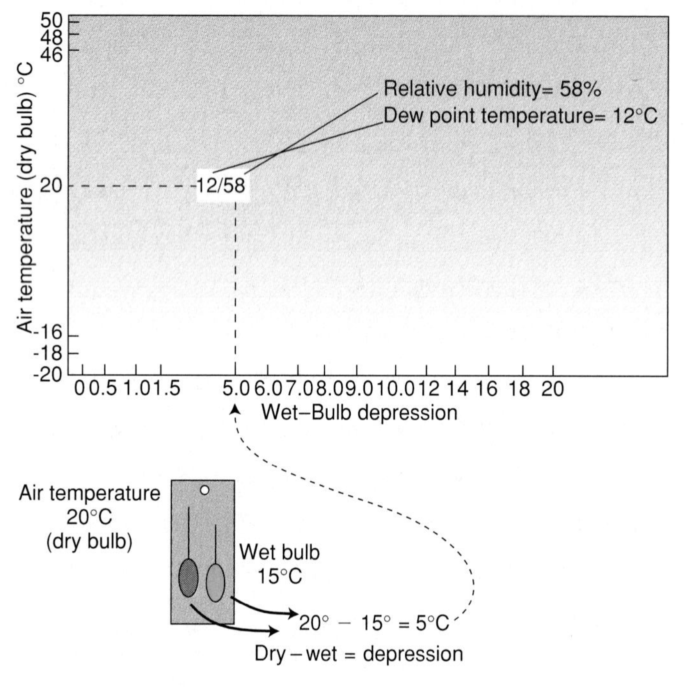


Figure J.1

## Dew Point/Relative Humidity (%)

| Temperature | Wetbulb Depression (°C) | | | | | | | | | | | | | | | | | | | | |
|---|---|---|---|---|---|---|---|---|---|---|---|---|---|---|---|---|---|---|---|---|---|
| (°C) | 0.0 | 1.0 | 2.0 | 3.0 | 4.0 | 5.0 | 6.0 | 7.0 | 8.0 | 9.0 | 10.0 | 11.0 | 12.0 | 13.0 | 14.0 | 15.0 | 16.0 | 17.0 | 18.0 | 19.0 | 20.0 |
| 40 | 40/100 | 39/94 | 38/88 | 36/82 | 35/77 | 34/71 | 33/66 | 31/62 | 30/57 | 29/52 | 27/48 | 26/44 | 24/40 | 22/36 | 21/33 | 19/29 | 17/26 | 15/23 | 13/20 | 10/17 | 7/14 |
| 38 | 38/100 | 37/94 | 36/88 | 34/82 | 33/76 | 32/71 | 31/65 | 29/60 | 28/56 | 26/51 | 25/47 | 23/43 | 22/38 | 20/35 | 18/31 | 16/27 | 14/24 | 12/20 | 9/17 | 6/14 | 3/11 |
| 36 | 36/100 | 35/93 | 34/87 | 32/81 | 31/75 | 30/70 | 28/64 | 27/59 | 26/54 | 24/50 | 22/45 | 21/41 | 19/37 | 17/33 | 15/29 | 13/25 | 11/21 | 8/18 | 5/15 | 2/12 | -3/8 |
| 34 | 34/100 | 33/93 | 32/87 | 30/81 | 29/75 | 28/69 | 26/63 | 25/58 | 23/53 | 22/48 | 20/43 | 18/39 | 16/35 | 14/30 | 12/26 | 10/23 | 7/19 | 4/15 | 1/12 | -4/9 | -10/ |
| 32 | 32/100 | 31/93 | 29/86 | 28/80 | 27/74 | 25/68 | 24/62 | 22/57 | 21/51 | 19/46 | 17/42 | 16/37 | 14/32 | 11/28 | 9/24 | 6/20 | 3/16 | -1/12 | -5/9 | -12/ | -24/ |
| 30 | 30/100 | 29/93 | 27/86 | 26/79 | 25/73 | 23/67 | 22/61 | 20/55 | 18/50 | 17/44 | 15/39 | 13/35 | 11/30 | 8/25 | 5/21 | 2/17 | -2/13 | -6/9 | -13/ | -28/ | |
| 28 | 28/100 | 27/93 | 25/85 | 24/78 | 22/72 | 21/65 | 19/59 | 18/53 | 16/48 | 14/42 | 12/37 | 10/32 | 7/27 | 5/22 | 1/18 | -2/14 | -7/9 | -15/ | -31/ | | |
| 26 | 26/100 | 25/92 | 23/85 | 22/78 | 20/71 | 19/64 | 17/58 | 15/52 | 13/46 | 11/40 | 9/34 | 7/29 | 4/24 | 1/19 | -3/14 | -8/10 | -16/ | -35/ | | | |
| 24 | 24/100 | 23/92 | 21/84 | 20/77 | 18/69 | 16/63 | 15/56 | 13/49 | 11/43 | 9/37 | 6/32 | 3/26 | 0/21 | -4/15 | -9/10 | -17/ | -38/ | | | | |
| 22 | 22/100 | 21/92 | 19/83 | 18/76 | 16/68 | 14/61 | 12/54 | 10/47 | 8/41 | 6/34 | 3/28 | -1/22 | -5/17 | -10/11 | -18/ | -40/ | | | | | |
| 20 | 20/100 | 19/91 | 17/83 | 15/74 | 14/66 | 12/59 | 10/51 | 8/44 | 5/38 | 2/31 | -1/24 | -5/18 | -10/12 | -18/ | -41/ | | | | | | |
| 18 | 18/100 | 16/91 | 15/82 | 13/73 | 11/65 | 9/57 | 7/49 | 5/41 | 2/34 | -1/27 | -5/20 | -10/14 | -18/ | -39/ | | | | | | | |
| 16 | 16/100 | 14/90 | 13/81 | 11/72 | 9/63 | 7/54 | 4/46 | 2/38 | -2/30 | -5/23 | -10/15 | -18/ | -37/ | | | | | | | | |
| 14 | 14/100 | 12/90 | 11/80 | 9/70 | 6/60 | 4/51 | 1/43 | -2/34 | -5/26 | -10/18 | -18/10 | -33/ | | | | | | | | | |
| 12 | 12/100 | 10/89 | 8/78 | 6/68 | 4/58 | 1/48 | -2/39 | -5/30 | -10/21 | -17/12 | -30/ | | | | | | | | | | |
| 10 | 10/100 | 8/88 | 6/77 | 4/66 | 1/55 | -2/45 | -5/34 | -10/25 | -16/15 | -27/ | | | | | | | | | | | |
| 8 | 8/100 | 6/87 | 4/75 | 1/63 | -1/52 | -5/40 | -9/29 | -15/19 | -24/8 | | | | | | | | | | | | |
| 6 | 6/100 | 4/86 | 2/73 | -1/60 | -4/48 | -8/36 | -14/24 | -22/12 | | | | | | | | | | | | | |
| 4 | 4/100 | 2/85 | -1/71 | -4/57 | -8/43 | -12/30 | -19/17 | -35/ | | | | | | | | | | | | | |
| 2 | 2/100 | -0/84 | -3/68 | -7/53 | -11/38 | -17/24 | -28/9 | | | | | | | | | | | | | | |
| 0 | 0/100 | -3/82 | -6/65 | -10/48 | -15/32 | -23/16 | | | | | | | | | | | | | | | |
| -2 | -2/100 | -5/81 | -8/62 | -13/43 | -20/25 | -34/7 | | | | | | | | | | | | | | | |
| -4 | -4/100 | -7/79 | -11/58 | -17/37 | -26/17 | | | | | | | | | | | | | | | | |
| -6 | -6/100 | -10/76 | -14/53 | -21/30 | -36/8 | | | | | | | | | | | | | | | | |
| -8 | -8/100 | -12/73 | -17/47 | -27/22 | | | | | | | | | | | | | | | | | |
| -10 | -10/100 | -15/70 | -21/41 | -35/12 | | | | | | | | | | | | | | | | | |
| -12 | -12/100 | -17/66 | -25/33 | | | | | | | | | | | | | | | | | | |
| -14 | -14/100 | -20/62 | -31/24 | | | | | | | | | | | | | | | | | | |
| -16 | -16/100 | -23/56 | -39/13 | | | | | | | | | | | | | | | | | | |
| -18 | -18/100 | -26/50 | | | | | | | | | | | | | | | | | | | |
| -20 | -20/100 | -30/42 | | | | | | | | | | | | | | | | | | | |
| -22 | -22/100 | -35/ | | | | | | | | | | | | | | | | | | | |
| -24 | -24/100 | -41/ | | | | | | | | | | | | | | | | | | | |

# Appendix K

# Saturation Mixing Ratio Diagram

**Saturation Mixing Ratio over Liquid Water and over Ice, at Air Pressure = 1000 mb**

| Temperature (°C) | Ratio over Water (g/kg) | Ratio over Ice (g/kg) |
|---|---|---|
| 40 | 49.8 | |
| 38 | 44.4 | |
| 36 | 39.5 | |
| 34 | 35.1 | |
| 32 | 31.2 | |
| 30 | 27.7 | |
| 28 | 24.6 | |
| 26 | 21.7 | |
| 24 | 19.2 | |
| 22 | 17.0 | |
| 20 | 15.0 | |
| 18 | 13.2 | |
| 16 | 11.6 | |
| 14 | 10.1 | |
| 12 | 8.9 | |
| 10 | 7.8 | |
| 8 | 6.8 | |
| 6 | 5.9 | |
| 4 | 5.1 | |
| 2 | 4.4 | |
| 0 | 3.84 | 3.84 |
| −2 | 3.31 | 3.25 |
| −4 | 2.85 | 2.74 |
| −6 | 2.45 | 2.31 |
| −8 | 2.10 | 1.94 |
| −10 | 1.79 | 1.63 |
| −12 | 1.53 | 1.36 |
| −14 | 1.30 | 1.13 |
| −16 | 1.10 | 0.94 |
| −18 | 0.93 | 0.78 |
| −20 | 0.78 | 0.65 |
| −22 | 0.66 | 0.53 |
| −24 | 0.55 | 0.44 |
| −26 | 0.46 | 0.36 |

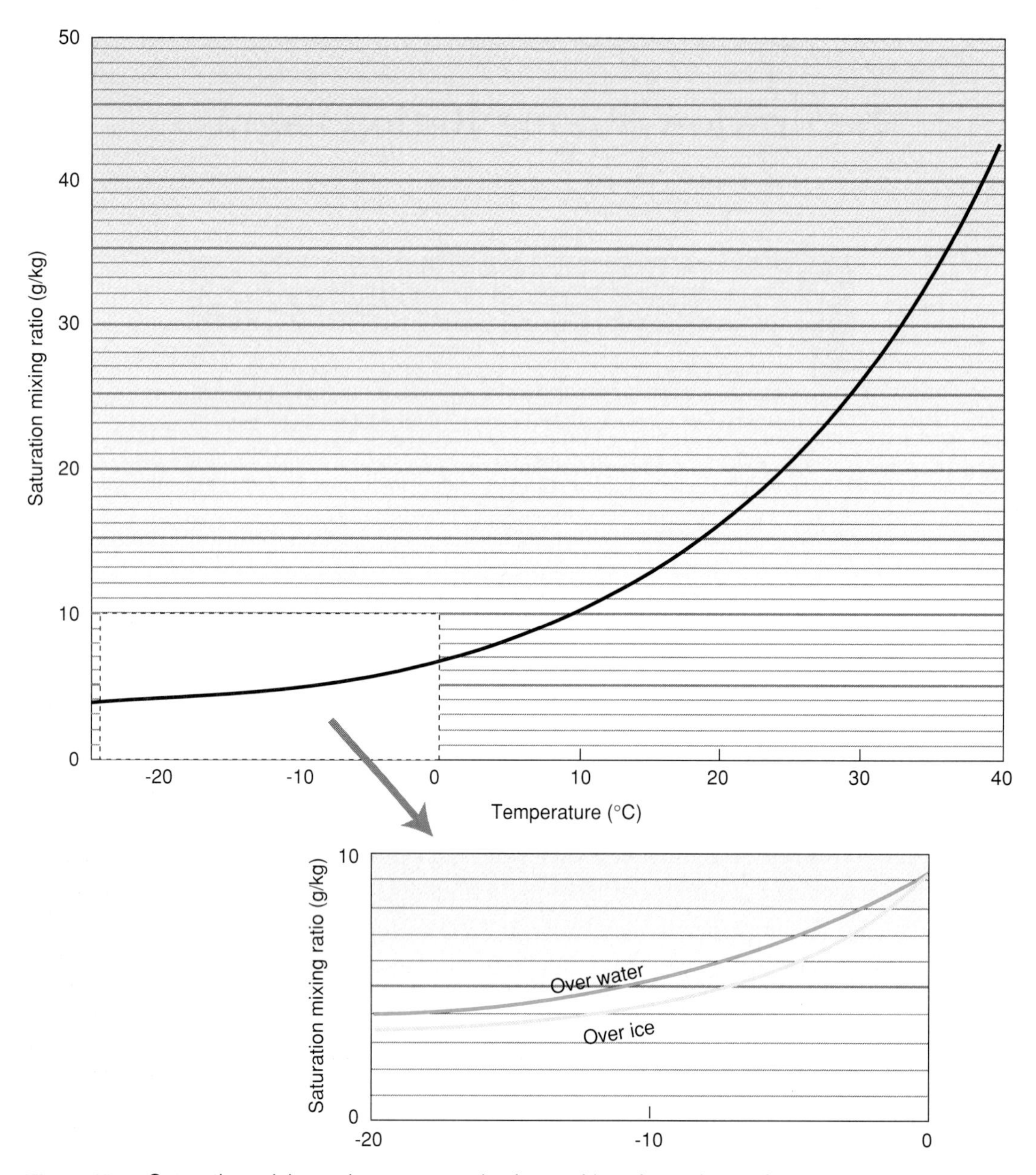


Figure K.1 Saturation mixing ratio over water (main graph) and over ice and water (insert graph) for air with pressure equal to 1,000 millibars. Units are grams of water vapor per kilogram of dry air. (For air pressures other than 1,000 millibars, use the thermodynamic diagram in Appendix L.)

# Appendix L

# Thermodynamic Diagram

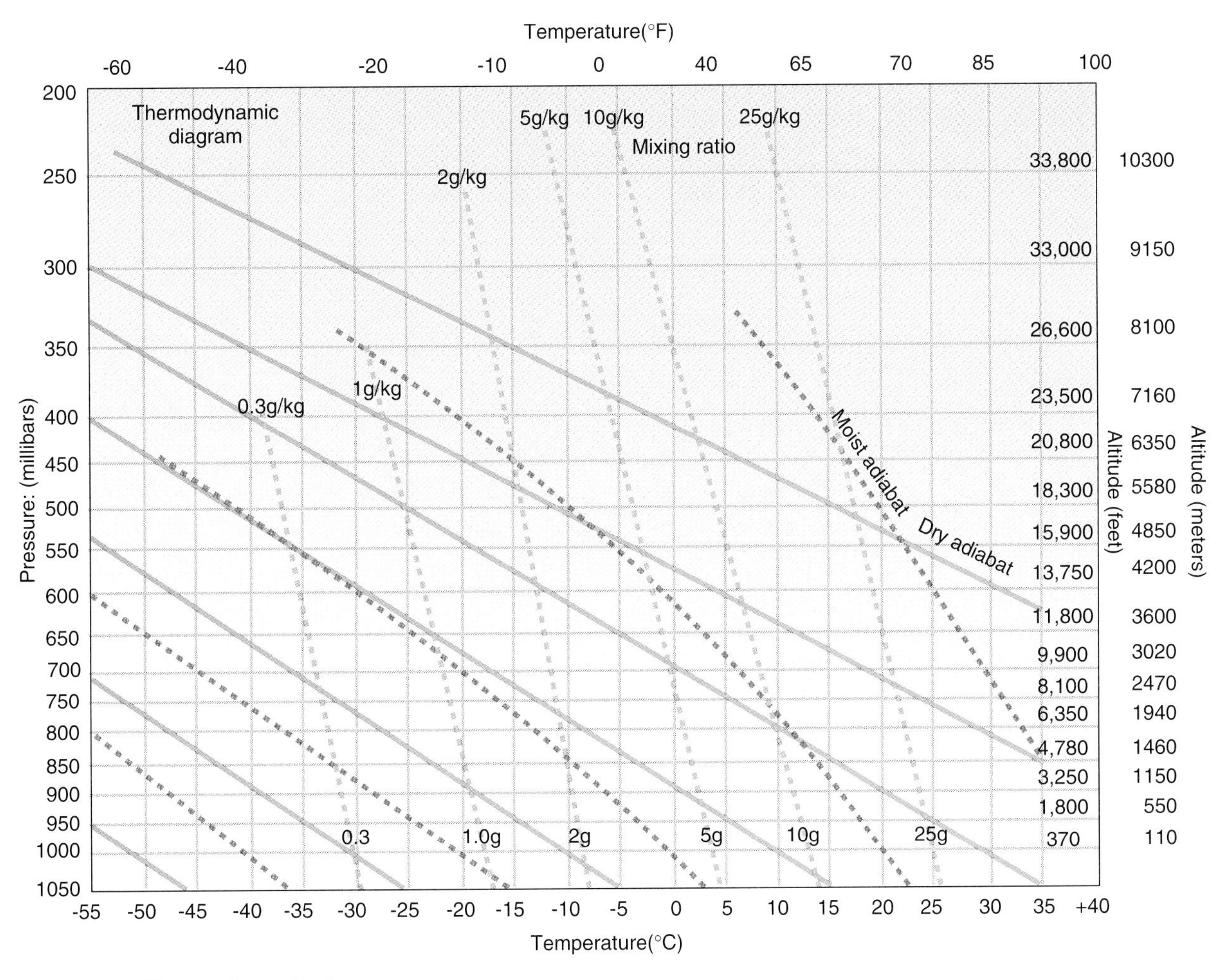


Figure L.1 Thermodynamic diagram.

Appendix M

# Climatological Data

## Climatological Data

| Location | Latitude | Longitude | Elevation (meters) | Jan. | Feb. | March | April | May | June | July | Aug. | Sept. | Oct. | Nov. | Dec. | Annual Total Precipitation, Average Temperature |
|---|---|---|---|---|---|---|---|---|---|---|---|---|---|---|---|---|
| Addis Abba, Ethiopia | | | | | | | | | | | | | | | | |
| | 9.0 | 38.8 E | 2360 | | | | | | | | | | | | | |
| P | | | | 17 | 38 | 68 | 86 | 86 | 132 | 268 | 281 | 186 | 28 | 11 | 10 | 1210 mm |
| T | | | | 16.2 | 17.1 | 18.1 | 18.1 | 18.4 | 17.0 | 15.4 | 15.3 | 16.0 | 16.0 | 15.8 | 15.7 | 16.6 C |
| Algiers, Algeria | | | | | | | | | | | | | | | | |
| | 36.7 | 3.3 E | 25 | | | | | | | | | | | | | |
| P | | | | 89 | 78 | 67 | 68 | 36 | 19 | 3 | 4 | 32 | 93 | 92 | 121 | 702 mm |
| T | | | | 10.8 | 11.3 | 12.9 | 14.8 | 17.9 | 21.4 | 24.4 | 25.2 | 22.9 | 19.1 | 15.0 | 11.9 | 17.3 C |
| Alice Springs, Australia | | | | | | | | | | | | | | | | |
| | −23.8 | 133.9 E | 537 | | | | | | | | | | | | | |
| P | | | | 41 | 42 | 35 | 17 | 17 | 17 | 12 | 10 | 9 | 20 | 25 | 37 | 281 mm |
| T | | | | 28.5 | 27.7 | 24.8 | 20.1 | 15.5 | 12.4 | 11.6 | 14.4 | 18.3 | 22.8 | 25.8 | 27.7 | 20.8 C |
| Barrow, Alaska | | | | | | | | | | | | | | | | |
| | 71.3 | 156.8 W | 13 | | | | | | | | | | | | | |
| P | | | | 4 | 4 | 4 | 5 | 4 | 9 | 23 | 25 | 16 | 12 | 6 | 4 | 116 mm |
| T | | | | −25.6 | −28.0 | −26.1 | −18.6 | −7.2 | 1.0 | 4.1 | 3.3 | −0.8 | −9.9 | −18.6 | −24.3 | −12.6 C |
| Billings, Montana | | | | | | | | | | | | | | | | |
| | 45.8 | 108.5 | 1099 | | | | | | | | | | | | | |
| P | | | | 18 | 14 | 23 | 41 | 63 | 54 | 23 | 28 | 32 | 27 | 18 | 17 | 357 mm |
| T | | | | −5.5 | −1.0 | 2.1 | 8.2 | 13.6 | 18.2 | 22.2 | 20.9 | 15.3 | 10.0 | 2.3 | −2.0 | 8.7 C |
| Bilma, Niger | | | | | | | | | | | | | | | | |
| | 18.7 | 12.9 E | 355 | | | | | | | | | | | | | |
| P | | | | 0 | 0 | 0 | 0 | 0 | 1 | 3 | 9 | 2 | 1 | 0 | 0 | 17 mm |
| T | | | | 16.9 | 20.3 | 24.1 | 28.8 | 32.3 | 33.5 | 32.8 | 32.3 | 31.0 | 27.2 | 22.0 | 17.8 | 26.6 C |
| Birmingham, Alabama | | | | | | | | | | | | | | | | |
| | 33.6 | 86.8 W | 196 | | | | | | | | | | | | | |
| P | | | | 126 | 119 | 151 | 117 | 114 | 92 | 132 | 93 | 99 | 70 | 100 | 126 | 1339 mm |
| T | | | | 6.1 | 8.1 | 12.2 | 16.8 | 21.1 | 25.0 | 26.7 | 26.4 | 23.3 | 17.1 | 11.5 | 7.4 | 16.8 C |
| Brookings, Oregon | | | | | | | | | | | | | | | | |
| | 42.1 | 124.3 | 21 | | | | | | | | | | | | | |
| P | | | | 314 | 247 | 244 | 130 | 95 | 40 | 12 | 29 | 58 | 159 | 284 | 325 | 1937 mm |
| T | | | | 8.605 | 9.397 | 9.458 | 10.43 | 12.46 | 14.29 | 15.01 | 15.26 | 15.38 | 13.38 | 10.87 | 9.113 | 12.0 C |
| Bulawayo, Zimbabwe | | | | | | | | | | | | | | | | |
| | −20.2 | 28.6 E | 1343 | | | | | | | | | | | | | |
| P | | | | 134 | 107 | 69 | 25 | 9 | 2 | 1 | 1 | 5 | 27 | 87 | 127 | 595 mm |
| T | | | | 22.8 | 21.4 | 20.6 | 19.7 | 16.0 | 14.1 | 14.2 | 16.9 | 20.1 | 21.9 | 21.5 | 21.4 | 19.2 C |
| Cairo, Egypt | | | | | | | | | | | | | | | | |
| | 30.0 | 31.4 E | 64 | | | | | | | | | | | | | |
| P | | | | 5 | 4 | 4 | 2 | 1 | 0 | 0 | 0 | 0 | 1 | 3 | 6 | 25 mm |
| T | | | | 13.8 | 15.3 | 17.6 | 21.4 | 24.8 | 27.2 | 27.9 | 27.8 | 26.2 | 23.8 | 19.3 | 15.2 | 21.7 C |

## Climatological Data *continued*

| Location | Latitude | Longitude | Elevation (meters) | Jan. | Feb. | March | April | May | June | July | Aug. | Sept. | Oct. | Nov. | Dec. | Annual Total Precipitation, Average Temperature |
|---|---|---|---|---|---|---|---|---|---|---|---|---|---|---|---|---|
| Carrasco, Uruguay | | | | | | | | | | | | | | | | |
| | −34.8 | 56.0 W | 32 | | | | | | | | | | | | | |
| P | | | | 95 | 100 | 111 | 83 | 76 | 74 | 86 | 84 | 90 | 98 | 78 | 84 | 1059 mm |
| T | | | | 22.9 | 22.3 | 20.5 | 17.2 | 14.0 | 11.1 | 10.8 | 11.5 | 13.2 | 15.8 | 18.4 | 21.2 | 16.6 C |
| Chengdu, China | | | | | | | | | | | | | | | | |
| | 30.7 | 104.0 E | 498 | | | | | | | | | | | | | |
| P | | | | 7 | 11 | 20 | 48 | 93 | 110 | 236 | 231 | 122 | 41 | 17 | 6 | 941 mm |
| T | | | | 5.6 | 7.4 | 11.8 | 16.7 | 21.0 | 23.6 | 25.4 | 25.0 | 21.2 | 16.8 | 11.9 | 7.1 | 16.1 C |
| Christchurch, New Zealand | | | | | | | | | | | | | | | | |
| | −43.5 | 172.6 E | 34 | | | | | | | | | | | | | |
| P | | | | 49 | 43 | 58 | 54 | 63 | 61 | 65 | 60 | 39 | 47 | 49 | 54 | 642 mm |
| T | | | | 16.9 | 16.6 | 14.9 | 12.1 | 8.9 | 6.3 | 5.8 | 7.1 | 9.3 | 11.6 | 13.8 | 15.5 | 11.6 C |
| Cokurdah, Russia | | | | | | | | | | | | | | | | |
| | 70.6 | 147.9 E | 48 | | | | | | | | | | | | | |
| P | | | | 14 | 11 | 8 | 7 | 10 | 25 | 34 | 33 | 24 | 21 | 16 | 12 | 214 mm |
| T | | | | −34.3 | −33.7 | −28.0 | −19.3 | −6.1 | 6.0 | 9.4 | 6.5 | 0.5 | −12.9 | −26.8 | −33.2 | −14.3 C |
| Comodoro Rivadavia, Argentina | | | | | | | | | | | | | | | | |
| | −45.8 | 67.5 W | 58 | | | | | | | | | | | | | |
| P | | | | 16 | 11 | 21 | 21 | 34 | 21 | 25 | 22 | 13 | 13 | 13 | 15 | 224 mm |
| T | | | | 18.9 | 18.3 | 16.0 | 12.8 | 9.4 | 6.8 | 6.7 | 7.7 | 9.8 | 12.7 | 15.7 | 17.7 | 12.7 C |
| Cuiaba, Brazil | | | | | | | | | | | | | | | | |
| | −15.6 | 56.1 W | 160 | | | | | | | | | | | | | |
| P | | | | 221 | 201 | 208 | 113 | 49 | 13 | 9 | 17 | 49 | 126 | 163 | 198 | 1367 mm |
| T | | | | 26.6 | 26.5 | 26.4 | 26.0 | 24.5 | 23.3 | 23.1 | 25.0 | 26.9 | 27.3 | 27.1 | 26.8 | 25.8 C |
| Des Moines, Iowa | | | | | | | | | | | | | | | | |
| | 41.5 | 93.7 W | 292 | | | | | | | | | | | | | |
| P | | | | 25 | 31 | 57 | 82 | 98 | 112 | 91 | 107 | 81 | 62 | 43 | 29 | 818 mm |
| T | | | | −6.8 | −3.8 | 2.3 | 10.3 | 16.7 | 22.0 | 24.6 | 23.2 | 18.3 | 12.0 | 3.5 | −3.8 | 9.9 C |
| Ghanzi, Botswanna | | | | | | | | | | | | | | | | |
| | −21.7 | 21.7 E | 1131 | | | | | | | | | | | | | |
| P | | | | 95 | 81 | 68 | 35 | 8 | 1 | 0 | 1 | 3 | 22 | 45 | 68 | 426 mm |
| T | | | | 26.7 | 25.4 | 24.3 | 21.5 | 17.2 | 14.1 | 14.0 | 16.9 | 21.6 | 24.9 | 25.4 | 25.7 | 21.5 C |
| Harbin, China | | | | | | | | | | | | | | | | |
| | 45.8 | 126.8 E | 143 | | | | | | | | | | | | | |
| P | | | | 4 | 5 | 11 | 23 | 37 | 81 | 157 | 118 | 67 | 25 | 8 | 6 | 542 mm |
| T | | | | −19.3 | −15.2 | −4.6 | 6.4 | 14.4 | 20.0 | 22.8 | 21.1 | 14.3 | 5.5 | −5.8 | −15.4 | 3.7 C |
| Houlton, Maine | | | | | | | | | | | | | | | | |
| | 46.1 | 67.8 W | 125 | | | | | | | | | | | | | |
| P | | | | 72 | 63 | 66 | 72 | 70 | 80 | 86 | 86 | 80 | 78 | 101 | 95 | 951 mm |
| T | | | | −10.9 | −9.3 | −3.3 | 3.7 | 10.6 | 16.0 | 19.1 | 17.5 | 12.6 | 6.7 | 0.2 | −7.8 | 4.6 C |
| Howard Air Force Base, Canal Zone | | | | | | | | | | | | | | | | |
| | 8.9 | 79.6 W | 16 | | | | | | | | | | | | | |
| P | | | | 33 | 18 | 16 | 67 | 175 | 178 | 155 | 175 | 175 | 238 | 231 | 119 | 1579 mm |
| T | | | | 28.2 | 28.6 | 28.9 | 28.9 | 28.1 | 27.7 | 27.7 | 27.7 | 27.5 | 27.0 | 27.5 | 27.8 | 28.0 C |
| Hyderabad, India | | | | | | | | | | | | | | | | |
| | 17.5 | 78.5 E | 530 | | | | | | | | | | | | | |
| P | | | | 6 | 10 | 15 | 22 | 29 | 104 | 171 | 151 | 169 | 85 | 28 | 5 | 795 mm |
| T | | | | 22.2 | 24.8 | 28.3 | 31.2 | 33.0 | 29.5 | 26.7 | 26.2 | 26.2 | 25.7 | 23.0 | 21.3 | 26.5 C |

## Climatological Data *continued*

| Location | Latitude | Longitude | Elevation (meters) | Jan. | Feb. | March | April | May | June | July | Aug. | Sept. | Oct. | Nov. | Dec. | Annual Total Precipitation, Average Temperature |
|---|---|---|---|---|---|---|---|---|---|---|---|---|---|---|---|---|
| Iquitos, Peru | | | | | | | | | | | | | | | | |
| | −3.8 | 73.3 W | 125 | | | | | | | | | | | | | |
| P | | | | 268 | 254 | 323 | 301 | 267 | 208 | 163 | 166 | 190 | 231 | 249 | 258 | 2878 mm |
| T | | | | 26.4 | 26.4 | 26.4 | 26.0 | 25.9 | 25.5 | 25.3 | 25.9 | 26.4 | 26.6 | 26.6 | 26.5 | 26.2 C |
| Kap Tobin, Greenland | | | | | | | | | | | | | | | | |
| | 70.4 | 22.0 W | 41 | | | | | | | | | | | | | |
| P | | | | 35 | 31 | 33 | 27 | 21 | 23 | 31 | 42 | 46 | 58 | 36 | 45 | 428 mm |
| T | | | | −16.2 | −17.3 | −16.5 | −11.7 | −3.7 | 0.9 | 3.2 | 3.1 | −0.1 | −6.1 | −11.5 | −14.2 | −7.5 C |
| Kazalinsk, Kazakhstan | | | | | | | | | | | | | | | | |
| | 45.8 | 62.1 E | 65 | | | | | | | | | | | | | |
| P | | | | 13 | 13 | 18 | 15 | 12 | 6 | 6 | 6 | 8 | 15 | 13 | 16 | 141 mm |
| T | | | | −11 | −10 | −2 | 10 | 19 | 24 | 26 | 24 | 17 | 8 | 0 | −7 | 8 C |
| Key West, Florida | | | | | | | | | | | | | | | | |
| | 24.6 | 81.8 W | 2 | | | | | | | | | | | | | |
| P | | | | 48 | 51 | 45 | 45 | 81 | 122 | 88 | 120 | 162 | 115 | 77 | 50 | 1002 mm |
| T | | | | 20.9 | 21.4 | 23.2 | 25.1 | 27.0 | 28.4 | 29.1 | 29.1 | 28.4 | 26.6 | 24.1 | 21.8 | 25.4 C |
| Kisangani, Zaire | | | | | | | | | | | | | | | | |
| | 0.5 | 25.2 E | 415 | | | | | | | | | | | | | |
| P | | | | 97 | 107 | 172 | 190 | 162 | 128 | 114 | 178 | 164 | 233 | 207 | 105 | 1858 mm |
| T | | | | 24.8 | 25.2 | 25.2 | 25.0 | 25.0 | 24.3 | 23.7 | 23.7 | 24.2 | 24.5 | 24.5 | 24.6 | 24.6 C |
| Lima, Peru | | | | | | | | | | | | | | | | |
| | −12.0 | 77.1 W | 13 | | | | | | | | | | | | | |
| P | | | | 1 | 1 | 1 | 0 | 1 | 2 | 4 | 3 | 3 | 2 | 1 | 1 | 20 mm |
| T | | | | 22.2 | 22.8 | 22.3 | 20.6 | 18.5 | 17.1 | 16.3 | 16.1 | 16.2 | 17.1 | 18.6 | 20.6 | 19.0 C |
| Los Angeles, California | | | | | | | | | | | | | | | | |
| | 33.9 | 118.4 W | 38 | | | | | | | | | | | | | |
| P | | | | 69 | 61 | 46 | 22 | 3 | 1 | 0 | 3 | 5 | 8 | 41 | 39 | 297 mm |
| T | | | | 13.9 | 14.9 | 15.4 | 16.6 | 18.2 | 20.5 | 23.2 | 23.5 | 22.8 | 20.4 | 17.1 | 14.7 | 18.4 C |
| Madrid, Spain | | | | | | | | | | | | | | | | |
| | 40.4 | 3.7 W | 657 | | | | | | | | | | | | | |
| P | | | | 30 | 31 | 37 | 37 | 37 | 37 | 33 | 43 | 45 | 47 | 42 | 29 | 447 mm |
| T | | | | 5.8 | 7.1 | 9.9 | 12.3 | 16.0 | 20.4 | 24.2 | 23.6 | 20.1 | 14.4 | 9.3 | 6.3 | 14.1 C |
| Moscow, Russia | | | | | | | | | | | | | | | | |
| | 55.8 | 37.6 E | 190 | | | | | | | | | | | | | |
| P | | | | 43 | 35 | 34 | 39 | 55 | 73 | 86 | 76 | 60 | 56 | 52 | 51 | 661 mm |
| T | | | | −9.5 | −8.4 | −3.3 | 5.7 | 12.8 | 16.8 | 18.2 | 16.5 | 10.9 | 4.9 | −1.8 | −6.2 | 4.7 C |
| Paris, France | | | | | | | | | | | | | | | | |
| | 49.0 | 2.45 E | 66 | | | | | | | | | | | | | |
| P | | | | 31 | 26 | 27 | 29 | 28 | 33 | 39 | 39 | 36 | 36 | 36 | 33 | 392 mm |
| T | | | | 3.4 | 4.0 | 6.7 | 9.6 | 13.3 | 16.4 | 18.4 | 17.9 | 15.3 | 11.3 | 6.7 | 4.4 | 10.6 C |
| Port Moresby, Papua New Guinea | | | | | | | | | | | | | | | | |
| | −9.4 | 142.2 E | 48 | | | | | | | | | | | | | |
| P | | | | 28 | 27 | 27 | 27 | 27 | 26 | 26 | 26 | 26 | 27 | 28 | 28 | 323 mm |
| T | | | | 178.8 | 195.9 | 190.1 | 119.7 | 65.3 | 39.4 | 26.5 | 25.5 | 32.9 | 34.9 | 55.5 | 121.4 | 90.5 C |
| Resolute, Northwest Territories, Canada | | | | | | | | | | | | | | | | |
| | 74.7 | 95.0 W | 67 | | | | | | | | | | | | | |
| P | | | | 3 | 3 | 4 | 6 | 8 | 12 | 23 | 32 | 21 | 14 | 6 | 4 | 138 mm |
| T | | | | −32.0 | −33.2 | −31.2 | −23.2 | −10.8 | −0.3 | 4.2 | 2.2 | −4.8 | −15.1 | −24.4 | −28.9 | −16.5 C |

## Climatological Data *continued*

| Location | Latitude | Longitude | Elevation (meters) | Jan. | Feb. | March | April | May | June | July | Aug. | Sept. | Oct. | Nov. | Dec. | Annual Total Precipitation, Average Temperature |
|---|---|---|---|---|---|---|---|---|---|---|---|---|---|---|---|---|
| Roma, Australia | | | | | | | | | | | | | | | | |
| | −26.6 | 148.8 | 299 | | | | | | | | | | | | | |
| P | | | | 82 | 75 | 65 | 33 | 37 | 36 | 38 | 27 | 32 | 50 | 57 | 67 | 598 mm |
| T | | | | 27.3 | 26.6 | 24.5 | 20.5 | 16.1 | 12.7 | 11.9 | 13.7 | 17.4 | 21.6 | 24.7 | 26.7 | 20.3 C |
| Rome, Italy | | | | | | | | | | | | | | | | |
| | 41.8 | 12.2 E | 2 | | | | | | | | | | | | | |
| P | | | | 80 | 71 | 61 | 55 | 38 | 23 | 15 | 27 | 68 | 88 | 104 | 93 | 723 mm |
| T | | | | 32.1 | 28.4 | 24.7 | 18.3 | 17.6 | 16.8 | 18.4 | 20.3 | 26.9 | 31.7 | 33.3 | 32.5 | 25.1 C |
| Schefferville, Quebec, Canada | | | | | | | | | | | | | | | | |
| | 54.8 | 66.8 W | 522 | | | | | | | | | | | | | |
| P | | | | 46 | 41 | 45 | 49 | 48 | 76 | 100 | 92 | 92 | 77 | 67 | 49 | 783 mm |
| T | | | | −23.2 | −21.3 | −15.2 | −6.9 | 1.2 | 8.5 | 12.5 | 11.0 | 5.4 | −1.4 | −9.1 | −19.2 | −4.8 C |
| Sonnblick, Austria | | | | | | | | | | | | | | | | |
| | 47.1 | 13.0 W | 3106 | | | | | | | | | | | | | |
| P | | | | 132 | 106 | 134 | 165 | 148 | 149 | 156 | 156 | 112 | 107 | 126 | 131 | 1622 mm |
| T | | | | −12.6 | −12.7 | −11.1 | −8.2 | −3.8 | −0.6 | 1.6 | 1.6 | −0.4 | −3.4 | −7.8 | −10.7 | −5.7 C |
| Stockholm, Sweden | | | | | | | | | | | | | | | | |
| | 59.4 | 18.0 E | 15 | | | | | | | | | | | | | |
| P | | | | 40 | 28 | 25 | 31 | 31 | 47 | 69 | 67 | 54 | 51 | 51 | 47 | 541 mm |
| T | | | | −3.1 | −3.6 | −0.4 | 4.2 | 10.2 | 15.4 | 17.1 | 16.1 | 11.9 | 7.4 | 2.5 | −0.9 | 6.4 C |
| Tomsk, Russia | | | | | | | | | | | | | | | | |
| | 56.4 | 85.0 E | 122 | | | | | | | | | | | | | |
| P | | | | 27 | 18 | 22 | 26 | 45 | 61 | 74 | 68 | 44 | 48 | 46 | 34 | 512 mm |
| T | | | | −18.8 | −16.5 | −10.0 | −0.3 | 8.4 | 15.3 | 18.2 | 15.3 | 9.1 | 0.8 | −10.6 | −17.1 | −0.5 C |
| Tucson, Arizona | | | | | | | | | | | | | | | | |
| | 32.1 | 110.9 W | 802 | | | | | | | | | | | | | |
| P | | | | 23 | 20 | 21 | 11 | 4 | 7 | 58 | 50 | 28 | 25 | 17 | 27 | 292 mm |
| T | | | | 10.8 | 12.4 | 14.7 | 18.8 | 23.2 | 28.6 | 30.1 | 29.0 | 26.9 | 21.4 | 15.0 | 11.2 | 20.2 C |
| Tulsa, Oklahoma | | | | | | | | | | | | | | | | |
| | 36.2 | 95.9 W | 206 | | | | | | | | | | | | | |
| P | | | | 37 | 49 | 84 | 97 | 139 | 107 | 82 | 71 | 108 | 92 | 70 | 50 | 985 mm |
| T | | | | 2.1 | 4.9 | 9.9 | 16.1 | 20.7 | 25.5 | 28.4 | 27.7 | 23.2 | 16.9 | 9.7 | 4.1 | 15.8 C |
| Turuhansk, Russia | | | | | | | | | | | | | | | | |
| | 65.7 | 87.9 E | 32 | | | | | | | | | | | | | |
| P | | | | 32 | 20 | 28 | 27 | 29 | 49 | 62 | 73 | 69 | 62 | 42 | 40 | 533 mm |
| T | | | | −26.8 | −24.8 | −17.0 | −8.4 | 0.2 | 10.3 | 16.8 | 12.4 | 5.7 | −5.8 | −19.3 | −24.0 | −6.7 C |
| Ulan-Bator, Mongolia | | | | | | | | | | | | | | | | |
| | 47.9 | 107.0 E | 1337 | | | | | | | | | | | | | |
| P | | | | 2 | 4 | 6 | 13 | 22 | 64 | 79 | 85 | 42 | 12 | 8 | 4 | 341 mm |
| T | | | | −19.5 | −15.6 | −9.8 | −0.6 | 7.9 | 13.3 | 15.2 | 13.4 | 7.9 | −0.8 | −9.1 | −18.2 | −1.3 C |
| Udon Thani, Thailand | | | | | | | | | | | | | | | | |
| | 17.4 | 102.8 E | 177 | | | | | | | | | | | | | |
| P | | | | 7 | 18 | 41 | 88 | 212 | 216 | 209 | 254 | 244 | 81 | 16 | 5 | 1391 mm |
| T | | | | 22.5 | 25.1 | 28.0 | 29.9 | 29.4 | 28.8 | 28.5 | 28.0 | 27.7 | 27.0 | 24.9 | 22.2 | 26.8 C |
| Vostok, Antarctica | | | | | | | | | | | | | | | | |
| | −78.5 | 106.9 E | 3488 | | | | | | | | | | | | | |
| P | | | | 0 | 0 | 1 | 1 | 0 | 1 | 1 | 1 | 0 | 0 | 0 | 0 | 4 mm |
| T | | | | −32.1 | −44.3 | −58.0 | −64.8 | −65.6 | −65.3 | −66.9 | −67.7 | −66.0 | −57.2 | −43.4 | −32.2 | −55.3 C |

## Climatological Data *continued*

| Location | Latitude | Longitude | Elevation (meters) | Jan. | Feb. | March | April | May | June | July | Aug. | Sept. | Oct. | Nov. | Dec. | Annual Total Precipitation, Average Temperature |
|---|---|---|---|---|---|---|---|---|---|---|---|---|---|---|---|---|
| Whitehorse, Yukon Territory, Canada | | | | | | | | | | | | | | | | |
| | 60.7 | 135.1 W | 703 | | | | | | | | | | | | | |
| P | | | | 18 | 13 | 13 | 9 | 14 | 30 | 36 | 38 | 31 | 22 | 19 | 19 | 262 mm |
| T | | | | −19.2 | −13.3 | −7.5 | 0.3 | 6.7 | 11.9 | 14.1 | 12.4 | 7.4 | 0.5 | −9.2 | −15.5 | −0.9 C |

Source: Derived from GHCN Global Climate Dataset, distributed by EarthInfo, Inc., Boulder, Colorado.

T = Mean temperature, °C.

P = Mean precipitation (melted), in millimeters.

# Glossary

## A

**absolute vorticity** The total amount of spin (vorticity) an air parcel possesses; the sum of the parcel's relative vorticity plus the earth's rotation.

**absolutely stable** Pertains to any portion of an atmospheric sounding whose lapse rate is less than the moist adiabatic rate.

**absolutely unstable** Pertains to any portion of an atmospheric sounding whose lapse rate is greater than the dry adiabatic rate.

**acceleration** The change in an object's velocity vector, to include a change in speed or direction of motion or both.

**acid rain and fog** Precipitation and fog made acidic through the addition of atmospheric pollutants such as sulfuric and nitric acid.

**adiabatic** A process involving an air parcel (for example, when the parcel moves vertically) during which no exchange of heat energy occurs between the parcel and its surroundings.

**advection** The horizontal transport of some property (heat, for example) by the winds or ocean currents.

**advection fog** Fog formed when moist air moves over a surface whose temperature is colder than the dew point of the moist air blowing across it.

**aerosol particles** Tiny, airborne liquid or solid particles.

**AgI** See **silver iodide**.

**air mass thunderstorm** Also called an ordinary thunderstorm; a thunderstorm that forms individually, due to local heating, and not due to large-scale lifting mechanisms, such as fronts or orography.

**air parcel** A small sample of air under study, often in comparison with its surrounding environment.

**air pollutant** A substance present in the air in amounts sufficient to cause injury to life or the environment.

**air pressure** The force exerted by the air on a given area.

**albedo** The percentage of reflected sunlight compared to the amount incoming.

**altocumulus** A middle cloud possessing cumuliform structure.

**altostratus** A middle cloud possessing stratiform structure.

**anemometer** An instrument for measuring wind speed.

**anthropogenic** Caused by human activity.

**anticyclone** A high pressure system and its attendant circulation.

**Arctic (air mass)** An air mass type that forms over land in high latitudes. Arctic air is colder than cP air and is marked by extremely sharp, low-level temperature inversions.

**atmosphere** The layer of gases that surrounds the earth; known informally as air.

**atom** A particle of matter consisting of a nucleus of protons and (usually) neutrons surrounded by a region occupied generally by one or more electrons.

**atomic mass number** The number of protons plus neutrons in an atomic nucleus.

**atomic number** The number of protons residing in the nucleus of an atom.

**aurora** Light emissions occurring in the upper atmosphere, caused by interactions of fast-moving, charged particles, mostly from the sun, with the ionospheric gas.

**backdoor cold front** A cold front that moves from the east or northeast, bringing cold, moist (mP) air to the Atlantic seaboard of North America.

**backing wind** Wind that changes in a counterclockwise direction (for example, from south the southeast).

**baroclinic** Pertains to any region of the atmosphere in which temperature advection is occurring.

**baroclinic instability** Growth of an atmospheric disturbance due to feedback between baroclinic flow (warm and cold air advection) within the disturbance and upper-level wind flow patterns.

**barometer** An instrument that measures air pressure.

**barotropic** A part of the atmosphere in which no temperature advection is present.

**Beaufort wind scale** A system of estimating wind speeds by observing the wind's effect on the sea surface (originally) or (now, commonly) on objects such as smoke and trees.

**Bergeron process** The production of precipitation-sized particles in a cloud containing ice crystals and supercooled water droplets through the evaporation of the droplets and deposition of moisture onto the ice crystals.

**blackbody radiation** Radiation emitted by a perfectly efficient radiator; the Planck radiation curve depicts blackbody radiation.

**boundary layer** The surface layer of air, typically a kilometer or so in depth, in which the atmosphere's kinetic energy is moved downward and out to the earth by frictional dissipation.

**bounded weak echo region** Abbreviated BWER; within a severe thunderstorm, a region of updrafts so strong that precipitation particles move through too fast to grow into effective radar targets.

**broken clouds** Sky condition when cloud cover is in the range of five- through nine-tenths.

**butterfly effect** A hypothetical example of a system's sensitivity to initial conditions: the flap of a butterfly's wings might eventually affect wind flow and other weather patterns worldwide.

**calorie** The quantity of energy required to raise 1 gram of water by 1°C (from 14.5° to 15.5°C).

**causal link** A cause-and-effect relation between two or more variables.

**Celsius** A temperature scale in which the freezing and boiling points of water at standard air pressure are 0 and 100, respectively.

**chaos** Behavior that is nonperiodic and highly sensitive to initial conditions but deterministic nonetheless.

**chemical bond** A force that joins atoms in ways that do not change the nuclear makeup of the atoms involved.

**Chinook wind** A warm, dry downslope wind that forms on the east slope of the Rocky Mountains.

**chlorofluorocarbons** Abbreviated CFC, compounds composed of chlorine, fluorine, and carbon atoms.
**cirrocumulus** A high cloud possessing cumuliform structure.
**cirrostratus** A high cloud possessing stratiform structure.
**cirrus** A high cloud, composed of ice crystals, possessing a wispy, filamentlike structure.
**clear** Sky condition when sky is cloudless or less than one-tenth covered (to the nearest tenth).
**clear air turbulence** Abbreviated CAT; microscale turbulence not associated with clouds; severe CAT can be a hazard to aircraft operations.
**climate** The sum of all weather events affecting a region over an extended period of time (typically 30 years).
**climograph** A graph that depicts a locale's monthly normal and extreme temperatures and monthly mean precipitation data.
**coalescence** The joining together of two or more drops to form a larger drop.
**coalescence efficiency** The ratio of the number of raindrops or droplets actually captured, compared to the number that collide with a given drop or droplet.
**cold air damming** The restriction of the advance of cold air by a range of mountains.
**cold front** A front in which colder air advances to replace warmer air.
**cold occlusion** An occlusion in which the advancing cold air is densest and therefore pushes forward beneath the warm front.
**collection efficiency** The ratio of the number of raindrops or droplets coalescing compared to the number in a droplet's path; equal in value to the product of the collision efficiency times the collection efficiency.
**collision efficiency** The ratio of the number of raindrops or droplets actually colliding with a given drop or droplet, compared to the number in its path.
**compound** A substance composed of a single kind of molecule.
**concept map** A chart that indicates relations among a group of processes, concepts, facts, and so on.
**condensation** The process in which a substance changes from gas to liquid phase.
**condensation level** Abbreviated CL, the altitude at which a rising air parcel's temperature and dew point become equal.
**condensation nuclei** Small particles in the atmosphere that provide sites for condensation of water vapor.
**conditionally unstable** Any portion of an atmospheric sounding whose lapse rate lies between the dry adiabatic and moist adiabatic rates.
**conduction** Heat transfer due to molecular collisions.
**conservation of energy (law of)** Within a closed system, the total energy remains constant; that is, energy is neither created nor destroyed.
**constant height map** A weather map that depicts variations in air pressure at one particular altitude in the atmosphere.
**constant pressure map** A weather map that depicts variations in altitude of a specific pressure surface (for example, the 500-millibar surface).
**contact nucleation** The freezing of supercooled droplets as a result of collision with other particles in the atmosphere.
**continental polar** Abbreviated cP; an air mass type that forms over land in high latitudes, especially in winter. The air is typically cold and dry, with a surface-based inversion.
**continental tropical** Abbreviated cT; an air mass type that forms over tropical and subtropical land masses, particularly deserts. The air is hot and dry.
**continuous spectrum** A spectrum in which all wavelengths of radiation are present, with no gaps in the sequence.
**convection** Heat transfer due to the motion of matter from one location to another.
**convection hypothesis** A hypothesis that convective air currents are responsible for electric charge separation within clouds.
**convergence** The contraction of a given parcel of air; the opposite of divergence.
**Coriolis force** An apparent force, arising from the earth's rotation, that causes moving objects to travel in curved paths.
**corona** A ring of light surrounding the sun or moon, caused by diffraction of light by cloud (liquid) droplets in altostratus clouds.
**correlation** A trend that relates values of one variable (for example, school grade) and values of another variable (for example, students' height).
**cumulonimbus** A cumulus cloud that generates precipitation.
**cumulus stage** The initial stage of air mass thunderstorm development, characterized by updrafts and warm air but no precipitation.
**cumulus** A cloud of vertical development, its base typically lying within 2,000 meters of the earth's surface.
**curvature** Vorticity of an air parcel manifested through changes over time in the parcel's direction of motion.
**curvature effect** The increase in equilibrium vapor pressure at small cloud droplet diameters.
**cyclone** A low pressure system and its attendant circulation.

**density** The amount of mass contained in a given volume.
**deposition** The process in which a substance changes from gas to solid phase.
**dew point** The temperature at which atmospheric water vapor begins to condense into dew (or fog) as the air is cooled.
**dew-point temperature** The temperature reached by cooling an air parcel to saturation without changing its mixing ratio.
**diatomic** Composed of molecules bonded in pairs (such as most oxygen in the lower atmosphere, $O_2$).
**dissipating stage** The final stage of thunderstorm development, characterized by weak downdrafts throughout the cloud and light precipitation.
**divergence** The spreading out of a given parcel of air; the opposite of convergence.
**doldrums** The ITCZ; the term often refers specifically to the light and variable winds within the ITCZ.
**Doppler effect** The shift in wavelength of radiation emitted or reflected from an object moving toward or away from the observer.
**Doppler radar** A radar that uses the Doppler effect to measure motions of objects toward or away from the radar instrument, in addition to the object's position.
**drizzle** Liquid precipitation having a drop diameter of less than 0.5 millimeters.

**dry adiabatic** Pertains to any adiabatic process during which no phase change of water within the parcel occurs.

**dry adiabatic lapse rate** The rate at which an unsaturated air parcel changes temperature due to adiabatic lifting. (When applied to vertical motion, the rate is 9.8°C per kilometer, or 5.4°F per 1000 feet.)

**dry ice** Carbon dioxide in solid (frozen) form.

**dry line** A front separating moist and dry air masses (typically mT and cT air). Dry lines are marked by substantial dew-point contrasts.

**dynamic effects** In cloud seeding operations, actions that cause or alter air currents within the cloud.

**easterly wave** A synoptic scale, westward-moving, wave-shaped disturbance embedded in the tropical trade winds.

**eddy viscosity** Frictional resistance due to eddy motions within a fluid experiencing turbulent flow.

**El Niño** A warming in coastal surface waters of Peru and Ecuador that occurs around Christmas time (El Niño is Spanish for the Christ Child) every few years as part of the southern oscillation.

**electric discharge current** An electric spark; a momentary flow of electrically charged particles between oppositely charged objects.

**electromagnetic spectrum** The range of all forms of electromagnetic radiation, from those of very short wavelength such as X-rays through visible light, extending to those of very long wavelength such as radio waves.

**electron** A negatively charged subatomic particle.

**element** A substance composed of atoms of the same atomic number.

**energy budget** An accounting of all fluxes of energy into and out of a particular object (such as the atmosphere).

**ENSO** El Niño Southern Oscillation; the warming of surface and near surface waters in the eastern tropical Pacific Ocean due to a reversal in east-to-west pressure patterns in the tropical Pacific.

**entrainment** The mixing of environmental air with that of an air current such as a cloud updraft or downdraft.

**environment** In atmospheric physics, a large volume of air, such as the air over a city extending upwards for several miles. Compare with **air parcel**.

**equilibrium vapor pressure** The value of vapor pressure at which the rate of evaporation from a plane water surface equals the rate of condensation.

**equinox** Point on sun's apparent annual path when it crosses the equator heading northward (spring equinox, approximately March 21) or southward (autumnal equinox, September 23).

**evaporation** The process in which a substance changes from liquid to gas phase.

**excited state** An electron energy level greater than the ground state.

**extended forecast** A forecast of conditions expected up to five days in the future.

**extrapolation** The process of predicting future positions of weather systems solely on their past motions.

**extreme value** The highest or lowest value in a given set of data.

**eye (of hurricane)** A distinctive region of clear or nearly clear skies, subsiding air, and light winds at the center of a hurricane.

**eyewall** A ring of towering thunderstorm clouds that surrounds the eye of a hurricane.

**Fahrenheit** A temperature scale in which the freezing and boiling points of water at standard air pressure are 32 and 212, respectively.

**flux** The rate of flow of a substance, such as the flux of water vapor due to evaporation.

**fog** A cloud whose base is at the earth's surface.

**force** A push or a pull.

**forecast verification** Evaluation of a forecast's accuracy by comparing the forecasted conditions with the weather that actually occurred.

**free atmosphere** Air in which frictional effects associated with the earth's surface are absent; thus, the air lying above the atmospheric boundary layer.

**freezing rain** Liquid precipitation that freezes upon striking the ground, trees, or other structures.

**front** A narrow zone of division between air masses of different characteristics—especially, of different densities resulting from temperature contrasts.

**frontal cyclone** A synoptic scale low pressure system containing one or more fronts.

**frontal thunderstorm** A thunderstorm that forms along a front due to forced lifting of air by the front.

**frontogenesis** The formation and intensification of a front.

**frontolysis** The dissipation of a front.

**frost point** A "dew-point temperature" whose value is lower than 0°C.

**Fujita scale** A tornado classification scale designed by Theodore Fujita and based on maximum wind speeds and the nature of the damage caused.

**funnel cloud** A vortex of spinning air extending downward from a thunderstorm (typically from a wall cloud) but not reaching the ground.

**gas law** Pressure = constant × density × temperature; also known as the ideal gas law, or the equation of state.

**general circulation** The set of global scale, more or less permanent wind flow regimes such as the trade winds, prevailing westerlies, and so on, that collectively describe the atmosphere's circulation as a whole.

**general circulation model** Abbreviated GCM, a mathematical description of atmospheric conditions and the natural laws that affect those conditions, generally expressed as a computer program.

**geostrophic balance** A condition in which pressure gradient and Coriolis forces are the only horizontal forces acting on an air parcel, and they are in balance (equal in magnitude, opposite in direction). Geostrophic balance is often a close approximation of actual conditions.

**geostrophic wind** Wind that occurs under conditions of geostrophic balance. Geostrophic winds blow parallel to isobars, with lower pressure to the left of motion (Northern Hemisphere).

**global scale** The largest scale of atmospheric phenomena, having dimensions of several thousand or more kilometers and life spans of weeks or longer. Example: Rossby waves.

**gradient flow or wind** Wind flow determined by pressure gradient, Coriolis, and centripetal forces; in gradient flow, winds blow parallel to curved isobars.

**graupel** See **snow pellet**.

**greenhouse effect** The warming effect caused by the action of greenhouse gases.

**greenhouse gases** Gases such as water vapor, carbon dioxide, and methane, which are relatively transparent to sunlight but absorb and emit longwave radiation.

**ground fog** See **radiation fog**.

**ground state** The lowest possible energy level occupied by an electron.

**gust front** The boundary between warm ascending air and cold descending air in a thunderstorm circulation.

**gyre** A large scale (synoptic or global scale) oceanic circulation cell, commonly associated with an atmospheric subtropical high pressure cell.

## H

**Hadley cell** A global scale, thermal circulation feature in which the trade winds carry air from the subtropical high pressure belt to the ITCZ; poleward wind flow aloft completes the cell.

**hail** Precipitation in the form of spherical or irregular chunks of ice 5 millimeters or more in diameter, formed in cumulonimbus clouds.

**halo** A ring of light surrounding the sun or moon, caused by refraction of light by ice crystals in cirrostratus clouds.

**heat index** A quantity calculated by combining temperature and humidity values, which gives a measure of human discomfort in hot weather.

**heterogeneous nucleation** The nucleation of water vapor onto microscopic, nonwater particles in the air.

**high cloud** A cloud whose base typically is higher than 7000 meters above earth's surface; high clouds are composed of ice crystals.

**high** A shorthand term for "high pressure area."

**highly significant** An event whose probability of occurrence by chance is 2% or less.

**homogeneous nucleation** Nucleation of water vapor to liquid or solid phase without the aid of a seed nucleus.

**hook echo** A small, highly reflecting radar image shaped like the numeral 6, often associated with tornado formation.

**horse latitudes** The latitudes along the axis of the subtropical high pressure belt. So named for horses that had starved to death and were thrown overboard by early explorers when their ships were becalmed in the region.

**humidity** A general term referring to the amount of water vapor present in the atmosphere.

**hurricane** A severe tropical cyclone on the order of 500 kilometers (300 miles) in diameter, with winds in excess of 64 knots (74 mph), that forms over warm ocean waters in maritime tropical air.

**hurricane warning** A bulletin stating that hurricane conditions are expected and that residents should begin immediately to take precautions, including evacuating low-lying areas subject to flooding.

**hurricane watch** A bulletin stating that a particular region faces the threat of hurricane conditions within 24 to 36 hours. It is not a call to evacuate.

**hydrocarbon** A compound formed from hydrogen and carbon atoms.

**hydrologic cycle** The processes of evaporation, precipitation, runoff, and so on that characterize the transport of water among the atmosphere, oceans, and land.

**hydrophobic** A substance that repels water vapor; condensation occurs only at relative humidities greater than 100%, compared to a plane water surface.

**hydrostatic balance** The balance of the upward pressure gradient force and the downward force of gravity.

**hygrometer** Any of several types of instruments that measure relative humidity.

**hygroscopic** A substance that attracts water vapor and on which condensation occurs at relative humidity less than 100%, compared to a plane water surface.

**ice nucleus** A particle whose molecular structure is similar to the crystal structure of ice. Ice crystals in clouds form on ice nuclei.

**ice pellet** A frozen raindrop. Also known as sleet.

**induction** Creation of electric charge in an object due to the proximity of another, previously charged object.

**infrared radiation** Electromagnetic radiation of wavelengths somewhat longer than those of visible light.

**insolation** Abbreviation of *in*coming *sol*ar radi*ation*. The quantity of sunlight incident onto a horizontal surface at the earth's distance from the sun.

**interpolation** The process of estimating conditions at some locale or at some time, when conditions surrounding the locale or time in question are known.

**intertropical convergence zone** Abbreviated ITCZ; the global scale, equatorial region of low surface air pressure along which the trade winds converge.

**inversion** An increase in temperature with altitude.

**ion** An atom or molecule that has become electrically charged, typically by loss of one or more electrons.

**ionosphere** A region within the upper mesosphere and the thermosphere that is occupied by large numbers of ions.

**isobar** A line of equal or constant air pressure on a weather chart.

**isothermal expansion** The expansion of air that occurs at constant temperature; can occur when air near the ocean surface is drawn to the center of a hurricane.

**isotope** Atoms of a particular element, all of which possess the same number of neutrons.

## J

**jet streak** A narrow zone of very strong winds located within a jet stream.

**jet stream** A narrow zone of strong winds, generally in the upper atmosphere.

## K

**Köppen classification** A worldwide climate classification system. Each Köppen climate zone, defined by temperature and precipitation means, contains certain characteristic forms of vegetation.

**kinetic theory of gases** A theory based on the assumption that the physical behavior of gases (such as the atmosphere) can be understood as an assembly of innumerable tiny particles in constant and rapid collisional motion.

**knot** One nautical mile per hour. One knot is equivalent to 1.15 mph.

## L

**lake effect snowstorms** Mesoscale snowstorms that occur most notably in the lee of the Great Lakes, due to cold winds gaining heat and moisture as they traverse open water and subsequently releasing their moisture in the form of heavy snow squalls.

**laminar flow** Fluid flow that is smooth and nonturbulent, progressing in parallel lines or sheets.

**land and sea breeze circulation** A thermally driven, coastal mesoscale circulation in which surface air moves inland during the day and seaward at night.

**latent heat flux** Energy transfer conveyed through phase changes of matter (for example, evaporation or condensation).

**law of conservation of angular momentum** In a rotating object, the product of its mass (m), its rotational speed (v), and its distance from the center of rotation (r) tends to remain constant. Thus, $mvr = \text{constant}$.

**lee trough** A horizontal low pressure trough caused by cross-mountain wind flow.

**lee waves** Waves in wind flow that develop downstream from an obstruction such as a mountain.

**lifted index** A stability measure, based on temperature and humidity values at different atmospheric levels, that indicates the probability of convective activity such as thunderstorms at a given locale.

**line spectrum** A spectrum in which only certain discrete photon energies and wavelengths are represented (compare with continuous spectrum).

**linear** In a straight line.

**long wave** Large, slowly moving or stationary horizontal waves (on the order of 5,000 kilometers in wavelength) in upper-level wind flow. Also known as Rossby waves.

**longwave radiation** Infrared radiation emitted and absorbed by the earth and atmosphere.

**low** A shorthand term for "low pressure area."

**low cloud** A cloud whose base typically is within 2000 meters (6500 feet) of the earth's surface.

## M

**maritime polar** Abbreviated mP; an air mass type that forms over oceans in high latitudes. It is characterized by cold, moist air.

**maritime tropical** Abbreviated mT; an air mass type that forms over tropical and subtropical oceans. The air is warm or hot (not as hot as cT air) and humid.

**mass** A measure of an object's resistance to a change of motion.

**mature stage** The middle stage of thunderstorm development; precipitation, updrafts, downdrafts, thunder, and lightning are present.

**mechanical turbulence** Turbulence caused by the obstruction of air flow due to the presence of objects such as buildings, trees, and mountains.

**mesocyclone** A small-scale, cyclonically rotating, low pressure center within a supercell thunderstorm; may be a precursor of a tornado.

**mesopause** The atmospheric boundary between the mesosphere and the thermosphere, found typically at an altitude of 80 to 90 kilometers.

**mesoscale** Atmospheric phenomena with dimensions from 1 or 2 to 200 kilometers and life spans of a few minutes to several hours. Example: a thunderstorm.

**mesoscale convective complex** Abbreviated MCC; a slow-moving region of concentrated thunderstorm activity, typically 400 kilometers or so in diameter and composed of dozens of cells, often most active at night.

**mesosphere** The atmospheric layer lying between the stratosphere below and the thermosphere above, typically from 50 to 85 kilometers.

**meteorology** The study of the atmosphere and the phenomena that occur in it.

**microclimate** Climate on microscale dimensions. Microclimate often varies significantly over small distances due to variations in surface conditions such as slope and vegetation.

**micrometer** One millionth of a meter, or one-one thousandth of a millimeter.

**microscale** Atmospheric phenomena with dimensions from millimeters to a few kilometers and lifetimes of a second to a few minutes. Example: a wind gust.

**middle cloud** A cloud whose base typically is between 2000 and 7000 meters of the earth's surface; middle clouds may be composed of liquid droplets, ice crystals, or a mixture of both.

**Milankovitch cycles** Periodic variations in the earth's orbit over periods of thousands of years, which are suspected of causing changes in the amount of solar energy reaching the earth.

**millibar** A unit of pressure measurement, abbreviated mb. Standard air pressure at sea level is equal to 1013.25 mb.

**mixing ratio** The number of grams of water vapor present per kilogram of dry air.

**mixture** A substance composed of individual atoms not bonded to each other.

**moist adiabatic** Pertains to any adiabatic process involving a saturated air parcel.

**moist adiabatic lapse rate** The rate of temperature decrease in a rising, saturated air parcel. The rate varies with air temperature and is less than the dry adiabatic rate due to release of latent heat. Typical values range from 4 to 7°C per kilometer (2 to 4°F per 1000 feet).

**molecular scale** The smallest scale of atmospheric phenomena, composed of features of microscopic or smaller size and lifetimes of a small fraction of a second. Example: heat conduction.

**molecular viscosity** Frictional resistance within a fluid, caused by random molecular motions.

**molecule** A particle formed by the bonding of two or more atoms.

**monatomic** Composed of individual atoms not bonded to each other.

**monsoon** A seasonal, thermal circulation caused by temperature contrasts from ocean to continent. Surface winds generally blow inland during summer and off the land during winter.

**mountain and valley winds** A mesoscale circulation that develops in hilly or mountainous terrain when diurnal heating and cooling combine with topography to create a circulation system that reverses direction from day to night or from morning to afternoon.

**multicell thunderstorm** A cluster of several thunderstorms in various stages of development.

## N

**negative feedback** Any process in which a system (the atmosphere, for example) responds to some forcing in a way that diminishes the original force.

**neutron** An uncharged subatomic particle, found in the nucleus of all atoms except those of hydrogen.

**Newton's first law of motion** An object at rest remains at rest; an object in motion moves in a straight line at constant speed unless acted on by an unbalanced external force.

**Newton's second law of motion** $F=ma$, where F is a net external force applied to an object of mass m. The resulting acceleration, a, is in the same direction as the applied force.

**Newton's third law of motion** Every force acting on an object is accompanied by an equal and opposite force exerted by the object.

**nimbostratus** A stratiform cloud that yields precipitation.

**noise** Unwanted, often random, variations in a data set, which tend to mask any signal that might be present.

**nor'easter** A wintertime frontal cyclone that moves northeastward along or just off the northeastern U.S. coast, bringing strong northeast winds and heavy precipitation to the region.

**normal** A climatological mean value of a variable such as temperature or precipitation.

**Norwegian cyclone model** A theory of frontal cyclone formation, development, and decay proposed by Jacob Bjerknes and others in the 1920s.

**nowcast** A very short-term weather forecast (for a period of a few hours).

**nucleation** The condensation, freezing, or deposition of water vapor in the free air.

**nucleus (atomic)** The core of an atom, composed of protons and neutrons.

**obscured** Sky condition when the sky is hidden due to surface-based phenomena such as fog, haze, or smoke.

**occluded front** A composite of two fronts, the result of a cold front overtaking a warm front.

**ocean conveyor belt** A global scale oceanic circulation system, driven primarily by the sinking of cold, salty water in the North Atlantic Ocean.

**orographic lifting** Upward air motion caused by wind flow over sloping terrain.

**outgoing long wave radiation** The infrared radiation loss to space from the earth's surface and atmosphere.

**overcast** Sky condition when cloud cover is complete (10/10, to the nearest tenth).

**overrunning** The process in which warm air glides upward over a surface-based cold air mass, sometimes producing extended periods of widespread precipitation.

**ozone** A substance composed of triatomic oxygen ($O_3$).

**persistence forecasting** A forecasting strategy based on the supposition that the present weather will remain unchanged.

**photochemical dissociation** The breakdown of a molecule (such as atmospheric $O_3$ dissociating to $O_2$ and O) due to the input of solar energy.

**photochemical smog** A visible haze or fog (*smog* derives from the words *smoke* and *fog*) polluted with ozone.

**photon** A particle of radiant energy.

**photosynthesis** The process by which a plant's leaves combine $CO_2$ from the air, water from the soil, and sunlight to make plant tissue.

**polar easterlies** Global scale, low-level easterly winds that flow from a polar high pressure cell toward the polar front.

**polar front** A more or less continuous boundary (front) between polar and tropical air that encircles the globe, according to the Norwegian cyclone model.

**polar high pressure cell** A cold, high pressure cell that forms in polar latitudes or in subpolar latitudes over continents in winter.

**positive feedback** Any process in which a system (the atmosphere, for example) responds to some forcing in a way that amplifies the original force.

**power** The rate of energy flow in unit time.

**precession** The change in the orientation of the earth's axis in space over a cycle of 22,000 years.

**precipitation** Any form of water (raindrops, snowflakes, etc.) that falls from the atmosphere and reaches the ground.

**precipitation hypothesis** A hypothesis that collisions among precipitation particles in convective clouds are responsible for electric charge separation.

**prefrontal squall line** A line of thunderstorms generally caused directly or indirectly by a front and lying 100 to 500 kilometers ahead of it.

**present weather** A description of (generally) visible events occurring in the atmosphere; see Appendix H.

**pressure gradient** A vector quantity describing the rate of pressure change over distance in a direction at right angles to the isobars on a weather chart.

**pressure gradient force** The net pressure force on an air parcel due to an existing pressure gradient.

**pressure tendency** The change in air pressure over a specified period of time (typically three hours).

**prevailing visibility** A measure of the atmosphere's horizontal transparency. Specifically, the greatest distance you can see along at least half of the horizon.

**prevailing westerlies** Global scale, middle latitude winds that blow from the subtropical high pressure belt toward the polar front.

**proton** A positively charged subatomic particle, found in the nuclei of all atoms.

**proxy data** Data from climate-sensitive phenomena such as glacier lengths, pollen deposits, tree ring spacings, and so forth that can be used to reconstruct estimates of past climate conditions.

**psychrometer** A humidity-measuring instrument consisting of two thermometers, one of which has a moistened bulb that indicates the wet-bulb temperature.

**radar** An instrument that detects objects by transmitting and then receiving reflections of radiation in microwave wavelengths.

**radiation budget** The sum of all radiative energy fluxes across a surface (such as the earth's surface) or into and out of an object (such as the atmosphere).

**radiation fog** Fog formed in air cooled to its dew point due to surface cooling from radiative heat loss at night; also known as ground fog.

**radiative forcing formula** (Solar radiation absorbed by the earth and atmosphere) − (outgoing longwave radiation from the earth and atmosphere) = the energy storage change by the earth and atmosphere.

**radiative forcing** The energy stimulus provided to the earth-atmosphere system by the net flux of energy through the top of the atmosphere. (See **radiative forcing formula**.)

**radioactive decay** The transmutation of a nuclear particle into one or more other particles plus energy.

**range** The span of values from highest to lowest in a set of data.

**recurvature** The change in a tropical cyclone's direction of motion from westward to northwest-, north-, and northeastward (in the Northern Hemisphere).

**relative humidity** A measure of the quantity of water vapor in the air. Specifically, the mixing ratio of an air parcel compared to (divided by) its saturation mixing ratio; generally expressed as a percentage.

**relative vorticity** The amount of rotation (vorticity) an air parcel possesses with respect to the earth's surface.

**return stroke** A positive electric current in a lightning bolt, typically moving from ground to cloud and carrying the main flow of electricity along a channel established by a stepped leader.

**rime ice** The freezing of supercooled droplets on an object, causing a coating of ice.

**roll or rotor** A turbulent eddy that typically forms in the lee of a mountain or other obstruction and rotates on a horizontal axis.

**Rossby wave** See long wave.

**running mean** A "smoothed" data set, obtained by averaging each data point in a sequence with one or more points within some set interval adjacent to it.

**Saffir-Simpson scale** A classification system for hurricanes based on their central pressure, maximum sustained winds, storm surge, and attendant damage.

**Santa Ana wind** A hot wind that carries desert air southwestward to portions of the southern California coast in the vicinity of the city of Santa Ana.

**saturated** An air parcel is saturated if, when brought into contact with a plane water surface, the rate of evaporation of water into the parcel equals the rate of water loss from the parcel by condensation.

**saturation mixing ratio** An air parcel's mixing ratio when the parcel is saturated.

**savanna** A tropical grassland regime, predominant in tropical wet-and-dry climates (Aw climates in the Köppen system).

**scalar** Any quantity, such as temperature, that has only a magnitude. (Compare with **vector**.)

**scatter** The amount by which individual data points (typically on a graph) differ from the mean value.

**scattered clouds** Sky condition when cloud cover is in the range of one-tenth to five-tenths.

**scattering** Deflection of sunlight in all directions by small particles in the atmosphere.

**selective emitters and absorbers** Substances that emit and absorb photons of only certain energies or wavelengths.

**sensible heat flux** Heat transfer through the processes of conduction and convection.

**severe storm or tornado watch** A bulletin stating that conditions are such that, although severe weather has not been reported, there is a significant chance of it occurring within the watch region.

**severe thunderstorm** A thunderstorm capable of producing heavy downpours, flash flooding, strong and gusty winds, frequent lightning, hail, and tornadoes.

**severe weather warning** A bulletin stating that visual or radar data indicate that severe weather is occurring or imminent within the region.

**shear** Vorticity of an air parcel manifested through variation in wind speeds perpendicular to the direction of flow.

**short wave** A horizontal wave in upper-level wind flow associated with an individual synoptic scale cyclone, having a wavelength on the same scale (1000 to 2000 kilometers).

**short-term forecast** A forecast of conditions expected up to 48 or so hours in the future.

**signal** The message or information contained in a set of data.

**significance level** A statistical gauge of the probability that a particular event might have occurred by chance.

**silver iodide** A compound whose molecules consist of one atom of silver (Ag) and one of iodine (I); hence, the formula AgI. AgI crystals resemble ice crystals and therefore are used in cloud seeding.

**sleet** Precipitation in the form of ice pellets.

**snow pellet** Soft, crunchy, roughly spherical frozen precipitation, formed by the freezing of water onto falling snowflakes. Also known as graupel.

**snowflake** A clump of a few to as many as a hundred or so individual snow crystals.

**solar constant** The amount of solar radiation incident per minute on a 1 square-centimeter surface oriented perpendicular to the beam and positioned at the top of the atmosphere at earth's mean distance from the sun.

**solstice** Points on the sun's apparent annual path when it reaches the greatest distance north (summer solstice) or south (winter solstice) of the equator; dates are approximately June 21 and December 21, respectively.

**solute effect** The decrease in equilibrium vapor pressure in small cloud droplets due to the presence of dissolved impurities ("solutes").

**source and sink** Agents or processes that act to increase some quantity (if a source) or decrease it (if a sink) in a cycle. (Thus, evaporation represents a source to the atmosphere and a sink to the oceans.)

**southern oscillation** A cycle of changing air pressure and wind regimes from east to west across the tropical Pacific Ocean, leading to El Niño episodes.

**special statement** An informational or advisory notice that describes an unusual weather event whose potential to cause damage or injury is less than that of events in the watch or warning categories.

**specific heat** The amount of energy required to raise the temperature of 1 gram of a substance by 1°C.

**spontaneous nucleation** The freezing of liquid droplets without the presence of ice nuclei.

**static stability** A measure of the restoring force an air parcel experiences upon undergoing a small vertical displacement.

**stationary front** A front showing little or no movement.

**statistically significant** An event whose probability of occurrence by chance is 5% or less.

**steering** A forecasting technique based on the fiction that upper-level wind flow "steers" surface weather patterns.

**Stefan-Boltzmann formula** A formula giving the quantity of blackbody radiation emitted per unit area of emitting surface, per unit time: $E = \sigma T^4$, where E is the emitted energy, $\sigma$ a constant, and T the temperature in Kelvins.

**steppe** A semiarid region characterized by short, sparse grasses, typical of conditions in Köppen Bs climate zones.

**stepped leader** The first stage of a lightning bolt, in which a flood of electrons travels earthward in a series of segments.

**storage (in budgets)** In a cycle, the amount of the substance in a given locale at a given time. For example, the total quantity of water in the atmosphere is the atmosphere's storage of water in the hydrologic cycle.

**storm surge** A rise of the sea surface along a storm's path due to winds and low pressure at the storm center.

**stratocumulus** A low cloud possessing an overall sheetlike stratiform structure within which roll or cellular structures typical of cumulus clouds are visible.

**stratopause** The atmospheric boundary between the stratosphere below and the mesosphere above, typically at an altitude of approximately 50 kilometers.

**stratosphere** The atmospheric layer bounded below by the tropopause and above by the stratopause; typically, between 15 and 50 kilometers above sea level.

**stratus** A low cloud of stratiform structure.

**sublimation** The process in which a substance changes from solid to gas phase.

**subsidence** The sinking or descending of air.

**subtropical high pressure belt** A global scale zone of high pressure lying at approximately 30 degrees north and south latitude. Within the zone are several distinct cells, such as the Bermuda High.

**subtropical jet stream** A zone of strong winds aloft located near the poleward boundary of the Hadley cell.

**suction vortex** A small, intense circulation within a tornado. Suction vortexes contain the strongest winds and most severe pressure changes, and therefore cause the greatest tornado-related damage.

**supercell thunderstorm** A thunderstorm of exceptional size and intensity that persists in the mature stage for considerably longer than most thunderstorms.

**supercooled water** Water that remains in the liquid phase at temperatures colder than 0°C.

**supersaturated** An air parcel is supersaturated if, when brought into contact with a plane water surface, the rate of evaporation of water into the parcel is less than the rate of water loss from the parcel by condensation.

**synoptic meteorology** The study of weather systems on the synoptic scale, which are on the order of 1500 kilometers in diameter and have lifetimes of approximately one week.

**synoptic scale** A scale of weather systems whose horizontal dimensions are on the order of 1000 miles or 1500 kilometers.

**terminal velocity** The maximum speed reached by a freely falling object; frictional forces prevent it from attaining greater speeds.

**thermal turbulence** Turbulence caused by the obstruction of air flow due to convective air motions.

**thermodynamic diagram** A pressure-temperature diagram on which are shown isotherms, isobars, dry and moist adiabats, and saturation mixing ratio lines; used for plotting atmospheric soundings and determining changes in vertically moving air parcels.

**thermosphere** The uppermost of earth's four atmospheric layers, extending upward from an altitude of around 85 kilometers.

**time step** A fixed interval by which a program advances time in a general circulation model.

**tilting** The sloping with height of a pressure system; as a result, the pressure center at upper levels does not lie directly above the surface center.

**tornado** A funnel cloud that makes contact with the earth's surface.

**trade winds** A global scale band of northeast (Northern Hemisphere) or southeast (Southern Hemisphere) winds in low latitudes (typically, between 5 and 20 degrees of latitude).

**transmissometer** An instrument that measures visibility along an atmospheric path (typically along airport runways).

**transpiration** The transfer of water (typically via latent heat flux) from plant surfaces to the atmosphere.

**trend forecasting** A forecasting strategy based on the assumption that weather features such as fronts and their associated weather will persist and will move in the future as they have in the past

**triatomic** Composed of molecules bonded in threes (such as $O_3$, ozone).

**triple point** On a weather map, the point at which warm, cold, and occluded fronts intersect.

**tropical cyclone** Any member of the hurricane family (hurricane, typhoon, tropical storm, etc.), whatever its location or strength.

**tropical depression** A tropical cyclone with maximum sustained winds between 20 and 34 knots (23 to 39 mph).

**tropical storm** A tropical cyclone with maximum sustained winds between 35 and 64 knots (40 to 74 mph).

**Tropics of Cancer and Capricorn** Latitudes +23.5 and −23.5, respectively, the northern and southernmost points reached by the sun during its apparent annual path (i.e., at the solstices).

**tropopause** The boundary between the troposphere and the next higher layer, called the stratosphere.

**troposphere** The lowest atmospheric layer, extending from the earth's surface to an altitude of 10 to 15 kilometers.

**turbulent flow or turbulence** Fluid flow characterized by complex patterns that change rapidly with time.

**typhoon** A hurricane in the Western Pacific Ocean.

## U

**ultraviolet radiation** Electromagnetic radiation of shorter wavelength and greater energy than visible light.

**unsaturated** An air parcel is unsaturated if, when brought into contact with a plane water surface, the rate of evaporation of water into the parcel is greater than the rate of water loss from the parcel by condensation.

**unstable layer** An atmospheric layer whose temperature decreases rapidly with altitude, which results in the development of vertical air motions (warm air rising, cold air sinking).

**upwelling** The rising of cold, deep water to the surface, to replace surface water moved offshore by the wind.

**urban heat island** A region of warmer air temperatures in a metropolitan area, compared with temperatures in the surrounding countryside.

**vapor pressure** That portion of the air pressure caused by water vapor molecules.

**variables of state** Pressure, density, and temperature of a gas.

**vector** A quantity, such as force or velocity, that possesses both a magnitude and a direction.

**veering wind** Wind that changes in a clockwise direction (for example, from east to south).

**velocity** A vector quantity that describes an object's motion; velocity includes both the object's speed and its direction of movement.

**virga** Precipitation that falls from a cloud base but evaporates in drier air between the cloud base and the ground.

**viscous forces** Frictional forces within a fluid.

**volume** The amount of space occupied by an object.

**vorticity** The amount of rotation present in a fluid, such as an air parcel.

## W

**wake capture** The process in which a cloud droplet is swept behind a larger particle and collides with the larger particle from the rear.

**wall cloud** A dark, slowly rotating cloud extending below the mesocyclone circulation in a potentially tornado-bearing thunderstorm.

**warm cloud** A cloud in which the temperature everywhere is greater than freezing (0°C).

**warm front** A front in which colder air recedes, allowing warmer air to replace it.

**warm occlusion** An occlusion in which the retreating cold air is the densest; as a result, the advancing cold air glides up the warm frontal surface.

**warm rain process** The production of precipitation-sized particles through the collision and coalescence of cloud particles.

**warning** A notice that severe weather is imminent or occurring in a given area.

**watch** An alert; a notice that conditions are ripe for the occurrence of severe weather in a specified area during a certain time period, but such weather has not yet occurred there.

**watt** Unit of power.

**wavelength** The distance from a given point on one wave to the same point on the next wave in the same train.

**weather types forecasting** A forecasting technique based on the idea that certain weather patterns recur and tend to behave similarly each time they recur.

**wet-bulb temperature** The temperature measured by a psychrometer's wet-bulb thermometer. The wet-bulb temperature lies between the air temperature and the dew-point temperature.

**wind vane** An instrument for measuring the direction from which the wind is blowing.

# Credits

## Permission Credits

### Chapter 1

Figure 1.9: From Donald J. Conte, et al., *Earth Science,* 2nd edition. Copyright © 1997 McGraw-Hill Company, Inc., Dubuque, Iowa. All Rights Reserved. Reprinted by permission.

### Chapter 2

Figure 2.15b: Updated figure from personal communication with C.D. Keeling and T.P. Whorf. Used by permission of Scripps Institution of Oceanography. The measurements were obtained in a cooperative program of NOAA and the Scripps Institution of Oceanography.

### Chapter 3

Figure 3.32: From John J. Houghton, *The Physics of Atmosphere,* 1977. Copyright © 1977 Cambridge University Press, New York, NY. Reprinted by permission.

### Chapter 4

Figure 4.3: From *Climate Change,* The IPCC Scientific Assessment Report prepared by Working Group 1, edited by J.T. Houghton, et al., page xvi, 1991, Cambridge University Press. Reprinted by permission of the Intergovernmental Panel on Climate Change, Switzerland.

Figure 4.4: From *Climate Change,* The IPCC Scientific Assessment Report prepared by Working Group 1, edited by J.T. Houghton, et al., page 11, 1991, Cambridge University Press. Reprinted by permission of the Intergovernmental Panel on Climate Change, Switzerland.

Figure 4.5a: From *Climate Change,* The IPCC Scientific Assessment Report prepared by Working Group 1, edited by J.T. Houghton, et al., page 62, 1991, Cambridge University Press. Reprinted by permission of the Intergovernmental Panel on Climate Change, Switzerland.

Figure 4.10: From *Climate Change,* The IPCC Scientific Assessment Report prepared by Working Group 1, edited by J.T. Houghton, et al., page 143, 1991, Cambridge University Press. Reprinted by permission of the Intergovernmental Panel on Climate Change, Switzerland.

Figure 4.11: From "Greenhouse Gas Induced Climate Changes Simulated with the CCC Second Generation General Circulation Model" in *Journal of Climate,* Vol. 5, No. 10, pp. 1045–1077, October, 1992. Reprinted by permission of the American Meteorological Society, Boston, MA.

Figure 4.13: Reprinted with permission from *Policy Implications of Greenhouse Warming.* Copyright 1991 by the National Academy of Sciences. Courtesy of the National Academy Press, Washington, D.C.

Figure 4.15c: From *Climate Change,* The IPCC Scientific Assessment Report prepared by Working Group 1, edited by J.T. Houghton, et al., page 269, 1991, Cambridge University Press. Reprinted by permission of the Intergovernmental Panel on Climate Change, Switzerland.

Figure 4.17: From *Climate Change,* The IPCC Scientific Assessment Report prepared by Working Group 1, edited by J.T. Houghton, et al., page 264, 1991, Cambridge University Press. Reprinted by permission of the Intergovernmental Panel on Climate Change, Switzerland.

Figure 4.19: From *Climate Change,* The IPCC Scientific Assessment Report prepared by Working Group 1, edited by J.T. Houghton, et al., page 202, 1991, Cambridge University Press. Reprinted by permission of the Intergovernmental Panel on Climate Change, Switzerland.

Figure 4.20: From *Climate Change,* The IPCC Scientific Assessment Report prepared by Working Group 1, edited by J.T. Houghton, et al., page 250, 1991, Cambridge University Press. Reprinted by permission of the Intergovernmental Panel on Climate Change, Switzerland.

Figure 4.21: From *Climate Change,* The IPCC Scientific Assessment Report prepared by Working Group 1, edited by J.T. Houghton, et al., page 246, 1991, Cambridge University Press. Reprinted by permission of the Intergovernmental Panel on Climate Change, Switzerland.

Figure 4.22: From *Climate Change,* The IPCC Scientific Assessment Report prepared by Working Group 1, edited by J.T. Houghton, et al., page xx and Appendix I, 1991, Cambridge University Press. Reprinted by permission of the Intergovernmental Panel on Climate Change, Switzerland.

Figure 4.23: From *Climate Change,* The IPCC Scientific Assessment Report prepared by Working Group 1, edited by J.T. Houghton, et al., page xxiii, 1991, Cambridge University Press. Reprinted by permission of the Intergovernmental Panel on Climate Change, Switzerland.

Figure 4.25: From *Climate Change,* The IPCC Scientific Assessment Report prepared by Working Group 1, edited by J.T. Houghton, et al., page 298, 1991, Cambridge University Press. Reprinted by permission of the Intergovernmental Panel on Climate Change, Switzerland.

Figure 4.28: From *Weatherwise,* December, 1964, Volume 17, Number 6, p. 265. Reprinted with permission of the Helen Dwight Reid Educational Foundation. Published by Heldref Publications, 1319 Eighteenth Street, NW, Washington, D.C. 20036-1802. Copyright © 1964.

Figure 11.30: From Alison B. Duxbury and Alyn C. Duxbury, *Fundamentals of Oceanography,* 2nd edition. Copyright © 1996 McGraw-Hill Company, Inc., Dubuque, Iowa. All Rights Reserved. Reprinted by permission.

### Chapter 11

Figure 11.34: Information from National Geographic Society as shown in Michael Bradshaw and Ruth Weaver, *Foundations of Physical Geography.* Copyright © 1995 McGraw-Hill Company, Inc., Dubuque, Iowa. All Rights Reserved. Reprinted by permission.

Figure 11.36: From *Science News,* May 5, 1990, by Richard Monastersky, Vol. 187, p. 280. Reprinted by permission of Edward Lorenz.

### Chapter 12

Figure 12.28: From G. Klazura and D. Imy, "A Description of the Initial Set of Analysis Products Available From the NEXRAD WSR-88D System" in *Bulletin of the American Meteorological Society,* Vol. 74, No. &, July 1993. Reprinted by permission of the American Meteorological Society, Boston, MA.

### Chapter 14

Figure 14.3: From Arthur Getis, et al., *Introduction to Geography,* 5th edition. Copyright © 1996 McGraw-Hill Company, Inc., Dubuque, Iowa. All Rights Reserved. Reprinted by permission.

Figure 14.10: From Michael Bradshaw and Ruth Weaver, *Foundations of Physical Geography.* Copyright © 1995 McGraw-Hill Company, Inc., Dubuque, Iowa. All Rights Reserved. Reprinted by permission.

Figure 14.25: From Alison B. Duxbury and Alyn C. Duxbury, *Fundamentals of Oceanography,* 2nd edition. Copyright © 1996 McGraw-Hill



450

Company, Inc., Dubuque, Iowa. All Rights Reserved. Reprinted by permission.

## Appendices

Figure D.1: From Donald J. Conte, et al., *Earth Science,* 2nd edition. Copyright © 1997 McGraw-Hill Company, Inc., Dubuque, Iowa. All Rights Reserved. Reprinted by permission.

Figure F.1 B,C: From Alison B. Duxbury and Alyn C. Duxbury, *Fundamentals of Oceanography,* 2nd edition. Copyright © 1996 McGraw-Hill Company, Inc., Dubuque, Iowa. All Rights Reserved. Reprinted by permission.

Figure F.1 D: From Jerome Fellmann, et al., *Human Geography,* 5th edition. Copyright © 1997 McGraw-Hill Company, Inc., Dubuque, Iowa. All Rights Reserved. Reprinted by permission.

# Photographs

## Chapter 1

UO1a,b: Northwind Picture Archives CO1: © Michael Agliolo/International Stock Fig 1.2a: Ronald & Shirley Holle Fig 1.2b–d: National Center for Atmospheric Research Fig 1.5a: © John D. Cunningham/Visuals Unlimited Fig 1.5b: © Van D. Bucher/Photo Researchers, Inc. Fig 1.5c: Weatherstock Fig 1.5d: © John D. Cunningham/Visuals Unlimited Fig 1.17a,b: Bill Danielson

## Chapter 2

CO2: © Corbis/Bettmann Fig 2.7a: Bill Danielson Fig 2.13: © Bert Gildart/Peter Arnold, Inc. Fig 2.15a: © Wilson North/International Stock Fig 2.16: © Will McIntyre/Photo Researchers, Inc. Fig 2.20: © David Baird/Tom Stack & Associates Fig 2.22: © The McGraw-Hill Companies, Inc./Bob Coyle, photographer. Fig 2.23: Richard Emanuel/Alaska Volcano Observatory Fig 2.24: © James L. Shaffer Fig 2.26a: Courtesy NOAA Fig 2.31b: NASA Fig 2.33: Carla Montgomery Fig 2.34a: NASA Fig 2.34b: NSSDC/Goddard Space Flight Center

## Chapter 3

UO2: © Scott Berner/Visuals Unlimited CO3: ©Cliff Hollenbeck/International Stock Fig 3.11: National Aeronautics and Space Administration, Langley Research Center. Fig 3.14: © David Parker/SPL/Science Source/Photo Researchers, Inc. Fig 3.25: T.D. Bess/NASA

## Chapter 4

CO4: © The McGraw-Hill Companies, Inc./Bob Coyle, photographer Fig 4.5b: The Observatories of the Carnegie Institution of Washington Fig 4.6: David Longstreath/AP Wide World Photos Fig 4.15a,b: NCAR/National Science Foundation Fig 4.16: © Charles Preitner/Visuals Unlimited Fig 4.24: © Jack S. Grove/Tom Stack & Associates Fig 4.26: Dr Cynthia Rosenzweig/NASA Fig 4.27a: Bill W. Tillery Fig 4.27b: Cameramann International, Ltd. Fig 4.27c: © The McGraw-Hill Companies, Inc./Doug Sherman, photographer UO3a: © Norbert Wu/Peter Arnold, Inc. UO3b: © Clyde H. Smith/Peter Arnold, Inc.

## Chapter 5

CO5: © G. Brad Lewis/Tony Stone Images Fig 5.11: Edge Tech Moisture & Humidity Systems Fig 5.12a,b: © McGraw-Hill/Photo by Bob Coyle Fig 5.18: © David Brownell Fig 5.20a: W. Wayne Lockwood, M.D./Corbis Fig 5.24: Eye Ubiquitous/Corbis Fig 5.29: Bill Danielson Fig 5.31a: © Richard Kaylin/Tony Stone Images Fig 5.31b: © Galen Rowell/Peter Arnold, Inc.

## Chapter 6

CO6: © James Steinberg/Photo Researchers, Inc. Fig 6.2a: © Glenn M. Oliver/Visuals Unlimited Fig 6.2b: Weatherstock Fig 6.2c: Papilio/Corbis Fig 6.2d: © Edna Bennett/Photo Researchers, Inc. Fig 6.2e: © W. Banaszewski/Visuals Unlimited Fig 6.2f: Scott Smith/Corbis Fig 6.2g: Ecoscene/Corbis Fig 6.2h: © Buff & Gerald Corsi/Visuals Unlimited Fig 6.2i: Scott Smith/Corbis Fig 6.2j: © Gregory K. Scott/Photo Researchers, Inc. Fig 6.2k: © John Lamb/Tony Stone Images Fig 6.19a: Paul Souders/Corbis Fig 6.19b: © Kent Wood/Photo Researchers, Inc. Fig 6.21: © Gary Ladd/Photo Researchers, Inc. Fig 6.22a: © Glenn Oliver/Visuals Unlimited Fig 6.25: © Val Corbett/Tony Stone Images Fig 6.27b: © John D. Cunningham/Visuals Unlimited Fig 6.28: © Mack Henley/Visuals Unlimited Fig 6.30b: © Howard Bluestein/Photo Researchers, Inc. Fig 6.30c: © McGraw-Hill/Bob Coyle, photographer Fig 6.30d: © J. Gerner/Visuals Unlimited Fig 6.32a: © David R. Frazier/Photo Researchers, Inc. Fig 6.32b: Weatherstock

## Chapter 7

CO7: © AP/Wide World Photos Fig 7.2: © The McGraw-Hill Companies, Inc./Bob Coyle, photographer Fig 7.4: AP/Wide World Photos Fig 7.7: Robert N. Wallen Fig 7.10a–d: Weather Graphics Courtesy of AccuWeather, Inc. Fig 7.15a: © L & D Klein/Photo Researchers, Inc. Fig 7.17a,b: National Center for Atmospheric Researcher /University Corporation for Atmospheric Research/National Science Foundation Figs 7.22 & 7.23: U.S. Signal Corps

## Chapter 8

UO4: National Aeronautics and Space Administration, Langley Research Center. CO8: © Armond Scavo Fig 8.1: © Bob Firth/International Stock Photo

## Chapter 9

CO9: © Armond Scavo Fig 9.19: From Sanders and Doswell; Bulletin of the AMS, Vol. 76, No. 4, April 1995, page 506 Fig 9.26a: © Beatriz Schiller/International Stock Photo Fig 9.26b: © David Brownell

## Chapter 10

CO10: © Armond Scavo Fig 10.18: Mark Elias/AP Wide World Photos Fig 10.19: U.S. Dept. of Commerce/NOAA Fig 10.21a,b: The McGraw-Hill Companies, Inc./Bob Coyle, photographer a: Journal Inquirer, Manchester, CT; b: The Boston Globe. Fig 10.22: © David Young-Wolff/PhotoEdit Fig 10.23: © John Feingersh/Tom Stack & Associates Fig 10.24: © Patrick Cone Photography

## Chapter 11

UO5: National Center for Atmospheric Research/University Corporation for Atmospheric Research/University Corporation for Atmospheric Research/National Science Foundation CO11: © TSADO/NCDC/NOAA/Tom Stack & Associates Fig 11.1a: NCAR; Fig 11.1b: NASA From Rainbow Fig 11.1c: International Stock Photo Fig 11.1c: © Dan McCoy/Rainbow Fig 11.1d: © Bob Firth Fig 11.1e: © Michael Newman/PhotoEdit Fig 11.1f–h: NASA Fig 11.8: Ronald & Shirley Holle Fig 11.9: Bettmann Fig 11.10a: © Beatriz Schiller/International Stock Photo Fig 11.14a–c: John Thorp, Pacific Northwest Laboratory, Bettelle Memorial Institute for the U.S. Department of Energy under contract No. DE-AC06-76LO-1830, previously published in Weatherwise, 39, #6, 1986. Fig 11.18: Buffalo Bill Historical Center, Cody, WY Gift of Charles Ulrick and Josephine Bay Foundation, Inc. Fig 11.20: Jeff & Alexa Henry Fig 11.23: Accu-Weather Fig 11.31: Courtesy Johns Hopkins University internet site/Analysis of imagery Janifer Clark's Gulfstream Fig 11.32: Images provided courtesy of G. Feldman, NASA/GSFC

## Chapter 12

CO12: © Warren Faidley/International Stock Fig 12.3: NOAA/National Climatic Data Center Fig 12.14: Courtesy of NOAA Fig 12.16a: © Shattil/Rozinski/Tom Stack & Associates Fig 12.20: NOAA/National Climatic Data Center Fig 12.21a: Peter Willing Fig 12.21b: Michael Mee, FEMA Fig 12.23: Steve Tegtmeier for Weather Data, Inc. Fig 12.25a–c: Arizona Daily Star Photo By Jack Sheafer Fig 12.27: Dave Williams/Wichita Eagle Fig 12.29: Reprinted with permission from Van Nostrand Reinhold Fig 12.30a–e: T. Fujita Fig 12.33a,b: Steve A. Tegtmeier for Weather Data, Inc.

## Chapter 13

CO13: Warren Faidley/International Stock Fig 13.7a–13.11a: NOAA/National Climatic Data Center Fig 13.11a: Courtesy of NOAA Fig 13.11b: Stephen Waide/Aurora Flight Sciences Corporation Fig 13.16c: NOAA/National Climatic Data Center Fig 13.18: NOAA Satellite Services Division Fig 13.21:

All Rights Reserved the Rhode Island Historical Society/RHI (X3) 1766 Fig 13.23: NOAA/ National Climatic Data Center Fig 13.24: Warren Faidley/International Stock Fig 13.26c: NOAA/National Climatic Data Center

## Chapter 14

CO14: © Norris Clarr/International Stock Photo Fig 14.2a: Brian Parker/Tom Stack & Associates Fig 14.2b: © Scott Blackman/Tom Stack & Associates Fig 14.4: © Brian Seed Fig 14.6: © Branson Reynolds Fig 14.8a: Raymond, W.H., R.M. Rabin, and H.G. Wade, 1994 Bulletin of the American Meteorology Society 75, 1019–1025 Fig 14.12a: © Richard Thom/Tom Stack & Associates Fig 14.12b–14.14a: Courtesy of Dr. Charles Hogue, Curator of Entomology, Los Angeles County Museum of Natural History Fig 14.14b: Brian Parker/Tom Stack & Associates Fig 14.17a: © J. Lotter/Tom Stack & Associates Fig 14.17b: Robert N. Wallen Fig 14.17c: © Terry Donnelly/Tom Stack & Associates Fig 14.17d: USDA Fig 14.19a,b: © John Cunningham/Visuals Unlimited Fig 14.20a: © William Ferguson Fig 14.21a: Seth Danielson

# Index

Page numbers followed by *t* and *f* indicate tables and figures, respectively.

## A

B



C

## D

## E



## F

## H



## I



## J



## K

## L



## M



## N

O



P



R

# T

# W



# Z